Springer Actuarial

This is a series on actuarial topics in a broad and interdisciplinary sense, aimed at students, academics and practitioners in the fields of insurance and finance.

Springer Actuarial informs timely on theoretical and practical aspects of topics like risk management, internal models, solvency, asset-liability management, market-consistent valuation, the actuarial control cycle, insurance and financial mathematics, and other related interdisciplinary areas.

The series aims to serve as a primary scientific reference for education, research, development and model validation.

The type of material considered for publication includes lecture notes, monographs and textbooks. All submissions will be peer-reviewed.

Mario V. Wüthrich • Michael Merz

Statistical Foundations of Actuarial Learning and its Applications

SAV Schweizerische Aktuarvereinigung
ASA Association Suisse des Actuaires
ASA Associazione Svizzera degli Attuari

Springer

Swiss Re

Mario V. Wüthrich
Department of Mathematics, RiskLab
Switzerland
ETH Zürich
Zürich, Switzerland

Michael Merz
Faculty of Business Administration
University of Hamburg
Hamburg, Germany

This work was supported by Schweizerische Aktuarvereinigung SAV and Swiss Re.

ISSN 2523-3262 ISSN 2523-3270 (electronic)
Springer Actuarial
ISBN 978-3-031-12408-2 ISBN 978-3-031-12409-9 (eBook)
https://doi.org/10.1007/978-3-031-12409-9

Mathematics Subject Classification: C13, C21/31, C24/34, G22, 62F10, 62F12, 62J07, 62J12, 62M45, 62P05, 68T01, 68T50

This Springer imprint is published by the registered company Springer Nature Switzerland AG
The registered company address is: Gewerbestrasse 11, 6330 Cham, Switzerland

Acknowledgments

We kindly thank our very generous sponsors, the Swiss Association of Actuaries (SAA) and Swiss Re, for financing the open access option of the electronic version of this book. Our special thanks go to Sabine Betz (President of SAA), Adrian Kolly (Swiss Re), and Holger Walz (SAA) who were very positive and interested in this book project from the very beginning, and who made this open access funding possible within their institutions.

A very special thank you goes to Hans Bühlmann who has been supporting us over the last 30 years. We have had so many inspiring discussions over these years, and we have greatly benefited and learned from Hans' incredible knowledge and intuition.

Jointly with Christoph Buser, we have started to teach the lecture "Data Analytics for Non-Life Insurance Pricing" at ETH Zurich in 2018. Our data analytics lecture focuses (only) on the Poisson claim counts case, but its lecture notes have provided a first draft for this book project. This draft has been developed and extended to the general case of the exponential family. Since our first lecture, we have greatly benefited from interactions with many colleagues and students. In particular, we would like to mention the data science initiative "Actuarial Data Science" of the Swiss Association of Actuaries (chaired by Jürg Schelldorfer), whose tutorials provided a great stimulus for this book. Moreover, we mention the annual Insurance Data Science Conference (chaired by Markus Gesmann and Andreas Tsanakas) and the ASTIN Reading Club (chaired by Ronald Richman and Dimitri Semenovich). Furthermore, we would like to kindly thank Ronald Richman who has always been a driving force behind learning and adapting new machine learning techniques, and we also kindly thank Simon Rentzmann for many interesting discussions on how to apply these techniques on real insurance problems.

We thank the following colleagues by name (in alphabetical order). We collaborated and had inspiring discussions in the field of statistical learning with the following colleagues: Johannes Abegglen, Hansjörg Albrecher, Davide Apolloni, Peter Bühlmann, Christoph Buser, Patrick Cheridito, Łukasz Delong, Paul Embrechts, Andrea Ferrario, Tobias Fissler, Luca Fontana, Daisuke Frei, Tsz Chai Fung, Guangyuan Gao, Yan-Xing Lan, Gee Lee, Mathias Lindholm, Christian

Lorentzen, Friedrich Loser, Michael Mayer, Daniel Meier, Alexander Noll, Gareth Peters, Jan Rabenseifner, Peter Reinhard, Simon Rentzmann, Ronald Richman, Ludger Rüschendorf, Robert Salzmann, Marc Sarbach, Jürg Schelldorfer, Pavel Shevchenko, Joël Thomann, Andreas Tsanakas, George Tzougas, Emiliano Valdez, Tim Verdonck, and Patrick Zöchbauer.

Contents

Chapter 1
Introduction

1.1 The Statistical Modeling Cycle

We consider statistical modeling of insurance problems. This comprises the process of data collection, data analysis and statistical model building to forecast insured events that (may) happen in the future. This problem is at the very heart of statistics and statistical modeling. Our goal here is to present and provide the statistical tools that are useful in daily actuarial practice, in particular, we aim at describing the mathematical foundation behind these statistical concepts and how they can be applied. Statistical modeling has a wide range of applications, and, depending on the application, the theoretical aspects may be weighted differently. In insurance pricing we are mainly interested in optimal predictions, whereas economists often use statistical tools to explain observations, and in medical fields one is interested in causal effects that medications have on patients. Therefore, statistical theory is wide ranging, and one should always keep the corresponding application in mind. Shmueli [338] nicely discusses the difference between prediction and explanation; our focus here is mainly on prediction.

Box–Jenkins [49] and McCullagh–Nelder [265] distinguish three processes in statistical modeling: (i) model identification/selection, (ii) estimation, and (iii) prediction. In our statistical modeling cycle these three points are slightly modified and extended:

(1) Data collection, cleaning and pre-processing:
This item takes at least 80% of the total time in statistical modeling. It includes exploratory data analysis, data visualization and data pre-processing. This part of the modeling cycle does not seem to be very scientific, however, it is a highly important step because only extended data analysis allows the modeler to fully understand the data. Based on this knowledge the modeler can formulate her/his research question, her/his model, etc.

M. V. Wüthrich, M. Merz, *Statistical Foundations of Actuarial Learning and its Applications*, Springer Actuarial, https://doi.org/10.1007/978-3-031-12409-9_1

(2) Selection of a model class:
Based on the knowledge collected in the first item, the modeler has to select a suitable model class that is able to answer her/his research question. This model class can be in the sense of a data model (proper stochastic model), but it can also be an algorithmic model; we refer to the discussion on the "two modeling cultures" by Breiman [53].
(3) Choice of an objective function:
Once the modeler has specified a model class, she/he needs to define a decision rule how a particular member of the model class is selected for the collected data. Often this is in terms of an objective function, e.g., a scoring rule or a loss function that quantifies misspecification.
(4) Solving a (non-convex) optimization problem:
Once the first three items are completed, one is left with an optimization problem that tries to find the best model within the selected model class w.r.t. the given objective function and the collected data. In simple cases this optimization problem is a convex minimization problem for which numerical tools are in place. In more complex cases the optimization problem is neither convex nor concave, and the 'best' solution can often not be found explicitly. In that case, also the meaning of solution needs to be discussed.
(5) Model validation:
In the final/next step, the selected and fitted model needs to be validated. That is, does the model fit to the data, does it serve at predicting new data, does it answer the research question adequately, is there any better model/process choice, etc.?
(6) Possibly go back to (1):
If the answers in item (5) are not satisfactory, one typically goes back to (1). For instance, data pre-processing needs to be done differently, etc.

Especially, the two modeling cultures discussion of Breiman [53], after the turn of the millennium, has shaken up the statistical community. Having predictive performance as the main criterion, the data modeling culture has gradually shifted to the algorithmic culture, where the model itself plays a secondary role as long as the prediction is accurate. The latter is often in the form of a point predictor which can come from an algorithm. Lifting this discussion to a more scientific level, providing prediction uncertainty will slowly merge the two modeling cultures. There is an other interesting discussion by Efron [116] on prediction, estimation (of model parameters) and attribution (predictor selection), that is very much at the core of statistical modeling. In these notes we want to especially emphasize the one modeling culture view of Yu–Barter [397] who expect the two modeling cultures of Breiman [53] to merge much closer than one would expect. Our goal is to demonstrate how all these different techniques and views can be seen as a unified modeling framework.

Concluding, the purpose of these notes is to discuss and illustrate how the different statistical techniques from the data modeling culture and the algorithmic modeling culture can be combined to solve actuarial questions in the best possible way. The main emphasis in this discussion lies on the statistical modeling tools, and we present these tools along with actuarial examples. In actuarial practice one often distinguishes between life and general insurance. This distinction is done for good reasons. There are legislative reasons that require to legally separate life from general insurance business, but there are also modeling reasons, because insurance products in life and general insurance can have rather different features. In this book, we do not make this distinction because the statistical methods presented here can be useful in both branches of insurance, and we are going to consider life and general insurance examples, e.g., the former considering mortality forecasting and the latter aiming at insurance claims prediction for pricing.

1.2 Preliminaries on Probability Theory

The modern axiomatic foundation of probability theory was introduced in 1933 by the famous mathematician Kolmogoroff [221] in his book called "Grundbegriffe der Wahrscheinlichkeitsrechnung". We give a brief introduction to probability theory and random variables; this introduction follows the lecture notes [387]. Throughout we assume to work on a sufficiently rich probability space $(\Omega, \mathcal{A}, \mathbb{P})$, meaning that this probability space should be able to carry all objects that we study. We denote (real-valued) random variables on this probability space by capital letters $Y, Z, \ldots$, and random vectors use boldface capital letters, e.g., we have a random vector $\boldsymbol{Y} = (Y_1, \ldots, Y_q)^\top$ of dimension $q \in \mathbb{N}$, where each component Y_k, $1 \le k \le q$, is a random variable. Random variables Y are characterized by (cumulative) distribution functions[1] $F : \mathbb{R} \to [0, 1]$, for $y \in \mathbb{R}$

$$F(y) = \mathbb{P}[Y \le y],$$

being the probability of the event that Y has a realization of less or equal to y. We write $Y \sim F$ for Y having distribution function F. Similarly random vectors $\boldsymbol{Y} \sim F$ are characterized by (cumulative) distribution functions $F : \mathbb{R}^q \to [0, 1]$ with

$$F(\boldsymbol{y}) = \mathbb{P}\left[Y_1 \le y_1, \ldots, Y_q \le y_q\right] \qquad \text{for } \boldsymbol{y} = (y_1, \ldots, y_q)^\top \in \mathbb{R}^q.$$

In insurance modeling, there are two important types of random variables, namely, discrete random variables and absolutely continuous random variables:

- The distribution function F of a discrete random variable Y is a step function with countably many steps in discrete points $k \in \mathfrak{N} \subset \mathbb{R}$. A discrete random variable has probability weights in these discrete points

$$f(k) = \mathbb{P}[Y = k] > 0 \qquad \text{for } k \in \mathfrak{N},$$

[1] Cumulative distribution functions F are right-continuous, non-decreasing with $\lim_{x\to-\infty} F(x) = 0$ and $\lim_{x\to\infty} F(x) = 1$.

satisfying $\sum_{k \in \mathfrak{N}} f(k) = 1$. If $\mathfrak{N} \subseteq \mathbb{N}_0$, the integer-valued random variable Y is called count random variable. Count random variables are used to model the number of claims in insurance. A similar situation occurs if Y models nominal outcomes, for instance, if Y models gender with female being encoded by 0 and male being encoded by 1, then $f(0)$ is the probability weight of having a female and $f(1) = 1 - f(0)$ the probability weight of having a male; in this case we identify the finite set $\mathfrak{N} = \{0, 1\} = \{\text{female}, \text{male}\}$.

- A random variable $Y \sim F$ is said to be absolutely continuous[2] if there exists a non-negative (measurable) function f, called density of Y, such that

$$F(y) = \int_{-\infty}^{y} f(x)\, dx \qquad \text{for all } y \in \mathbb{R}.$$

In that case we equivalently write $Y \sim f$ and $Y \sim F$. Absolutely continuous random variables are often used to model claim sizes in insurance.

More generally speaking, discrete and absolutely continuous random variables have densities $f(\cdot)$ w.r.t. a σ-finite measure ν on $\mathbb{R}$. In the former case, this σ-finite measure ν is the counting measure on $\mathfrak{N} \subset \mathbb{R}$, and in the latter case it is the Lebesgue measure on $\mathbb{R}$. In actuarial science we also consider mixed cases, for instance, Tweedie's compound Poisson random variable is absolutely continuous on $(0, \infty)$ having an additional point mass in 0; this model will be studied in Sect. 2.2.3, below.

Choose a random variable $Y \sim F$ and a measurable function $h : \mathbb{R} \to \mathbb{R}$. The expected value of $h(Y)$ is defined by (upon existence)

$$\mathbb{E}[h(Y)] = \int_{\mathbb{R}} h(y)\, dF(y).$$

We mainly focus on the following important examples of function h:

- expected value, mean or first moment of $Y \sim F$: for $h(y) = y$

$$\mu = \mathbb{E}[Y] = \int_{\mathbb{R}} y\, dF(y);$$

- k-th moment of $Y \sim F$ for $k \in \mathbb{N}$: for $h(y) = y^k$

$$\mathbb{E}\left[Y^k\right] = \int_{\mathbb{R}} y^k\, dF(y);$$

[2] Absolutely continuous is a stronger property than continuous.

- moment generating function of $Y \sim F$ in $r \in \mathbb{R}$: for $h(y) = e^{ry}$

$$M_Y(r) = \mathbb{E}\left[e^{rY}\right] = \int_{\mathbb{R}} e^{ry}\, dF(y);$$

always subject to existence.

The moment generating function $M_Y(\cdot)$ is sufficient for identifying distribution functions of random variables Y. The following statements are elementary and their proofs are based on Section 30 of Billingsley [34], for more details we also refer to Chapter 1 in the lecture notes [387]. Assume that the moment generating function of $Y \sim F$ has a strictly positive radius of convergence $\rho_0 > 0$ around the origin implying that $M_Y(r) < \infty$ for all $r \in (-\rho_0, \rho_0)$. In this case we can write $M_Y(r)$ as a power series expansion

$$M_Y(r) = \sum_{k=0}^{\infty} \frac{r^k}{k!}\, \mathbb{E}\left[Y^k\right] \qquad \text{for all } r \in (-\rho_0, \rho_0).$$

As a consequence we can differentiate $M_Y(\cdot)$ in the open interval $(-\rho_0, \rho_0)$ arbitrarily often, term by term under the sum. The derivatives in $r = 0$ provide the k-th moments (which all exist and are finite)

$$\frac{d^k}{dr^k} M_Y(r)|_{r=0} = \mathbb{E}\left[Y^k\right] \qquad \text{for all } k \in \mathbb{N}_0. \tag{1.1}$$

In particular, in this case we immediately know that all moments of Y exist, and these moments completely determine the moment generating function M_Y of Y. Another consequence is that for a random variable Y, whose moment generating function M_Y has a strictly positive radius of convergence around the origin, the distribution function F is fully determined by this moment generating function. That is, if we have two such random variables Y_1 and Y_2 with $M_{Y_1}(r) = M_{Y_2}(r)$ for all $r \in (-r_0, r_0)$, for some $r_0 > 0$, then $Y_1 \stackrel{(d)}{=} Y_2$.[3] Thus, these two random variables have the same distribution function. This statement carries over to the limit, i.e., if we have a sequence of random variables $(Y_n)_n$ whose moment generating functions converge on a common interval $(-r_0, r_0)$, for some $r_0 > 0$, to the moment generating function of Y, also being finite on $(-r_0, r_0)$, then $(Y_n)_n$ converges in distribution to Y; such an argument is used to prove the central limit theorem (CLT).

[3] The notation $Y_1 \stackrel{(d)}{=} Y_2$ is generally used for equality in distribution meaning that Y_1 and Y_2 have the same distribution function.

In insurance, we often deal with so-called positive random variables Y, meaning that $Y \geq 0$, almost surely (a.s.). In that case, the statements about moment generating functions and distributions hold true without the assumption of having a positive radius of convergence around the origin, see Theorem 22.2 in Billingsley [34]. Note that for positive random variables the moment generating function $M_Y(r)$ exists for all $r \leq 0$.

Existence of the moment generating function $M_Y(r)$ for some positive $r > 0$ can also be interpreted as having a light-tailed distribution function. Observe that if $M_Y(r)$ exists for some positive $r > 0$, then we can choose $s \in (0, r)$ and Chebychev's inequality gives us (we assume $Y \geq 0$, a.s., here)

$$\mathbb{P}[Y > y] = \mathbb{P}\left[\exp\{sY\} > \exp\{sy\}\right] \leq \exp\{-sy\} M_Y(s). \tag{1.2}$$

The latter tells us that the survival function $1 - F(y) = \mathbb{P}[Y > y]$ decays exponentially for $y \to \infty$. Heavy-tailed distribution functions do not have this property, but the survival function decays slower than exponentially as $y \to \infty$. This slower decay of the survival function is the case for so-called subexponential distribution functions (an example is the log-normal distribution, we refer to Rolski et al. [320]) and for regularly varying survival functions (an example is the Pareto distribution). Regularly varying survival functions $1 - F$ have the property

$$\lim_{y \to \infty} \frac{1 - F(ty)}{1 - F(y)} = t^{-\beta} \qquad \text{for all } t > 0 \text{ and some } \beta > 0. \tag{1.3}$$

These distribution functions have a polynomial tail (power tail) with tail index $\beta > 0$. In particular, if a positively supported distribution function F has a regularly varying survival function with tail index $\beta > 0$, then this distribution function is also subexponential, see Theorem 2.5.5 in Rolski et al. [320].

We are not going to specifically focus on heavy-tailed distribution functions, here, but we will explain how light-tailed random variables can be transformed to enjoy heavy-tailed properties. In these notes, we are mainly interested in studying different aspects of regression modeling. Regression modeling requires numerous observations to be able to successfully fit these models to the data. By definition, large claims are scarce, as they live in the tail of the distribution function and, thus, correspond to rare events. Therefore, it is often not possible to employ a regression model for scarce tail events. For this reason, extreme value analysis only plays a marginal role in these notes, though, it has a significant impact on insurance prices. For more on extreme value theory we refer to the relevant literature, see, e.g., Embrechts et al. [121], Rolski et al. [320], Mikosch [277] and Albrecher et al. [7].

1.3 Lab: Exploratory Data Analysis

Our theory is going to be supported by several data examples. These examples are mostly based on publicly available data. The different data sets are described in detail in Chap. 13. We highly recommend the reader to use these data sets to gain her/his own modeling experience.

We describe some tools here that allow for a descriptive and exploratory analysis of the available data; exploratory data analysis has been introduced and promoted by Tukey [357]. We consider the observed claim sizes of the Swedish motorcycle data set described in Sect. 13.2. This data set consists of 656 (positive) claim amounts y_i, $1 \le i \le n = 656$. These claim amounts are illustrated in the boxplots of Fig. 1.1.

Typically in insurance, there are large claims that dominate the picture, see Fig. 1.1 (lhs). This results in right-skewed distribution functions, and such data is better illustrated on the log scale, see Fig. 1.1 (rhs). The latter, of course, assumes that all claims are strictly positive.

Figure 1.2 (lhs) shows the empirical distribution function of the observations y_i, $1 \le i \le n$, which is obtained by

$$\widehat{F}_n(y) = \frac{1}{n} \sum_{i=1}^{n} \mathbb{1}_{\{y_i \le y\}} \qquad \text{for } y \in \mathbb{R}.$$

If this data set has been generated by i.i.d. random variables, then the Glivenko–Cantelli theorem [64, 159] tells us that this empirical distribution function $\widehat{F}_n$ converges uniformly to the (true) data generating distribution function, a.s., as the number n of observations converges to infinity, see Theorem 20.6 in Billingsley [34].

Figure 1.2 (rhs) shows the empirical density of the observations y_i, $1 \le i \le n$. This empirical density is obtained by considering a kernel smoother of a given

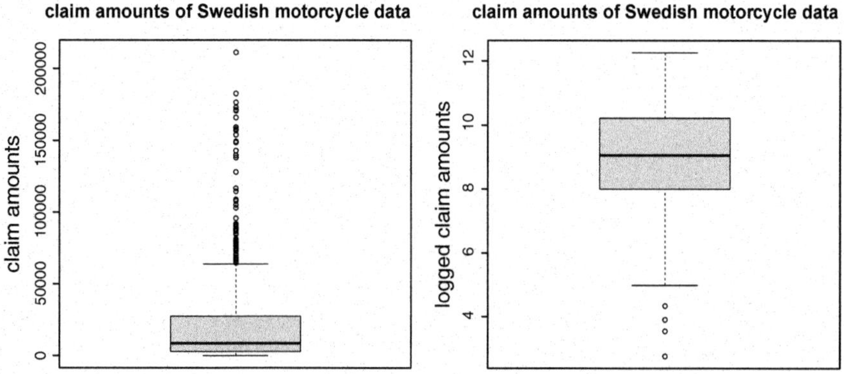

Fig. 1.1 Boxplot of the claim amounts of the Swedish motorcycle data set: (lhs) on the original scale and (rhs) on the log scale

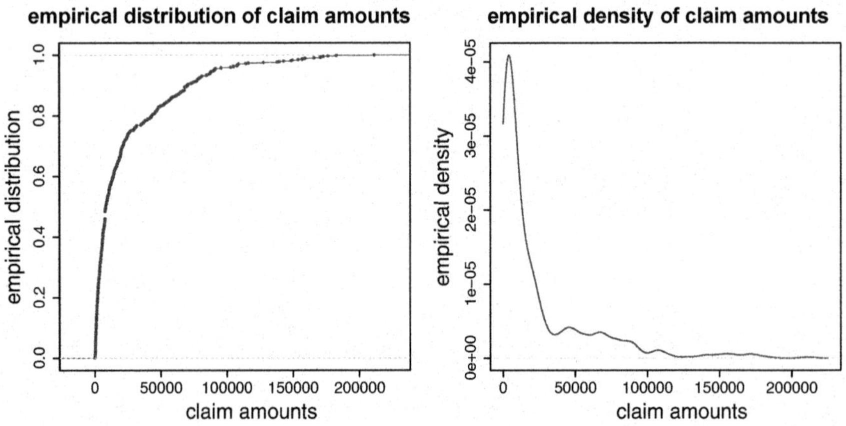

Fig. 1.2 (lhs) Empirical distribution and (rhs) empirical density of the observed claim amounts y_i, $1 \le i \le n$

bandwidth around each observation y_i. The standard choice is the Gaussian kernel, with the bandwidth determining the variance parameter $\sigma^2 > 0$ of the Gaussian density,

$$y \mapsto \widehat{f}_n(y) = \frac{1}{n} \sum_{i=1}^{n} \frac{1}{\sqrt{2\pi\sigma^2}} \exp\left\{ -\frac{1}{2} \frac{(y-y_i)^2}{\sigma^2} \right\}.$$

From the graph in Fig. 1.2 (rhs) we observe that the main body of the claim sizes is below an amount of 50'000, but the biggest claim exceeds 200'000. The latter motivates to study heavy-tailedness of the claim size data. Therefore, one usually benchmarks with a distribution function F that has a regularly varying survival function with a tail index $\beta > 0$, see (1.3). Asymptotically a regularly varying survival function behaves as $y^{-\beta}$; for this reason the log-log plot is a popular tool to identify regularly varying tails. The log-log plot of a distribution function F is obtained by considering

$$y > 0 \mapsto (\log y, \log(1 - F(y))) \in \mathbb{R}^2.$$

Figure 1.3 gives the log-log plot of the empirical distribution function $\widehat{F}_n$. If this plot looks asymptotically (for $y \to \infty$) like a straight line with a negative slope $-\beta$, then the data shows heavy-tailedness in the sense of regular variation. Such data cannot be modeled by a distribution function for which the moment generating function $M_Y(r)$ exists for some positive $r > 0$, see (1.2). Figure 1.3 does not suggest a regularly varying tail as we do not see an obvious asymptotic straight line for increasing claim sizes.

These graphs give us a first indication what the claim size data is about. Later on we are going to introduce explanatory variables that describe the insurance

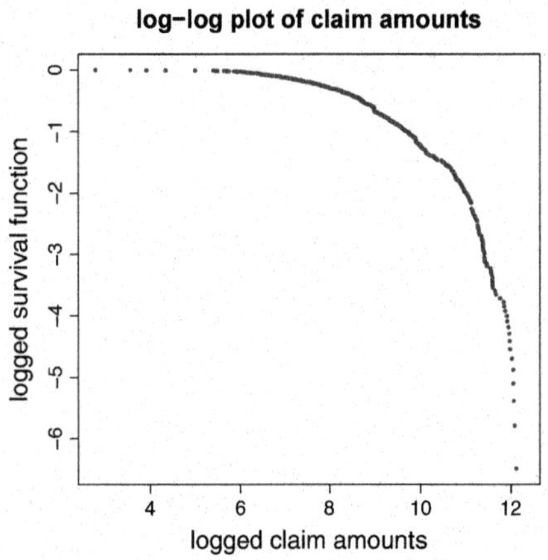

Fig. 1.3 Log-log plot of the empirical distribution function $\widehat{F}_n$

policyholders behind these claims. These explanatory variables characterize the policyholder and the general goal is to get a better description of the claim sizes as a function of these explanatory variables, e.g., older policyholders may cause larger claims than younger ones, etc. Such patterns are called *systematic effects* that can be explained by explanatory variables.

1.4 Outline of This Book

This book has eleven chapters (including the present one), and it has two appendices. We briefly describe the contents of these chapters and appendices.

In Chap. 2 we introduce and discuss the exponential family (EF) and the exponential dispersion family (EDF). The EF and the EDF are by far the most important classes of distribution functions for regression modeling. They include, among others, the Gaussian, the binomial, the Poisson, the gamma, the inverse Gaussian and Tweedie's models. We introduce these families of distribution functions, discuss their properties and provide several examples. Moreover, we introduce the Kullback–Leibler (KL) divergence and the Bregman divergence, which are important tools in model evaluation.

Chapter 3 is on classical statistical decision theory. This chapter is important for historical reasons, but it also provides the right mathematical grounding and intuition for more modern tools from data science and machine learning. In particular, we discuss maximum likelihood estimation (MLE), unbiasedness, consistency and asymptotic normality of MLEs in this chapter.

Chapter 4 is the core theoretical chapter on predictive modeling and forecast evaluation. The main problem in actuarial modeling is to forecast and price future claims. For this, we build predictive models, and this chapter deals with assessing and ranking these predictive models. We therefore introduce the mean squared error of prediction (MSEP) and, more generally, the generalization loss (GL) to assess predictive models. This chapter is complemented by a more decision-theoretic approach to forecast evaluation, it discusses deviance losses, proper scoring, elicitability, forecast dominance, cross-validation, Akaike's information criterion (AIC) and we give an introduction to the bootstrap simulation method.

Chapter 5 discusses state-of-the-art statistical modeling in insurance which is the generalized linear model (GLM). We discuss GLMs in the light of claim count and claim size modeling, we present feature engineering, model fitting, model selection, over-dispersion, zero-inflated claim counts problems, double GLMs, and insurance-specific issues such as the balance property for having unbiasedness.

Chapter 6 summarizes some techniques that use Bayes' theorem. These are classical Bayesian statistical models, e.g., using the Markov chain Monte Carlo (MCMC) method for model fitting. This chapter discusses regularization of regression models such as ridge and LASSO regularization, which has a Bayesian interpretation, and it concerns the Expectation-Maximization (EM) algorithm. The EM algorithm is a general purpose tool that can handle incomplete data settings. We illustrate this for different examples coming from mixture distributions, censored and truncated claims data.

The core of this book are deep learning methods and neural networks. Chapter 7 considers deep feed-forward neural (FN) networks. We introduce the generic architecture of deep FN networks, and we discuss universality theorems of FN networks. We present network fitting, back-propagation, embedding layers for categorical variables and insurance-specific issues such as the balance property in network fitting and network ensembling to reduce model uncertainty. This chapter is complemented by many examples on non-life insurance pricing, but also on mortality modeling, as well as tools that help to explain deep FN network regression results.

Chapters 8 and 9 consider recurrent neural (RN) networks and convolutional neural (CN) networks. These are special network architectures that are useful for time-series and spatial data modeling, e.g., applied to image recognition problems. Time-series and images have a natural topology, and RN and CN networks try to benefit from this additional structure (over tabular data). We introduce these network architectures and provide insurance-relevant examples.

Chapter 10 discusses natural language processing (NLP) which deals with regression modeling of non-tabular or unstructured text data. We explain how words can be embedded into low-dimension spaces that serve as numerical word encodings. These can then be used for text recognition, either using RN networks or attention layers. We give an example where we aim at predicting claim perils from claim descriptions.

Chapter 11 is a selection of different topics. We mention forecasting under model uncertainty, deep quantile regression, deep composite regression or the LocalGLMnet which is an interpretable FN network architecture. Moreover, we provide a bootstrap example to assess prediction uncertainty, and we discuss mixture density networks.

Chapter 12 (Appendix A) is a technical chapter that discusses universality theorems for networks and sieve estimators, which are useful for studying asymptotic normality within a network framework. Chapter 13 (Appendix B) illustrates the data used in this book.

Finally, we remark that the book is written in a typical mathematical style using the structure of Lemmas, Theorems, etc. Results and statements which are particularly important for applications are highlighted with gray boxes.

Chapter 2
Exponential Dispersion Family

We introduce the exponential family (EF) and the exponential dispersion family (EDF) in this chapter. The single-parameter EF has been introduced in 1934 by the British statistician Sir Fisher [128], and it has been extended to vector-valued parameters by Darmois [88], Koopman [223] and Pitman [306] between 1935 and 1936. It is the most commonly used family of distribution functions in statistical modeling; among others, it contains the Gaussian distribution, the gamma distribution, the binomial distribution and the Poisson distribution. Its parametrization is taken in a special form that is convenient for statistical modeling. The EF can be introduced in a constructive way providing the main properties of this family of distribution functions. In this chapter we follow Jørgensen [201–203] and Barndorff-Nielsen [23], and we state the most important results based on this constructive introduction. This gives us a unified notation which is going to be useful for our purposes.

2.1 Exponential Family

2.1.1 Definition and Properties

We define the EF w.r.t. a σ-finite measure ν on $\mathbb{R}$. The results in this section can be generalized to σ-finite measures on $\mathbb{R}^m$, but such an extension is not necessary for our purposes. Select an integer $k \in \mathbb{N}$, and choose measurable functions $a : \mathbb{R} \to \mathbb{R}$ and $T : \mathbb{R} \to \mathbb{R}^k$.[1] Consider for a *canonical parameter* $\boldsymbol{\theta} \in \mathbb{R}^k$ the Laplace

[1] We could also use boldface notation for T because $T(y) \in \mathbb{R}^k$ is vector-valued, but we prefer to not use boldface notation for (vector-valued) functions.

M. V. Wüthrich, M. Merz, *Statistical Foundations of Actuarial Learning and its Applications*, Springer Actuarial, https://doi.org/10.1007/978-3-031-12409-9_2

transform

$$\mathfrak{L}(\boldsymbol{\theta}) = \int_{\mathbb{R}} \exp\left\{\boldsymbol{\theta}^\top T(y) + a(y)\right\} d\nu(y).$$

Assume that this Laplace transform is not identically equal to $+\infty$. The *effective domain* is defined by

$$\boldsymbol{\Theta} = \left\{\boldsymbol{\theta} \in \mathbb{R}^k;\ \mathfrak{L}(\boldsymbol{\theta}) < \infty\right\} \subseteq \mathbb{R}^k. \tag{2.1}$$

Lemma 2.1 *The effective domain $\boldsymbol{\Theta} \subseteq \mathbb{R}^k$ is a convex set.*

The effective domain $\boldsymbol{\Theta}$ is not necessarily an open set, but in many applications it is open. Counterexamples are given in Problem 4.1 of Chapter 1 in Lehmann [244], and in the inverse Gaussian example in Sect. 2.1.3, below.

Proof of Lemma 2.1 Choose $\boldsymbol{\theta}_i \in \mathbb{R}^k$, $i = 1, 2$, with $\mathfrak{L}(\boldsymbol{\theta}_i) < \infty$. Set $\boldsymbol{\theta} = c\boldsymbol{\theta}_1 + (1-c)\boldsymbol{\theta}_2$ for $c \in (0, 1)$. We use Hölder's inequality, applied to the norms $p = 1/c$ and $q = 1/(1-c)$,

$$\begin{aligned}
\mathfrak{L}(\boldsymbol{\theta}) &= \int_{\mathbb{R}} \exp\left\{(c\boldsymbol{\theta}_1 + (1-c)\boldsymbol{\theta}_2)^\top T(y) + a(y)\right\} d\nu(y) \\
&= \int_{\mathbb{R}} \exp\left\{\boldsymbol{\theta}_1^\top T(y) + a(y)\right\}^c \exp\left\{\boldsymbol{\theta}_2^\top T(y) + a(y)\right\}^{1-c} d\nu(y) \\
&\le \mathfrak{L}(\boldsymbol{\theta}_1)^c \mathfrak{L}(\boldsymbol{\theta}_2)^{1-c} < \infty.
\end{aligned}$$

This implies $\boldsymbol{\theta} \in \boldsymbol{\Theta}$ and proves the claim. □

We define *the cumulant function* on the effective domain $\boldsymbol{\Theta}$

$$\kappa : \boldsymbol{\Theta} \to \mathbb{R}, \qquad \boldsymbol{\theta} \mapsto \kappa(\boldsymbol{\theta}) = \log \mathfrak{L}(\boldsymbol{\theta}).$$

Definition 2.2 The EF with σ-finite measure ν on $\mathbb{R}$ and cumulant function $\kappa : \boldsymbol{\Theta} \to \mathbb{R}$ is given by the distribution functions F on $\mathbb{R}$ with

$$dF(y; \boldsymbol{\theta}) = f(y; \boldsymbol{\theta}) d\nu(y) = \exp\left\{\boldsymbol{\theta}^\top T(y) - \kappa(\boldsymbol{\theta}) + a(y)\right\} d\nu(y), \tag{2.2}$$

for canonical parameters $\boldsymbol{\theta} \in \boldsymbol{\Theta} \subseteq \mathbb{R}^k$.

Remarks 2.3

- The definition of the EF (2.2) assumes that the effective domain $\boldsymbol{\Theta} \subseteq \mathbb{R}^k$ has been constructed from the choices $a : \mathbb{R} \to \mathbb{R}$ and $T : \mathbb{R} \to \mathbb{R}^k$ as described in (2.1). This is not explicitly stated in the surrounding text of (2.2).
- The support of any random variable $Y \sim F(\cdot; \boldsymbol{\theta})$ of this EF does *not* depend on the explicit choice of the canonical parameter $\boldsymbol{\theta} \in \boldsymbol{\Theta}$, but solely on the choice of the σ-finite measure ν on $\mathbb{R}$, and the distribution functions $F(\cdot; \boldsymbol{\theta})$ are mutually absolutely continuous (equivalent) w.r.t. ν.
- In statistics, the main object of interest is the canonical parameter $\boldsymbol{\theta}$. Importantly for parameter estimation, the function $a(\cdot)$ does not involve the canonical parameter. Therefore, it is irrelevant for parameter estimation and (only) serves as a normalization so that F in (2.2) is a proper distribution function. In fact, this is the way how the EF is often introduced in the statistical and actuarial literature, but in this latter introduction we lose the deeper interpretation of the cumulant function κ, nor is it immediately clear what properties it possesses.
- The case $k \geq 2$ gives a *vector-valued* canonical parameter $\boldsymbol{\theta}$. The case $k = 1$ gives a *single-parameter* EF, and, if additionally $T(y) = y$, it is called a *single-parameter linear EF*.

Theorem 2.4 *Assume the effective domain* $\boldsymbol{\Theta}$ *has a non-empty interior* $\mathring{\boldsymbol{\Theta}}$. *Choose* $Y \sim F(\cdot; \boldsymbol{\theta})$ *for fixed* $\boldsymbol{\theta} \in \mathring{\boldsymbol{\Theta}}$. *The moment generating function of* $T(Y)$ *for sufficiently small* $\boldsymbol{r} \in \mathbb{R}^k$ *is given by*

$$M_{T(Y)}(\boldsymbol{r}) = \mathbb{E}_{\boldsymbol{\theta}}\left[\exp\left\{\boldsymbol{r}^\top T(Y)\right\}\right] = \exp\left\{\kappa(\boldsymbol{\theta}+\boldsymbol{r}) - \kappa(\boldsymbol{\theta})\right\},$$

where the expectation operator $\mathbb{E}_{\boldsymbol{\theta}}$ *illustrates the selected canonical parameter* $\boldsymbol{\theta}$ *for* Y.

Proof Choose $\boldsymbol{\theta} \in \mathring{\boldsymbol{\Theta}}$ and $\boldsymbol{r} \in \mathbb{R}^k$ so small that $\boldsymbol{\theta} + \boldsymbol{r} \in \mathring{\boldsymbol{\Theta}}$. We receive

$$\begin{aligned} M_{T(Y)}(\boldsymbol{r}) &= \int_{\mathbb{R}} \exp\left\{(\boldsymbol{\theta}+\boldsymbol{r})^\top T(y) - \kappa(\boldsymbol{\theta}) + a(y)\right\} d\nu(y) \\ &= \exp\left\{\kappa(\boldsymbol{\theta}+\boldsymbol{r}) - \kappa(\boldsymbol{\theta})\right\} \int_{\mathbb{R}} \exp\left\{(\boldsymbol{\theta}+\boldsymbol{r})^\top T(y) - \kappa(\boldsymbol{\theta}+\boldsymbol{r}) + a(y)\right\} d\nu(y) \\ &= \exp\left\{\kappa(\boldsymbol{\theta}+\boldsymbol{r}) - \kappa(\boldsymbol{\theta})\right\}, \end{aligned}$$

where the last identity follows from the fact that the support of the EF does not depend on the explicit choice of the canonical parameter. □

Theorem 2.4 has a couple of immediate implications. First, in any interior point $\boldsymbol{\theta} \in \mathring{\boldsymbol{\Theta}}$ both the moment generating function $\boldsymbol{r} \mapsto M_{T(Y)}(\boldsymbol{r})$ (in the neighborhood of the origin) and the cumulant function $\boldsymbol{\theta} \mapsto \kappa(\boldsymbol{\theta})$ have derivatives of all orders, and, similarly to Sect. 1.2, moments of all orders of $T(Y)$ exist, see also (1.1). Existence of moments of all orders implies that the distribution function of $T(Y)$ cannot have a regularly varying tails.

Corollary 2.5 *Assume $\mathring{\boldsymbol{\Theta}}$ is non-empty. The cumulant function $\boldsymbol{\theta} \mapsto \kappa(\boldsymbol{\theta})$ is convex, and for $Y \sim F(\cdot; \boldsymbol{\theta})$ with $\boldsymbol{\theta} \in \mathring{\boldsymbol{\Theta}}$*

$$\mu = \mathbb{E}_{\boldsymbol{\theta}}[T(Y)] = \nabla_{\boldsymbol{\theta}} \kappa(\boldsymbol{\theta}) \qquad \textit{and} \qquad Var_{\boldsymbol{\theta}}(T(Y)) = \nabla^2_{\boldsymbol{\theta}} \kappa(\boldsymbol{\theta}),$$

where $\nabla_{\boldsymbol{\theta}}$ is the gradient and $\nabla^2_{\boldsymbol{\theta}}$ the Hessian w.r.t. vector $\boldsymbol{\theta}$.

Similarly to $T : \mathbb{R} \to \mathbb{R}^k$, we will not use boldface notation for the (multi-dimensional) mean because later on we will understand the mean $\mu = \mu(\boldsymbol{\theta}) \in \mathbb{R}^k$ as a function of the canonical parameter $\boldsymbol{\theta}$; see Footnote 1 on page 13 on boldface notation.

Proof Existence of the moment generating function for all sufficiently small $\boldsymbol{r} \in \mathbb{R}^k$ (around the origin) implies that we have first and second moments. For the first moment we receive

$$\mu = \mathbb{E}_{\boldsymbol{\theta}}[T(Y)] = \nabla_{\boldsymbol{r}} M_{T(Y)}(\boldsymbol{r})\big|_{\boldsymbol{r}=0} = \exp\{\kappa(\boldsymbol{\theta}+\boldsymbol{r}) - \kappa(\boldsymbol{\theta})\} \nabla_{\boldsymbol{r}} \kappa(\boldsymbol{\theta}+\boldsymbol{r})|_{\boldsymbol{r}=0} = \nabla_{\boldsymbol{\theta}} \kappa(\boldsymbol{\theta}).$$

Denote component j of $T(Y) \in \mathbb{R}^k$ by $T_j(Y)$. We have for $1 \le j, l \le k$

$$\begin{aligned}
\mathbb{E}_{\boldsymbol{\theta}}\left[T_j(Y)T_l(Y)\right] &= \frac{\partial^2}{\partial r_j \partial r_l} M_{T(Y)}(\boldsymbol{r})\bigg|_{\boldsymbol{r}=0} \\
&= \exp\{\kappa(\boldsymbol{\theta}+\boldsymbol{r}) - \kappa(\boldsymbol{\theta})\} \left(\frac{\partial^2}{\partial r_j \partial r_l}\kappa(\boldsymbol{\theta}+\boldsymbol{r}) + \frac{\partial}{\partial r_j}\kappa(\boldsymbol{\theta}+\boldsymbol{r})\frac{\partial}{\partial r_l}\kappa(\boldsymbol{\theta}+\boldsymbol{r})\right)\bigg|_{\boldsymbol{r}=0} \\
&= \left(\frac{\partial^2}{\partial \theta_j \partial \theta_l}\kappa(\boldsymbol{\theta}) + \frac{\partial}{\partial \theta_j}\kappa(\boldsymbol{\theta})\frac{\partial}{\partial \theta_l}\kappa(\boldsymbol{\theta})\right).
\end{aligned}$$

This implies for the covariance

$$\mathrm{Cov}_{\boldsymbol{\theta}}(T_j(Y), T_l(Y)) = \frac{\partial^2}{\partial \theta_j \partial \theta_l}\kappa(\boldsymbol{\theta}).$$

The convexity of κ follows because $\nabla^2_{\boldsymbol{\theta}}\kappa(\boldsymbol{\theta})$ is the positive semi-definite covariance matrix of $T(Y)$, for all $\boldsymbol{\theta} \in \mathring{\boldsymbol{\Theta}}$. This finishes the proof. □

Assumption 2.6 (Minimal Representation) *We assume that the interior $\mathring{\boldsymbol{\Theta}}$ of the effective domain $\boldsymbol{\Theta}$ is non-empty and that the cumulant function κ is strictly convex on this interior $\mathring{\boldsymbol{\Theta}}$.*

Remarks 2.7

- Throughout these notes we will work under Assumption 2.6 without making explicit reference. This assumption strengthens the properties of the cumulant function κ from being convex, see Corollary 2.5, to being strictly convex. This strengthening implies that the mean function $\boldsymbol{\theta} \mapsto \mu = \mu(\boldsymbol{\theta}) = \nabla_{\boldsymbol{\theta}}\kappa(\boldsymbol{\theta})$ can be inverted; this is needed for the canonical link, see Definition 2.8, below.
- The strict convexity of κ means that the covariance matrix $\nabla^2_{\boldsymbol{\theta}}\kappa(\boldsymbol{\theta})$ of $T(Y)$ is positive definite and has full rank k for all $\boldsymbol{\theta} \in \mathring{\boldsymbol{\Theta}}$, see Corollary 2.5. This property is important, otherwise we do not have identifiability in the canonical parameter $\boldsymbol{\theta}$ because we have a linear dependence between the components of $T(Y)$.
- Mathematically, this strict convexity is not a restriction because it can be obtained by working under a so-called *minimal representation*. If the covariance matrix $\nabla^2_{\boldsymbol{\theta}}\kappa(\boldsymbol{\theta})$ does not have full rank k, the choice k is "non-optimal" because the problem lives in a smaller dimension. Thus, w.l.o.g., we may and will assume to work in this smaller dimension, called minimal representation; for a rigorous derivation of a minimal representation we refer to Section 8.1 in Barndorff-Nielsen [23].

Definition 2.8 The canonical link is defined by $h = (\nabla_{\boldsymbol{\theta}}\kappa)^{-1}$.

The application of the canonical link h to the mean implies under Assumption 2.6

$$h(\mu) = h\left(\mathbb{E}_{\boldsymbol{\theta}}\left[T(Y)\right]\right) = \boldsymbol{\theta},$$

for mean $\mu = \mathbb{E}_{\boldsymbol{\theta}}[T(Y)]$ of $Y \sim F(\cdot; \boldsymbol{\theta})$ with $\boldsymbol{\theta} \in \mathring{\boldsymbol{\Theta}}$.

Remarks 2.9 (Dual Parameter Space) Assumption 2.6 provides that the canonical link h is well-defined, and we can either work with the canonical parameter representation $\boldsymbol{\theta} \in \mathring{\boldsymbol{\Theta}} \subseteq \mathbb{R}^k$ or with its dual (mean) parameter representation $\mu = \mathbb{E}_{\boldsymbol{\theta}}\left[T(Y)\right] \in \mathcal{M}$ with

$$\mathcal{M} \stackrel{\text{def.}}{=} \nabla_{\boldsymbol{\theta}}\kappa(\mathring{\boldsymbol{\Theta}}) = \{\nabla_{\boldsymbol{\theta}}\kappa(\boldsymbol{\theta});\ \boldsymbol{\theta} \in \mathring{\boldsymbol{\Theta}}\} \subseteq \mathbb{R}^k. \tag{2.3}$$

Strict convexity of κ implies that there is a one-to-one correspondence between these two parametrizations. $\boldsymbol{\Theta}$ is called the *effective domain* and $\mathcal{M}$ is called the *dual parameter space* or the *mean parameter space*.

In Sect. 2.2.4, below, we introduce one more property called *steepness* that the cumulant function κ should satisfy. This additional property gives a relationship between the support $\mathfrak{T}$ of the random variables $T(Y)$ of the given EF and the boundary of the dual parameter space $\mathcal{M}$. This steepness property is important for parameter estimation.

2.1.2 Single-Parameter Linear EF: Count Variable Examples

We start by giving single-parameter discrete linear EF examples based on counting measures on $\mathbb{N}_0$. Since we work in one dimension $k = 1$, we replace boldface $\boldsymbol{\theta}$ by scalar $\theta \in \boldsymbol{\Theta} \subseteq \mathbb{R}$ in this section.

Bernoulli Distribution as a Single-Parameter Linear EF

For the Bernoulli distribution with parameter $p \in (0, 1)$ we choose as ν the counting measure on $\{0, 1\}$. We make the following choices: $T(y) = y$,

$$a(y) = 0, \quad \kappa(\theta) = \log(1 + e^\theta), \quad p = \kappa'(\theta) = \frac{e^\theta}{1 + e^\theta}, \quad \theta = h(p) = \log\left(\frac{p}{1 - p}\right),$$

for effective domain $\boldsymbol{\Theta} = \mathbb{R}$, dual parameter space $\mathcal{M} = (0, 1)$ and support $\mathfrak{T} = \{0, 1\}$ of $Y = T(Y)$. With these choices we have

$$dF(y; \theta) = \exp\left\{\theta y - \log(1 + e^\theta)\right\} d\nu(y) = \left(\frac{e^\theta}{1 + e^\theta}\right)^y \left(\frac{1}{1 + e^\theta}\right)^{1-y} d\nu(y).$$

$\theta \mapsto \kappa'(\theta)$ is the logistic or sigmoid function, and the canonical link $p \mapsto h(p)$ is the logit function. Mean and variance are given by

$$\mu = \mathbb{E}_\theta [Y] = \kappa'(\theta) = p \quad \text{and} \quad \mathrm{Var}_\theta (Y) = \kappa''(\theta) = \frac{e^\theta}{(1 + e^\theta)^2} = p(1 - p),$$

and the probability weights satisfy for $y \in \mathfrak{T} = \{0, 1\}$

$$\mathbb{P}_\theta[Y = y] = p^y (1 - p)^{1-y}.$$

Binomial Distribution as a Single-Parameter Linear EF

For the binomial distribution with parameters $n \in \mathbb{N}$ and $p \in (0, 1)$ we choose as ν the counting measure on $\{0, \ldots, n\}$. We make the following choices: $T(y) = y$,

$$a(y) = \log\binom{n}{y}, \quad \kappa(\theta) = n\log(1+e^\theta), \quad \mu = \kappa'(\theta) = \frac{ne^\theta}{1 + e^\theta}, \quad \theta = h(\mu) = \log\left(\frac{\mu}{n - \mu}\right),$$

for effective domain $\boldsymbol{\Theta} = \mathbb{R}$, dual parameter space $\mathcal{M} = (0, n)$ and support $\mathfrak{T} = \{0, \ldots, n\}$ of $Y = T(Y)$. With these choices we have

$$dF(y; \theta) = \binom{n}{y} \exp\left\{\theta y - n\log(1 + e^\theta)\right\} d\nu(y) = \binom{n}{y} \left(\frac{e^\theta}{1 + e^\theta}\right)^y \left(\frac{1}{1 + e^\theta}\right)^{n-y} d\nu(y).$$

Mean and variance are given by

$$\mu = \mathbb{E}_\theta[Y] = \kappa'(\theta) = np \qquad \text{and} \qquad \mathrm{Var}_\theta(Y) = \kappa''(\theta) = n\frac{e^\theta}{(1+e^\theta)^2} = np(1-p),$$

where we set $p = e^\theta/(1+e^\theta)$. The probability weights satisfy for $y \in \mathfrak{T} = \{0,\ldots,n\}$

$$\mathbb{P}_\theta[Y=y] = \binom{n}{y} p^y(1-p)^{n-y}.$$

Poisson Distribution as a Single-Parameter Linear EF

For the Poisson distribution with parameter $\lambda > 0$ we choose as ν the counting measure on $\mathbb{N}_0$. We make the following choices: $T(y) = y$,

$$a(y) = \log\left(\frac{1}{y!}\right), \quad \kappa(\theta) = e^\theta, \quad \mu = \kappa'(\theta) = e^\theta, \quad \theta = h(\mu) = \log(\mu),$$

for effective domain $\boldsymbol{\Theta} = \mathbb{R}$, dual parameter space $\mathcal{M} = (0,\infty)$ and support $\mathfrak{T} = \mathbb{N}_0$ of $Y = T(Y)$. With these choices we have

$$dF(y;\theta) = \frac{1}{y!}\exp\left\{\theta y - e^\theta\right\} d\nu(y) = e^{-\mu}\frac{\mu^y}{y!} d\nu(y). \tag{2.4}$$

The canonical link $\mu \mapsto h(\mu)$ is the log-link. Mean and variance are given by

$$\mu = \mathbb{E}_\theta[Y] = \kappa'(\theta) = \lambda \qquad \text{and} \qquad \mathrm{Var}_\theta(Y) = \kappa''(\theta) = \lambda = \mu = \mathbb{E}_\theta[Y],$$

where we set $\lambda = e^\theta$. The probability weights in the Poisson case satisfy for $y \in \mathfrak{T} = \mathbb{N}_0$

$$\mathbb{P}_\theta[Y=y] = e^{-\lambda}\frac{\lambda^y}{y!}.$$

Negative-Binomial (Pólya) Distribution as a Single-Parameter Linear EF

For the negative-binomial distribution with $\alpha > 0$ and $p \in (0,1)$ we choose as ν the counting measure on $\mathbb{N}_0$; α plays the role of a nuisance parameter or hyper-parameter. We make the following choices: $T(y) = y$,

$$a(y) = \log\binom{y+\alpha-1}{y}, \quad \kappa(\theta) = -\alpha\log(1-e^\theta),$$

$$\mu = \kappa'(\theta) = \alpha \frac{e^\theta}{1 - e^\theta}, \quad \theta = h(\mu) = \log\left(\frac{\mu}{\mu + \alpha}\right),$$

for effective domain $\boldsymbol{\Theta} = (-\infty, 0)$, dual parameter space $\mathcal{M} = (0, \infty)$ and support $\mathfrak{T} = \mathbb{N}_0$ of $Y = T(Y)$. With these choices we have

$$\begin{aligned} dF(y;\theta) &= \binom{y+\alpha-1}{y} \exp\left\{\theta y + \alpha \log(1 - e^\theta)\right\} d\nu(y) \\ &= \binom{y+\alpha-1}{y} p^y (1-p)^\alpha \, d\nu(y), \end{aligned}$$

with $p = e^\theta$. Parameter $\alpha > 0$ is treated as nuisance parameter, otherwise we drop out of the EF framework. We have first the two moments

$$\mu = \mathbb{E}_\theta[Y] = \alpha \frac{e^\theta}{1 - e^\theta} = \alpha \frac{p}{1-p} \quad \text{and} \quad \mathrm{Var}_\theta(Y) = \mathbb{E}_\theta[Y]\left(1 + \frac{e^\theta}{1-e^\theta}\right) > \mathbb{E}_\theta[Y].$$

This model allows us to model over-dispersion, in contrast to the Poisson model. In fact, the negative-binomial model is a mixed Poisson model with a gamma mixing distribution, for details see Sect. 5.3.5, below. Typically, one uses a different parametrization. Set $e^\theta = \lambda/(\alpha + \lambda)$, for $\lambda > 0$. This implies

$$\mu = \mathbb{E}_\theta[Y] = \lambda \qquad \text{and} \qquad \mathrm{Var}_\theta(Y) = \lambda\left(1 + \frac{\lambda}{\alpha}\right) > \lambda.$$

For $\alpha \in \mathbb{N}$ this model can also be interpreted as the waiting time until we observe α successful trials among i.i.d. trials, for instance, for $\alpha = 1$ we have the geometric distribution (with a small reparametrization).

The probability weights of the negative-binomial model satisfy for $y \in \mathfrak{T} = \mathbb{N}_0$

$$\mathbb{P}_\theta[Y = y] = \binom{y+\alpha-1}{y} p^y (1-p)^\alpha. \tag{2.5}$$

2.1.3 Vector-Valued Parameter EF: Absolutely Continuous Examples

We give vector-valued parameter absolutely continuous EF examples with $k = 2$, and being based on the Lebesgue measure on (subsets of) $\mathbb{R}$, in this section.

Gaussian Distribution as a Vector-Valued Parameter EF

For the Gaussian distribution with parameters $\mu \in \mathbb{R}$ and $\sigma^2 > 0$ we choose as ν the Lebesgue measure on $\mathbb{R}$, and we make the following choices: $T(y) = (y, y^2)^\top$,

$$a(y) = -\frac{1}{2}\log(2\pi), \qquad \kappa(\boldsymbol{\theta}) = -\frac{\theta_1^2}{4\theta_2} - \frac{1}{2}\log(-2\theta_2),$$

$$(\mu, \sigma^2 + \mu^2)^\top = \nabla_{\boldsymbol{\theta}}\kappa(\boldsymbol{\theta}) = \left(\frac{\theta_1}{-2\theta_2}, (-2\theta_2)^{-1} + \frac{\theta_1^2}{4\theta_2^2}\right)^\top,$$

for effective domain $\boldsymbol{\Theta} = \mathbb{R} \times (-\infty, 0)$, dual parameter space $\mathcal{M} = \mathbb{R} \times (0, \infty)$ and support $\mathfrak{T} = \mathbb{R} \times [0, \infty)$ of $T(Y) = (Y, Y^2)^\top$. With these choices we have

$$\begin{aligned} dF(y; \boldsymbol{\theta}) &= \frac{1}{\sqrt{2\pi}} \exp\left\{\boldsymbol{\theta}^\top T(y) + \frac{\theta_1^2}{4\theta_2} + \frac{1}{2}\log(-2\theta_2)\right\} d\nu(y) \\ &= \frac{1}{\sqrt{2\pi}(-2\theta_2)^{-1/2}} \exp\left\{-\frac{1}{2}\frac{1}{(-2\theta_2)^{-1}}\left(y - \frac{\theta_1}{-2\theta_2}\right)^2\right\} d\nu(y). \end{aligned}$$

This is the Gaussian model with mean $\mu = \theta_1/(-2\theta_2)$ and variance $\sigma^2 = (-2\theta_2)^{-1}$.

If we treat $\sigma > 0$ as a nuisance parameter, we obtain the Gaussian model as a single-parameter EF. This is the most common example of an EF. Set $T(y) = y/\sigma$ and

$$a(y) = -\frac{1}{2}\log(2\pi\sigma^2) - y^2/(2\sigma^2), \quad \kappa(\theta) = \theta^2/2, \quad \mu = \kappa'(\theta) = \theta, \quad \theta = h(\mu) = \mu,$$

for effective domain $\boldsymbol{\Theta} = \mathbb{R}$, dual parameter space $\mathcal{M} = \mathbb{R}$ and support $\mathfrak{T} = \mathbb{R}$ of $T(Y) = Y/\sigma$. With these choices we have

$$\begin{aligned} dF(y; \theta) &= \frac{1}{\sqrt{2\pi}\sigma} \exp\left\{\theta y/\sigma - y^2/(2\sigma^2) - \theta^2/2\right\} d\nu(y) \\ &= \frac{1}{\sqrt{2\pi}\sigma} \exp\left\{-\frac{1}{2\sigma^2}(y - \sigma\theta)^2\right\} d\nu(y), \end{aligned}$$

and, in particular, the canonical link is the *identity link* $\mu \mapsto \theta = h(\mu) = \mu$ in this single-parameter EF example.

Gamma Distribution as a Vector-Valued Parameter EF

For the gamma distribution with parameters $\alpha, \beta > 0$ we choose as ν the Lebesgue measure on $\mathbb{R}_+$. Then we make the following choices: $T(y) = (y, \log y)^\top$,

$$a(y) = -\log y, \qquad \kappa(\boldsymbol{\theta}) = \log\Gamma(\theta_2) - \theta_2 \log(-\theta_1),$$

$$\left(\alpha/\beta, \frac{\Gamma'(\alpha)}{\Gamma(\alpha)} - \log(\beta)\right)^\top = \nabla_{\boldsymbol{\theta}}\kappa(\boldsymbol{\theta}) = \left(\frac{\theta_2}{-\theta_1}, \frac{\Gamma'(\theta_2)}{\Gamma(\theta_2)} - \log(-\theta_1)\right)^\top,$$

for effective domain $\boldsymbol{\Theta} = (-\infty, 0) \times (0, \infty)$, and setting $\beta = -\theta_1 > 0$ and $\alpha = \theta_2 > 0$. The dual parameter space is $\mathcal{M} = (0, \infty) \times \mathbb{R}$, and we have support $\mathfrak{T} = (0, \infty) \times \mathbb{R}$ of $T(Y) = (Y, \log Y)^\top$. With these choices we obtain

$$\begin{aligned} dF(y; \boldsymbol{\theta}) &= \exp\left\{\boldsymbol{\theta}^\top T(y) - \log\Gamma(\theta_2) + \theta_2\log(-\theta_1) - \log y\right\} d\nu(y) \\ &= \frac{(-\theta_1)^{\theta_2}}{\Gamma(\theta_2)} y^{\theta_2 - 1} \exp\{-(-\theta_1)y\}\, d\nu(y) \\ &= \frac{\beta^\alpha}{\Gamma(\alpha)} y^{\alpha-1} \exp\{-\beta y\}\, d\nu(y). \end{aligned}$$

This is a vector-valued parameter EF with $k = 2$, and the first moment is given by

$$\mathbb{E}_{\boldsymbol{\theta}}\left[(Y, \log Y)^\top\right] = \nabla_{\boldsymbol{\theta}}\kappa(\boldsymbol{\theta}) = \left(\alpha/\beta, \frac{\Gamma'(\alpha)}{\Gamma(\alpha)} - \log(\beta)\right)^\top.$$

Parameter α is called *shape parameter* and parameter β is called *scale parameter*.[2]

If we treat the shape parameter $\alpha > 0$ as a nuisance parameter we can turn the gamma distribution into a single-parameter linear EF. Set $T(y) = y$ and

$$a(y) = (\alpha - 1)\log y - \log\Gamma(\alpha),\ \kappa(\theta) = -\alpha\log(-\theta),\ \mu = \kappa'(\theta) = \frac{\alpha}{-\theta},\ \theta = h(\mu) = -\frac{\alpha}{\mu},$$

for effective domain $\Theta = (-\infty, 0)$, dual parameter space $\mathcal{M} = (0, \infty)$ and support $\mathfrak{T} = (0, \infty)$. With these choices we have for $\beta = -\theta > 0$

$$dF(y; \theta) = \frac{(-\theta)^\alpha}{\Gamma(\alpha)} y^{\alpha-1} \exp\{-(-\theta)y\}\, d\nu(y). \tag{2.6}$$

This provides us with mean and variance

$$\mu = \mathbb{E}_\theta[Y] = \frac{\alpha}{\beta} \qquad \text{and} \qquad \sigma^2 = \mathrm{Var}_\theta(Y) = \frac{\alpha}{\beta^2} = \frac{1}{\alpha}\mu^2.$$

---

[2] The function $\Psi(x) = \frac{d}{dx}\log\Gamma(x) = \Gamma'(x)/\Gamma(x)$ is called digamma function.

For parameter estimation one often needs to invert these identities which gives us

$$\alpha = \frac{\mu^2}{\sigma^2} \qquad \text{and} \qquad \beta = \frac{\mu}{\sigma^2}.$$

Remarks 2.10

- The gamma distribution contains as special cases the exponential distribution for $\alpha = \theta_2 = 1$ and $\beta = -\theta_1 > 0$, and the χ_r^2-distribution with r degrees of freedom for $\alpha = \theta_2 = r/2$ and $\beta = -\theta_1 = 1/2$.
- The distributions of the EF are all light-tailed in the sense that all moments of $T(Y)$ exist. Therefore, the EF does not allow for regularly varying survival functions, see (1.3). If Y is gamma distributed, then $Z = \exp\{Y\}$ is log-gamma distributed (with the special case of the Pareto distribution for the exponential case $\alpha = \theta_2 = 1$). For an example we refer to Sect. 2.2.5. However, this log-transformation is not always recommended because it may provide accurate models on the transformed log-scale, but back-transformation to the original scale may not necessarily provide a good predictive model on that original scale.
- The gamma density (2.6) may be a bit tricky in applications because the effective domain $\boldsymbol{\Theta} = (-\infty, 0)$ is one-sided bounded (we come back to this below). For this reason, in practice, one often uses links different from the canonical link $h(\mu) = -\alpha/\mu$. For instance, a parametrization $\theta = -\exp\{-\vartheta\}$ for $\vartheta \in \mathbb{R}$, see Ohlsson–Johansson [290], leads to the following model

$$dF(y; \vartheta) = \frac{y^{\alpha-1}}{\Gamma(\alpha)} \exp\left\{-e^{-\vartheta} y - \alpha\vartheta\right\} d\nu(y). \tag{2.7}$$

  We will study the gamma model in more depth below, and parametrization (2.7) will correspond to the log-link choice, see Example 5.5, below.

Figure 2.1 gives examples of gamma densities for shape parameters $\alpha \in \{1/2, 1, 3/2, 2\}$ and scale parameters $\beta \in \{1/2, 1, 3/2, 2\}$ with $\alpha = \beta$ all providing the same mean $\mu = \mathbb{E}_\theta[Y] = \alpha/\beta = 1$. The crucial observation is that these gamma densities can have two different shapes, for $\alpha \leq 1$ we have a strictly decreasing shape and for $\alpha > 1$ we have a unimodal density with mode in $(\alpha - 1)/\beta$.

Inverse Gaussian Distribution as a Vector-Valued Parameter EF

For the inverse Gaussian distribution with parameters $\alpha, \beta > 0$ we choose as ν the Lebesgue measure on $\mathbb{R}_+$. Then we make the following choices: $T(y) = (y, 1/y)^\top$,

$$a(y) = -\frac{1}{2}\log(2\pi y^3), \quad \kappa(\boldsymbol{\theta}) \; = \; -2(\theta_1\theta_2)^{1/2} - \frac{1}{2}\log(-2\theta_2),$$

$$\left(\alpha/\beta, \beta/\alpha + 1/\alpha^2\right)^\top = \nabla_{\boldsymbol{\theta}}\kappa(\boldsymbol{\theta}) \; = \; \left(\left(\frac{-2\theta_2}{-2\theta_1}\right)^{1/2}, \left(\frac{-2\theta_1}{-2\theta_2}\right)^{1/2} + \frac{1}{-2\theta_2}\right)^\top,$$

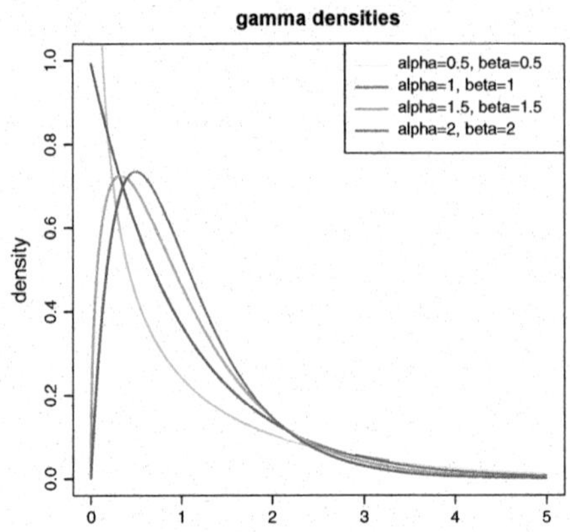

Fig. 2.1 Gamma densities for shape parameters $\alpha \in \{1/2, 1, 3/2, 2\}$ and scale parameters $\beta \in \{1/2, 1, 3/2, 2\}$ all providing the same mean $\mu = \alpha/\beta = 1$

for $\boldsymbol{\theta} = (\theta_1, \theta_2)^\top \in (-\infty, 0)^2$, and setting $\beta = (-2\theta_1)^{1/2}$ and $\alpha = (-2\theta_2)^{1/2}$. The dual parameter space is $\mathcal{M} = (0, \infty)^2$, and we have support $\mathfrak{T} = (0, \infty)^2$ of $T(Y) = (Y, 1/Y)^\top$. With these choices we obtain

$$\begin{aligned} dF(y;\boldsymbol{\theta}) &= \exp\left\{\boldsymbol{\theta}^\top T(y) + 2(\theta_1\theta_2)^{1/2} + \frac{1}{2}\log(-2\theta_2) - \frac{1}{2}\log(2\pi y^3)\right\} d\nu(y) \\ &= \frac{1}{(2\pi y^3)^{1/2}}(-2\theta_2)^{1/2}\exp\left\{-\frac{1}{2y}\left((-2\theta_1)y^2 + (-2\theta_2) - 4(\theta_1\theta_2)^{1/2}y\right)\right\} d\nu(y) \\ &= \frac{\alpha}{(2\pi y^3)^{1/2}}\exp\left\{-\frac{\alpha^2}{2y}\left(1 - \frac{\beta}{\alpha}y\right)^2\right\} d\nu(y). \end{aligned} \tag{2.8}$$

This is a vector-valued parameter EF with $k = 2$ and with first moment

$$\mathbb{E}_{\boldsymbol{\theta}}\left[(Y, 1/Y)^\top\right] = \nabla_{\boldsymbol{\theta}}\kappa(\boldsymbol{\theta}) = \left(\alpha/\beta, \beta/\alpha + 1/\alpha^2\right)^\top.$$

For receiving (2.8) we have chosen canonical parameter $\boldsymbol{\theta} = (\theta_1, \theta_2)^\top \in (-\infty, 0)^2$. Interestingly, we can close this parameter space for $\theta_1 = 0$, i.e., the effective domain $\boldsymbol{\Theta}$ is not open in this example. The choice $\theta_1 = 0$ gives us cumulant function $\kappa(\boldsymbol{\theta}) = -\frac{1}{2}\log(-2\theta_2)$ and boundary case

$$\begin{aligned} dF(y;\boldsymbol{\theta}) &= \exp\left\{\boldsymbol{\theta}^\top T(y) + \frac{1}{2}\log(-2\theta_2) - \frac{1}{2}\log(2\pi y^3)\right\} d\nu(y) \\ &= \frac{1}{(2\pi y^3)^{1/2}}(-2\theta_2)^{1/2}\exp\left\{-\frac{-2\theta_2}{2y}\right\} d\nu(y) \\ &= \frac{\alpha}{(2\pi y^3)^{1/2}}\exp\left\{-\frac{\alpha^2}{2y}\right\} d\nu(y). \end{aligned} \tag{2.9}$$

This is the distribution of the first-passage time of level $\alpha > 0$ of a standard Brownian motion, see Bachelier [20]; this distribution is also known as Lévy distribution.

If we treat $\alpha > 0$ as a nuisance parameter, we can turn the inverse Gaussian distribution into a single-parameter linear EF by setting $T(y) = y$,

$$a(y) = \log\left(\frac{\alpha}{(2\pi y^3)^{1/2}}\right) - \frac{\alpha^2}{2y}, \quad \kappa(\theta) = -\alpha(-2\theta)^{1/2},$$

$$\mu = \kappa'(\theta) = \frac{\alpha}{(-2\theta)^{1/2}}, \quad \theta = h(\mu) = -\frac{1}{2}\frac{\alpha^2}{\mu^2},$$

for $\theta \in (-\infty, 0)$, dual parameter space $\mathcal{M} = (0, \infty)$ and support $\mathfrak{T} = (0, \infty)$. With these choices we have the inverse Gaussian model for $\beta = (-2\theta)^{1/2} > 0$

$$\begin{aligned} dF(y;\theta) &= \exp\{a(y)\}\exp\left\{-\frac{1}{2y}\left((-2\theta)y^2 - 2\alpha(-2\theta)^{1/2}y\right)\right\}d\nu(y) \\ &= \frac{\alpha}{(2\pi y^3)^{1/2}}\exp\left\{-\frac{\alpha^2}{2y}\left(1 - \frac{\beta}{\alpha}y\right)^2\right\}d\nu(y). \end{aligned}$$

This provides us with mean and variance

$$\mu = \mathbb{E}_\theta[Y] = \frac{\alpha}{\beta} \qquad \text{and} \qquad \sigma^2 = \mathrm{Var}_\theta(Y) = \frac{\alpha}{\beta^3} = \frac{1}{\alpha^2}\mu^3.$$

For parameter estimation one often needs to invert these identities, which gives us

$$\alpha = \frac{\mu^{3/2}}{\sigma} \qquad \text{and} \qquad \beta = \frac{\mu^{1/2}}{\sigma}.$$

Figure 2.2 gives examples of inverse Gaussian densities for parameter choices $\alpha = \beta \in \{1/2, 1, 3/2, 2\}$ all providing the same mean $\mu = \mathbb{E}_\theta[Y] = \alpha/\beta = 1$.

Generalized Inverse Gaussian Distribution as a Vector-Valued Parameter EF

For the generalized inverse Gaussian distribution with parameters $\alpha, \beta > 0$ and $\gamma \in \mathbb{R}$ we choose as ν the Lebesgue measure on $\mathbb{R}_+$. We combine the terms of the gamma and the inverse Gaussian models to the vector-valued choice: $T(y) = (y, \log y, 1/y)^\top$ with $k = 3$. Moreover, we choose $a(y) = -\log y$ and cumulant function

$$\kappa(\boldsymbol{\theta}) = \log\left(2K_{\theta_2}(2\sqrt{\theta_1\theta_3})\right) - \frac{\theta_2}{2}\log(\theta_1/\theta_3),$$

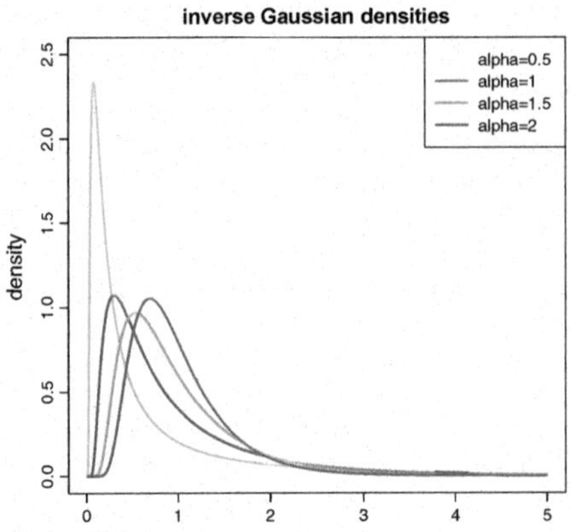

Fig. 2.2 Inverse Gaussian densities for parameters $\alpha = \beta \in \{1/2, 1, 3/2, 2\}$ all providing the same mean $\mu = \alpha/\beta = 1$

for $\boldsymbol{\theta} = (\theta_1, \theta_2, \theta_3)^\top \in (-\infty, 0) \times \mathbb{R} \times (-\infty, 0)$, and where K_{θ_2} denotes the modified Bessel function of the second kind with index $\gamma = \theta_2 \in \mathbb{R}$. With these choices we obtain generalized inverse Gaussian density

$$dF(y;\boldsymbol{\theta}) = \exp\left\{\boldsymbol{\theta}^\top T(y) - \log\left(2K_{\theta_2}(2\sqrt{\theta_1\theta_3})\right) + \frac{\theta_2}{2}\log(\theta_1/\theta_3) - \log y\right\} d\nu(y)$$

$$= \frac{(\alpha/\beta)^{\gamma/2}}{2K_\gamma(\sqrt{\alpha\beta})}\, y^{\gamma-1} \exp\left\{-\frac{1}{2}\left(\alpha y + \beta y^{-1}\right)\right\} d\nu(y), \qquad (2.10)$$

setting $\alpha = -2\theta_1$ and $\beta = -2\theta_3$. This is a vector-valued parameter EF with $k = 3$, and the first moment is given by

$$\mathbb{E}_{\boldsymbol{\theta}}\left[\left(Y, \log Y, \frac{1}{Y}\right)^\top\right] = \nabla_{\boldsymbol{\theta}}\kappa(\boldsymbol{\theta})$$

$$= \left(\frac{K_{\gamma+1}(\sqrt{\alpha\beta})}{K_\gamma(\sqrt{\alpha\beta})}\sqrt{\frac{\beta}{\alpha}},\ \log\sqrt{\frac{\beta}{\alpha}} + \frac{\partial}{\partial\gamma}\log K_\gamma(\sqrt{\alpha\beta}),\ \frac{K_{\gamma+1}(\sqrt{\alpha\beta})}{K_\gamma(\sqrt{\alpha\beta})}\sqrt{\frac{\alpha}{\beta}} - \frac{2\gamma}{\beta}\right)^\top.$$

The effective domain $\boldsymbol{\Theta}$ is a bit complicated because the possible choices of (θ_1, θ_3) depend on $\theta_2 \in \mathbb{R}$, namely, for $\theta_2 < 0$ the negative half-line $(-\infty, 0]$ can be closed at the origin for θ_1, and for $\theta_2 > 0$ it can be closed at the origin for θ_3. The inverse Gaussian model is obtained for $\theta_2 = -1/2$ and the gamma model is obtained for $\theta_3 = 0$. For further properties of the generalized inverse Gaussian distribution we refer to the textbook of Jørgensen [200].

2.1.4 Vector-Valued Parameter EF: Count Variable Example

We close our EF examples by giving a discrete example with a vector-valued parameter.

Categorical Distribution as a Vector-Valued Parameter EF

For the categorical distribution with $k \in \mathbb{N}$ and $\boldsymbol{p} \in (0,1)^k$ such that $\sum_{i=1}^k p_i < 1$, we choose as ν the counting measure on the finite set $\{1, \ldots, k+1\}$. Then we make the following choices: $T(y) = (\mathbb{1}_{\{y=1\}}, \ldots, \mathbb{1}_{\{y=k\}})^\top \in \mathbb{R}^k$, $\boldsymbol{\theta} = (\theta_1, \ldots, \theta_k)^\top$, $e^{\boldsymbol{\theta}} = (e^{\theta_1}, \ldots, e^{\theta_k})^\top$ and

$$a(y) = 0, \qquad \kappa(\boldsymbol{\theta}) = \log\left(1 + \sum_{i=1}^k e^{\theta_i}\right), \qquad \boldsymbol{p} = \nabla_{\boldsymbol{\theta}} \kappa(\boldsymbol{\theta}) = \frac{e^{\boldsymbol{\theta}}}{1 + \sum_{i=1}^k e^{\theta_i}},$$

for effective domain $\boldsymbol{\Theta} = \mathbb{R}^k$, dual parameter space $\mathcal{M} = (0,1)^k$, and the support $\mathfrak{T}$ of $T(Y)$ are the $k+1$ corners of the unit simplex in $\mathbb{R}^k$. This representation is minimal, see Assumption 2.6. With these choices we have (set $\theta_{k+1} = 0$)

$$dF(y; \boldsymbol{\theta}) = \exp\left\{\boldsymbol{\theta}^\top T(y) - \log\left(1 + \sum_{i=1}^k e^{\theta_i}\right)\right\} d\nu(y) = \prod_{j=1}^{k+1} \left(\frac{e^{\theta_j}}{\sum_{i=1}^{k+1} e^{\theta_i}}\right)^{\mathbb{1}_{\{y=j\}}} d\nu(y).$$

This is a vector-valued parameter EF with $k \in \mathbb{N}$. The canonical link is slightly more complicated. Set vectors $\boldsymbol{v} = \exp\{\boldsymbol{\theta}\} \in \mathbb{R}^k$ and $\boldsymbol{w} = (1, \ldots, 1)^\top \in \mathbb{R}^k$. This provides $\boldsymbol{p} = \nabla_{\boldsymbol{\theta}} \kappa(\boldsymbol{\theta}) = \frac{1}{1+\boldsymbol{w}^\top \boldsymbol{v}} \boldsymbol{v} \in \mathbb{R}^k$. Set matrix $A_{\boldsymbol{p}} = \mathbb{1} - \boldsymbol{p}\boldsymbol{w}^\top \in \mathbb{R}^{k \times k}$, the latter gives us $\boldsymbol{p} = A_{\boldsymbol{p}} \boldsymbol{v}$, and since $A_{\boldsymbol{p}}$ has full rank k, we obtain canonical link

$$\boldsymbol{p} \mapsto \boldsymbol{\theta} = h(\boldsymbol{p}) = \log\left(A_{\boldsymbol{p}}^{-1} \boldsymbol{p}\right) = \log\left(\frac{\boldsymbol{p}}{1 - \boldsymbol{w}^\top \boldsymbol{p}}\right).$$

The last identity can be verified by explicit calculation

$$\log\left(\frac{\boldsymbol{p}}{1 - \boldsymbol{w}^\top \boldsymbol{p}}\right) = \log\left(\frac{e^{\boldsymbol{\theta}}/(1 + \sum_{j=1}^k e^{\theta_j})}{1 - \sum_{i=1}^k e^{\theta_i}/(1 + \sum_{j=1}^k e^{\theta_j})}\right) = \log\left(e^{\boldsymbol{\theta}}\right) = \boldsymbol{\theta}.$$

Remarks 2.11

- There are many more examples that belong to the EF. From Theorem 2.4, we know that all examples of the EF are light-tailed in the sense that all moments of $T(Y)$ exist. If we want to model heavy-tailed distributions within the EF, we first need to apply a suitable transformation. We could model the Pareto distribution using transformation $T(y) = \log y$, and assuming that the transformed random variable has an exponential distribution. Different light-tailed examples are obtained by, e.g., using transformation $T(y) = y^\tau$ for the Weibull distribution or $T(y) = (\log y, \log(1-y))^\top$ for the beta distribution. We refrain from giving explicit formulas for these or other examples.

- Observe that in all examples above we have $\mathfrak{T} \subset \overline{\mathcal{M}}$, i.e., the support of $T(Y)$ is contained in the closure of the dual parameter space $\mathcal{M}$, we come back to this observation in Sect. 2.2.4, below.

2.2 Exponential Dispersion Family

In the previous section we have introduced the EF, and we have explicitly studied the vector-valued parameter EF examples of the Gaussian, the gamma and the inverse Gaussian models. We have highlighted that these three vector-valued parameter EFs can be turned into single-parameter EFs by declaring one parameter to be a nuisance parameter that is not modeled (and acts as a hyper-parameter). This changes these three models into single-parameter EFs. These three single-parameter EFs with nuisance parameter can also be interpreted as EDF models. In this section we discuss the single-parameter EDF; this is sufficient for our purposes, and vector-valued parameter extensions can be obtained in a canonical way.

2.2.1 Definition and Properties

The EFs of Sect. 2.1 can be extended to EDFs. In the single-parameter case this is achieved by a transformation $Y = X/\omega$, where $\omega > 0$ is a scaling and where X belongs to a single-parameter linear EF, i.e., with $T(x) = x$. We restrict ourselves to the single-parameter case $k = 1$ throughout this section. Choose a σ-finite measure ν_1 on $\mathbb{R}$ and a measurable function $a_1 : \mathbb{R} \to \mathbb{R}$. These choices give a single-parameter linear EF, directly modeling a real-valued random variable $T(X) = X$. By (2.2) we have distribution for the single-parameter linear EF random variable X

$$dF(x;\theta,1) = f(x;\theta,1)d\nu_1(x) = \exp\Big\{\theta x - \kappa(\theta) + a_1(x)\Big\}d\nu_1(x),$$

on the effective domain

$$\boldsymbol{\Theta} = \left\{\theta \in \mathbb{R};\ \int_{\mathbb{R}} \exp\{\theta x + a_1(x)\}\, d\nu_1(x) < \infty\right\}, \tag{2.11}$$

and with cumulant function

$$\theta \in \boldsymbol{\Theta} \;\mapsto\; \kappa(\theta) = \log\left(\int_{\mathbb{R}} \exp\{\theta x + a_1(x)\}\, d\nu_1(x)\right). \tag{2.12}$$

Throughout, we assume that the effective domain $\boldsymbol{\Theta}$ has a non-empty interior $\mathring{\boldsymbol{\Theta}}$. Thus, since $\boldsymbol{\Theta}$ is convex, we assume that $\mathring{\boldsymbol{\Theta}}$ is a non-empty (possibly infinite) open interval in $\mathbb{R}$.

Following Jørgensen [201, 202], we extend this linear EF to an EDF as follows. Choose a family of σ-finite measures ν_ω on $\mathbb{R}$ and measurable functions $a_\omega : \mathbb{R} \to \mathbb{R}$ for a given index set $\mathcal{W} \ni \omega$ with $\{1\} \subset \mathcal{W} \subset \mathbb{R}_+$. Assume that we have an ω-independent scaled cumulant function κ on this index set $\mathcal{W}$, that is,

$$\theta \in \boldsymbol{\Theta} \;\mapsto\; \kappa(\theta) = \frac{1}{\omega}\left(\log \int_{\mathbb{R}} \exp\{\theta x + a_\omega(x)\}\, d\nu_\omega(x)\right) \qquad \text{for all } \omega \in \mathcal{W},$$

with effective domain $\boldsymbol{\Theta}$ defined by (2.11), i.e., for $\omega = 1$. This allows us to consider the distribution functions

$$\begin{aligned} dF(x;\theta,\omega) = f(x;\theta,\omega)d\nu_\omega(x) &= \exp\Big\{\theta x - \omega\kappa(\theta) + a_\omega(x)\Big\} d\nu_\omega(x) \\ &= \exp\Big\{\omega\left(\theta y - \kappa(\theta)\right) + a_\omega(\omega y)\Big\} d\nu_\omega(\omega y), \end{aligned} \tag{2.13}$$

in the third identity we did a change of variable $x \mapsto y = x/\omega$. By reparametrizing the function $a_\omega(\omega\ \cdot)$ and the σ-finite measures $\nu_\omega(\omega\ \cdot)$ slightly differently, depending on the particular structure of the chosen σ-finite measures, we arrive at the following single-parameter EDF.

Definition 2.12 The (single-parameter) EDF is given by densities of the form

$$Y \sim f(y;\theta, v/\varphi) = \exp\left\{\frac{y\theta - \kappa(\theta)}{\varphi/v} + a(y; v/\varphi)\right\}, \tag{2.14}$$

with

$\kappa : \boldsymbol{\Theta} \to \mathbb{R}$	is the cumulant function (2.12),
$\theta \in \boldsymbol{\Theta}$	is the canonical parameter in the effective domain (2.11),
$v > 0$	is a given weight (exposure, volume),
$\varphi > 0$	is the dispersion parameter,
$a(\cdot;\cdot)$	is the normalization, *not* depending on the canonical parameter θ.

Remarks 2.13

- Exposure $v > 0$ and dispersion parameter $\varphi > 0$ provide the parametrization usually used for $\omega = v/\varphi \in \mathcal{W}$. Their meaning and interpretation will become clear below, and they will always appear as a ratio $\omega = v/\varphi$.
- The support of these EDF distributions does not depend on the explicit choice of the canonical parameter $\theta \in \boldsymbol{\Theta}$, but it may depend on $\omega = v/\varphi \in \mathcal{W}$ through the choices of the σ-finite measures ν_ω, for $\omega \in \mathcal{W}$. Consequently, $a(y;\omega)$ is a normalization such that $f(y;\theta,\omega)$ integrates to 1 w.r.t. the chosen σ-finite measure ν_ω to receive a proper distributional model.
- The transformation $x \mapsto y = x/\omega$ in (2.13) is called duality transformation, see Section 3.1 in Jørgensen [203]. It provides the duality between the *additive form* (in variable x in (2.13)) and the *reproductive form* (in variable y in (2.13)) of the EDF; Definition 2.12 is the reproductive form.
- Lemma 2.1 tells us that $\boldsymbol{\Theta}$ is convex, thus, it is a possibly infinite interval in $\mathbb{R}$. To exclude trivial cases we will always assume that the σ-finite measure ν_1 is not concentrated in one single point (this relates to the minimal representation for $k = 1$ in the linear EF case, see Assumption 2.6), and that the interior $\mathring{\boldsymbol{\Theta}}$ of the effective domain $\boldsymbol{\Theta}$ is non-empty.

Corollary 2.14 *Assume $\mathring{\boldsymbol{\Theta}}$ is non-empty and that ν_1 is not concentrated in one single point. Choose $Y \sim F(\cdot;\theta, v/\varphi)$ for fixed $\theta \in \mathring{\boldsymbol{\Theta}}$. The moment generating function of Y for small $r \in \mathbb{R}$ satisfies*

$$M_Y(r) = \mathbb{E}_\theta\left[\exp\{rY\}\right] = \exp\left\{\frac{v}{\varphi}\left[\kappa(\theta + r\varphi/v) - \kappa(\theta)\right]\right\}.$$

The first two moments of Y are given by

$$\mu = \mathbb{E}_\theta[Y] = \kappa'(\theta) \qquad \text{and} \qquad \mathrm{Var}_\theta(Y) = \frac{\varphi}{v}\kappa''(\theta) > 0.$$

The cumulant function κ is smooth and strictly convex on $\mathring{\boldsymbol{\Theta}}$ with canonical link $h = (\kappa')^{-1}$. The variance function is defined by $\mu \mapsto V(\mu) = (\kappa'' \circ h)(\mu)$ and, consequently, for the variance of Y we have $\mathrm{Var}_\mu(Y) = \frac{\varphi}{v}V(\mu)$ for $\mu \in \mathcal{M}$.

Proof This follows analogously to Theorem 2.4. The linear case $T(y) = y$ with ν_1 not being concentrated in one single point guarantees that the minimal dimension is $k = 1$, providing a minimal representation in this dimension, see Assumption 2.6. □

Before giving explicit examples we state the so-called convolution formula.

Corollary 2.15 (Convolution Formula) *Assume $\mathring{\boldsymbol{\Theta}}$ is non-empty and that ν_1 is not concentrated in one single point. Assume that $Y_i \sim F(\cdot; \theta, v_i/\varphi)$ are independent, for $1 \le i \le n$, with fixed $\theta \in \mathring{\boldsymbol{\Theta}}$. Set $v_+ = \sum_{i=1}^n v_i$. Then*

$$Y_+ = \frac{1}{v_+} \sum_{i=1}^{n} v_i Y_i \sim F(\cdot; \theta, v_+/\varphi).$$

Proof The proof immediately follows from calculating the moment generating function $M_{Y_+}(r)$ and from using the independence between the Y_i's. □

2.2.2 *Exponential Dispersion Family Examples*

The single-parameter linear EF examples introduced above can be reformulated as EDF examples.

Binomial Distribution as a Single-Parameter EDF

For the binomial distribution with parameters $p \in (0, 1)$ and $n \in \mathbb{N}$ we choose the counting measure on $\{0, 1/n, \ldots, 1\}$ with $\omega = n$. Then we make the following choices

$$a(y) = \log\binom{n}{ny}, \quad \kappa(\theta) = \log(1+e^\theta), \quad p = \kappa'(\theta) = \frac{e^\theta}{1+e^\theta}, \quad \theta = h(p) = \log\left(\frac{p}{1-p}\right),$$

for effective domain $\boldsymbol{\Theta} = \mathbb{R}$ and dual parameter space $\mathcal{M} = (0, 1)$. With these choices we have

$$f(y; \theta, n) = \binom{n}{ny} \exp\left\{n\left(\theta y - \log(1+e^\theta)\right)\right\} = \binom{n}{ny} \left(\frac{e^\theta}{1+e^\theta}\right)^{ny} \left(\frac{1}{1+e^\theta}\right)^{n-ny}.$$

This is a single-parameter EDF. The canonical link $p \mapsto h(p)$ gives the logit function. Mean and variance are given by

$$p = \mathbb{E}_\theta[Y] = \kappa'(\theta) = \frac{e^\theta}{1+e^\theta} \quad \text{and} \quad \mathrm{Var}_\theta(Y) = \frac{1}{n}\kappa''(\theta) = \frac{1}{n}\frac{e^\theta}{(1+e^\theta)^2} = \frac{1}{n}p(1-p),$$

and the variance function is given by $V(\mu) = \mu(1 - \mu)$. The binomial random variable is obtained by setting $X = nY \sim \mathrm{Binom}(n, p)$.

Poisson Distribution as a Single-Parameter EDF

For the Poisson distribution with parameters $\lambda > 0$ and $v > 0$ we choose the counting measure on $\mathbb{N}_0/v$ for exposure $\omega = v$. Then we make the following choices

$$a(y) = \log\left(\frac{v^{vy}}{(vy)!}\right), \quad \kappa(\theta) = e^{\theta}, \quad \lambda = \kappa'(\theta) = e^{\theta}, \quad \theta = h(\lambda) = \log(\lambda),$$

for effective domain $\boldsymbol{\Theta} = \mathbb{R}$ and dual parameter space $\mathcal{M} = (0, \infty)$. With these choices we have

$$f(y; \theta, v) = \frac{v^{vy}}{(vy)!} \exp\left\{v\left(\theta y - e^{\theta}\right)\right\} = e^{-v\lambda} \frac{(v\lambda)^{vy}}{(vy)!}. \tag{2.15}$$

This is a single-parameter EDF. The canonical link $\lambda \mapsto h(\lambda)$ is the log-link. Mean and variance are given by

$$\lambda = \mathbb{E}_{\theta}[Y] = \kappa'(\theta) = e^{\theta} \qquad \text{and} \qquad \mathrm{Var}_{\theta}(Y) = \frac{1}{v}\kappa''(\theta) = \frac{1}{v}e^{\theta} = \frac{1}{v}\lambda,$$

and the variance function is given by $V(\lambda) = \lambda$, that is, the variance function is linear in the mean parameter λ. The Poisson random variable is obtained by setting $X = vY \sim \mathrm{Poi}(v\lambda)$. We choose $\varphi = 1$, here, meaning that we have neither under- nor over-dispersion. Thus, the choices v and φ in $\omega = v/\varphi$ have the interpretation of an exposure and a dispersion parameter, respectively. This interpretation is going to be important in claim counts modeling, below.

Gamma Distribution as a Single-Parameter EDF

For the gamma distribution with parameters $\alpha, \beta > 0$ we choose the Lebesgue measure on $\mathbb{R}_+$ and shape parameter $\omega = v/\varphi = \alpha$. We make the following choices

$$a(y) = (\alpha - 1)\log y + \alpha \log \alpha - \log \Gamma(\alpha), \quad \kappa(\theta) = -\log(-\theta),$$
$$\mu = \kappa'(\theta) = -1/\theta, \quad \theta = h(\mu) = -1/\mu,$$

for effective domain $\boldsymbol{\Theta} = (-\infty, 0)$ and dual parameter space $\mathcal{M} = (0, \infty)$. With these choices we have

$$f(y; \theta, \alpha) = \frac{\alpha^{\alpha}}{\Gamma(\alpha)} y^{\alpha-1} \exp\left\{\alpha\left(y\theta + \log(-\theta)\right)\right\} = \frac{(-\theta\alpha)^{\alpha}}{\Gamma(\alpha)} y^{\alpha-1} \exp\{-(-\theta\alpha)y\}.$$

This is analogous to (2.6) with shape parameter $\alpha > 0$ and scale parameter $\beta = -\theta > 0$. Mean and variance are given by

$$\mu = \mathbb{E}_\theta[Y] = \kappa'(\theta) = -\theta^{-1} \qquad \text{and} \qquad \mathrm{Var}_\theta(Y) = \frac{1}{\alpha}\kappa''(\theta) = \frac{1}{\alpha}\theta^{-2},$$

and the variance function is given by $V(\mu) = \mu^2$, that is, the variance function is quadratic in the mean parameter μ. The gamma random variable is obtained by setting $X = \alpha Y \sim \Gamma(\alpha, \beta)$. This gives us for the first two moments of X

$$\mu_X = \mathbb{E}_\theta[X] = \frac{\alpha}{\beta} \qquad \text{and} \qquad \mathrm{Var}_\theta(X) = \frac{\alpha}{\beta^2} = \frac{1}{\alpha}\mu_X^2.$$

Suppose $v = 1$, for shape parameter $\alpha > 1$, we have under-dispersion $\varphi = 1/\alpha < 1$ and the gamma density is unimodal; for shape parameter $\alpha < 1$, we have over-dispersion $\varphi = 1/\alpha > 1$ and the gamma density is strictly decreasing, we refer to Fig. 2.1.

Inverse Gaussian Distribution as a Single-Parameter EDF

For the inverse Gaussian distribution with parameters $\alpha, \beta > 0$ we choose the Lebesgue measure on $\mathbb{R}_+$ and we set $\omega = v/\varphi = \alpha$. We make the following choices

$$a(y) = \log\left(\frac{\alpha^{1/2}}{(2\pi y^3)^{1/2}}\right) - \frac{\alpha}{2y}, \quad \kappa(\theta) = -(-2\theta)^{1/2},$$

$$\mu = \kappa'(\theta) = \frac{1}{(-2\theta)^{1/2}}, \quad \theta = h(\mu) = -\frac{1}{2\mu^2},$$

for $\theta \in (-\infty, 0)$ and dual parameter space $\mathcal{M} = (0, \infty)$. With these choices we have

$$\begin{aligned} f(y; \theta, \alpha)dy &= \frac{\alpha^{1/2}}{(2\pi y^3)^{1/2}} \exp\left\{\alpha\left(\theta y + (-2\theta)^{1/2}\right) - \frac{\alpha}{2y}\right\} dy \\ &= \frac{\alpha^{1/2}}{(2\pi y^3)^{1/2}} \exp\left\{-\frac{\alpha}{2y}\left(1 - (-2\theta)^{1/2} y\right)^2\right\} dy \\ &= \frac{\alpha}{(2\pi x^3)^{1/2}} \exp\left\{-\frac{\alpha^2}{2x}\left(1 - \frac{(-2\theta)^{1/2}}{\alpha} x\right)^2\right\} dx, \end{aligned}$$

where in the last step we did a change of variable $y \mapsto x = \alpha y$. This is exactly (2.8). Mean and variance are given by

$$\mu = \mathbb{E}_\theta[Y] = \kappa'(\theta) = (-2\theta)^{-1/2} \quad \text{and} \quad \mathrm{Var}_\theta(Y) = \frac{1}{\alpha}\kappa''(\theta) = \frac{1}{\alpha}(-2\theta)^{-3/2},$$

and the variance function is given by $V(\mu) = \mu^3$, that is, the variance function is cubic in the mean parameter μ. The inverse Gaussian random variable is obtained by setting $X = \alpha Y$. The mean and variance of X are given by, set $\beta = (-2\theta)^{1/2} > 0$,

$$\mu_X = \mathbb{E}_\theta[X] = \frac{\alpha}{\beta} \qquad \text{and} \qquad \mathrm{Var}_\theta(X) = \frac{\alpha}{\beta^3} = \frac{1}{\alpha^2}\mu_X^3.$$

This inverse Gaussian density is illustrated in Fig. 2.2.

Similarly to (2.9), we can extend the inverse Gaussian model to the boundary case $\theta = 0$, i.e., the effective domain $\boldsymbol{\Theta} = (-\infty, 0]$ is not open. This provides us with density

$$f(y; \theta = 0, \alpha)dy = \frac{\alpha}{(2\pi x^3)^{1/2}} \exp\left\{-\frac{\alpha^2}{2x}\right\} dx, \tag{2.16}$$

using, as above, the change of variable $y \mapsto x = \alpha y$. An additional transformation $x \mapsto 1/x$ gives a gamma distribution with shape parameter 1/2 and scale parameter $\alpha^2/2$.

Remark 2.16 The inverse Gaussian case gives an example of a non-open effective domain $\boldsymbol{\Theta} = (-\infty, 0]$. It is worth noting that for the boundary parameter $\theta = 0$, the first moment does not exist, i.e., Corollary 2.14 only makes statements in the interior $\mathring{\boldsymbol{\Theta}}$ of the effective domain $\boldsymbol{\Theta}$. This also relates to Remarks 2.9 on the dual parameter space $\mathcal{M}$.

2.2.3 *Tweedie's Distributions*

Tweedie's compound Poisson (CP) model was introduced in 1984 by Tweedie [358], and it has been studied in detail in Jørgensen [202], Jørgensen–de Souza [204], Smyth–Jørgensen [342] and in the review paper of Delong et al. [94]. Tweedie's CP model belongs to the EDF. We spend more time on explaining Tweedie's CP model because it plays an important role in actuarial modeling.

Tweedie's CP model is received by choosing as σ-finite measure ν_1 a mixture of the Lebesgue measure on $(0, \infty)$ and a point measure in 0. Furthermore, we choose *power variance parameter* $p \in (1, 2)$ and cumulant function

$$\kappa(\theta) = \kappa_p(\theta) = \frac{1}{2-p}\left((1-p)\theta\right)^{\frac{2-p}{1-p}}, \tag{2.17}$$

on the effective domain $\theta \in \boldsymbol{\Theta} = (-\infty, 0)$. This provides us with Tweedie's CP model

$$Y \sim f(y; \theta, v/\varphi) = \exp\left\{\frac{y\theta - \kappa_p(\theta)}{\varphi/v} + a(y; v/\varphi)\right\},$$

with exposure $v > 0$ and dispersion parameter $\varphi > 0$; the normalizing function $a(\cdot; v/\varphi)$ does not have any simple closed form, we refer to Section 2.1 in Jørgensen–de Souza [204] and Section 4.2 in Jørgensen [203].

The first two moments of Tweedie's CP random variable Y are given by

$$\mu = \mathbb{E}_\theta[Y] = \kappa_p'(\theta) = ((1-p)\theta)^{\frac{1}{1-p}} \in \mathcal{M} = (0, \infty), \tag{2.18}$$

$$\mathrm{Var}_\theta(Y) = \frac{\varphi}{v}\kappa_p''(\theta) = \frac{\varphi}{v}((1-p)\theta)^{\frac{p}{1-p}} = \frac{\varphi}{v}\mu^p > 0. \tag{2.19}$$

The parameter $p \in (1, 2)$ determines the power variance functions $V(\mu) = \mu^p$ between the Poisson $p = 1$ and the gamma $p = 2$ cases, see Sect. 2.2.2.

The moment generating function of Tweedie's CP random variable $X = vY/\varphi = \omega Y$ in its additive form is given by, we use Corollary 2.14,

$$M_X(r) = M_{vY/\varphi}(r) = \exp\left\{\frac{v}{\varphi}\kappa_p(\theta)\left(\left(\frac{-\theta}{-\theta - r}\right)^{\frac{2-p}{p-1}} - 1\right)\right\} \qquad \text{for } r < -\theta.$$

Some readers will notice that this is the moment generating function of a CP distribution having i.i.d. gamma claim sizes. This is exactly the statement of the next proposition which is found, e.g., in Smyth–Jørgensen [342].

Proposition 2.17 *Assume $S = \sum_{i=1}^N Z_i$ is CP distributed with Poisson claim counts $N \sim \mathrm{Poi}(\lambda v)$ and i.i.d. gamma claim sizes $Z_i \sim \Gamma(\alpha, \beta)$ being independent of N. We have $S \stackrel{(d)}{=} vY/\varphi$ by identifying the parameters as follows*

$$p = \frac{\alpha + 2}{\alpha + 1} \in (1, 2), \qquad \beta = -\theta > 0 \qquad \textit{and} \qquad \lambda = \frac{1}{\varphi}\kappa_p(\theta) > 0.$$

Proof of Proposition 2.17 Assume S is CP distributed with i.i.d. gamma claim sizes. From Proposition 2.11 and Section 3.2.1 in Wüthrich [387] we receive that the moment generating function of S is given by

$$M_S(r) = \exp\left\{\lambda v\left(\left(\frac{\beta}{\beta - r}\right)^\alpha - 1\right)\right\} \qquad \text{for } r < \beta.$$

Using the proposed parameter identification, the claim immediately follows. □

Proposition 2.17 gives us a second interpretation of Tweedie's CP model which was introduced in an EDF fashion, above. This second interpretation explains the name of this EDF model, it explains the mixture of the Lebesgue measure and the point measure in 0, and it also highlights why the Poisson model and the gamma model are the boundary cases in terms of power variance functions.

An interesting question is whether the EDF can be extended beyond power variance functions $V(\mu) = \mu^p$ with $p \in [1, 2]$. The answer to this question is yes, and the full answer is provided in Theorem 2 of Jørgensen [202]:

Theorem 2.18 (Jørgensen [202], Without Proof) *Only power variance parameters $p \in (0, 1)$ do not allow for EDF models.*

Table 2.1 gives the EDF distributions that have a power variance function. These distributions are called *Tweedie's distributions*, with the special case of Tweedie's CP distributions for $p \in (1, 2)$. The densities for $p \in \{0, 1, 2, 3\}$ have a closed form, but the other Tweedie's distributions do not have a closed-form density. Thus, they cannot explicitly be constructed as suggested in Sect. 2.2.1. Besides the constructive approach presented above, there is a uniqueness theorem saying that the variance function $V(\cdot)$ on the domain $\mathcal{M}$ characterizes the single-parameter linear EF, see Theorem 2.11 in Jørgensen [203]. This uniqueness theorem is the basis of the proof of Theorem 2.18. Tweedie's distributions for $p \notin [0, 1] \cup \{2, 3\}$ involve infinite sums for the normalization $\exp\{a(\cdot, \cdot)\}$, we refer to formulas (4.19), (4.20) and (4.31) in Jørgensen [203], this is the reason that one has to go via the uniqueness theorem to prove Theorem 2.18. Dunn–Smyth [112] provide methods of fast calculation of some of these infinite sums; in Sect. 5.5.2, below, we present an approximation (saddlepoint approximation). The uniqueness theorem is also useful to construct new examples within the EF, see, e.g., Section 2 of Awad et al. [15].

Table 2.1 Power variance function models $V(\mu) = \mu^p$ within the EDF (taken from Table 4.1 in Jørgensen [203])

p	Distribution	Support of Y	$\boldsymbol{\Theta}$	$\mathcal{M}$
$p < 0$	Generated by extreme stable distributions	$\mathbb{R}$	$[0, \infty)$	$(0, \infty)$
$p = 0$	Gaussian distribution	$\mathbb{R}$	$\mathbb{R}$	$\mathbb{R}$
$p = 1$	Poisson distribution	$\mathbb{N}_0$	$\mathbb{R}$	$(0, \infty)$
$1 < p < 2$	Tweedie's CP distribution	$[0, \infty)$	$(-\infty, 0)$	$(0, \infty)$
$p = 2$	Gamma distribution	$(0, \infty)$	$(-\infty, 0)$	$(0, \infty)$
$p > 2$	Generated by positive stable distributions	$(0, \infty)$	$(-\infty, 0]$	$(0, \infty)$
$p = 3$	Inverse Gaussian distribution	$(0, \infty)$	$(-\infty, 0]$	$(0, \infty)$

2.2.4 Steepness of the Cumulant Function

Assume we have a fixed EF satisfying Assumption 2.6. All random variables $T(Y)$ belonging to this EF have the same support, not depending on the particular choice of the canonical parameter $\boldsymbol{\theta} \in \boldsymbol{\Theta}$. We denote this support of $T(Y)$ by $\mathfrak{T}$.

Below, we are going to estimate the canonical parameter $\boldsymbol{\theta} \in \boldsymbol{\Theta}$ from data using maximum likelihood estimation. For this it is advantageous to have the property $\mathfrak{T} \subset \mathcal{M}$, because, intuitively, this allows us to directly select $\widehat{\mu} = T(Y)$ as the parameter estimate in the dual parameter space $\mathcal{M}$, for a given observation $T(Y) \in \mathfrak{T}$. This then translates to a canonical parameter $\widehat{\boldsymbol{\theta}} = h(\widehat{\mu}) = h(T(Y)) \in \boldsymbol{\Theta}$, using the canonical link h; this estimation approach will be better motivated in Chap. 3, below. Unfortunately, many examples of the EF do not satisfy this property $\mathfrak{T} \subset \mathcal{M}$. For instance, in the Poisson model the observation $T(Y) = Y = 0$ is not included in $\mathcal{M}$, see Table 2.1. This poses some challenges in parameter estimation, and the purpose of this small discussion is to be prepared for these challenges.

A cumulant function κ is called *steep* if for all $\boldsymbol{\theta} \in \mathring{\boldsymbol{\Theta}}$ and all $\widetilde{\boldsymbol{\theta}}$ in the boundary of $\boldsymbol{\Theta}$

$$\left(\widetilde{\boldsymbol{\theta}} - \boldsymbol{\theta}\right)^{\top} \nabla_{\boldsymbol{\theta}} \kappa\left(\alpha\boldsymbol{\theta} + (1-\alpha)\widetilde{\boldsymbol{\theta}}\right) \to \infty \qquad \text{for } \alpha \downarrow 0, \tag{2.20}$$

we refer to Formula (20) in Section 8.1 of Barndorff-Nielsen [23]. Define the convex closure of the support $\mathfrak{T}$ by $\mathfrak{C} = \overline{\text{conv}}(\mathfrak{T})$.

Theorem 2.19 (Theorem 9.2 in Barndorff-Nielsen [23], Without Proof) *Assume we have a fixed EF satisfying Assumption 2.6. The cumulant function κ is steep if and only if $\mathring{\mathfrak{C}} = \mathcal{M} = \nabla_{\boldsymbol{\theta}}\kappa(\mathring{\boldsymbol{\Theta}})$.*

Theorem 2.19 tells us that for a steep cumulant function we have $\mathfrak{C} = \overline{\mathcal{M}} = \overline{\nabla_{\boldsymbol{\theta}}\kappa(\mathring{\boldsymbol{\Theta}})}$. In this case parameter estimation can be extended to observations $T(Y) \notin \mathcal{M}$ such that we may obtain a degenerate model at the boundary of $\mathcal{M}$. Coming back to our Poisson example from above, in this case we set $\widehat{\mu} = 0$, which gives a degenerate Poisson model.

Throughout this book we will work under the assumption that κ is steep. The classical examples satisfy this assumption: the examples with power variance parameter p in $\{0\} \cup [1, \infty)$ satisfy Theorem 2.19; this includes the Gaussian, the Poisson, the gamma, the inverse Gaussian and Tweedie's CP models, see Table 2.1. Moreover, the examples we have met in Sect. 2.1 fulfill this assumption; these are the single-parameter linear EF models of the Bernoulli, the binomial and the negative binomial distributions, as well as the vector-valued parameter examples of the Gaussian, the gamma and the inverse Gaussian models and of the categorical distribution. The only models we have seen that do not have a steep cumulant function are the power variance models with $p < 0$, see Table 2.1.

Remark 2.20 Working within the EDF needs some additional thoughts because the support $\mathfrak{T} = \mathfrak{T}_\omega$ of the single-parameter linear EDF random variable $Y = T(Y)$ may depend on the specific choice of the dispersion parameter $\omega \in \mathcal{W} \supset \{1\}$ through the σ-finite measure $d\nu_\omega(\omega \cdot)$, see (2.13). For instance, in the binomial case the support of Y is given by $\mathfrak{T}_\omega = \{0, 1/n, \ldots, 1\}$ with $\omega = n$, see Sect. 2.2.2.

Assume that the cumulant function κ is steep for the single-parameter linear EF that corresponds to the single-parameter EDF with $\omega = 1$. Theorem 2.19 then implies that for this choice we have $\mathring{\mathfrak{C}}_{\omega=1} = \nabla_{\boldsymbol{\theta}} \kappa(\mathring{\boldsymbol{\Theta}})$ with convex closure $\mathfrak{C}_{\omega=1} = \overline{\text{conv}}(\mathfrak{T}_{\omega=1})$.

Consider $\omega \in \mathcal{W} \backslash \{1\}$ which corresponds to the choice ν_ω of the σ-finite measure on $\mathbb{R}$. This choice belongs to the cumulant function $\theta \mapsto \omega\kappa(\theta)$ in the additive form (x-parametrization in (2.13)). Since steepness (2.20) holds for any $\omega > 0$ we receive that the convex closure of the support of this distribution in the x-parametrization in (2.13) is given by $\overline{\nabla_{\boldsymbol{\theta}} \omega\kappa(\mathring{\boldsymbol{\Theta}})} = \omega \overline{\nabla_{\boldsymbol{\theta}} \kappa(\mathring{\boldsymbol{\Theta}})}$. The duality transformation $x \mapsto y = x/\omega$ leads to the change of measure $d\nu_\omega(x) \mapsto d\nu_\omega(\omega y)$ and to the corresponding change of support, see (2.13). The latter implies that in the reproductive form (y-parametrization) the convex closure of the support does not depend on the specific choice of $\omega \in \mathcal{W}$. Since the EDF representation given in (2.14) corresponds to the y-parametrization (reproductive form), we can use Theorem 2.19 without limitation also for the single-parameter linear EDF given by (2.14), and $\mathfrak{C}$ does not depend on $\omega \in \mathcal{W}$.

2.2.5 *Lab: Large Claims Modeling*

From Corollary 2.14 we know that the moment generating function exists around the origin for all examples belonging to the EDF. This implies that the moments of all orders exist, and that we have an exponentially decaying survival function $\mathbb{P}_\theta[Y > y] = 1 - F(y; \theta, \omega) \sim \exp\{-\varrho y\}$ for some $\varrho > 0$ as $y \to \infty$, see (1.2). In many applied situations the data is more heavy-tailed and, thus, cannot be modeled by such an exponentially decaying survival function. In such cases one often chooses a distribution function with a regularly varying survival function; regular variation with tail index $\beta > 0$ has been introduced in (1.3). A popular choice is a log-gamma distribution which can be obtained from the gamma distribution (belonging to the EDF). We briefly explain how this is done and how it relates to the Pareto and the Lomax [256] distributions.

We start from the gamma density (2.6). The random variable Z has a log-gamma distribution with shape parameter $\alpha > 0$ and scale parameter $\beta = -\theta > 0$ if $\log(Z) = Y$ has a gamma distribution with these parameters. Thus, the gamma density of $Y = \log(Z)$ is given by

$$f(y; \beta, \alpha)dy = \frac{\beta^\alpha}{\Gamma(\alpha)} y^{\alpha-1} \exp\{-\beta y\} dy \qquad \text{for } y > 0.$$

We do a change of variable $y \mapsto z = \exp\{y\}$ to receive the density of the log-gamma distributed random variable $Z = \exp\{Y\}$

$$f(z; \beta, \alpha)dz = \frac{\beta^\alpha}{\Gamma(\alpha)}(\log z)^{\alpha-1} z^{-(\beta+1)} dz \qquad \text{for } z > 1.$$

This log-gamma density has support $(1, \infty)$. The distribution function of this log-gamma distributed random variable needs to be calculated numerically, and its survival function is regularly varying with tail index $\beta > 0$.

A special case of the log-gamma distribution is the Pareto distribution. The Pareto distribution is more tractable and it is obtained by setting shape parameter $\alpha = 1$ in the log-gamma density. This gives us the Pareto density

$$f(z; \beta)dz = f(z; \beta, \alpha = 1)dz = \beta z^{-(\beta+1)} dz \qquad \text{for } z > 1.$$

The distribution function in this Pareto case is for $z \geq 1$ given by

$$F(z; \beta) = 1 - z^{-\beta}.$$

Obviously, this provides a regularly varying survival function with tail index $\beta > 0$; in fact, in this case we do not need to go over to the limit in (1.3) because we have an exact identity. The Pareto distribution has the nice property that it is closed under thresholding (lower-truncation) with M, that is, we remain within the family of Pareto distributions with the same tail index β by considering lower-truncated claims: for $1 \leq M \leq z$ we have

$$F(z; \beta, M) = \mathbb{P}[Z \leq z | Z > M] = \frac{\mathbb{P}[M < Z \leq z]}{\mathbb{P}[Z > M]} = 1 - \left(\frac{z}{M}\right)^{-\beta}.$$

This is the classical definition of the Pareto distribution, and it allows to preserve full flexibility in the choice of the threshold $M > 0$.

The disadvantage of the Pareto distribution is that it does not provide a continuous density on $\mathbb{R}_+$ as there is a discontinuity in threshold M. For this reason, one sometimes explores another change of variable $Z \mapsto X = Z - M$ for a Pareto distributed random variable $Z \sim F(\cdot; \beta, M)$. This provides the Lomax distribution, also called Pareto Type II distribution. X has the following distribution function on $(0, \infty)$

$$\mathbb{P}[X \leq x] = 1 - \left(\frac{x + M}{M}\right)^{-\beta} \qquad \text{for } x \geq 0.$$

This distribution has again a regularly varying survival function with tail index $\beta > 0$. Moreover, we have

$$\lim_{x\to\infty} \frac{\left(\frac{x+M}{M}\right)^{-\beta}}{\left(\frac{x}{M}\right)^{-\beta}} = \lim_{x\to\infty} \left(1 + \frac{M}{x}\right)^{-\beta} = 1.$$

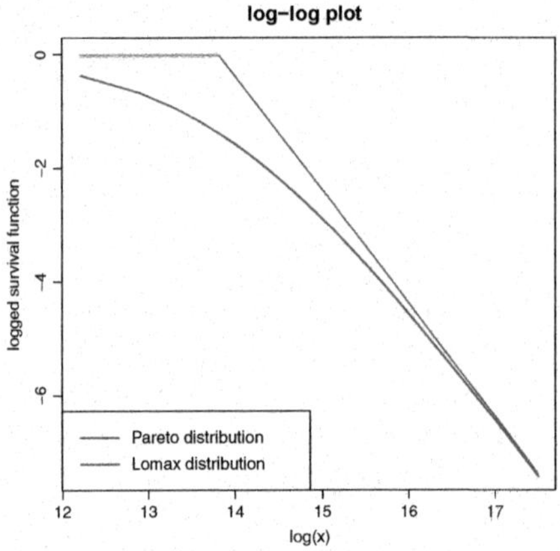

Fig. 2.3 Log-log plot of a Pareto and a Lomax distribution with tail index $\beta = 2$ and threshold $M = 1'000'000$

This says that we should choose the same threshold $M > 0$ for both the Pareto and the Lomax distribution to receive the same asymptotic tail behavior, and this also quantifies the rate of convergence between the two survival functions. Figure 2.3 illustrates this convergence in a log-log plot choosing tail index $\beta = 2$ and threshold $M = 1'000'000$.

For completeness we provide the density of the Pareto distribution

$$f(z; \beta, M) = \frac{\beta}{M}\left(\frac{z}{M}\right)^{-(\beta+1)} \qquad \text{for } z \geq M,$$

and of the Lomax distribution

$$f(x; \beta, M) = \frac{\beta}{M}\left(\frac{x+M}{M}\right)^{-(\beta+1)} \qquad \text{for } x \geq 0.$$

2.3 Information Geometry in Exponential Families

We do a short excursion to information geometry. This excursion may look a bit disconnected from what we have done so far, but it provides us with important background information for the chapter on forecast evaluation, see Chap. 4, below.

2.3.1 Kullback–Leibler Divergence

There is literature in information geometry which uses techniques from differential geometry to study EFs as Riemannian manifolds with points corresponding to EF densities parametrized by their canonical parameters $\boldsymbol{\theta} \in \boldsymbol{\Theta}$, we refer to Amari [10],

Ay et al. [16] and Nielsen [285] for an extended treatment of these mathematical concepts.

Choose a fixed EF (2.2) with cumulant function κ on the effective domain $\boldsymbol{\Theta} \subseteq \mathbb{R}^k$ and with σ-finite measure ν on $\mathbb{R}$. We define the Kullback–Leibler (KL) divergence (relative entropy) from model $\boldsymbol{\theta}_1 \in \boldsymbol{\Theta}$ to model $\boldsymbol{\theta}_0 \in \boldsymbol{\Theta}$ within this EF by

$$D_{\mathrm{KL}}(f(\cdot;\boldsymbol{\theta}_0)||f(\cdot;\boldsymbol{\theta}_1)) = \int_{\mathbb{R}} f(y;\boldsymbol{\theta}_0)\log\left(\frac{f(y;\boldsymbol{\theta}_0)}{f(y;\boldsymbol{\theta}_1)}\right) d\nu(y) \;\geq\; 0.$$

Recall that the support of the EF does not depend on the specific choice of the canonical parameter $\boldsymbol{\theta}$ in $\boldsymbol{\Theta}$, see Remarks 2.3; this implies that the KL divergence is well-defined, here. The positivity of the KL divergence is obtained from Jensen's inequality; this is proved in Lemma 2.21, below.

The KL divergence has the interpretation of having a data model that is characterized by the distribution $f(\cdot;\boldsymbol{\theta}_0)$, and we would like to measure how close another model $f(\cdot;\boldsymbol{\theta}_1)$ is to the data model. Note that the KL divergence is not a distance function because it is neither symmetric nor does it satisfy the triangle inequality.

We calculate the KL divergence within the chosen EF

$$\begin{aligned} D_{\mathrm{KL}}(f(\cdot;\boldsymbol{\theta}_0)||f(\cdot;\boldsymbol{\theta}_1)) &= \int_{\mathbb{R}} f(y;\boldsymbol{\theta}_0)\left[(\boldsymbol{\theta}_0-\boldsymbol{\theta}_1)^\top T(y) - \kappa(\boldsymbol{\theta}_0) + \kappa(\boldsymbol{\theta}_1)\right] d\nu(y) \\ &= (\boldsymbol{\theta}_0-\boldsymbol{\theta}_1)^\top \nabla_{\boldsymbol{\theta}}\kappa(\boldsymbol{\theta}_0) - \kappa(\boldsymbol{\theta}_0) + \kappa(\boldsymbol{\theta}_1) \;\geq\; 0, \qquad (2.21) \end{aligned}$$

where we have used Corollary 2.5, and the positivity of the KL divergence can be seen from the convexity of κ. This allows us to consider the following (Taylor) expansion

$$\kappa(\boldsymbol{\theta}_1) = \kappa(\boldsymbol{\theta}_0) + \nabla_{\boldsymbol{\theta}}\kappa(\boldsymbol{\theta}_0)^\top (\boldsymbol{\theta}_1-\boldsymbol{\theta}_0) + D_{\mathrm{KL}}(f(\cdot;\boldsymbol{\theta}_0)||f(\cdot;\boldsymbol{\theta}_1)). \qquad (2.22)$$

This illustrates that the KL divergence corresponds to second and higher order differences between the cumulant value $\kappa(\boldsymbol{\theta}_0)$ and another cumulant value $\kappa(\boldsymbol{\theta}_1)$. The gradients of the KL divergence w.r.t. $\boldsymbol{\theta}_1$ in $\boldsymbol{\theta}_1 = \boldsymbol{\theta}_0$ and w.r.t. $\boldsymbol{\theta}_0$ in $\boldsymbol{\theta}_0 = \boldsymbol{\theta}_1$ are given by

$$\begin{aligned} &\nabla_{\boldsymbol{\theta}_1} D_{\mathrm{KL}}(f(\cdot;\boldsymbol{\theta}_0)||f(\cdot;\boldsymbol{\theta}_1))\big|_{\boldsymbol{\theta}_1=\boldsymbol{\theta}_0} \qquad (2.23) \\ &\quad = \nabla_{\boldsymbol{\theta}_0} D_{\mathrm{KL}}(f(\cdot;\boldsymbol{\theta}_0)||f(\cdot;\boldsymbol{\theta}_1))\big|_{\boldsymbol{\theta}_0=\boldsymbol{\theta}_1} \;=\; \mathbf{0}. \end{aligned}$$

This emphasizes that the KL divergence reflects second and higher-order terms in cumulant function κ; and that the data model $\boldsymbol{\theta}_0$ forms the minimum of this KL divergence (as a function of $\boldsymbol{\theta}_1$) as we will just see. We calculate the Hessian (second order term) w.r.t. $\boldsymbol{\theta}_1$ in $\boldsymbol{\theta}_1 = \boldsymbol{\theta}_0$

$$\nabla^2_{\boldsymbol{\theta}_1} D_{\mathrm{KL}}(f(\cdot;\boldsymbol{\theta}_0)||f(\cdot;\boldsymbol{\theta}_1))\Big|_{\boldsymbol{\theta}_1=\boldsymbol{\theta}_0} = \nabla^2_{\boldsymbol{\theta}}\kappa(\boldsymbol{\theta})\Big|_{\boldsymbol{\theta}=\boldsymbol{\theta}_0} \stackrel{\text{def.}}{=} \mathcal{I}(\boldsymbol{\theta}_0).$$

The positive definite matrix $\mathcal{I}(\boldsymbol{\theta}_0)$ (in a minimal representation) is called *Fisher's information*. Fisher's information is an important tool in statistics that we will meet in Theorem 3.13 of Sect. 3.3, below. A function satisfying (2.21) (with being zero if and only if $\boldsymbol{\theta}_0 = \boldsymbol{\theta}_1$), fulfilling (2.23) and having positive definite Fisher's information is called *divergence*, see Definition 5 in Nielsen [285]. Fisher's information $\mathcal{I}(\boldsymbol{\theta}_0)$ measures the curvature of the KL divergence in $\boldsymbol{\theta}_0$ and we have the second order Taylor approximation

$$\kappa(\boldsymbol{\theta}_1) \approx \kappa(\boldsymbol{\theta}_0) + \nabla_{\boldsymbol{\theta}}\kappa(\boldsymbol{\theta}_0)^\top(\boldsymbol{\theta}_1 - \boldsymbol{\theta}_0) + \frac{1}{2}(\boldsymbol{\theta}_1 - \boldsymbol{\theta}_0)^\top \mathcal{I}(\boldsymbol{\theta}_0)(\boldsymbol{\theta}_1 - \boldsymbol{\theta}_0).$$

Next-order terms are obtained from the so-called Amari–Chentsov tensor, see Amari [10] and Section 4.2 in Ay et al. [16]. In information geometry one studies the (possibly degenerate) Riemannian metric on the effective domain $\boldsymbol{\Theta}$ induced by Fisher's information; we refer to Section 3.7 in Nielsen [285].

Lemma 2.21 *Consider two densities p and q w.r.t. a given σ-finite measure ν. We have $D_{KL}(p||q) \geq 0$, and $D_{KL}(p||q) = 0$ if and only if $p = q$, ν-a.s.*

Proof Assume $Y \sim p d\nu$, then we can rewrite the KL divergence, using Jensen's inequality,

$$\begin{aligned} D_{\mathrm{KL}}(p||q) = \int p(y)\log\left(\frac{p(y)}{q(y)}\right)d\nu(y) &= -\mathbb{E}_p\left[\log\left(\frac{q(Y)}{p(Y)}\right)\right] \\ \geq -\log\mathbb{E}_p\left[\frac{q(Y)}{p(Y)}\right] &= -\log\int q(y)d\nu(y) \geq 0. \end{aligned} \tag{2.24}$$

Equality holds if and only if $p = q$, ν-a.s. The last inequality of (2.24) considers that q does not necessarily need to be a density w.r.t. ν, i.e., we can also have $\int q(y)d\nu(y) < 1$. □

2.3.2 *Unit Deviance and Bregman Divergence*

In the next chapter we are going to introduce maximum likelihood estimation for parameters, see Definition 3.4, below. Maximum likelihood estimators are obtained by maximizing likelihood functions (evaluated in the observations). Maximizing likelihood functions within the EDF is equivalent to minimizing deviance loss functions. Deviance loss functions are based on unit deviances, which, in turn, correspond to KL divergences. The purpose of this small section is to discuss this relation. This should be viewed as a preparation for Chap. 4.

Assume we work within a single-parameter linear EDF, i.e., $T(y) = y$. Using the canonical link h we obtain the canonical parameter $\theta = h(\mu) \in \boldsymbol{\Theta} \subseteq \mathbb{R}$ from the mean parameter $\mu \in \mathcal{M}$. If we replace the (typically unknown) mean parameter μ by an observation Y, supposed $Y \in \mathcal{M}$, we get the specific model that is exactly calibrated to this observation. This provides us with the canonical parameter estimate $\widehat{\theta}_Y = h(Y)$ for θ. We can now measure the KL divergence from any model represented by θ to the observation calibrated model $\widehat{\theta}_Y = h(Y)$. This KL divergence is given by (we use (2.21) and we set $\omega = v/\varphi = 1$)

$$\begin{aligned} D_{\mathrm{KL}}\left(f(\cdot; h(Y), 1) || f(\cdot; \theta, 1)\right) &= \int_{\mathbb{R}} f(y; \widehat{\theta}_Y, 1) \log\left(\frac{f(y; \widehat{\theta}_Y, 1)}{f(y; \theta, 1)}\right) d\nu(y) \\ &= (h(Y) - \theta)\, Y - \kappa(h(Y)) + \kappa(\theta) \;\geq\; 0. \end{aligned}$$

This latter object is the unit deviance (up to factor 2) of the chosen EDF. It plays a crucial role in predictive modeling.

We define the *unit deviance* under the assumption that κ is steep as follows:

$$\begin{aligned} \mathfrak{d} : \mathring{\mathfrak{C}} \times \mathcal{M} &\to \mathbb{R}_+ \qquad\qquad (2.25) \\ (y, \mu) &\mapsto \mathfrak{d}(y, \mu) = 2\left(yh(y) - \kappa\left(h(y)\right) - yh(\mu) + \kappa\left(h(\mu)\right)\right) \;\geq\; 0, \end{aligned}$$

where $\mathfrak{C}$ is the convex closure of the support $\mathfrak{T}$ of Y and $\mathcal{M}$ is the dual parameter space of the chosen EDF. Steepness of κ implies $\mathring{\mathfrak{C}} = \mathcal{M}$, see Theorem 2.19.

This unit deviance $\mathfrak{d}$ is received from the KL divergence, and it is (twice) the difference of two log-likelihood functions, one using canonical parameter $h(y)$ and the other one having any canonical parameter $\theta = h(\mu) \in \mathring{\boldsymbol{\Theta}}$. That is, for $\mu = \kappa'(\theta)$,

$$\begin{aligned} \mathfrak{d}(y, \mu) &= 2\, D_{\mathrm{KL}}(f(\cdot; h(y), 1) || f(\cdot; \theta, 1)) \qquad (2.26) \\ &= 2\, \frac{\varphi}{v}\, \left(\log f(y; h(y), v/\varphi) - \log f(y; \theta, v/\varphi)\right), \end{aligned}$$

for general $\omega = v/\varphi \in \mathcal{W}$. The latter can be rewritten as

$$f(y; \theta, v/\varphi) = f(y; h(y), v/\varphi)\, \exp\left\{-\frac{1}{2\varphi/v}\mathfrak{d}(y, \kappa'(\theta))\right\}. \qquad (2.27)$$

This looks like a generalization of the Gaussian distribution, where the square difference $(y - \mu)^2$ in the exponent is replaced by the unit deviance $\mathfrak{d}(y, \mu)$ with $\mu = \kappa'(\theta)$. This interpretation gets further support by the following lemma.

Lemma 2.22 *Under Assumption 2.6 and the assumption that the cumulant function κ is steep, the unit deviance $\mathfrak{d}(y, \mu) \geq 0$ of the chosen EDF is zero if and only if $y = \mu$. Moreover, the unit deviance $\mathfrak{d}(y, \mu)$ is twice continuously differentiable w.r.t. (y, μ) in $\mathring{\mathfrak{C}} \times \mathcal{M}$, and*

$$\left.\frac{\partial^2 \mathfrak{d}(y, \mu)}{\partial \mu^2}\right|_{y=\mu} = \left.\frac{\partial^2 \mathfrak{d}(y, \mu)}{\partial y^2}\right|_{y=\mu} = -\left.\frac{\partial^2 \mathfrak{d}(y, \mu)}{\partial \mu \partial y}\right|_{y=\mu} = 2/V(\mu) > 0.$$

Proof The positivity and the if and only if statement follows from Lemma 2.21 and the strict convexity of κ. Continuous differentiability follows from the smoothness of κ in the interior of $\boldsymbol{\Theta}$. Moreover we have

$$\left.\frac{\partial^2 \mathfrak{d}(y, \mu)}{\partial \mu^2}\right|_{y=\mu} = \left.\frac{\partial}{\partial \mu} 2\left(-yh'(\mu) + \mu h'(\mu)\right)\right|_{y=\mu} = 2h'(\mu) = 2/\kappa''(h(\mu)) = 2/V(\mu) > 0,$$

where $V(\mu)$ is the variance function of the chosen EDF introduced in Corollary 2.14. The remaining second derivatives are received by similar (straightforward) calculations. □

Remarks 2.23

- Lemma 2.22 shows that the unit deviance definition of $\mathfrak{d}(y, \mu)$ provides a so-called regular unit deviance according to Definition 1.1 in Jørgensen [203]. Moreover, any model that can be brought into the form (2.27) for a (regular) unit deviance is called (regular) reproductive dispersion model, see Definition 1.2 of Jørgensen [203].
- In general the unit deviance $\mathfrak{d}(y, \mu)$ is not symmetric in its two arguments y and μ, we come back to this in Fig. 11.1, below.

More generally, the KL divergence and the unit deviance can be embedded into the framework of Bregman loss functions [50]. We restrict to the single-parameter EDF case. Assume that $\psi : \mathring{\mathfrak{C}} \to \mathbb{R}$ is a strictly convex function. The *Bregman divergence* w.r.t. ψ between y and μ is defined by

$$D_\psi(y, \mu) = \psi(y) - \psi(\mu) - \psi'(\mu)(y - \mu) \geq 0, \tag{2.28}$$

where ψ' is a (sub-)gradient of ψ. The lower bound holds because of convexity of ψ. Consider the specific choice $\psi(\mu) = \mu h(\mu) - \kappa(h(\mu))$ for the chosen EDF. Similar to Lemma 2.22 we have $\psi''(\mu) = h'(\mu) = 1/V(\mu) > 0$, which says that this choice is strictly convex. Using this choice for ψ gives us unit deviance (up to factor 1/2)

$$D_\psi(y, \mu) = yh(y) - \kappa(h(y)) + \kappa(h(\mu)) - h(\mu)y = \frac{1}{2}\mathfrak{d}(y, \mu). \tag{2.29}$$

Thus, the unit deviance $\mathfrak{d}$ can be understood as a difference of log-likelihoods (2.26), as a KL divergence D_{KL} and as a Bregman divergence D_ψ.

Example 2.24 (Poisson Model) We start with a single-parameter EF example. Consider cumulant function $\kappa(\theta) = \exp\{\theta\}$ for canonical parameter $\theta \in \boldsymbol{\Theta} = \mathbb{R}$, this gives us the Poisson model. For the KL divergence from model θ_1 to model θ_0 we receive

$$D_{\mathrm{KL}}(f(\cdot;\theta_0)||f(\cdot;\theta_1)) = \exp\{\theta_1\} - \exp\{\theta_0\} - (\theta_1 - \theta_0)\exp\{\theta_0\} \ \geq\ 0,$$

which is zero if and only if $\theta_0 = \theta_1$. Fisher's information is given by

$$\mathcal{I}(\theta) = \kappa''(\theta) = \exp\{\theta\} > 0.$$

If we have observation $Y > 0$ we receive a model described by canonical parameter $\widehat{\theta}_Y = h(Y) = \log(Y)$. This gives us unit deviance, see (2.26),

$$\begin{aligned}\mathfrak{d}(Y,\mu) &= 2D_{\mathrm{KL}}(f(\cdot;h(Y),1)||f(\cdot;\theta,1))\\ &= 2\left(e^{\theta} - Y - (\theta - \log(Y))Y\right)\\ &= 2\left(\mu - Y - Y\log\left(\frac{\mu}{Y}\right)\right) \ \geq\ 0,\end{aligned}$$

with $\mu = \kappa'(\theta) = \exp\{\theta\}$. This Poisson unit deviance will commonly be used for model fitting and forecast evaluation, see, e.g., (5.28). ■

Example 2.25 (Gamma Model) The second example considers a vector-valued parameter EF example. We consider the cumulant function $\kappa(\boldsymbol{\theta}) = \log\Gamma(\theta_2) - \theta_2\log(-\theta_1)$ for $\boldsymbol{\theta} = (\theta_1,\theta_2)^\top \in \boldsymbol{\Theta} = (-\infty,0)\times(0,\infty)$; this gives us the gamma model, see Sect. 2.1.3. For the KL divergence from model $\boldsymbol{\theta}_1$ to model $\boldsymbol{\theta}_0$ we receive

$$\begin{aligned}D_{\mathrm{KL}}(f(\cdot;\boldsymbol{\theta}_0)||f(\cdot;\boldsymbol{\theta}_1)) &= \left(\theta_{0,2} - \theta_{1,2}\right)\frac{\Gamma'(\theta_{0,2})}{\Gamma(\theta_{0,2})} - \log\left(\frac{\Gamma(\theta_{0,2})}{\Gamma(\theta_{1,2})}\right)\\ &\quad + \theta_{1,2}\log\left(\frac{-\theta_{0,1}}{-\theta_{1,1}}\right) + \theta_{0,2}\left(\frac{-\theta_{1,1}}{-\theta_{0,1}} - 1\right) \ \geq\ 0.\end{aligned}$$

Fisher's information matrix is given by

$$\mathcal{I}(\boldsymbol{\theta}) = \nabla^2_{\boldsymbol{\theta}}\kappa(\boldsymbol{\theta}) = \begin{pmatrix} \frac{\theta_2}{(-\theta_1)^2} & \frac{1}{-\theta_1} \\ \frac{1}{-\theta_1} & \frac{\Gamma''(\theta_2)\Gamma(\theta_2) - \Gamma'(\theta_2)^2}{\Gamma(\theta_2)^2} \end{pmatrix}.$$

The off-diagonal terms in Fisher's information matrix $\mathcal{I}(\boldsymbol{\theta})$ are non-zero which means that the two components of the canonical parameter $\boldsymbol{\theta}$ interact. Choosing a different parametrization $\mu = \theta_2/(-\theta_1)$ (dual mean parametrization) and $\alpha = \theta_2$ we receive diagonal Fisher's information in (μ, α)

$$\mathcal{I}(\mu, \alpha) = \begin{pmatrix} \frac{\alpha}{\mu^2} & 0 \\ 0 & \frac{\Gamma''(\alpha)\Gamma(\alpha)-\Gamma'(\alpha)^2}{\Gamma(\alpha)^2} - \frac{1}{\alpha} \end{pmatrix} = \begin{pmatrix} \frac{\alpha}{\mu^2} & 0 \\ 0 & \Psi'(\alpha) - \frac{1}{\alpha} \end{pmatrix}, \tag{2.30}$$

where Ψ is the digamma function, see Footnote 2 on page 22. This transformation is obtained by using the corresponding Jacobian matrix for variable transformation; more details are provided in (3.16) below. In this new representation, the parameters μ and α are orthogonal; the term $\Psi'(\alpha) - \frac{1}{\alpha}$ is further discussed in Remarks 5.26 and Remarks 5.28, below.

Using this second parametrization based on mean μ and dispersion $1/\alpha$, we arrive at the EDF representation of the gamma model. This allows us to calculate the corresponding unit deviance (within the EDF), which in the gamma case is given by

$$\mathfrak{d}(Y, \mu) = 2\left(\frac{Y}{\mu} - 1 + \log\left(\frac{\mu}{Y}\right)\right) \geq 0.$$

■

Example 2.26 (Inverse Gaussian Model) Our final example considers the inverse Gaussian vector-valued parameter EF case. We consider the cumulant function $\kappa(\boldsymbol{\theta}) = -2(\theta_1\theta_2)^{1/2} - \frac{1}{2}\log(-2\theta_2)$ for $\boldsymbol{\theta} = (\theta_1, \theta_2)^\top \in \boldsymbol{\Theta} = (-\infty, 0] \times (-\infty, 0)$, see Sect. 2.1.3. For the KL divergence from model $\boldsymbol{\theta}_1$ to model $\boldsymbol{\theta}_0$ we receive

$$\begin{aligned} D_{\mathrm{KL}}(f(\cdot; \boldsymbol{\theta}_0)||f(\cdot; \boldsymbol{\theta}_1)) = &-\theta_{1,1}\sqrt{\frac{-\theta_{0,2}}{-\theta_{0,1}}} - \theta_{1,2}\sqrt{\frac{-\theta_{0,1}}{-\theta_{0,2}}} - 2\sqrt{\theta_{1,1}\theta_{1,2}} \\ &+ \frac{\theta_{0,2} - \theta_{1,2}}{-2\theta_{0,2}} + \frac{1}{2}\log\left(\frac{-\theta_{0,2}}{-\theta_{1,2}}\right) \geq 0. \end{aligned}$$

Fisher's information matrix is given by

$$\mathcal{I}(\boldsymbol{\theta}) = \nabla^2_{\boldsymbol{\theta}}\kappa(\boldsymbol{\theta}) = \begin{pmatrix} \frac{(-2\theta_2)^{1/2}}{(-2\theta_1)^{3/2}} & -\frac{1}{2(\theta_1\theta_2)^{1/2}} \\ -\frac{1}{2(\theta_1\theta_2)^{1/2}} & \frac{(-2\theta_1)^{1/2}}{(-2\theta_2)^{3/2}} + \frac{2}{(-2\theta_2)^2} \end{pmatrix}.$$

Again the off-diagonal terms in Fisher's information matrix $\mathcal{I}(\boldsymbol{\theta})$ are non-zero in the canonical parametrization. We switch to the mean parametrization by setting

$\mu = (-2\theta_2/(-2\theta_1))^{1/2}$ and $\alpha = -2\theta_2$. This provides us with diagonal Fisher's information

$$\mathcal{I}(\mu, \alpha) = \begin{pmatrix} \frac{\alpha}{\mu^3} & 0 \\ 0 & \frac{1}{2\alpha^2} \end{pmatrix}. \tag{2.31}$$

This transformation is again obtained by using the corresponding Jacobian matrix for variable transformation, see (3.16), below. We compare the lower-right entries of (2.30) and (2.31). Remark that we have first order approximation of the digamma function

$$\Psi(\alpha) \approx \log\alpha - \frac{1}{2\alpha},$$

and taking derivatives says that these entries of Fisher's information are first order equivalent; this is also used in the saddlepoint approximation in Sect. 5.5.2, below. Using this second parametrization based on mean μ and dispersion $1/\alpha$, we arrive at the EDF representation of the inverse Gaussian model with unit deviance

$$\mathfrak{d}(Y, \mu) = \frac{(Y-\mu)^2}{\mu^2 Y} \geq 0.$$

■

More examples will be given in Chap. 4, below.

Chapter 3
Estimation Theory

This chapter gives an introduction to decision and estimation theory. This introduction is based on the books of Lehmann [243, 244], the lecture notes of Künsch [229] and the book of Van der Vaart [363]. This chapter presents classical statistical estimation theory, it embeds estimation into a historical context, and it provides important aspects and intuition for modern data science and predictive modeling. For further reading we recommend the books of Barndorff-Nielsen [23], Berger [31], Bickel–Doksum [33] and Efron–Hastie [117].

3.1 Introduction to Decision Theory

We start from an observation vector $\boldsymbol{Y} = (Y_1, \ldots, Y_n)^\top$ taking values in a measurable space $\mathbb{Y} \subset \mathbb{R}^n$, where $n \in \mathbb{N}$ denotes the number of components Y_i, $1 \le i \le n$, in $\boldsymbol{Y}$. Assume that this observation vector $\boldsymbol{Y}$ has been generated by a distribution belonging to the family $\mathcal{P} = \{P(\cdot; \theta); \theta \in \boldsymbol{\Theta}\}$ being parametrized by a parameter set $\boldsymbol{\Theta}$.

Remarks 3.1 There are some subtle points in the notation that we are going to use. We use $P(\cdot; \theta)$ for the distribution of the observation vector $\boldsymbol{Y}$, and if we consider a specific component Y_i of $\boldsymbol{Y}$ we will use the notation $Y_i \sim F(\cdot; \theta)$. We make this distinction as in estimation theory one often considers i.i.d. observations $Y_i \sim F(\cdot; \theta)$, $1 \le i \le n$, with (in this case) joint product distribution $\boldsymbol{Y} \sim P(\cdot; \theta)$. This latter distribution is then used for purposes of maximum likelihood estimation, etc. The family $\mathcal{P}$ is parametrized by $\theta \in \boldsymbol{\Theta}$, and if we want to emphasize that this parameter is a k-dimensional vector we use boldface notation $\boldsymbol{\theta}$, this is similar to the EFs introduced in Chap. 2, but in this chapter we do not restrict to EFs. Finally, we assume identifiability meaning that different parameters θ give different distributions $P(\cdot; \theta) \in \mathcal{P}$.

M. V. Wüthrich, M. Merz, *Statistical Foundations of Actuarial Learning and its Applications*, Springer Actuarial, https://doi.org/10.1007/978-3-031-12409-9_3

To fix ideas, assume we want to determine $\gamma(\theta)$ of a given functional $\gamma(\cdot)$ on $\boldsymbol{\Theta}$. Typically, the true value $\theta \in \boldsymbol{\Theta}$ is not known, and we are not able to determine $\gamma(\theta)$ explicitly. Therefore, we try to *estimate* $\gamma(\theta)$ from data $\boldsymbol{Y} \sim P(\cdot;\theta)$ that belongs to the same $\theta \in \boldsymbol{\Theta}$. As an example we may think of working in the EDF of Chap. 2, and we are interested in the mean $\mu = \mathbb{E}_\theta[Y] = \kappa'(\theta)$ of Y. Thus, we aim at determining $\gamma(\theta) = \kappa'(\theta)$. If the true θ is unknown, and if we have an observation Y from this model, we can try to estimate $\gamma(\theta) = \kappa'(\theta)$ from Y. This motivation is based on estimation of $\gamma(\theta)$, but the following framework of decision making is more general, for instance, it may also be used for statistical hypothesis testing.

Denote the *action space* of possible decisions (actions) by $\mathbb{A}$. In decision theory we are looking for a *decision rule* (*action rule*)

$$A : \mathbb{Y} \to \mathbb{A}, \qquad \boldsymbol{Y} \mapsto A(\boldsymbol{Y}), \tag{3.1}$$

which should be understood as an educated guess for $\gamma(\theta)$ based on observation $\boldsymbol{Y}$. A decision rule is evaluated in terms of a (given) *loss function*

$$L : \boldsymbol{\Theta} \times \mathbb{A} \to \mathbb{R}_+, \qquad (\theta, a) \mapsto L(\theta, a) \geq 0. \tag{3.2}$$

$L(\theta, a)$ describes the loss of an action $a \in \mathbb{A}$ w.r.t. a true parameter choice $\theta \in \boldsymbol{\Theta}$. The *risk function* of decision rule A for data generated by $\boldsymbol{Y} \sim P(\cdot;\theta)$ is defined by

$$\theta \mapsto \mathcal{R}(\theta, A) = \mathbb{E}_\theta[L(\theta, A(\boldsymbol{Y}))] = \int_{\mathbb{Y}} L\left(\theta, A(\boldsymbol{y})\right) dP(\boldsymbol{y};\theta), \tag{3.3}$$

where $\mathbb{E}_\theta$ is the expectation w.r.t. the probability distribution $P(\cdot;\theta)$. Risk function (3.3) describes the long-term average loss of using decision rule A. As an example we may think of estimating $\gamma(\theta)$ for unknown (true) parameter θ by a decision rule $\boldsymbol{Y} \mapsto A(\boldsymbol{Y})$. Then, the loss function $L(\theta, A(\boldsymbol{Y}))$ should describe the *estimation loss* if we consider the discrepancy between $\gamma(\theta)$ and its estimate $A(\boldsymbol{Y})$, and the risk function $\mathcal{R}(\theta, A)$ is the *average estimation loss* in that case.

Good decision rules A should provide a small risk $\mathcal{R}(\theta, A)$. Unfortunately, this statement is of rather theoretical nature because, in general, the true data generating parameter θ is not known and the goodness of a decision rule for the true parameter cannot be evaluated explicitly, but the risk can only be estimated (for instance, using a bootstrap approach). Moreover, typically, there does not exist a uniformly best decision rule A over all $\theta \in \boldsymbol{\Theta}$. For these reasons we may (just) try to eliminate decision rules that are obviously not good. We give two introductory examples.

Example 3.2 (Minimax Decision Rule) Decision rule A is called minimax if for all alternative decision rules $\widetilde{A} : \mathbb{Y} \to \mathbb{A}$ we have

$$\sup_{\theta \in \boldsymbol{\Theta}} \mathcal{R}(\theta, A) \leq \sup_{\theta \in \boldsymbol{\Theta}} \mathcal{R}(\theta, \widetilde{A}).$$

A minimax decision rule is the best choice in the worst case of the true θ, i.e., it minimizes the worst case risk. ■

Example 3.3 (Bayesian Decision Rule) Assume we are given a distribution π on $\boldsymbol{\Theta}$. Decision rule A is called Bayesian w.r.t. π if it satisfies

$$A = \arg\min_{\widetilde{A}} \int_{\boldsymbol{\Theta}} \mathcal{R}(\theta, \widetilde{A}) d\pi(\theta).$$

Distribution π is called *prior distribution* on $\boldsymbol{\Theta}$. ■

The above examples give two possible choices of decision rules. The first one tries to minimize the worst case risk, whereas the second one uses additional knowledge in terms of a prior distribution π on $\boldsymbol{\Theta}$. This means that we impose stronger assumptions in the second case to get stronger conclusions. The difficult part in practice is to justify these stronger assumptions in order to validate the stronger conclusions. Below, we are going to introduce other criteria that should be satisfied by good decision rules, an important one in estimation will be unbiasedness.

3.2 Parameter Estimation

This section focuses on estimating the (unknown) parameter $\theta \in \boldsymbol{\Theta}$ from observation $\boldsymbol{Y} \sim P(\cdot; \theta)$. For this we consider decision rules $A : \mathbb{Y} \to \mathbb{A} = \boldsymbol{\Theta}$ with $A(\boldsymbol{Y})$ estimating θ. We assume there exist densities $p(\cdot; \theta)$ w.r.t. a fixed σ-finite measure ν on $\mathbb{Y} \subset \mathbb{R}^n$,

$$dP(\boldsymbol{y}; \theta) = p(\boldsymbol{y}; \theta) d\nu(\boldsymbol{y}),$$

for all distributions $P(\cdot; \theta) \in \mathcal{P}$, i.e., all $\theta \in \boldsymbol{\Theta}$.

Definition 3.4 (Maximum Likelihood Estimator, MLE) The maximum likelihood estimator (MLE) of θ for a given observation $\boldsymbol{Y} \in \mathbb{Y}$ is given by (subject to existence and uniqueness)

$$\widehat{\theta}^{\mathrm{MLE}} = \arg\max_{\widetilde{\theta} \in \boldsymbol{\Theta}} p(\boldsymbol{Y}; \widetilde{\theta}) = \arg\max_{\widetilde{\theta} \in \boldsymbol{\Theta}} \ell_{\boldsymbol{Y}}(\widetilde{\theta}),$$

where the log-likelihood function of $p(\boldsymbol{Y}; \theta)$ is defined by $\theta \mapsto \ell_{\boldsymbol{Y}}(\theta) = \log p(\boldsymbol{Y}; \theta)$.

The MLE $\boldsymbol{Y} \mapsto \widehat{\theta}^{\mathrm{MLE}} = \widehat{\theta}^{\mathrm{MLE}}(\boldsymbol{Y}) = A(\boldsymbol{Y})$ is nothing else than a specific decision rule with action space $\mathbb{A} = \boldsymbol{\Theta}$ for estimating θ. We can now start to explore the risk function $\mathcal{R}(\theta, \widehat{\theta}^{\mathrm{MLE}})$ of that decision rule for a given loss function L.

Example 3.5 (MLE within the EDF) We emphasize that this example is used throughout these notes. Assume that the (independent) components of $\boldsymbol{Y} = (Y_1, \ldots, Y_n)^\top \sim P(\cdot; \theta)$ follow a given EDF distribution. That is, we assume that $Y_1, \ldots, Y_n$ are independent and have densities w.r.t. σ-finite measures on $\mathbb{R}$ given by, see (2.14),

$$Y_i \sim f(y_i; \theta, v_i/\varphi) = \exp\left\{ \frac{y_i\theta - \kappa(\theta)}{\varphi/v_i} + a(y_i; v_i/\varphi) \right\},$$

for $1 \leq i \leq n$. Note that these random variables are not i.i.d. because they may differ in exposures $v_i > 0$. Throughout, we assume that Assumption 2.6 is fulfilled and that the cumulant function κ is steep, see Theorem 2.19. For the latter we also refer to Remark 2.20: the supports $\mathfrak{T}_{v_i/\varphi}$ of Y_i may differ; however, these supports share the same convex closure.

Independence between the Y_i's implies that the joint probability $P(\cdot; \theta)$ is the product distribution of the individual distributions $F(\cdot; \theta, v_i/\varphi)$, $1 \leq i \leq n$. Therefore, the MLE of θ in the EDF is found by solving

$$\widehat{\theta}^{\mathrm{MLE}} = \underset{\widetilde{\theta} \in \boldsymbol{\Theta}}{\arg\max}\ \ell_{\boldsymbol{Y}}(\widetilde{\theta}) = \underset{\widetilde{\theta} \in \boldsymbol{\Theta}}{\arg\max} \sum_{i=1}^{n} \frac{Y_i\widetilde{\theta} - \kappa(\widetilde{\theta})}{\varphi/v_i}.$$

Since the cumulant function κ is strictly convex we receive the MLE (subject to existence)

$$\widehat{\theta}^{\mathrm{MLE}} = \widehat{\theta}^{\mathrm{MLE}}(\boldsymbol{Y}) = (\kappa')^{-1}\left(\frac{\sum_{i=1}^n v_i Y_i}{\sum_{i=1}^n v_i}\right) = h\left(\frac{\sum_{i=1}^n v_i Y_i}{\sum_{i=1}^n v_i}\right).$$

Thus, the MLE is received by applying the canonical link $h = (\kappa')^{-1}$, see Definition 2.8, and strict convexity of κ implies that the MLE is unique. However, existence needs to be analyzed more carefully! It may happen that the MLE $\widehat{\theta}^{\mathrm{MLE}}$ is a boundary point of the effective domain $\boldsymbol{\Theta}$ which may not exist (if $\boldsymbol{\Theta}$ is open). We give an example. Assume we work in the Poisson model presented in Sect. 2.1.2. The canonical link in the Poisson model is the log-link $\mu \mapsto h(\mu) = \log(\mu)$, for $\mu > 0$. With positive probability we have in the Poisson case $\sum_{i=1}^n v_i Y_i = 0$.

Therefore, with positive probability the MLE $\widehat{\theta}^{\text{MLE}}$ does not exist (we have a degenerate Poisson model in that case).

Since the canonical link is strictly increasing we can also perform MLE in the dual (mean) parametrization. The dual parameter space is given by $\mathcal{M} = \kappa'(\mathring{\boldsymbol{\Theta}})$, see Remarks 2.9, with mean parameters $\mu = \kappa'(\theta) \in \mathcal{M}$. This motivates

$$\widehat{\mu}^{\text{MLE}} = \underset{\widetilde{\mu}\in\mathcal{M}}{\arg\max}\ \ell_{\boldsymbol{Y}}(h(\widetilde{\mu})) = \underset{\widetilde{\mu}\in\mathcal{M}}{\arg\max} \sum_{i=1}^{n} \frac{Y_i h(\widetilde{\mu}) - \kappa(h(\widetilde{\mu}))}{\varphi/v_i}. \tag{3.4}$$

Subject to existence, this provides the unique MLE

$$\widehat{\mu}^{\text{MLE}} = \widehat{\mu}^{\text{MLE}}(\boldsymbol{Y}) = \frac{\sum_{i=1}^{n} v_i Y_i}{\sum_{i=1}^{n} v_i}. \tag{3.5}$$

Also this dual MLE does not need to exist (in the dual parameter space $\mathcal{M}$). Under the assumption that the cumulant function κ is steep, we know that the closure of the dual parameter space $\overline{\mathcal{M}}$ contains the supports $\mathfrak{T}_{v_i/\varphi}$ of Y_i, see Theorem 2.19 and Remark 2.20. Thus, in that case we can close the dual parameter space and receive MLE $\widehat{\mu}^{\text{MLE}} \in \overline{\mathcal{M}}$ (in a possibly degenerate model). In the aforementioned degenerate Poisson situation we receive $\widehat{\mu}^{\text{MLE}} = 0$ which is in the boundary $\partial\mathcal{M}$ of the dual parameter space. ■

Definition 3.6 (Bayesian Estimator) The Bayesian estimator of θ for a given observation $\boldsymbol{Y} \in \mathbb{Y}$ and a given prior distribution π on $\boldsymbol{\Theta}$ is given by (subject to existence)

$$\widehat{\theta}^{\text{Bayes}} = \widehat{\theta}^{\text{Bayes}}(\boldsymbol{Y}) = \mathbb{E}_{\pi}[\theta|\boldsymbol{Y}],$$

where the conditional expectation on the right-hand side is calculated under the posterior distribution $\pi(\theta|\boldsymbol{y}) \propto p(\boldsymbol{y};\theta)\pi(\theta)$ for a given observation $\boldsymbol{Y} = \boldsymbol{y}$.

Example 3.7 (Bayesian Estimator) Assume that $\mathbb{A} = \boldsymbol{\Theta} = \mathbb{R}$ and choose the square loss function $L(\theta, a) = (\theta - a)^2$. Assume that for ν-a.e. $\boldsymbol{y} \in \mathbb{Y}$ the following decision rule $A : \mathbb{Y} \to \mathbb{A}$ exists

$$A(\boldsymbol{y}) = \underset{a\in\mathbb{A}}{\arg\min}\ \mathbb{E}_{\pi}[(\theta - a)^2|\boldsymbol{Y} = \boldsymbol{y}], \tag{3.6}$$

where the expectation is calculated w.r.t. the posterior distribution $\pi(\theta|\boldsymbol{y})$. In this case, A is a Bayesian decision rule w.r.t. π and $L(\theta, a) = (\theta - a)^2$: by assumption (3.6) we have for any other decision rule $\widetilde{A} : \mathbb{Y} \to \mathbb{A}$, ν-a.s.,

$$\mathbb{E}_\pi[(\theta - A(\boldsymbol{Y}))^2|\boldsymbol{Y} = \boldsymbol{y}] \leq \mathbb{E}_\pi[(\theta - \widetilde{A}(\boldsymbol{Y}))^2|\boldsymbol{Y} = \boldsymbol{y}].$$

Applying the tower property we receive for any other decision rule $\widetilde{A}$

$$\int_{\Theta} \mathcal{R}(\theta, A) d\pi(\theta) = \mathbb{E}[(\theta - A(\boldsymbol{Y}))^2] \leq \mathbb{E}[(\theta - \widetilde{A}(\boldsymbol{Y}))^2] = \int_{\Theta} \mathcal{R}(\theta, \widetilde{A}) d\pi(\theta),$$

where the expectation $\mathbb{E}$ is calculated over the joint distribution of $\boldsymbol{Y}$ and θ. This proves that A is a Bayesian decision rule w.r.t. π and $L(\theta, a) = (\theta - a)^2$, see Example 3.3. Finally, note that the conditional expectation given in Definition 3.6 is the minimizer of (3.6). This justifies the name Bayesian estimator in Definition 3.6 (for the square loss function). The case of the Bayesian estimator for a general loss function L is considered in Theorem 4.1.1 of Lehmann [244]. ■

Definition 3.8 (Method of Moments Estimator) Assume that $\boldsymbol{\Theta} \subseteq \mathbb{R}^k$ and that the components Y_i of $\boldsymbol{Y}$ are i.i.d. $F(\cdot; \boldsymbol{\theta})$ distributed with finite k-th moments for all $\boldsymbol{\theta} \in \boldsymbol{\Theta}$. The law of large numbers provides, a.s., for all $1 \leq l \leq k$,

$$\lim_{n\to\infty} \frac{1}{n} \sum_{i=1}^{n} Y_i^l = \mathbb{E}_{\boldsymbol{\theta}}[Y_1^l].$$

Assume that the following map is invertible (on suitable range definitions for (3.7)–(3.8))

$$\gamma : \boldsymbol{\Theta} \to \mathbb{R}^k, \qquad \boldsymbol{\theta} \mapsto \gamma(\boldsymbol{\theta}) = (\mathbb{E}_{\boldsymbol{\theta}}[Y_1], \ldots, \mathbb{E}_{\boldsymbol{\theta}}[Y_1^k])^\top. \tag{3.7}$$

The method of moments estimator of $\boldsymbol{\theta}$ is defined by

$$\widehat{\boldsymbol{\theta}}^{\mathrm{MM}} = \widehat{\boldsymbol{\theta}}^{\mathrm{MM}}(\boldsymbol{Y}) = \gamma^{-1}\left(\frac{1}{n}\sum_{i=1}^{n} Y_i, \ldots, \frac{1}{n}\sum_{i=1}^{n} Y_i^k\right)^\top. \tag{3.8}$$

The MLE, the Bayesian estimator and the method of moments estimator are the most commonly used parameter estimators. They may have additional properties (under certain assumptions) that we are going to explore below. In the remainder of this section we give an additional view on estimators which is based on the empirical distribution of the observation $\boldsymbol{Y}$.

Assume that the components Y_i of $\boldsymbol{Y}$ are real-valued and i.i.d. F distributed. The empirical distribution induced by the observation $\boldsymbol{Y} = (Y_1, \ldots, Y_n)^\top$ is given by

$$\widehat{F}_n(y) = \frac{1}{n}\sum_{i=1}^{n} \mathbb{1}_{\{Y_i \le y\}} \qquad \text{for } y \in \mathbb{R}, \tag{3.9}$$

we also refer to Fig. 1.2 (lhs). The Glivenko–Cantelli theorem [64, 159] tells us that the empirical distribution $\widehat{F}_n$ converges uniformly to F, a.s., for $n \to \infty$.

Definition 3.9 (Fisher-Consistency) Denote by $\mathfrak{P}$ the set of all distribution functions on the given probability space. Let $Q : \mathfrak{P} \to \boldsymbol{\Theta}$ be a functional with the property

$$Q(F(\cdot;\theta)) = \theta \qquad \text{for all } F(\cdot;\theta) \in \mathcal{F} = \{F(\cdot;\theta); \theta \in \boldsymbol{\Theta}\} \subset \mathfrak{P}.$$

Such a functional is called *Fisher-consistent* for $\mathcal{F}$ and $\theta \in \boldsymbol{\Theta}$, respectively.

A given Fisher-consistent functional Q motivates the estimator $\widehat{\theta} = Q(\widehat{F}_n) \in \boldsymbol{\Theta}$. This is exactly what we have applied for the method of moments estimator (3.8) with Fisher-consistent functional induced by the inverse of (3.7). The next example shows that this also works for MLE.

Example 3.10 (MLE and Kullback–Leibler (KL) Divergence) The MLE can be received from a Fisher-consistent functional. Consider for $F \in \mathfrak{P}$ the functional

$$Q(F) = \underset{\widetilde{\theta}}{\arg\max} \int \log f(y;\widetilde{\theta})dF(y),$$

assuming that $f(\cdot;\widetilde{\theta})$ are densities w.r.t. a σ-finite measure on $\mathbb{R}$. Assume that F has density f w.r.t. the σ-finite measure ν on $\mathbb{R}$. Then, we can rewrite the above as

$$Q(F) = \underset{\widetilde{\theta}}{\arg\min} \int \log\left(\frac{f(y)}{f(y;\widetilde{\theta})}\right) f(y)d\nu(y) = \underset{\widetilde{\theta}}{\arg\min}\; D_{\mathrm{KL}}(f||f(\cdot;\widetilde{\theta})).$$

The latter is the Kullback–Leibler (KL) divergence which we have met in Sect. 2.3. Lemma 2.21 states that the KL divergence is non-negative, and it is zero if and only if the two densities f and $f(\cdot;\widetilde{\theta})$ are identical, ν-a.s. This implies that $Q(F(\cdot;\theta)) = \theta$. Thus, Q is Fisher-consistent for $\theta \in \boldsymbol{\Theta}$, assuming identifiability, see Remarks 3.1.

Next, we use this Fisher-consistent functional (KL divergence) to receive the MLE. Replace the unknown distribution F by the empirical one to receive

$$\begin{aligned} Q(\widehat{F}_n) &= \underset{\widetilde{\theta}}{\arg\min}\; D_{\mathrm{KL}}(\widehat{f}_n||f(\cdot;\widetilde{\theta})) \\ &= \underset{\widetilde{\theta}}{\arg\max}\; \frac{1}{n}\sum_{i=1}^{n} \log f(Y_i;\widetilde{\theta}) \;=\; \widehat{\theta}^{\mathrm{MLE}}, \end{aligned}$$

where we have used that the empirical density $\widehat{f}_n$ allocates point masses of size $1/n$ to the i.i.d. observations $Y_1, \ldots, Y_n$. Thus, the MLE $\widehat{\theta}^{\mathrm{MLE}}$ of θ can be obtained by choosing the model $f(\cdot; \widetilde{\theta})$, $\widetilde{\theta} \in \boldsymbol{\Theta}$, that is closest in KL divergence to the empirical distribution $\widehat{F}_n$ of i.i.d. observations $Y_i \sim F$. Note that in this construction we do not assume that the true distribution F is in $\mathcal{F}$, see Definition 3.9. ■

Remarks 3.11

- Many properties of estimators of θ are based on properties of Fisher-consistent functionals Q (in cases where they exist). For instance, asymptotic properties as $n \to \infty$ are obtained from smoothness properties of Fisher-consistent functionals Q, or using the influence function we can analyze the impact of individual observations Y_i on decision rules $\widehat{\theta} = \widehat{\theta}(\boldsymbol{Y}) = Q(\widehat{F}_n)$. The latter is the basis of robust statistics, see Huber [194] and Hampel et al. [180]. Since Fisher-consistent functionals do not require that the true distribution belongs to $\mathcal{F}$ it requires a careful consideration of the quantity to be estimated.
- The discussion on parameter estimation has implicitly assumed that the true data generating model belongs to the family $\mathcal{P} = \{P(\cdot; \theta); \theta \in \boldsymbol{\Theta}\}$, and the only problem was to find the true parameter in $\boldsymbol{\Theta}$. More generally, one should also consider model uncertainty w.r.t. the chosen family $\mathcal{P}$, i.e., the data generating model may not belong to this family. Of course, this problem is by far more difficult. We explore this in more detail in Sect. 11.1.4, below.

3.3 Unbiased Estimators

We introduce the property of uniformly minimum variance unbiased (UMVU) for decision rules in this section. This is a very attractive property in insurance pricing because it gives a quality statement to decision rules (and to the resulting prices). At the current stage it is not clear how unbiasedness is related, e.g., to the MLE of θ.

3.3.1 Cramér–Rao Information Bound

Above we have stated some quality criteria for decision rules like the minimax property. A crucial property in financial applications is the so-called *unbiasedness* (for mean estimates) because this guarantees that the overall (price) levels are correctly specified.

Definition 3.12 (Uniformly Minimum Variance Unbiased, UMVU) A decision rule $A : \mathbb{Y} \to \mathbb{A} = \mathbb{R}$ is unbiased for $\gamma : \boldsymbol{\Theta} \to \mathbb{R}$ if for all $\boldsymbol{Y} \sim P(\cdot; \theta), \theta \in \boldsymbol{\Theta}$, we have

$$\mathbb{E}_\theta[A(\boldsymbol{Y})] = \gamma(\theta). \tag{3.10}$$

The decision rule A is called UMVU for γ if additionally to the unbiasedness (3.10) we have

$$\mathrm{Var}_\theta(A(\boldsymbol{Y})) \leq \mathrm{Var}_\theta(\widetilde{A}(\boldsymbol{Y})),$$

for all $\theta \in \boldsymbol{\Theta}$ and for any other decision rule $\widetilde{A} : \mathbb{Y} \to \mathbb{R}$ that is unbiased for γ.

Note that unbiasedness is not invariant under transformations, i.e., if $A(\boldsymbol{Y})$ is unbiased for $\gamma(\theta)$, then, in general, $b(A(\boldsymbol{Y}))$ is not unbiased for $b(\gamma(\theta))$. For instance, if b is strictly convex then we get a counterexample by simply applying Jensen's inequality.

Our first step is to derive a general lower bound for $\mathrm{Var}_\theta(A(\boldsymbol{Y}))$. If this general lower bound is met for an unbiased decision rule A for γ, then we know that it is UMVU for γ. We start with the one-dimensional case given in Section 2.6 of Lehmann [244].

Theorem 3.13 (Cramér–Rao Information Bound) *Assume that the distributions $P(\cdot; \theta)$, $\theta \in \boldsymbol{\Theta}$, have densities $p(\cdot; \theta)$ for a given σ-finite measure ν on $\mathbb{Y}$, and that $\boldsymbol{\Theta} \subset \mathbb{R}$ is an open interval such that the set $\{\boldsymbol{y};\ p(\boldsymbol{y}; \theta) > 0\}$ does not depend on $\theta \in \boldsymbol{\Theta}$. Let $A(\boldsymbol{Y})$ be unbiased for $\gamma : \boldsymbol{\Theta} \to \mathbb{R}$ having finite second moment. If the limit*

$$\frac{\partial}{\partial \theta} \log p(\boldsymbol{y}; \theta) = \lim_{\Delta \to 0} \frac{1}{\Delta} \frac{p(\boldsymbol{y}; \theta + \Delta) - p(\boldsymbol{y}; \theta)}{p(\boldsymbol{y}; \theta)}$$

exists in $L^2(P(\cdot; \theta))$ and if

$$\mathcal{I}(\theta) = \mathbb{E}_\theta\left[\left(\frac{\partial}{\partial \theta} \log p(\boldsymbol{Y}; \theta)\right)^2\right] \in (0, \infty),$$

then the function $\theta \mapsto \gamma(\theta)$ is differentiable, $\mathbb{E}_\theta[\frac{\partial}{\partial \theta} \log p(\boldsymbol{Y}; \theta)] = 0$ and we have information bound

$$\mathrm{Var}_\theta(A(\boldsymbol{Y})) \geq \frac{\gamma'(\theta)^2}{\mathcal{I}(\theta)}.$$

Proof We start from an arbitrary function $\psi : \mathbf{\Theta} \times \mathbb{Y} \to \mathbb{R}$ with finite variance $\mathrm{Var}_\theta(\psi(\theta, \boldsymbol{Y})) \in (0, \infty)$ for all $\theta \in \mathbf{\Theta}$. The Cauchy–Schwarz inequality implies

$$\mathrm{Var}_\theta(A(\boldsymbol{Y})) \geq \frac{\mathrm{Cov}_\theta(A(\boldsymbol{Y}), \psi(\theta, \boldsymbol{Y}))^2}{\mathrm{Var}_\theta(\psi(\theta, \boldsymbol{Y}))}. \tag{3.11}$$

If we manage to make the right-hand side of (3.11) independent of decision rule $A(\cdot)$ we have a general lower bound, we also refer to Theorem 2.6.1 in Lehmann [244].

The Cauchy–Schwarz inequality implies that for any $U \in L^2(P(\cdot; \theta))$ the following limit exists and is equal to

$$\lim_{\Delta \to 0} \mathbb{E}_\theta \left[\frac{1}{\Delta} \frac{p(\boldsymbol{Y}; \theta + \Delta) - p(\boldsymbol{Y}; \theta)}{p(\boldsymbol{Y}; \theta)} U \right] = \mathbb{E}_\theta \left[\frac{\partial}{\partial \theta} \log p(\boldsymbol{Y}; \theta) U \right]. \tag{3.12}$$

Setting $U \equiv 1$ gives average score $\mathbb{E}_\theta[\frac{\partial}{\partial \theta} \log p(\boldsymbol{Y}; \theta)] = 0$ because for sufficiently small Δ

$$\mathbb{E}_\theta \left[\frac{p(\boldsymbol{Y}; \theta + \Delta) - p(\boldsymbol{Y}; \theta)}{p(\boldsymbol{Y}; \theta)} \right] = \int_{\mathbb{Y}} \frac{p(\boldsymbol{y}; \theta + \Delta) - p(\boldsymbol{y}; \theta)}{p(\boldsymbol{y}; \theta)} p(\boldsymbol{y}; \theta) d\nu(\boldsymbol{y}) = 0,$$

where we have used that the support of the random variables does not depend on θ and that the domain $\mathbf{\Theta}$ of θ is open.

Secondly, we set $U = A(\boldsymbol{Y})$ in (3.12). We have similarly to above using unbiasedness w.r.t. γ

$$\begin{aligned}\mathrm{Cov}_\theta \left(A(\boldsymbol{Y}), \frac{p(\boldsymbol{Y}; \theta + \Delta) - p(\boldsymbol{Y}; \theta)}{p(\boldsymbol{Y}; \theta)} \right) &= \int_{\mathbb{Y}} A(\boldsymbol{y}) \frac{p(\boldsymbol{y}; \theta + \Delta) - p(\boldsymbol{y}; \theta)}{p(\boldsymbol{y}; \theta)} p(\boldsymbol{y}; \theta) d\nu(\boldsymbol{y}) \\ &= \gamma(\theta + \Delta) - \gamma(\theta).\end{aligned}$$

Existence of limit (3.12) provides the differentiability of γ. Finally, from (3.11) we have

$$\mathrm{Var}_\theta(A(\boldsymbol{Y})) \geq \lim_{\Delta \to 0} \frac{\mathrm{Cov}_\theta \left(A(\boldsymbol{Y}), \frac{p(\boldsymbol{Y}; \theta + \Delta) - p(\boldsymbol{Y}; \theta)}{p(\boldsymbol{Y}; \theta)} \right)^2}{\mathrm{Var}_\theta \left(\frac{p(\boldsymbol{Y}; \theta + \Delta) - p(\boldsymbol{Y}; \theta)}{p(\boldsymbol{Y}; \theta)} \right)} = \frac{\gamma'(\theta)^2}{\mathcal{I}(\theta)}. \tag{3.13}$$

This completes the proof. □

Remarks 3.14 (Fisher's Information and Score)

- $\mathcal{I}(\theta)$ is called *Fisher's information* or *Fisher metric*.
- $s(\theta, \boldsymbol{Y}) = \frac{\partial}{\partial \theta} \log p(\boldsymbol{Y}; \theta)$ is called *score*, and $\mathbb{E}_\theta[s(\boldsymbol{Y}; \theta)] = 0$ in Theorem 3.13 expresses that the average score is zero under the assumptions of that theorem.
- Under the regularity conditions of Lemma 6.1 in Section 2.6 of Lehmann [244]

$$\mathcal{I}(\theta) = \mathbb{E}_\theta\left[\left(\frac{\partial}{\partial\theta}\log p(\boldsymbol{Y};\theta)\right)^2\right] = -\mathbb{E}_\theta\left[\frac{\partial^2}{\partial\theta^2}\log p(\boldsymbol{Y};\theta)\right]. \qquad (3.14)$$

Fisher's information $\mathcal{I}(\theta)$ expresses the variance of the score $s(\theta, \boldsymbol{Y})$. Identity (3.14) justifies the notion Fisher's information in Sect. 2.3 for the EF.

- In order to determine the Cramér–Rao information bound for unknown θ we need to estimate Fisher's information $\mathcal{I}(\theta)$ from the available data. There are two different ways to do so, either we choose

$$\mathcal{I}(\widehat{\theta}) = \mathbb{E}_{\widehat{\theta}}\left[\left(\frac{\partial}{\partial\theta}\log p(\boldsymbol{Y};\theta)\right)^2\right],$$

or we choose the *observed Fisher's information*

$$\widehat{\mathcal{I}}(\widehat{\theta}) = \left(\frac{\partial}{\partial\theta}\log p(\boldsymbol{Y};\theta)\right)^2\Bigg|_{\theta=\widehat{\theta}},$$

for given data $\boldsymbol{Y}$ and where $\widehat{\theta} = \widehat{\theta}(\boldsymbol{Y})$. Both estimated Fisher's information $\mathcal{I}(\widehat{\theta})$ and $\widehat{\mathcal{I}}(\widehat{\theta})$ play a central role in MLE of generalized linear models (GLMs). They are used in Fisher's scoring method, the iterated re-weighted least squares (IRLS) algorithm and the Newton–Raphson algorithm to determine the MLE.

- The Cramér–Rao information bound in Theorem 3.13 is stated in terms of the observation $\boldsymbol{Y} \sim p(\cdot;\theta)$. Assume that the components Y_i of $\boldsymbol{Y}$ are i.i.d. $f(\cdot;\theta)$ distributed. In this case, Fisher's information scales as

$$\mathcal{I}(\theta) = \mathcal{I}_n(\theta) = n\mathcal{I}_1(\theta), \qquad (3.15)$$

with single risk's Fisher's information (contribution)

$$\mathcal{I}_1(\theta) = \mathbb{E}_\theta\left[\left(\frac{\partial}{\partial\theta}\log f(Y_1;\theta)\right)^2\right].$$

In general, Fisher's information is additive in independent random variables, because the product of densities is additive after applying the logarithm, and because the average score is zero.

Proposition 3.15 *The unbiased decision rule A for γ attains the Cramér–Rao information bound if and only if the density is of the form $p(\boldsymbol{y};\theta) = \exp\{\delta(\theta)T(\boldsymbol{y}) - \beta(\theta) + a(\boldsymbol{y})\}$ with $T = A$. In that case we have $\gamma(\theta) = \beta'(\theta)/\delta'(\theta)$.*

Proof of Proposition 3.15 The Cauchy–Schwarz inequality provides equality in (3.13) if and only if $\frac{\partial}{\partial\theta}\log p(\boldsymbol{y};\theta) = \delta'(\theta)A(\boldsymbol{y}) - \beta'(\theta)$, ν-a.s, for some functions $\delta'(\theta)$ and $\beta'(\theta)$ on $\boldsymbol{\Theta}$. Integration and the fact that $p(\cdot;\theta)$ is a density whose support does not depend on the explicit choice of $\theta \in \boldsymbol{\Theta}$ provide the implication "$\Rightarrow$". For the implication "$\Leftarrow$" we study for $A = T$

$$0 = \mathbb{E}_\theta\left[\frac{\partial}{\partial\theta}\log p(\boldsymbol{Y};\theta)\right] = \int_{\mathbb{Y}} (\delta'(\theta)A(\boldsymbol{y}) - \beta'(\theta))p(\boldsymbol{y};\theta)d\nu(\boldsymbol{y}) = \delta'(\theta)\mathbb{E}_\theta[A(\boldsymbol{Y})] - \beta'(\theta).$$

In that case we have $\gamma(\theta) = \mathbb{E}_\theta[A(\boldsymbol{Y})] = \beta'(\theta)/\delta'(\theta)$. Moreover, we have equality in the Cauchy–Schwarz inequality. This finishes the proof. □

The single-parameter EF fulfills the properties of Proposition 3.15 with $\delta(\theta) = \theta$ and $\beta(\theta) = \kappa(\theta)$, and decision rule $A(y) = T(y)$ attains the Cramér–Rao information bound for $\gamma(\theta) = \kappa'(\theta)$.

We give a multi-dimensional version of the Cramér–Rao information bound.

Theorem 3.16 (Multi-Dimensional Version of the Cramér–Rao Information Bound, Without Proof) *Assume that the distributions $P(\cdot;\boldsymbol{\theta})$, $\boldsymbol{\theta} \in \boldsymbol{\Theta}$, have densities $p(\cdot;\boldsymbol{\theta})$ for a given σ-finite measure ν on $\mathbb{Y}$, and that $\boldsymbol{\Theta} \subseteq \mathbb{R}^k$ is an open convex set such that the set $\{\boldsymbol{y};\ p(\boldsymbol{y};\boldsymbol{\theta}) > 0\}$ does not depend on $\boldsymbol{\theta} \in \boldsymbol{\Theta}$. Let $A(\boldsymbol{Y})$ be unbiased for $\gamma : \boldsymbol{\Theta} \to \mathbb{R}$ having finite second moment. Under additional regularity conditions, see Theorem 7.3 in Section 2.7 of Lehmann [244], we have*

$$\mathrm{Var}_{\boldsymbol{\theta}}(A(\boldsymbol{Y})) \geq (\nabla_{\boldsymbol{\theta}}\gamma(\boldsymbol{\theta}))^\top \mathcal{I}(\boldsymbol{\theta})^{-1}\nabla_{\boldsymbol{\theta}}\gamma(\boldsymbol{\theta}),$$

with (positive definite) Fisher's information matrix $\mathcal{I}(\boldsymbol{\theta}) = (\mathcal{I}_{l,j}(\boldsymbol{\theta}))_{1\le l,j\le k}$ given by

$$\mathcal{I}_{l,j}(\boldsymbol{\theta}) = \mathbb{E}_{\boldsymbol{\theta}}\left[\frac{\partial}{\partial\theta_l}\log p(\boldsymbol{Y};\boldsymbol{\theta})\frac{\partial}{\partial\theta_j}\log p(\boldsymbol{Y};\boldsymbol{\theta})\right],$$

for $1 \le l, j \le k$.

Remarks 3.17

- Whenever an unbiased decision rule $A(\boldsymbol{Y})$ for $\gamma(\boldsymbol{\theta})$ meets the Cramér–Rao information bound it is UMVU. Thus, it minimizes the risk function $\mathcal{R}(\boldsymbol{\theta}, A)$ being based on the square loss $L(\boldsymbol{\theta}, a) = (\gamma(\boldsymbol{\theta}) - a)^2$ among all unbiased decision rules, because unbiasedness for $\gamma(\boldsymbol{\theta})$ gives $\mathcal{R}(\boldsymbol{\theta}, A) = \mathrm{Var}_{\boldsymbol{\theta}}(A(\boldsymbol{Y}))$.
- The regularity conditions in Theorem 3.16 include that Fisher's information matrix $\mathcal{I}(\boldsymbol{\theta})$ is positive definite.
- Under additional regularity conditions we have the following identity for Fisher's information matrix

$$\mathcal{I}(\boldsymbol{\theta}) = \mathbb{E}_{\boldsymbol{\theta}}\left[(\nabla_{\boldsymbol{\theta}} \log p(\boldsymbol{Y}; \boldsymbol{\theta}))\,(\nabla_{\boldsymbol{\theta}} \log p(\boldsymbol{Y}; \boldsymbol{\theta}))^\top\right] = -\mathbb{E}_{\boldsymbol{\theta}}\left[\nabla_{\boldsymbol{\theta}}^2 \log p(\boldsymbol{Y}; \boldsymbol{\theta})\right] \in \mathbb{R}^{k\times k}.$$

  Thus, Fisher's information matrix can either be calculated from a quadratic form of the score $s(\boldsymbol{\theta}, \boldsymbol{Y}) = \nabla_{\boldsymbol{\theta}} \log p(\boldsymbol{Y}; \boldsymbol{\theta})$ or from the Hessian $\nabla_{\boldsymbol{\theta}}^2$ of the log-likelihood $\ell_{\boldsymbol{Y}}(\boldsymbol{\theta}) = \log p(\boldsymbol{Y}; \boldsymbol{\theta})$. Since the score has mean zero, Fisher's information matrix is equal to the covariance matrix of the score $s(\boldsymbol{\theta}, \boldsymbol{Y})$.

In many situations we do not work under the canonical parametrization $\boldsymbol{\theta}$. Considerations then require a change of variable. Assume that

$$\boldsymbol{\zeta} \in \mathbb{R}^r \;\mapsto\; \boldsymbol{\theta} = \boldsymbol{\theta}(\boldsymbol{\zeta}) \in \mathbb{R}^k,$$

such that all derivatives $\partial\theta_l(\boldsymbol{\zeta})/\partial\zeta_j$ exist for $1 \le l \le k$ and $1 \le j \le r$. The Jacobian matrix is given by

$$J(\boldsymbol{\zeta}) = \left(\frac{\partial}{\partial\zeta_j}\theta_l(\boldsymbol{\zeta})\right)_{1\le l\le k, 1\le j\le r} \in \mathbb{R}^{k\times r}.$$

Fisher's information matrix w.r.t. $\boldsymbol{\zeta}$ is given by

$$\mathcal{I}^*(\boldsymbol{\zeta}) = \left(\mathbb{E}_{\boldsymbol{\theta}(\boldsymbol{\zeta})}\left[\frac{\partial}{\partial\zeta_l}\log p(\boldsymbol{Y}; \boldsymbol{\theta}(\boldsymbol{\zeta}))\frac{\partial}{\partial\zeta_j}\log p(\boldsymbol{Y}; \boldsymbol{\theta}(\boldsymbol{\zeta}))\right]\right)_{1\le l, j\le r} \in \mathbb{R}^{r\times r},$$

and we have the identity

$$\mathcal{I}^*(\boldsymbol{\zeta}) = J(\boldsymbol{\zeta})^\top\, \mathcal{I}(\boldsymbol{\theta}(\boldsymbol{\zeta}))\, J(\boldsymbol{\zeta}). \tag{3.16}$$

This formula is used quite frequently, e.g., in generalized linear models when changing the parametrization of the models.

3.3.2 *Information Bound in the Exponential Family Case*

The purpose of this section is to summarize the Cramér–Rao information bound results for the EF and the EDF, since these families play a distinguished role in statistical and actuarial modeling.

Cramér–Rao Information Bound in the EF Case

We start with the EF case. Assume we have i.i.d. observations $Y_1, \ldots, Y_n$ having densities w.r.t. a σ-finite measure ν on $\mathbb{R}$ given by the EF, see (2.2),

$$dF(y;\boldsymbol{\theta}) \;=\; f(y;\boldsymbol{\theta})d\nu(y) \;=\; \exp\left\{\boldsymbol{\theta}^\top T(y) - \kappa(\boldsymbol{\theta}) + a(y)\right\} d\nu(y),$$

for canonical parameter $\boldsymbol{\theta} \in \boldsymbol{\Theta} \subseteq \mathbb{R}^k$. We assume to work under a minimal representation implying that the cumulant function κ is strictly convex on the interior $\mathring{\boldsymbol{\Theta}}$, see Assumption 2.6. Moreover, we assume that the cumulant function κ is steep in the sense of Theorem 2.19. Consider the (aggregated) *statistics* of the joint EF $\mathcal{P} = \{P(\cdot;\theta); \theta \in \boldsymbol{\Theta}\}$

$$\boldsymbol{y} \mapsto S(\boldsymbol{y}) \stackrel{\text{def.}}{=} \left(\sum_{i=1}^n T_1(y_i), \ldots, \sum_{i=1}^n T_k(y_i)\right)^\top \in \mathbb{R}^k. \tag{3.17}$$

We calculate the score of this EF

$$s(\boldsymbol{\theta}, \boldsymbol{Y}) = \nabla_{\boldsymbol{\theta}} \log p(\boldsymbol{Y};\boldsymbol{\theta}) = \nabla_{\boldsymbol{\theta}} \left(\boldsymbol{\theta}^\top \sum_{i=1}^n T(Y_i) - n\kappa(\boldsymbol{\theta})\right) = S(\boldsymbol{Y}) - n\nabla_{\boldsymbol{\theta}}\kappa(\boldsymbol{\theta}).$$

An immediate consequence of Corollary 2.5 is that the expected value of the score is zero for any $\boldsymbol{\theta} \in \mathring{\boldsymbol{\Theta}}$. This then reads as

$$\mu = \mathbb{E}_\theta\left[T(Y_1)\right] = \mathbb{E}_\theta\left[S(\boldsymbol{Y})/n\right] = \nabla_{\boldsymbol{\theta}}\kappa(\boldsymbol{\theta}) \in \mathbb{R}^k. \tag{3.18}$$

Thus, the statistics $S(\boldsymbol{Y})/n$ is an unbiased decision rule for the mean $\mu = \nabla_{\boldsymbol{\theta}}\kappa(\boldsymbol{\theta})$, and we can study its Cramér–Rao information bound. Fisher's information matrix is given by the positive definite matrix

$$\mathcal{I}(\boldsymbol{\theta}) = \mathcal{I}_n(\boldsymbol{\theta}) = \mathbb{E}_{\boldsymbol{\theta}}\left[s(\boldsymbol{\theta}, \boldsymbol{Y})s(\boldsymbol{\theta}, \boldsymbol{Y})^\top\right] = -\mathbb{E}_{\boldsymbol{\theta}}\left[\nabla^2_{\boldsymbol{\theta}} \log p(\boldsymbol{Y};\boldsymbol{\theta})\right] = n\nabla^2_{\boldsymbol{\theta}}\kappa(\boldsymbol{\theta}) \in \mathbb{R}^{k\times k}.$$

Note that the multi-dimensionally extended Cramér–Rao information bound in Theorem 3.16 applies to the individual components of vector $\mu = \nabla_{\boldsymbol{\theta}}\kappa(\boldsymbol{\theta}) \in \mathbb{R}^k$. Assume we would like to estimate its j-th component, set $\gamma_j(\boldsymbol{\theta}) = \mu_j =$

$(\nabla_{\boldsymbol{\theta}}\kappa(\boldsymbol{\theta}))_j = \partial\kappa(\boldsymbol{\theta})/\partial\theta_j$, for $1 \le j \le k$. This corresponds to the j-th component $S_j(\boldsymbol{Y})$ of the statistics $S(\boldsymbol{Y})$. We have unbiasedness of $S_j(\boldsymbol{Y})/n$ for $\gamma_j(\boldsymbol{\theta}) = \mu_j = (\nabla_{\boldsymbol{\theta}}\kappa(\boldsymbol{\theta}))_j$, and this unbiased statistics attains the Cramér–Rao information bound

$$\mathrm{Var}_{\boldsymbol{\theta}}(S_j(\boldsymbol{Y})/n) = \frac{1}{n}\left(\nabla^2_{\boldsymbol{\theta}}\kappa(\boldsymbol{\theta})\right)_{j,j} = (\nabla_{\boldsymbol{\theta}}\gamma_j(\boldsymbol{\theta}))^\top \mathcal{I}(\boldsymbol{\theta})^{-1}(\nabla_{\boldsymbol{\theta}}\gamma_j(\boldsymbol{\theta})). \tag{3.19}$$

Recall that $\mathcal{I}(\boldsymbol{\theta})^{-1}$ scales as n^{-1}, see (3.15). This provides us with the following corollary.

Corollary 3.18 *Assume $Y_1, \ldots, Y_n$ are i.i.d. and follow an EF (under a minimal representation). The components of the statistics $S(\boldsymbol{Y})/n$ are UMVU for $\gamma_j(\boldsymbol{\theta}) = \partial\kappa(\boldsymbol{\theta})/\partial\theta_j$, $1 \le j \le k$ and $\boldsymbol{\theta} \in \mathring{\boldsymbol{\Theta}}$, with*

$$\mathrm{Var}_{\boldsymbol{\theta}}\left(\frac{1}{n}S_j(\boldsymbol{Y})\right) = \frac{1}{n}\frac{\partial^2}{\partial\theta_j^2}\kappa(\boldsymbol{\theta}).$$

The corresponding covariance terms are for $1 \le j, l \le k$ given by

$$\mathrm{Cov}_{\boldsymbol{\theta}}\left(\frac{1}{n}S_j(\boldsymbol{Y}), \frac{1}{n}S_l(\boldsymbol{Y})\right) = \frac{1}{n}\frac{\partial^2}{\partial\theta_j\partial\theta_l}\kappa(\boldsymbol{\theta}).$$

The UMVU property stated in Corollary 3.18 is, in general, not related to MLE, but within the EF there is the following link. We have (subject to existence)

$$\widehat{\boldsymbol{\theta}}^{\mathrm{MLE}} = \underset{\widetilde{\boldsymbol{\theta}}\in\boldsymbol{\Theta}}{\arg\max}\ p(\boldsymbol{Y};\widetilde{\boldsymbol{\theta}}) = \underset{\widetilde{\boldsymbol{\theta}}\in\boldsymbol{\Theta}}{\arg\max}\ \left(\widetilde{\boldsymbol{\theta}}^\top S(\boldsymbol{Y}) - n\kappa(\widetilde{\boldsymbol{\theta}})\right) = h\left(\frac{1}{n}S(\boldsymbol{Y})\right), \tag{3.20}$$

where $h = (\nabla_{\boldsymbol{\theta}}\kappa)^{-1}$ is the canonical link of this EF, see Definition 2.8; and where we need to ensure that a solution to (3.20) exists; e.g., the solution to (3.20) might be at the boundary of $\boldsymbol{\Theta}$ which may cause problems, see Example 3.5.[1] Because the cumulant function κ is strictly convex (in a minimal representation), we receive the

[1] Another example where there does not exist a proper solution to the MLE problem (3.20) is, for instance, obtained within the 2-dimensional Gaussian EF if we have only one single observation Y_1. Intuitively this is clear because we cannot estimate two parameters from one observation $T(Y_1) = (Y_1, Y_1^2)$.

MLE for the mean parameter $\mu = \mathbb{E}_\theta [T(Y_1)]$

$$\widehat{\mu}^{\text{MLE}} = \underset{\widetilde{\mu} \in \overline{\mathcal{M}}}{\arg\max} \left(h(\widetilde{\mu})^\top S(\boldsymbol{Y}) - n\kappa(h(\widetilde{\mu})) \right) = \frac{1}{n} S(\boldsymbol{Y}),$$

the dual parameter space $\mathcal{M} = \nabla_{\boldsymbol{\theta}} \kappa(\boldsymbol{\Theta}) \subseteq \mathbb{R}^k$ has been introduced in Remarks 2.9. If $S(\boldsymbol{Y})/n$ is contained in $\mathcal{M}$, then this MLE is a proper solution; otherwise, because we have assumed that the cumulant function κ is steep, the MLE exists in the closure $\overline{\mathcal{M}}$, see Theorem 2.19, and it is UMVU for μ, see Corollary 3.18.

Corollary 3.19 (Balance Property) *Assume $Y_1, \ldots, Y_n$ are i.i.d. and follow an EF with $\boldsymbol{\theta} \in \mathring{\boldsymbol{\Theta}}$ and $T(Y_i) \in \overline{\mathcal{M}}$, a.s. The MLE $\widehat{\mu}^{\text{MLE}} \in \overline{\mathcal{M}}$ is UMVU for μ, and it fulfills the balance property on portfolio level, i.e.,*

$$\sum_{i=1}^n \mathbb{E}_{\widehat{\mu}^{\text{MLE}}} [T(Y_i)] = n\widehat{\mu}^{\text{MLE}} = S(\boldsymbol{Y}).$$

Remarks 3.20

- The balance property is a very important property in insurance pricing because it implies that the portfolio is priced on the right level: we have unbiasedness

$$\mathbb{E}_{\boldsymbol{\theta}} \left[\sum_{i=1}^n \mathbb{E}_{\widehat{\mu}^{\text{MLE}}} [T(Y_i)] \right] = \mathbb{E}_{\boldsymbol{\theta}} [S(\boldsymbol{Y})] = n\mu. \tag{3.21}$$

- We emphasize that the balance property is much stronger than unbiasedness (3.21), note that the balance property provides unbiasedness even if $\boldsymbol{Y}$ follows a completely different model, i.e., even if the chosen EF $\mathcal{P}$ is completely misspecified.
- In general, the MLE $\widehat{\boldsymbol{\theta}}^{\text{MLE}}$ is not unbiased for $\boldsymbol{\theta}$. E.g., if the canonical link $h = (\nabla_{\boldsymbol{\theta}} \kappa)^{-1}$ is strictly concave, we have from Jensen's inequality, subject to existence at the boundary of $\boldsymbol{\Theta}$,

$$\mathbb{E}_{\boldsymbol{\theta}} \left[\widehat{\boldsymbol{\theta}}^{\text{MLE}} \right] = \mathbb{E}_{\boldsymbol{\theta}} \left[h\left(\frac{1}{n} S(\boldsymbol{Y}_n) \right) \right] < h\left(\mathbb{E}_{\boldsymbol{\theta}} \left[\frac{1}{n} S(\boldsymbol{Y}_n) \right] \right) = h(\mu) = \boldsymbol{\theta}. \tag{3.22}$$

- The statistics $S(\boldsymbol{Y})$ is a sufficient statistics of $\boldsymbol{Y}$, this follows from the factorization criterion; see Theorem 1.5.2 of Lehmann [244].

Cramér–Rao Information Bound in the EDF Case

The single-parameter linear EDF case is very similar to the above vector-valued parameter EF case. We briefly summarize the main results in the EDF case.

Recall Example 3.5: assume that $Y_1, \ldots, Y_n$ are independent having densities w.r.t. a σ-finite measures on $\mathbb{R}$ (not being concentrated in a single point) given by, see (2.14),

$$Y_i \sim f(y_i; \theta, v_i/\varphi) = \exp\left\{\frac{y_i\theta - \kappa(\theta)}{\varphi/v_i} + a(y_i; v_i/\varphi)\right\}, \tag{3.23}$$

for $1 \leq i \leq n$. Note that these random variables are not i.i.d. because they may differ in the exposures $v_i > 0$. The MLE of $\mu = \kappa'(\theta)$, $\theta \in \mathring{\boldsymbol{\Theta}}$, is found by, see (3.5),

$$\widehat{\mu}^{\mathrm{MLE}} = \underset{\widetilde{\mu} \in \overline{\mathcal{M}}}{\arg\max} \sum_{i=1}^{n} \frac{Y_i h(\widetilde{\mu}) - \kappa(h(\widetilde{\mu}))}{\varphi/v_i} = \frac{\sum_{i=1}^{n} v_i Y_i}{\sum_{i=1}^{n} v_i}, \tag{3.24}$$

we assume that κ is steep to ensure $\widehat{\mu}^{\mathrm{MLE}} \in \overline{\mathcal{M}}$. The convolution formula of Corollary 2.15 says that the MLE $\widehat{\mu}^{\mathrm{MLE}} = Y_+$ belongs to the same EDF with the same canonical parameter θ and the same dispersion φ, only the weight changes to $v_+ = \sum_{i=1}^{n} v_i$.

Corollary 3.21 (Balance Property) *Assume $Y_1, \ldots, Y_n$ are independent with EDF distribution (3.23) for $\theta \in \mathring{\boldsymbol{\Theta}}$ and $Y_i \in \overline{\mathcal{M}}$, a.s. The MLE $\widehat{\mu}^{\mathrm{MLE}} \in \overline{\mathcal{M}}$ is UMVU for $\mu = \kappa'(\theta)$, and it fulfills the balance property on portfolio level, i.e.,*

$$\sum_{i=1}^{n} \mathbb{E}_{\widehat{\mu}^{\mathrm{MLE}}}\left[v_i Y_i\right] = \sum_{i=1}^{n} v_i \widehat{\mu}^{\mathrm{MLE}} = \sum_{i=1}^{n} v_i Y_i.$$

The score in this EDF is given by

$$s(\theta, \boldsymbol{Y}) = \frac{\partial}{\partial\theta} \log p(\boldsymbol{Y}; \theta) = \frac{\partial}{\partial\theta} \sum_{i=1}^{n} \frac{v_i}{\varphi} \left(\theta Y_i - \kappa(\theta)\right) = \sum_{i=1}^{n} \frac{v_i}{\varphi} \left(Y_i - \kappa'(\theta)\right).$$

Of course, we have $\mathbb{E}_\theta[s(\theta, \boldsymbol{Y})] = 0$ and we receive Fisher's information for $\theta \in \mathring{\boldsymbol{\Theta}}$

$$\mathcal{I}(\theta) = -\mathbb{E}_\theta\left[\frac{\partial^2}{\partial\theta^2} \log p(\boldsymbol{Y}; \theta)\right] = \sum_{i=1}^{n} \frac{v_i}{\varphi} \kappa''(\theta) > 0. \tag{3.25}$$

Corollary 2.15 gives for the variance of the MLE

$$\mathrm{Var}_\theta\left(\widehat{\mu}^{\mathrm{MLE}}\right) = \frac{\varphi}{\sum_{i=1}^n v_i}\kappa''(\theta) = \frac{(\kappa''(\theta))^2}{\mathcal{I}(\theta)} = \frac{(\partial\mu(\theta)/\partial\theta)^2}{\mathcal{I}(\theta)}.$$

This verifies that $\widehat{\mu}^{\mathrm{MLE}}$ meets the Cramér–Rao information bound and is UMVU for the mean $\mu = \kappa'(\theta)$.

Example 3.22 (Poisson Case) For this example, we consider independent Poisson random variables $N_i \sim \mathrm{Poi}(v_i\lambda)$. In Sect. 2.2.2 we have seen that $Y_i = N_i/v_i$ can be modeled within the single-parameter linear EDF framework using as cumulant function the exponential function $\kappa(\theta) = e^\theta$, and setting $\omega_i = v_i$ and $\varphi = 1$. Thus, the probability weights of a single observation Y_i are given by, see (2.15),

$$f(y_i;\theta, v_i) = \exp\left\{v_i\left(\theta y_i - e^\theta\right) + a(y_i; v_i)\right\},$$

with canonical parameter $\theta = \log(\lambda) \in \boldsymbol{\Theta} = \mathbb{R}$. The MLE in the mean parametrization is given by, see (3.24),

$$\widehat{\lambda}^{\mathrm{MLE}} = \frac{\sum_{i=1}^n v_i Y_i}{\sum_{i=1}^n v_i} = \frac{\sum_{i=1}^n N_i}{\sum_{i=1}^n v_i} \in \overline{\mathcal{M}} = [0,\infty).$$

This estimator is unbiased for λ. Having independent Poisson random variables we can calculate the variance of this estimator as

$$\mathrm{Var}\left(\widehat{\lambda}^{\mathrm{MLE}}\right) = \frac{\lambda}{\sum_{i=1}^n v_i}.$$

Moreover, from Corollary 3.21 we know that this estimator is UMVU for λ, which can easily be seen, and uses Fisher's information (3.25) with dispersion parameter $\varphi = 1$

$$\mathcal{I}(\theta) = -\mathbb{E}_\theta\left[\frac{\partial^2}{\partial\theta^2}\log p(\boldsymbol{Y};\theta)\right] = \sum_{i=1}^n v_i\kappa''(\theta) = \lambda\sum_{i=1}^n v_i.$$

■

One could study many other properties of decision rules (and corresponding estimators), for instance, admissibility or uniformly minimum risk equivariance (UMRE), and we could also study other families of distribution functions such as group families. We refrain from doing so because we will not need this for our purposes.

3.4 Asymptotic Behavior of Estimators

All results above have been based on a finite sample $\boldsymbol{Y}_n = (Y_1, \ldots, Y_n)^\top$, we add a lower index n to $\boldsymbol{Y}_n$ to indicate the finite sample size $n \in \mathbb{N}$. The aim of this section is to analyze properties of decision rules when the sample size n tends to infinity.

3.4.1 Consistency

Assume we have an infinite sequence of observations Y_i, $i \geq 1$, which allows us to construct an infinite sequence of decision rules $A_n = A_n(\boldsymbol{Y}_n)$, $n \geq 1$, where A_n always considers the first n observations $\boldsymbol{Y}_n = (Y_1, \ldots, Y_n)^\top \sim P_n(\cdot; \theta)$, for $\theta \in \boldsymbol{\Theta}$ not depending on n. To fix ideas, one may think of i.i.d. random variables Y_i.

Definition 3.23 (Consistency) The sequence $A_n = A_n(\boldsymbol{Y}_n) \in \mathbb{R}^r$, $n \geq 1$, is consistent for $\gamma : \boldsymbol{\Theta} \to \mathbb{R}^r$ if for all $\theta \in \boldsymbol{\Theta}$ and for all $\varepsilon > 0$ we have

$$\lim_{n\to\infty} \mathbb{P}_\theta \left[\|A_n(\boldsymbol{Y}_n) - \gamma(\theta)\|_2 > \varepsilon \right] = 0.$$

Definition 3.23 says that $A_n(\boldsymbol{Y}_n)$ converges in probability to $\gamma(\theta)$ as $n \to \infty$. If we (even) have a.s. convergence, we call $A_n, n \geq 1$, *strongly consistent* for $\gamma : \boldsymbol{\Theta} \to \mathbb{R}^r$. Consistency is a minimal property that decision rules should fulfill. Typically, in applications, this is not enough, and we are interested in (fast) rates of convergence, i.e., we would like to know the error rates between $A_n(\boldsymbol{Y}_n)$ and $\gamma(\theta)$ for $n \to \infty$.

Example 3.24 (Consistency of the MLE in the EF) We revisit Corollary 3.19 and consider an i.i.d. sequence of random variables Y_i, $i \geq 1$, belonging to an EF, and we assume to work under a minimal representation and to have a steep cumulant function κ. The MLE for μ is given by the statistics

$$\widehat{\mu}_n^{\mathrm{MLE}} = \frac{1}{n} S(\boldsymbol{Y}_n) = \frac{1}{n} \sum_{i=1}^n (T_1(Y_i), \ldots, T_k(Y_i))^\top \in \overline{\mathcal{M}}.$$

We add a lower index n to the MLE to indicate the sample size. The i.i.d. property of Y_i, $i \geq 1$, implies that we can apply the strong law of large numbers which tells us that we have $\lim_{n\to\infty} \widehat{\mu}_n^{\mathrm{MLE}} = \mathbb{E}_{\boldsymbol{\theta}}[T(Y_1)] = \nabla_{\boldsymbol{\theta}} \kappa(\boldsymbol{\theta}) = \mu$, a.s., for all $\boldsymbol{\theta} \in \boldsymbol{\Theta}$. This implies strong consistency of the sequence of MLEs $\widehat{\mu}_n^{\mathrm{MLE}}$, $n \geq 1$, for μ.

We have seen that these MLEs are also UMVU for μ, but if we transform them to the canonical scale $\widehat{\boldsymbol{\theta}}_n^{\mathrm{MLE}}$ they are, in general, biased for $\boldsymbol{\theta}$, see (3.22). However, since the cumulant function κ is strictly convex (under a minimal representation) we receive $\lim_{n\to\infty} \widehat{\boldsymbol{\theta}}_n^{\mathrm{MLE}} = \boldsymbol{\theta}$, a.s., which provides strong consistency also on the canonical scale. ■

Proposition 3.25 *Assume the real-valued random variables* Y_i, $i \geq 1$, *are i.i.d.* $F(\cdot;\theta)$ *distributed with fixed* $\theta \in \boldsymbol{\Theta}$. *The resulting empirical distributions* $\widehat{F}_n$, $n \geq 1$, *are given by* (3.9). *Assume* Q *is a Fisher-consistent functional for* $\gamma(\theta)$, *i.e.,* $Q(F(\cdot;\theta)) = \gamma(\theta)$ *for all* $\theta \in \boldsymbol{\Theta}$. *Moreover, assume that* Q *is continuous in* $F(\cdot;\theta)$, *for all* $\theta \in \boldsymbol{\Theta}$, *w.r.t. the supremum norm. The functionals* $Q(\widehat{F}_n)$, $n \geq 1$, *are consistent for* $\gamma(\theta)$.

Sketch of Proof The Glivenko–Cantelli theorem [64, 159] says that the empirical distribution $\widehat{F}_n$ converges uniformly to $F(\cdot;\theta)$, a.s., for $n \to \infty$. Using the assumptions made, we are allowed to exchange the corresponding limits, which provides consistency. □

In view of Proposition 3.25, we discuss the case of the MLE of $\theta \in \boldsymbol{\Theta}$. In Example 3.10 we have seen that the MLE of $\theta \in \boldsymbol{\Theta}$ is obtained from a Fisher-consistent functional Q for θ on the set of probability distributions $\mathfrak{P}$ given by

$$Q(F) = \underset{\widetilde{\theta}}{\arg\max} \int \log f(y;\widetilde{\theta}) dF(y) = \underset{\widetilde{\theta}}{\arg\min}\, D_{\mathrm{KL}}(f || f(\cdot;\widetilde{\theta})),$$

in the second step we assumed that F has a density f w.r.t. a σ-finite measure ν on $\mathbb{R}$.

Assume we have i.i.d. data $Y_i \sim f(\cdot;\theta)$, $i \geq 1$. Thus, the true data generating distribution is described by the parameter $\theta \in \boldsymbol{\Theta}$. MLE requires the study of the log-likelihood function (we scale with the sample size n)

$$\widetilde{\theta} \mapsto \frac{1}{n}\ell_{\boldsymbol{Y}_n}(\widetilde{\theta}) = \frac{1}{n}\sum_{i=1}^{n} \log f(Y_i;\widetilde{\theta}).$$

The law of large numbers gives us, a.s.,

$$\lim_{n\to\infty} \frac{1}{n}\sum_{i=1}^{n} \log f(Y_i;\widetilde{\theta}) = \mathbb{E}_\theta\left[\log f(Y;\widetilde{\theta})\right]. \tag{3.26}$$

Thus, *if* we are allowed to exchange the arg max operation and the limit in $n \to \infty$ we receive, a.s.,

$$\begin{aligned} \lim_{n\to\infty} \widehat{\theta}_n^{\mathrm{MLE}} &= \lim_{n\to\infty} \left(\underset{\widetilde{\theta}}{\arg\max} \frac{1}{n}\sum_{i=1}^{n} \log f(Y_i;\widetilde{\theta}) \right) \\ &\overset{?}{=} \underset{\widetilde{\theta}}{\arg\max} \left(\lim_{n\to\infty} \frac{1}{n}\sum_{i=1}^{n} \log f(Y_i;\widetilde{\theta}) \right) \\ &= \underset{\widetilde{\theta}}{\arg\max}\, \mathbb{E}_\theta\left[\log f(Y;\widetilde{\theta})\right] = Q(F(\cdot;\theta)) = \theta. \end{aligned} \tag{3.27}$$

That is, we receive consistency of the MLE for θ if we are allowed to exchange the arg max operation and the limit in $n \to \infty$. This requires regularity conditions on the considered family of distributions $\mathcal{F} = \{F(\cdot;\theta); \theta \in \boldsymbol{\Theta}\}$. The case of a finite parameter space $\boldsymbol{\Theta} = \{\theta_1, \ldots, \theta_J\}$ is easy, this is a simplified version of Wald's [374] consistency proof,

$$\mathbb{P}_{\theta_j}\left[\theta_j \notin \underset{\theta_k}{\arg\max} \frac{1}{n}\sum_{i=1}^n \log f(Y_i;\theta_k)\right] \le \sum_{k\neq j} \mathbb{P}_{\theta_j}\left[\frac{1}{n}\sum_{i=1}^n \log f(Y_i;\theta_k) > \frac{1}{n}\sum_{i=1}^n \log f(Y_i;\theta_j)\right].$$

The right-hand side converges to 0 as $n \to \infty$ for all $\theta_k \neq \theta_j$, which gives consistency. For regularity conditions on more general parameter spaces we refer to Section 5.2 in Van der Vaart [363]. Basically, one needs that the arg max of the limiting function given on the right-hand side of (3.26) is well-separated from other large values of that function, see Theorem 5.7 in Van der Vaart [363].

Remarks 3.26

- The estimator from the arg max operation in (3.27) is also called M-estimator, and $(y, a) \mapsto \log(f(y; a))$ plays the role of a scoring function (similar to a loss function). The the last line of (3.27) says that this scoring function is strictly consistent for the functional $Q : \mathcal{F} \to \boldsymbol{\Theta}$, and Fisher-consistency of this functional Q implies

$$\mathbb{E}_\theta\left[\log f(Y;\widetilde{\theta})\right] \le \mathbb{E}_\theta\left[\log f(Y; Q(F(\cdot;\theta)))\right] = \mathbb{E}_\theta\left[\log f(Y;\theta)\right],$$

  for all $\widetilde{\theta} \in \boldsymbol{\Theta}$. Strict consistency of loss and scoring functions is going to be defined formally in Sect. 4.1.3, below, and we have just seen that this plays an important role for the consistency of M-estimators in the sense of Definition 3.23.
- Consistency (3.27) assumes that the data generating model $Y \sim F$ belongs to the specified family $\mathcal{F} = \{F(\cdot;\theta); \theta \in \boldsymbol{\Theta}\}$. Model uncertainty may imply that the data generating model does not belong to $\mathcal{F}$. In this situation, and if we are allowed to exchange the arg max operation and the limit in n in (3.27), the MLE will provide the model in $\mathcal{F}$ that is closest in KL divergence to the true model F. We come back to this in Sect. 11.1.4, below.

3.4.2 Asymptotic Normality

As mentioned above, typically, we would like to have stronger results than just consistency. We give an introductory example based on the EF.

Example 3.27 (Asymptotic Normality of the MLE in the EF) We work under the same EF as in Example 3.24. This example has provided consistency of the sequence of MLEs $\widehat{\mu}_n^{\text{MLE}}$, $n \ge 1$, for μ. Note that the i.i.d. property together with the finite variance property immediately implies the following convergence in distribution

$$\sqrt{n}\left(\widehat{\mu}_n^{\mathrm{MLE}} - \mu\right) \Rightarrow \mathcal{N}(0, \nabla_{\boldsymbol{\theta}}^2 \kappa(\boldsymbol{\theta})) \stackrel{\text{(d)}}{=} \mathcal{N}(0, \mathcal{I}_1(\boldsymbol{\theta})) \qquad \text{as } n \to \infty,$$

where $\boldsymbol{\theta} = \boldsymbol{\theta}(\mu) = (\nabla_{\boldsymbol{\theta}}\kappa)^{-1}(\mu) \in \boldsymbol{\Theta}$ for $\mu \in \mathcal{M}$, and $\mathcal{N}$ denotes the Gaussian distribution. This is the multivariate version of the central limit theorem (CLT), and it tells us that the rate of convergence is $1/\sqrt{n}$. This asymptotic result is stated in terms of Fisher's information matrix under parametrization $\boldsymbol{\theta}$. We transform this to the dual mean parametrization and call Fisher's information matrix under the dual mean parametrization $\mathcal{I}_1^*(\mu)$. This involves the change of variable $\mu \mapsto \boldsymbol{\theta} = \boldsymbol{\theta}(\mu) = (\nabla_{\boldsymbol{\theta}}\kappa)^{-1}(\mu)$. The Jacobian matrix of this change of variable is given by $J(\mu) = \mathcal{I}_1(\boldsymbol{\theta}(\mu))^{-1}$ and, thus, the transformation of Fisher's information matrix gives, see also (3.16),

$$\mu \mapsto \mathcal{I}_1^*(\mu) = J(\mu)^\top \mathcal{I}_1(\boldsymbol{\theta}(\mu))\, J(\mu) = \mathcal{I}_1(\boldsymbol{\theta}(\mu))^{-1}.$$

This allows us to express the above CLT w.r.t. Fisher's information matrix corresponding to μ and it gives us

$$\sqrt{n}\left(\widehat{\mu}_n^{\mathrm{MLE}} - \mu\right) \Rightarrow \mathcal{N}\left(0, \mathcal{I}_1^*(\mu)^{-1}\right) \qquad \text{as } n \to \infty. \tag{3.28}$$

We conclude that the appropriately normalized MLE $\widehat{\mu}_n^{\mathrm{MLE}}$ converges in distribution to the centered Gaussian distribution having as covariance matrix the *inverse of Fisher's information matrix* $\mathcal{I}_1^*(\mu)$, and the *rate of convergence* is $1/\sqrt{n}$.

Assume that the effective domain $\boldsymbol{\Theta}$ is open, and that $\boldsymbol{\theta} = \boldsymbol{\theta}(\mu) \in \boldsymbol{\Theta}$. This allows us to transform asymptotic normality (3.28) to the canonical scale. Consider again the change of variable $\mu \mapsto \boldsymbol{\theta} = \boldsymbol{\theta}(\mu) = (\nabla_{\boldsymbol{\theta}}\kappa)^{-1}(\mu)$ with Jacobian matrix $J(\mu) = \mathcal{I}_1(\boldsymbol{\theta}(\mu))^{-1} = \mathcal{I}_1^*(\mu)$. Theorem 1.9 in Section 5.2 of Lehmann [244] tells us how the CLT transforms under such a change of variable, namely,

$$\begin{aligned}\sqrt{n}\left(\widehat{\boldsymbol{\theta}}_n^{\mathrm{MLE}} - \boldsymbol{\theta}\right) &= \sqrt{n}\left((\nabla_{\boldsymbol{\theta}}\kappa)^{-1}\left(\widehat{\mu}_n^{\mathrm{MLE}}\right) - (\nabla_{\boldsymbol{\theta}}\kappa)^{-1}(\mu)\right) \\ &\Rightarrow \mathcal{N}\left(0, J(\mu)\mathcal{I}_1^*(\mu)^{-1}J(\mu)\right) \stackrel{\text{(d)}}{=} \mathcal{N}\left(0, \mathcal{I}_1(\boldsymbol{\theta})^{-1}\right) \qquad \text{as } n \to \infty.\end{aligned} \tag{3.29}$$

We have exactly the same structural form in the two asymptotic results (3.28) and (3.29). There is a main difference, $\widehat{\mu}_n^{\mathrm{MLE}}$ is unbiased for μ whereas, in general, $\widehat{\boldsymbol{\theta}}_n^{\mathrm{MLE}}$ is not unbiased for $\boldsymbol{\theta}$, but we receive the same asymptotic behavior. ∎

There are many different versions of asymptotic normality results similar to (3.28) and (3.29), and the main difficulty often is to verify the assumptions made. For instance, one can prove asymptotic normality based on a Fisher-consistent functional Q. The assumptions made are, among others, that Q needs to be Fréchet differentiable in $P(\cdot;\theta)$ which, unfortunately, is rather difficult to verify. We make a list of assumptions here that are easier to check and then we give a version of the asymptotic normality result which is stated in the book of Lehmann [244]. This list of assumptions in the one-dimensional case $\boldsymbol{\Theta} \subseteq \mathbb{R}$ reads as follows:

(i) $\boldsymbol{\Theta} \subseteq \mathbb{R}$ is an open interval (possibly infinite).
(ii) The real-valued random variables $Y_i \sim F(\cdot;\theta)$, $i \geq 1$, have common support $\mathfrak{T} = \{y \in \mathbb{R};\ f(y;\theta) > 0\}$ which is independent of $\theta \in \boldsymbol{\Theta}$.
(iii) For every $y \in \mathfrak{T}$, the density $f(y;\theta)$ is three times continuously differentiable in θ.
(iv) The integral $\int f(y;\theta)d\nu(y)$ is twice differentiable under the integral sign.
(v) Fisher's information satisfies $\mathcal{I}_1(\theta) = \mathbb{E}_\theta[(\partial \log f(Y_1;\theta)/\partial\theta)^2] \in (0,\infty)$.
(vi) For every $\theta_0 \in \boldsymbol{\Theta}$ there exist a positive constant c and a function $M(y)$ (both may depend on θ_0) such that $\mathbb{E}_{\theta_0}[M(Y_1)] < \infty$ and

$$\left|\frac{\partial^3}{\partial\theta^3}\log f(y;\theta)\right| \leq M(y) \qquad \text{for all } y \in \mathfrak{T} \text{ and } \theta \in (\theta_0 - c, \theta_0 + c).$$

Theorem 3.28 (Theorem 2.3 in Section 6.2 of Lehmann [244]) *Assume Y_i, $i \geq 1$, are i.i.d. $F(\cdot;\theta)$ distributed satisfying* (i)–(vi) *from above. Assume that $\widehat{\theta}_n = \widehat{\theta}_n(\boldsymbol{Y}_n)$, $n \geq 1$, is a sequence of roots that solves the score equations*

$$\frac{\partial}{\partial\widetilde{\theta}}\sum_{i=1}^n \log f(Y_i;\widetilde{\theta}) = \frac{\partial}{\partial\widetilde{\theta}}\ell_{\boldsymbol{Y}_n}(\widetilde{\theta}) = 0,$$

and which is consistent for θ, i.e. this sequence of roots $\widehat{\theta}_n(\boldsymbol{Y}_n)$ converges in probability to the true parameter θ. Then we have asymptotic normality

$$\sqrt{n}\left(\widehat{\theta}_n - \theta\right) \Rightarrow \mathcal{N}\left(0, \mathcal{I}_1(\theta)^{-1}\right) \qquad \text{as } n \to \infty. \tag{3.30}$$

Sketch of Proof Fix $\theta \in \boldsymbol{\Theta}$ and consider a Taylor expansion of the score $\ell'_{\boldsymbol{Y}_n}(\cdot)$ in θ for $\widehat{\theta}_n$. It is given by

$$\ell'_{\boldsymbol{Y}_n}(\widehat{\theta}_n) = \ell'_{\boldsymbol{Y}_n}(\theta) + \ell''_{\boldsymbol{Y}_n}(\theta)\left(\widehat{\theta}_n - \theta\right) + \frac{1}{2}\ell'''_{\boldsymbol{Y}_n}(\theta_n)\left(\widehat{\theta}_n - \theta\right)^2,$$

for $\theta_n \in [\theta, \widehat{\theta}_n]$. Since $\widehat{\theta}_n$ is a root of the score, the left-hand side is equal to zero. This allows us to re-arrange the above Taylor expansion as follows

$$\sqrt{n}\left(\widehat{\theta}_n - \theta\right) = \frac{\frac{1}{\sqrt{n}}\ell'_{\mathbf{Y}_n}(\theta)}{-\frac{1}{n}\ell''_{\mathbf{Y}_n}(\theta) - \frac{1}{2n}\ell'''_{\mathbf{Y}_n}(\theta_n)\left(\widehat{\theta}_n - \theta\right)}.$$

The enumerator on the right-hand side converges in distribution to $\mathcal{N}(0, \mathcal{I}_1(\theta))$, see (18) in Section 6.2 of [244], the first term in the denominator converges in probability to $\mathcal{I}_1(\theta)$, see (19) in Section 6.2 of [244], and in the second term of the denominator we have $\frac{1}{2n}\ell'''_{\mathbf{Y}_n}(\theta_n)$ which is bounded in probability, see (20) in Section 6.2 of [244]. The claim then follows from Slutsky's theorem.

□

Remarks 3.29

- A sequence $(\widehat{\theta}_n)_{n\geq 1}$ satisfying Theorem 3.28 is called *efficient likelihood estimator* (ELE) of θ. Typically, the sequence of MLEs $\widehat{\theta}_n^{\mathrm{MLE}}$ gives such an ELE sequence, but there are counterexamples where this is not the case, see Example 3.1 in Section 6.2 of Lehmann [244]. In that example $\widehat{\theta}_n^{\mathrm{MLE}}$ exists for all $n \geq 1$, but it converges in probability to ∞, regardless of the value of the true parameter θ.
- Any sequence of estimators that fulfills (3.30) is called *asymptotically efficient*, because, similarly to the Cramér–Rao information bound of Theorem 3.13, it attains $\mathcal{I}_1(\theta)^{-1}$ (which under certain assumptions is a lower variance bound except on Lebesgue measure zero, see Theorem 1.1 in Section 6.1 of Lehmann [244]). However, there are two important differences here: (1) the Cramér–Rao information bound statement needs unbiasedness of the decision rule, whereas (3.30) only requires consistency (but not unbiasedness nor asymptotically vanishing bias); and (2) the lower bound in the Cramér–Rao statement is an effective variance (on a finite sample), whereas the quantity in (3.30) is only an asymptotic variance. Moreover, any other sequence that differs in probability from an asymptotically efficient one less than $o(1/\sqrt{n})$ is asymptotically efficient, too.
- If we consider a differentiable function $\theta \mapsto \gamma(\theta)$, then Theorem 3.28 implies

$$\sqrt{n}\left(\gamma\left(\widehat{\theta}_n\right) - \gamma(\theta)\right) \Rightarrow \mathcal{N}\left(0, \frac{(\gamma'(\theta))^2}{\mathcal{I}_1(\theta)}\right) \qquad \text{as } n \to \infty. \tag{3.31}$$

  This follows from asymptotic normality, consistency and considering a Taylor expansion around θ.
- We were starting from the MLE problem

$$\widehat{\theta}_n^{\mathrm{MLE}} = \underset{\widetilde{\theta}}{\arg\max} \frac{1}{n}\sum_{i=1}^{n} \log f(Y_i; \widetilde{\theta}). \tag{3.32}$$

In statistical theory a parameter estimator that is obtained through a maximization operation is called M-estimator (for maximizing or minimizing), see also Remarks 3.26. If the log-likelihood is differentiable in $\widetilde{\theta}$ we can turn the above problem into a root search problem for $\widetilde{\theta}$

$$\frac{1}{n}\sum_{i=1}^{n}\frac{\partial}{\partial\widetilde{\theta}}\log f(Y_i;\widetilde{\theta})=0. \tag{3.33}$$

If a parameter estimator is obtained through a root search problem it is called Z-estimator (for equating to zero). The Z-estimator (3.33) does not require a maximum of the original function, but only a critical point; this is exactly what we have been exploring in Theorem 3.28. More generally, for a sufficiently nice function $\psi(\cdot;\theta)$ a Z-estimator $\widehat{\theta}_n^{\mathrm{Z}}$ for θ is obtained by solving the following equation for $\widetilde{\theta}$

$$\frac{1}{n}\sum_{i=1}^{n}\psi(Y_i;\widetilde{\theta})=0, \tag{3.34}$$

for i.i.d. data $Y_i \sim F(\cdot;\theta)$. Suppose that the first moment of $\psi(Y_i;\widetilde{\theta})$ exists. The law of large numbers gives us, a.s., see also (3.26),

$$\lim_{n\to\infty}\frac{1}{n}\sum_{i=1}^{n}\psi(Y_i;\widetilde{\theta})=\mathbb{E}_\theta\left[\psi(Y;\widetilde{\theta})\right]. \tag{3.35}$$

Consistency of the Z-estimator $\widehat{\theta}_n^{\mathrm{Z}}$, $n\ge 1$, for θ is related to the right-hand side of (3.35) being zero for $\widetilde{\theta}=\theta$. Under additional regularity conditions (and consistency) it then holds asymptotic normality

$$\sqrt{n}\left(\widehat{\theta}_n^{\mathrm{Z}}-\theta\right)\Rightarrow\mathcal{N}\left(0,\frac{\mathbb{E}_\theta\left[\psi(Y;\theta)^2\right]}{\mathbb{E}_\theta\left[\frac{\partial}{\partial\theta}\psi(Y;\theta)\right]^2}\right)\qquad\text{as } n\to\infty. \tag{3.36}$$

For rigorous statements we refer to Theorems 5.21 and 5.41 in Van der Vaart [363]. A modification to the regression case is given in Theorem 11.6 below.

Example 3.30 We consider the single-parameter linear EF for given strictly convex and steep cumulant function κ and w.r.t. a σ-finite measure ν on $\mathbb{R}$. The score equation gives requirement

$$\frac{1}{n}S(\boldsymbol{Y}_n)\overset{!}{=}\kappa'(\theta)=\mathbb{E}_\theta[Y_1]. \tag{3.37}$$

Strict convexity implies that the right-hand side strictly increases in θ. Therefore, we have at most one solution of the score equation here. We assume that the effective domain $\boldsymbol{\Theta} \subseteq \mathbb{R}$ is open. It is easily verified that assumptions (ii)–(vi) hold, in particular, (vi) saying that the third derivative should have a uniformly bounded integrable bound holds because the third derivative is independent of y and continuous in θ. With probability converging to 1, (3.37) has a solution $\widehat{\theta}_n$ which is unique, consistent and Theorem 3.28 holds. Note that in Example 3.5 we have mentioned the Poisson case which can be degenerate. For the asymptotic normality result we use here that this degeneracy asymptotically vanishes with probability converging to one. ■

Remark 3.31 (Multi-Dimensional Extension) For an extension of Theorem 3.28 to the multi-dimensional case $\boldsymbol{\Theta} \subseteq \mathbb{R}^k$ we refer to Section 6.4 in Lehmann [244]. The assumptions made in the multi-dimensional case do not essentially differ from the ones in the 1-dimensional case.

Chapter 4
Predictive Modeling and Forecast Evaluation

In the previous chapter, we have fully focused on parameter estimation $\theta \in \boldsymbol{\Theta}$ and the estimation of functions $\theta \mapsto \gamma(\theta)$ by exploiting decision rules A for estimating $\boldsymbol{Y}_n \mapsto \widehat{\theta} = A(\boldsymbol{Y}_n)$ or $\boldsymbol{Y}_n \mapsto \widehat{\gamma}(\theta) = A(\boldsymbol{Y}_n)$, respectively. The derivations in that chapter analyzed the quality of decision rules in terms of loss functions which compare, e.g., the action $\widehat{\theta} = A(\boldsymbol{Y}_n)$ to the true parameter θ. The Cramér–Rao information bound considers this in terms of a square loss function. In actuarial modeling, parameter estimation is only part of the problem, and the second part is to predict new random variables Y. These new random variables should be thought as claims in the future that we try to predict (and price) using decision rules being developed based on past information $\boldsymbol{Y}_n = (Y_1, \ldots, Y_n)^\top$. In this case, we would like to study how a decision rule $A(\boldsymbol{Y}_n)$ *generalizes* to new data Y, and we then call the decision rule rather a *predictor* for Y. This capability of suitable decision rules to generalize to new (unseen) data is analyzed in Sect. 4.1. Such an analysis often relies on (numerical) techniques such as cross-validation, which is examined in Sect. 4.2, or the bootstrap technique, being presented in Sect. 4.3, below. In this chapter, we denote past observations by $\boldsymbol{Y}_n = (Y_1, \ldots, Y_n)^\top$ supported on $\mathbb{Y}$, and the (real-valued) random variables to be predicted are denoted by Y with support $\mathcal{Y} \subset \mathbb{R}$. Often we have $\mathbb{Y} = \mathcal{Y} \times \cdots \times \mathcal{Y}$.

4.1 Generalization Loss

We start by considering the most commonly used *expected generalization loss* (GL) which is the *mean squared error of prediction* (MSEP). The MSEP is based on the square loss function, and it can be seen as a distribution-free approach to measure expected GL. In subsequent sections we will study distribution-adapted GL approaches. Expected GL measurement with MSEP is considered to be general knowledge and we do not give a specific reference in this section. Distribution-

M. V. Wüthrich, M. Merz, *Statistical Foundations of Actuarial Learning and its Applications*, Springer Actuarial, https://doi.org/10.1007/978-3-031-12409-9_4

adapted versions are mainly based on the strictly consistent scoring framework of Gneiting–Raftery [163] and Gneiting [162]. In particular, we will discuss *deviance losses* in Sect. 4.1.2 that are strictly consistent scoring functions for mean estimation and, hence, provide proper scoring rules.

4.1.1 Mean Squared Error of Prediction

We denote by $\boldsymbol{Y}_n = (Y_1, \ldots, Y_n)^\top$ (past) observations on which predictors and decision rules $A : \mathbb{Y} \to \mathbb{A}$ are based on. The new observation that we would like to predict is denoted by Y having support $\mathcal{Y} \subset \mathbb{R}$. In the previous chapter we have used decision rule the $A(\boldsymbol{Y}_n)$ to estimate an unknown quantity $\gamma(\theta)$. In this section we will use this decision rule to directly predict the new (unseen) observation Y.

Theorem 4.1 (Mean Squared Error of Prediction, MSEP) *Assume that $\boldsymbol{Y}_n$ and Y are independent. Assume that the predictor $A : \mathbb{Y} \to \mathbb{A} \subseteq \mathbb{R}$, $\boldsymbol{Y}_n \mapsto A(\boldsymbol{Y}_n)$ has finite second moment, and that the real-valued random variable Y has finite second moment, too. The MSEP of predictor A to predict Y is given by*

$$\mathbb{E}\left[(Y - A(\boldsymbol{Y}_n))^2\right] = (\mathbb{E}[Y] - \mathbb{E}[A(\boldsymbol{Y}_n)])^2 + Var(A(\boldsymbol{Y}_n)) + Var(Y). \tag{4.1}$$

Proof of Theorem 4.1 We compute

$$\begin{aligned}
\mathbb{E}\left[(A(\boldsymbol{Y}_n) - Y)^2\right] &= \mathbb{E}\left[(A(\boldsymbol{Y}_n) - \mathbb{E}[Y] + \mathbb{E}[Y] - Y)^2\right] \\
&= \mathbb{E}\left[(A(\boldsymbol{Y}_n) - \mathbb{E}[Y])^2\right] + \mathbb{E}\left[(\mathbb{E}[Y] - Y)^2\right] \\
&\quad +2\, \mathbb{E}\left[(A(\boldsymbol{Y}_n) - \mathbb{E}[Y])\, (\mathbb{E}[Y] - Y)\right] \\
&= \mathbb{E}\left[(\mathbb{E}[Y] - \mathbb{E}[A(\boldsymbol{Y}_n)] + \mathbb{E}[A(\boldsymbol{Y}_n)] - A(\boldsymbol{Y}_n))^2\right] + \mathrm{Var}(Y) \\
&= (\mathbb{E}[Y] - \mathbb{E}[A(\boldsymbol{Y}_n)])^2 + \mathrm{Var}(A(\boldsymbol{Y}_n)) + \mathrm{Var}(Y),
\end{aligned}$$

where on the second last line we use the independence between $\boldsymbol{Y}_n$ and Y. This finishes the proof. □

Remarks 4.2 (Expected Generalization Loss)

- The quantity $\mathbb{E}[(Y - A(\boldsymbol{Y}_n))^2]$ is an expected GL because it measures how well the decision rule (predictor) $A(\boldsymbol{Y}_n)$ generalizes to new (unseen) data Y. As loss function we use the square loss function

$$L : \mathcal{Y} \times \mathbb{A} \to \mathbb{R}_+, \qquad (y, a) \mapsto L(y, a) = (y - a)^2. \tag{4.2}$$

Therefore, this expected GL is called MSEP.

- MSEP (4.1) is called *expected* GL. If we condition on $\boldsymbol{Y}_n$, then we call it GL. For the square loss function the GL (conditional MSEP) is given by

$$\mathbb{E}\left[(Y - A(\boldsymbol{Y}_n))^2 \middle| \boldsymbol{Y}_n \right] = (\mathbb{E}[Y] - A(\boldsymbol{Y}_n))^2 + \mathrm{Var}(Y), \tag{4.3}$$

  where we have used independence between Y and $\boldsymbol{Y}_n$.
- We do not distinguish the terms 'prediction' and 'forecast'. Sometimes the literature makes a subtle difference between the two, the latter involving a temporal component and the former not. In the context of prediction/forecasting a loss function (4.2) is also called *scoring function*. We also use these two terms interchangeably in the context of prediction/forecasting.
- The MSEP in Theorem 4.1 decouples into three terms:
  - The first term $(\mathbb{E}[Y] - \mathbb{E}[A(\boldsymbol{Y}_n)])^2$ is the (squared) *bias*. Obviously, good decision rules $A(\boldsymbol{Y}_n)$ under the MSEP should be unbiased for $\mathbb{E}[Y]$. If we compare this to the previous chapter, we note that now the bias is measured w.r.t. the mean of the new observation Y. Additionally, there might be a slight difference to the previous chapter if $\boldsymbol{Y}_n$ and Y do not belong to the same parameter $\theta \in \boldsymbol{\Theta}$ (if we work in a parametrized family): the risk function in (3.3) considers $\mathcal{R}(\theta, A) = \mathbb{E}_\theta[L(\theta, A(\boldsymbol{Y}_n))]$ with both components of the loss function L belonging to the same parameter value θ. For the MSEP we replace θ in $L(\theta, A(\boldsymbol{Y}_n))$ by the new observation Y that might originate from a different distribution (or from a randomized θ in a Bayesian case).
  - The second term $\mathrm{Var}(A(\boldsymbol{Y}_n))$ is called *estimation variance* or *statistical error*.
  - The last term $\mathrm{Var}(Y)$ is called *process variance* or *irreducible risk*. It reflects the pure randomness received from the fact that we try to predict random variables Y with deterministic means $\mathbb{E}[Y]$.
- All three terms on the right-hand side of (4.1) are non-negative. The *MSEP optimal predictor* for Y is its expected value $\mathbb{E}[Y]$. For this choice, the first two terms (squared bias and estimation variance) vanish, and we are only left with the irreducible risk. Since this MSEP optimal predictor is typically unknown it is replaced by a decision rule $A(\boldsymbol{Y}_n)$ that is based on past experience $\boldsymbol{Y}_n$. This decision rule is used to predict Y, but it can also be seen as an *estimator* for $\mathbb{E}[Y]$. A good decision rule $A(\boldsymbol{Y}_n)$ is unbiased for $\mathbb{E}[Y]$, making the first term on the right-hand side of (4.1) equal to zero, and at the same time trying to make the estimation variance small. Typically, this cannot be achieved simultaneously and, therefore, there is a trade-off between bias and estimation variance in most applied statistical problems.

- We emphasize that in financial applications we typically aim for unbiased estimators for $\mathbb{E}[Y]$, we especially refer to Sect. 7.4.2 that studies the balance property in network regression models under a stationary portfolio assumption. Here, this stationarity may, e.g., translate into a (stronger) i.i.d. assumption on $Y_1, \ldots, Y_n, Y$. Unbiasedness then implies that the predictor $A(\boldsymbol{Y}_n)$ is optimal in (4.1) if it meets the Cramér–Rao information bound, see Theorem 3.13.

Theorem 4.1 considers the MSEP which implicitly assumes that the square loss function is the objective (scoring) function of interest. The square loss function may be considered as being distribution-free, but it is motivated by a Gaussian model for $\boldsymbol{Y}_n$ and Y, respectively; this will be justified in Remarks 4.6, below. If we use the square loss function for observations different from Gaussian ones it might under- or over-weigh particular characteristics in these observations because they may not look very Gaussian (e.g. more heavy-tailed). Therefore, we should always choose a scoring function that fits the problem considered, for instance, a square loss function is not appropriate if we model claim counts following a Poisson distribution. We close this section with the example of the EDF.

Example 4.3 (MSEP Within the EDF) We choose a fixed single-parameter linear EDF satisfying Assumption 2.6 and having a steep cumulant function κ, see Theorem 2.19 and Remark 2.20. Assume we have independent random variables $Y_1, \ldots, Y_n, Y$ belonging to this EDF having densities, see Example 3.5,

$$Y_i \sim f(y_i; \theta, v_i/\varphi) = \exp\left\{ \frac{y_i\theta - \kappa(\theta)}{\varphi/v_i} + a(y_i; v_i/\varphi) \right\}, \tag{4.4}$$

and similarly for $Y \sim f(y; \theta, v/\varphi)$. Note that all random variables share the same canonical parameter $\theta \in \mathring{\boldsymbol{\Theta}}$. The MLE of $\mu \in \mathcal{M}$ based on $\boldsymbol{Y}_n = (Y_1, \ldots, Y_n)^\top$ is found by solving, see (3.4)–(3.5),

$$\begin{aligned} \widehat{\mu}^{\mathrm{MLE}} &= \widehat{\mu}^{\mathrm{MLE}}(\boldsymbol{Y}_n) = \underset{\widetilde{\mu} \in \overline{\mathcal{M}}}{\arg\max}\ \ell_{\boldsymbol{Y}_n}(\widetilde{\mu}) \qquad (4.5) \\ &= \underset{\widetilde{\mu} \in \overline{\mathcal{M}}}{\arg\max} \sum_{i=1}^n \frac{Y_i h(\widetilde{\mu}) - \kappa(h(\widetilde{\mu}))}{\varphi/v_i}, \end{aligned}$$

with canonical link $h = (\kappa')^{-1}$. Since the cumulant function κ is strictly convex and assumed to be steep, there exists a unique solution $\widehat{\mu}^{\mathrm{MLE}} \in \overline{\mathcal{M}}$. If $\widehat{\mu}^{\mathrm{MLE}} \in \mathcal{M}$ we have a proper solution providing $\widehat{\theta}^{\mathrm{MLE}} = h(\widehat{\mu}^{\mathrm{MLE}}) \in \boldsymbol{\Theta}$, otherwise $\widehat{\mu}^{\mathrm{MLE}}$ provides a degenerate model. This decision rule $\boldsymbol{Y}_n \mapsto \widehat{\mu}^{\mathrm{MLE}} = \widehat{\mu}^{\mathrm{MLE}}(\boldsymbol{Y}_n)$ is now used to predict the (independent) new random variable Y and to estimate the unknown parameters θ and μ, respectively. That is, we use the following predictor for Y

$$\boldsymbol{Y}_n \mapsto \widehat{Y} = \widehat{\mathbb{E}}_\theta[Y] = \mathbb{E}_{\widehat{\theta}^{\mathrm{MLE}}}[Y] = \widehat{\mu}^{\mathrm{MLE}} = \widehat{\mu}^{\mathrm{MLE}}(\boldsymbol{Y}_n).$$

Note that this predictor $\widehat{Y}$ is used to predict an unobserved (new) random variable Y, and it is itself a random variable as a function of (independent) past observations $\boldsymbol{Y}_n$. We calculate the MSEP in this model. Using Theorem 4.1 we obtain

$$\begin{aligned}\mathbb{E}_\theta\left[\left(Y-\widehat{\mu}^{\mathrm{MLE}}\right)^2\right] &= \left(\mathbb{E}_\theta\left[Y\right]-\mathbb{E}_\theta\left[\widehat{\mu}^{\mathrm{MLE}}\right]\right)^2+\mathrm{Var}_\theta\left(\widehat{\mu}^{\mathrm{MLE}}\right)+\mathrm{Var}_\theta(Y)\\ &= \left(\kappa'(\theta)-\kappa'(\theta)\right)^2+\frac{\varphi\kappa''(\theta)}{\sum_{i=1}^n v_i}+\frac{\varphi\kappa''(\theta)}{v} \qquad (4.6)\\ &= \frac{(\kappa''(\theta))^2}{\mathcal{I}(\theta)}+\frac{\varphi\kappa''(\theta)}{v},\end{aligned}$$

see (3.25) for Fisher's information $\mathcal{I}(\theta)$. In this calculation we have used that the MLE $\widehat{\mu}^{\mathrm{MLE}}$ is UMVU for $\mu = \kappa'(\theta)$ and that $\boldsymbol{Y}_n$ and Y come from the same EDF with the same canonical parameter $\theta \in \mathring{\boldsymbol{\Theta}}$. As a result, we are only left with estimation variance and process variance, moreover, the estimation variance asymptotically vanishes as $\sum_{i=1}^n v_i \to \infty$. ■

4.1.2 Unit Deviances and Deviance Generalization Loss

The main estimation technique used in these notes is MLE introduced in Definition 3.4. At this stage, MLE is un-related to any specific scoring function L because it has been received by maximizing the log-likelihood function. In this section we discuss the deviance loss function (as a scoring function) and we highlight its connection to the Bregman divergence introduced in Sect. 2.3. Based on the deviance loss function choice we rephrase Theorem 4.1 in terms of this scoring function. A theoretical foundation to these considerations will be given in Sect. 4.1.3, below.

For the derivations in this section we rely on the same single-parameter linear EDF as in Example 4.3, having a steep cumulant function κ. The MLE of $\mu = \kappa(\theta)$ is found by solving, see (4.5),

$$\widehat{\mu}^{\mathrm{MLE}} = \widehat{\mu}^{\mathrm{MLE}}(\boldsymbol{Y}_n) = \underset{\widetilde{\mu}\in\overline{\mathcal{M}}}{\arg\max} \sum_{i=1}^n \frac{Y_i h(\widetilde{\mu})-\kappa(h(\widetilde{\mu}))}{\varphi/v_i} \in \overline{\mathcal{M}},$$

with canonical link $h = (\kappa')^{-1}$. This decision rule $\boldsymbol{Y}_n \mapsto \widehat{\mu}^{\mathrm{MLE}} = \widehat{\mu}^{\mathrm{MLE}}(\boldsymbol{Y}_n)$ is now used to predict the (new) random variable Y and to estimate the unknown parameters θ and μ, respectively. We aim at studying the expected GL under a distribution-adapted loss function choice potentially different from the square loss function. Below we will justify this second choice more extensively.

For the *saturated model* the *common* canonical parameter θ of the independent random variables $Y_1, \ldots, Y_n$ in (4.4) is replaced by *individual* canonical parameters θ_i, $1 \le i \le n$. These individual canonical parameters are estimated with individual MLEs. The individual MLEs are given by, respectively,

$$\widehat{\theta}_i^{\mathrm{MLE}} = (\kappa')^{-1}(Y_i) = h(Y_i) \qquad \text{and} \qquad \widehat{\mu}_i^{\mathrm{MLE}} = Y_i \in \overline{\mathcal{M}},$$

the latter always exists because of strict convexity and steepness of κ. Since the MLE $\widehat{\mu}_i^{\mathrm{MLE}} = Y_i$ maximizes the log-likelihood, we receive for any $\mu \in \mathcal{M}$ the inequality

$$\begin{aligned} 0 &\le 2\Big(\log f(Y_i; h(Y_i), v_i/\varphi) - \log f(Y_i; h(\mu), v_i/\varphi)\Big) \\ &= 2\,\frac{v_i}{\varphi}\Big(Y_i h(Y_i) - \kappa(h(Y_i)) - Y_i h(\mu) + \kappa(h(\mu))\Big) \qquad (4.7) \\ &= \frac{v_i}{\varphi}\,\mathfrak{d}(Y_i, \mu). \end{aligned}$$

The function $(y, \mu) \mapsto \mathfrak{d}(y, \mu) \ge 0$ is the unit deviance introduced in (2.25), extended to $\mathfrak{C}$, and it is zero if and only if $y = \mu$, see Lemma 2.22. The latter is also an immediate consequence of the fact that the MLE is unique within EDFs.

Remark 4.4 The unit deviance $\mathfrak{d}(y, \mu)$ has only been considered on $\mathring{\mathfrak{C}} \times \mathcal{M}$ in (2.25). Having steepness of cumulant function κ implies $\mathring{\mathfrak{C}} = \mathcal{M}$, see Theorem 2.19, and in the absolutely continuous EDF case, we always have $Y_i \in \mathcal{M}$, a.s., which makes (4.7) well-defined for all observations Y_i, a.s. In the discrete or the mixed EDF case, an observation Y_i can be at the boundary of $\mathcal{M}$. In that case (4.7) must be calculated from

$$\mathfrak{d}(Y_i, \mu) = 2\left(\sup_{\widetilde{\theta} \in \boldsymbol{\Theta}}\left[Y_i\widetilde{\theta} - \kappa\big(\widetilde{\theta}\big)\right] - Y_i h(\mu) + \kappa(h(\mu))\right). \qquad (4.8)$$

This applies, e.g., to the Poisson or Bernoulli cases for observation $Y_i = 0$, in these cases we obtain unit deviances 2μ and $-2\log(1-\mu)$, respectively.

The previous considerations (4.7)–(4.8) have been studying one single observation Y_i of $\boldsymbol{Y}_n$. Aggregating over all observations in $\boldsymbol{Y}_n$ (and additionally using independence between the individual components of $\boldsymbol{Y}_n$) we arrive at the so-called *deviance loss function*

$$\mathfrak{D}(\boldsymbol{Y}_n, \mu) \stackrel{\text{def.}}{=} \frac{1}{n} \sum_{i=1}^{n} \frac{v_i}{\varphi} \mathfrak{d}\left(Y_i, \mu\right) \tag{4.9}$$

$$= \frac{2}{n} \sum_{i=1}^{n} \frac{v_i}{\varphi} \Big(Y_i h\left(Y_i\right) - \kappa\left(h\left(Y_i\right)\right) - Y_i h\left(\mu\right) + \kappa\left(h\left(\mu\right)\right) \Big) \geq 0.$$

The deviance loss function $\mathfrak{D}(\boldsymbol{Y}_n, \mu)$ subtracts twice the log-likelihood $\ell_{\boldsymbol{Y}_n}(\mu)$ from the one of the saturated model. Thus, it introduces a sign flip compared to (4.5). This immediately gives us the following corollary.

Corollary 4.5 (Deviance Loss Function) *The MLE problem* (4.5) *is equivalent to solving*

$$\widehat{\mu}^{MLE} = \underset{\widetilde{\mu} \in \overline{\mathcal{M}}}{\arg\max}\ \ell_{\boldsymbol{Y}_n}(\widetilde{\mu}) = \underset{\widetilde{\mu} \in \overline{\mathcal{M}}}{\arg\min}\ \mathfrak{D}(\boldsymbol{Y}_n, \widetilde{\mu}). \tag{4.10}$$

Remarks 4.6

- Formula (4.10) replaces a maximization problem by a minimization problem with objective function $\mathfrak{D}(\boldsymbol{Y}_n, \mu)$ being bounded below by zero. We can use this deviance loss function as a loss function not only for parameter estimation, but also as a scoring function for analyzing GLs within the EDF (similarly to Theorem 4.1).
- We draw the link to the KL divergence discussed in Sect. 2.3. In formula (2.26) we have shown that the unit deviance is equal to the KL divergence (up to scaling with factor 2), thus, equivalently, MLE aims at minimizing the average KL divergence over all observations $\boldsymbol{Y}_n$

$$\widehat{\theta}^{\text{MLE}} = \underset{\widetilde{\theta} \in \boldsymbol{\Theta}}{\arg\min}\ \frac{1}{n} \sum_{i=1}^{n} D_{\text{KL}}\Big(f(\cdot; h(Y_i), v_i/\varphi) \Big|\Big| f(\cdot; \widetilde{\theta}, v_i/\varphi) \Big),$$

by finding an optimal parameter $\widehat{\theta}^{\rm MLE}$ somewhere 'in the middle' of the observation $\widehat{\theta}_1^{\rm MLE} = h(Y_1), \ldots, \widehat{\theta}_n^{\rm MLE} = h(Y_n)$. This then provides us with, see (2.27),

$$\prod_{i=1}^{n} f\left(Y_i; \widetilde{\theta}, v_i/\varphi\right) = \left[\prod_{i=1}^{n} f\left(Y_i; h\left(Y_i\right), v_i/\varphi\right)\right] e^{-\frac{1}{2}\sum_{i=1}^{n} \frac{v_i}{\varphi} \mathfrak{d}(Y_i, \kappa'(\widetilde{\theta}))} \tag{4.11}$$

$$\propto \exp\left\{-\sum_{i=1}^{n} D_{\rm KL}\Big(f(\cdot; h(Y_i), v_i/\varphi)\Big|\Big|f(\cdot; \widetilde{\theta}, v_i/\varphi)\Big)\right\},$$

where $\propto$ highlights that we drop all terms that do not involve $\widetilde{\theta}$. This describes the change in joint likelihood by varying the canonical parameter $\widetilde{\theta}$ over its domain $\boldsymbol{\Theta}$. The first line of (4.11) is in the spirit of minimizing a weighted square loss, but the Gaussian square is replaced by the unit deviance $\mathfrak{d}$. The second line of (4.11) is in the spirit of information geometry considered in Sect. 2.3, where we try to find a canonical parameter $\widetilde{\theta}$ that has a small KL divergence to the n individual models being parametrized by $h(Y_1), \ldots, h(Y_n)$, thus, the MLE $\widehat{\theta}^{\rm MLE}$ provides an optimal balance over the entire set of (independent) observations $Y_1, \ldots, Y_n$ w.r.t. the KL divergence.

- In contrast to the square loss function, the deviance loss function $\mathfrak{D}(\boldsymbol{Y}_n, \mu)$ respects the distributional properties of $\boldsymbol{Y}_n$, see (4.11). That is, if the underlying distribution allows for larger or smaller claims, this fact is appropriately valued in the deviance loss function (supposed that we have chosen the right family of distributions; model uncertainty will be studied in Sect. 11.1, below).
- Assume we work in the Gaussian model. In this model we have $\kappa(\theta) = \theta^2/2$ and canonical link $h(\mu) = \mu$, see Sect. 2.1.3. This provides unit deviance in the Gaussian case $\mathfrak{d}\,(y, \mu) = (y - \mu)^2$, which is exactly the square loss function for action space $\mathbb{A} = \overline{\mathcal{M}}$. Thus, the square loss function is most appropriate in the Gaussian case.
- As explained above, we use unit deviances $\mathfrak{d}(y, \mu)$ as a measure of discrepancy. Alternatively, as in the introduction to this section, see (4.6), we can consider Pearson's χ^2-statistic which corresponds to the weighted square loss function

$$X^2(y, \mu) = \frac{(y - \mu)^2}{V(\mu)}, \tag{4.12}$$

where $\mu \mapsto V(\mu)$ is the variance function of the chosen EDF. Similarly, to the deviance loss function (4.9), we can aggregate these Pearson's χ^2-statistics $X^2(Y_i, \mu)$ over all observations Y_i in $\boldsymbol{Y}_n$ to receive a second overall measure of discrepancy. In the Gaussian case the deviance loss and Pearson's χ^2-statistic coincide and have a χ^2-distribution, for other distributions asymptotic results are available.

In the non-Gaussian case, (4.12) is not always robust. For instance, if we work in the Poisson model, we have variance function $V(\mu) = \mu$. Our examples below will have low claim frequencies which implies that μ will be small. The appearance of a small μ in the denominator of (4.12) will imply that Pearson's χ^2-statistic is not very robust in small frequency applications, in particular, if we need to estimate this μ from $\boldsymbol{Y}_n$. Therefore, we refrain from using (4.12).

Naturally, in analogy to Theorem 4.1 and derivation (4.6), the above considerations motivate us to consider expected GLs under unit deviances within the EDF. We use the decision rule $\widehat{\mu}^{\mathrm{MLE}}(\boldsymbol{Y}_n) \in \mathbb{A} = \overline{\mathcal{M}}$ to predict a new observation Y.

The expected *deviance GL* is defined and given by

$$\begin{aligned}
&\mathbb{E}_\theta\left[\mathfrak{d}\left(Y, \widehat{\mu}^{\mathrm{MLE}}(\boldsymbol{Y}_n)\right)\right] \\
&= \mathbb{E}_\theta\left[\mathfrak{d}\left(Y, \mu\right)\right] + 2\,\mathbb{E}_\theta\left[Yh(\mu) - \kappa\left(h(\mu)\right) - Yh(\widehat{\mu}^{\mathrm{MLE}}(\boldsymbol{Y}_n)) + \kappa\left(h(\widehat{\mu}^{\mathrm{MLE}}(\boldsymbol{Y}_n))\right)\right] \\
&= \mathbb{E}_\theta\left[\mathfrak{d}\left(Y, \mu\right)\right] + \mathcal{E}\left(\mu, \widehat{\mu}^{\mathrm{MLE}}(\boldsymbol{Y}_n)\right), \qquad (4.13)
\end{aligned}$$

the last identity uses independence between $\boldsymbol{Y}_n$ and Y, and with *estimation risk function*

$$\mathcal{E}\left(\mu, \widehat{\mu}^{\mathrm{MLE}}(\boldsymbol{Y}_n)\right) = \mathbb{E}_\theta\left[\mathfrak{d}\left(\mu, \widehat{\mu}^{\mathrm{MLE}}(\boldsymbol{Y}_n)\right)\right] > 0, \qquad (4.14)$$

we use steepness of the cumulant function, $\mathfrak{C} = \overline{\mathrm{conv}(\mathfrak{T})} = \overline{\mathcal{M}}$, and Lemma 2.22 for the strict positivity of the estimation risk function. Thus, for the estimation risk function $\mathcal{E}$ we replace Y by μ in the unit deviance and the expectation $\mathbb{E}_\theta$ is only over the observations $\boldsymbol{Y}_n$. This looks like a very convincing generalization of the MSEP, however, one needs to ensure that all terms in (4.13) exist.

Theorem 4.7 (Expected Deviance Generalization Loss) *Assume that $\boldsymbol{Y}_n$ and Y are independent and belong to the same linear EDF having the same canonical parameter $\theta \in \mathring{\boldsymbol{\Theta}}$ and having strictly convex and steep cumulant function κ. Choose a predictor $A : \mathbb{Y} \to \mathbb{A} = \overline{\mathcal{M}}$, $\boldsymbol{Y}_n \mapsto A(\boldsymbol{Y}_n)$ and assume that all expectations in the following formula exist. The expected deviance GL of predictor A to predict Y is given by*

$$\mathbb{E}_\theta\left[\mathfrak{d}\left(Y, A(\boldsymbol{Y}_n)\right)\right] = \mathbb{E}_\theta\left[\mathfrak{d}\left(Y, \mu\right)\right] + \mathcal{E}\left(\mu, A(\boldsymbol{Y}_n)\right) \geq \mathbb{E}_\theta\left[\mathfrak{d}\left(Y, \mu\right)\right].$$

Remarks 4.8

- $\mathbb{E}_\theta[\mathfrak{d}(Y, \mu)]$ plays the role of the pure process variance (irreducible risk) of Theorem 4.1. This term does not involve any parameter estimation bias and uncertainty because it is based on the true parameter θ and $\mu = \kappa'(\theta)$, respectively. In Sect. 4.1.3, below, we are going to justify the appropriateness of this object as a tool for forecast evaluation. In particular, because the unit deviance is strictly consistent for the mean functional, the true mean $\mu = \mu(\theta)$ minimizes $\mathbb{E}_\theta[\mathfrak{d}(Y, \mu)]$, see (4.28), below.
- The second term $\mathcal{E}(\mu, A(\boldsymbol{Y}_n))$ measures parameter estimation bias and uncertainty of decision rule $A(\boldsymbol{Y}_n)$ versus the true parameter $\mu = \kappa'(\theta)$. The first remark is that we can do this for any decision rule A, i.e., we do not necessarily need to consider the MLE. The second remark is that we can no longer get a clear cut differentiation between a bias term and a parameter estimation uncertainty term for deviance loss functions not coming from the Gaussian distribution. We come back to this in Remarks 7.17, below, where we give more characterization to the individual terms of the expected deviance GL.
- An issue in applying Theorem 4.7 to the MLE decision rule $A(\boldsymbol{Y}_n) = \widehat{\mu}^{\mathrm{MLE}}(\boldsymbol{Y}_n)$ is that, in general, it does not lead to a finite estimation risk function. For instance, in the Poisson case we have with positive probability $\widehat{\mu}^{\mathrm{MLE}}(\boldsymbol{Y}_n) = 0$, which results in an infinite estimation risk. In order to avoid this, we need to bound away the decision rule form the boundary of $\mathcal{M}$ and $\boldsymbol{\Theta}$, respectively. In the Poisson case this can be achieved by considering a decision rule $A(\boldsymbol{Y}_n) = \max\{\widehat{\mu}^{\mathrm{MLE}}(\boldsymbol{Y}_n), \epsilon\}$ for a fixed given $\epsilon \in (0, \mu = \kappa'(\theta))$. This decision rule has a bias which asymptotically vanishes as $n \to \infty$. Moreover, consistency and asymptotic normality tells us that this lower bound does not affect prediction for large sample sizes n (with large probability).
- Similar to (4.3), we can also consider the deviance GL, given $\boldsymbol{Y}_n$. Under independence of $\boldsymbol{Y}_n$ and Y we have deviance GL

$$\begin{aligned}\mathbb{E}_\theta\left[\mathfrak{d}\left(Y, A(\boldsymbol{Y}_n)\right)\middle|\,\boldsymbol{Y}_n\right] &= \mathbb{E}_\theta\left[\mathfrak{d}\left(Y, \mu\right)\middle|\,\boldsymbol{Y}_n\right] + \mathfrak{d}(\mu, A(\boldsymbol{Y}_n)) \qquad (4.15)\\ &\geq \mathbb{E}_\theta\left[\mathfrak{d}\left(Y, \mu\right)\right].\end{aligned}$$

Thus, here we directly compare $A(\boldsymbol{Y}_n)$ to the true parameter μ.

Example 4.9 (Estimation Risk Function in the Gaussian Case) We consider the Gaussian case with cumulant function $\kappa(\theta) = \theta^2/2$ and canonical link $h(\mu) = \mu$.

The estimation risk function is in the Gaussian case for a square integrable predictor $A(\boldsymbol{Y}_n)$ given by

$$\begin{aligned}
\mathcal{E}(\mu, A(\boldsymbol{Y}_n)) &= \mathbb{E}_\theta\left[\mathfrak{d}(\mu, A(\boldsymbol{Y}_n))\right] \\
&= 2\Big(\mu h(\mu) - \kappa(h(\mu)) - \mu\mathbb{E}_\theta\left[h(A(\boldsymbol{Y}_n))\right] + \mathbb{E}_\theta\left[\kappa(h(A(\boldsymbol{Y}_n)))\right]\Big) \\
&= \mu^2 - 2\mu\mathbb{E}_\theta\left[A(\boldsymbol{Y}_n)\right] + \mathbb{E}_\theta\left[(A(\boldsymbol{Y}_n))^2\right] \\
&= (\mu - \mathbb{E}_\theta\left[A(\boldsymbol{Y}_n)\right])^2 + \mathrm{Var}_\theta(A(\boldsymbol{Y}_n)).
\end{aligned}$$

These are exactly the squared bias and the estimation variance, see (4.1). Thus, in the Gaussian case, the MSEP and the expected deviance GL coincide. Moreover, adding a deterministic bias $c \in \mathbb{R}$ to $A(\boldsymbol{Y}_n)$ increases the estimation risk function, supposed that $A(\boldsymbol{Y}_n)$ is unbiased for μ. We emphasize the latter as this is an important property to have, and we refer to the next Example 4.10 for an example where this property fails to hold. ■

Example 4.10 (Estimation Risk Function in the Poisson Case) We consider the Poisson case with cumulant function $\kappa(\theta) = e^\theta$ and canonical link $h(\mu) = \log\mu$. The estimation risk function is given by (subject to existence)

$$\mathcal{E}(\mu, A(\boldsymbol{Y}_n)) = 2\Big(\mu\log(\mu) - \mu - \mu\mathbb{E}_\theta\left[\log(A(\boldsymbol{Y}_n))\right] + \mathbb{E}_\theta\left[A(\boldsymbol{Y}_n)\right]\Big). \tag{4.16}$$

Assume that decision rule $A(\boldsymbol{Y}_n)$ is non-deterministic and unbiased for μ. Using Jensen's inequality these assumptions imply for the estimation risk function

$$\mathcal{E}(\mu, A(\boldsymbol{Y}_n)) = 2\mu\Big(\log(\mu) - \mathbb{E}_\theta\left[\log(A(\boldsymbol{Y}_n))\right]\Big) > 0.$$

We now add a small deterministic bias $c \in \mathbb{R}$ to the unbiased estimator $A(\boldsymbol{Y}_n)$ for μ. This gives us estimation risk function, see (4.16) and subject to existence,

$$\mathcal{E}(\mu, A(\boldsymbol{Y}_n) + c) = 2\Big(\mu\log(\mu) - \mu\mathbb{E}_\theta\left[\log(A(\boldsymbol{Y}_n) + c)\right] + c\Big).$$

Consider the derivative w.r.t. bias c in 0, we use Jensen's inequality on the last line,

$$\begin{aligned}
\left.\frac{\partial}{\partial c}\mathcal{E}(\mu, A(\boldsymbol{Y}_n) + c)\right|_{c=0} &= \left. 2\left(- \mu\mathbb{E}_\theta\left[\frac{1}{A(\boldsymbol{Y}_n) + c}\right] + 1\right)\right|_{c=0} \\
&= -2\mu\mathbb{E}_\theta\left[\frac{1}{A(\boldsymbol{Y}_n)}\right] + 2 \\
&< -2\mu\frac{1}{\mathbb{E}_\theta\left[A(\boldsymbol{Y}_n)\right]} + 2 = 0.
\end{aligned} \tag{4.17}$$

Thus, the estimation risk becomes smaller if we add a small bias to the (non-deterministic) unbiased predictor $A(\boldsymbol{Y}_n)$. This issue has been raised in Denuit et al. [97]. Of course, this is a very unfavorable property, and it is rather different from the Gaussian case in Example 4.9. It is essentially driven by the fact that parameter estimation is based on a finite sample, which implies a strict inequality in (4.17) for the finite sample estimate $A(\boldsymbol{Y}_n)$. A conclusion of this example is that if we use expected deviance GLs for forecast evaluation we need to insist on having unbiased predictors. This will become especially important for more complex regression models, see Sect. 7.4.2, below.

More generally, one can prove this result of a smaller estimation risk function for a small positive bias for any EDF member with power variance function $V(\mu) = \mu^p$ with $p \geq 1$, see also (4.18) below. The proof uses the Fortuin–Kasteleyn–Ginibre (FKG) inequality [133] providing $\mathbb{E}_\theta[A(\boldsymbol{Y}_n)^{1-p}] < \mathbb{E}_\theta[A(\boldsymbol{Y}_n)]\mathbb{E}_\theta[A(\boldsymbol{Y}_n)^{-p}] = \mu\mathbb{E}_\theta[A(\boldsymbol{Y}_n)^{-p}]$ to receive (4.17) for power variance parameters $p \geq 1$. ■

Remarks 4.11 (Conclusion from Examples 4.9 and 4.10 and a Further Remark)

- Working with expected deviance GLs for evaluating forecasts requires some care because a bigger bias in the (finite sample) estimate $A(\boldsymbol{Y}_n)$ may provide a smaller estimation risk function $\mathcal{E}(\mu, A(\boldsymbol{Y}_n))$. For this reason, we typically insist on having unbiased predictors/forecasts. The latter is also an important requirement in financial applications to guarantee that the overall price is set to the right level, we refer to the balance property in Corollary 3.19 and to Sect. 7.4.2, below.
- In Theorems 4.1 and 4.7 we use independence between the predictor $A(\boldsymbol{Y}_n)$ and the random variable Y to receive the split of the expected deviance GL into irreducible risk and estimation risk function. In regression models, this independence between the predictor $A(\boldsymbol{Y}_n)$ and the random variable Y may no longer hold. In that case we will still work with the expected deviance GL $\mathbb{E}_\theta[\mathfrak{d}(Y, A(\boldsymbol{Y}_n))]$, but a clear split between estimation and forecasting will no longer be possible, see Sect. 4.2, below.

The next example gives the most important unit deviances in actuarial modeling.

Example 4.12 (Unit Deviances) We give the most prominent examples of unit deviances within the single-parameter linear EDF. We recall unit deviance (2.25)

$$\mathfrak{d}(y, \mu) = 2\left(yh(y) - \kappa\left(h(y)\right) - yh(\mu) + \kappa\left(h(\mu)\right)\right) \geq 0.$$

In Sect. 2.2 we have met the examples given in Table 4.1.

Table 4.1 Unit deviances of selected distributions commonly used in actuarial science

Distribution	Cumulant function $\kappa(\theta)$	Unit deviance $\mathfrak{d}(y, \mu)$
Gaussian	$\theta^2/2$	$(y-\mu)^2$
Gamma	$-\log(-\theta)$	$2\left((y-\mu)/\mu + \log(\mu/y)\right)$
Inverse Gaussian	$-\sqrt{-2\theta}$	$(y-\mu)^2/(\mu^2 y)$
Poisson	e^θ	$2\left(\mu - y - y\log(\mu/y)\right)$
Negative-binomial	$-\log(1-e^\theta)$	$2\left(y\log\left(\frac{y}{\mu}\right) - (y+1)\log\left(\frac{y+1}{\mu+1}\right)\right)$
Tweedie's CP	$\frac{((1-p)\theta)^{\frac{2-p}{1-p}}}{2-p}$, $p \in (1,2)$	$2\left(y\frac{y^{1-p}-\mu^{1-p}}{1-p} - \frac{y^{2-p}-\mu^{2-p}}{2-p}\right)$
Bernoulli	$\log(1+e^\theta)$	$2\left(-y\log\mu - (1-y)\log(1-\mu)\right)$

If we focus on Tweedie's distributions having power variance functions $V(\mu) = \mu^p$, see Table 2.1, we get a unified expression for the unit deviances for $p \in \{0\} \cup (1,2) \cup (2,\infty)$

$$\mathfrak{d}(y,\mu) = 2\left(y\frac{y^{1-p}-\mu^{1-p}}{1-p} - \frac{y^{2-p}-\mu^{2-p}}{2-p}\right) \tag{4.18}$$
$$= 2\left(\frac{y^{2-p}}{(1-p)(2-p)} - \frac{y\mu^{1-p}}{1-p} + \frac{\mu^{2-p}}{2-p}\right).$$

For the remaining power variance cases we have: $p = 1$ corresponds to the Poisson case, $p = 2$ gives the gamma case, the cases $p < 0$ do not have a steep cumulant function, and, moreover, there are no EDF models for $p \in (0,1)$, see Theorem 2.18.

The unit deviance in the Bernoulli case is also called *binary cross-entropy*. This binary cross-entropy has a categorical generalization, called *multi-class cross-entropy*. Assume we have a categorical EF with levels $\{1,\ldots,k+1\}$ and corresponding probabilities $p_1,\ldots,p_{k+1} \in (0,1)$ summing up to 1, see Sect. 2.1.4. We denote by $\boldsymbol{Y} = (\mathbb{1}_{\{Y=1\}},\ldots,\mathbb{1}_{\{Y=k+1\}})^\top \in \mathbb{R}^{k+1}$ the indicator variable that shows which level the categorical random variable Y takes; $\boldsymbol{Y}$ is called one-hot encoding of the categorical random variable Y. Assume $\boldsymbol{y}$ is a realization of $\boldsymbol{Y}$ and set $\boldsymbol{\mu} = \boldsymbol{p} = (p_1,\ldots,p_{k+1})^\top$. The categorical (multi-class) cross-entropy loss function is given by

$$\mathfrak{d}(\boldsymbol{y},\boldsymbol{\mu}) = \mathfrak{d}(\boldsymbol{y},\boldsymbol{p}) = -2\sum_{j=1}^{k+1} y_j \log p_j \;\geq\; 0. \tag{4.19}$$

This cross-entropy is closely related to the KL divergence between two categorical distributions $\boldsymbol{p}$ and $\boldsymbol{q}$ on $\{1,\ldots,k+1\}$. The KL divergence from $\boldsymbol{p}$ to $\boldsymbol{q}$ is given by

$$D_{\mathrm{KL}}(\boldsymbol{q}||\boldsymbol{p}) = \sum_{j=1}^{k+1} q_j \log\left(\frac{q_j}{p_j}\right) = \sum_{j=1}^{k+1} q_j \log q_j - \sum_{j=1}^{k+1} q_j \log p_j.$$

If we replace the true (but unknown) distribution $\boldsymbol{q}$ by observation $\boldsymbol{Y} = \boldsymbol{y}$ we receive unit deviance (4.19) (scaled by 2), and the MLE is obtained by minimizing this KL divergence, see also Example 3.10. ■

Outlook 4.13 In the regression modeling, below, each response Y_i will have its own mean parameter $\mu_i = \mu(\boldsymbol{\beta}, \boldsymbol{x}_i)$ which will be a function of its covariate information $\boldsymbol{x}_i$, and $\boldsymbol{\beta}$ denotes a regression parameter to be estimated with MLE. In that case, we modify the deviance loss function (4.9) to

$$\boldsymbol{\beta} \mapsto \mathfrak{D}(\boldsymbol{Y}_n, \boldsymbol{\beta}) = \frac{1}{n} \sum_{i=1}^{n} \frac{v_i}{\varphi} \mathfrak{d}\left(Y_i, \mu_i\right) = \frac{1}{n} \sum_{i=1}^{n} \frac{v_i}{\varphi} \mathfrak{d}\left(Y_i, \mu(\boldsymbol{\beta}, \boldsymbol{x}_i)\right), \tag{4.20}$$

and the MLE of $\boldsymbol{\beta}$ can be found by solving

$$\widehat{\boldsymbol{\beta}}^{\text{MLE}} = \underset{\boldsymbol{\beta}}{\arg\min}\ \mathfrak{D}(\boldsymbol{Y}_n, \boldsymbol{\beta}). \tag{4.21}$$

If Y is a new response with covariate information $\boldsymbol{x}$ and following the same EDF as $\boldsymbol{Y}_n$, we will evaluate the corresponding expected scaled deviance GL given by

$$\mathbb{E}_{\boldsymbol{\beta}}\left[\frac{v}{\varphi} \mathfrak{d}\left(Y, \mu(\widehat{\boldsymbol{\beta}}^{\text{MLE}}, \boldsymbol{x})\right)\right], \tag{4.22}$$

where $\mathbb{E}_{\boldsymbol{\beta}}$ is the expectation under the true regression parameter $\boldsymbol{\beta}$ for $\boldsymbol{Y}_n$ and Y. This will be discussed in Sect. 5.1.7, below. If we interpret $(Y, \boldsymbol{x}, v)$ as a random vector describing a randomly selected insurance policy from our portfolio, and being independent of $\boldsymbol{Y}_n$ (and the corresponding covariate information $\boldsymbol{x}_i$, $1 \leq i \leq n$), then $\widehat{\boldsymbol{\beta}}^{\text{MLE}}$ will be independent of $(Y, \boldsymbol{x}, v)$. Nevertheless, the predictor $\mu(\widehat{\boldsymbol{\beta}}^{\text{MLE}}, \boldsymbol{x})$ will introduce dependence between the chosen decision rule and Y through $\boldsymbol{x}$, and we no longer receive the split of the expected deviance GL as stated in Theorem 4.7, for a related discussion we also refer to Remarks 7.17, below.

If we interpret $(Y, \boldsymbol{x}, v)$ as a randomly selected insurance policy, then the expected GL (4.22) is evaluated under the joint (portfolio) distribution of $(Y, \boldsymbol{x}, v)$, and the deviance loss $\mathfrak{D}(\boldsymbol{Y}_n, \widehat{\boldsymbol{\beta}}^{\text{MLE}})$ is an (in-sample) empirical version of (4.22). ■

4.1.3 A Decision-Theoretic Approach to Forecast Evaluation

We present an excursion to a decision-theoretic approach to forecast evaluation. This excursion gives the theoretical foundation to the unit deviance considerations from above. This section follows Gneiting [162], Krüger–Ziegel [227] and Denuit et al. [97], and we refrain from giving complete proofs in this section. Forecast evaluation should involve consistent loss/scoring functions and proper scoring rules to encourage the forecaster to make careful assessments and honest forecasts. Consistent loss functions are also a necessary tool to receive consistency of M-estimators, we refer to Remarks 3.26.

Consistency and Proper Scoring Rules

Denote by $\mathfrak{C} \subseteq \mathbb{R}$ the convex closure of the support of a real-valued random variable Y, and let the action space be $\mathbb{A} = \mathfrak{C}$, see also (3.1). Predictions are evaluated in terms of a loss/scoring function

$$L : \mathfrak{C} \times \mathbb{A} \to \mathbb{R}_+, \qquad (y, a) \mapsto L(y, a) \geq 0. \tag{4.23}$$

Remark 4.14 In (4.23) we assume that the loss function L is bounded below by zero. This can be an advantage in applications because it gives a calibration to the loss function. In general, this lower bound is not a necessary condition for forecast evaluation. If we drop this lower bound property, we rather call L (only) a scoring function. For instance, the log-likelihood $\log(f(y, a))$ in (3.27) plays the role of a scoring function.

The forecaster can take the position of minimizing the expected loss to choose her/his action rule. That is, subject to existence, an optimal action w.r.t. L is received by

$$\widehat{a} = \widehat{a}(F) = \underset{a \in \mathbb{A}}{\arg\min}\, \mathbb{E}_F\left[L(Y, a)\right] = \underset{a \in \mathbb{A}}{\arg\min} \int_{\mathfrak{C}} L(y, a) dF(y). \tag{4.24}$$

In this setup the scoring function $L(y, a)$ describes the loss that the forecaster suffers if she/he uses action $a \in \mathbb{A}$ and observation $y \in \mathfrak{C}$ materializes. Since we do not want to insist on uniqueness in (4.24) we rather think of set-valued functionals in this section, which may provide solutions to problems like (4.24).[1]

We now reverse the line of arguments, and we start from a general set-valued functional. Denote by $\mathcal{F}$ the family of distribution functions of interest supported on $\mathfrak{C}$. Consider the set-valued functional

$$\mathfrak{A} : \mathcal{F} \to \mathcal{P}(\mathbb{A}), \qquad F \mapsto \mathfrak{A}(F) \subset \mathbb{A}, \tag{4.25}$$

that maps each distribution $F \in \mathcal{F}$ to a subset $\mathfrak{A}(F)$ of the action space $\mathbb{A} = \mathfrak{C}$, that is, an element of the power set $\mathcal{P}(\mathbb{A})$. The main question that we want to study in this section is the following: can we find a loss function L so that the set-valued

[1] In fact, also for the MLE in Definition 3.4 we should consider a set-valued functional. We have decided to skip this distinction to avoid any kind of complication and to not disturb the flow of reading.

functional $\mathfrak{A}$ is obtained by a loss minimization (4.24)? This motivates the following definition.

Definition 4.15 (Strict Consistency) The loss function $L : \mathfrak{C} \times \mathbb{A} \to \mathbb{R}_+$ is consistent for the functional $\mathfrak{A} : \mathcal{F} \to \mathcal{P}(\mathbb{A})$ relative to the class $\mathcal{F}$ if

$$\mathbb{E}_F\left[L(Y, \widehat{a})\right] \leq \mathbb{E}_F\left[L(Y, a)\right], \tag{4.26}$$

for all $F \in \mathcal{F}$, $\widehat{a} \in \mathfrak{A}(F)$ and $a \in \mathbb{A}$. It is strictly consistent if it is consistent and equality in (4.26) implies that $a \in \mathfrak{A}(F)$.

As stated in Theorem 1 of Gneiting [162], a loss function L is consistent for the functional $\mathfrak{A}$ relative to the class $\mathcal{F}$ if and only if, given any $F \in \mathcal{F}$, every $\widehat{a} \in \mathfrak{A}(F)$ is an optimal action under L in the sense of (4.24).

We give an example. Assume we start from the functional $F \mapsto \mathfrak{A}(F) = \mathbb{E}_F[Y]$ that maps each distribution F to its expected value. In this case we do not need to consider a set-valued functional because the expected value is a singleton (we assume that $\mathcal{F}$ only contains distributions with a finite first moment). The question then is whether we can find a loss function L such that this mean can be received by a minimization (4.24). This question is answered in Theorem 4.19, below.

Next we relate a consistent loss function L to a *proper scoring rule*. A proper scoring rule is a function $R : \mathfrak{C} \times \mathcal{F} \to \mathbb{R}$ such that

$$\mathbb{E}_F\left[R(Y, F)\right] \leq \mathbb{E}_F\left[R(Y, G)\right], \tag{4.27}$$

for all $F, G \in \mathcal{F}$, supposed that the expectations are well-defined. A scoring rule R analyzes the penalty $R(y, G)$ if the forecaster works with a distribution G and an observation y of $Y \sim F$ materializes. Proper scoring rules have been promoted in Gneiting–Raftery [163] and Gneiting [162]. They are important because they encourage the forecaster to make honest forecasts, i.e., it gives the forecaster the incentive to minimize the expected score by following his true belief about the true distribution, because only this minimizes the expected penalty in (4.27).

Theorem 4.16 (Gneiting [162, Theorem 3]) *Assume that L is a consistent loss function for the functional $\mathfrak{A}$ relative to the class $\mathcal{F}$. For each $F \in \mathcal{F}$, let $a_F \in \mathfrak{A}(F)$. The scoring rule*

$$R : \mathfrak{C} \times \mathcal{F} \to \mathbb{R}, \qquad (y, F) \mapsto R(y, F) = L(y, a_F),$$

is a proper scoring rule.

Example 4.17 Consider the unit deviance $\mathfrak{d}(\cdot, \cdot) : \mathfrak{C} \times \mathcal{M} \to \mathbb{R}_+$ for a given EDF $\mathcal{F} = \{F(\cdot; \theta, v/\varphi);\ \theta \in \mathring{\boldsymbol{\Theta}}\}$ with cumulant function κ. Lemma 2.22 says that under suitable assumptions this unit deviance $\mathfrak{d}(y, \mu)$ is zero if and only if $y = \mu$. We consider the mean functional on $\mathcal{F}$

$$\mathfrak{A} : \mathcal{F} \to \mathbb{A} = \mathcal{M}, \qquad F_\theta = F(\cdot; \theta, v/\varphi) \mapsto \mathfrak{A}(F_\theta) = \mu(\theta),$$

where $\mu = \mu(\theta) = \kappa'(\theta)$ is the mean of the chosen EDF. Choosing the unit deviance as loss function we receive for any action $a \in \mathbb{A}$, see (4.13),

$$\begin{aligned}\mathbb{E}_\theta\left[\mathfrak{d}(Y,a)\right] &= \mathbb{E}_\theta\left[\mathfrak{d}(Y,\mu)\right] + 2\,\mathbb{E}_\theta\left[Yh(\mu) - \kappa\left(h(\mu)\right) - Yh(a) + \kappa\left(h(a)\right)\right] \\ &= \mathbb{E}_\theta\left[\mathfrak{d}(Y,\mu)\right] + 2\left(\mu h(\mu) - \kappa\left(h(\mu)\right) - \mu h(a) + \kappa\left(h(a)\right)\right) \\ &= \mathbb{E}_\theta\left[\mathfrak{d}(Y,\mu)\right] + \mathfrak{d}(\mu, a).\end{aligned}$$

This is minimized for $a = \mu$ and it proves that the unit deviance is strictly consistent for the mean functional $\mathfrak{A} : F_\theta \mapsto \mathfrak{A}(F_\theta) = \mu(\theta)$ relative to the chosen EDF $\mathcal{F} = \{F(\cdot;\theta, v/\varphi);\ \theta \in \mathring{\boldsymbol{\Theta}}\}$. Using Theorem 4.16, the scoring rule

$$R : \mathfrak{C} \times \mathcal{F} \to \mathbb{R}, \qquad (y, F_\theta) \mapsto R(y, F_\theta) = \mathfrak{d}(y, \mu(\theta)),$$

is a strictly proper scoring rule, that is,

$$\mathbb{E}_\theta\left[R(Y, F_\theta)\right] = \mathbb{E}_\theta\left[\mathfrak{d}(Y, \mu(\theta))\right] \ < \ \mathbb{E}_\theta\left[\mathfrak{d}(Y, \mu(\widetilde{\theta}))\right] = \mathbb{E}_\theta\left[R(Y, F_{\widetilde{\theta}})\right],$$

for any $\widetilde{\theta} \neq \theta$. We conclude from this small example that the unit deviance is a strictly consistent loss function for the mean functional on the chosen EDF, and this provides us with a strictly proper scoring rule. ∎

In the above Example 4.17 we have chosen the mean functional

$$\mathfrak{A} : \mathcal{F} \to \mathbb{A} = \mathcal{M}, \qquad F_\theta = F(\cdot;\theta, v/\varphi) \mapsto \mathfrak{A}(F_\theta) = \mu(\theta),$$

within a given EDF $\mathcal{F} = \{F(\cdot;\theta, v/\varphi);\ \theta \in \mathring{\boldsymbol{\Theta}}\}$. We have seen that

- the unit deviance $\mathfrak{d}(\cdot,\cdot)$ is a strictly consistent loss function for the mean functional $\mathfrak{A}$ relative to the EDF $\mathcal{F}$;
- the function $(y, F_\theta) \mapsto R(y, F_\theta) = \mathfrak{d}(y, \mu(\theta))$ is a strictly proper scoring rule for the EDF $\mathcal{F}$, i.e.,

$$\mathbb{E}_\theta\left[\mathfrak{d}(Y, \mu(\theta))\right] \ < \ \mathbb{E}_\theta\left[\mathfrak{d}(Y, \mu(\widetilde{\theta}))\right],$$

for any $\widetilde{\theta} \neq \theta$.

The consideration of the mean functional $F \mapsto \mathfrak{A}(F) = \mathbb{E}_F[Y]$ in Example 4.17 is motivated by the fact that we typically forecast random variables by their means. However, more generally, we may ask the question for which functionals $\mathfrak{A} : \mathcal{F} \to \mathcal{P}(\mathbb{A})$, relative to a given set of distributions $\mathcal{F}$, there exists a loss function L that is strictly consistent.

Definition 4.18 (Elicitable) The functional $\mathfrak{A}$ is elicitable relative to a given set of distributions $\mathcal{F}$ if there exists a loss function L that is strictly consistent for $\mathfrak{A}$ and $\mathcal{F}$.

Above we have seen that the mean functional is elicitable relative to the EDF using the unit deviance loss; expected values relative to $\mathcal{F}$ with finite second moments are also elicitable using the square loss function. Savage [327] more generally identifies the Bregman divergences as being the only consistent scoring functions for the mean functional; recall that the unit deviance is a special case of a Bregman divergence, see (2.29). We are going to state the corresponding result.

For a general loss function L we make the following (standard) assumptions:

(L0) $L(y, a) \geq 0$ and we have an equality if and only if $y = a$;
(L1) $L(y, a)$ is measurable in y and continuous in a;
(L2) the partial derivative $\partial L(y, a)/\partial a$ exists and is continuous in a whenever $a \neq y$.

This then allows us to cite the following theorem.

Theorem 4.19 (Gneiting [162, Theorem 7]) *Let $\mathcal{F}$ be the class of distributions on an interval $\mathfrak{C} \subseteq \mathbb{R}$ having finite first moments.*

- *Assume the loss function $L : \mathfrak{C} \times \mathbb{A} \to \mathbb{R}$ satisfies (L0)–(L2) for interval $\mathfrak{C} = \mathbb{A} \subseteq \mathbb{R}$. L is consistent for the mean functional relative to the class $\mathcal{F}$ of compactly supported distributions on $\mathfrak{C}$ if and only if the loss function L is of Bregman divergence form*

$$D_\psi(y, a) = \psi(y) - \psi(a) - \psi'(a)(y - a),$$

  for a convex function ψ with (sub-)gradient ψ' on $\mathfrak{C}$.
- *If ψ is strictly convex on $\mathfrak{C}$, then the Bregman divergence D_ψ is strictly consistent for the mean functional relative to the class $\mathcal{F}$ on $\mathfrak{C}$ for which both $\mathbb{E}_F[Y]$ and $\mathbb{E}_F[\psi(Y)]$ exist and are finite.*

Theorem 4.19 tells us that Bregman divergences are the only consistent loss functions for the mean functional (under some additional assumptions). Consider the specific choice $\psi(a) = a^2/2$ which is a strictly convex function. For this choice, the Bregman divergence is the square loss function $D_\psi(y, a) = (y - a)^2/2$, which is strictly consistent for the mean functional relative to the class $\mathcal{F} \subset L^2(\mathbb{P})$. We remark that also quantiles are elicitable, the corresponding result is going to be stated in Theorem 5.33, below.

The second bullet point of Theorem 4.19 immediately implies that the unit deviance $\mathfrak{d}(\cdot, \cdot)$ is a strictly consistent loss function for the mean functional within the chosen EDF, see also (2.29) and Example 4.17. In particular, for $\theta \in \mathring{\boldsymbol{\Theta}}$

$$\mu = \mu(\theta) = \underset{a \in \mathcal{M}}{\arg\min}\, \mathbb{E}_\theta \left[\mathfrak{d}(Y, a)\right]. \tag{4.28}$$

Explicit evaluation of (4.28) requires that the true distribution F_θ of Y is known. Since, typically, this is not the case, we need to evaluate it empirically. Assume that the random variables Y_i are independent and F_θ distributed, with F_θ belonging to the fixed EDF providing the corresponding unit deviance $\mathfrak{d}$. Then, the objective function in (4.28) is approximated by, a.s.,

$$\mathfrak{D}(\boldsymbol{Y}_n, a) = \frac{1}{n}\sum_{i=1}^{n}\frac{v_i}{\varphi}\mathfrak{d}(Y_i, a) \;\rightarrow\; \mathbb{E}_\theta\left[\frac{v}{\varphi}\mathfrak{d}(Y, a)\right] \qquad \text{as } n \to \infty. \tag{4.29}$$

The convergence statement follows from the strong law of large numbers applied to the i.i.d. random variables (Y_i, v_i), $i \geq 1$, and supposed that the right-hand side of (4.29) exists. Thus, the deviance loss function (4.9) is an empirical version of the expected deviance loss function, and this approach is successful if we can exchange the 'argmin' operator of (4.28) and the limit $n \to \infty$ in (4.29). This closes the circle and brings us back to the M-estimator considered in Remarks 3.26 and 3.29, and which also links forecast evaluation and M-estimation.

Forecast Dominance

A consequence of Theorem 4.19 is that there are infinitely many strictly consistent loss functions for the mean functional, and, in principle, we could choose any of these for forecast evaluation. Choosing the unit deviance $\mathfrak{d}$ that matches the distribution F_θ of the observations $\boldsymbol{Y}_n$ and Y, respectively, gives us the MLE $\widehat{\mu}^{\text{MLE}}$, and we have seen that the MLE $\widehat{\mu}^{\text{MLE}}$ is not only unbiased for $\mu = \kappa'(\theta)$, but it also meets the Cramér–Rao information bound. That is, it is UMVU within the data generating model reflected by the true unit deviance $\mathfrak{d}$. This provides us (in the finite sample case) with a natural candidate for $\mathfrak{d}$ in (4.29) and, thus, a canonical proper scoring rule for (out-of-sample) forecast evaluation.

The previous statements have all been done under the assumption that there is no uncertainty about the underlying family of distribution functions that generates Y and $\boldsymbol{Y}_n$, respectively. Uncertainty was limited to the true canonical parameter θ and the true mean $\mu(\theta)$. This situation changes under model uncertainty. Krüger–Ziegel [227] study the question of having multiple strictly consistent loss functions in the situation where there is no natural candidate choice. Different choices may give different rankings to different (finite sample) predictors. Assume we have two predictors $\widehat{\mu}_1$ and $\widehat{\mu}_2$ for a random variable Y. Similarly to the definition of the expected deviance GL, we understand these predictors $\widehat{\mu}_1$ and $\widehat{\mu}_2$ as random variables, and we assume that all considered random variables have a finite first moment. Importantly, we do not assume independence between $\widehat{\mu}_1$, $\widehat{\mu}_2$ and Y, and in regression models we typically receive dependence between predictors $\widehat{\mu}$ and random variables Y through the features (covariates) $\boldsymbol{x}$, see also Outlook 4.13. Following Krüger–Ziegel [227] and Ehm et al. [119] we define *forecast dominance* as follows.

Definition 4.20 (Forecast Dominance) Predictor $\widehat{\mu}_1$ dominates predictor $\widehat{\mu}_2$ if

$$\mathbb{E}\left[D_\psi(Y,\widehat{\mu}_1)\right] \le \mathbb{E}\left[D_\psi(Y,\widehat{\mu}_2)\right],$$

for all Bregman divergences D_ψ with (convex) ψ supported on $\mathfrak{C}$, the latter being the convex closure of the supports of Y, $\widehat{\mu}_1$ and $\widehat{\mu}_2$.

If we work with a fixed member of the EDF, e.g., the gamma distribution, then we typically study the corresponding expected deviance GL for forecast evaluation in one single model, see Theorem 4.7 and (4.29). This evaluation may involve model risk in the decision making process, and forecast dominance provides a robust selection criterion.

Krüger–Ziegel [227] build on Theorem 1b and Corollary 1b of Ehm et al. [119] to prove the following theorem (which prevents from considering all convex functions ψ).

Theorem 4.21 (Theorem 2.1 of Krüger–Ziegel [227]) *Predictor $\widehat{\mu}_1$ dominates predictor $\widehat{\mu}_2$ if and only if for all $\tau \in \mathfrak{C}$*

$$\mathbb{E}\left[(Y-\tau)\,\mathbb{1}_{\{\widehat{\mu}_1>\tau\}}\right] \ge \mathbb{E}\left[(Y-\tau)\,\mathbb{1}_{\{\widehat{\mu}_2>\tau\}}\right]. \tag{4.30}$$

Denuit et al. [97] argue that in insurance one typically works with Tweedie's distributions having power variances $V(\mu)=\mu^p$ with power variance parameters $p \ge 1$. This motivates the following weaker form of forecast dominance.

Definition 4.22 (Tweedie's Forecast Dominance) Predictor $\widehat{\mu}_1$ Tweedie-dominates predictor $\widehat{\mu}_2$ if

$$\mathbb{E}\left[\mathfrak{d}_p(Y,\widehat{\mu}_1)\right] \le \mathbb{E}\left[\mathfrak{d}_p(Y,\widehat{\mu}_2)\right],$$

for all Tweedie's unit deviances $\mathfrak{d}_p$ with power variance parameters $p \ge 1$, we refer to (4.18) for $p \in (1,\infty)\setminus\{2\}$ and Table 4.1 for the Poisson and gamma cases $p \in \{1,2\}$.

Recall that Tweedie's unit deviances $\mathfrak{d}_p$ are a subclass of Bregman divergences, see (2.29). Define the following function for power variance parameters $p \ge 1$

$$\Upsilon_p(\mu) = \begin{cases} \log\mu & \text{for } p=2, \\ \frac{\mu^{2-p}}{2-p} & \text{otherwise.} \end{cases}$$

Denuit et al. [97] prove the following proposition.

Proposition 4.23 (Proposition 4.1 of Denuit et al. [97]) *Predictor $\widehat{\mu}_1$ Tweedie-dominates predictor $\widehat{\mu}_2$ if*

$$\mathbb{E}\left[\Upsilon_p(\widehat{\mu}_1)\right] \le \mathbb{E}\left[\Upsilon_p(\widehat{\mu}_2)\right] \qquad \textit{for all } p \ge 1,$$

and

$$\mathbb{E}\left[Y\mathbb{1}_{\{\widehat{\mu}_1>\tau\}}\right] \geq \mathbb{E}\left[Y\mathbb{1}_{\{\widehat{\mu}_2>\tau\}}\right] \qquad \textit{for all } \tau \in \mathfrak{C}.$$

Theorem 4.21 gives necessary and sufficient conditions to have forecast dominance, Proposition 4.23 gives sufficient conditions to have the weaker Tweedie's forecast dominance. In Theorem 7.15, below, we give another characterization of forecast dominance in terms of convex orders, under the additional assumption that the predictors are so-called auto-calibrated.

4.2 Cross-Validation

This section focuses on estimating the expected deviance GL (4.13) in cases where the canonical parameter θ is not known. Of course, the same concepts apply to the MSEP. In the remainder of this section we scale the unit deviances with v/φ, to bring them in line with the deviance loss (4.9).

4.2.1 In-Sample and Out-of-Sample Losses

The general aim in predictive modeling is to predict an unobserved random variable Y as good as possible based on past information $\boldsymbol{Y}_n$. Within the EDF, the predictive performance is then evaluated under an empirical version of the expected deviance GL

$$\mathbb{E}_\theta\left[\frac{v}{\varphi}\mathfrak{d}\left(Y, A(\boldsymbol{Y}_n)\right)\right] = 2\mathbb{E}_\theta\left[\frac{v}{\varphi}\Big(Yh(Y) - \kappa\left(h(Y)\right) - Yh(A(\boldsymbol{Y}_n)) + \kappa\left(h(A(\boldsymbol{Y}_n))\right)\Big)\right]. \tag{4.31}$$

Here, we no longer assume that Y and $A(\boldsymbol{Y}_n)$ are independent, and in the dependent case Theorem 4.7 does not apply. The reason for dropping the independence assumption is that below we consider regression models of a similar type as in Outlook 4.13. The expected deviance GL (4.31) as such is not directly useful because it cannot be calculated if the true canonical parameter θ is not known. Therefore, we are going to explain how it can be estimated empirically.

We start from the expected deviance GL in the EDF applied to the MLE decision rule $\widehat{\mu}^{\mathrm{MLE}}(\boldsymbol{Y}_n)$. It can be rewritten as

$$\mathbb{E}_\theta\left[\frac{v}{\varphi}\mathfrak{d}\left(Y, \widehat{\mu}^{\mathrm{MLE}}(\boldsymbol{Y}_n)\right)\right] = \int \mathbb{E}_\theta\left[\frac{v}{\varphi}\mathfrak{d}\left(Y, \widehat{\mu}^{\mathrm{MLE}}(\boldsymbol{Y}_n)\right)\bigg| \boldsymbol{Y}_n = \boldsymbol{y}_n\right] dP(\boldsymbol{y}_n;\theta), \tag{4.32}$$

where we use the tower property for conditional expectations. In view of (4.32), there are two things to be done:

(1) For given observations $\boldsymbol{Y}_n = \boldsymbol{y}_n$, we need to estimate the deviance GL, see also (4.15),

$$\mathbb{E}_\theta\left[\frac{v}{\varphi}\mathfrak{d}\left(Y, \widehat{\mu}^{\text{MLE}}(\boldsymbol{Y}_n)\right)\middle|\boldsymbol{Y}_n = \boldsymbol{y}_n\right] = \mathbb{E}_\theta\left[\frac{v}{\varphi}\mathfrak{d}\left(Y, \widehat{\mu}^{\text{MLE}}(\boldsymbol{y}_n)\right)\middle|\boldsymbol{Y}_n = \boldsymbol{y}_n\right]. \tag{4.33}$$

This is the part that we are going to solve empirically in the this section. Typically, we assume that Y and $\boldsymbol{Y}_n$ are independent, nevertheless, Y and its MLE predictor may still be dependent because we may have a predictor $\widehat{\mu}^{\text{MLE}}(\boldsymbol{Y}_n) = \widehat{\mu}^{\text{MLE}}(\boldsymbol{Y}_n, \boldsymbol{x})$. That is, this predictor often depends on covariate information $\boldsymbol{x}$ that describes Y, an example is provided in (4.22) of Outlook 4.13 and this is different from (4.15). In that case, the decision rule $A : \mathbb{Y} \times \mathcal{X} \to \mathbb{A}$ is extended by an additional covariate component $\boldsymbol{x} \in \mathcal{X}$, we refer to Sect. 5.1.1, where $\mathcal{X}$ is introduced and discussed.

(2) We have to find a way to generate more observations $\boldsymbol{Y}_n$ from $P(\boldsymbol{y}_n; \theta)$ in order to evaluate the outer integral in (4.32) empirically. One way to do so is the bootstrap method that is going to be discussed in Sect. 4.3, below.

We address the first problem of estimating the deviance GL given in (4.33). We do this under the assumption that $\boldsymbol{Y}_n$ and Y are independent. In order to estimate (4.33) we need observations for Y. However, typically, there are no observations available for this random variable because it is only going to be observed in the future. For this reason, one uses past observations for both, model fitting and the GL analysis. In order to perform this analysis in a proper way, the general paradigm is to partition the entire data into two *disjoint* data sets, a so-called *learning data set* $\mathcal{L} = \{Y_1, \ldots, Y_n\}$ and a *test data set* $\mathcal{T} = \{Y_1^\dagger, \ldots, Y_T^\dagger\}$. If we assume that all observations in $\mathcal{L} \cup \mathcal{T}$ are independent, then we receive a suitable observation $\boldsymbol{Y}_n$ from the learning data set $\mathcal{L}$ that can be used for model fitting. The test sample $\mathcal{T}$ can then play the role of the unobserved random variable Y (by assumption being independent of $\boldsymbol{Y}_n$). Note that $\mathcal{L}$ is *only* used for model fitting and $\mathcal{T}$ is *only* used for the deviance GL evaluation, see Fig. 4.1.

This setup motivates to estimate the mean parameter μ with MLE $\widehat{\mu}_{\mathcal{L}}^{\text{MLE}} = \widehat{\mu}^{\text{MLE}}(\boldsymbol{Y}_n)$ from the learning data $\mathcal{L}$ and $\boldsymbol{Y}_n$, respectively, by minimizing the deviance loss function $\mu \mapsto \mathfrak{D}(\boldsymbol{Y}_n, \mu)$ on the learning data $\mathcal{L}$, according to Corollary 4.5. Then we use this predictor $\widehat{\mu}_{\mathcal{L}}^{\text{MLE}}$ to empirically evaluate the conditional expectation in (4.33) on $\mathcal{T}$. The perception used is that we *(in-sample) learn a model* on $\mathcal{L}$ and we *out-of-sample test this model* on $\mathcal{T}$ to see how it generalizes to unobserved variables $Y_t^\dagger$, $1 \le t \le T$, that are of a similar nature as Y.

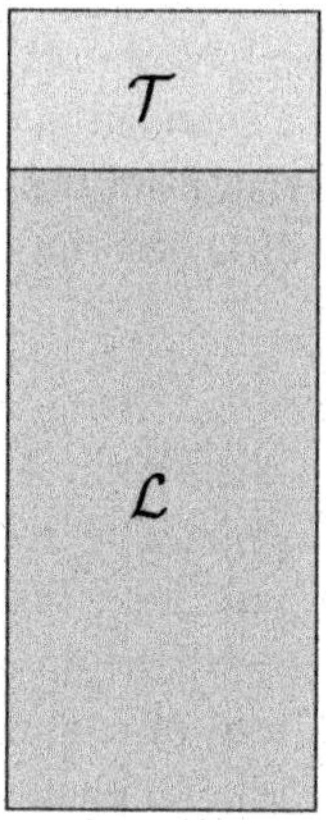

Fig. 4.1 Partition of entire data into learning data set $\mathcal{L}$ and test data set $\mathcal{T}$

Definition 4.24 (In-Sample and Out-of-Sample Losses) The *in-sample deviance loss* on the learning data $\mathcal{L} = \{Y_1, \ldots, Y_n\}$ is given by

$$\mathfrak{D}(\mathcal{L}, \widehat{\mu}_{\mathcal{L}}^{\mathrm{MLE}}) = \frac{2}{n} \sum_{i=1}^{n} \frac{v_i}{\varphi} \Big(Y_i h(Y_i) - \kappa(h(Y_i)) - Y_i h(\widehat{\mu}_{\mathcal{L}}^{\mathrm{MLE}}) + \kappa\left(h(\widehat{\mu}_{\mathcal{L}}^{\mathrm{MLE}})\right) \Big),$$

with MLE $\widehat{\mu}_{\mathcal{L}}^{\mathrm{MLE}} = \widehat{\mu}^{\mathrm{MLE}}(\boldsymbol{Y}_n)$ on $\mathcal{L}$.

The out-of-sample deviance loss on the test data $\mathcal{T} = \{Y_1^\dagger, \ldots, Y_T^\dagger\}$ of predictor $\widehat{\mu}_{\mathcal{L}}^{\mathrm{MLE}}$ is

$$\mathfrak{D}(\mathcal{T}, \widehat{\mu}_{\mathcal{L}}^{\mathrm{MLE}}) = \frac{2}{T} \sum_{t=1}^{T} \frac{v_t^\dagger}{\varphi} \Big(Y_t^\dagger h\left(Y_t^\dagger\right) - \kappa\left(h\left(Y_t^\dagger\right)\right) - Y_t^\dagger h(\widehat{\mu}_{\mathcal{L}}^{\mathrm{MLE}}) + \kappa\left(h(\widehat{\mu}_{\mathcal{L}}^{\mathrm{MLE}})\right) \Big),$$

where the sum runs over the test sample $\mathcal{T}$ having exposures $v_1^\dagger, \ldots, v_T^\dagger > 0$.

For MLE we minimize the objective function (4.9), therefore, the in-sample deviance loss $\mathfrak{D}(\mathcal{L}, \widehat{\mu}_{\mathcal{L}}^{\mathrm{MLE}}) = \mathfrak{D}(\boldsymbol{Y}_n, \widehat{\mu}^{\mathrm{MLE}}(\boldsymbol{Y}_n))$ exactly corresponds to the minimal deviance loss (4.9) achieved on the learning data $\mathcal{L}$, i.e., when using MLE $\widehat{\mu}_{\mathcal{L}}^{\mathrm{MLE}} = \widehat{\mu}^{\mathrm{MLE}}(\boldsymbol{Y}_n)$. We call this *in-sample* because the *same* data $\mathcal{L}$ is used for parameter estimation and deviance loss calculation. Typically, this loss is biased because it uses the optimal (in-sample) parameter estimate, we also refer to Sect. 4.2.3, below.

The out-of-sample loss $\mathfrak{D}(\mathcal{T}, \widehat{\mu}_{\mathcal{L}}^{\mathrm{MLE}})$ then empirically estimates the inner expectation in (4.32). This is a proper out-of-sample analysis because the test data $\mathcal{T}$ is disjoint from the learning data $\mathcal{L}$ on which the decision rule $\widehat{\mu}_{\mathcal{L}}^{\mathrm{MLE}}$ has been trained. Note that this out-of-sample figure reflects (4.33) in the following sense.

We have a portfolio of risks $(Y_t^\dagger, v_t^\dagger)$, $1 \le t \le T$, and (4.33) does not only reflect the calculation of the deviance GL of a given risk, but also the random selection of a risk from the portfolio. In this sense, (4.33) is an average over a given portfolio whose description is also included in the probability $\mathbb{P}_\theta$.

Summary 4.25 Definition 4.24 gives the general principle in predictive modeling according to which model learning and the generalization analysis are done. Namely, based on two disjoint and independent data sets $\mathcal{L}$ and $\mathcal{T}$, we perform model calibration on $\mathcal{L}$, and we analyze (conditional) GLs (using out-of-sample losses) on $\mathcal{T}$, respectively. For this concept to be useful, the learning data $\mathcal{L}$ and the test data $\mathcal{T}$ have to be sufficiently similar, i.e., ideally coming from the same model.

This approach does not estimate the outer expectation in the expected deviance GL (4.32), i.e., it is only an estimate for the deviance GL, given $\boldsymbol{Y}_n$, see (4.33).

4.2.2 Cross-Validation Techniques

In many applications one is not in the comfortable situation of having two sufficiently large data sets $\mathcal{L}$ and $\mathcal{T}$ available to support model learning and an out-of-sample generalization analysis. That is, we are usually equipped with only one data set of average size, let us call it $\mathcal{D}$. In order to calculate the objects in Definition 4.24 we could partition this data set (at random) into two data sets and then calculate in-sample and out-of-sample deviance losses on this partition. The disadvantage of this approach is that it is an inefficient use of information if only little data is available. In that case we require (almost) all data for learning. However, we still need a sufficiently large share of data for testing, to receive reliable deviance GL estimates for (4.33). The classical approach in this situation is to use cross-validation for estimating out-of-sample losses. The concept works as follows:

1. Perform model learning and in-sample loss calculation $\mathfrak{D}(\mathcal{L}, \widehat{\mu}_{\mathcal{L}}^{\text{MLE}})$ on all available data $\mathcal{L} = \mathcal{D}$, i.e., this part is not affected by selecting test data $\mathcal{T}$ and it is not touched by cross-validation.
2. For out-of-sample deviance loss calculation use the data $\mathcal{D}$ iteratively in an efficient way such that part of the data is used for model learning and the other part for the out-of-sample generalization analysis. This second step

(continued)

is (only) done for *estimating* the deviance GL of the model learned on all data. I.e. for prediction we work with MLE $\widehat{\mu}^{\text{MLE}}_{\mathcal{L}=\mathcal{D}}$, but the out-of-sample deviance loss is estimated using this data in a different way.

The three most commonly used methods are leave-one-out, K-fold and stratified K-fold cross-validation. We briefly describe these three cross-validation methods.

Leave-One-Out Cross-Validation

Denote all available data by $\mathcal{D} = \{Y_1, \ldots, Y_n\}$, and assume independence between the components. For leave-one-out (loo) cross-validation we select $1 \le i \le n$ and define the partition $\mathcal{L}_{(-i)} = \mathcal{D} \setminus \{Y_i\}$ for the learning data and $\mathcal{T}_i = \{Y_i\}$ for the test data. Based on the learning data $\mathcal{L}_{(-i)}$ we calculate the MLE

$$\widehat{\mu}^{(-i)} \stackrel{\text{def.}}{=} \widehat{\mu}^{\text{MLE}}_{\mathcal{L}_{(-i)}},$$

which is based on all data except observation Y_i. This observation is now used to do an out-of-sample analysis, and averaging this over all $1 \le i \le n$ we receive the *leave-one-out cross-validation loss*

$$\begin{aligned}\widehat{\mathfrak{D}}^{\text{loo}} &= \frac{1}{n}\sum_{i=1}^{n} \frac{v_i}{\varphi}\,\mathfrak{d}\left(Y_i, \widehat{\mu}^{(-i)}\right) = \frac{1}{n}\sum_{i=1}^{n} \mathfrak{D}\left(\mathcal{T}_i, \widehat{\mu}^{(-i)}\right) \qquad (4.34)\\ &= \frac{2}{n}\sum_{i=1}^{n} \frac{v_i}{\varphi}\left(Y_i h\left(Y_i\right) - \kappa\left(h\left(Y_i\right)\right) - Y_i h\left(\widehat{\mu}^{(-i)}\right) + \kappa\left(h\left(\widehat{\mu}^{(-i)}\right)\right)\right),\end{aligned}$$

where $\mathfrak{D}(\mathcal{T}_i, \widehat{\mu}^{(-i)})$ is the (out-of-sample) *cross-validation loss* on $\mathcal{T}_i = \{Y_i\}$ using the predictor $\widehat{\mu}^{(-i)}$. This leave-one-out cross-validation loss $\widehat{\mathfrak{D}}^{\text{loo}}$ is now used as estimate for the out-of-sample deviance loss $\mathfrak{D}(\mathcal{T}, \widehat{\mu}^{\text{MLE}}_{\mathcal{L}})$. Leave-one-out cross-validation uses all data $\mathcal{D}$ for learning and testing, namely, the data $\mathcal{D}$ is partitioned into a learning set $\mathcal{L}_{(-i)}$ for (partial) learning and a test set $\mathcal{T}_i = \{Y_i\}$ for an out-of-sample generalization analysis. This is done for all instances $1 \le i \le n$, and the out-of-sample loss is estimated by the resulting average cross-validation loss. This averaging allows us to not only understand (4.34) as a conditional out-of-sample loss in the spirit of Definition 4.24. The outer empirical average in (4.34) also makes it suitable for an expected deviance GL estimate according to (4.32).

The variance of this empirical deviance GL is given by (subject to existence)

$$\text{Var}_\theta\left(\widehat{\mathfrak{D}}^{\text{loo}}\right) = \frac{1}{n^2}\sum_{i=1}^{n}\sum_{j=1}^{n} \text{Cov}_\theta\left(\frac{v_i}{\varphi}\mathfrak{d}\left(Y_i, \widehat{\mu}^{(-i)}\right), \frac{v_j}{\varphi}\mathfrak{d}\left(Y_j, \widehat{\mu}^{(-j)}\right)\right).$$

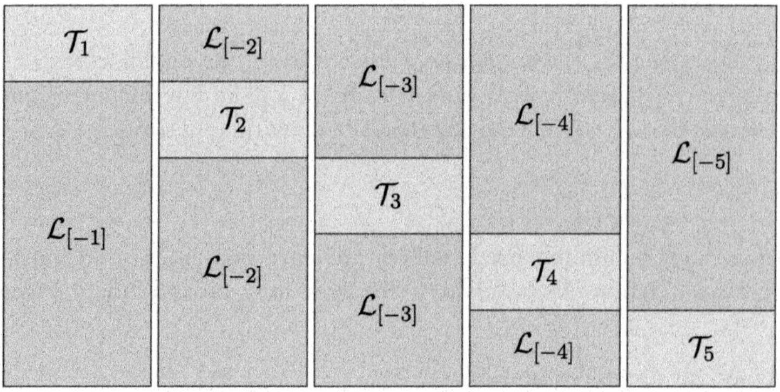

Fig. 4.2 Partitions of K-fold cross-validation for $K = 5$

These covariances use exactly the same observations on $\mathcal{D} \setminus \{Y_i, Y_j\}$, therefore, there are strong correlations between the estimators $\widehat{\mu}^{(-i)}$ and $\widehat{\mu}^{(-j)}$. In addition, the leave-one-out cross-validation is often computationally not feasible because it requires fitting the model n times, which in the situation of complex models and of large insurance portfolios can be too demanding. We come back to this in Sect. 5.6 where we provide the generalized cross-validation (GCV) loss approximation within generalized linear models (GLMs).

K-Fold Cross-Validation

Choose a fixed integer $K \geq 2$ and partition the entire data $\mathcal{D}$ at random into K disjoint subsets (called folds) $\mathcal{L}_1, \ldots, \mathcal{L}_K$ of approximately the same size. The learning data for fixed $1 \leq k \leq K$ is then defined by $\mathcal{L}_{[-k]} = \mathcal{D} \setminus \mathcal{L}_k$ and the test data by $\mathcal{T}_k = \mathcal{L}_k$, see Fig. 4.2. Based on learning data $\mathcal{L}_{[-k]}$ we calculate the MLE

$$\widehat{\mu}^{[-k]} \stackrel{\text{def.}}{=} \widehat{\mu}^{\text{MLE}}_{\mathcal{L}_{[-k]}},$$

which is based on all data except $\mathcal{T}_k$.

These observations are now used to do an (out-of-sample) cross-validation analysis, and averaging this over all $1 \leq k \leq K$ we receive the *K-fold cross-validation (CV) loss*.

$$
\begin{aligned}
\widehat{\mathfrak{D}}^{\mathrm{CV}} &= \frac{1}{K}\sum_{k=1}^{K} \mathfrak{D}\left(\mathcal{T}_k, \widehat{\mu}^{[-k]}\right) \\
&= \frac{1}{K}\sum_{k=1}^{K} \frac{1}{|\mathcal{T}_k|} \sum_{Y_i\in\mathcal{T}_k} \frac{v_i}{\varphi}\,\mathfrak{d}\left(Y_i, \widehat{\mu}^{[-k]}\right) \qquad (4.35)\\
&\approx \frac{1}{n}\sum_{k=1}^{K} \sum_{Y_i\in\mathcal{T}_k} \frac{v_i}{\varphi}\,\mathfrak{d}\left(Y_i, \widehat{\mu}^{[-k]}\right).
\end{aligned}
$$

The last step is an approximation because not all $\mathcal{T}_k$ may have exactly the same sample size if n is not a multiple of K. We can understand (4.35) not only as a conditional out-of-sample loss estimate in the spirit of Definition 4.24. The outer empirical average in (4.35) also makes it suitable for an expected deviance GL estimate according to (4.32). The variance of this empirical deviance GL is given by (subject to existence)

$$
\mathrm{Var}_\theta\left(\widehat{\mathfrak{D}}^{\mathrm{CV}}\right) \approx \frac{1}{n^2}\sum_{k,l=1}^{K} \sum_{Y_i\in\mathcal{T}_k} \sum_{Y_j\in\mathcal{T}_l} \mathrm{Cov}_\theta\left(\frac{v_i}{\varphi}\mathfrak{d}\left(Y_i, \widehat{\mu}^{[-k]}\right), \frac{v_j}{\varphi}\mathfrak{d}\left(Y_j, \widehat{\mu}^{[-l]}\right)\right).
$$

Typically, in applications, one uses K-fold cross-validation with $K = 10$.

Stratified K-Fold Cross-Validation

A disadvantage of the above K-fold cross-validation is that it may happen that there are two outliers in the data, and there is a positive probability that these two outliers belong to the same subset $\mathcal{L}_k$. This may substantially distort K-fold cross-validation because in that case the subsets $\mathcal{L}_k$, $1 \le k \le K$, are of different quality. Stratified K-fold cross-validation aims at distributing outliers more equally across the partition. Order the observations Y_i, $1 \le i \le n$, as follows

$$
Y_{(1)} \ge Y_{(2)} \ge \ldots \ge Y_{(n)}.
$$

For stratified K-fold cross-validation, we randomly distribute (partition) the K biggest claims $Y_{(1)}, \ldots, Y_{(K)}$ to the subsets $\mathcal{L}_k$, $1 \le k \le K$, then we randomly partition the next K biggest claims $Y_{(K+1)}, \ldots, Y_{(2K)}$ to the subsets $\mathcal{L}_k$, $1 \le k \le K$, and so forth. This implies, e.g., that the two biggest claims cannot fall into the same set $\mathcal{L}_k$. This stratified partition $\mathcal{L}_k$, $1 \le k \le K$, is then used for K-fold cross-validation.

Summary 4.26 (Cross-Validation)

- A model is calibrated on the learning data set $\mathcal{L}$ by minimizing the in-sample deviance loss $\mathfrak{D}(\mathcal{L}, \mu)$ in μ. This provides MLE $\widehat{\mu}_{\mathcal{L}}^{\mathrm{MLE}}$.
- The quality of this model is assessed on test data $\mathcal{T}$ being disjoint of $\mathcal{L}$ considering the corresponding out-of-sample deviance loss $\mathfrak{D}(\mathcal{T}, \widehat{\mu}_{\mathcal{L}}^{\mathrm{MLE}})$.
- If there is no test data set $\mathcal{T}$ available we perform (stratified) K-fold cross-validation. This provides the (stratified) K-fold cross-validation loss $\widehat{\mathfrak{D}}^{\mathrm{CV}}$ which is an estimate for the out-of-sample deviance loss and for the expected deviance GL (4.32).

Example 4.27 (Out-of-Sample Deviance Loss Estimation) We consider a claim counts example using the Poisson EDF model. The claim counts N_i and exposures $v_i > 0$ used come from the French motor insurance data given in Listing 13.2 of Chap. 13.1. We model the claim frequencies $Y_i = N_i/v_i$ with the Poisson EDF model having cumulant function $\kappa(\theta) = \exp\{\theta\}$ and dispersion parameter $\varphi = 1$ for all $1 \le i \le n$. The expected frequency is given by $\mu = \mathbb{E}_\theta[Y_i] = \kappa'(\theta)$. Moreover, we assume that all claim counts N_i, $1 \le i \le n$, are independent. This provides us with the Poisson deviance loss function for observations $\boldsymbol{Y}_n = (Y_1, \ldots, Y_n)^\top$, see Example 4.12,

$$\begin{aligned}\mathfrak{D}(\boldsymbol{Y}_n, \mu) &= \frac{1}{n}\sum_{i=1}^n v_i \mathfrak{d}(Y_i, \mu) = \frac{1}{n}\sum_{i=1}^n 2v_i \left(\mu - Y_i - Y_i \log\left(\frac{\mu}{Y_i}\right)\right) \\ &= \frac{1}{n}\sum_{i=1}^n 2\left(v_i\mu - N_i - N_i \log\left(\frac{v_i\mu}{N_i}\right)\right) \ \ge\ 0,\end{aligned}$$

where, for $Y_i = 0$, we set $\mathfrak{d}(Y_i = 0, \mu) = 2\mu$. Minimizing the Poisson deviance loss function $\mathfrak{D}(\boldsymbol{Y}_n, \mu)$ in μ gives us the MLE for μ and $\theta = h(\mu)$, respectively. It is given by, see (3.24),

$$\widehat{\mu}^{\mathrm{MLE}} = \widehat{\mu}_{\mathcal{L}}^{\mathrm{MLE}} = \frac{\sum_{i=1}^n N_i}{\sum_{i=1}^n v_i} = 7.36\%,$$

for learning data set $\mathcal{L} = \{Y_1, \ldots, Y_n\}$. This provides us with an in-sample Poisson deviance loss of $\mathfrak{D}(\boldsymbol{Y}_n, \widehat{\mu}_{\mathcal{L}}^{\mathrm{MLE}}) = \mathfrak{D}(\mathcal{L}, \widehat{\mu}_{\mathcal{L}}^{\mathrm{MLE}}) = 25.213 \cdot 10^{-2}$.

Since we do not have test data $\mathcal{T}$, we explore tenfold cross-validation. We therefore partition the entire data at random into $K = 10$ disjoint sets $\mathcal{L}_1, \ldots, \mathcal{L}_{10}$, and compute the tenfold cross-validation loss as described in (4.35). This gives us $\widehat{\mathfrak{D}}^{\mathrm{CV}} = 25.213 \cdot 10^{-2}$, thus, we receive the same value as for the in-sample loss which says that we do not have in-sample over-fitting, here. This is not surprising in the homogeneous model $\lambda = \mathbb{E}_\theta[Y_i]$. We can also quantify the uncertainty in this estimate by the corresponding empirical standard deviation for $\mathcal{T}_k = \mathcal{L}_k$

$$\sqrt{\frac{1}{K-1}\sum_{k=1}^{K}\left(\mathfrak{D}\left(\mathcal{T}_k, \widehat{\mu}^{[-k]}\right) - \widehat{\mathfrak{D}}^{\mathrm{CV}}\right)^2} = 0.234 \cdot 10^{-2}. \tag{4.36}$$

This says that there is quite some fluctuation in the data because uncertainty in estimate $\widehat{\mathfrak{D}}^{\mathrm{CV}} = 25.213 \cdot 10^{-2}$ is roughly 1%. This finishes this example, and we will come back to it in Sect. 5.2.4, below. ■

4.2.3 Akaike's Information Criterion

The out-of-sample analysis in terms of GLs and cross-validation evaluates the predictive performance on unseen data. Another way of model selection is to study in-sample losses instead, but penalize model complexity. Akaike's information criterion (AIC), see Akaike [5], is the most popular tool that follows such a model selection methodology. AIC is based on a set of assumptions which should be fulfilled to apply, this is going to be discussed in this section; we therefore follow the lecture notes of Künsch [229].

Assume we have independent random variables Y_i from some (unknown) density f. Assume we have two candidate models with densities h_θ and g_ϑ from which we would like to select the preferred one for the given data $\boldsymbol{Y}_n = (Y_1, \ldots, Y_n)$. The two unknown parameters in these densities h_θ and g_ϑ are called θ and ϑ, respectively. We neither assume that one of the two models h_θ and g_ϑ contains the true model f, nor that the two models are nested. That is, f, h_θ and g_ϑ are quite general densities w.r.t. a given σ-finite measure ν.

Assume that both models under consideration have a unique MLE $\widehat{\theta}^{\mathrm{MLE}} = \widehat{\theta}^{\mathrm{MLE}}(\boldsymbol{Y}_n)$ and $\widehat{\vartheta}^{\mathrm{MLE}} = \widehat{\vartheta}^{\mathrm{MLE}}(\boldsymbol{Y}_n)$ which is based on the same observations $\boldsymbol{Y}_n$. AIC [5] says that model $h_{\widehat{\theta}^{\mathrm{MLE}}}$ should be preferred over model $g_{\widehat{\vartheta}^{\mathrm{MLE}}}$ if

$$-2\sum_{i=1}^{n}\log\left(h_{\widehat{\theta}^{\mathrm{MLE}}}(Y_i)\right) + 2\dim(\theta) \;<\; -2\sum_{i=1}^{n}\log\left(g_{\widehat{\vartheta}^{\mathrm{MLE}}}(Y_i)\right) + 2\dim(\vartheta), \tag{4.37}$$

where $\dim(\cdot)$ denotes the dimension of the corresponding parameter. Thus, we compute the log-likelihoods of the data $\boldsymbol{Y}_n$ in the corresponding MLEs $\widehat{\theta}^{\mathrm{MLE}}$ and $\widehat{\vartheta}^{\mathrm{MLE}}$, and we penalize the resulting values with the number of parameters to correct for model complexity. We give some remarks.

Remarks 4.28

- AIC is neither an in-sample loss nor an out-of-sample loss to measure generalization accuracy, but it considers penalized log-likelihoods. Under certain assumptions one can prove that asymptotically minimizing AICs is equivalent to minimizing leave-one-out cross-validation mean squared errors.
- The two penalized log-likelihoods have to be evaluated on the *same* data $\boldsymbol{Y}_n$ and they need to consider the *MLEs* $\widehat{\theta}^{\mathrm{MLE}}$ and $\widehat{\vartheta}^{\mathrm{MLE}}$ because the justification of AIC is based on the asymptotic normality of MLEs, otherwise there is no mathematical justification why (4.37) should be a reasonable model selection tool.
- AIC does not require (but allows for) nested models h_θ and g_ϑ nor need they be Gaussian, it is only based on asymptotic normality. We give a heuristic argument below.
- Evaluation of (4.37) involves all terms of the log-likelihoods, also those that do not depend on the parameters θ and ϑ.
- Both models should consider the data $\boldsymbol{Y}_n$ in the same units, i.e., AIC does not apply if h_θ is a density for Y_i and g_ϑ is a density for cY_i. In that case, one has to perform a transformation of variables to ensure that both densities consider the data in the same units. We briefly highlight this by considering a Gaussian example. We choose i.i.d. observations $Y_i \sim \mathcal{N}(\theta, \sigma^2)$ for known variance $\sigma^2 > 0$. Choose $c > 0$, we have $cY_i \sim \mathcal{N}(\vartheta = c\theta, c^2\sigma^2)$. We obtain MLE $\widehat{\theta}^{\mathrm{MLE}} = \sum_{i=1}^n Y_i/n$ and log-likelihood in MLE $\widehat{\theta}^{\mathrm{MLE}}$

$$\sum_{i=1}^n \log\left(h_{\widehat{\theta}^{\mathrm{MLE}}}(Y_i)\right) = -\frac{n}{2}\log(2\pi\sigma^2) - \sum_{i=1}^n \frac{1}{2\sigma^2}\left(Y_i - \widehat{\theta}^{\mathrm{MLE}}\right)^2.$$

  On the transformed scale we have MLE $\widehat{\vartheta}^{\mathrm{MLE}} = \sum_{i=1}^n cY_i/n = c\widehat{\theta}^{\mathrm{MLE}}$ and log-likelihood in MLE $\widehat{\vartheta}^{\mathrm{MLE}}$

$$\sum_{i=1}^n \log\left(g_{\widehat{\vartheta}^{\mathrm{MLE}}}(cY_i)\right) = -\frac{n}{2}\log(2\pi c^2\sigma^2) - \sum_{i=1}^n \frac{1}{2c^2\sigma^2}\left(cY_i - c\widehat{\theta}^{\mathrm{MLE}}\right)^2.$$

  Thus, find that the two log-likelihoods differ by $-n\log(c)$, but we consider the same model only under different measurement units of the data. The same applies when we work, e.g., with a log-normal model or logged data in a Gaussian model.

We give a heuristic justification of AIC. In Example 3.10 we have seen that the MLE is obtained by minimizing the KL divergence from h_θ to the empirical distribution $\widehat{f}_n$ of $\boldsymbol{Y}_n$. This motivates to use the KL divergence also for comparing the MLE estimated models to the true model, i.e., we consider the difference (supposed the densities are defined on the same domain)

$$\begin{aligned}
&D_{\mathrm{KL}}\left(f \,\|h_{\widehat{\theta}^{\mathrm{MLE}}}(\cdot)\right) - D_{\mathrm{KL}}\left(f \,\|g_{\widehat{\vartheta}^{\mathrm{MLE}}}(\cdot)\right) \\
&= \int \log\left(\frac{f(y)}{h_{\widehat{\theta}^{\mathrm{MLE}}}(y)}\right) f(y)d\nu(y) - \int \log\left(\frac{f(y)}{g_{\widehat{\vartheta}^{\mathrm{MLE}}}(y)}\right) f(y)d\nu(y) \\
&= \int \log\left(g_{\widehat{\vartheta}^{\mathrm{MLE}}}(y)\right) f(y)d\nu(y) - \int \log\left(h_{\widehat{\theta}^{\mathrm{MLE}}}(y)\right) f(y)d\nu(y). \qquad (4.38)
\end{aligned}$$

If this difference is negative, model $h_{\widehat{\theta}^{\mathrm{MLE}}}$ should be preferred over model $g_{\widehat{\vartheta}^{\mathrm{MLE}}}$ because it is closer to the true model f w.r.t. the KL divergence. Thus, we need to calculate the two integrals in (4.38). Since the true density f is not known, these two integrals need to be estimated.

As a first idea we estimate the integrals on the right-hand side empirically using the observations $\boldsymbol{Y}_n$, say, the first integral is estimated by

$$\frac{1}{n}\sum_{i=1}^{n} \log\left(g_{\widehat{\vartheta}^{\mathrm{MLE}}}(Y_i)\right).$$

However, this will lead to a biased estimate because the MLE $\widehat{\vartheta}^{\mathrm{MLE}}$ exactly maximizes this empirical estimate (as a function of ϑ). The integrals in (4.38), on the other hand, can be interpreted as an out-of-sample calculation between independent random variables $\boldsymbol{Y}_n$ (used for MLE) and $Y \sim f d\nu$ used in the integral. The bias results from the fact that in the empirical estimate the independence gets lost. Therefore, we need to correct this estimate for the bias in order to obtain a reasonable estimate for the difference of the KL divergences. Under the following assumptions this bias correction is asymptotically given by $-\dim(\vartheta)/n$: (1) $\sqrt{n}(\widehat{\vartheta}^{\mathrm{MLE}}(\boldsymbol{Y}_n) - \vartheta_0)$ is asymptotically normally distributed $\mathcal{N}(0, \Sigma(\vartheta_0)^{-1})$ as $n \to \infty$, where ϑ_0 is the parameter that minimizes the KL divergence from g_ϑ to f; we also refer to Remarks 3.26. (2) The true f is sufficiently close to g_{ϑ_0} such that the $\mathbb{E}_f$-covariance matrix of the score $\nabla_\vartheta \log g_{\vartheta_0}$ is close to the negative $\mathbb{E}_f$-expected Hessian $\nabla^2_\vartheta \log g_{\vartheta_0}$; see also (3.36) and Sect. 11.1.4, below. In that case, $\Sigma(\vartheta_0)$ approximately corresponds to Fisher's information matrix $\mathcal{I}_1(\vartheta_0)$ and AIC is justified.

This shows that AIC applies if both models are evaluated under the same observations $\boldsymbol{Y}_n$, the models need to use the MLEs, and asymptotic normality needs to hold with limits such that the true model is close to a member of the selected model classes $\{h_\theta; \theta\}$ and $\{g_\vartheta; \vartheta\}$. We remark that this is not the only set-up under which AIC can be justified, but other set-ups do not essentially differ.

The Bayesian information criterion (BIC) is similar to AIC but in a Bayesian context. The BIC says that model $h_{\widehat{\theta}^{\mathrm{MLE}}}$ should be preferred over model $g_{\widehat{\vartheta}^{\mathrm{MLE}}}$ if

$$-2\sum_{i=1}^{n} \log\left(h_{\widehat{\theta}^{\mathrm{MLE}}}(Y_i)\right) + \log(n)\dim(\theta) \;<\; -2\sum_{i=1}^{n} \log\left(g_{\widehat{\vartheta}^{\mathrm{MLE}}}(Y_i)\right) + \log(n)\dim(\vartheta),$$

where n is the sample size of $\boldsymbol{Y}_n$ used for model fitting. The BIC has been derived by Schwarz [331]. Therefore, it is also called Schwarz' information criterion (SIC).

4.3 Bootstrap

The bootstrap method has been invented by Efron [115] and Efron–Tibshirani [118]. The bootstrap is used to simulate new data from either the empirical distribution $\widehat{F}_n$ or from an estimated model $F(\cdot;\widehat{\theta})$. This allows, for instance, to evaluate the outer expectation in the expected deviance GL (4.32) which requires a data model for $\boldsymbol{Y}_n$. The presentation in this section is based on the lecture notes of Bühlmann–Mächler [59, Chapter 5].

4.3.1 Non-parametric Bootstrap Simulation

Assume we have i.i.d. observations $Y_1, \ldots, Y_n$ from an unknown distribution function $F(\cdot;\theta)$. Based on these observations $\boldsymbol{Y} = (Y_1, \ldots, Y_n)$ we choose a decision rule $A : \mathbb{Y} \to \mathbb{A} = \boldsymbol{\Theta} \subseteq \mathbb{R}$ which provides us with an estimator for θ

$$\boldsymbol{Y} \mapsto \widehat{\theta} = A(\boldsymbol{Y}). \tag{4.39}$$

Typically, the decision rule $A(\cdot)$ is a known function and we would like to determine the distributional properties of parameter estimator (4.39) as a function of the (random) observations $\boldsymbol{Y}$. E.g., for any measurable set C, we might want to compute

$$\mathbb{P}_\theta\left[\widehat{\theta} \in C\right] = \mathbb{P}_\theta\left[A(\boldsymbol{Y}) \in C\right] = \int \mathbb{1}_{\{A(\boldsymbol{y}) \in C\}}\, dP(\boldsymbol{y};\theta). \tag{4.40}$$

Since, typically, the true data generating distribution $Y_i \sim F(\cdot;\theta)$ is not known, the distributional properties of $\widehat{\theta}$ cannot be determined, also not by Monte Carlo simulation. The idea behind bootstrap is to approximate $F(\cdot;\theta)$. Choose as approximation to $F(\cdot;\theta)$ the empirical distribution of the i.i.d. observations $\boldsymbol{Y}$ given by, see (3.9),

$$\widehat{F}_n(y) = \frac{1}{n}\sum_{i=1}^{n} \mathbb{1}_{\{Y_i \le y\}} \qquad \text{for } y \in \mathbb{R}.$$

The Glivenko–Cantelli theorem [64, 159] tells us that the empirical distribution $\widehat{F}_n$ converges uniformly to $F(\cdot;\theta)$, a.s., for $n \to \infty$, so it should be a good approximation to $F(\cdot;\theta)$ for large n. The idea now is to simulate from the empirical distribution $\widehat{F}_n$.

(Non-parametric) bootstrap algorithm

(1) Repeat for $m = 1, \ldots, M$

(a) simulate i.i.d. observations $Y_1^*, \ldots, Y_n^*$ from $\widehat{F}_n$ (these are obtained by random drawings with replacements from the observations $Y_1, \ldots, Y_n$; we denote this resampling distribution of $\boldsymbol{Y}^* = (Y_1^*, \ldots, Y_n^*)$ by $\mathbb{P}^* = \mathbb{P}_{\boldsymbol{Y}}^*$);

(b) calculate the estimator $\widehat{\theta}^{(m*)} = A(\boldsymbol{Y}^*)$.

(2) Return $\widehat{\theta}^{(1*)}, \ldots, \widehat{\theta}^{(M*)}$ and the resulting empirical bootstrap distribution

$$\widehat{F}_M^*(\vartheta) = \frac{1}{M} \sum_{m=1}^{M} \mathbb{1}_{\{\widehat{\theta}^{(m*)} \leq \vartheta\}},$$

for the estimated distribution of $\widehat{\theta}$.

We can use the *empirical bootstrap distribution* $\widehat{F}_M^*$ as an estimate of the true distribution of $\widehat{\theta}$, that is, we estimate and approximate

$$\mathbb{P}_\theta \left[\widehat{\theta} \in C \right] \approx \mathbb{P}_{\widehat{\theta}} \left[\widehat{\theta} \in C \right] \stackrel{\text{def.}}{=} \mathbb{P}_{\boldsymbol{Y}}^* \left[\widehat{\theta}^* \in C \right] \approx \frac{1}{M} \sum_{m=1}^{M} \mathbb{1}_{\{\widehat{\theta}^{(m*)} \in C\}}, \tag{4.41}$$

where $\mathbb{P}_{\boldsymbol{Y}}^*$ corresponds to the *bootstrap distribution* of Step (1a) of the above algorithm, and where we set $\widehat{\theta}^* = A(\boldsymbol{Y}^*)$. This bootstrap distribution $\mathbb{P}_{\boldsymbol{Y}}^*$ is empirically approximated by the empirical bootstrap distribution $\widehat{F}_M^*$ for studying $\widehat{\theta}^*$.

Remarks 4.29

- The quality of the approximations in (4.41) depend on the richness of the observation $\boldsymbol{Y} = (Y_1, \ldots, Y_n)$, because the bootstrap distribution

$$\mathbb{P}_{\boldsymbol{Y}}^* \left[\widehat{\theta}^* \in C \right] = \mathbb{P}_{\boldsymbol{Y}=\boldsymbol{y}}^* \left[\widehat{\theta}^* \in C \right],$$

depends on the realization $\boldsymbol{y}$ of the data $\boldsymbol{Y}$ from which we generate the bootstrap sample $\boldsymbol{Y}^*$. It also depends on M and the explicit random drawings Y_i^* providing the empirical bootstrap distribution $\widehat{F}_M^*$. The latter uncertainty can be controlled since the bootstrap distribution $\mathbb{P}_{\boldsymbol{Y}}^*$ corresponds to a multinomial distribution, and the Glivenko–Cantelli theorem [64, 159] applies to $\widehat{F}_M^*$ and $\mathbb{P}_{\boldsymbol{Y}}^*$ for $M \to \infty$. The former uncertainty inherited from the realization $\boldsymbol{Y} = \boldsymbol{y}$ cannot be diminished because we cannot enrich the observation $\boldsymbol{Y}$.

- The empirical bootstrap distribution $\widehat{F}^*_M$ can be used to estimate the mean of the estimator $\widehat{\theta}$ given in (4.39)

$$\widehat{\mathbb{E}}_\theta\left[\widehat{\theta}\right] = \mathbb{E}^*_{\boldsymbol{Y}}\left[\widehat{\theta}^*\right] \approx \frac{1}{M}\sum_{m=1}^{M}\widehat{\theta}^{(m*)},$$

  and its variance

$$\widehat{\mathrm{Var}}_\theta\left(\widehat{\theta}\right) = \mathrm{Var}_{\mathbb{P}^*_{\boldsymbol{Y}}}\left(\widehat{\theta}^*\right) \approx \frac{1}{M-1}\sum_{m=1}^{M}\left(\widehat{\theta}^{(m*)} - \frac{1}{M}\sum_{k=1}^{M}\widehat{\theta}^{(k*)}\right)^2.$$

- The previous item discusses the approximation of the bootstrap mean and variance, respectively. Bootstrap intervals for coverage ratios need some care, and there are different versions. The naive way of just calculating quantiles from $\widehat{F}^*_M$ often does not work well, and methods like a double bootstrap may need to be considered.
- In (4.39) we have assumed that the quantity of interest is the parameter θ, but similar considerations also apply to general decision rules estimating $\gamma(\theta)$.
- The bootstrap as defined above directly acts on the observations $Y_1, \ldots, Y_n$, and the basic assumption is that these observations are i.i.d. If this is not the case, one may first need to transform the observations, for instance, one can calculate residuals and assume that these residuals are i.i.d. In more complicated cases, one even drops the i.i.d. assumption and replaces it by an identical mean and variance assumption, that is, that all residuals are assumed to be independent, centered and with unit variance. This is sometimes also called *residual bootstrap* and it may be suitable in regression models as will be introduced below. Thus, in this latter case we estimate for each observation Y_i its mean $\widehat{\mu}_i$ and its standard deviation $\widehat{\sigma}_i$, for instance, using the variance function of the chosen EDF. This then allows for calculating the residuals $\widehat{\varepsilon}_i = (Y_i - \widehat{\mu}_i)/\widehat{\sigma}_i$. For the residual bootstrap we resample the residuals $\widehat{\varepsilon}^*_i$ from $\widehat{\varepsilon}_1, \ldots, \widehat{\varepsilon}_n$. This provides bootstrap observations

$$Y_i^* = \widehat{\mu}_i + \widehat{\sigma}_i\widehat{\varepsilon}^*_i.$$

  The *wild bootstrap* proposed by Wu [386] additionally uses a centered and normalized i.i.d. random variable V_i (also being independent of $\widehat{\varepsilon}^*_i$) to modify the residual bootstrap observations to

$$Y_i^* = \widehat{\mu}_i + \widehat{\sigma}_i V_i\widehat{\varepsilon}^*_i.$$

The bootstrap is called *consistent* for $\widehat{\theta}$ if we have for all $z \in \mathbb{R}$ the following convergence in probability as $n \to \infty$

$$\mathbb{P}_\theta\left[\sqrt{n}\left(\widehat{\theta}-\theta\right) \le z\right] - \mathbb{P}^*_{\boldsymbol{Y}}\left[\sqrt{n}\left(\widehat{\theta}^*-\widehat{\theta}\right) \le z\right] \stackrel{\text{prob.}}{\to} 0,$$

the quantities $\widehat{\theta} = \widehat{\theta}_n$ and $\widehat{\theta}^* = \widehat{\theta}^*_n$ depend on (the size n of) the observation $\boldsymbol{Y} = \boldsymbol{Y}_n$; the convergence in probability is needed because $\boldsymbol{Y} = \boldsymbol{Y}_n$ are random vectors. Assume that $\widehat{\theta}^{\text{MLE}} = \widehat{\theta}$ is the MLE of θ satisfying the assumptions of Theorem 3.28. Then we have asymptotic normality, see (3.30),

$$\sqrt{n}\left(\widehat{\theta}-\theta\right) \implies \mathcal{N}\left(0, \mathcal{I}_1(\theta)^{-1}\right) \qquad \text{as } n \to \infty,$$

with Fisher's information $\mathcal{I}_1(\theta)$. Bootstrap consistency then requires

$$\sqrt{n}\left(\widehat{\theta}^*-\widehat{\theta}\right) \stackrel{\mathbb{P}^*_{\boldsymbol{Y}}}{\implies} \mathcal{N}\left(0, \mathcal{I}_1(\theta)^{-1}\right) \qquad \text{in probability as } n \to \infty.$$

Bootstrap consistency typically holds if $\widehat{\theta}$ is asymptotically normal (as $n \to \infty$) and if the underlying data Y_i is i.i.d. Moreover, bootstrap consistency usually implies consistent variance and bias estimation

$$\frac{\text{Var}_{\mathbb{P}^*_{\boldsymbol{Y}}}\left(\widehat{\theta}^*\right)}{\text{Var}_\theta\left(\widehat{\theta}\right)} \stackrel{\text{prob.}}{\to} 1 \qquad \text{and} \qquad \frac{\mathbb{E}^*_{\boldsymbol{Y}}\left[\widehat{\theta}^*\right]-\widehat{\theta}}{\mathbb{E}_\theta\left[\widehat{\theta}\right]-\theta} \stackrel{\text{prob.}}{\to} 1 \qquad \text{as } n \to \infty.$$

For more information and bootstrap confidence intervals we refer to Chapter 5 in the lecture notes of Bühlmann–Mächler [59].

4.3.2 *Parametric Bootstrap Simulation*

For the parametric bootstrap we assume to know the parametric family $\mathcal{F} = \{F(\cdot;\theta); \theta \in \boldsymbol{\Theta}\}$ from which the i.i.d. observations $Y_1, \ldots, Y_n \sim F(\cdot;\theta)$ have been generated from, and only the explicit choice of the parameter $\theta \in \boldsymbol{\Theta}$ is not known. Based on these observations we construct an estimator $\widehat{\theta} = A(\boldsymbol{Y})$, for the unknown parameter $\theta \in \boldsymbol{\Theta}$.

(Parametric) bootstrap algorithm

---

(1) Repeat for $m = 1, \ldots, M$

(a) simulate i.i.d. observations $Y_1^*, \ldots, Y_n^*$ from $F(\cdot;\widehat{\theta})$ (we denote the resampling distribution of $\boldsymbol{Y}^* = (Y_1^*, \ldots, Y_n^*)$ by $\mathbb{P}^* = \mathbb{P}^*_{\boldsymbol{Y}}$);

(b) calculate the estimator $\widehat{\theta}^{(m*)} = A(\boldsymbol{Y}^*)$.

(2) Return $\widehat{\theta}^{(1*)}, \ldots, \widehat{\theta}^{(M*)}$ and the resulting empirical bootstrap distribution

$$\widehat{F}^*_M(\vartheta) = \frac{1}{M} \sum_{m=1}^{M} \mathbb{1}_{\{\widehat{\theta}^{(m*)} \le \vartheta\}}.$$

We then estimate and approximate the distribution of $\widehat{\theta}$ analogously to (4.41), and the same remarks apply as for the non-parametric bootstrap. The parametric bootstrap has the advantage that it can enrich the data by sampling new observations from the distribution $F(\cdot; \widehat{\theta})$. A shortfall of the parametric bootstrap will occur if the family $\mathcal{F}$ is misspecified, then the bootstrap sample $\boldsymbol{Y}^*$ will only poorly describe the true data $\boldsymbol{Y}$, e.g., if the data shows over-dispersion but the select family $\mathcal{F}$ does not allow to model such over-dispersion.

Chapter 5
Generalized Linear Models

Most of the theory in the previous chapters has been based on the assumption of having similarity (or homogeneity) between the different observations. This was expressed by making an i.i.d. assumption on the observations, see, e.g., Sect. 3.3.2. In many practical applications such a homogeneity assumption is not reasonable, one may for example think of car insurance pricing where different car drivers have different driving experience and they drive different cars, or of health insurance where policyholders may have different genders and ages. Figure 5.1 shows a health insurance example where the claim sizes depend on the gender and the age of the policyholders. The most popular statistical models that are able to cope with such heterogeneous data are the *generalized linear models* (GLMs). The notion of GLMs has been introduced in the seminal work of Nelder–Wedderburn [283] in 1972. Their work has introduced a unified procedure for modeling and fitting distributions within the EDF to data having systematic differences (effects) that can be described by explanatory variables. Today, GLMs are the state-of-the-art statistical models in many applied fields including statistics, actuarial science and economics. However, the specific use of GLMs in the different fields may substantially differ. In fields like actuarial science these models are mainly used for *predictive* modeling, in other fields like economics or social sciences GLMs have become the main tool in exploring and *explaining* (hopefully) causal relations. For a discussion on "predicting" versus "explaining" we refer to Shmueli [338].

It is difficult to give a good list of references for GLMs, since GLMs and their offsprings are present in almost every statistical modeling publication and in every lecture on statistics. Classical statistical references are the books of McCullagh–Nelder [265], Fahrmeir–Tutz [123] and Dobson [107], in the actuarial literature we mention the textbooks (in alphabetical order) of Charpentier [67], De Jong–Heller [89], Denuit et al. [99–101], Frees [134] and Ohlsson–Johansson [290], but this list is far from being complete.

M. V. Wüthrich, M. Merz, *Statistical Foundations of Actuarial Learning and its Applications*, Springer Actuarial, https://doi.org/10.1007/978-3-031-12409-9_5

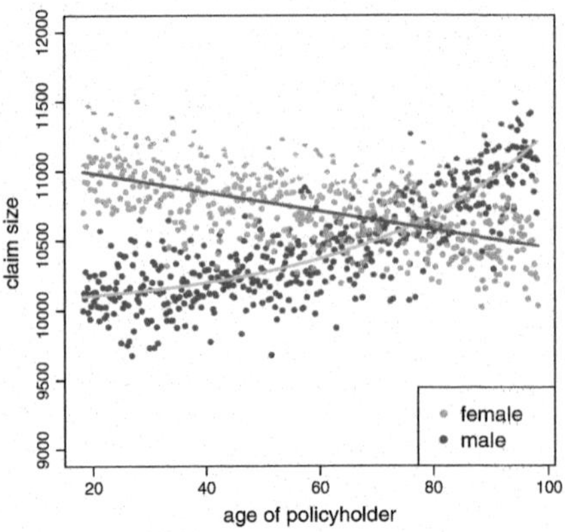

Fig. 5.1 Claim sizes in health insurance as a function of the age of the policyholder, and split by gender

In this chapter we introduce and discuss GLMs in the context of actuarial modeling. We do this in such a way that GLMs can be seen as a building block of network regression models which will be the main topic of Chap. 7 on deep learning.

5.1 Generalized Linear Models and Log-Likelihoods

5.1.1 Regression Modeling

We start by assuming of having independent random variables $Y_1, \ldots, Y_n$ which are described by a fixed member of the EDF. That is, we assume that all Y_i are independent and have densities w.r.t. a σ-finite measure ν on $\mathbb{R}$ given by

$$Y_i \sim f(y_i; \theta_i, v_i/\varphi) = \exp\left\{ \frac{y_i\theta_i - \kappa(\theta_i)}{\varphi/v_i} + a(y_i; v_i/\varphi) \right\} \qquad \text{for } 1 \le i \le n, \tag{5.1}$$

with canonical parameters $\theta_i \in \mathring{\boldsymbol{\Theta}}$, exposures $v_i > 0$ and dispersion parameter $\varphi > 0$. Throughout, we assume that the effective domain $\boldsymbol{\Theta}$ has a non-empty interior. There is a fundamental difference between (5.1) and Example 3.5. We now allow every random variable Y_i to have its own canonical parameter $\theta_i \in \mathring{\boldsymbol{\Theta}}$. We call this a *heterogeneous* situation because the observations are allowed to differ in a systematic way expressed by different canonical parameters. This is highlighted by the lines in the health insurance example of Fig. 5.1 where (expected) claim sizes differ by gender and age of policyholder.

In Sect. 4.1.2 we have introduced the *saturated model* where every observation Y_i has its own parameter θ_i. In general, if we have n observations $\boldsymbol{Y} = (Y_1, \ldots, Y_n)^\top$ we can estimate at most n parameters. The other extreme case is the homogeneous one, meaning that $\theta_i = \theta \in \mathring{\boldsymbol{\Theta}}$ for all $1 \le i \le n$. In this latter case we have exactly one parameter to estimate, and we call this model *null model*, *intercept model* or *homogeneous model*, because all components of $\boldsymbol{Y}$ are assumed to follow the same law expressed in a single common parameter θ. Both the saturated model and the null model may behave very poorly in predicting new observations. Typically, the saturated model fully reflects the data $\boldsymbol{Y}$ including the noisy part (random component, irreducible risk, see Remarks 4.2) and, therefore, it is not useful for prediction. We also say that this model (in-sample) over-fits to the data $\boldsymbol{Y}$ and does not generalize (out-of-sample) to new data. The null model often has a poor predictive performance because if the data has systematic effects these cannot be captured by a null model. GLMs try to find a good balance between these two extreme cases, by trying to extract (only) the systematic effects from noisy data $\boldsymbol{Y}$. We therefore model the canonical parameters θ_i as a low-dimensional function of *explanatory variables* which capture the systematic effects in the data. In Fig. 5.1 gender and age of policyholder play the role of such explanatory variables.

Assume that each observation Y_i is equipped with a *feature* (explanatory variable, covariate) $\boldsymbol{x}_i$ that belongs to a fixed given *feature space* $\mathcal{X}$. These features $\boldsymbol{x}_i$ are assumed to describe the *systematic effects* in the observations Y_i, i.e., these features are assumed to be appropriate descriptions of the heterogeneity between the observations. In a nutshell, we then assume of having a suitable *regression function*

$$\theta : \mathcal{X} \to \mathring{\boldsymbol{\Theta}}, \qquad \boldsymbol{x} \mapsto \theta(\boldsymbol{x}),$$

such that we can appropriately describe the observations by

$$Y_i \overset{\text{ind.}}{\sim} f(y_i; \theta_i = \theta(\boldsymbol{x}_i), v_i/\varphi) = \exp\left\{ \frac{y_i \theta(\boldsymbol{x}_i) - \kappa(\theta(\boldsymbol{x}_i))}{\varphi/v_i} + a(y_i; v_i/\varphi) \right\}, \tag{5.2}$$

for $1 \le i \le n$. As a result we receive for the first moment of Y_i, see Corollary 2.14,

$$\mu_i = \mu(\boldsymbol{x}_i) = \mathbb{E}_{\theta(\boldsymbol{x}_i)}[Y_i] = \kappa'(\theta(\boldsymbol{x}_i)). \tag{5.3}$$

Thus, the regression function $\theta : \mathcal{X} \to \mathring{\boldsymbol{\Theta}}$ is assumed to describe the systematic differences (effects) between the random variables $Y_1, \ldots, Y_n$ being expressed by the means $\mu(\boldsymbol{x}_i)$ for features $\boldsymbol{x}_1, \ldots, \boldsymbol{x}_n$. In GLMs this regression function takes a linear form after a suitable transformation, which exactly motivates the terminology *generalized linear model*.

5.1.2 Definition of Generalized Linear Models

We start with the discussion of the features $\boldsymbol{x} \in \mathcal{X}$. Features are also called explanatory variables, covariates, independent variables or regressors. Throughout, we assume that the features $\boldsymbol{x} = (x_0, x_1, \ldots, x_q)^\top$ include a first component $x_0 = 1$, and we choose feature space $\mathcal{X} \subset \{1\} \times \mathbb{R}^q$. The inclusion of this first component $x_0 = 1$ is useful in what follows. We call this first component *intercept* or *bias component* because it will be modeling an intercept of a regression model. The null model (homogeneous model) has features that only consist of this intercept component. For later purposes it will be useful to introduce the *design matrix* $\mathfrak{X}$ which collects the features $\boldsymbol{x}_1, \ldots, \boldsymbol{x}_n \in \mathcal{X}$ of all responses $Y_1, \ldots, Y_n$. The design matrix is defined by

$$\mathfrak{X} = (\boldsymbol{x}_1, \ldots, \boldsymbol{x}_n)^\top = \begin{pmatrix} 1 & x_{1,1} & \cdots & x_{1,q} \\ \vdots & \vdots & \ddots & \vdots \\ 1 & x_{n,1} & \cdots & x_{n,q} \end{pmatrix} \in \mathbb{R}^{n \times (q+1)}. \tag{5.4}$$

Based on these choices we assume existence of a *regression parameter* $\boldsymbol{\beta} \in \mathbb{R}^{q+1}$ and of a strictly monotone and smooth *link function* $g : \mathcal{M} \to \mathbb{R}$ such that we can express (5.3) by the following function (we drop index i)

$$\boldsymbol{x} \mapsto g(\mu(\boldsymbol{x})) = g\left(\mathbb{E}_{\theta(\boldsymbol{x})}[Y]\right) = \eta(\boldsymbol{x}) = \langle \boldsymbol{\beta}, \boldsymbol{x} \rangle = \beta_0 + \sum_{j=1}^{q} \beta_j x_j. \tag{5.5}$$

Here, $\langle \cdot, \cdot \rangle$ describes the scalar product in the Euclidean space $\mathbb{R}^{q+1}$, $\theta(\boldsymbol{x}) = h(\mu(\boldsymbol{x}))$ is the resulting canonical parameter (using canonical link $h = (\kappa')^{-1}$), and $\eta(\boldsymbol{x})$ is the so-called *linear predictor*. After applying a suitable link function g, the systematic effects of the random variable Y with features $\boldsymbol{x}$ can be described by a linear predictor $\eta(\boldsymbol{x}) = \langle \boldsymbol{\beta}, \boldsymbol{x} \rangle$, linear in the components of $\boldsymbol{x} \in \mathcal{X}$. This gives a particular functional form to (5.3), and the random variables $Y_1, \ldots, Y_n$ share a common regression parameter $\boldsymbol{\beta} \in \mathbb{R}^{q+1}$. Remark that the link function g used in (5.5) can be different from the canonical link h used to calculate $\theta(\boldsymbol{x}) = h(\mu(\boldsymbol{x}))$. We come back to this distinction below.

Summary of (5.5)

1. The independent random variables Y_i follow a fixed member of the EDF (5.1) with individual canonical parameters $\theta_i \in \mathring{\boldsymbol{\Theta}}$, for all $1 \le i \le n$.
2. The canonical parameters θ_i and the corresponding mean parameters μ_i are related by the canonical link $h = (\kappa')^{-1}$ as follows $h(\mu_i) = \theta_i$, where κ is the cumulant function of the chosen EDF, see Corollary 2.14.
3. We assume that the systematic effects in the random variables Y_i can be described by linear predictors $\eta_i = \eta(\boldsymbol{x}_i) = \langle \boldsymbol{\beta}, \boldsymbol{x}_i \rangle$ and a strictly monotone and smooth link function g such that we have $g(\mu_i) = \eta_i = \langle \boldsymbol{\beta}, \boldsymbol{x}_i \rangle$, for all $1 \le i \le n$, with common regression parameter $\boldsymbol{\beta} \in \mathbb{R}^{q+1}$.

We can either express this GLM regression structure in the dual (mean) parameter space $\mathcal{M}$ or in the effective domain $\boldsymbol{\Theta}$, see Remarks 2.9,

$$x \mapsto \mu(x) = g^{-1}(\eta(x)) = g^{-1}\langle \beta, x \rangle \in \mathcal{M} \qquad \text{or}$$
$$x \mapsto \theta(x) = (h \circ g^{-1})(\eta(x)) = (h \circ g^{-1})\langle \beta, x \rangle \in \mathring{\Theta},$$

where $(h \circ g^{-1})$ is the composition of the inverse link g^{-1} and the canonical link h. For the moment, the link function g is quite general. In practice, the explicit choice needs some care. The right-hand side of (5.5) is defined on the whole real line if at least one component of x is both-sided unbounded. On the other hand, $\mathcal{M}$ and $\mathring{\Theta}$ may be bounded sets. Therefore, the link function g may require some restrictions such that the domain and the range fulfill the necessary constraints. The dimension of β should satisfy $1 \leq 1 + q \leq n$, the lower bound will provide a null model and the upper bound a saturated model.

5.1.3 Link Functions and Feature Engineering

As link function we choose a strictly monotone and smooth function $g : \mathcal{M} \to \mathbb{R}$ such that we do not have any conflicts in domains and ranges. Beside these requirements, we may want further properties for the link function g and the features x. From (5.5) we have

$$\mu(x) = \mathbb{E}_{\theta(x)}[Y] = g^{-1}\langle \beta, x \rangle. \tag{5.6}$$

Of course, a basic requirement is that the selected features x can appropriately describe the mean of Y by the function in (5.6), see also Fig. 5.1. This may require so-called *feature engineering* of x, for instance, we may want to replace the first component x_1 of the *raw features* x by, say, x_1^2 in the *pre-processed features*. For example, if this first component describes the age of the insurance policyholder, then, in some regression problems, it might be more appropriate to consider $\texttt{age}^2$ instead of $\texttt{age}$ to bring the predictive problem into structure (5.6). It may also be that we would like to enforce a certain type of *interaction* between the components of the raw features. For instance, we may include in a pre-processed feature a component x_1/x_2^2 which might correspond to $\texttt{weight}/\texttt{height}^2$ if the policyholder has body weight x_1 and body height x_2. In fact, this pre-processed feature is exactly the body mass index of the policyholder. We will come back to feature engineering in Sect. 5.2.2, below.

Another important requirement is the ability of model interpretation. In insurance pricing problems, one often prefers additive and multiplicative effects in feature components. Choosing the identity link $g(m) = m$ we receive a model with additive effects

$$\mu(\boldsymbol{x}) = \mathbb{E}_{\theta(\boldsymbol{x})}[Y] = \langle \boldsymbol{\beta}, \boldsymbol{x} \rangle = \beta_0 + \sum_{j=1}^{q} \beta_j x_j,$$

and choosing the log-link $g(m) = \log(m)$ we receive a model with multiplicative effects

$$\mu(\boldsymbol{x}) = \mathbb{E}_{\theta(\boldsymbol{x})}[Y] = \exp\langle \boldsymbol{\beta}, \boldsymbol{x} \rangle = e^{\beta_0} \prod_{j=1}^{q} e^{\beta_j x_j}.$$

The latter is probably the most commonly used GLM in insurance pricing because it leads to explainable tariffs where feature values directly relate to price de- and increases in percentages of a base premium $\exp\{\beta_0\}$.

Another very popular choice is the canonical (natural) link, i.e., $g = h = (\kappa')^{-1}$. The canonical link substantially simplifies the analysis and it has very favorable statistical properties (as we will see below). However, in some applications practical needs overrule good statistical properties. Under the canonical link $g = h$ we have in the dual mean parameter space $\mathcal{M}$ and in the effective domain $\boldsymbol{\Theta}$, respectively,

$$\boldsymbol{x} \mapsto \mu(\boldsymbol{x}) = \kappa'(\eta(\boldsymbol{x})) = \kappa'\langle \boldsymbol{\beta}, \boldsymbol{x} \rangle \qquad \text{and} \qquad \boldsymbol{x} \mapsto \theta(\boldsymbol{x}) = \eta(\boldsymbol{x}) = \langle \boldsymbol{\beta}, \boldsymbol{x} \rangle.$$

Thus, the linear predictor η and the canonical parameter θ coincide under the canonical link choice $g = h = (\kappa')^{-1}$.

5.1.4 Log-Likelihood Function and Maximum Likelihood Estimation

After having a fully specified GLM within the EDF, there remains estimation of the regression parameter $\boldsymbol{\beta} \in \mathbb{R}^{q+1}$. This is done within the framework of MLE.

The log-likelihood function of $\boldsymbol{Y} = (Y_1, \ldots, Y_n)^\top$ for regression parameter $\boldsymbol{\beta} \in \mathbb{R}^{q+1}$ is given by, see (5.2) and we use the independence between the Y_i's,

(continued)

$$\boldsymbol{\beta} \mapsto \ell_{\boldsymbol{Y}}(\boldsymbol{\beta}) = \sum_{i=1}^{n} \frac{v_i}{\varphi}\Big[Y_i h(\mu(\boldsymbol{x}_i)) - \kappa\left(h(\mu(\boldsymbol{x}_i))\right)\Big] + a(Y_i; v_i/\varphi), \tag{5.7}$$

where we set $\mu(\boldsymbol{x}_i) = g^{-1}\langle\boldsymbol{\beta}, \boldsymbol{x}_i\rangle$. For the canonical link $g = h = (\kappa')^{-1}$ this simplifies to

$$\boldsymbol{\beta} \mapsto \ell_{\boldsymbol{Y}}(\boldsymbol{\beta}) = \sum_{i=1}^{n} \frac{v_i}{\varphi}\Big[Y_i\langle\boldsymbol{\beta}, \boldsymbol{x}_i\rangle - \kappa\langle\boldsymbol{\beta}, \boldsymbol{x}_i\rangle\Big] + a(Y_i; v_i/\varphi). \tag{5.8}$$

MLE of $\boldsymbol{\beta}$ needs maximization of log-likelihoods (5.7) and (5.8), respectively; these are the GLM counterparts to the homogeneous case treated in Section 3.3.2. We calculate the score, we set $\eta_i = \langle\boldsymbol{\beta}, \boldsymbol{x}_i\rangle$ and $\mu_i = \mu(\boldsymbol{x}_i) = g^{-1}\langle\boldsymbol{\beta}, \boldsymbol{x}_i\rangle$,

$$\begin{aligned} s(\boldsymbol{\beta}, \boldsymbol{Y}) &= \nabla_{\boldsymbol{\beta}}\ell_{\boldsymbol{Y}}(\boldsymbol{\beta}) = \sum_{i=1}^{n} \frac{v_i}{\varphi}\left[Y_i - \mu_i\right] \nabla_{\boldsymbol{\beta}} h(\mu(\boldsymbol{x}_i)) \\ &= \sum_{i=1}^{n} \frac{v_i}{\varphi}\left[Y_i - \mu_i\right] \frac{\partial h(\mu_i)}{\partial \mu_i} \frac{\partial \mu_i}{\partial \eta_i} \nabla_{\boldsymbol{\beta}} \eta(\boldsymbol{x}_i) \\ &= \sum_{i=1}^{n} \frac{v_i}{\varphi} \frac{Y_i - \mu_i}{V(\mu_i)} \left(\frac{\partial g(\mu_i)}{\partial \mu_i}\right)^{-1} \boldsymbol{x}_i, \end{aligned} \tag{5.9}$$

where we use the definition of the variance function $V(\mu) = (\kappa'' \circ h)(\mu)$, see Corollary 2.14. We define the diagonal working weight matrix, which in general depends on $\boldsymbol{\beta}$ through the means $\mu_i = g^{-1}\langle\boldsymbol{\beta}, \boldsymbol{x}_i\rangle$,

$$W(\boldsymbol{\beta}) = \operatorname{diag}\left(\left(\frac{\partial g(\mu_i)}{\partial \mu_i}\right)^{-2} \frac{v_i}{\varphi} \frac{1}{V(\mu_i)}\right)_{1\le i\le n} \in \mathbb{R}^{n\times n},$$

and the working residuals

$$\boldsymbol{R} = \boldsymbol{R}(\boldsymbol{Y}, \boldsymbol{\beta}) = \left(\frac{\partial g(\mu_i)}{\partial \mu_i}(Y_i - \mu_i)\right)^{\top}_{1\le i\le n} \in \mathbb{R}^n.$$

This allows us to write the score equations in a compact form, which provides the following proposition.

Proposition 5.1 *The MLE for $\boldsymbol{\beta}$ is found by solving the score equations*

$$s(\boldsymbol{\beta}, \boldsymbol{Y}) = \nabla_{\boldsymbol{\beta}} \ell_{\boldsymbol{Y}}(\boldsymbol{\beta}) = \mathfrak{X}^\top W(\boldsymbol{\beta}) \boldsymbol{R}(\boldsymbol{Y}, \boldsymbol{\beta}) = 0.$$

For the canonical link $g = h = (\kappa')^{-1}$ the score equations simplify to

$$s(\boldsymbol{\beta}, \boldsymbol{Y}) = \nabla_{\boldsymbol{\beta}} \ell_{\boldsymbol{Y}}(\boldsymbol{\beta}) = \mathfrak{X}^\top \mathrm{diag}\left(\frac{v_i}{\varphi}\right)_{1 \le i \le n} \left(\boldsymbol{Y} - \kappa'(\mathfrak{X}\boldsymbol{\beta})\right) = 0,$$

where $\kappa'(\mathfrak{X}\boldsymbol{\beta}) \in \mathbb{R}^n$ is understood element-wise.

Remarks 5.2

- In general, the MLE of $\boldsymbol{\beta}$ is not calculated by maximizing the log-likelihood function $\ell_{\boldsymbol{Y}}(\boldsymbol{\beta})$, but rather by solving the score equations $s(\boldsymbol{\beta}, \boldsymbol{Y}) = 0$; we also refer to Remarks 3.29 on M- and Z-estimators. The score equations provide the critical points for $\boldsymbol{\beta}$, from which the global maximum of the log-likelihood function can be determined, supposed it exists.
- Existence of a MLE of $\boldsymbol{\beta}$ is not always given, similarly to Example 3.5, we may face the problem that the solution lies at the boundary of the parameter space (which itself may be an open set).
- If the log-likelihood function $\boldsymbol{\beta} \mapsto \ell_{\boldsymbol{Y}}(\boldsymbol{\beta})$ is strictly concave, then the critical point of the score equations $s(\boldsymbol{\beta}, \boldsymbol{Y}) = 0$ is unique, supposed it exists, and, henceforth, we have a unique MLE $\widehat{\boldsymbol{\beta}}^{\mathrm{MLE}}$ for $\boldsymbol{\beta}$. Below, we give cases where the strict concavity of the log-likelihood holds.
- In general, there is no closed from solution for the MLE of $\boldsymbol{\beta}$, except in the Gaussian case with canonical link, thus, we need to solve the score equations numerically.

Similarly to Remarks 3.17 we can calculate Fisher's information matrix w.r.t. $\boldsymbol{\beta}$ through the negative expected Hessian of $\ell_{\boldsymbol{Y}}(\boldsymbol{\beta})$.

We get Fisher's information matrix w.r.t. $\boldsymbol{\beta}$

$$\mathcal{I}(\boldsymbol{\beta}) = \mathbb{E}_{\boldsymbol{\beta}}\left[\nabla_{\boldsymbol{\beta}} \ell_{\boldsymbol{Y}}(\boldsymbol{\beta}) \left(\nabla_{\boldsymbol{\beta}} \ell_{\boldsymbol{Y}}(\boldsymbol{\beta})\right)^\top\right] = -\mathbb{E}_{\boldsymbol{\beta}}\left[\nabla_{\boldsymbol{\beta}}^2 \ell_{\boldsymbol{Y}}(\boldsymbol{\beta})\right] = \mathfrak{X}^\top W(\boldsymbol{\beta}) \mathfrak{X}. \tag{5.10}$$

If the design matrix $\mathfrak{X} \in \mathbb{R}^{n \times (q+1)}$ has full rank $q + 1 \le n$, Fisher's information matrix $\mathcal{I}(\boldsymbol{\beta})$ is positive definite.

Dispersion parameter $\varphi > 0$ has been treated as a nuisance parameter above. Its explicit specification does not influence the MLE of $\boldsymbol{\beta}$ because it cancels in the score equations. If necessary, we can also estimate this dispersion parameter with MLE. This requires solving the additional score equation

$$\frac{\partial}{\partial \varphi} \ell_{\boldsymbol{Y}}(\boldsymbol{\beta}, \varphi) = \sum_{i=1}^{n} -\frac{v_i}{\varphi^2} \Big[Y_i h(\mu(\boldsymbol{x}_i)) - \kappa \left(h(\mu(\boldsymbol{x}_i)) \right) \Big] + \frac{\partial}{\partial \varphi} a(Y_i; v_i/\varphi) = 0, \tag{5.11}$$

and we can plug in the MLE of $\boldsymbol{\beta}$ (which can be estimated independently of φ). Fisher's information matrix is in this extended framework given by

$$\mathcal{I}(\boldsymbol{\beta}, \varphi) = -\mathbb{E}_{\boldsymbol{\beta}} \left[\nabla^2_{(\boldsymbol{\beta},\varphi)} \ell_{\boldsymbol{Y}}(\boldsymbol{\beta}, \varphi) \right] = \begin{pmatrix} \mathfrak{X}^\top W(\boldsymbol{\beta}) \mathfrak{X} & 0 \\ 0 & -\mathbb{E}_{\boldsymbol{\beta}} \left[\partial^2 \ell_{\boldsymbol{Y}}(\boldsymbol{\beta}, \varphi)/\partial \varphi^2 \right] \end{pmatrix},$$

that is, the off-diagonal terms between $\boldsymbol{\beta}$ and φ are zero.

In view of Proposition 5.1 we need a root search algorithm to obtain the MLE of $\boldsymbol{\beta}$. Typically, one uses Fisher's scoring method or the iterative re-weighted least squares (IRLS) algorithm to solve this root search problem. This is a main result derived in the seminal work of Nelder–Wedderburn [283] and it explains the popularity of GLMs, namely, GLMs can be solved efficiently by this algorithm. Fisher's scoring method/IRLS algorithm explore the updates for $t \geq 0$ until convergence

$$\widehat{\boldsymbol{\beta}}^{(t)} \mapsto \widehat{\boldsymbol{\beta}}^{(t+1)} = \left(\mathfrak{X}^\top W(\widehat{\boldsymbol{\beta}}^{(t)}) \mathfrak{X} \right)^{-1} \mathfrak{X}^\top W(\widehat{\boldsymbol{\beta}}^{(t)}) \left(\mathfrak{X} \widehat{\boldsymbol{\beta}}^{(t)} + \boldsymbol{R}(\boldsymbol{Y}, \widehat{\boldsymbol{\beta}}^{(t)}) \right), \tag{5.12}$$

where all terms on the right-hand side are evaluated at algorithmic time t. If we have n observations $\boldsymbol{Y} = (Y_1, \ldots, Y_n)^\top$ we can estimate at most n parameters. Therefore, in our GLM we assume to have a regression parameter $\boldsymbol{\beta} \in \mathbb{R}^{q+1}$ of dimension $q + 1 \leq n$. Moreover, we require that the design matrix $\mathfrak{X}$ has full rank $q + 1 \leq n$. Otherwise the regression parameter is not uniquely identifiable since linear dependence in the columns of $\mathfrak{X}$ allows us to reduce the dimension of the parameter space to a smaller representation. This is also needed to calculate the inverse matrix in (5.12). This motivates the following assumption.

Assumption 5.3 *Throughout, we assume that the design matrix $\mathfrak{X} \in \mathbb{R}^{n\times(q+1)}$ has full rank $q+1 \le n$.*

Remarks 5.4 (Justification of Fisher's Scoring Method/IRLS Algorithm)

- We give a short justification of Fisher's scoring method/IRLS algorithm, for a more detailed treatment we refer to Section 2.5 in McCullagh–Nelder [265] and Section 2.2 in Fahrmeir–Tutz [123].

  The Newton–Raphson algorithm provides a numerical scheme to find solutions to the score equations. It requires to iterate for $t \ge 0$

$$\widehat{\boldsymbol{\beta}}^{(t)} \mapsto \widehat{\boldsymbol{\beta}}^{(t+1)} = \widehat{\boldsymbol{\beta}}^{(t)} + \widehat{\mathcal{I}}(\widehat{\boldsymbol{\beta}}^{(t)})^{-1}\, s(\widehat{\boldsymbol{\beta}}^{(t)}, \boldsymbol{Y}),$$

  where $\widehat{\mathcal{I}}(\boldsymbol{\beta}) = -\nabla^2_{\boldsymbol{\beta}} \ell_{\boldsymbol{Y}}(\boldsymbol{\beta})$ denotes the observed information matrix in $\boldsymbol{\beta} \in \mathbb{R}^{q+1}$. The calculation of the inverse of the observed information matrix $(\widehat{\mathcal{I}}(\widehat{\boldsymbol{\beta}}^{(t)}))^{-1}$ can be time consuming and unstable because we need to calculate second derivatives and the eigenvalues of the observed information matrix can be close to zero. A stable scheme is obtained by replacing the observed information matrix $\widehat{\mathcal{I}}(\boldsymbol{\beta})$ by Fisher's information matrix $\mathcal{I}(\boldsymbol{\beta}) = \mathbb{E}_{\boldsymbol{\beta}}[\widehat{\mathcal{I}}(\boldsymbol{\beta})]$ being positive definite under Assumption 5.3; this provides a quasi-Newton method. Thus, for Fisher's scoring method we iterate for $t \ge 0$

$$\widehat{\boldsymbol{\beta}}^{(t)} \mapsto \widehat{\boldsymbol{\beta}}^{(t+1)} = \widehat{\boldsymbol{\beta}}^{(t)} + \mathcal{I}(\widehat{\boldsymbol{\beta}}^{(t)})^{-1}\, s(\widehat{\boldsymbol{\beta}}^{(t)}, \boldsymbol{Y}), \tag{5.13}$$

  and rewriting this provides us exactly with (5.12). The latter can also be interpreted as an IRLS scheme where the response $g(Y_i)$ is replaced by an adjusted linearized version $Z_i = g(\mu_i) + \frac{\partial g(\mu_i)}{\partial \mu_i}(Y_i - \mu_i)$. This corresponds to the last bracket in (5.12), and with corresponding weights.
- Under the canonical link choice, Fisher's information matrix and the observed information matrix coincide, i.e. $\mathcal{I}(\boldsymbol{\beta}) = \widehat{\mathcal{I}}(\boldsymbol{\beta})$, and the Newton–Raphson algorithm, Fisher's scoring method and the IRLS algorithm are identical. This can easily be seen from Proposition 5.1. We receive under the canonical link choice

$$\begin{aligned}\nabla^2_{\boldsymbol{\beta}} \ell_{\boldsymbol{Y}}(\boldsymbol{\beta}) &= -\widehat{\mathcal{I}}(\boldsymbol{\beta}) = -\mathfrak{X}^\top \mathrm{diag}\left(\frac{v_i}{\varphi} V(\mu_i)\right)_{1\le i\le n} \mathfrak{X} \\ &= -\mathfrak{X}^\top W(\boldsymbol{\beta})\mathfrak{X} = -\mathcal{I}(\boldsymbol{\beta}).\end{aligned} \tag{5.14}$$

The full rank assumption $q + 1 \leq n$ on the design matrix $\mathfrak{X}$ implies that Fisher's information matrix $\mathcal{I}(\boldsymbol{\beta})$ is positive definite. This in turn implies that the log-likelihood function $\ell_{\boldsymbol{Y}}(\boldsymbol{\beta})$ is strictly concave, providing uniqueness of a critical point (supposed it exists). This indicates that the canonical link has very favorable properties for MLE. Examples 5.5 and 5.6 give two examples not using the canonical link, the first one is a concave maximization problem, the second one is not for $p > 2$.

Example 5.5 (Gamma Model with Log-Link) We study the gamma distribution as a single-parameter EDF model, choosing the shape parameter $\alpha = 1/\varphi$ as the inverse of the dispersion parameter, see Sect. 2.2.2. Cumulant function $\kappa(\theta) = -\log(-\theta)$ gives us the canonical link $\theta = h(\mu) = -1/\mu$. Moreover, we choose the log-link $\eta = g(\mu) = \log(\mu)$ for the GLM. This gives a canonical parameter $\theta = -\exp\{-\eta\}$. We receive the score

$$s(\boldsymbol{\beta}, \boldsymbol{Y}) = \nabla_{\boldsymbol{\beta}} \ell_{\boldsymbol{Y}}(\boldsymbol{\beta}) = \sum_{i=1}^{n} \frac{v_i}{\varphi} \left[\frac{Y_i}{\mu_i} - 1 \right] \boldsymbol{x}_i = \mathfrak{X}^\top \operatorname{diag}\left(\frac{v_i}{\varphi} \right)_{1 \leq i \leq n} \boldsymbol{R}(\boldsymbol{Y}, \boldsymbol{\beta}).$$

Unlike in other examples with non-canonical links, we receive a favorable expression here because only one term in the square bracket depends on the regression parameter $\boldsymbol{\beta}$, or equivalently, the working weight matrix W does not dependent on $\boldsymbol{\beta}$. We calculate the negative Hessian (observed information matrix)

$$\widehat{\mathcal{I}}(\boldsymbol{\beta}) = -\nabla^2_{\boldsymbol{\beta}} \ell_{\boldsymbol{Y}}(\boldsymbol{\beta}) = \mathfrak{X}^\top \operatorname{diag}\left(\frac{v_i}{\varphi} \frac{Y_i}{\mu_i} \right)_{1 \leq i \leq n} \mathfrak{X}.$$

In the gamma model all observations Y_i are strictly positive, a.s., and under the full rank assumption $q + 1 \leq n$, the observed information matrix $\widehat{\mathcal{I}}(\boldsymbol{\beta})$ is positive definite, thus, we have a strictly concave log-likelihood function in the gamma case with log-link. ■

Example 5.6 (Tweedie's Models with Log-Link) We study Tweedie's models for power variance parameters $p > 1$ as a single-parameter EDF model, see Sect. 2.2.3. The cumulant function κ_p is given in Table 4.1. This gives us the canonical link $\theta = h_p(\mu) = \mu^{1-p}/(1-p) < 0$ for $\mu > 0$ and $p > 1$. Moreover, we choose the log-link $\eta = g(\mu) = \log(\mu)$ for the GLM. This implies $\theta = \exp\{(1-p)\eta\}/(1-p) < 0$ for $p > 1$. We receive the score

$$s(\boldsymbol{\beta}, \boldsymbol{Y}) = \nabla_{\boldsymbol{\beta}} \ell_{\boldsymbol{Y}}(\boldsymbol{\beta}) = \sum_{i=1}^{n} \frac{v_i}{\varphi} \frac{Y_i - \mu_i}{\mu_i^{p-1}} \boldsymbol{x}_i = \mathfrak{X}^\top \operatorname{diag}\left(\frac{v_i}{\varphi} \frac{1}{\mu_i^{p-2}} \right)_{1 \leq i \leq n} \boldsymbol{R}(\boldsymbol{Y}, \boldsymbol{\beta}).$$

We calculate the negative Hessian (observed information matrix) for $\mu_i > 0$

$$\widehat{\mathcal{I}}(\boldsymbol{\beta}) = -\nabla^2_{\boldsymbol{\beta}} \ell_{\boldsymbol{Y}}(\boldsymbol{\beta}) = \mathfrak{X}^\top \mathrm{diag}\left(\frac{v_i}{\varphi} \frac{(p-1)Y_i - (p-2)\mu_i}{\mu_i^{p-1}}\right)_{1 \le i \le n} \mathfrak{X}.$$

This matrix is positive definite for $p \in [1, 2]$, and for $p > 2$ it is not positive definite because $(p-1)Y_i-(p-2)\mu_i$ may have positive or negative values if we vary $\mu_i > 0$ over its domain $\mathcal{M}$. Thus, we do not have concavity of the optimization problem under the log-link choice in Tweedie's GLMs for power variance parameters $p > 2$. This in particular applies to the inverse Gaussian GLM with log-link. ■

5.1.5 Balance Property Under the Canonical Link Choice

Throughout this section we work under the canonical link choice $g = h = (\kappa')^{-1}$. This choice has very favorable statistical properties. We have already seen in Remarks 5.4 that the derivation of the MLE of $\boldsymbol{\beta}$ becomes particularly easy under the canonical link choice and the observed information matrix $\widehat{\mathcal{I}}(\boldsymbol{\beta})$ coincides with Fisher's information matrix $\mathcal{I}(\boldsymbol{\beta})$ in this case, see (5.14).

For insurance pricing, canonical links have another very remarkable property, namely, that the estimated model automatically fulfills the balance property and, henceforth, is unbiased. This is particularly important in insurance pricing because it tells us that the insurance prices (over the entire portfolio) are on the right level. We have already met the balance property in Corollary 3.19.

Corollary 5.7 (Balance Property) *Assume that $\boldsymbol{Y}$ has independent components being modeled by a GLM under the canonical link choice $g = h = (\kappa')^{-1}$. Assume that the MLE of regression parameter $\boldsymbol{\beta} \in \mathbb{R}^{q+1}$ exists and denote it by $\widehat{\boldsymbol{\beta}}^{\mathrm{MLE}}$. We have balance property on portfolio level (for constant dispersion φ)*

$$\sum_{i=1}^n \mathbb{E}_{\widehat{\boldsymbol{\beta}}^{\mathrm{MLE}}}[v_i Y_i] = \sum_{i=1}^n v_i \kappa'\langle \widehat{\boldsymbol{\beta}}^{\mathrm{MLE}}, \boldsymbol{x}_i\rangle = \sum_{i=1}^n v_i Y_i.$$

Proof The first column of the design matrix $\mathfrak{X}$ is identically equal to 1 representing the intercept, see (5.4). The second part of Proposition 5.1 then provides for this first column of $\mathfrak{X}$, we cancel the (constant) dispersion φ,

$$(1, \ldots, 1)\,\mathrm{diag}(v_1, \ldots, v_n)\,\kappa'(\mathfrak{X}\widehat{\boldsymbol{\beta}}^{\mathrm{MLE}}) = (1, \ldots, 1)\,\mathrm{diag}(v_1, \ldots, v_n)\,\boldsymbol{Y}.$$

This proves the claim. □

Remark 5.8 We mention once more that this balance property is very strong and useful, see also Remarks 3.20. In particular, the balance property holds, even though the chosen GLM might be completely misspecified. Misspecification may include an incorrect distributional model, not the right link function choice, or if we have not pre-processed features appropriately, etc. Such misspecification will imply that we have a poor model on an insurance policy level (observation level). However, the total premium charged over the entire portfolio will be on the right level (supposed that the structure of the portfolio does not change) because it matches the observations, and henceforth, we have unbiasedness for the portfolio mean.

From the log-likelihood function (5.8) we see that under the canonical link choice we consider the statistics $S(\boldsymbol{Y}) = \mathfrak{X}^\top \text{diag}(v_i/\varphi)_{1\le i\le n}\boldsymbol{Y} \in \mathbb{R}^{q+1}$, and to prove the balance property we have used the first component of this statistics. Considering all components, $S(\boldsymbol{Y})$ is an unbiased estimator (decision rule) for

$$\mathbb{E}_{\boldsymbol{\beta}}\left[S(\boldsymbol{Y})\right] = \mathfrak{X}^\top \text{diag}(v_i/\varphi)_{1\le i\le n}\kappa'(\mathfrak{X}\boldsymbol{\beta}) = \left(\sum_{i=1}^n \frac{v_i}{\varphi}\kappa'\langle\boldsymbol{\beta},\boldsymbol{x}_i\rangle x_{i,j}\right)^\top_{0\le j\le q}. \tag{5.15}$$

This unbiased estimator $S(\boldsymbol{Y})$ meets the Cramér–Rao information bound, hence it is UMVU: taking the partial derivatives of the previous expression gives $\nabla_{\boldsymbol{\beta}}\mathbb{E}_{\boldsymbol{\beta}}\left[S(\boldsymbol{Y})\right] = \mathcal{I}(\boldsymbol{\beta})$, the latter also being the multivariate Cramér–Rao information bound for the unbiased decision rule $S(\boldsymbol{Y})$ for (5.15). Focusing on the first component we have

$$\text{Var}_{\boldsymbol{\beta}}\left(\sum_{i=1}^n \mathbb{E}_{\widehat{\boldsymbol{\beta}}^{\text{MLE}}}\left[v_i Y_i\right]\right) = \text{Var}_{\boldsymbol{\beta}}\left(\sum_{i=1}^n v_i Y_i\right) = \sum_{i=1}^n \varphi v_i V(\mu_i) = \varphi^2\left(\mathcal{I}(\boldsymbol{\beta})\right)_{0,0}, \tag{5.16}$$

where the component $(0,0)$ in the last expression is the top-left entry of Fisher's information matrix $\mathcal{I}(\boldsymbol{\beta})$ under the canonical link choice.

5.1.6 Asymptotic Normality

Formula (5.16) quantifies the uncertainty in the premium calculation of the insurance policies if we use the MLE estimated model (under the canonical link choice). That is, this quantifies the uncertainty in the dual mean parametrization in terms of the resulting variance. We could also focus on the MLE $\widehat{\boldsymbol{\beta}}^{\text{MLE}}$ itself (for general link function g). In general, this MLE is not unbiased but we have consistency and asymptotic normality similar to Theorem 3.28. Under "certain regularity conditions"[1] we have for n large

$$\widehat{\boldsymbol{\beta}}_n^{\mathrm{MLE}} \overset{(\mathrm{d})}{\approx} \mathcal{N}\left(\boldsymbol{\beta}, \mathcal{I}_n(\boldsymbol{\beta})^{-1}\right), \tag{5.17}$$

where $\widehat{\boldsymbol{\beta}}_n^{\mathrm{MLE}}$ is the MLE based on the observations $\boldsymbol{Y}_n = (Y_1, \ldots, Y_n)^\top$, and $\mathcal{I}_n(\boldsymbol{\beta})$ is Fisher's information matrix of $\boldsymbol{Y}_n$, which scales linearly in n in the homogeneous EF case, see Remarks 3.14, and in the homogeneous EDF case it scales as $\sum_{i=1}^n v_i$, see (3.25).

5.1.7 Maximum Likelihood Estimation and Unit Deviances

From formula (5.7) we conclude that the MLE $\widehat{\boldsymbol{\beta}}^{\mathrm{MLE}}$ of $\boldsymbol{\beta} \in \mathbb{R}^{q+1}$ is found by the solution of (subject to existence)

$$\widehat{\boldsymbol{\beta}}^{\mathrm{MLE}} = \underset{\boldsymbol{\beta}}{\arg\max}\, \ell_{\boldsymbol{Y}}(\boldsymbol{\beta}) = \underset{\boldsymbol{\beta}}{\arg\max} \sum_{i=1}^n \frac{v_i}{\varphi}\Big[Y_i h(\mu(\boldsymbol{x}_i)) - \kappa\left(h(\mu(\boldsymbol{x}_i))\right)\Big],$$

with $\mu_i = \mu(\boldsymbol{x}_i) = \mathbb{E}_{\theta(\boldsymbol{x}_i)}[Y] = g^{-1}\langle \boldsymbol{\beta}, \boldsymbol{x}_i\rangle$ under the link choice g. If we prefer to work with an objective function that reflects the notion of a loss function, we can work under the unit deviances $\mathfrak{d}(Y_i, \mu_i)$ studied in Sect. 4.1.2. The MLE is then obtained by, see (4.20)–(4.21),

$$\widehat{\boldsymbol{\beta}}^{\mathrm{MLE}} = \underset{\boldsymbol{\beta}}{\arg\max}\, \ell_{\boldsymbol{Y}}(\boldsymbol{\beta}) = \underset{\boldsymbol{\beta}}{\arg\min} \sum_{i=1}^n \frac{v_i}{\varphi}\, \mathfrak{d}(Y_i, \mu_i), \tag{5.18}$$

the latter satisfying $\mathfrak{d}(Y_i, \mu_i) \geq 0$ for all $1 \leq i \leq n$, and being zero if and only if $Y_i = \mu_i$, see Lemma 2.22. Thus, using the unit deviances we have a loss function that is bounded below by zero, and we determine the regression parameter $\boldsymbol{\beta}$ such that this loss is (in-sample) minimized. This can also be interpreted in a more geometric way. Consider the $(q+1)$-dimensional manifold $\mathfrak{M} \subset \mathbb{R}^n$ spanned by the GLM function

$$\boldsymbol{\beta} \mapsto \boldsymbol{\mu}(\boldsymbol{\beta}) = g^{-1}(\mathfrak{X}\boldsymbol{\beta}) = (g^{-1}\langle \boldsymbol{\beta}, \boldsymbol{x}_1\rangle, \ldots, g^{-1}\langle \boldsymbol{\beta}, \boldsymbol{x}_n\rangle)^\top \in \mathbb{R}^n. \tag{5.19}$$

[1] The regularity conditions for asymptotic normality results will depend on the particular regression problem studied, we refer to pages 43–44 in Fahrmeir–Tutz [123].

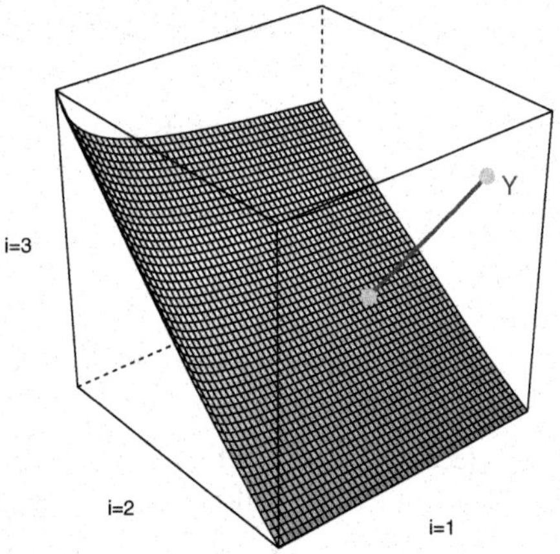

Fig. 5.2 2-dimensional manifold $\mathfrak{M} \subset \mathbb{R}^3$ for observation $\boldsymbol{Y} = (Y_1, Y_2, Y_3)^\top \in \mathbb{R}^3$, the straight line illustrates the projection (w.r.t. the unit deviance distances $\mathfrak{d}$) of $\boldsymbol{Y}$ onto $\mathfrak{M}$ which gives MLE $\widehat{\boldsymbol{\beta}}^{\text{MLE}}$ satisfying $\boldsymbol{\mu}(\widehat{\boldsymbol{\beta}}^{\text{MLE}}) \in \mathfrak{M}$

Minimization (5.18) then tries to find the point $\boldsymbol{\mu}(\boldsymbol{\beta})$ in this manifold $\mathfrak{M} \subset \mathbb{R}^n$ that minimizes simultaneously all unit deviances $\mathfrak{d}(Y_i, \cdot)$ w.r.t. the observation $\boldsymbol{Y} = (Y_1, \ldots, Y_n)^\top \in \mathbb{R}^n$. Or in other words, the optimal parameter $\boldsymbol{\beta}$ is obtained by "projecting" observation $\boldsymbol{Y}$ onto this manifold $\mathfrak{M}$, where "projection" is understood as a simultaneous minimization of loss function $\sum_{i=1}^n \frac{v_i}{\varphi} \mathfrak{d}(Y_i, \mu_i)$, see Fig. 5.2. In the un-weighted Gaussian case, this corresponds to the usual orthogonal projection as the next example shows, and in the non-Gaussian case it is understood in the KL divergence minimization sense as displayed in formula (4.11).

Example 5.9 (Gaussian Case) Assume we have the Gaussian EDF case $\kappa(\theta) = \theta^2/2$ with canonical link $g(\mu) = h(\mu) = \mu$. In this case, the manifold (5.19) is the linear space spanned by the columns of the design matrix $\mathfrak{X}$

$$\boldsymbol{\beta} \mapsto \boldsymbol{\mu}(\boldsymbol{\beta}) = \mathfrak{X}\boldsymbol{\beta} = (\langle \boldsymbol{\beta}, \boldsymbol{x}_1 \rangle, \ldots, \langle \boldsymbol{\beta}, \boldsymbol{x}_n \rangle)^\top \in \mathbb{R}^n.$$

If additionally we assume $v_i/\varphi = c > 0$ for all $1 \le i \le n$, the minimization problem (5.18) reads as

$$\widehat{\boldsymbol{\beta}}^{\text{MLE}} = \underset{\boldsymbol{\beta}}{\arg\min} \sum_{i=1}^n \frac{v_i}{\varphi} \mathfrak{d}(Y_i, \mu_i) = \underset{\boldsymbol{\beta}}{\arg\min} \|\boldsymbol{Y} - \mathfrak{X}\boldsymbol{\beta}\|_2^2,$$

where we have used that the unit deviances in the Gaussian case are given by the square loss function, see Example 4.12. As a consequence, the MLE $\widehat{\boldsymbol{\beta}}^{\text{MLE}}$ is found by orthogonally projecting $\boldsymbol{Y}$ onto $\mathfrak{M} = \{\mathfrak{X}\boldsymbol{\beta} | \, \boldsymbol{\beta} \in \mathbb{R}^{q+1}\} \subset \mathbb{R}^n$, and this orthogonal projection is given by $\mathfrak{X}\widehat{\boldsymbol{\beta}}^{\text{MLE}} \in \mathfrak{M}$. ■

5.2 Actuarial Applications of Generalized Linear Models

The purpose of this section is to illustrate how the concept of GLMs is used in actuarial modeling. We therefore explore the typical actuarial examples of claim counts and claim size modeling.

5.2.1 *Selection of a Generalized Linear Model*

The selection of a predictive model within GLMs for solving an applied actuarial problem requires the following choices.

Choice of the Member of the EDF Select a member of the EDF that fits the modeling problem. In a first step, we should try to understand the properties of the data $\boldsymbol{Y}$ before doing this selection, for instance, do we have count data, do we have a classification problem, do we have continuous observations?

All members of the EDF are light-tailed because the moment generating function exists around the origin, see Corollary 2.14, and the EDF is not suited to model heavy-tailed data, for instance, having a regularly varying tail. Therefore, a datum Y is sometimes first transformed before being modeled by a member of the EDF. A popular transformation is the logarithm for positive observations. After this transformation a member of the EDF can be chosen to model $\log(Y)$. For instance, if we choose the Gaussian distribution for $\log(Y)$, then Y will be log-normally distributed, or if we choose the exponential distribution for $\log(Y)$, then Y will be Pareto distributed, see Sect. 2.2.5. One can then model the transformed datum with a GLM. Often this provides very accurate models, say, on the log scale for the log-transformed data. There is one issue with this approach, namely, if a model is unbiased on the transformed scale then it is typically biased on the original observation scale; if the transformation is concave this easily follows from Jensen's inequality. The problematic part now is that the bias correction itself often has systematic effects which means that the transformation (or the involved nuisance parameters) should be modeled with a regression model, too, see Sect. 5.3.9. In many cases this will not easily work, unfortunately. Therefore, if possible, clear preference should be given to modeling the data on the original observation scale (if unbiasedness is a central requirement).

Choice of Link Function From a statistical point of view we should choose the canonical link $g = h$ to connect the mean μ of the model to the linear predictor η because this implies many favorable mathematical properties. However, as seen, sometimes we have different needs. Practical reasons may require that we have a model with additive or multiplicative effects, which favors the identity or the log-link, respectively. Another requirement is that the resulting canonical parameter $\theta = (h \circ g^{-1})(\eta)$ needs to be within the effective domain $\boldsymbol{\Theta}$. If this effective domain is bounded, for instance, if it covers the negative real line as for the gamma model, a (transformation of the) log-link might be more suitable than the canonical link because $g^{-1}(\cdot) = -\exp(\cdot)$ has a strictly negative range, see Example 5.5.

Choice of Features and Feature Engineering Assume we have selected the member of the EDF and the link function g. This gives us the relationship between the mean μ and the linear predictor η, see (5.5),

$$\mu(\boldsymbol{x}) = \mathbb{E}_{\theta(\boldsymbol{x})}[Y] = g^{-1}(\eta(\boldsymbol{x})) = g^{-1}\langle \boldsymbol{\beta}, \boldsymbol{x}\rangle. \tag{5.20}$$

Thus, the features $\boldsymbol{x} \in \mathcal{X} \subset \mathbb{R}^{q+1}$ need to be in the right functional form so that they can appropriately describe the systematic effect via the function (5.20). We distinguish the following feature types:

- *Continuous real-valued feature components*, examples are age of policyholder, weight of car, body mass index, etc.
- *Ordinal categorical feature components*, examples are ratings like good-medium-bad or A-B-C-D-E.
- *Nominal categorical feature components*, examples are vehicle brands, occupation of policyholders, provinces of living places of policyholders, etc. The values that the categorical feature components can take are called *levels*.
- *Binary feature components* are special categorical features that only have two levels, e.g. female-male, open-closed. Because binary variables often play a distinguished role in modeling they are separated from categorical variables which are typically assumed to have more than two levels.

All these components need to be brought into a suitable form so that they can be used in a linear predictor $\eta(\boldsymbol{x}) = \langle \boldsymbol{\beta}, \boldsymbol{x}\rangle$, see (5.20). This requires the consideration of the following points (1) transformation of continuous components so that they can describe the systematic effects in a linear form, (2) transformation of categorical components to real-valued components, (3) interaction of components beyond an additive structure in the linear predictor, and (4) the resulting design matrix $\mathfrak{X}$ should have full rank $q + 1 \leq n$. We are going to describe these points (1)–(4) in the next section.

5.2.2 *Feature Engineering*

Categorical Feature Components: Dummy Coding

Categorical variables need to be embedded into a Euclidean space. This embedding needs to be done such that the resulting design matrix $\mathfrak{X}$ has full rank $q + 1 \leq n$. There are many different ways to do so, and the particular choice depends on the modeling purpose. The most popular way is *dummy coding*. We only describe dummy coding here because it is sufficient for our purposes, but we mention that

Table 5.1 Dummy coding example that maps the $K = 11$ levels (colors) to the unit vectors of the 10-dimensional Euclidean space $\mathbb{R}^{10}$ selecting the last level a_{11} (brown color) as reference level, and showing the resulting dummy vectors $\boldsymbol{x}_j^\top$ as row vectors

$a_1 =$ white	1	0	0	0	0	0	0	0	0	0
$a_2 =$ yellow	0	1	0	0	0	0	0	0	0	0
$a_3 =$ orange	0	0	1	0	0	0	0	0	0	0
$a_4 =$ red	0	0	0	1	0	0	0	0	0	0
$a_5 =$ magenta	0	0	0	0	1	0	0	0	0	0
$a_6 =$ violet	0	0	0	0	0	1	0	0	0	0
$a_7 =$ blue	0	0	0	0	0	0	1	0	0	0
$a_8 =$ cyan	0	0	0	0	0	0	0	1	0	0
$a_9 =$ green	0	0	0	0	0	0	0	0	1	0
$a_{10} =$ beige	0	0	0	0	0	0	0	0	0	1
$a_{11} =$ brown	0	0	0	0	0	0	0	0	0	0

there are also other codings like effects coding or Helmert's contrast coding.[2] The choice of the coding will not influence the predictive model (if we work with a full rank design matrix), but it may influence parameter selection, parameter reduction and model interpretation. For instance, the choice of the coding is (more) important in medical studies where one tries to understand the effects between certain therapies.

Assume that the raw feature component $\widetilde{x}_j$ is a categorical variable taking K different levels $\{a_1, \ldots, a_K\}$. For dummy coding we declare one level, say a_K, to be the reference level and all other levels are described relative to that reference level. Formally, this can be described by an embedding map

$$\widetilde{x}_j \mapsto \boldsymbol{x}_j = \left(\mathbb{1}_{\{\widetilde{x}_j = a_1\}}, \ldots, \mathbb{1}_{\{\widetilde{x}_j = a_{K-1}\}}\right)^\top \in \mathbb{R}^{K-1}. \tag{5.21}$$

This is closely related to the categorical distribution in Sect. 2.1.4. An explicit example is given in Table 5.1.

Example 5.10 (Multiplicative Model) If we choose the log-link function $\eta = g(\mu) = \log(\mu)$, we receive the regression function for the categorical example of Table 5.1

$$\widetilde{x}_j \mapsto \exp\langle \boldsymbol{\beta}, \boldsymbol{x}_j \rangle = \exp\{\beta_0\} \prod_{k=1}^{K-1} \exp\left\{\beta_k \mathbb{1}_{\{\widetilde{x}_j = a_k\}}\right\}, \tag{5.22}$$

including an intercept component. Thus, the base value $\exp\{\beta_0\}$ is determined by the reference level $a_{11} =$ brown, and any color different from brown has a deviation from the base value described by the multiplicative correction term $\exp\{\beta_k \mathbb{1}_{\{\widetilde{x}_j = a_k\}}\}$. ■

[2] There is an example of Helmert's contrast coding in Remarks 2.7 of lecture notes [392], and for more examples we refer to the UCLA statistical consulting website: https://stats.idre.ucla.edu/r/library/r-library-contrast-coding-systems-for-categorical-variables/.

Remarks 5.11

- Importantly, dummy coding leads to full rank design matrices $\mathfrak{X}$ and, henceforth, Assumption 5.3 is fulfilled.
- Dummy coding is different from one-hot encoding which is going to be introduced in Sect. 7.3.1, below.
- Dummy coding needs some care if we have categorical feature components with many levels, for instance, considering car brands and car models we can get hundreds of levels. In that case we will have sparsity in the resulting design matrix. This may cause computational issues, and, as the following example will show, it may lead to high uncertainty in parameter estimation. In particular, the columns of the design matrix $\mathfrak{X}$ of very rare levels will be almost collinear which implies that we do not receive very well-conditioned matrices in Fisher's scoring method (5.12). For this reason, it is recommended to merge levels to bigger classes. In Sect. 7.3.1, below, we are going to present a different treatment. Categorical variables are embedded into low-dimensional spaces, so that proximity in these spaces has a reasonable meaning for the regression task at hand.

Example 5.12 (Balance Property and Dummy Coding) A main argument for the use of the canonical link function has been the fulfillment of the balance property, see Corollary 5.7. If we have categorical feature components and if we apply dummy coding to those, then the balance property is projected down to the individual levels of that categorical variable. Assume that columns 2 to K of design matrix $\mathfrak{X}$ are used to model a raw categorical feature $\widetilde{x}_1$ with K levels according to (5.21). In that case, columns $2 \leq k \leq K$ will indicate all observations Y_i which belong to levels a_{k-1}. Analogously to the proof of Corollary 5.7, we receive (summation i runs over the different instances/policies)

$$\sum_{i:\, \widetilde{x}_{i,1}=a_{k-1}} \mathbb{E}_{\widehat{\beta}^{\mathrm{MLE}}}\left[v_i Y_i\right] = \sum_{i=1}^{n} x_{i,k} \mathbb{E}_{\widehat{\beta}^{\mathrm{MLE}}}\left[v_i Y_i\right] = \sum_{i=1}^{n} x_{i,k} v_i Y_i = \sum_{i:\, \widetilde{x}_{i,1}=a_{k-1}} v_i Y_i. \tag{5.23}$$

Thus, we receive the balance property for all policies $1 \leq i \leq n$ that belong to level a_{k-1}.

If we have many levels, then it will happen that some levels have only very few observations, and the above summation (5.23) only runs over very few insurance policies with $\widetilde{x}_{i,1} = a_{k-1}$. Suppose additionally the volumes v_i are small. This can lead to considerable estimation uncertainty, because the estimated prices on the left-hand side of (5.23) will be based too much on individual observations Y_i having the corresponding level, and we are not in the regime of a law of large numbers that balances these observations.

Thus, this balance property from dummy coding is a natural property under the canonical link choice. Actuarial pricing is very familiar with such a property. Early distribution-free approaches have postulated this property resulting in the method of the total marginal sums, see Bailey and Jung [22, 206], where the balance property is enforced for marginal sums of all categorical levels in parameter estimation. However, if we have scarce levels in categorical variables, this approach needs careful consideration. ■

Binary Feature Components

Binary feature components do not need a treatment different from the categorical ones, they are Bernoulli variables which can be encoded as 0 or 1. This is exactly dummy coding for $K = 2$ levels.

Continuous Feature Components

Continuous feature components are already real-valued. Therefore, from the viewpoint of 'variable types', the continuous feature components do not need any pre-processing because they are already in the right format to be included in scalar products.

Nevertheless, in many cases, also continuous feature components need feature engineering because only in rare cases they directly fit the functional form (5.20). We give an example. Consider car drivers that have different driving experience and different driving skills. To explain experience and skills we typically choose the age of driver as explanatory variable. Modeling the claim frequency as a function of the age of driver, we often observe a U-shaped function, thus, a function that is non-monotone in the age of driver variable. Since the link function g needs to be strictly monotone, this regression problem cannot be modeled by (5.20), directly including the age of driver as a feature because this leads to monotonicity of the regression function in the age of driver variable.

Typically, in such situations, the continuous variable is discretized to categorical classes. In the driver's age example, we build age classes. These age classes are then treated as categorical variables using dummy coding (5.21). We will give examples below. These age classes should fulfill the requirement of being sufficiently homogeneous in the sense that insurance policies that fall into the same class should have a similar propensity to claims. This implies that we would like to have many small homogeneous classes. However, the classes should be sufficiently large, otherwise parameter estimation involves high uncertainty, see also Example 5.12. Thus, there is a trade-off to sufficiently meet both of these two requirements.

A disadvantage of this discretization approach is that neighboring age classes will not be recognized by the regression function because, per se, dummy coding is based on nominal variables not having any topology. This is also illustrated by the fact, that all categorical levels (excluding the reference level) have, in view of embedding (5.21), the same mutual Euclidean distance. Therefore, in some applications, one prefers a different approach by rather trying to find an appropriate functional form. For instance, we can pre-process a strictly positive raw feature component $\widetilde{x}_l$ to a higher-dimensional functional form

$$\widetilde{x}_l \mapsto \beta_1 \widetilde{x}_l + \beta_2 \widetilde{x}_l^2 + \beta_3 \widetilde{x}_l^3 + \beta_4 \log(\widetilde{x}_l), \tag{5.24}$$

with regression parameter $(\beta_1, \ldots, \beta_4)^\top$, i.e., we have a polynomial function of degree 3 plus a logarithmic term in this choice. If one does not want to choose a specific functional form, one often chooses natural cubic splines. This, together with regularization, leads to the framework of generalized additive models (GAMs), which is popular family of regression models besides GLMs; for literature on GAMs we refer to Hastie–Tibshirani [182], Wood [384], Ohlsson–Johansson [290], Denuit et al. [99] and Wüthrich–Buser [392]. In these notes we will not further pursue GAMs.

Example 5.13 (Multiplicative Model) If we choose the log-link function $\eta = g(\mu) = \log(\mu)$ we receive a multiplicative regression function

$$\boldsymbol{x} \mapsto \mu(\boldsymbol{x}) = \exp\langle \boldsymbol{\beta}, \boldsymbol{x} \rangle = \exp\{\beta_0\} \prod_{j=1}^{q} \exp\left\{\beta_j x_j\right\}.$$

That is, all feature components x_j enter the regression function in an exponential form. In general insurance, one may have specific variables for which it is explicitly known that they should enter the regression function as a power function. Having a raw feature $\widetilde{x}_l$ we can pre-process it as $\widetilde{x}_l \mapsto x_l = \log(\widetilde{x}_l)$. This implies

$$\mu(\boldsymbol{x}) = \exp\langle \boldsymbol{\beta}, \boldsymbol{x} \rangle = \exp\{\beta_0\}\, \widetilde{x}_l^{\beta_l} \prod_{j=1, j\neq l}^{q} \exp\left\{\beta_j x_j\right\},$$

which gives a power term of order β_l. The GLM estimates in this case the power parameter that should be used for $\widetilde{x}_l$. If the power parameter is known, then one can even include this component as an offset; offsets are discussed in Sect. 5.2.3, below. ■

Interactions

Naturally, GLMs only allow for an additive structure in the linear predictor. Similar to continuous feature components, such an additive structure may not always be suitable and one wants to model more complex interaction terms. Such interactions need to be added manually by the modeler, for instance, if we have two raw feature components $\widetilde{x}_l$ and $\widetilde{x}_k$, we may want to consider a functional form

$$(\widetilde{x}_l, \widetilde{x}_k) \mapsto \beta_1 \widetilde{x}_l + \beta_2 \widetilde{x}_k + \beta_3 \widetilde{x}_l \widetilde{x}_k + \beta_4 \widetilde{x}_l^2 \widetilde{x}_k,$$

with regression parameter $(\beta_1, \ldots, \beta_4)^\top$.

More generally, this manual feature engineering of adding interactions and of specifying functional forms (5.24) can be understood as a new representation of raw features. Representation learning in relation to deep learning is going to be discussed in Sect. 7.1, and this discussion is also related to Mercer's kernels.

5.2.3 *Offsets*

In many heterogeneous portfolio problems with observations $\boldsymbol{Y} = (Y_1, \ldots, Y_n)^\top$, there are known prior differences between the individual risks Y_i, for instance, the time exposure varies between the different policies i. Such known prior differences can be integrated into the predictors, and this integration typically does not involve any additional model parameters. A simple way is to use an *offset* (constant) in the linear predictor of a GLM. Assume that each observation Y_i is equipped with a feature $\boldsymbol{x}_i \in \mathcal{X}$ and a known offset $o_i \in \mathbb{R}$ such that the linear predictor η_i takes the form

$$(\boldsymbol{x}_i, o_i) \mapsto g(\mu_i) = \eta_i = \eta(\boldsymbol{x}_i, o_i) = o_i + \langle \boldsymbol{\beta}, \boldsymbol{x}_i \rangle, \tag{5.25}$$

for all $1 \le i \le n$. An offset o_i does not change anything from a structural viewpoint, in fact, it could be integrated into the feature $\boldsymbol{x}_i$ with a regression parameter that is identically equal to 1.

Offsets are frequently used in Poisson models with the (canonical) log-link choice to model multiplicative time exposures in claim frequency modeling. Under the log-link choice we receive from (5.25) the following mean function

$$(\boldsymbol{x}_i, o_i) \mapsto \mu(\boldsymbol{x}_i, o_i) = \exp\{\eta(\boldsymbol{x}_i, o_i)\} = \exp\{o_i + \langle \boldsymbol{\beta}, \boldsymbol{x}_i \rangle\} = \exp\{o_i\} \exp\langle \boldsymbol{\beta}, \boldsymbol{x}_i \rangle.$$

In this version, the offset o_i provides us with an exposure $\exp\{o_i\}$ that acts multiplicatively on the regression function. If $w_i = \exp\{o_i\}$ measures time, then w_i is a so-called pro-rata temporis (proportional in time) exposure.

Remark 5.14 (Boosting) A popular machine learning technique in statistical modeling is boosting. Boosting tries to step-wise adaptively improve a regression model. Offsets (5.25) are a simple way of constructing boosted models. Assume we have constructed a predictive model using any statistical model, and denote the resulting estimated means of Y_i by $\widehat{\mu}_i^{(0)}$. The idea of boosting is that we select another statistical model and we try to see whether this second model can still find systematic structure in the data which has not been found by the first model. In view of (5.25), we include the first model into the offset and we build a second model around this offset, that is, we may explore a GLM

$$\widehat{\mu}_i^{(1)} = g^{-1}\left(g(\widehat{\mu}_i^{(0)}) + \langle \boldsymbol{\beta}, \boldsymbol{x}_i\rangle\right).$$

If the first model is perfect we come up with a regression parameter $\boldsymbol{\beta} = 0$, otherwise the linear predictor $\langle \boldsymbol{\beta}, \boldsymbol{x}_i\rangle$ of the second model starts to compensate for weaknesses in $\widehat{\mu}_i^{(0)}$. Of course, this boosting procedure can then be iterated and one should stop boosting before the resulting model starts to over-fit to the data. Typically, this approach is applied to regression trees instead of GLMs, see Ferrario–Hämmerli [125], Section 7.4 in Wüthrich–Buser [392], Lee–Lin [241] and Denuit et al. [100].

5.2.4 *Lab: Poisson GLM for Car Insurance Frequencies*

We present a first GLM example. This example is based on French motor third party liability (MTPL) insurance claim counts data. The data is described in detail in Chap. 13.1; an excerpt of the available MTPL data is given in Listing 13.2. For the moment we only consider claim frequency modeling. We use the following data: N_i describes the number of claims, $v_i \in (0, 1]$ describes the duration of the insurance policy, and $\widetilde{\boldsymbol{x}}_i$ describes the available raw feature information of insurance policy i, see Listing 13.2.

We are going to model the claim counts N_i with a Poisson GLM using the canonical link function of the Poisson model. In the Poisson approach there are two different ways to account for the duration of the insurance policy. Either we model $Y_i = N_i/v_i$ with the Poisson model of the EDF, see Sect. 2.2.2 and Remarks 2.13 (reproductive form), or we directly model N_i with the Poisson distribution from the EF and treat the log-duration as an offset variable $o_i = \log v_i$. In the first approach we have for the log-link choice $g(\cdot) = h(\cdot) = \log(\cdot)$ and dispersion $\varphi = 1$

$$Y_i = N_i/v_i \sim f(y_i; \theta_i, v_i) = \exp\left\{\frac{y_i\langle \boldsymbol{\beta}, \boldsymbol{x}_i\rangle - e^{\langle \boldsymbol{\beta}, \boldsymbol{x}_i\rangle}}{1/v_i} + a(y_i; v_i)\right\}, \tag{5.26}$$

where $\boldsymbol{x}_i \in \mathcal{X}$ is the suitably pre-processed feature information of insurance policy i, and with canonical parameter $\theta_i = \eta(\boldsymbol{x}_i) = \langle \boldsymbol{\beta}, \boldsymbol{x}_i\rangle$. In the second approach we include the log-duration as offset into the regression function and model N_i with the Poisson distribution from the EF. Using notation (2.2) this gives us

$$N_i \sim f(n_i;\theta_i) = \exp\left\{n_i\left(\log v_i + \langle \boldsymbol{\beta}, \boldsymbol{x}_i\rangle\right) - e^{\log v_i + \langle \boldsymbol{\beta}, \boldsymbol{x}_i\rangle} + a(n_i)\right\} \quad (5.27)$$

$$= \exp\left\{\frac{\frac{n_i}{v_i}\langle \boldsymbol{\beta}, \boldsymbol{x}_i\rangle - e^{\langle \boldsymbol{\beta}, \boldsymbol{x}_i\rangle}}{1/v_i} + a(n_i) + n_i \log v_i\right\},$$

with canonical parameter $\theta_i = \eta(\boldsymbol{x}_i, o_i) = o_i + \langle \boldsymbol{\beta}, \boldsymbol{x}_i\rangle = \log v_i + \langle \boldsymbol{\beta}, \boldsymbol{x}_i\rangle$ for observation $n_i = v_i y_i$. That is, we receive the same model in both cases (5.26) and (5.27) under the canonical log-link choice for the Poisson GLM.

Finally, we make the assumption that all observations N_i are independent. There remains the pre-processing of the raw features $\widetilde{\boldsymbol{x}}_i$ to features $\boldsymbol{x}_i$ so that they can be used in a sensible way in the linear predictors $\eta_i = \eta(\boldsymbol{x}_i, o_i) = o_i + \langle \boldsymbol{\beta}, \boldsymbol{x}_i\rangle$.

Feature Engineering

Categorical and Binary Variables: Dummy Coding

For categorical and binary variables we use dummy coding as described in Sect. 5.2.2. We have two categorical variables `VehBrand` and `Region`, as well as a binary variable `VehGas`, see Listing 13.2. We choose the first level as reference level, and the remaining levels are characterized by $(K-1)$-dimensional embeddings (5.21). This provides us with $K-1 = 10$ parameters for `VehBrand`, $K-1 = 21$ parameters for `Region` and $K-1 = 1$ parameter for `VehGas`.

Figure 5.3 shows the empirical marginal frequencies $\bar{\lambda} = \sum N_i / \sum v_i$ on all levels of the categorical feature components `VehBrand`, `Region` and `VehGas`. Moreover, the blue areas (in the colored version) give confidence bounds of $\pm 2\sqrt{\bar{\lambda}/\sum v_i}$ (under a Poisson assumption), see Example 3.22. The more narrow these confidence bounds, the bigger the volumes $\sum v_i$ behind these empirical marginal estimates.

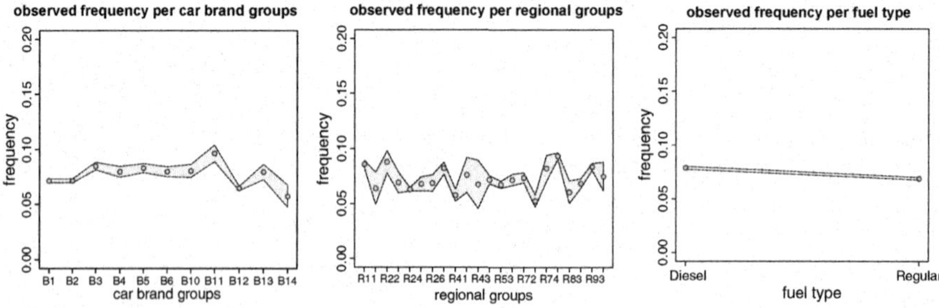

Fig. 5.3 Empirical marginal frequencies on each level of the categorical variables (lhs) `VehBrand`, (middle) `Region`, and (rhs) `VehGas`

Continuous Variables

We consider feature engineering of the continuous variables `Area`, `VehPower`, `VehAge`, `DrivAge`, `BonusMalus` and log-`Density` (`Density` on the log scale); note that we map the `Area` codes $(A, \ldots, F) \mapsto (1, \ldots, 6)$. Some of these variables do not show any monotonicity nor log-linearity in the empirical marginal frequency plots, see Fig. 5.4.

These non-monotonicity and non-log-linearity suggest in a first step to build homogeneous classes for these feature components and use dummy coding for the resulting classes. We make the following choices here (motivated by the marginal graphs of Fig. 5.4):

- `Area`: continuous log-linear feature component for $\{A, \ldots, F\} \mapsto \{1, \ldots, 6\}$;
- `VehPower`: discretize into categorical classes where we merge vehicle power groups bigger and equal to 9 (totally $K = 6$ levels);
- `VehAge`: we build categorical classes $[0, 6)$, $[6, 13)$, $[13, \infty)$ (totally $K = 3$ levels);
- `DrivAge`: we build categorical classes $[18, 21)$, $[21, 26)$, $[26, 31)$, $[31, 41)$, $[41, 51)$, $[51, 71)$, $[71, \infty)$ (totally $K = 7$ levels);
- `BonusMalus`: continuous log-linear feature component (we censor at 150);
- `Density`: log-density is chosen as continuous log-linear feature component.

This encoding is slightly different from Noll et al. [287] because of different data cleaning. The discretization has been chosen quite ad-hoc by just looking at the empirical plots; as illustrated in Section 6.1.6 of Wüthrich–Buser [392] regression trees may provide an algorithmic way of choosing homogeneous classes of sufficient volume. This provides us with a feature space (the initial component stands for the intercept $x_{i,0} = 1$ and the order of the terms is the same as in Listing 13.2)

$$\mathcal{X} \subset \{1\} \times \mathbb{R} \times \{0,1\}^5 \times \{0,1\}^2 \times \{0,1\}^6 \times \mathbb{R} \times \{0,1\}^{10} \times \{0,1\} \times \mathbb{R} \times \{0,1\}^{21},$$

of dimension $q+1 = 1+1+5+2+6+1+10+1+1+21 = 49$. The R code [307] for this pre-processing of continuous variables is shown in Listing 5.1, categorical variables do not need any special treatment because variables of `factor` type are consider internally in R by dummy coding; we call this model Poisson GLM1.

Choice of Learning and Test Samples

To measure predictive performance we follow the generalization approach as proposed in Chap. 4. This requires that we partition our entire data into learning sample $\mathcal{L}$ and test sample $\mathcal{T}$, see Fig. 4.1. Model selection and model fitting will be done on the learning sample $\mathcal{L}$, only, and the test sample $\mathcal{T}$ is used to analyze the generalization of the fitted models to unseen data. We partition the data at random (non-stratified) in a ratio of $9:1$, and we are going to hold on to the same partitioning throughout this monograph whenever we study this example. The R code used is given in Listing 5.2.

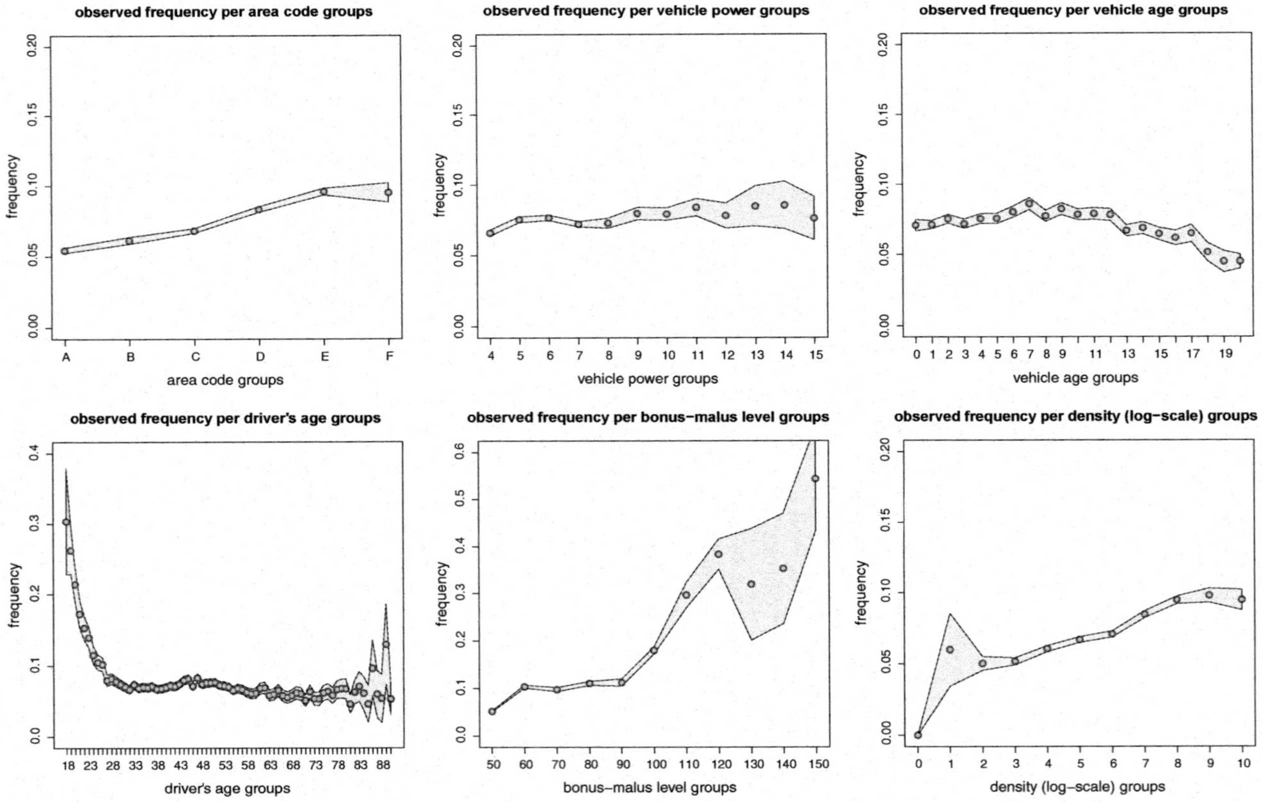

Fig. 5.4 Empirical marginal frequencies of the continuous variables: top row (lhs) `Area`, (middle) `VehPower`, (rhs) `VehAge`, and bottom row (lhs) `DrivAge`, (middle) `BonusMalus`, (rhs) log-`Density`, i.e., `Density` on the log scale; note that `DrivAge` and `BonusMalus` have a different y-scale in these plots

Listing 5.1 Pre-processing of features for model Poisson GLM1 in R

```
dat$AreaGLM       <- as.integer(dat$Area)
dat$VehPowerGLM <- as.factor(pmin(dat$VehPower, 9))
dat$VehAgeGLM     <- as.factor(cut(dat$VehAge, c(0,5,12,101),
                        labels = c("0-5","6-12","12+"),
                        include.lowest = TRUE))
dat$DrivAgeGLM    <- as.factor(cut(dat$DrivAge, c(18,20,25,30,40,50,70,101),
                        labels = c("18-20","21-25","26-30","31-40","41-50",
                                   "51-70","71+"), include.lowest = TRUE))
dat$BonusMalusGLM <- pmin(dat$BonusMalus, 150)
dat$DensityGLM    <- log(dat$Density)
```

Table 5.2 shows the summary of the chosen partition into learning and test samples

$$\mathcal{L} = \left\{(Y_i = N_i/v_i, \boldsymbol{x}_i, v_i) : i = 1, \ldots, n = 610'206\right\},$$

and

$$\mathcal{T} = \left\{(Y_t^\dagger = N_t^\dagger/v_t^\dagger, \boldsymbol{x}_t^\dagger, v_t^\dagger) : t = 1, \ldots, T = 67'801\right\}.$$

In contrast to Sect. 4.2 we also include feature information and exposure information to $\mathcal{L}$ and $\mathcal{T}$.

Listing 5.2 Partition of the data to learning sample $\mathcal{L}$ and test sample $\mathcal{T}$

```
RNGversion("3.5.0")     # we use R version 3.5.0 for this partition
set.seed(500)
ll     <- sample(c(1:nrow(dat)), round(0.9*nrow(dat)), replace = FALSE)
learn <- dat[ll,]
test   <- dat[-ll,]
```

Table 5.2 Choice of learning data set $\mathcal{L}$ and test data set $\mathcal{T}$; the empirical frequency on both data sets is similar (last column), and the split of the policies w.r.t. the numbers of claims is also rather similar

	Numbers of observed claims						Empirical
	0	1	2	3	4	5	frequency
Learning sample $\mathcal{L}$	96.32%	3.47%	0.19%	0.01%	0.0006%	0.0002%	7.36%
Test sample $\mathcal{T}$	96.31%	3.50%	0.18%	0.01%	0.0015%	0.0015%	7.35%

Maximum-Likelihood Estimation and Results

The remaining step is to perform MLE to estimate regression parameter $\boldsymbol{\beta} \in \mathbb{R}^{q+1}$. This can be done either by maximizing the Poisson log-likelihood function or by minimizing the Poisson deviance loss. In view of (4.9) and Example 4.27, the Poisson deviance loss on the learning data $\mathcal{L}$ is given by

$$\boldsymbol{\beta} \mapsto \mathfrak{D}(\mathcal{L}, \boldsymbol{\beta}) = \frac{2}{n} \sum_{i=1}^{n} v_i \left(\mu(\boldsymbol{x}_i) - Y_i - Y_i \log \left(\frac{\mu(\boldsymbol{x}_i)}{Y_i} \right) \right) \geq 0, \tag{5.28}$$

where the terms under the summation are set equal to $v_i \mu(\boldsymbol{x}_i)$ for $Y_i = 0$, see (4.8), and we have GLM regression function

$$\boldsymbol{x} \mapsto \mu(\boldsymbol{x}) = \mu_{\boldsymbol{\beta}}(\boldsymbol{x}) = \exp\langle \boldsymbol{\beta}, \boldsymbol{x} \rangle.$$

That is, we work under the canonical link with the canonical parameter being equal to the linear predictor. The MLE of $\boldsymbol{\beta}$ is found by minimizing (5.28). This is done with Fisher's scoring method. In order to receive a non-degenerate solution we need to ensure that we have sufficiently many claims $Y_i > 0$, otherwise it might happen that the MLE provides a (degenerate) solution at the boundary of the effective domain $\boldsymbol{\Theta}$. We denote the MLE by $\widehat{\boldsymbol{\beta}}_{\mathcal{L}}^{\text{MLE}} = \widehat{\boldsymbol{\beta}}^{\text{MLE}}$, because it has been estimated on the learning data $\mathcal{L}$, only. This gives us estimated regression function

$$\boldsymbol{x} \mapsto \widehat{\mu}(\boldsymbol{x}) = \mu_{\widehat{\boldsymbol{\beta}}_{\mathcal{L}}^{\text{MLE}}}(\boldsymbol{x}) = \exp\langle \widehat{\boldsymbol{\beta}}_{\mathcal{L}}^{\text{MLE}}, \boldsymbol{x} \rangle.$$

We emphasize that we only use the learning data $\mathcal{L}$ for this model fitting. In view of Definition 4.24 we receive in-sample and out-of-sample Poisson deviance losses

$$\mathfrak{D}(\mathcal{L}, \widehat{\boldsymbol{\beta}}_{\mathcal{L}}^{\text{MLE}}) = \frac{2}{n} \sum_{i=1}^{n} v_i \left(\widehat{\mu}(\boldsymbol{x}_i) - Y_i - Y_i \log \left(\frac{\widehat{\mu}(\boldsymbol{x}_i)}{Y_i} \right) \right) \geq 0,$$

$$\mathfrak{D}(\mathcal{T}, \widehat{\boldsymbol{\beta}}_{\mathcal{L}}^{\text{MLE}}) = \frac{2}{T} \sum_{t=1}^{T} v_t^\dagger \left(\widehat{\mu}(\boldsymbol{x}_t^\dagger) - Y_t^\dagger - Y_t^\dagger \log \left(\frac{\widehat{\mu}(\boldsymbol{x}_t^\dagger)}{Y_t^\dagger} \right) \right) \geq 0.$$

We implement this GLM on the data of Listing 5.1 (and including the categorical features) in R using the function `glm` [307], a short overview of the results is presented in Listing 5.3. This overview presents the regression model implemented, an excerpt of the parameter estimates $\widehat{\boldsymbol{\beta}}_{\mathcal{L}}^{\text{MLE}}$, standard errors which are received from the square-rooted diagonal entries of the inverse of the estimated Fisher's information matrix $\mathcal{I}_n(\widehat{\boldsymbol{\beta}}_{\mathcal{L}}^{\text{MLE}})$, see (5.17); the remaining columns will be described in Sect. 5.3.2 on the Wald test (5.33). The bottom line of the output says that Fisher's scoring algorithm has converged in 6 iterations, it gives the in-sample deviance loss $n\mathfrak{D}(\mathcal{L}, \widehat{\boldsymbol{\beta}}_{\mathcal{L}}^{\text{MLE}})$ called `Residual deviance` (not being scaled by the number of

Listing 5.3 Results in model Poisson GLM1 using the R command `glm`

```
1  Call:
2  glm(formula = ClaimNb ~ VehPowerGLM + VehAgeGLM + DrivAgeGLM +
3                BonusMalusGLM + VehBrand + VehGas + DensityGLM + Region +
4                AreaGLM, family = poisson(), data = learn, offset = log(Exposure))
5
6  Deviance Residuals:
7      Min       1Q    Median       3Q       Max
8  -1.4728  -0.3256  -0.2456  -0.1383   7.7971
9
10 Coefficients:
11                  Estimate Std. Error z value Pr(>|z|)
12 (Intercept)     -4.8175439  0.0579296 -83.162  < 2e-16 ***
13 VehPowerGLM5     0.0604293  0.0229841   2.629 0.008559 **
14 VehPowerGLM6     0.0868252  0.0225509   3.850 0.000118 ***
15 .                    .                    .
16 .                    .                    .
17 RegionR93        0.1388160  0.0294901   4.707 2.51e-06 ***
18 RegionR94        0.1918538  0.0938250   2.045 0.040874 *
19 AreaGLM          0.0407973  0.0200818   2.032 0.042199 *
20 ---
21 Signif. codes:  0 '***' 0.001 '**' 0.01 '*' 0.05 '.' 0.1 ' ' 1
22
23 (Dispersion parameter for poisson family taken to be 1)
24
25     Null deviance: 153852  on 610205  degrees of freedom
26 Residual deviance: 147069  on 610157  degrees of freedom
27 AIC: 192818
28
29 Number of Fisher Scoring iterations: 6
```

Table 5.3 Run times, number of parameters, AICs, in-sample and out-of-sample deviance losses, tenfold cross-validation losses with empirical standard deviation in brackets, see also (4.36), (units are in 10^{-2}) and the in-sample average frequency of the null model (Poisson intercept model, see Example 4.27) and of model Poisson GLM1

	Run time	# Param.	AIC	In-sample loss on $\mathcal{L}$	Out-of-sample loss on $\mathcal{T}$	Tenfold CV loss $\widehat{\mathfrak{D}}^{\mathrm{CV}}$	Aver. freq.
Poisson null	–	1	199'506	25.213	25.445	25.213(0.234)	7.36%
Poisson GLM1	16 s	49	192'818	24.101	24.146	24.121(0.245)	7.36%

observations), as well as Akaike's Information Criterion (AIC), see Sect. 4.2.3 for AIC. Note that we have implemented Poisson version (5.27) with the exposures entering the offset, see lines 2–4 of Listing 5.3; this is important for understanding AIC being calculated on the (unscaled) claim counts N_i.

Table 5.3 summarizes the results of model Poisson GLM1 and it compares the figures to the null model (only having an intercept β_0); the null model has already been introduced in Example 4.27. We present the run time needed to fit the model,[3] the number of regression parameters $q+1$ in $\boldsymbol{\beta} \in \mathbb{R}^{q+1}$, AIC, in-sample and out-of-sample deviance losses, as well as tenfold cross-validation losses on the

[3] All run times are measured on a personal laptop Intel(R) Core(TM) i7-8550U CPU @ 1.80 GHz 1.99 GHz with 16 GB RAM, and they only correspond to fitting the model (or the corresponding step) once, i.e., they do not account for multiple runs, for instance, for K-fold cross-validation.

learning data $\mathcal{L}$. For tenfold cross-validation we always use the same (non-stratified) partition of $\mathcal{L}$ (in all examples in this monograph), and in bracket we show the empirical standard deviation received by (4.36). Tenfold cross-validation would not be necessary in this case because we have test data $\mathcal{T}$ on which we can evaluate the out-of-sample deviance GL. We present both figures to back-test whether tenfold cross-validation works properly in our example. We observe that the out-of-sample deviance losses $\mathfrak{D}(\mathcal{T}, \widehat{\boldsymbol{\beta}}^{\text{MLE}}_{\mathcal{L}})$ are within one empirical standard deviation of the tenfold cross-validation losses $\widehat{\mathfrak{D}}^{\text{CV}}$, which supports this methodology of model comparison.

From Table 5.3 we conclude that we should prefer model Poisson GLM1 over the null model, this decision is supported by a smaller AIC, a smaller out-of-sample deviance loss $\mathfrak{D}(\mathcal{T}, \widehat{\boldsymbol{\beta}}^{\text{MLE}}_{\mathcal{L}})$ as well as a smaller cross-validation loss $\widehat{\mathfrak{D}}^{\text{CV}}$. The last column of Table 5.3 confirms that the estimated model meets the balance property (we work with the canonical link here). Note that this balance property should be fulfilled for two reasons. Firstly, we would like to have the overall portfolio price on the right level, and secondly, deviance losses should only be compared on the same overall frequency, see Example 4.10.

Before we continue to introduce more models to challenge model Poisson GLM1, we are going to discuss statistical tools for model evaluation. Of course, we would like to know whether model Poisson GLM1 is a good model for this data or whether it is just the better model of two bad options.

Remark 5.15 (Prior and Posterior Information) Pricing literature distinguishes between prior feature information and posterior feature information, see Verschuren [372]. Prior feature information is available at the inception of the (new) insurance contract before having any claims history. This includes, for instance, age of driver, vehicle brand, etc. For policy renewals, past claims history is available and prices of policy renewals can also be based on such posterior information. Past claims history has led to the development of so-called bonus-malus systems (BMS) which often are in the form of multiplicative factors to the base premium to reward and punish good and bad past experience, respectively. One stream of literature studies optimal designs of BMS, we refer to Loimaranta [255], De Pril [91], Lemaire [245], Denuit et al. [102], Brouhns et al. [57] Pinquet [304], Pinquet et al. [305], Tzougas et al. [360] or Ágoston–Gyetvai [4]. Another stream of literature studies how one can optimally extract predictive information from an existing BMS, see Boucher–Inoussa [46], Boucher–Pigeon [47] and Verschuren [372].

The latter is basically what we also do in the above example: note that we include the variable `BonusMalus` into the feature information and, thus, we use past claims information to predict future claims. For new policies, the bonus-malus level is at 100%, and our information does not allow to clearly distinguish between new policies and policy renewals for drivers that have posterior information reflected by a bonus-malus level of 100%. Since young drivers are more likely new customers we expect interactions between the driver's age variable and the bonus-malus level, this intuition is supported by Fig. 13.12 (lhs). In order to improve our model, we would require more detailed information about past claims history. Remark that we do not strictly distinguish between prior and posterior information, here. If we go over to a time-series consideration, where more and more claims experience becomes available of an individual driver, we should clearly distinguish the different sets of information, because otherwise it may happen that in prior and posterior pricing factors we correct twice for the same factor; an interesting paper is Corradin et al. [82].

We also mention that a new source of posterior information is emerging through the collection of telematics car driving data. Telematics car driving data leads to a completely new way of posterior information rate making (experience rating), we refer to Ayuso et al. [17–19], Boucher et al. [42], Lemaire et al. [246] and Denuit et al. [98]. We mention the papers of Gao et al. [152, 154] and Meng et al. [271] who directly extract posterior feature information from telematics car driving data in order to improve rate making. This approach combines a Poisson GLM with a network extractor for the telematics car driving data.

5.3 Model Validation

One of the purposes of Chap. 4 has been to describe measures to analyze how well a fitted model generalizes to unseen data. In a proper generalization analysis this requires learning data $\mathcal{L}$ for in-sample model fitting and a test sample $\mathcal{T}$ for an out-of-sample generalization analysis. In many cases, one is not in the comfortable situation of having a test sample. In such situations one can use AIC that tries to correct the in-sample figure for model complexity or, alternatively, K-fold cross-validation as used in Table 5.3.

The purpose of this section is to introduce diagnostic tools for fitted models; these are often based on unit deviances $\mathfrak{d}(Y_i, \mu_i)$, which play the role of squared residuals in classical linear regression. Moreover, we discuss parameter and model selection, for instance, by step-wise backward elimination or forward selection using the analysis of variance (ANOVA) or the likelihood ratio test (LRT).

5.3.1 Residuals and Dispersion

Within the EDF we distinguish two different types of residuals. The first type of residuals are based on the unit deviances $\mathfrak{d}(Y_i, \mu_i)$ studied in (4.7). The *deviance residuals* are given by

$$r_i^{\mathrm{D}} = \mathrm{sign}(Y_i - \mu_i)\sqrt{\frac{v_i}{\varphi}\,\mathfrak{d}\,(Y_i, \mu_i)}.$$

Secondly, *Pearson's residuals* are given by, see also (4.12),

$$r_i^{\mathrm{P}} = \sqrt{\frac{v_i}{\varphi}}\frac{Y_i - \mu_i}{\sqrt{V(\mu_i)}}.$$

In the Gaussian case the two residuals coincide. This indicates that Pearson's residuals are most appropriate in the Gaussian case because they respect the distributional properties in that case. For other distributions, Pearson's residuals can be markedly skewed, as stated in Section 2.4.2 of McCullagh–Nelder [265], and therefore may fail to have properties similar to Gaussian residuals. An other issue occurs in Pearson's residuals when the denominator involves an estimated standard deviation $\sqrt{V(\widehat{\mu}_i)}$, for instance, if we work in a small frequency Poisson problem. Estimation uncertainty in small denominators of Pearson's residuals may substantially distort the estimated residuals. For this reason, we typically work with (the more robust) deviance residuals; this is related to the discussion in Chap. 4 on MSEPs versus expected deviance GLs, see Remarks 4.6.

The squared residuals provide unit deviance and weighted square loss, respectively,

$$(r_i^{\mathrm{D}})^2 = \frac{v_i}{\varphi}\,\mathfrak{d}\,(Y_i, \mu_i) \qquad \text{and} \qquad (r_i^{\mathrm{P}})^2 = \frac{v_i}{\varphi}\frac{(Y_i - \mu_i)^2}{V(\mu_i)},$$

the latter corresponds to Pearson's χ^2-statistic, see (4.12).

Example 5.16 (Residuals in the Poisson Case) In the Poisson case, Pearson's χ^2-statistic is for $v_i = \varphi = 1$ given by

$$(r_i^{\mathrm{P}})^2 = \frac{(Y_i - \mu_i)^2}{\mu_i},$$

because we have variance function $V(\mu) = \mu$. A second order Taylor expansion around Y_i on the scale $\mu_i^{1/3}$ (for μ_i) provides approximation to the unit deviances in the Poisson case, see formula (6.4) and Figure 6.2 in McCullagh–Nelder [265],

$$\mathfrak{d}\,(Y_i, \mu_i) \;\approx\; 9Y_i^{1/3}\left(Y_i^{1/3} - \mu_i^{1/3}\right)^2. \tag{5.29}$$

This emphasizes the different behaviors around the observation Y_i of the two types of residuals in the Poisson case. The scale $\mu_i^{1/3}$ has been motivated in McCullagh–

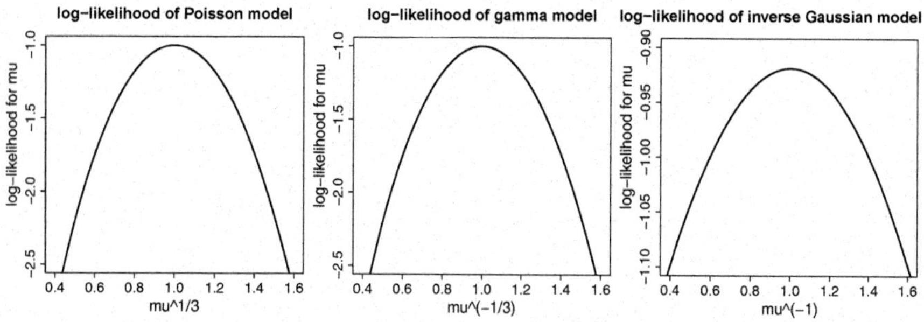

Fig. 5.5 Log-likelihoods $\ell_Y(\mu)$ in $Y = 1$ as a function of μ plotted against (lhs) $\mu^{1/3}$ in the Poisson case, (middle) $\mu^{-1/3}$ in the gamma case with shape parameter $\alpha = 1$, and (rhs) μ^{-1} in the inverse Gaussian case with $\alpha = 1$

Nelder [265] by providing a symmetric behavior around the mode in $Y_i = 1$ of the resulting log-likelihood function, see Fig. 5.5 (lhs).

■

The explicit calculation of the residuals requires knowledge of the dispersion parameter $\varphi > 0$. In the Poisson Example 5.16 this dispersion parameter has been set equal to 1 because the Poisson model does neither allow for under- nor for over-dispersion. Typically, this is not the case for other models, and this requires determination of the dispersion parameter if we want to simulate from these other models. So far, this dispersion parameter has been treated as a nuisance parameter and, in fact, it canceled in MLE (because it was assumed to be constant), see Proposition 5.1.

If we need to estimate the dispersion parameter, we can either do this within MLE, see Remarks 5.2, or we can use Pearson's or the deviance estimates, respectively,

$$\widehat{\varphi}^{\mathrm{P}} = \frac{1}{n-(q+1)} \sum_{i=1}^{n} \frac{(Y_i - \widehat{\mu}_i)^2}{V(\widehat{\mu}_i)/v_i} \quad \text{and} \quad \widehat{\varphi}^{\mathrm{D}} = \frac{1}{n-(q+1)} \sum_{i=1}^{n} v_i \mathfrak{d}\left(Y_i, \widehat{\mu}_i\right), \tag{5.30}$$

where $\widehat{\mu}_i = \widehat{\mu}(\boldsymbol{x}_i)$ are the MLE estimated means involving $q+1$ estimated parameters $\widehat{\boldsymbol{\beta}}^{\mathrm{MLE}} \in \mathbb{R}^{q+1}$. We briefly motivate these choices. Firstly, Pearson's estimate $\widehat{\varphi}^{\mathrm{P}}$ is consistent for φ. Note that in the Gaussian case this is just the standard estimate for the variance parameter. Justification of the deviance dispersion estimate is more challenging. Consider the unscaled deviance with $\widehat{\boldsymbol{\mu}}_n = (\widehat{\mu}_1, \ldots, \widehat{\mu}_n)^\top$, see (4.9),

$$n\varphi \mathfrak{D}(\boldsymbol{Y}_n, \widehat{\boldsymbol{\mu}}_n) = \sum_{i=1}^{n} v_i \mathfrak{d}\left(Y_i, \widehat{\mu}_i\right).$$

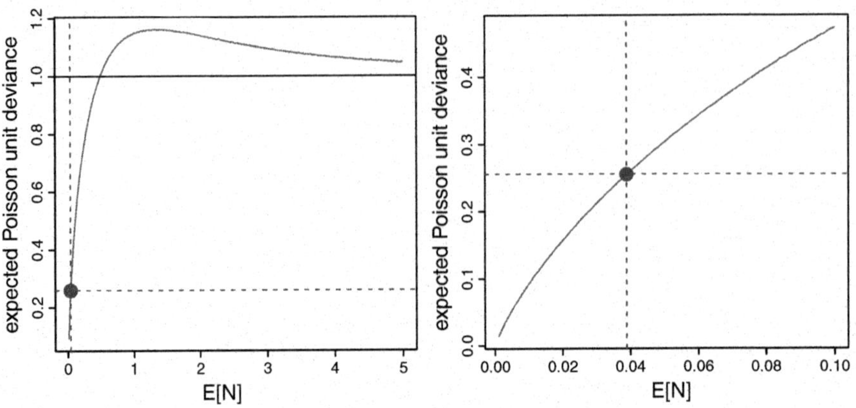

Fig. 5.6 Expected unit deviance $v\mathbb{E}_\mu[\mathfrak{d}(Y, \mu)]$ in the Poisson case as a function of $\mathbb{E}[N] = \mathbb{E}[vY] = v\mu$; the two plots only differ in the scale on the x-axis

This statistic is under *certain* assumptions asymptotically $\varphi\chi^2_{n-(q+1)}$-distributed, where $\chi^2_{n-(q+1)}$ denotes a χ^2-distribution with $n-(q+1)$ degrees of freedom. Thus, this approximation gives us an expected value of $\varphi(n-(q+1))$. This exactly justifies the deviance dispersion estimate (5.30) in these cases. However, as stated in the last paragraph of Section 2.3 of McCullagh–Nelder [265], often a χ^2-approximation is not suitable even as $n \to \infty$. We give an example.

Example 5.17 (Poisson Unit Deviances) The deviance statistics in the Poisson model with means $\boldsymbol{\mu}_n = (\mu_1, \ldots, \mu_n)^\top$ is given by

$$\mathfrak{D}(\boldsymbol{Y}_n, \boldsymbol{\mu}_n) = \frac{1}{n}\sum_{i=1}^n v_i \mathfrak{d}(Y_i, \mu_i) = \frac{1}{n}\sum_{i=1}^n 2v_i \left(\mu_i - Y_i - Y_i \log\left(\frac{\mu_i}{Y_i}\right)\right),$$

note that in the Poisson model we have (by definition) $\varphi = 1$. We evaluate the expected value of this deviance statistics. It is given by

$$\mathbb{E}_{\boldsymbol{\mu}_n}\left[\mathfrak{D}(\boldsymbol{Y}_n, \boldsymbol{\mu}_n)\right] = \frac{1}{n}\sum_{i=1}^n 2v_i \mathbb{E}_{\mu_i}\left[\mu_i - Y_i - Y_i \log\left(\frac{\mu_i}{Y_i}\right)\right] = \frac{1}{n}\sum_{i=1}^n 2\mathbb{E}_{\mu_i}\left[N_i \log\left(\frac{N_i}{v_i\mu_i}\right)\right],$$

with $N_i \stackrel{\text{ind.}}{\sim} \text{Poi}(v_i\mu_i)$.

In Fig. 5.6 we plot the expected unit deviance $v\mu \mapsto v\mathbb{E}_\mu[\mathfrak{d}(Y, \mu)]$ in the Poisson model. In our example of Table 5.3, we have $\mathbb{E}_\mu[vY] = v\mu \approx 3.89\%$, which results in an expected unit deviance of $v\mathbb{E}_\mu[\mathfrak{d}(Y, \mu)] \approx 25.52 \cdot 10^{-2} < 1$. This is in line with the losses in Table 5.3. Thus, the expected deviance $n\mathbb{E}_{\boldsymbol{\mu}_n}\left[\mathfrak{D}(\boldsymbol{Y}_n, \boldsymbol{\mu}_n)\right] \approx n/4 < n$. Therefore it is substantially smaller than n. But this implies that $n\mathfrak{D}(\boldsymbol{Y}_n, \boldsymbol{\mu}_n)$ cannot be asymptotically $\chi^2_{n-(q+1)}$-distributed because the latter has an expected of value $n-(q+1) \approx n$ for $n \to \infty$. In fact, the deviance dispersion estimate is not consistent in this example, and for a consistent estimate one should rely on Pearson's deviance estimate.

In order to have an asymptotic χ^2-distribution we need to have large volumes v because then a saddlepoint approximation holds that allows to approximate the (scaled) unit deviances by χ^2-distributions, see Sect. 5.5.2, below. ■

5.3.2 *Hypothesis Testing*

Consider a sub-vector $\boldsymbol{\beta}_r \in \mathbb{R}^r$ of the GLM parameter $\boldsymbol{\beta} \in \mathbb{R}^{q+1}$, for $r < q+1$. We would like to understand if we can set this sub-vector $\boldsymbol{\beta}_r = 0$, and at the same time we do not lose any generalization power. Thus, we investigate whether there is a simpler *nested* GLM that provides a similar prediction accuracy. If this is the case, preference should be given to the simpler model because the bigger model seems over-parametrized (has redundancy, is not parsimonious). This section is based on Section 2.2.2 of Fahrmeir–Tutz [123].

Geometric Interpretation We begin by giving a geometric interpretation. We start from the full model being expressed by the design matrix $\mathfrak{X} \in \mathbb{R}^{n\times(q+1)}$. This design matrix together with the link function g generates a $(q+1)$-dimensional manifold $\mathfrak{M} \subset \mathbb{R}^n$ given by, see (5.19) and Fig. 5.2,

$$\mathfrak{M} = \left\{ \boldsymbol{\mu} = g^{-1}(\mathfrak{X}\boldsymbol{\beta}) = (g^{-1}\langle\boldsymbol{\beta}, \boldsymbol{x}_1\rangle, \ldots, g^{-1}\langle\boldsymbol{\beta}, \boldsymbol{x}_n\rangle)^\top \in \mathbb{R}^n \middle|\ \boldsymbol{\beta} \in \mathbb{R}^{q+1} \right\} \subset \mathbb{R}^n.$$

The MLE $\widehat{\boldsymbol{\beta}}^{\mathrm{MLE}}$ is determined by the point in $\mathfrak{M}$ that minimizes the distance to $\boldsymbol{Y}$, where distance between $\boldsymbol{Y}$ and $\mathfrak{M}$ is measured component-wise by $\frac{v_i}{\varphi}\mathfrak{d}(Y_i, \mu_i)$ with $\boldsymbol{\mu} \in \mathfrak{M}$, i.e., w.r.t. the KL divergence.

Assume, now, that we want to drop the components $\boldsymbol{\beta}_r$ in $\boldsymbol{\beta}$, i.e., we want to drop these columns from the design matrix resulting in a smaller design matrix $\mathfrak{X}_r \in \mathbb{R}^{n\times(q+1-r)}$. This generates a $(q+1-r)$-dimensional *nested* manifold $\mathfrak{M}_r \subset \mathfrak{M}$ described by

$$\mathfrak{M}_r = \left\{ \boldsymbol{\mu} = g^{-1}(\mathfrak{X}_r\boldsymbol{\beta}) \in \mathbb{R}^n \middle|\ \boldsymbol{\beta} \in \mathbb{R}^{q+1-r} \right\} \subset \mathfrak{M}.$$

If the distance of $\boldsymbol{Y}$ to $\mathfrak{M}_r$ and $\mathfrak{M}$ is roughly the same, we should go for the smaller model. In the Gaussian case of Example 5.9 this can be explained by the Pythagorean theorem applied to successive orthogonal projections. In the general unit deviance case, this has to be studied in terms of information geometry considering the KL divergence, see Sect. 2.3.

Likelihood Ratio Test (LRT) We consider the testing problem of the null hypothesis H_0 against the alternative hypothesis H_1

$$H_0 : \boldsymbol{\beta}_r = 0 \qquad \text{against} \qquad H_1 : \boldsymbol{\beta}_r \neq 0. \tag{5.31}$$

Denote by $\widehat{\boldsymbol{\beta}}^{\text{MLE}}$ the MLE under the full model and by $\widehat{\boldsymbol{\beta}}^{\text{MLE}}_{(-r)}$ the MLE under the null hypothesis model. Define the (log-)*likelihood ratio test (LRT) statistics*

$$\Lambda = -2\left(\ell_{\boldsymbol{Y}}(\widehat{\boldsymbol{\beta}}^{\text{MLE}}_{(-r)}) - \ell_{\boldsymbol{Y}}(\widehat{\boldsymbol{\beta}}^{\text{MLE}})\right) \geq 0.$$

The inequality holds because the null hypothesis model is nested in the full model, henceforth, the latter needs to have a bigger log-likelihood value in the MLE. If the LRT statistics Λ is large, the null hypothesis should be rejected because the reduced model is not competitive compared to the full model. More mathematically, under similar conditions as for the asymptotic normality results of the MLE of $\boldsymbol{\beta}$ in (5.17), we have that under the null hypothesis H_0 the LRT statistics Λ is asymptotically χ^2-distributed with r degrees of freedom. Therefore, we should reject the null hypothesis in favor of the full model if the resulting p-value of Λ under the χ^2_r-distribution is too small. These results remain true if the unknown dispersion parameter φ is replaced by a consistent estimator $\widehat{\varphi}$, e.g., Pearson's dispersion estimate $\widehat{\varphi}^{\text{P}}$ (from the bigger model).

The LRT statistics Λ may not be properly defined in over-dispersed situations where the distributional assumptions are not fully specified, for instance, in an over-dispersed Poisson model. In such situations, one usually divides the log-likelihood (of the Poisson model) by the estimated over-dispersion and then uses the resulting scaled LRT statistics Λ as an approximation to the unspecified model.

Wald Test Alternatively, we can use the Wald statistics. The Wald statistics uses a second order approximation to the log-likelihood and, therefore, is only based on the first two moments (and not on the entire distribution). Define the matrix $I_r \in \mathbb{R}^{r\times(q+1)}$ such that $\boldsymbol{\beta}_r = I_r\boldsymbol{\beta}$, i.e., matrix I_r selects exactly the components of $\boldsymbol{\beta}$ that are included in $\boldsymbol{\beta}_r$ (and which are set to 0 under the null hypothesis H_0 given in (5.31)).

Asymptotic normality (5.17) motivates consideration of the Wald statistics

$$W = (I_r\widehat{\boldsymbol{\beta}}^{\text{MLE}} - 0)^\top \left(I_r\mathcal{I}(\widehat{\boldsymbol{\beta}}^{\text{MLE}})^{-1}I_r^\top\right)^{-1}(I_r\widehat{\boldsymbol{\beta}}^{\text{MLE}} - 0). \tag{5.32}$$

The Wald statistics measures the distance between the MLE in the full model $I_r\widehat{\boldsymbol{\beta}}^{\text{MLE}}$ restricted to the components of $\boldsymbol{\beta}_r$ and the null hypothesis H_0 (being $\boldsymbol{\beta}_r = 0$). The estimated Fisher's information matrix $\mathcal{I}(\widehat{\boldsymbol{\beta}}^{\text{MLE}})$ is used to bring all components onto the same unit scale (and to account for collinearity). The Wald statistics W is asymptotically χ^2_r-distributed under the same assumptions as for (5.17) to hold. Thus, the null hypothesis H_0 should be rejected if the resulting p-

value of W under the χ_r^2-distribution is too small. Note that this test does not require calculation of the MLE in the null hypothesis model, i.e., this test is computationally more attractive than the LRT because we only need to fit one model. Again, an unknown dispersion parameter φ in Fisher's information matrix $\mathcal{I}(\boldsymbol{\beta})$ is replaced by a consistent estimator $\widehat{\varphi}$ (from the bigger model).

In the special case of considering only one component of $\boldsymbol{\beta}$, i.e., if $\boldsymbol{\beta}_r = \beta_k$ with $r = 1$ and for one selected component $0 \le k \le q$, the Wald statistics reduces to

$$W_k = \frac{(\widehat{\beta}_k^{\text{MLE}})^2}{\widehat{\sigma}_k^2} \qquad \text{or} \qquad T_k = W_k^{1/2} = \frac{\widehat{\beta}_k^{\text{MLE}}}{\widehat{\sigma}_k}, \tag{5.33}$$

with diagonal entries of the inverse of the estimated Fisher's information matrix given by $\widehat{\sigma}_k^2 = (\mathcal{I}(\widehat{\boldsymbol{\beta}}^{\text{MLE}})^{-1})_{k,k}$, $0 \le k \le q$. The square-roots of these estimates are provided in column `Std. Error` of the R output in Listing 5.3.

In this case the Wald statistics W_k is equal to the square of the t-statistics T_k; this t-statistics is provided in column `z value` of the R output of Listing 5.3. Remark that Fisher's information matrix involves the dispersion parameter φ. If this dispersion parameter is estimated with a consistent estimator $\widehat{\varphi}$ we have a t-statistics. For known dispersion parameter the t-statistics reduces to a z-statistics, i.e., the corresponding p-values can be calculated from a normal distribution instead of a t-distribution. In the Poisson case, the dispersion $\varphi = 1$ is known, and for this reason, we perform a z-test (and not a t-test) in the last column of Listing 5.3; and we call T_k a z-statistics in that case.

5.3.3 *Analysis of Variance*

In the previous section, we have presented tests that allow for model selection in the case of nested models. More generally, if we have a full model, say, based on regression parameter $\boldsymbol{\beta} \in \mathbb{R}^{q+1}$ we would like to select the "best" sub-model according to some selection criterion. In most cases, it is computationally not feasible to fit all sub-models if q is large, therefore, this is not a practical solution. For large models and data sets step-wise procedures are a feasible tool. *Backward elimination* starts from the full model, and then recursively drops feature components which have high p-values in the corresponding Wald statistics (5.32) and (5.33). Performing this recursively will provide us with hierarchy of nested models. *Forward selection* works just in the opposite direction, that is, we start with the null model and we include feature components one after the other that have a low p-value in the corresponding Wald statistics.

Remarks 5.18

- The order of the inclusion/exclusion of the feature components matters in this selection algorithms because we do not have additivity in this selection process. For this reason, often backward elimination and forward selection is combined in an alternating way.
- This process as well as the tests from Sect. 5.3.2 are based on a fixed pre-processing of features. If the feature pre-processing is done differently, all analysis needs to be repeated for this new model. Moreover, between two different models we need to apply different tools for model selection (if they are not nested), for instance, AIC, cross-validation or an out-of-sample generalization analysis.
- For categorical variables with dummy coding we should apply the forward selection or the backward elimination simultaneously on the entire dummy coded vector of a categorical variable. This will include or exclude this variable; if we only apply the Wald test to one component of the dummy vector, then we test whether this level should be merged with the reference level.

Typically, in practice, a so-called analysis of variance (ANOVA) table is studied. The ANOVA table is mainly motivated by the Gaussian model with orthogonal data. The Gaussian assumption implies that the deviance loss is equal to the square loss and the orthogonality implies that the square loss decouples in an additive way w.r.t. the feature components. This implies that one can explicitly study the contribution of each feature component to the decrease in square loss; an example is given in Section 2.3.2 of McCullagh–Nelder [265]. In non-Gaussian and non-orthogonal situations one loses this additivity property and, as mentioned in Remarks 5.18, the order of inclusion matters. Therefore, for the ANOVA table we pre-specify the order in which the components are included and then we analyze the decrease of deviance loss by the inclusion of additional components.

Example 5.19 (Poisson GLM1, Revisited) We revisit the MTPL claim frequency example of Sect. 5.2.4 to illustrate the variable selection procedures. Based on the model presented in Listing 5.3 we run an ANOVA analysis using the R command `anova`, the results are presented in Listing 5.4.

Listing 5.4 shows the hierarchy of models starting from the null model by sequentially including feature components one by one. The column `Df` gives the number of regression parameters involved and the column `Deviance` the decrease of deviance loss by the inclusion of this feature component. The biggest model improvements are provided by the bonus-malus level and driver's age, this is not surprising in view of the empirical analysis in Figs. 5.3 and 5.4, and in Chap. 13.1. At the other end we have the `Area` code which only seems to improve the model marginally. However, this does not imply, yet, that this variable should be dropped. There are two points that need to be considered: (1) maybe feature pre-processing of `Area` has not been done in an appropriate way and the variable is not in the right functional form for the chosen link function; and (2) `Area` is the last variable included in the model in Listing 5.4 and, maybe, there are already other variables

Listing 5.4 ANOVA table of model Poisson GLM1

```
Analysis of Deviance Table

Model: poisson, link: log

Response: ClaimNb

Terms added sequentially (first to last)


              Df Deviance Resid. Df Resid. Dev
NULL                         610205     153852
VehPowerGLM    5     73.7    610200     153779
VehAgeGLM      2    179.7    610198     153599
DrivAgeGLM     6   1199.4    610192     152400
BonusMalusGLM  1   4300.6    610191     148099
VehBrand      10    240.3    610181     147859
VehGas         1     82.4    610180     147776
DensityGLM     1    512.1    610179     147264
Region        21    191.3    610158     147073
AreaGLM        1      4.1    610157     147069
```

that take over the role of `Area` in smaller models which is possible if we have correlations between the feature components. In our data, `Area` and `Density` are highly correlated. For this reason, we exchange the order of these two components and run the same analysis again, we call this model Poisson GLM1B (which of course provides the same predictive model as Poisson GLM1).

Listing 5.5 ANOVA table of model Poisson GLM1B

```
Analysis of Deviance Table

Model: poisson, link: log

Response: ClaimNb

Terms added sequentially (first to last)


              Df Deviance Resid. Df Resid. Dev
NULL                         610205     153852
VehPowerGLM    5     73.7    610200     153779
VehAgeGLM      2    179.7    610198     153599
DrivAgeGLM     6   1199.4    610192     152400
BonusMalusGLM  1   4300.6    610191     148099
VehBrand      10    240.3    610181     147859
VehGas         1     82.4    610180     147776
AreaGLM        1    505.0    610179     147271
Region        21    192.4    610158     147079
DensityGLM     1     10.1    610157     147069
```

Listing 5.5 shows the ANOVA table if we exchange the order of these two variables. We observe that the magnitudes of the decrease of the deviance loss has switched between the two variables. Overall, `Density` seems slightly more predictive, and we may consider dropping `Area` from the model, also because the correlation between `Density` and `Area` is very high.

If we want to perform backward elimination (sequentially drop one variable after the other) we can use the R command `drop1`. For small models this is doable, for larger models it is computationally demanding.

Listing 5.6 `drop1` analysis of model Poisson GLM1

```
Single term deletions

Model:
ClaimNb ~ VehPowerGLM + VehAgeGLM + DrivAgeGLM + BonusMalusGLM +
    VehBrand + VehGas + DensityGLM + Region + AreaGLM
              Df Deviance    AIC    LRT  Pr(>Chi)
<none>             147069 192818
VehPowerGLM    5   147152 192892   83.4 < 2.2e-16 ***
VehAgeGLM      2   147283 193028  214.1 < 2.2e-16 ***
DrivAgeGLM     6   147603 193341  534.5 < 2.2e-16 ***
BonusMalusGLM  1   150970 196718 3901.5 < 2.2e-16 ***
VehBrand      10   147298 193027  228.9 < 2.2e-16 ***
VehGas         1   147213 192961  144.5 < 2.2e-16 ***
DensityGLM     1   147079 192826   10.1  0.001459 **
Region        21   147259 192967  190.7 < 2.2e-16 ***
AreaGLM        1   147073 192820    4.1  0.042180 *
---
Signif. codes:  0 '***' 0.001 '**' 0.01 '*' 0.05 '.' 0.1 ' ' 1
```

In Listing 5.6 we present the results of this `drop1` analysis. Both, according to AIC and according to the LRT, we should keep all variables in the model. Again, `Area` and `Density` provide the smallest LRT statistics Λ which illustrates the high collinearity between these two variables (note that the values in Listing 5.6 are identical to the ones in Listings 5.4 and 5.5, respectively).

We conclude that in model Poisson GLM1 we should keep all feature components, and a model improvement can only be obtained by a different feature pre-processing, by a different regression function or by a different distributional model. ■

5.3.4 Lab: Poisson GLM for Car Insurance Frequencies, Revisited

Continuous Coding of Non-monotone Feature Components

We revisit model Poisson GLM1 studied in Sect. 5.2.4 for MTPL claim frequency modeling, and we consider additional competing models by using different feature pre-processing. From Example 5.19, above, we conclude that we should keep all variables in the model if we work with model Poisson GLM1.

Table 5.4 Contingency table of observed number of policies against predicted number of policies with given claim counts `ClaimNb`

	Numbers of claims `ClaimNb`					
	0	1	2	3	4	5
Observed number of policies	587'772	21'198	1'174	57	4	1
Predicted number of policies	587'325	22'064	779	34	3	0.3

We calculate Pearson's dispersion estimate which provides $\widehat{\varphi}^{\mathrm{P}} = 1.6697 > 1$. This indicates that the model is not fully suitable for our data because in a Poisson model the dispersion parameter should be equal to 1. There may be two reasons for this over-dispersion: (1) the Poisson assumption is not appropriate because, for instance, the tail of the observations is more heavy-tailed, or (2) the Poisson assumption is appropriate but the regression function has not been chosen in a fully suitable way (maybe also due to missing feature information).

We believe that in our example the observed over-dispersion is a mixture of the two reasons (1) and (2). Surely, the regression structure can be improved since our feature pre-processing is non-optimal and since the chosen regression function only considers multiplicative interactions between the feature components (we have chosen the log-link regression function without adding interaction terms to the regression function).

Table 5.4 gives a contingency table. We observe that we have much more policies with more than 1 claim compared to what is predicted by the fitted model. As a result, a χ^2-test rejects this Poisson model because the resulting p-value is close to 0.

In our data, we have a rather large number of policies with short exposures v_i, and further analysis suggests that these short exposures are not suitably modeled. We will not invest more time into improving the exposure modeling. As mentioned in the appendix, there seem to be a couple of issues how the exposures are displayed and how policy renewals are accounted for in this data. However, it is difficult (almost impossible) to clean the data for better exposure measures without more detailed information about the data collection process.

Our next aim is to model continuous feature components differently, if their raw form does not match the linear predictor assumption. In Poisson GLM1 we have categorized such components and then used dummy coding for the resulting classes, see Sect. 5.2.4. Alternatively, we can use different functional forms, for instance, we can use for `DrivAge` the following pre-processing

$$\texttt{DrivAge} \mapsto \beta_l\, \texttt{DrivAge} + \beta_{l+1} \log(\texttt{DrivAge}) + \sum_{j=2}^{4} \beta_{l+j} (\texttt{DrivAge})^j. \tag{5.34}$$

Table 5.5 Run times, number of parameters, AICs, in-sample and out-of-sample deviance losses, tenfold cross-validation losses (units are in 10^{-2}) and in-sample average frequency of the null model (intercept model) and of different Poisson GLMs

	Run time	# Param.	AIC	In-sample loss on $\mathcal{L}$	Out-of-sample loss on $\mathcal{T}$	Tenfold CV loss $\widehat{\mathfrak{D}}^{\text{CV}}$	Aver. freq.
Poisson null	–	1	199'506	25.213	25.445	25.213	7.36%
Poisson GLM1	16s	49	192'818	24.101	24.146	24.121	7.36%
Poisson GLM2	15s	48	192'753	24.091	24.113	24.110	7.36%
Poisson GLM3	15s	50	192'716	24.084	24.102	24.104	7.36%

This replaces the $K = 7$ categorical age classes of model Poisson GLM1 by 5 continuous functions of the variable `DrivAge`, and the number of regression parameters is reduced from $K - 1 = 6$ to 5. We call this model Poisson GLM2.

Besides improving the modeling of the feature components we can also start to add interactions beyond the multiplicative ones. For instance, Fig. 13.12 in Chap. 13 may indicate that there is an interaction term between `BonusMalus` and `DrivAge`. New young drivers enter the bonus-malus system at level 100, and it takes some years free of accidents to reach the lowest bonus-malus level of 50. Whereas for senior drivers a bonus-malus level of 100 may indicate that they have had a bad claim experience because otherwise they would be on the lowest bonus-malus level, see also Remark 5.15. We are adding the following interaction to Poisson GLM2 and we call the resulting model Poisson GLM3

$$\beta_{l'}\,\texttt{BonusMalus} \cdot \texttt{DrivAge} + \beta_{l'+1}\texttt{BonusMalus} \cdot (\texttt{DrivAge})^2. \tag{5.35}$$

From Table 5.5 we observe that this leads to a further small model improvement. We mention that this model improvement can also be observed in a decrease of Pearson's dispersion estimate to $\widehat{\varphi}^P = 1.6644$. Noteworthy, all model selection criteria AIC, out-of-sample generalization loss and cross-validation come to the same conclusion in this example.

The tedious task of the modeler now is to find all these systematic effects and bring them in an appropriate form into the model. Here, this is still possible because we have a comparably small model. However, if we have hundreds of feature components, such a manual analysis becomes intractable. Other regression models such as network regression models should be preferred, or at least should be used to find systematic effects. But, one should also keep in mind that the (final) chosen model should be as simple as possible (parsimonious).

Remarks 5.20

- An advantage of GLMs is that these regression models can deal with collinearity in feature components. Nevertheless, the results should be carefully checked if the collinearity in feature components is very high. If we have a high collinearity between two feature components then we may observe large values with opposite signs in the corresponding regression parameters compensating each other. The

Listing 5.7 `drop1` analysis of model Poisson GLM2

```
Single term deletions

Model:
ClaimNb ~ VehPowerGLM + VehAgeGLM + DrivAge + log(DrivAge) +
    I(DrivAge^2) + I(DrivAge^3) + I(DrivAge^4) + BonusMalusGLM +
    VehBrand + VehGas + DensityGLM + Region + AreaGLM
              Df Deviance    AIC    LRT  Pr(>Chi)
<none>             147005 192753
VehPowerGLM    5   147087 192825   82.4 2.671e-16 ***
VehAgeGLM      2   147225 192969  220.3 < 2.2e-16 ***
DrivAge        1   147157 192902  151.9 < 2.2e-16 ***
log(DrivAge)   1   147190 192935  184.8 < 2.2e-16 ***
I(DrivAge^2)   1   147123 192869  118.1 < 2.2e-16 ***
I(DrivAge^3)   1   147094 192840   89.0 < 2.2e-16 ***
I(DrivAge^4)   1   147071 192816   65.5 5.687e-16 ***
BonusMalusGLM  1   150907 196653 3902.0 < 2.2e-16 ***
VehBrand      10   147232 192959  226.5 < 2.2e-16 ***
VehGas         1   147148 192893  142.8 < 2.2e-16 ***
DensityGLM     1   147015 192761   10.1  0.001498 **
Region        21   147193 192899  188.0 < 2.2e-16 ***
AreaGLM        1   147009 192755    4.1  0.043123 *
---
Signif. codes:  0 '***' 0.001 '**' 0.01 '*' 0.05 '.' 0.1 ' ' 1
```

resulting GLM will not be very robust, and a slight change in the observations may change these regression parameters completely. In this case one should drop one of the two highly collinear feature components. This problem may also occur if we include too many terms in functional forms like in (5.34).

- A tool to find suitable functional forms of regression functions in continuous feature components are the partial residual plots of Cook–Croos-Dabrera [80]. If we want to analyze the first feature component x_1 of $\boldsymbol{x}$, we can fit a GLM to the data using the entire feature vector $\boldsymbol{x}$. The partial residuals for component x_1 are defined by, see formula (8) in Cook–Croos-Dabrera [80],

$$r_i^{\text{partial}} = (Y_i - \mu(\boldsymbol{x}_i))g'(\mu(\boldsymbol{x}_i)) + \beta_1 x_{i,1} \qquad \text{for } 1 \le i \le n,$$

where g is the chosen link function and $g(\mu(\boldsymbol{x}_i)) = \langle \boldsymbol{\beta}, \boldsymbol{x}_i \rangle$. These partial residuals offset the effect of feature component x_1. The partial residual plot shows r_i^{partial} against $x_{i,1}$. If this plot shows a linear structure then including x_1 linearly is justified, and any other functional form may be detected from that plot.

Under-Sampling and Over-Sampling

Often run times are an issue in model fitting, in particular, if we want to experiment with different models, different feature codings, etc. Under-sampling is an interesting approach that can be applied in imbalanced situations (like in our claim frequency data situation) to speed up calculations, and still receiving accurate approximations. We briefly describe under-sampling in this subsection.

Under-sampling is based on the idea that we do not need to consider all $n = 610'206$ insurance policies for model fitting, and we can still receive accurate results. For this we select all insurance policies that have at least 1 claim; in our data these are 22'434 insurance policies, we call this data set $\mathcal{L}^*_{\geq 1}$. The motivation for selecting these insurance policies is that these are exactly the policies that have information about the drivers causing claims. These selected insurance policies need to be complemented with policies that do not cause any claims. We select at random (under-sample) 22'434 insurance policies of drivers without claims, we call this data set $\mathcal{L}^*_0$. Merging the two sets we receive data $\mathcal{L}^* = \mathcal{L}^*_0 \cup \mathcal{L}^*_{\geq 1}$ comprising 44'868 insurance policies. This data is balanced from the viewpoint of claim causing policies because exactly half of the policies in $\mathcal{L}^*$ suffers a claim and the other half does not. The idea now is to fit a GLM only on this learning data $\mathcal{L}^*$, and because we only consider 44'868 insurance policies the fitting should be fast.

There is still one point to be considered, namely, in the new learning data $\mathcal{L}^*$ policies with claims are over-represented (because we work in a low frequency problem). This motivates that we adjust the time exposures v_i in $\mathcal{L}^*_0$ accordingly by multiplying as follows

$$v_i \mapsto v_i^* = v_i \frac{\sum_{j=1}^n v_j \mathbb{1}_{\{N_j=0\}}}{\sum_{v_j \in \mathcal{L}^*_0} v_j}.$$

Thus, we stretch the exposures of the policies without claims in $\mathcal{L}^*$; for our data this factor is 26.17. This then provides us with an empirical frequency on $\mathcal{L}^*$ of 7.36% which is identical to the observed frequency on the entire learning data $\mathcal{L}$.

We fit model Poisson GLM3 on this reduced (and exposure adjusted) learning data $\mathcal{L}^*$, the results are presented on the last line of Table 5.6. This model can be fitted in 1s, and by construction it fulfills the balance property. The resulting in-sample and out-of-sample losses (evaluated on the entire data $\mathcal{L}$ and $\mathcal{T}$) are very close to model Poisson GLM3 which verifies that the model fitted only on the learning data $\mathcal{L}^*$ gives a good approximation. We do not provide AIC because the data used is not identical to the data used to fit the other models. The tenfold cross-

Table 5.6 Run times, number of parameters, AICs, in-sample and out-of-sample deviance losses, tenfold cross-validation losses (units are in 10^{-2}) and in-sample average frequency of the null model (intercept model) and of different Poisson GLMs, the last row uses under-sampling in model Poisson GLM3

	Run time	# param.	AIC	In-sample loss on $\mathcal{L}$	Out-of-sample loss on $\mathcal{T}$	Tenfold CV loss $\widehat{\mathfrak{D}}^{\text{CV}}$	Aver. freq.
Poisson null	–	1	199'506	25.213	25.445	25.213	7.36%
Poisson GLM1	16 s	49	192'818	24.101	24.146	24.121	7.36%
Poisson GLM2	15 s	48	192'753	24.091	24.113	24.110	7.36%
Poisson GLM3	15 s	50	192'716	24.084	24.102	24.104	7.36%
under-sampling	1 s	50	–	24.098	24.108	24.120	7.36%

validation loss is a little bit bigger which seems to be a consequence of applying the non-stratified version to only 44'868 insurance policies, i.e., this higher cross-validation loss shows that we fit the model on less data which provides higher uncertainty in model fitting. This finishes this example.

The presented method is called under-sampling because we under-sample from the insurance policies without claims to make both classes (policies with claims and policies without claims) equally large. Alternatively, to achieve a class balance we could also over-sample from the minority class by duplicating policies. This has a similar effect, but it increases run times. Importantly, if we under- or over-sample we *have* to adjust the exposures correspondingly. Otherwise we obtain a biased model that is not useful for pricing, the same applies to methods such as the synthetic minority oversampling technique (SMOTE) and similar techniques.

Alternatively, to under-sampling we could also fit a so-called zero-truncated Poisson (ZTP) model to the data by only fitting a model on the insurance policies that suffer at least one claim, and adjusting the distribution to the observations $N_i|_{\{N_i \geq 1\}}$. This is rather similar to a hurdle Poisson model and we come back to this in Example 6.19, below.

5.3.5 Over-Dispersion in Claim Counts Modeling

Mixed Poisson Distribution

In the previous example we have seen that the considered Poisson GLMs do not fully fit our data, at least not with the chosen feature engineering, because there is over-dispersion in the data (relative to the chosen models). This may give rise to consider models that allow for over-dispersion. Typically, such over-dispersed models are constructed starting from the Poisson model, because the Poisson model enjoys many nice properties as we have seen above. A natural extension is to introduce the family of mixed Poisson models, where the frequency is not modeled with a single parameter but rather with a whole family of parameters described by an underlying mixing distribution.

In the dual mean parametrization the Poisson distribution for $Y = N/v$ reads as

$$Y \sim f(y; \lambda, v) = e^{-v\lambda} \frac{(v\lambda)^{vy}}{(vy)!} \qquad \text{for } y \in \mathbb{N}_0/v,$$

where the mean parameter is given by $\lambda = \kappa'(\theta) = \exp\{\theta\}$, and θ denotes the canonical parameter; on purpose we use for the mean notation λ instead of μ, here, the reason will become clear below. This model satisfies for the first two moments of $N = vY$

$$\mathbb{E}_\lambda [N] = v\kappa'(\theta) = v\lambda \qquad \text{and} \qquad \text{Var}_\lambda (N) = v\kappa''(\theta) = v\lambda = \mathbb{E}_\lambda [N],$$

with dispersion parameter $\varphi = 1$. A mixed Poisson distribution is obtained by mixing/integrating over different frequency parameters $\lambda > 0$. We choose a distribution π on $\mathbb{R}_+$ (strictly positively supported), and define the new distribution

$$Y = N/v \sim f_\pi(y; v) = \int_{\mathbb{R}_+} f(y; \lambda, v)\, d\pi(\lambda) = \int_{\mathbb{R}_+} e^{-v\lambda} \frac{(v\lambda)^{vy}}{(vy)!}\, d\pi(\lambda). \tag{5.36}$$

If π is not concentrated in a single point, the tower property immediately implies

$$\mathbb{E}_\pi [N] < \mathrm{Var}_\pi (N), \tag{5.37}$$

supposed that the moments exist, we refer to Lemma 2.18 in Wüthrich [387]. Hence, mixing over different frequency parameters allows us to receive over-dispersion. Of course, this concept can also be applied to mixing over the canonical parameter θ in the EF (instead of the mean parameter).

This leads to the framework of Bayesian credibility models which are widely used and studied in actuarial science, we refer to the textbook of Bühlmann–Gisler [58]. We have already met this idea in the Bayesian decision rule of Example 3.3 which has led to the Bayesian estimator in Definition 3.6.

Negative-Binomial Model

In the case of the Poisson model, the gamma distribution is a particularly attractive mixing distribution for λ because it allows for a closed-form solution in (5.36), and $f_{\pi=\Gamma}(y; v)$ will be a negative-binomial distribution.[4] One can choose different parametrizations of this mixing distribution, and they will provide different scalings in the resulting negative-binomial distribution. We choose the following parametrization $\pi(\lambda) \stackrel{(d)}{=} \Gamma(v\alpha, v\alpha/\mu)$ for mean parameter $\mu > 0$ and shape parameter $v\alpha > 0$. This implies, see (5.36),

$$\begin{aligned} f_{\mathrm{NB}}(y; \mu, v, \alpha) &= \int_{\mathbb{R}_+} e^{-v\lambda} \frac{(v\lambda)^{vy}}{(vy)!} \frac{(v\alpha/\mu)^{v\alpha}}{\Gamma(v\alpha)} \lambda^{v\alpha-1} e^{-v\alpha\lambda/\mu} d\lambda \\ &= \frac{\Gamma(vy + v\alpha)}{(vy)!\Gamma(v\alpha)} \frac{v^{vy}(v\alpha/\mu)^{v\alpha}}{(v + v\alpha/\mu)^{vy+v\alpha}} \\ &= \binom{vy + v\alpha - 1}{vy} \left(e^\theta\right)^{vy} \left(1 - e^\theta\right)^{v\alpha}, \end{aligned}$$

[4] The gamma distribution is the conjugate prior to the Poisson distribution. As a result, the posterior distribution, given observations, will again be a gamma distribution with posterior parameters, see Section 8.1 of Wüthrich [387]. This Bayesian model has been introduced to the actuarial literature by Bichsel [32].

setting for canonical parameter $\theta = \log(\mu/(\mu+\alpha)) < 0$. This is the negative-binomial distribution we have already met in (2.5). A single-parameter linear EDF representation is given by, we set unit dispersion parameter $\varphi = 1$,

$$Y \sim f_{\mathrm{NB}}(y;\theta,v,\alpha) = \exp\left\{\frac{y\theta + \alpha\log(1-e^{\theta})}{1/v} + \log\binom{vy+v\alpha-1}{vy}\right\}, \tag{5.38}$$

where this is a density w.r.t. the counting measure on $\mathbb{N}_0/v$. The cumulant function and the canonical link, respectively, are given by

$$\kappa(\theta) = -\alpha\log(1-e^{\theta}) \quad \text{and} \quad \theta = h(\mu) = \log\left(\frac{\mu}{\mu+\alpha}\right) \in \boldsymbol{\Theta} = (-\infty, 0).$$

Note that $\alpha > 0$ is treated as nuisance parameter (which is a fixed part of the cumulant function, here). The first two moments of the claim count $N = vY$ are given by

$$v\mu = \mathbb{E}_{\theta}[N] = v\alpha\frac{e^{\theta}}{1-e^{\theta}}, \tag{5.39}$$

$$\mathrm{Var}_{\theta}(N) = \mathbb{E}_{\theta}[N]\left(1+\frac{e^{\theta}}{1-e^{\theta}}\right) = \mathbb{E}_{\theta}[N]\left(1+\frac{\mu}{\alpha}\right) > \mathbb{E}_{\theta}[N]. \tag{5.40}$$

This shows that we receive a fixed over-dispersion of size μ/α, which (in this parametrization) does not depend on the exposure v; this is the reason for choosing a mixing distribution $\pi(\lambda) \stackrel{(d)}{=} \Gamma(v\alpha, v\alpha/\mu)$. This parametrization is called NB2 parametrization.

Remarks 5.21

- We emphasize that the effective domain $\boldsymbol{\Theta} = (-\infty, 0)$ is one-sided bounded. Therefore, the canonical link for the linear predictor will not work in general because the linear predictor $\boldsymbol{x} \mapsto \eta(\boldsymbol{x})$ can be both-sided unbounded in a GLM setting. Instead, we use the log-link for $g(\cdot)$ in our example below, with the downside that one loses the balance property.
- The unit deviance in this negative-binomial EDF model is given by

$$(y,\mu) \mapsto \mathfrak{d}(y,\mu) = 2\left[y\log\left(\frac{y}{\mu}\right) - (y+\alpha)\log\left(\frac{y+\alpha}{\mu+\alpha}\right)\right],$$

  we also refer to Table 4.1 for $\alpha = 1$. We emphasize that this is the unit deviance in a single-parameter linear EDF, and we only aim at estimating canonical parameter $\theta \in \boldsymbol{\Theta}$ and mean parameter $\mu \in \mathcal{M}$, respectively, whereas $\alpha > 0$ is treated as a given nuisance parameter. This is important because the unit deviance relies on the saturated model which, in general, estimates a one-dimensional parameter θ and μ, respectively, from the one-dimensional observation Y. The nuisance parameter is not affected by the consideration of the saturated model, and it is treated as a fixed part of the cumulant function, which is not estimated at this stage. An important consequence of this is that model comparison using deviance residuals only works for identical nuisance parameters.

- We mention that we receive over-dispersion in (5.40) though we have dispersion parameter $\varphi = 1$ in (5.38). Alternatively, we could do the duality transformation $y \mapsto \widetilde{y} = y/\alpha$ for nuisance parameter $\alpha > 0$; this gives the reproductive form of the negative-binomial model NB2, see also Remarks 2.13. This provides us with a density on $\mathbb{N}_0/(v\alpha)$, set $\widetilde{\varphi} = 1/\alpha$,

$$\widetilde{Y} \sim f_{\mathrm{NB}}(\widetilde{y}; \theta, v/\widetilde{\varphi}) = \exp\left\{ \frac{\widetilde{y}\theta + \log(1 - e^{\theta})}{1/(v\alpha)} + \log\binom{v\alpha\widetilde{y} + v\alpha - 1}{v\alpha\widetilde{y}} \right\}.$$

The cumulant function and the canonical link, respectively, are now given by

$$\kappa(\theta) = -\log(1 - e^{\theta}) \quad \text{and} \quad \theta = h(\widetilde{\mu}) = \log\left(\frac{\widetilde{\mu}}{\widetilde{\mu} + 1}\right) \in \boldsymbol{\Theta} = (-\infty, 0).$$

The first two moments are for $\theta \in \boldsymbol{\Theta}$ given by

$$\widetilde{\mu} = \mathbb{E}_\theta[\widetilde{Y}] = \frac{e^{\theta}}{1 - e^{\theta}},$$

$$\mathrm{Var}_\theta(\widetilde{Y}) = \frac{\widetilde{\varphi}}{v}\kappa''(\theta) = \frac{1}{v\alpha}\widetilde{\mu}(1 + \widetilde{\mu}).$$

Thus, we receive the reproductive EDF representation with dispersion parameter $\widetilde{\varphi} = 1/\alpha$ and variance function $V(\widetilde{\mu}) = \widetilde{\mu}(1 + \widetilde{\mu})$. Moreover, $N = vY = v\alpha\widetilde{Y}$.

- The negative-binomial model with the NB1 parametrization uses the mixing distribution $\pi(\lambda) \stackrel{(d)}{=} \Gamma(\mu v/\alpha, v/\alpha)$. This leads to mean $\mathbb{E}_\theta[N] = v\mu$ and variance $\mathrm{Var}_\theta(N) = \mathbb{E}_\theta[N](1 + \alpha)$. In this parametrization, μ enters the gamma function as $\Gamma(\mu v/\alpha)$ in the gamma density which does not allow for an EDF representation. This parametrization has been called NB1 by Cameron–Trivedi [63] because both terms in the variance $\mathrm{Var}_\theta(N) = v\mu + v\mu\alpha$ are linear in μ. In contrast, in the NB2 parametrization the second term has a square $v\mu^2/\alpha$ in μ, see (5.40). Further discussion is provided in Greene [171].

Nuisance Parameter Estimation

All previous statements have been based on the assumption that $\alpha > 0$ is a *given* nuisance parameter. If α needs to be estimated too, then, we drop out of the EF. In this case, an iterative estimation procedure is applied to the EDF representation (5.38). One starts with a fixed nuisance parameter $\alpha^{(0)}$ and fits the negative-binomial GLM with MLE which provides a first set of MLE $\widehat{\boldsymbol{\beta}}^{(1)} = \widehat{\boldsymbol{\beta}}^{(1)}(\alpha^{(0)})$. Based on this estimate the nuisance parameter is updated $\alpha^{(0)} \mapsto \alpha^{(1)}$ by maximizing the log-likelihood in α for given $\widehat{\boldsymbol{\beta}}^{(1)}$. Iteration of this procedure then leads to a joint estimation of regression parameter $\boldsymbol{\beta}$ and nuisance parameter α. Both MLE steps in this algorithm increase the joint log-likelihood.

Remark 5.22 (Implementation of the Negative-Binomial GLM in R) Implementation of the negative-binomial model needs some care. There are two R procedures `glm` and `glm.nb` that can be used to fit negative-binomial GLMs, the latter being built on the former. The procedure `glm` is just the classical R procedure [307] that is usually used to fit GLMs within the EDF, it requires to set

```
family=negative.binomial(theta, link="log").
```

This parametrization considers the single-parameter linear EF on $\mathbb{N}$ (for mean $\mu \in \mathcal{M}$)

$$f_{\mathrm{NB}}(n; \mu, \mathtt{theta}) = \binom{n + \mathtt{theta} - 1}{n} \left(\frac{\mu}{\mu + \mathtt{theta}} \right)^{n} \left(1 - \frac{\mu}{\mu + \mathtt{theta}} \right)^{\mathtt{theta}},$$

where $\mathtt{theta} > 0$ denotes the nuisance parameter. The tricky part now is that we have to bring in the different exposures v_i of all policies $1 \leq i \leq n$. That is, we would like to have for claim counts $n_i = v_i y_i$, see (5.38),

$$\begin{aligned} f_{\mathrm{NB}}(y_i; \mu_i, v_i, \alpha) &= \binom{v_i y_i + v_i\alpha - 1}{v_i y_i} \left(\frac{v_i \mu_i}{v_i \mu_i + v_i \alpha} \right)^{v_i y_i} \left(1 - \frac{v_i \mu_i}{v_i \mu_i + v_i \alpha} \right)^{v_i \alpha} \\ &= \binom{v_i y_i + v_i\alpha - 1}{v_i y_i} \left[\left(\frac{\mu_i}{\mu_i + \alpha} \right)^{y_i} \left(1 - \frac{\mu_i}{\mu_i + \alpha} \right)^{\alpha} \right]^{v_i}. \end{aligned}$$

The square bracket can be implemented in `glm` as a scaled and weighted regression problem, see Listing 5.8 with $\mathtt{theta} = \alpha$. This approach provides the correct GLM parameter estimates $\widehat{\boldsymbol{\beta}}^{\mathrm{MLE}}$ for given α, however, the outputted AIC values cannot be compared to the Poisson case. Note that the Poisson case of Table 5.5 considers observations N_i whereas Listing 5.8 uses $Y_i = N_i/v_i$. For this reason we calculate the log-likelihood and AIC by an own implementation.

The same remark applies to `glm.nb`, and also nuisance parameter estimation cannot be performed by that routine under different exposures v_i. Therefore, we have implemented an iterative estimation algorithm ourselves, alternating `glm` of Listing 5.8 for given α and a maximization routine `optimize` to find the optimal α for given $\boldsymbol{\beta}$ using (5.38). We have applied this iteration in Example 5.23, below, and it has converged in 5 iterations.

Example 5.23 (Negative-Binomial Distribution for Claim Counts) We revisit the MTPL claim frequency GLM example of Sect. 5.3.4, but we replace the Poisson distribution by the negative-binomial one. We start with the negative-binomial (NB)

Listing 5.8 Implementation of model NB GLM3

```
d.glmnb <- glm(ClaimNb/Exposure ~ VehPowerGLM + VehAgeGLM
                    + log(DrivAge) + I(DrivAge^3) + I(DrivAge^4)
                    + BonusMalusGLM*DrivAge + BonusMalusGLM*I(DrivAge^2)
                    + VehBrand + VehGas + DensityGLM + Region + AreaGLM,
                    data=learn, weights=Exposure,
                    family=negative.binomial(alpha, link="log"))
```

Table 5.7 Run times, number of parameters, AICs, in-sample and out-of-sample deviance losses (units are in 10^{-2}) and in-sample average frequency of the null models (Poisson and negative-binomial) and the Poisson and negative-binomial GLMs. The optimal model is highlighted in boldface

	Run time	# Param.	AIC	In-sample loss on $\mathcal{L}$	Out-of-sample loss on $\mathcal{T}$	Aver. freq.
Poisson null	–	1	199'506	25.213	25.445	7.36%
Poisson GLM3	15 s	50	192'716	24.084	24.102	7.36%
NB null $\widehat{\alpha}^{\rm MLE}_{\rm null} = 1.059$	–	2	198'466	20.357	20.489	7.36%
NB null $\widehat{\alpha}^{\rm MLE}_{\rm NB} = 1.810$	–	1	198'564	21.796	21.948	7.36%
NB GLM3 $\widehat{\alpha}^{\rm MLE}_{\rm NB} = 1.810$	85s	51	**192'113**	20.722	20.674	7.38%

null model. The NB null model has two parameters, the homogeneous (overall) frequency and the nuisance parameter. MLE of the homogeneous overall frequency is identical to the one in the Poisson null model, and MLE of the nuisance parameter provides $\widehat{\alpha}^{\rm MLE}_{\rm null} = 1.059$. This is substantially smaller than infinity and suggests over-dispersion. The results are presented on the third line of Table 5.7. We observe a smaller AIC of the NB null model against the Poisson null model which says that we should allow for over-dispersion.

We now focus on the NB GLM. The feature pre-processing is done exactly as in model Poisson GLM3, and we choose the log-link for g. We call this model NB GLM3. The iterative estimation procedure outlined above provides a nuisance parameter estimate $\widehat{\alpha}^{\rm MLE}_{\rm NB} = 1.810$. This is bigger than in the NB null model because the regression structure explains some part of the over-dispersion, however, it is still substantially smaller than infinity which justifies the inclusion of this over-dispersion parameter.

The last line of Table 5.7 gives the result of model NB GLM3. From AIC we conclude that we favor the negative-binomial GLM over the Poisson GLM since AIC decreases from 192'716 to 192'113. The in-sample and out-of-sample deviance losses can only be compared within the same models, i.e., the models that have the same cumulant function. This also applies to the negative-binomial models which have cumulant function $\kappa(\theta) = -\alpha \log(1 - e^{\theta})$. Thus, to compare the NB null model and model NB GLM3, we need to choose the same nuisance parameter α. For this reason we added this second NB null model to Table 5.7. This second NB null model no longer uses the MLE $\widehat{\alpha}^{\rm MLE}_{\rm null}$, therefore, the corresponding AIC only includes one estimated parameter.

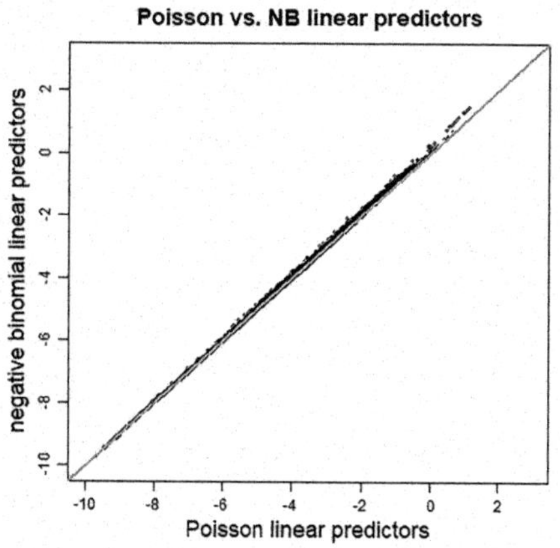

Fig. 5.7 Poisson logged predictors vs. negative-binomial logged predictors

Table 5.8 Out-of-sample deviance losses: forecast dominance. The optimal model is highlighted in boldface

Model	Poisson deviance	NB deviance $\widehat{\alpha}_{\text{null}}^{\text{MLE}} = 1.059$	NB deviance $\widehat{\alpha}_{\text{NB}}^{\text{MLE}} = 1.810$
Null model	25.445	20.489	21.948
Poisson GLM3	24.102	19.266	20.678
NB GLM3 $\widehat{\alpha}_{\text{NB}}^{\text{MLE}} = 1.810$	**24.100**	**19.262**	**20.674**

As mentioned above, deviance losses can only be compared under exactly the same cumulant function (including the same nuisance parameters). If we want to have a more robust model selection, we can consider forecast dominance according to Definition 4.20. Being less ambitious, here, we consider forecast dominance only for the three considered cumulant functions Poisson, negative-binomial with $\widehat{\alpha}_{\text{null}}^{\text{MLE}} = 1.059$ and negative-binomial with $\widehat{\alpha}_{\text{NB}}^{\text{MLE}} = 1.810$. The out-of-sample deviance losses are given in Table 5.8 in the different columns. According to this forecast dominance analysis we also give preference to model NB GLM3, but model Poisson GLM3 is pretty close.

Figure 5.7 compares the logged predictors $\log(\widehat{\mu}_i)$, $1 \leq i \leq n$, of the models Poisson GLM3 and NB GLM3. We see a huge similarity in these predictors, only high frequency policies are judged slightly differently by the NB model compared to the Poisson model.

Table 5.9 gives the predicted number of claims against the observed ones. We observe that model NB GLM3 predicts more accurately the number of policies with 2 or less claims, but it over-estimates the number of policies with more than 2 claims. This may also be related to the fact that the estimated in-sample frequency has a

Table 5.9 Contingency table of observed number of policies against predicted number of policies with given claim counts `ClaimNb`

	Numbers of claims `ClaimNb`					
	0	1	2	3	4	5
Observed number of policies	587'772	21'198	1'174	57	4	1
Poisson predicted number of policies	587'325	22'064	779	34	3	0.3
NB predicted number of policies	587'902	20'982	1'200	100	15	4

positive bias in model NB GLM3, see Table 5.7. That is, since we do not work with the canonical link, we do not have the balance property.

Listing 5.9 `drop1` analysis of model NB GLM3

```
Single term deletions

Model:
ClaimNb/Exposure ~ VehPowerGLM + VehAgeGLM + DrivAge + log(DrivAge) +
    I(DrivAge^2) + I(DrivAge^3) + I(DrivAge^4) + BonusMalusGLM *
    DrivAge + BonusMalusGLM * I(DrivAge^2) + BonusMalusGLM +
    VehBrand + VehGas + DensityGLM + Region + AreaGLM
                              Df Deviance    AIC scaled dev.  Pr(>Chi)
<none>                           126446 171064
VehPowerGLM                    5 126524 171102      48.266 3.134e-09 ***
VehAgeGLM                      2 126655 171190     130.070 < 2.2e-16 ***
log(DrivAge)                   1 126592 171153      91.057 < 2.2e-16 ***
I(DrivAge^3)                   1 126527 171112      50.483 1.202e-12 ***
I(DrivAge^4)                   1 126508 171100      38.381 5.820e-10 ***
VehBrand                      10 126658 171176     132.098 < 2.2e-16 ***
VehGas                         1 126583 171147      85.232 < 2.2e-16 ***
DensityGLM                     1 126456 171068       6.137   0.01324 *
Region                        21 126622 171132     109.838 5.042e-14 ***
AreaGLM                        1 126450 171064       2.411   0.12049
DrivAge:BonusMalusGLM          1 126484 171085      23.481 1.262e-06 ***
I(DrivAge^2):BonusMalusGLM     1 126490 171089      27.199 1.836e-07 ***
---
Signif. codes:  0 *** 0.001 ** 0.01 * 0.05 . 0.1   1
```

We close this example by providing the `drop1` analysis in Listing 5.9. From this analysis we conclude that the feature component `Area` should be dropped. Of course, this confirms the high collinearity between `Density` and `Area` which implies that we do not need both variables in the model. We remark that the AIC values in Listing 5.9 are not on our scale, as stated in Remark 5.22. ■

5.3.6 *Zero-Inflated Poisson Model*

In many applications it is the case that the Poisson distribution does not fully fit the claim counts data because there are too many policies with zero claims, i.e., policies with $Y = 0$, compared to a Poisson assumption. This topic has attracted some attention in the recent actuarial literature, see, e.g., Boucher et al. [43–45], Frees et al. [137], Calderín-Ojeda et al. [62] and Lee [239]. An obvious solution to this problem is to 'artificially' increase the probability of a zero claim compared to a Poisson model, this is the proposal introduced by Lambert [232]. Y has a zero-inflated Poisson (ZIP) distribution if the probability weights of Y are given by (set $v = 1$)

$$f_{\mathrm{ZIP}}(y; \theta, \pi_0) = \begin{cases} \pi_0 + (1 - \pi_0)e^{-\mu} & \text{for } y = 0, \\ (1 - \pi_0)e^{-\mu}\frac{\mu^y}{y!} & \text{for } y \in \mathbb{N}, \end{cases}$$

for $\pi_0 \in (0, 1)$, $\mu = e^{\theta} > 0$, and for the Poisson probability weights we refer to (2.4). For $\pi_0 > 0$ the weight of a zero claim $Y = 0$ is increased (inflated) compared to the original Poisson distribution.

Remarks 5.24

- The ZIP distribution has different interpretations. It can be interpreted as a hierarchical model where we have a latent variable Z which indicates with probability π_0 that we have an excess zero, and with probability $1 - \pi_0$ we have an ordinary Poisson distribution, i.e. for $y \in \mathbb{N}_0$

$$\mathbb{P}_{\theta}\left[Y = y \middle| Z = z\right] = \begin{cases} \mathbb{1}_{\{y=0\}} & \text{for } z = 0, \\ e^{-\mu}\frac{\mu^y}{y!} & \text{for } z = 1, \end{cases} \tag{5.41}$$

  with $\mathbb{P}[Z = 0] = 1 - \mathbb{P}[Z = 1] = \pi_0$.

  The latter shows that we can also understand it as a mixture of two distributions, namely, of the Poisson distribution and of a single point measure in $y = 0$ with mixing probability π_0. Mixture distributions are going to be discussed in Sect. 6.3.1, below. In this sense, we can also interpret the model as a mixed Poisson model with mixing distribution $\pi(\lambda)$ being a Bernoulli distribution taking values 0 and μ with probability π_0 and $1 - \pi_0$, respectively, see (5.36), and the former parameter $\lambda = 0$ leads to a degenerate Poisson model.
- We have introduced the ZIP model, but this approach is neither limited to the Poisson model nor the zeros. For instance, we could also consider an inflated negative-binomial model where both the zeros and the ones are inflated with probabilities $\pi_0, \pi_1 > 0$ such that $\pi_0 + \pi_1 < 1$.
- Hurdle models are an alternative way to model excess zeros. Hurdle models have been introduced by Cragg [83], and they also allow for too little zeros. A hurdle (Poisson) model mixes a lower-truncated (Poisson) count distribution with a point mass in zero

$$f_{\text{hurdle Poisson}}(y; \theta, \pi_0) = \begin{cases} \pi_0 & \text{for } y = 0, \\ (1 - \pi_0)\frac{e^{-\mu}\frac{\mu^y}{y!}}{1-e^{-\mu}} & \text{for } y \in \mathbb{N}, \end{cases} \tag{5.42}$$

for $\pi_0 \in (0, 1)$ and $\mu > 0$. For $\pi_0 > e^{-\mu}$ the weight of a zero claim is increased and for $\pi_0 < e^{-\mu}$ it is decreased. This distribution is called a hurdle distribution, because we first need to overcome the hurdle at zero to come to the Poisson model. Lower-truncated distributions are studied in Sect. 6.4, below, and mixture distributions are discussed in Sect. 6.3.1. In general, fitting lower-truncated distributions is challenging because the density and the distribution function should both have tractable forms to perform MLE for truncated distributions. The Expectation-Maximization (EM) algorithm is a useful tool to perform model fitting under truncation. We come back to the hurdle Poisson model in Example 6.19, below, and it is also closely related to the zero-truncated Poisson (ZTP) model discussed in Remarks 6.20.

The first two moments of a ZIP random variable $Y \sim f_{\text{ZIP}}(\cdot; \theta, \pi_0)$ are given by

$$\begin{aligned}\mathbb{E}_{\theta,\pi_0}[Y] &= (1-\pi_0)\mu,\\ \text{Var}_{\theta,\pi_0}(Y) &= (1-\pi_0)\mu + (\pi_0 - \pi_0^2)\mu^2 \;=\; \mathbb{E}_{\theta,\pi_0}[Y]\,(1+\pi_0\mu),\end{aligned}$$

these calculations easily follow with the latent variable Z interpretation from above. As a consequence, we receive an over-dispersed model with over-dispersion $\pi_0\mu$ (the latter also follows from the fact that we consider a mixed Poisson distribution with a Bernoulli mixing distribution having weights π_0 in 0 and $1-\pi_0$ in $\mu > 0$, see (5.37)).

Unfortunately, MLE does not allow for explicit solutions in this model. The score equations of $Y_i \overset{\text{i.i.d.}}{\sim} f_{\text{ZIP}}(\cdot; \theta, \pi_0)$ are given by

$$\begin{aligned}\nabla_{(\pi_0,\mu)}\ell_{\boldsymbol{Y}}(\pi_0,\mu) &= \nabla_{(\pi_0,\mu)}\sum_{i=1}^n \log\left(\pi_0 + (1-\pi_0)e^{-\mu}\right)\mathbb{1}_{\{Y_i=0\}}\\ &\quad + \nabla_{(\pi_0,\mu)}\sum_{i=1}^n \log\left((1-\pi_0)e^{-\mu}\frac{\mu^y}{y!}\right)\mathbb{1}_{\{Y_i>0\}} \;=\; 0.\end{aligned}$$

The R package `pscl` [401] has a function called `zeroinfl` which uses the general purpose optimizer `optim` to find the MLEs in the ZIP model. Alternatively, we could explore the EM algorithm for mixture distributions presented in Sect. 6.3, below.

In insurance applications, the ZIP application can be problematic if we have different exposures $v_i > 0$ for different insurance policies i. In the Poisson GLM case with canonical link choice we typically integrate the different exposures into the offset, see (5.27). However, it is not clear whether and how we should integrate the different exposures into the zero-inflation probability π_0. It seems natural to believe that shorter exposures should increase π_0, but the explicit functional form of this increase can be debated, some options are discussed in Section 5 of Lee [239].

Listing 5.10 Implementation of model ZIP GLM3

```
1 d.ZIP <- zeroinfl(ClaimNb ~ VehPowerGLM + VehAgeGLM
2                  + log(DrivAge) + I(DrivAge^3) + I(DrivAge^4)
3                  + BonusMalusGLM*DrivAge + BonusMalusGLM*I(DrivAge^2)
4                  + VehBrand + VehGas + DensityGLM + Region
5                  + AreaGLM | 1,
6                  data=learn, offset=log(Exposure), dist='poisson', link='logit',
7                  start=list(count=glm3$coefficients, zero=c(-0.4153)) )
```

Table 5.10 Run times, number of parameters, AICs, in-sample and out-of-sample deviance losses (units are in 10^{-2}) and in-sample average frequency of the null models (Poisson, negative-binomial and ZIP) and the Poisson, negative-binomial and ZIP GLMs. The optimal model is highlighted in boldface

	Run time	# Param.	AIC	In-sample loss on $\mathcal{L}$	Out-of-sample loss on $\mathcal{T}$	Aver. freq.
Poisson null	–	1	199'506	25.213	25.445	7.36%
Poisson GLM3	15 s	50	192'716	24.084	24.102	7.36%
NB null $\widehat{\alpha}^{\text{MLE}}_{\text{null}} = 1.059$	–	2	198'466	20.357	20.489	7.36%
NB null $\widehat{\alpha}^{\text{MLE}}_{\text{NB}} = 1.810$	–	1	198'564	21.796	21.948	7.36%
NB GLM3 $\widehat{\alpha}^{\text{MLE}}_{\text{NB}} = 1.810$	85 s	51	**192'113**	20.722	20.674	7.38%
ZIP null	20 s	2	198'638	–	–	7.43%
ZIP GLM3 (null π_0)	270 s	51	192'393	–	–	7.37%

In the following application, we simply choose π_0 independent of the exposures, but certainly this is not the best modeling choice.

Example 5.25 (ZIP Model for Claim Counts) We revisit the MTPL claim frequency example of Sect. 5.3.4, but this time we fit a ZIP model. For the Poisson part we use exactly the same GLM regression function as in model Poisson GLM3 and, in particular, we use for the different exposures v_i of the insurance policies the offset term $o_i = \log v_i$, see line 6 of Listing 5.10. This offset only acts on the Poisson part of the ZIP GLM. The zero-inflating probability π_0 is modeled with a logistic Bernoulli model, see Sect. 2.1.2. For computational reasons, we choose the null model for the Bernoulli part modeling the zero-inflation π_0. This is indicated by the "1" on line 5 of Listing 5.10. This 1 should be expanded if we also want to consider a regression model for the zero-inflating probability π_0 and, in particular, if we want to integrate an offset term for the exposure. We can set this term to `offset(f)`, where `f` is a suitable transformation of the exposure. Furthermore, successful calibration requires meaningful starting values, otherwise `zeroinfl` will not find the MLEs. We start the algorithm in the parameters of model Poisson GLM3, see line 7 of Listing 5.10. The results are presented in Table 5.10.

Firstly, we see that the run times are not fully competitive in this implementation, even if we choose the null model for the zero-inflating probability π_0, i.e., only

Table 5.11 Out-of-sample deviance losses: forecast dominance. The optimal model is highlighted in boldface

Model	Poisson deviance	NB deviance $\widehat{\alpha}^{\text{MLE}}_{\text{null}} = 1.059$	NB deviance $\widehat{\alpha}^{\text{MLE}}_{\text{NB}} = 1.810$
Null model	25.445	20.489	21.948
Poisson GLM3	24.102	19.266	20.678
NB GLM3 $\widehat{\alpha}^{\text{MLE}}_{\text{NB}} = 1.810$	**24.100**	**19.262**	**20.674**
ZIP null model	25.446	20.490	21.949
ZIP GLM3	24.103	19.267	20.679

Table 5.12 Contingency table of observed numbers of policies against predicted numbers of policies with given claim counts `ClaimNb`

	Numbers of claims `ClaimNb`					
	0	1	2	3	4	5
Observed number of policies	587'772	21'198	1'174	57	4	1
Poisson predicted number of policies	587'325	22'064	779	34	3	0.3
NB predicted number of policies	587'902	20'982	1'200	100	15	4
ZIP predicted number of policies	587'829	21'094	1'191	79	9	4

one intercept parameter is involved for determining π_0. Secondly, in this model we cannot calculate deviance losses because the saturated model has two parameters for each observation. Thirdly, the model does not satisfy the balance property though we work with the canonical links for the Poisson part and the Bernoulli part, however, this property gets lost under the combination of these two model parts.

Most interesting are the AIC values. We observe that the ZIP GLM improves the Poisson GLM, but it has a bigger AIC value than the negative-binomial GLM. From this we conclude that we give preference to the negative-binomial model in our case.

Considering forecast dominance according to Definition 4.20, but restricted to the three deviance losses studied in Example 5.23, we receive Table 5.11. Also this table gives preference to the negative-binomial GLM. However, if we consider the table of the observed numbers of policies against the predicted numbers of claims, see Table 5.12, we give preference to the ZIP GLM because it has the lowest χ^2-value, i.e., it reflects best (in-sample) our observations.

Figure 5.8 compares the resulting predictors on the log-scale. From this plot we conclude that in our example the predictors of the ZIP GLM are closer to the Poisson ones than the NB GLM predictors. In a next step, one could refine the zero-inflating probability π_0 modeling by integrating the exposure and further feature information. This would lead to a further model improvement. We refrain here from doing so and close this example; in Example 6.19, below, we study the hurdle Poisson model. ■

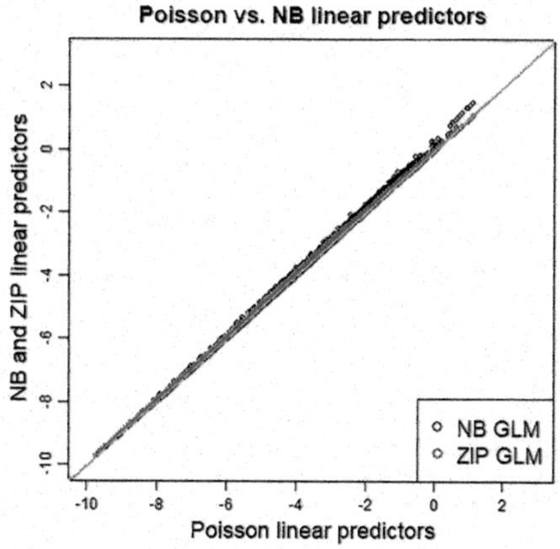

Fig. 5.8 Comparison linear predictors of the NB and ZIP GLMs against the ones of the Poisson GLM

5.3.7 Lab: Gamma GLM for Claim Sizes

As a second example we consider claim size modeling within GLMs. For this example we do not use the French MTPL claims data because the empirical density plot in Fig. 13.15 indicates that a GLM will not fit to that data. The French MTPL data seems to have three distinct modes, which suggests to use a mixture distribution. Moreover, the log-log plot indicates a regularly varying tail, which cannot be captured by the EDF on the original observation scale; we are going to study this data in Example 6.14, below. Here, we use the Swedish motorcycle data, previously used in the textbook of Ohlsson–Johansson [290] and described in Chap. 13.2. From Fig. 5.9 we see that the empirical density has one mode, and the log-log plot supports light tails, i.e., the gamma model might be a suitable choice for this data. Therefore, we choose a gamma GLM with log-link g. As described above, the log-link is not the canonical link for the gamma EDF distribution but it ensures the right sign w.r.t. the linear predictor $\eta_i = \langle \boldsymbol{\beta}, \boldsymbol{x}_i \rangle$. Working with the log-link in the gamma model will imply that the balance property is not fulfilled.

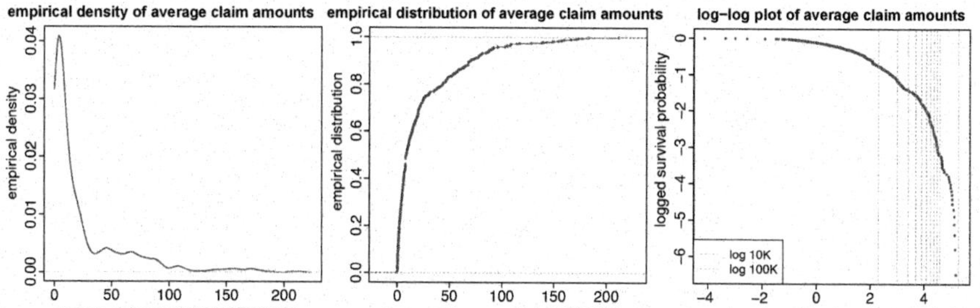

Fig. 5.9 (lhs) Empirical density, (middle) empirical distribution and (rhs) log-log plot of claim amounts of the Swedish motorcycle data presented in Chap. 13.2

Feature Engineering

We have 4 continuous feature components `OwnerAge`, `RiskClass`, `VehAge` and `BonusClass`, one binary feature component `Gender` and a categorical component `Area`, see Listing 13.4. We have decided for a minimal feature engineering; we refer to Figs. 13.19 (rhs) and 13.20 (rhs) for descriptive plots. We use the continuous variables directly in a log-linear fashion, we add quadratic terms for `OwnerAge` and `VehAge`, we merge `RiskClass` 6 and 7, and we censor `VehAge` at 20. `Area` is categorical, but we may interpret the `Zone` levels as ordinal categorical, and mapping them to integers allows us to use them in a continuous fashion; Fig. 13.19 (middle row, rhs) shows that this is a reasonable choice. Moreover, we merge `Zone` 5, 6 and 7 due to small volumes and their similar behavior.

Gamma Generalized Linear Model

The Swedish motorcycle claim amount data poses the special difficulty that we do not have individual claim observations $Z_{i,j}$, but we only know the total claim amounts $S_i = \sum_{j=1}^{N_i} Z_{i,j}$ and the number of claims N_i on each insurance policy; Fig. 5.9 shows average claims S_i/N_i of insurance policies i with $N_i > 0$. In general, this poses a problem in statistical modeling, but in the gamma model this problem can be handled because the gamma distribution is closed under aggregation of i.i.d. gamma claims $Z_{i,j}$. In all what follows in this section, we only study insurance policies with $N_i > 0$, and we label these insurance policies i accordingly.

Assume that $Z_{i,j}$ are i.i.d. gamma distributed with shape parameter α_i and scale parameter c_i, we refer to (2.6). The mean, the variance and the moment generating function of $Z_{i,j}$ are given by

$$\mathbb{E}[Z_{i,j}] = \frac{\alpha_i}{c_i}, \qquad \text{Var}(Z_{i,j}) = \frac{\alpha_i}{c_i^2} \qquad \text{and} \qquad M_{Z_{i,j}}(r) = \left(\frac{c_i}{c_i - r}\right)^{\alpha_i}, \tag{5.43}$$

where the moment generating function requires $r < c_i$ to be finite. Assuming that the number of claims N_i is a known positive integer $n_i \in \mathbb{N}$, we see from the moment generating function that $S_i = \sum_{j=1}^{n_i} Z_{i,j}$ is again gamma distributed with shape parameter $n_i\alpha_i$ and scale parameter c_i. We change the notation from N_i to n_i to emphasize that the number of claims is treated as a known constant (and also to avoid using the notation of conditional probabilities, here). Finally, we scale $Y_i = S_i/(n_i\alpha_i) \sim \Gamma(n_i\alpha_i, n_i\alpha_i c_i)$. This random variable Y_i has a single-parameter EDF gamma distribution with weight $v_i = n_i$, dispersion $\varphi_i = 1/\alpha_i$ and cumulant function $\kappa(\theta_i) = -\log(-\theta_i)$, for $\theta_i \in \boldsymbol{\Theta} = (-\infty, 0)$,

$$\begin{aligned} Y_i \sim f(y; \theta_i, v_i/\varphi_i) &= \exp\left\{\frac{y\theta_i - \kappa(\theta_i)}{\varphi_i/v_i} + a(y; v_i/\varphi_i)\right\} \\ &= \frac{(-\theta_i\alpha_i v_i)^{v_i\alpha_i}}{\Gamma(v_i\alpha_i)} y^{v_i\alpha_i - 1} \exp\{-(-\theta_i\alpha_i v_i)y\}, \end{aligned} \tag{5.44}$$

and the canonical parameter is $\theta_i = -c_i$. For our GLM analysis we treat the shape parameter $\alpha_i \equiv \alpha > 0$ as a nuisance parameter that does not depend on the specific policy i, i.e., we set constant dispersion $\varphi = 1/\alpha$, and only the scale parameter c_i is chosen policy dependent through $\theta_i = -c_i$.

Random variable $Y_i = S_i/(n_i\alpha) \sim \Gamma(n_i\alpha, n_i\alpha c_i)$ gives the reproductive form of the gamma EDF, see Remarks 2.13. In applications, this form is not directly useful because under unknown shape parameter α, we cannot calculate observations $Y_i = S_i/(n_i\alpha)$. For this reason, we parametrize the model differently, here. We consider instead

$$Y_i = S_i/n_i \sim \Gamma(n_i\alpha, n_i c_i). \tag{5.45}$$

This (new) random variable has the same gamma EDF (5.44), we only need to reinterpret the canonical parameter as $\theta_i = -c_i/\alpha$. Then, we choose the log-link for g which implies

$$\mu_i = \mathbb{E}_{\theta_i}[Y_i] = \kappa'(\theta_i) = -\frac{1}{\theta_i} = \exp\{\eta_i\} = \exp\langle \boldsymbol{\beta}, \boldsymbol{x}_i\rangle,$$

if $\boldsymbol{x}_i \in \mathcal{X} \subset \mathbb{R}^{q+1}$ describes the pre-processed features of policy i. The gamma GLM is now fully specified and can be fitted to the data; from Example 5.5 we know that we have a concave maximization problem. We call this model Gamma GLM1 (with the feature pre-processing as described above). Note that the (constant) dispersion parameter φ cancels in the score equations, thus, we do not need to explicitly specify the nuisance parameter α to estimate regression parameter $\boldsymbol{\beta} \in \mathbb{R}^{q+1}$.

Maximum Likelihood Estimation and Model Selection

Because we have only few claims data in this Swedish motorcycle example (only $m = 656$ insurance policies suffer claims), we do not perform a generalization analysis with learning and test samples. In this situation we need all data for model fitting, and model performance is analyzed with AIC and with tenfold cross-validation.

The in-sample deviance loss in the gamma GLM is given by

$$\mathfrak{D}(\mathcal{L}, \widehat{\mu}(\cdot)) = \frac{2}{m}\sum_{i=1}^{m}\frac{n_i}{\varphi}\left(\frac{Y_i - \widehat{\mu}(\boldsymbol{x}_i)}{\widehat{\mu}(\boldsymbol{x}_i)} - \log\left(\frac{Y_i}{\widehat{\mu}(\boldsymbol{x}_i)}\right)\right), \tag{5.46}$$

where i runs over the policies $i = 1, \ldots, m$ with positive claims $Y_i = S_i/n_i > 0$, and $\widehat{\mu}(\boldsymbol{x}_i) = \exp\langle\widehat{\boldsymbol{\beta}}^{\text{MLE}}, \boldsymbol{x}_i\rangle$ is the MLE estimated regression function. Similar to the Poisson case (5.29), McCullagh–Nelder [265] derive the following behavior

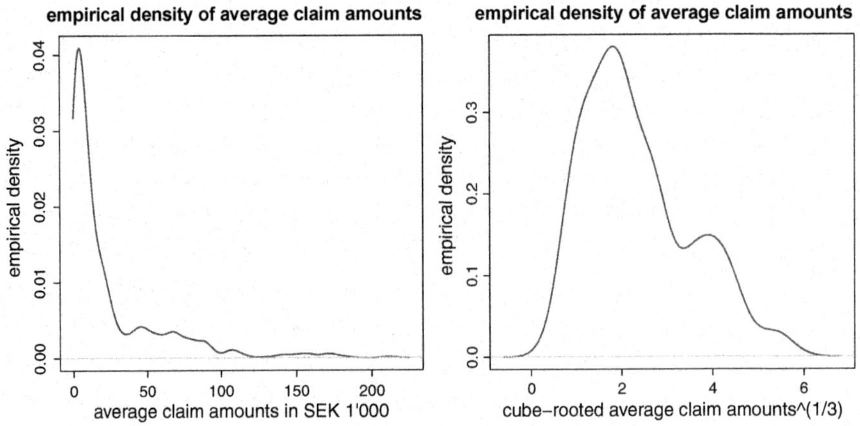

Fig. 5.10 (lhs) Empirical density of Y_i and (rhs) empirical density of $Y_i^{1/3}$

for the gamma unit deviance around its mode, see Section 7.2 and Figure 7.2 in McCullagh–Nelder [265],

$$\mathfrak{d}(Y_i, \mu_i) \approx 9Y_i^{2/3}\left(Y_i^{-1/3} - \mu_i^{-1/3}\right)^2, \tag{5.47}$$

this uses that the log-likelihood is symmetric around its mode for scale $\mu_i^{-1/3}$, see Fig. 5.5 (middle). This shows that the gamma deviance scales differently around Y_i compared to the square loss function. From this we receive an approximation to the deviance residuals (for $v/\varphi = 1$)

$$r_i^{\mathrm{D}} = \mathrm{sign}(Y_i - \mu_i)\sqrt{\mathfrak{d}(Y_i, \mu_i)} \approx 3\left(\left(\frac{Y_i}{\mu_i}\right)^{1/3} - 1\right) = 3\,\frac{Y_i^{1/3} - \mu_i^{1/3}}{\mu_i^{1/3}}. \tag{5.48}$$

This is the cube-root transformation derived by Wilson–Hilferty [383]. This suggests that if the empirical distribution of $Y_i^{1/3}$ looks roughly Gaussian we can use a gamma distribution. Figure 5.10 gives the empirical densities of Y_i on the left-hand side and of $Y_i^{1/3}$ on the right-hand side. The latter looks roughly Gaussian (except of the second mode close to 4), this supports the use of a gamma model.

Listing 5.11 provides the summary statistics of the fitted model Gamma GLM1; note that we integrate the number of claims n_i through scaling into the `weights`. We have $q + 1 = 9$ regression parameters, and from this summary statistics we observe that not all variables should be kept in the model. If we perform backward elimination using `drop1` in each step, see Sect. 5.3.3, we first drop `BonusClass` and then `Gender`, resulting in a reduced model with 7 parameters. We call this reduced model Gamma GLM2.

Listing 5.11 Results in model Gamma GLM1 using the R command `glm`

```
Call:
glm(formula = ClaimAmount/ClaimNb ~ OwnerAge + I(OwnerAge^2) +
    AreaGLM + RiskClass + VehAge + I(VehAge^2) + Gender + BonusClass,
    family = Gamma(link = "log"), data = mcdata0, weights = ClaimNb)

Deviance Residuals:
    Min       1Q   Median       3Q      Max
-3.3683  -1.4585  -0.5979   0.4354   3.4763

Coefficients:
                Estimate Std. Error t value Pr(>|t|)
(Intercept)    8.9737854  0.5532821  16.219  < 2e-16 ***
OwnerAge       0.1072781  0.0280862   3.820 0.000147 ***
I(OwnerAge^2) -0.0014508  0.0003489  -4.158 3.65e-05 ***
AreaGLM       -0.0768512  0.0368284  -2.087 0.037303 *
RiskClass      0.0615575  0.0327553   1.879 0.060651 .
VehAge        -0.2051148  0.0296184  -6.925 1.05e-11 ***
I(VehAge^2)    0.0062649  0.0015946   3.929 9.45e-05 ***
GenderMale     0.1085538  0.1673443   0.649 0.516772
BonusClass     0.0089004  0.0225371   0.395 0.693029
---
Signif. codes:  0 '***' 0.001 '**' 0.01 '*' 0.05 '.' 0.1 ' ' 1

(Dispersion parameter for Gamma family taken to be 1.536577)

    Null deviance: 1368.0  on 655  degrees of freedom
Residual deviance: 1126.5  on 647  degrees of freedom
AIC: 14922

Number of Fisher Scoring iterations: 11
```

Table 5.13 Run times, number of parameters, AICs, Pearson's dispersion estimate, in-sample losses, tenfold cross-validation losses and the in-sample average claim amounts of the null model (gamma intercept model) and the gamma GLMs

	Run time	# Param.	AIC	Dispersion est. $\widehat{\varphi}^{P}$	In-sample loss on $\mathcal{L}$	Tenfold CV loss $\widehat{\mathfrak{D}}^{CV}$	Average amount
Gamma null	–	$1+1$	14'416	2.057	2.085	2.091	24'641
Gamma GLM1	1s	$9+1$	14'277	1.537	1.717	1.752	25'105
Gamma GLM2	1s	$7+1$	14'274	1.544	1.719	1.747	25'130

The results of models Gamma GLM1 and Gamma GLM2 are presented in Table 5.13. We show AICs, Pearson's dispersion estimate, the in-sample deviance losses on all available data, the corresponding tenfold cross-validation losses, and the average claim amounts.

Firstly, we observe that the GLMs do not meet the balance property. This is implied by the fact that we do not use the canonical link to avoid any sort of difficulty of dealing with the one-sided bounded effective domain $\boldsymbol{\Theta} = (-\infty, 0)$. For pricing, the intercept parameter $\widehat{\beta}_0^{\mathrm{MLE}}$ should be shifted to eliminate this bias, i.e, we need to shift this parameter under the log-link by $-\log(25'130/24'641)$ for model Gamma GLM2.

Secondly, the in-sample and tenfold cross-validation losses are not directly comparable to AIC. Observe that we need to know the dispersion parameter φ in order to calculate both of these statistics. For the in-sample and cross-validation losses we have set $\varphi = 1$, thus, all these figures are directly comparable. For AIC we have estimated the dispersion parameter φ with MLE. This is the reason for increasing the number of parameters in Table 5.13 by +1. Moreover, the resulting AICs differ from the ones received from the R command `glm`, see, for instance, Listing 5.11. The AIC value in Listing 5.11 does not consider all terms appropriately due to the inclusion of `weights`, this is similar to Remark 5.22, it uses the deviance dispersion estimate $\widehat{\varphi}^{\mathrm{D}}$, i.e., not the MLE and (still) increases the number of parameters by 1 because the dispersion is estimated. For these reasons, we have implemented our own code for calculating AIC. Both AIC and the tenfold cross-validation losses say that we should give preference to model Gamma GLM2.

The dispersion estimate in Listing 5.11 corresponds to Pearson's estimate

$$\widehat{\varphi}^{\mathrm{P}} = \frac{1}{m-(q+1)} \sum_{i=1}^{m} n_i \frac{(Y_i - \widehat{\mu}_i)^2}{\widehat{\mu}_i^2}. \tag{5.49}$$

We observe that the dispersion estimate is roughly 1.5 which gives an estimate of the shape parameter $\alpha = 1/\varphi$ of 2/3. A shape parameter less than 1 implies that the density of the gamma distribution is strictly decreasing, see Fig. 2.1. Often this is a sign that the model does not fully fit the data, and if we use this model for simulation we may receive too many observations close to zero compared to the true data. A shape parameter less than 1 may be implied by more heterogeneity in the data compared to what the chosen gamma GLM allows for or by large claims that cannot be explained by the present gamma density structure. Thus, there is some sign here that the data is more heavy-tailed than our model choice suggests. Alternatively, there might be some need to also model the shape parameter with a regression model; this could be done using the vector-valued parameter EF representation of the gamma model, see Sect. 2.1.3. In view of Fig. 5.10 (rhs) it may also be that the feature information is not sufficient to describe the second mode in 4, thus, we probably need more explanatory information to reduce dispersion.

In Fig. 5.11 we give the Tukey–Anscombe plot and a QQ plot. Note that the observations for $n_i = 1$ follow a gamma distribution with shape parameter α and scale parameter $c_i = \alpha/\mu_i = -\alpha\theta_i$. Thus, if we scale Y_i/μ_i, we receive i.i.d. gamma random variables with shape and scale parameters equal to α. This then allows us for $n_i = 1$ to plot the empirical distribution of $Y_i/\widehat{\mu}_i$ against $\Gamma(\alpha, \alpha)$ in a QQ plot where we estimate $1/\alpha$ by Pearson's dispersion estimate. The Tukey–Anscombe plot looks reasonable, but the QQ plot shows that the gamma model does not entirely fit the data. From this plot we cannot conclude whether the gamma distribution is causing the problem or whether it is a missing term in the regression structure. We only see that the data is over-dispersed, resulting in more heavy-tailed observations than the theoretical gamma model can explain, and a compensation by too many small observations (which is induced by over-dispersion, i.e., a shape parameter smaller than one). In the network chapter we will refine the regression function, keeping the gamma assumption, to understand which modeling part is causing the difficulty.

Remark 5.26 For the calculation of AIC in Table 5.13 we have used the MLE of the dispersion parameter φ. This is obtained by solving the score equation (5.11) for the

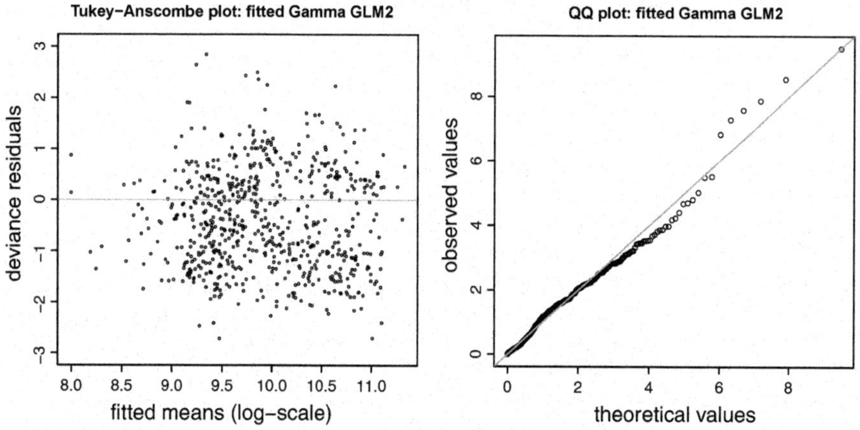

Fig. 5.11 (lhs) Tukey–Anscombe plot of the fitted model Gamma GLM2, and (rhs) QQ plot of the fitted model Gamma GLM2

gamma case. It is given by, we set $\alpha = 1/\varphi$ and we calculate the MLE of α instead,

$$\frac{\partial}{\partial \alpha}\ell_Y(\boldsymbol{\beta}, \alpha) = \sum_{i=1}^{n} v_i \Big[Y_i h(\mu(\boldsymbol{x}_i)) - \kappa\left(h(\mu(\boldsymbol{x}_i))\right) + \log Y_i + \log(\alpha v_i) + 1 - \Psi(\alpha v_i) \Big] = 0,$$

where $\Psi(\alpha) = \Gamma'(\alpha)/\Gamma(\alpha)$ is the digamma function. We calculate the second derivative w.r.t. α, see also (2.30),

$$\frac{\partial^2}{\partial \alpha^2}\ell_Y(\boldsymbol{\beta}, \alpha) = \sum_{i=1}^{n} v_i \left[\frac{1}{\alpha} - v_i \Psi'(\alpha v_i) \right] = \sum_{i=1}^{n} v_i^2 \left[\frac{1}{\alpha v_i} - \Psi'(\alpha v_i) \right] < 0 \qquad \text{for } \alpha > 0,$$

the negativity follows from Theorem 1 in Alzner [9]. In fact, the function $\log \alpha - \Psi(\alpha)$ is strictly completely monotonic for $\alpha > 0$. This says that the log-likelihood $\ell_Y(\boldsymbol{\beta}, \alpha)$ is a concave function in $\alpha > 0$ and the solution to the score equation is unique, giving the MLE of α and φ, respectively.

5.3.8 Lab: Inverse Gaussian GLM for Claim Sizes

We present the inverse Gaussian GLM in this section as a competing model to the gamma GLM studied in the previous section.

Infinite Divisibility

In the gamma model above we have used that the total claim amount $S = \sum_{j=1}^{n} Z_j$ has a gamma distribution for given claim counts $N = n > 0$ and i.i.d. gamma claim sizes Z_j. This property is closely related to divisibility. A random variable S is called divisible by $n \in \mathbb{N}$ if there exist i.i.d. random variables $Z_1, \ldots, Z_n$ such that

$$S \overset{(d)}{=} \sum_{j=1}^{n} Z_j,$$

and S is called *infinitely divisible* if S is divisible by n for all $n \in \mathbb{N}$. The EDF is based on parameters $(\theta, \omega) \in \boldsymbol{\Theta} \times \mathcal{W}$. Jørgensen [203] gives the following interesting result.

Theorem 5.27 (Theorem 3.7 in Jørgensen [203], Without Proof) *Choose a member of the EDF with parameter set* $\boldsymbol{\Theta} \times \mathcal{W}$. *Then*

- *the index set* $\mathcal{W}$ *is an additive semi-group and* $\mathbb{N} \subseteq \mathcal{W} \subseteq \mathbb{R}_+$, *and*
- *the members of the chosen EDF are infinitely divisible if and only if* $\mathcal{W} = \mathbb{R}_+$.

This theorem tells us how to aggregate and disaggregate within EDFs, e.g., the Poisson, gamma and inverse Gaussian models are infinitely divisible, and the binomial distribution is divisible by n with the disaggregated random variables belonging to the same EDF and the same canonical parameter, see Sect. 2.2.2. In particular, we also refer to Corollary 2.15 on the convolution property.

Inverse Gaussian Generalized Linear Model

Alternatively to the gamma GLM one often explores an inverse Gaussian GLM which has a cubic variance function $V(\mu) = \mu^3$. We bring this inverse Gaussian model into the same form as the gamma model of Sect. 5.3.7, so that we can aggregate claims within insurance policies. The mean, the variance and the moment generating function of an inverse Gaussian random variable $Z_{i,j}$ with parameters $\alpha_i, c_i > 0$ are given by

$$\mathbb{E}[Z_{i,j}] = \frac{\alpha_i}{c_i}, \quad \mathrm{Var}(Z_{i,j}) = \frac{\alpha_i}{c_i^3} \quad \text{and} \quad M_{Z_{i,j}}(r) = \exp\left\{\alpha_i\left[c_i - \sqrt{c_i^2 - 2r}\right]\right\},$$

where the moment generating function requires $r < c_i^2/2$ to be finite. From the moment generating function we see that $S_i = \sum_{j=1}^{n_i} Z_{i,j}$ is inverse Gaussian distributed with parameters $n_i\alpha_i$ and c_i. Finally, we scale $Y_i = S_i/(n_i\alpha_i)$ which provides us with an inverse Gaussian distribution with parameters $n_i^{1/2}\alpha_i^{1/2}$ and $n_i^{1/2}\alpha_i^{1/2}c_i$. This random variable Y_i has a single-parameter EDF inverse Gaussian distribution in its reproductive form, namely,

$$\begin{aligned} Y_i \sim f(y; \theta_i, v_i/\varphi_i) &= \exp\left\{\frac{y\theta_i - \kappa(\theta_i)}{\varphi_i/v_i} + a(y; v_i/\varphi_i)\right\} \qquad (5.50) \\ &= \frac{\alpha_i^{1/2}}{\sqrt{\frac{2\pi}{v_i} y^3}} \exp\left\{-\frac{\alpha_i}{2y/v_i}\left(1 - \sqrt{-2\theta_i y}\right)^2\right\}, \end{aligned}$$

with cumulant function $\kappa(\theta) = -\sqrt{-2\theta}$ for $\theta \in \boldsymbol{\Theta} = (-\infty, 0]$, weight $v_i = n_i$, dispersion parameter $\varphi_i = 1/\alpha_i$ and canonical parameter $\theta_i = -c_i^2/2$.

Similarly to the gamma case, this representation is not directly useful if the parameter α_i is not known. Therefore, we parametrize this model differently. Namely, we consider

$$Y_i = S_i/n_i \sim \text{InvGauss}\left(n_i^{1/2}\alpha_i, n_i^{1/2}c_i\right). \tag{5.51}$$

This re-scaled random variable has that same inverse Gaussian EDF (5.50), but we need to re-interpret the parameters. We have dispersion parameter $\varphi_i = 1/\alpha_i^2$ and canonical parameter $\theta_i = -c_i^2/(2\alpha_i^2)$. For our GLM analysis we will treat the parameter $\alpha_i \equiv \alpha > 0$ as a nuisance parameter that does not depend on the specific policy i. Thus, we have constant dispersion $\varphi = 1/\alpha^2$ and only the scale parameter c_i is assumed to be policy dependent through the canonical parameter $\theta_i = -c_i^2/(2\alpha^2)$.

We are now in the same situation as in the gamma case in Sect. 5.3.7. We choose the log-link for g which implies

$$\mu_i = \mathbb{E}_{\theta_i}[Y_i] = \kappa'(\theta_i) = \frac{1}{\sqrt{-2\theta_i}} = \exp\{\eta_i\} = \exp\langle \boldsymbol{\beta}, \boldsymbol{x}_i \rangle,$$

for $\boldsymbol{x}_i \in \mathcal{X} \subset \mathbb{R}^{q+1}$ describing the pre-processed features of policy i. We use the same feature pre-processing as in model Gamma GLM2, and we call this resulting model IG GLM2. Again the constant dispersion parameter $\varphi = 1/\alpha^2$ cancels in the score equations, thus, we do not need to explicitly specify the nuisance parameter α to estimate the regression parameter $\boldsymbol{\beta} \in \mathbb{R}^{q+1}$. However, there is an important difference to the gamma GLM, namely, as stated in Example 5.6, we do not have a concave maximization problem and Fisher's scoring method needs a suitable initial value. We start the fitting algorithm in the parameters of model Gamma GLM2.

The in-sample deviance loss in the inverse Gaussian GLM is given by

$$\mathfrak{D}(\mathcal{L}, \widehat{\mu}(\cdot)) = \frac{1}{m} \sum_{i=1}^{m} \frac{n_i}{\varphi} \frac{(Y_i - \widehat{\mu}(\boldsymbol{x}_i))^2}{\widehat{\mu}(\boldsymbol{x}_i)^2\, Y_i}, \tag{5.52}$$

where i runs over the policies $i = 1, \ldots, m$ with positive claims $Y_i = S_i/n_i > 0$, and $\widehat{\mu}(\boldsymbol{x}_i) = \exp\langle \widehat{\boldsymbol{\beta}}^{\text{MLE}}, \boldsymbol{x}_i \rangle$ is the MLE estimated regression function. The unit deviances behave as

$$\mathfrak{d}(Y_i, \mu_i) = Y_i \left(Y_i^{-1} - \mu_i^{-1}\right)^2, \tag{5.53}$$

Table 5.14 Run times, number of parameters, AICs, in-sample losses, tenfold cross-validation losses and the in-sample average claim amounts of the null gamma model, model Gamma GLM2, the null inverse Gaussian model, and model inverse Gaussian GLM2; the deviance losses use unit dispersion $\varphi = 1$

	Run time	# Param.	AIC	In-sample loss on $\mathcal{L}$	Tenfold CV loss $\widehat{\mathfrak{D}}^{\text{CV}}$	Average amount
Gamma null	–	$1+1$	14'416	2.085	2.091	24'641
Gamma GLM2	1 s	$7+1$	14'274	1.719	1.747	25'130
IG null	–	$1+1$	14'715	$5.012 \cdot 10^{-4}$	$5.016 \cdot 10^{-4}$	24'641
IG GLM2	1 s	$7+1$	14'686	$4.793 \cdot 10^{-4}$	$4.820 \cdot 10^{-4}$	32'268

note that the log-likelihood is symmetric around its mode for scale μ_i^{-1}, see Fig. 5.5 (rhs). From this we receive deviance residuals (for $v/\varphi = 1$)

$$r_i^{\text{D}} = \text{sign}(Y_i - \mu_i)\sqrt{\mathfrak{d}\,(Y_i, \mu_i)} = Y_i^{1/2}\left(\mu_i^{-1} - Y_i^{-1}\right).$$

Thus, these residuals behave as $Y_i^{1/2}$ for $Y_i \to \infty$ (and fixed μ_i^{-1}), which is more heavy-tailed than the cube-root behavior $Y_i^{1/3}$ in the gamma case, see (5.48). Another difference to the gamma case is that the deviance loss (5.52) is not scale-invariant, see also (11.4), below.

We revisit the example of Table 5.13, but we replace the gamma distribution by the inverse Gaussian distribution. The results in Table 5.14 show that the inverse Gaussian model is not fully competitive on this data set. In view of (5.43) we observe that the coefficient of variation (standard deviation divided by mean) is in the gamma model given by $1/\sqrt{\alpha}$, thus, in the gamma model this coefficient of variation is independent of the expected claim size μ_i and only depends on the shape parameter α. In the inverse Gaussian model the coefficient of variation is given by

$$\text{Vco}(Z_{i,j}) = \frac{\sqrt{\text{Var}(Z_{i,j})}}{\mathbb{E}[Z_{i,j}]} = \frac{\sqrt{\mu_i}}{\alpha},$$

thus, it monotonically increases in the expected claim size μ_i. It seems that this structure is not fully suitable for this data set, i.e., there is no indication that the coefficient of variation increases in the expected claim size. We come back to a comparison of the gamma and the inverse Gaussian model in Sect. 11.1, below.

5.3.9 Log-Normal Model for Claim Sizes: A Short Discussion

Another way to improve the gamma model of Sect. 5.3.7 could be to use a log-normal distribution instead. In the above situation this does not work because the observations are not in the right format. If the claim observations $Z_{i,j}$ are log-normally distributed, then $\log(Z_{i,j})$ are normally distributed. Unfortunately, in our Swedish motorcycle data set we do not have individual claim observations $Z_{i,j}$, but the provided information is aggregated over all claims per insurance policy, i.e., $S_i = \sum_{j=1}^{N_i} Z_{i,j}$. Therefore, there is no possibility here to challenge the gamma framework of Sect. 5.3.7 with a corresponding log-normal framework, because the log-normal framework is not closed under summation of i.i.d. log-normally distributed random variables.

We would like to give some remarks that concern calculations on the log-scale (or any other strictly increasing and concave transformation of the original data). For the log-normal distribution, as well as in similar cases like the log-gamma distribution, one works with logged observations $Y_i = \log(Z_i)$. This is a strictly monotone transformation and the MLEs in the log-normal model based on observations Z_i and in the normal model based on observations $Y_i = \log(Z_i)$ coincide. This can be seen from the following calculation. We start from the log-normal density on $\mathbb{R}_+$, and we do a transformation of variable $z > 0 \mapsto y = \log(z) \in \mathbb{R}$ with $dy = dz/z$

$$
\begin{aligned}
f_{\mathrm{LN}}(z;\mu,\sigma^2)dz &= \frac{1}{\sqrt{2\pi\sigma^2}}\frac{1}{z}\exp\left\{-\frac{1}{2\sigma^2}\left(\log(z)-\mu\right)^2\right\}dz \\
&= \frac{1}{\sqrt{2\pi\sigma^2}}\exp\left\{-\frac{1}{2\sigma^2}\left(y-\mu\right)^2\right\}dy \;=\; f_{\Phi}(y;\mu,\sigma^2)dy.
\end{aligned}
$$

From this we see that the MLEs will coincide.

In many situations, one assumes that $\sigma^2 > 0$ is a given nuisance parameter, and one models $\boldsymbol{x} \mapsto \mu(\boldsymbol{x})$ with a GLM within the single-parameter EDF. In the log-normal/Gaussian case one typically chooses the canonical link on the log-scale which is the identity function. This then allows one to perform a classical linear regression for $\mu(\boldsymbol{x}) = \langle \boldsymbol{\beta}, \boldsymbol{x}\rangle$ using the logged observations $\boldsymbol{Y} = (Y_1,\ldots,Y_n)^\top = (\log(Z_1),\ldots,\log(Z_n))^\top$, and the corresponding MLE is given by

$$
\widehat{\boldsymbol{\beta}}^{\mathrm{MLE}} = (\mathfrak{X}^\top\mathfrak{X})^{-1}\mathfrak{X}^\top \boldsymbol{Y}, \tag{5.54}
$$

for full rank $q+1 \le n$ design matrix $\mathfrak{X}$. Note that in this case we have a closed-form solution for the MLE of $\boldsymbol{\beta}$. This is called the homoskedastic case because all observations Y_i are assumed to have the same variance σ^2, otherwise, in the heteroskedastic case, we would still have to include the covariance matrix.

Since we work with the canonical link on the log-scale we have the balance property on the log-scale, see Corollary 5.7. Thus, we receive unbiasedness

$$
\sum_{i=1}^n \mathbb{E}_{\boldsymbol{\beta}}\left[\mathbb{E}_{\widehat{\boldsymbol{\beta}}^{\mathrm{MLE}}}[Y_i]\right] = \sum_{i=1}^n \mathbb{E}_{\boldsymbol{\beta}}\left[\langle \widehat{\boldsymbol{\beta}}^{\mathrm{MLE}}, \boldsymbol{x}_i\rangle\right] = \sum_{i=1}^n \mathbb{E}_{\boldsymbol{\beta}}\left[Y_i\right] = \sum_{i=1}^n \mu(\boldsymbol{x}_i). \tag{5.55}
$$

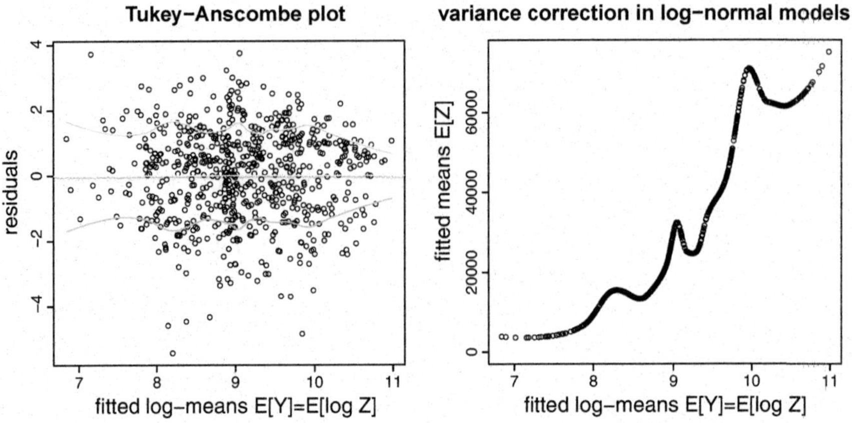

Fig. 5.12 (lhs) Tukey–Anscombe plot of the fitted Gaussian model $\widehat{\mu}(\boldsymbol{x}_i)$ on the logged claim sizes $Y_i = \log(Z_i)$, and (rhs) estimated means $\widehat{\mu}_{Z_i}$ as a function of $\widehat{\mu}(\boldsymbol{x}_i)$ considering heteroskedasticity $\widehat{\sigma}(\boldsymbol{x}_i)$

If we move back to the original scale of the observations Z_i we receive from the log-normal assumption

$$\mathbb{E}_{(\widehat{\boldsymbol{\beta}}^{\text{MLE}},\sigma^2)}[Z_i] = \exp\left\{\langle \widehat{\boldsymbol{\beta}}^{\text{MLE}}, \boldsymbol{x}_i\rangle + \sigma^2/2\right\}.$$

Therefore, we need to adjust with the nuisance parameter σ^2 for the back-transformation to the original observation scale. At this point, typically, the difficulties start. Often, a good back-transformation involves a feature dependent variance parameter $\sigma^2(\boldsymbol{x}_i)$, thus, in many practical applications the homoskedasticity assumption is not fulfilled, and a constant variance parameter choice leads to a poor model on the original observation scale.

A suitable estimation of $\sigma^2(\boldsymbol{x}_i)$ may turn out to be rather difficult. This is illustrated in Fig. 5.12. The left-hand side of this figure shows the Tukey–Anscombe plot of the homoskedastic case providing unscaled ($\sigma^2 \equiv 1$) (Pearson's) residuals on the log-scale

$$r_i^{\text{P}} = \log(Z_i) - \widehat{\mu}(\boldsymbol{x}_i) = Y_i - \widehat{\mu}(\boldsymbol{x}_i).$$

The light-blue color shows an insurance policy dependent standard deviation estimate $\widehat{\sigma}(\boldsymbol{x}_i)$. In our case this estimate is non-monotone in $\widehat{\mu}(\boldsymbol{x}_i)$ (which is quite common on real data). Using this estimate we can estimate the means of the log-normal random variables by

$$\widehat{\mu}_{Z_i} = \widehat{\mathbb{E}}[Z_i] = \exp\left\{\widehat{\mu}(\boldsymbol{x}_i) + \widehat{\sigma}(\boldsymbol{x}_i)^2/2\right\}.$$

The right-hand side of Fig. 5.12 plots these estimated means $\widehat{\mu}_{Z_i}$ against the estimated means $\widehat{\mu}(\boldsymbol{x}_i)$ on the log-scale. We observe a graph that is non-monotone, implied by the non-monotonicity of the standard deviation estimate $\widehat{\sigma}(\boldsymbol{x}_i)$ as a function of $\widehat{\mu}(\boldsymbol{x}_i)$. This non-monotonicity is not bad per se, as we still have a proper statistical model, however, it might be rather counter-intuitive and difficult to explain. For this reason it is advisable to directly model the expected value by one single function, and not to decompose it into different regression functions.

Another important point to be considered is that for model selection using AIC we have to work on the same scale for all models. Thus, if we use a gamma model to model Z_i, then for an AIC selection we need to evaluate also the log-normal model on that scale. This can be seen from the justification in Sect. 4.2.3.

Finally, we focus on unbiasedness. Note that on the log-scale we have unbiasedness (5.55) through the balance property. Unfortunately, this does not carry over to the original scale. We give a small example, where we assume that there is neither any uncertainty about the distributional model nor about the nuisance parameter. That is, we assume that Z_i are i.i.d. log-normally distributed with parameters μ and σ^2, where only μ is unknown. The MLE of μ is given by

$$\widehat{\mu}^{\mathrm{MLE}} = \frac{1}{n}\sum_{i=1}^{n} \log(Z_i) \sim \mathcal{N}(\mu, \sigma^2/n).$$

In this case we have

$$\begin{aligned}
\frac{1}{n}\sum_{i=1}^{n} \mathbb{E}_{(\mu,\sigma^2)}\left[\mathbb{E}_{(\widehat{\mu}^{\mathrm{MLE}},\sigma^2)}[Z_i]\right] &= \frac{1}{n}\sum_{i=1}^{n} \mathbb{E}_{(\mu,\sigma^2)}\left[\exp\{\widehat{\mu}^{\mathrm{MLE}}\}\right]\exp\{\sigma^2/2\} \\
&= \exp\left\{\mu + (1+n^{-1})\sigma^2/2\right\} \\
&> \exp\left\{\mu + \sigma^2/2\right\} = \frac{1}{n}\sum_{i=1}^{n} \mathbb{E}_{(\mu,\sigma^2)}\left[Z_i\right].
\end{aligned}$$

Volatility in parameter estimation $\widehat{\mu}^{\mathrm{MLE}}$ leads to a positive bias in this case. Note that we have assumed full knowledge of the distributional model (i.i.d. log-normal) and the nuisance parameter σ^2 in this calculation. If, for instance, we do not know the true nuisance parameter and we work with (deterministic) $\widetilde{\sigma}^2 \ll \sigma^2$ and $n > 1$, we can get a negative bias

$$\begin{aligned}
\frac{1}{n}\sum_{i=1}^{n} \mathbb{E}_{(\mu,\sigma^2)}\left[\mathbb{E}_{(\widehat{\mu}^{\mathrm{MLE}},\widetilde{\sigma}^2)}[Z_i]\right] &= \frac{1}{n}\sum_{i=1}^{n} \mathbb{E}_{(\mu,\sigma^2)}\left[\exp\{\widehat{\mu}^{\mathrm{MLE}}\}\right]\exp\{\widetilde{\sigma}^2/2\} \\
&= \exp\left\{\mu + \sigma^2/(2n) + \widetilde{\sigma}^2/2\right\} \\
&< \exp\left\{\mu + \sigma^2/2\right\} = \frac{1}{n}\sum_{i=1}^{n} \mathbb{E}_{(\mu,\sigma^2)}\left[Z_i\right].
\end{aligned}$$

This shows that working on the log-scale is rather difficult because the back-transformation is far from being trivial, and for unknown nuisance parameter not even the sign of the bias is clear. Similar considerations apply to the frequently used Box–Cox transformation [48] for $\chi \neq 1$

$$Z_i \mapsto Y_i = \frac{Z_i^{\chi} - 1}{\chi}.$$

For this reason, if unbiasedness is a central requirement (like in insurance pricing) non-linear transformations should only be used with great care (and only if necessary).

5.4 Quasi-Likelihoods

Above we have been mentioning the notion of over-dispersed Poisson models. This naturally leads to so-called quasi-Poisson models and quasi-likelihoods. The framework of quasi-likelihoods has been introduced by Wedderburn [376]. In this section we give the main idea behind quasi-likelihoods, and for a more detailed treatment and mathematical results we refer to Chapter 8 of McCullagh–Nelder [265].

In Sect. 5.1.4 we have discussed the estimation of GLMs. This has been based on the explicit knowledge of the full log-likelihood function $\ell_{\boldsymbol{Y}}(\boldsymbol{\beta})$ for given data $\boldsymbol{Y}$. This has allowed us to calculate the score equations $s(\boldsymbol{\beta}, \boldsymbol{Y}) = \nabla_{\boldsymbol{\beta}} \ell_{\boldsymbol{Y}}(\boldsymbol{\beta}) = 0$ whose solutions (Z-estimators) contain the MLE for $\boldsymbol{\beta}$. The solutions of the score equations themselves, using Fisher's scoring method, no longer need the explicit functional form of the log-likelihood, but they are only based on the first and second moments, see (5.9) and Remarks 5.4. Thus, all models where these first two moments coincide will provide the same MLE for the regression parameter $\boldsymbol{\beta}$; this is also the explanation behind the IRLS algorithm. Moreover, the first two moments are sufficient for prediction and uncertainty quantification based on mean squared errors, and they are also sufficient to quantify asymptotic normality. This is exactly what motivates the quasi-likelihood considerations, and these considerations are also related to the quasi-generalized pseudo maximum likelihood estimator (QPMLE) that we are going to discuss in Theorem 11.8, below.

Assume that $\boldsymbol{Y}$ is a random vector having first moment $\boldsymbol{\mu} \in \mathbb{R}^n$, positive definite variance function $V(\boldsymbol{\mu}) \in \mathbb{R}^{n \times n}$ and dispersion parameter φ. The quasi-(log-)likelihood function $\ell_{\boldsymbol{Y}}(\boldsymbol{\mu})$ assumes that its gradient is given by

$$\nabla_{\boldsymbol{\mu}} \ell_{\boldsymbol{Y}}(\boldsymbol{\mu}) = \frac{1}{\varphi} V(\boldsymbol{\mu})^{-1} (\boldsymbol{Y} - \boldsymbol{\mu}).$$

In case of a diagonal variance function $V(\boldsymbol{\mu})$ this relates to the score (5.9). The remaining step is to model the mean parameter $\boldsymbol{\mu} = \boldsymbol{\mu}(\boldsymbol{\beta}) \in \mathbb{R}^n$ as a function of a lower dimensional regression parameter $\boldsymbol{\beta} \in \mathbb{R}^{q+1}$, we also refer to Fig. 5.2. For this last step we assume that the Jacobian $B \in \mathbb{R}^{n\times(q+1)}$ of $d\boldsymbol{\mu}/d\boldsymbol{\beta}$ has full rank $q+1$. The score equations for $\boldsymbol{\beta}$ and given observations $\boldsymbol{Y}$ then read as

$$\frac{1}{\varphi} B^\top V(\boldsymbol{\mu}(\boldsymbol{\beta}))^{-1} (\boldsymbol{Y} - \boldsymbol{\mu}(\boldsymbol{\beta})) = 0.$$

This is of exactly the same structure as the score equations in Proposition 5.1, and the roots are found by using the IRLS algorithm for $t \geq 0$, see (5.12),

$$\widehat{\boldsymbol{\beta}}^{(t)} \mapsto \widehat{\boldsymbol{\beta}}^{(t+1)} = \left(B^\top V(\widehat{\boldsymbol{\mu}}^{(t)})^{-1} B\right)^{-1} B^\top V(\widehat{\boldsymbol{\mu}}^{(t)})^{-1} \left(B\widehat{\boldsymbol{\beta}}^{(t)} + \boldsymbol{Y} - \widehat{\boldsymbol{\mu}}^{(t)}\right),$$

where $\widehat{\boldsymbol{\mu}}^{(t)} = \boldsymbol{\mu}(\widehat{\boldsymbol{\beta}}^{(t)})$.

We conclude with the following points about quasi-likelihoods:

- For regression parameter estimation within the quasi-likelihood framework it is sufficient to know the structure of the first two moments $\boldsymbol{\mu}(\boldsymbol{\beta}) \in \mathbb{R}^n$ and $V(\boldsymbol{\mu}) \in \mathbb{R}^{n\times n}$ as well as the score equations. Thus, we do not need to explicitly specify a distributional family for the observations $\boldsymbol{Y}$. This structure of the first two moments is then sufficient for their estimation using the IRLS algorithm, i.e., we receive the predictors within this framework.
- Since we do not specify the full distribution of $\boldsymbol{Y}$ we can neither simulate from this model nor can we calculate quantities where the full log-likelihood of the model needs to be known. For example, we cannot calculate AIC in a quasi-likelihood model.
- The quasi-likelihood model is characterized by the functional forms of $\boldsymbol{\mu}(\boldsymbol{\beta})$ and $V(\boldsymbol{\mu})$. The former plays the role of the link function and the linear predictor in the GLM, and the latter plays the role of the variance function within the EDF which is characterized through the cumulant function κ. For instance, if we assume to have a diagonal matrix

  $$V(\boldsymbol{\mu}) = \text{diag}(V(\mu_1), \ldots, V(\mu_n)),$$

  then, the choice of the variance function $\mu \mapsto V(\mu)$ describes the explicit selection of the quasi-likelihood model. If we choose the power variance function $V(\mu) = \mu^p$, $p \notin (0, 1)$, we have a quasi-Tweedie's model.
- For prediction uncertainty evaluation we also need an estimate of the dispersion parameter $\varphi > 0$. Since we do not know the full likelihood in this approach, Pearson's estimate $\widehat{\varphi}^{\mathrm{P}}$ is the only option we have to estimate φ.
- For asymptotic normality results and hypothesis testing within the quasi-likelihood framework we refer to Section 8.4 of McCullagh–Nelder [265].

5.5 Double Generalized Linear Model

In the derivations above we have treated the dispersion parameter φ in the GLM as a nuisance parameter. In the case of a homogeneous dispersion parameter it can be canceled in the score equations for MLE, see (5.9). Therefore, it does not influence MLE, and in a subsequent step this nuisance parameter can still be estimated using, e.g., Pearson's or deviance residuals, see Sect. 5.3.1 and Remark 5.26. In some examples we may have systematic effects in the dispersion parameter, too. In this case the above approach will not work because a heterogeneous dispersion parameter no longer cancels in the score equations. This has been considered in Smyth [341] and Smyth–Verbyla [343]. The heterogeneous dispersion situation is of general interest for GLMs, and it is of particular interest for Tweedie's CP GLM if we interpret Tweedie's distribution [358] as a CP model with i.i.d. gamma claim sizes, see Proposition 2.17; we also refer to Jørgensen–de Souza [204], Smyth–Jørgensen [342] and Delong et al. [94].

5.5.1 *The Dispersion Submodel*

We extend model assumption (5.1) by assuming that also the dispersion parameter φ_i is policy i dependent. Assume that all random variables Y_i are independent and have densities w.r.t. a σ-finite measure ν on $\mathbb{R}$ given by

$$Y_i \sim f(y_i; \theta_i, v_i/\varphi_i) = \exp\left\{ \frac{y_i\theta_i - \kappa(\theta_i)}{\varphi_i/v_i} + a(y_i; v_i/\varphi_i) \right\},$$

for $1 \le i \le n$, with canonical parameters $\theta_i \in \mathring{\boldsymbol{\Theta}}$, exposures $v_i > 0$ and dispersion parameters $\varphi_i > 0$. As in (5.5) we assume that every policy i is equipped with feature information $\boldsymbol{x}_i \in \mathcal{X}$ such that for a given link function $g : \mathcal{M} \to \mathbb{R}$ we can model its mean as

$$\boldsymbol{x}_i \mapsto g(\mu_i) = g(\mu(\boldsymbol{x}_i)) = g\left(\mathbb{E}_{\theta(\boldsymbol{x}_i)}[Y_i]\right) = \eta_i = \eta(\boldsymbol{x}_i) = \langle \boldsymbol{\beta}, \boldsymbol{x}_i \rangle. \tag{5.56}$$

This provides us with log-likelihood function for observation $\boldsymbol{Y} = (Y_1, \ldots, Y_n)^\top$

$$\boldsymbol{\beta} \mapsto \ell_{\boldsymbol{Y}}(\boldsymbol{\beta}) = \sum_{i=1}^{n} \frac{v_i}{\varphi_i}\Big[Y_i h(\mu(\boldsymbol{x}_i)) - \kappa\left(h(\mu(\boldsymbol{x}_i))\right) \Big] + a(Y_i; v_i/\varphi_i),$$

with canonical link $h = (\kappa')^{-1}$. The difference to (5.7) is that the dispersion parameter φ_i now depends on the insurance policy which requires additional modeling. We choose a second strictly monotone and smooth link function g_φ :

$\mathbb{R}_+ \to \mathbb{R}$, and we express the dispersion of policy $1 \le i \le n$ by

$$g_\varphi(\varphi_i) = g_\varphi(\varphi(z_i)) = \langle \boldsymbol{\gamma}, z_i \rangle, \tag{5.57}$$

where z_i is the feature of policy i, which may potentially differ from $\boldsymbol{x}_i$. The rationale behind this different feature is that different information might be relevant for modeling the dispersion parameter, or feature information might be differently pre-processed compared to the response Y_i. We now need to estimate two regression parameters $\boldsymbol{\beta}$ and $\boldsymbol{\gamma}$ in this approach on possibly differently pre-processed feature information $\boldsymbol{x}_i$ and z_i of policy i. In general, this is not easily doable because the term $a(Y_i; v_i/\varphi_i)$ of the log-likelihood of Y_i may have a complicated structure (or may not be available in closed form like in Tweedie's CP model).

5.5.2 *Saddlepoint Approximation*

We reformulate the EDF density using the unit deviance $\mathfrak{d}(Y, \mu)$ defined in (2.25); we drop the lower index i for the moment. Set $\theta = h(\mu) \in \mathring{\boldsymbol{\Theta}}$ for the canonical link h, then

$$\begin{aligned} f(y; \theta, v/\varphi) &= \exp\left\{\frac{v}{\varphi}\left[yh(\mu) - \kappa(h(\mu))\right] + a(y; v/\varphi)\right\} \\ &= \exp\left\{\frac{v}{\varphi}\left[yh(y) - \kappa(h(y))\right] + a(y; v/\varphi)\right\} \exp\left\{-\frac{1}{2\varphi/v}\mathfrak{d}(y, \mu)\right\} \\ &\stackrel{\text{def.}}{=} a^*(y; \omega) \exp\left\{-\frac{\omega}{2}\mathfrak{d}(y, \mu)\right\}, \end{aligned} \tag{5.58}$$

with $\omega = v/\varphi \in \mathcal{W}$. This corresponds to (2.27), and it brings the EDF density into a Gaussian-looking form. A general difficulty is that the term $a^*(y; \omega)$ may have a complicated structure or may not be given in closed form. Therefore, we consider its saddlepoint approximation; this is based on Section 3.5 of Jørgensen [203].

Suppose that we are in the absolutely continuous EDF case and that κ is steep. In that case $Y \in \mathcal{M}$, a.s., and the variance function $y \mapsto V(y)$ is well-defined for all observations $Y = y$, a.s. Based on Daniels [87], Barndorff-Nielsen–Cox [24] proved the following statement, see Theorem 3.10 in Jørgensen [203]: assume there exists $\omega_0 \in \mathcal{W}$ such that for all $\omega > \omega_0$ the density (5.58) is bounded. Then, the following saddlepoint approximation is uniform on compact subsets of the support $\mathfrak{T}$ of Y

$$f(y; \theta, v/\varphi) = \left(\frac{2\pi\varphi}{v}V(y)\right)^{-1/2} \exp\left\{-\frac{1}{2\varphi/v}\mathfrak{d}(y, \mu)\right\}(1 + O(\varphi/v)), \tag{5.59}$$

as $\varphi/v \to 0$. What makes this saddlepoint approximation attractive is that we can get rid of a complicated function $a^*(y;\omega)$ by a neat approximation $(\frac{2\pi\varphi}{v}V(y))^{-1/2}$ for sufficiently large volumes v, and at the same time, this does not affect the unit deviance $\mathfrak{d}(y,\mu)$, preserving the estimation properties of μ. The discrete counterpart is given in Theorem 3.11 of Jørgensen [203].

Using saddlepoint approximation (5.59) we receive an approximate log-likelihood function

$$\ell_Y(\mu,\varphi) \approx \frac{1}{2}\left[-\varphi^{-1}v\mathfrak{d}(Y,\mu) - \log(\varphi)\right] - \frac{1}{2}\log\left(\frac{2\pi}{v}V(Y)\right).$$

This approximation has an attractive form for dispersion estimation because it gives an approximate EDF for observation $\mathfrak{d} \stackrel{\text{def.}}{=} v\mathfrak{d}(Y,\mu)$, for given μ. Namely, for canonical parameter $\phi = -\varphi^{-1} < 0$ we have approximation

$$\ell_Y(\mu,\phi) \approx \frac{\mathfrak{d}\phi - (-\log(-\phi))}{2} - \frac{1}{2}\log\left(\frac{2\pi}{v}V(Y)\right). \tag{5.60}$$

The right-hand side has the structure of a gamma EDF for observation $\mathfrak{d}$ with canonical parameter $\phi < 0$, cumulant function $\kappa_\varphi(\phi) = -\log(-\phi)$ and dispersion parameter 2. Thus, we have the structure of an approximate gamma model on the right-hand side of (5.60) with, for given μ,

$$\mathbb{E}_\phi[\mathfrak{d}|\mu] \approx \kappa'_\varphi(\phi) = -\frac{1}{\phi} = \varphi, \tag{5.61}$$

$$\mathrm{Var}_\phi(\mathfrak{d}|\mu) \approx 2\kappa''_\varphi(\phi) = 2\frac{1}{\phi^2} = 2\varphi^2. \tag{5.62}$$

These statements say that for given μ and assuming that the saddlepoint approximation is sufficiently accurate, $\mathfrak{d}$ is approximately gamma distributed with shape parameter 1/2 and canonical parameter ϕ (which relates to the dispersion φ in the mean parametrization). Thus, we can estimate ϕ and φ, respectively, with a (second) GLM from (5.60), for given mean parameter μ.

Remarks 5.28

- The accuracy of the saddlepoint approximation is discussed in Section 3.2 of Smyth–Verbyla [343]. The saddlepoint approximation is exact in the Gaussian and the inverse Gaussian case. In the Gaussian case, we have log-likelihood

$$\ell_Y(\mu,\phi) = \frac{\mathfrak{d}\phi - (-\log(-\phi))}{2} - \frac{1}{2}\log\left(\frac{2\pi}{v}\right),$$

with variance function $V(Y) = 1$. In the inverse Gaussian case, we have log-likelihood

$$\ell_Y(\mu,\phi) = \frac{\mathfrak{d}\phi - (-\log(-\phi))}{2} - \frac{1}{2}\log\left(\frac{2\pi}{v}Y^3\right),$$

with variance function $V(Y) = Y^3$. Thus, in the Gaussian case and in the inverse Gaussian case we have a gamma model for $\mathfrak{d}$ with mean φ and shape parameter $1/2$, for given μ; for a related result we also refer to Theorem 3 of Blæsild–Jensen [38]. For Tweedie's models with $p \geq 1$, one can show that the relative error of the saddlepoint approximation is a non-increasing function of the squared coefficient of variation $\tau = \frac{\varphi}{v}V(y)/y^2 = \frac{\varphi}{v}y^{p-2}$, leading to small approximation errors if φ/v is sufficiently small; typically one requires $\tau < 1/3$, see Section 3.2 of Smyth–Verbyla [343].

- The saddlepoint approximation itself does not provide a density because in general the term $O(\varphi/v)$ in (5.59) is non-zero. Nelder–Pregibon [282] renormalized the saddlepoint approximation to a proper density and studied its properties.
- In the gamma EDF case, the saddlepoint approximation would not be necessary because this case can still be solved in closed form. In fact, in the gamma EDF case we have log-likelihood, set $\phi = -v/\varphi < 0$,

$$\ell_Y(\mu,\phi) = \frac{\phi\mathfrak{d}(Y,\mu) - \chi(\phi)}{2} - \log Y, \tag{5.63}$$

with $\chi(\phi) = 2(\log\Gamma(-\phi) + \phi\log(-\phi) - \phi)$. For given μ, this is an EDF for $\mathfrak{d}(Y,\mu)$ with cumulant function χ on the effective domain $(-\infty, 0)$. This provides us with expected value and variance

$$\mathbb{E}_\phi[\mathfrak{d}(Y,\mu)|\mu] = \chi'(\phi) = 2\left(-\Psi(-\phi) + \log(-\phi)\right) \approx -\frac{1}{\phi},$$

$$\mathrm{Var}_\phi(\mathfrak{d}(Y,\mu)|\mu) = 2\chi''(\phi) = 4\left(\Psi'(-\phi) - \frac{1}{-\phi}\right),$$

with digamma function Ψ and the approximation exactly refers to the saddlepoint approximation; for the variance statement we also refer to Fisher's information (2.30). For receiving more accurate mean approximations one can consider higher order terms, e.g., the second order approximation is $\chi'(\phi) \approx -1/\phi + 1/(6\phi^2)$. In fact, from the saddlepoint approximation (5.60) and from the exact formula (5.63) we receive in the gamma case Stirling's formula

$$\Gamma(\gamma) \approx \sqrt{2\pi}\,\gamma^{\gamma-1/2}e^{-\gamma}.$$

In the subsequent examples we will just use the saddlepoint approximation also in the gamma EDF case.

5.5.3 Residual Maximum Likelihood Estimation

The saddlepoint approximation (5.60) proposes to alternate MLE of $\boldsymbol{\beta}$ for the mean model (5.56) and of $\boldsymbol{\gamma}$ for the dispersion model (5.57). Fisher's information matrix of the saddlepoint approximation (5.60) w.r.t. the canonical parameters θ and ϕ is given by

$$\mathcal{I}(\theta,\phi) = -\mathbb{E}_{\theta,\phi}\begin{pmatrix} \phi v \kappa''(\theta) & -v\left(Y-\kappa'(\theta)\right) \\ -v\left(Y-\kappa'(\theta)\right) & -\frac{1}{2}\frac{1}{\phi^2}\end{pmatrix} = \begin{pmatrix} \frac{v}{\varphi(\phi)} V(\mu(\theta)) & 0 \\ 0 & \frac{1}{2}V_\varphi(\varphi(\phi))\end{pmatrix},$$

with variance function $V_\varphi(\varphi) = \varphi^2$, and emphasizing that we work in the canonical parametrization (θ, ϕ). This is a positive definite diagonal matrix which suggests that the algorithm alternating the $\boldsymbol{\beta}$ and $\boldsymbol{\gamma}$ estimations will have a fast convergence. For fixed estimate $\widehat{\boldsymbol{\gamma}}$ we calculate estimated dispersion parameters $\widehat{\varphi}_i = g_\varphi^{-1}\langle\widehat{\boldsymbol{\gamma}}, \boldsymbol{z}_i\rangle$ of policies $1 \le i \le n$, see (5.57). These then allow us to calculate diagonal working weight matrix

$$W(\boldsymbol{\beta}) = \operatorname{diag}\left(\left(\frac{\partial g(\mu_i)}{\partial \mu_i}\right)^{-2} \frac{v_i}{\widehat{\varphi}_i}\frac{1}{V(\mu_i)}\right)_{1\le i\le n} \in \mathbb{R}^{n\times n},$$

which is used in Fisher's scoring method/IRLS algorithm (5.12) to receive MLE $\widehat{\boldsymbol{\beta}}$, given the estimates $(\widehat{\varphi}_i)_i$. These MLEs allow us to estimate the mean parameters $\widehat{\mu}_i = g^{-1}\langle\widehat{\boldsymbol{\beta}}, \boldsymbol{x}_i\rangle$, and to calculate the deviances

$$\mathfrak{d}_i = v_i \mathfrak{d}\left(Y_i, \widehat{\mu}_i\right) = 2v_i\Big(Y_i h\left(Y_i\right) - \kappa\left(h\left(Y_i\right)\right) - Y_i h\left(\widehat{\mu}_i\right) + \kappa\left(h\left(\widehat{\mu}_i\right)\right)\Big) \ge 0.$$

Using (5.60) we know that these deviances can be approximated by gamma distributions $\Gamma(1/2, 1/(2\varphi_i))$. This is a single-parameter EDF with dispersion parameter 2 (as nuisance parameter) and mean parameter φ_i. This motivates the definition of the working weight matrix (based on the gamma EDF model)

$$W_\varphi(\boldsymbol{\gamma}) = \operatorname{diag}\left(\left(\frac{\partial g_\varphi(\varphi_i)}{\partial \varphi_i}\right)^{-2} \frac{1}{2}\frac{1}{V_\varphi(\varphi_i)}\right)_{1\le i\le n} \in \mathbb{R}^{n\times n},$$

and the working residuals

$$\boldsymbol{R}_\varphi(\mathfrak{d}, \boldsymbol{\gamma}) = \left(\frac{\partial g_\varphi(\varphi_i)}{\partial \varphi_i}(\mathfrak{d}_i - \varphi_i)\right)^\top_{1\le i\le n} \in \mathbb{R}^n.$$

Fisher's scoring method (5.12) iterates for $s \geq 0$ the following recursion to receive $\widehat{\gamma}$

$$\widehat{\gamma}^{(s)} \mapsto \widehat{\gamma}^{(s+1)} = \left(\mathfrak{Z}^\top W_\varphi(\widehat{\gamma}^{(s)})\mathfrak{Z}\right)^{-1} \mathfrak{Z}^\top W_\varphi(\widehat{\gamma}^{(s)}) \left(\mathfrak{Z}\widehat{\gamma}^{(s)} + \boldsymbol{R}_\varphi(\mathfrak{d}, \widehat{\gamma}^{(s)})\right), \tag{5.64}$$

where $\mathfrak{Z} = (z_1, \ldots, z_n)^\top$ is the design matrix used to estimate the dispersion parameters.

5.5.4 Lab: Double GLM Algorithm for Gamma Claim Sizes

We revisit the Swedish motorcycle claim size data studied in Sect. 5.3.7. We expand the gamma claim size GLM to a double GLM also modeling the systematic effects in the dispersion parameter. In a first step we need to change the parametrization of the gamma model of Sect. 5.3.7. In the former section we have modeled the average claim size $S_i/n_i \sim \Gamma(n_i\alpha_i, n_ic_i)$, but for applying the saddlepoint approximation we should use the reproductive form (5.44) of the gamma model. We therefore set

$$Y_i = S_i/(n_i\alpha_i) \sim \Gamma(n_i\alpha_i, n_i\alpha_ic_i). \tag{5.65}$$

The reason for the different parametrization in Sect. 5.3.7 has been that (5.65) is not directly useful if α_i is unknown because in that case the observations Y_i cannot be calculated. In this section we estimate $\varphi_i = 1/\alpha_i$ which allows us to model (5.65); a different treatment within Tweedie's family is presented in Sect. 11.1.3. The only difficulty is to initialize the double GLM algorithm. We proceed as follows.

(0) In an initial step we assume constant dispersion $\varphi_i = 1/\alpha_i \equiv 1/\alpha = 1$. This gives us exactly the mean estimates of Sect. 5.3.7 for $S_i/n_i \sim \Gamma(n_i\alpha, n_ic_i)$; note that for constant shape parameter α the mean of S_i/n_i can be estimated without explicit knowledge of α (because it cancels in the score equations). Using these mean estimates we calculate the MLE $\widehat{\alpha}^{(0)}$ of the (constant) shape parameter α, see Remark 5.26. This then allows us to determine the (scaled) observations $Y_i^{(1)} = S_i/(n_i\widehat{\alpha}^{(0)})$ and we initialize $\widehat{\varphi}_i^{(0)} = 1/\widehat{\alpha}^{(0)}$.

(1) Iterate for $t \geq 1$:

- estimate the mean μ_i of Y_i using the mean GLM (5.56) based on the observations $Y_i^{(t)}$ and the dispersion estimates $\widehat{\varphi}_i^{(t-1)}$. This provides us with $\widehat{\mu}_i^{(t)}$;
- based on the deviances $\mathfrak{d}_i^{(t)} = v_i\mathfrak{d}(Y_i^{(t)}, \widehat{\mu}_i^{(t)})$, calculate the updated dispersion estimates $\widehat{\varphi}_i^{(t)}$ using the dispersion GLM (5.57) and the residual MLE iteration (5.64) with the saddlepoint approximation. Set for the updated observations $Y_i^{(t+1)} = S_i\widehat{\varphi}_i^{(t)}/n_i$.

Table 5.15 Number of parameters, AICs, Pearson's dispersion estimate, in-sample losses, tenfold cross-validation losses and the in-sample average claim amounts of the null model (gamma intercept model) and the (double) gamma GLM

	# Param.	AIC	Dispersion est. $\widehat{\varphi}^{\mathrm{P}}$	In-sample loss on $\mathcal{L}$	Tenfold CV loss $\widehat{\mathfrak{D}}^{\mathrm{CV}}$	Average amount
Gamma null	$1+1$	14'416	2.057	2.085	2.091	24'641
Gamma GLM2	$7+1$	14'274	1.544	1.719	1.747	25'130
Double gamma GLM	$7+6$	14'258	–	(1.721)	–	26'413

In an initial double GLM analysis we use the feature information $z_i = \boldsymbol{x}_i$ for the dispersion φ_i modeling (5.57). We choose for both GLMs the log-link which leads to concave maximization problems, see Example 5.5. Running the above double GLM algorithm converges in 4 iterations, and analyzing the resulting model we observe that we should drop the variable `RiskClass` from the feature z_i. We then run the same double GLM algorithm with the feature information $\boldsymbol{x}_i$ and the new z_i again, and the results are presented in Table 5.15.

The considered double GLM has parameter dimensions $\boldsymbol{\beta} \in \mathbb{R}^7$ and $\boldsymbol{\gamma} \in \mathbb{R}^6$. To have comparability with AIC of Sect. 5.3.7, we evaluate AIC of the double GLM in the observations S_i/n_i (and not in Y_i; i.e., similar to the gamma GLM). We observe that it has an improved AIC value compared to model Gamma GLM2. Thus, indeed, dispersion modeling seems necessary in this example (under the GLM2 regression structure). We do not calculate in-sample and cross-validation losses in the double GLM because in the other two models of Table 5.15 we have set $\varphi = 1$ in these statistics. However, the in-sample loss of model Gamma GLM2 with $\varphi = 1$ corresponds to the (homogeneous) deviance dispersion estimate (up to scaling $n/(n-(q+1))$), and this in-sample loss of 1.719 can directly be compared to the average estimated dispersion $m^{-1}\sum_{i=1}^m \widehat{\varphi}_i = 1.721$ (in round brackets in Table 5.15). On the downside, the double GLM has a bigger bias which needs an adjustment.

In Fig. 5.13 (lhs) we give the normal plots of model Gamma GLM2 and the double gamma GLM model. This plot is received by transforming the observations to normal quantiles using the corresponding estimated gamma models. We see quite some similarity between the two estimated gamma models. Both models seem to have similar deficiencies, i.e., dispersion modeling improves explanation of observations, however, either the regression function or the gamma distributional assumption does not fully fit the data, especially for small claims. Finally, in Fig. 5.13 (rhs) we plot the estimated dispersion parameters $\widehat{\varphi}_i$ against the logged estimated means $\log(\widehat{\mu}_i)$ (linear predictors). We observe that the estimated dispersion has a (weak) U-shape as a function of the expected claim sizes which indicates that the tails cannot fully be captured by our model. This closes this example.

Remark 5.29 For the dispersion estimation $\widehat{\varphi}_i$ we use as observations the deviances $\mathfrak{d}_i = v_i \mathfrak{d}(Y_i, \widehat{\mu}_i)$, $1 \le i \le n$. On a finite sample, these deviances are typically biased due to the use of the estimated means $\widehat{\mu}_i$. Smyth–Verbyla [343] propose the

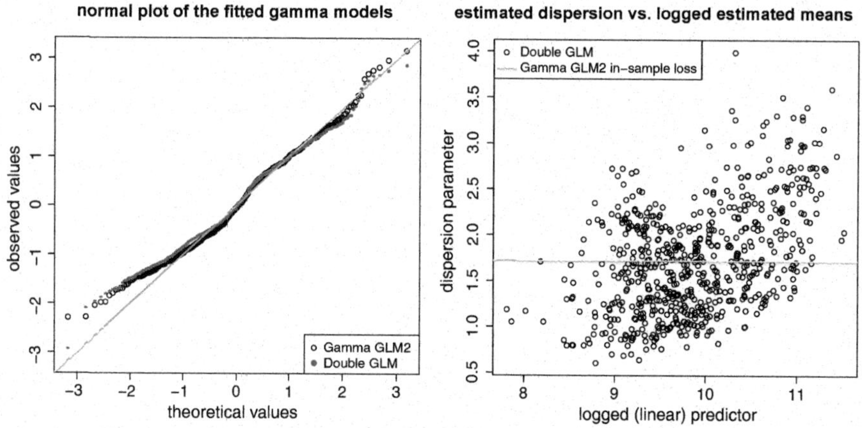

Fig. 5.13 (lhs) Normal plot of the fitted models Gamma GLM2 and double GLM, (rhs) estimated dispersion parameters $\widehat{\varphi}_i$ against the logged estimated means $\log(\widehat{\mu}_i)$ (the orange line gives the in-sample loss in model Gamma GLM2)

following bias correction. Consider the estimated hat matrix defined by

$$H = W(\widehat{\beta},\widehat{\gamma})^{1/2}\mathfrak{X}\left(\mathfrak{X}^\top W(\widehat{\beta},\widehat{\gamma})\,\mathfrak{X}\right)^{-1}\mathfrak{X}^\top W(\widehat{\beta},\widehat{\gamma})^{1/2},$$

with the diagonal work weight matrix $W(\widehat{\beta},\widehat{\gamma})$ depending on the estimated regression parameters $\widehat{\beta}$ and $\widehat{\gamma}$ through μ and φ. Denote the diagonal entries of the hat matrix by $(h_{i,i})_{1\le i\le n}$. A bias corrected version of the deviances is received by considering observations $(1-h_{i,i})^{-1}\mathfrak{d}_i = (1-h_{i,i})^{-1}v_i\mathfrak{d}(Y_i,\widehat{\mu}_i)$, $1\le i\le n$. We will come back to the hat matrix H in Sect. 5.6.1, below.

5.5.5 *Tweedie's Compound Poisson GLM*

A popular situation for applying the double GLM framework is Tweedie's CP model introduced in Sect. 2.2.3, in particular, we refer to Proposition 2.17 for the corresponding parametrization. Having claim frequency and claim sizes involved, such a model can hardly be calibrated with one single regression function and a constant dispersion parameter. An obvious choice is a double GLM, this is the proposal presented in Smyth–Jørgensen [342]. In most of the cases one chooses for both link functions g and g_φ the log-links because positivity needs to be guaranteed.

This implies for the two working weight matrices of the double GLM

$$W(\boldsymbol{\beta}) = \mathrm{diag}\left(\mu_i^2 \frac{v_i}{\varphi_i} \frac{1}{V(\mu_i)}\right)_{1\le i\le n} = \mathrm{diag}\left(\mu_i^{2-p} \frac{v_i}{\varphi_i}\right)_{1\le i\le n},$$

$$W_\varphi(\boldsymbol{\gamma}) = \mathrm{diag}\left(\varphi_i^2 \frac{1}{2} \frac{1}{V_\varphi(\varphi_i)}\right)_{1\le i\le n} = \mathrm{diag}(1/2, \ldots, 1/2).$$

The deviances in Tweedie's CP model are given by, see (4.18),

$$\mathfrak{d}_i = v_i \mathfrak{d}(Y_i, \widehat{\mu}_i) = 2v_i \left(Y_i \frac{Y_i^{1-p} - \widehat{\mu}_i^{1-p}}{1-p} - \frac{Y_i^{2-p} - \widehat{\mu}_i^{2-p}}{2-p} \right) \ge 0,$$

and these deviances could still be de-biased, see Remark 5.29. The working responses for the two GLMs are

$$\boldsymbol{R} = (Y_i/\mu_i - 1)_{1\le i\le n}^\top \qquad \text{and} \qquad \boldsymbol{R}_\varphi = (\mathfrak{d}_i/\varphi_i - 1)_{1\le i\le n}^\top.$$

The drawback of this approach is that it only considers the (scaled) total claim amounts $Y_i = S_i\varphi_i/v_i$ as observations, see Proposition 2.17. These total claim amounts consist of the number of claims N_i and i.i.d. individual claim sizes $Z_{i,j} \sim \Gamma(\alpha, c_i)$, supposed $N_i \ge 1$. Having observations of both claim amounts S_i and claim counts N_i allows one to build a Poisson GLM for claim counts and a gamma GLM for claim sizes which can be estimated separately. This has also been the reason of Smyth–Jørgensen [342] to enhance Tweedie's model estimation for known claim counts in their Section 4. Moreover, in Theorem 4 of Delong et al. [94] it is proved that the two GLM approaches can be identified under log-link choices.

5.6 Diagnostic Tools

In our examples we have studied several figures like AIC, cross-validation losses, etc., for model and parameter selection. Moreover, we have plotted the results, for instance, using the Tukey–Anscombe plot or the QQ plot. Of course, there are numerous other plots and tools that can help us to analyze the results and to improve the resulting models. We present some of these in this section.

5.6.1 The Hat Matrix

The MLE $\widehat{\boldsymbol{\beta}}^{\mathrm{MLE}}$ satisfies at convergence of the IRLS algorithm, see (5.12),

$$\widehat{\boldsymbol{\beta}}^{\mathrm{MLE}} = \left(\mathfrak{X}^\top W(\widehat{\boldsymbol{\beta}}^{\mathrm{MLE}})\mathfrak{X}\right)^{-1} \mathfrak{X}^\top W(\widehat{\boldsymbol{\beta}}^{\mathrm{MLE}}) \left(\mathfrak{X}\widehat{\boldsymbol{\beta}}^{\mathrm{MLE}} + \boldsymbol{R}(\boldsymbol{Y}, \widehat{\boldsymbol{\beta}}^{\mathrm{MLE}})\right),$$

with working residuals for $\boldsymbol{\beta} \in \mathbb{R}^{q+1}$

$$\boldsymbol{R}(\boldsymbol{Y}, \boldsymbol{\beta}) = \left(\left. \frac{\partial g(\mu_i)}{\partial \mu_i} \right|_{\mu_i = \mu_i(\boldsymbol{\beta})} (Y_i - \mu_i(\boldsymbol{\beta})) \right)^{\top}_{1 \le i \le n} \in \mathbb{R}^n.$$

Following Section 4.2.2 of Fahrmeir–Tutz [123], this allows us to define the so-called *hat matrix*, see also Remark 5.29,

$$H = H(\widehat{\boldsymbol{\beta}}^{\text{MLE}}) = W(\widehat{\boldsymbol{\beta}}^{\text{MLE}})^{1/2} \mathfrak{X} \left(\mathfrak{X}^{\top} W(\widehat{\boldsymbol{\beta}}^{\text{MLE}}) \mathfrak{X} \right)^{-1} \mathfrak{X}^{\top} W(\widehat{\boldsymbol{\beta}}^{\text{MLE}})^{1/2} \in \mathbb{R}^{n \times n}, \tag{5.66}$$

recall that the working weight matrix $W(\boldsymbol{\beta})$ is diagonal. The hat matrix H is symmetric and idempotent, i.e. $H^2 = H$, with $\text{trace}(H) = \text{rank}(H) = q + 1$. Therefore, H acts as a projection, mapping the observations $\widetilde{\boldsymbol{Y}}$ to the fitted values

$$\widetilde{\boldsymbol{Y}} \stackrel{\text{def.}}{=} W(\widehat{\boldsymbol{\beta}}^{\text{MLE}})^{1/2} \left(\mathfrak{X} \widehat{\boldsymbol{\beta}}^{\text{MLE}} + \boldsymbol{R}(\boldsymbol{Y}, \widehat{\boldsymbol{\beta}}^{\text{MLE}}) \right) \mapsto H \widetilde{\boldsymbol{Y}} = W(\widehat{\boldsymbol{\beta}}^{\text{MLE}})^{1/2} \mathfrak{X} \widehat{\boldsymbol{\beta}}^{\text{MLE}} = W(\widehat{\boldsymbol{\beta}}^{\text{MLE}})^{1/2} \widehat{\boldsymbol{\eta}},$$

the latter being the fitted linear predictors. The diagonal elements $h_{i,i}$ of this hat matrix H satisfy $0 \le h_{i,i} \le 1$, and values close to 1 correspond to extreme data points i, in particular, for $h_{i,i} = 1$ only observation $\widetilde{Y}_i$ influences $\widehat{\eta}_i$, whereas for $h_{i,i} = 0$ observation $\widetilde{Y}_i$ has no influence on $\widehat{\eta}_i$.

Figure 5.14 gives the resulting hat matrices of the double gamma GLM of Sect. 5.5.4. On the left-hand side we show the diagonal entries $h_{i,i}$ of the claim

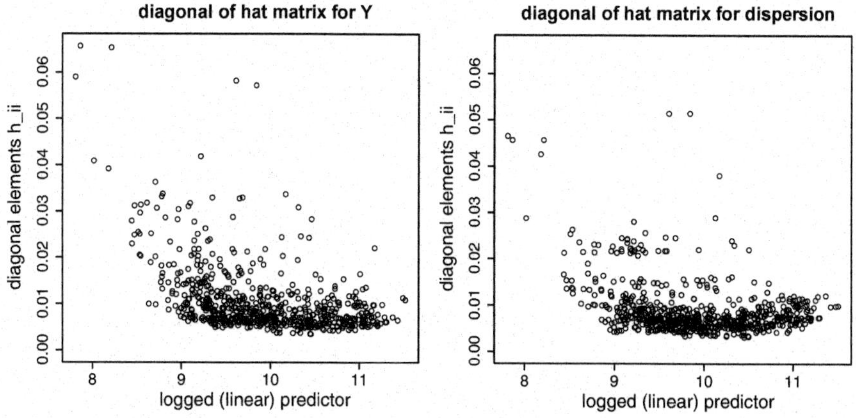

Fig. 5.14 Diagonal entries $h_{i,i}$ of the two hat matrices of the example in Sect. 5.5.4: (lhs) for means $\widehat{\mu}_i$ and responses Y_i, and (rhs) for dispersions $\widehat{\varphi}_i$ and responses $\mathfrak{d}_i$

amount responses Y_i (for the estimation of μ_i), and on the right-hand side the corresponding plots for the deviance responses $\mathfrak{d}_i$ (for the estimation of φ_i). These diagonal elements $h_{i,i}$ are ordered on the x-axis w.r.t. the linear predictors $\widehat{\eta}_i$. From this figure we conclude that the diagonal entries of the hat matrices are bigger for very small responses in our example, and the dispersion plot has a couple of more special observations that may require further analysis.

5.6.2 Case Deletion and Generalized Cross-Validation

As a continuation of the previous subsection we can analyze the influence of an individual observation Y_i on the estimation of regression parameter $\boldsymbol{\beta}$. This influence is naturally measured by fitting the regression parameter based on the full data $\mathcal{D}$ and based only on the observations $\mathcal{L}_{(-i)} = \mathcal{D} \setminus \{Y_i\}$, we also refer to leave-one-out cross-validation in Sect. 4.2.2. The influence of observation Y_i is then obtained by comparing $\widehat{\boldsymbol{\beta}}^{\text{MLE}}$ and $\widehat{\boldsymbol{\beta}}^{\text{MLE}}_{(-i)}$. Since fitting n different models by individually leaving out each observation Y_i is too costly, one only explores a one-step Fisher's scoring update starting from $\widehat{\boldsymbol{\beta}}^{\text{MLE}}$ that provides an approximation to $\widehat{\boldsymbol{\beta}}^{\text{MLE}}_{(-i)}$, that is,

$$\begin{aligned}
\widehat{\boldsymbol{\beta}}^{(1)}_{(-i)} &= \left(\mathfrak{X}^\top_{(-i)} W_{(-i)}(\widehat{\boldsymbol{\beta}}^{\text{MLE}})\mathfrak{X}_{(-i)}\right)^{-1} \mathfrak{X}^\top_{(-i)} W_{(-i)}(\widehat{\boldsymbol{\beta}}^{\text{MLE}}) \left(\mathfrak{X}\widehat{\boldsymbol{\beta}}^{\text{MLE}} + \boldsymbol{R}(\boldsymbol{Y}, \widehat{\boldsymbol{\beta}}^{\text{MLE}})\right)_{(-i)} \\
&= \left(\mathfrak{X}^\top_{(-i)} W_{(-i)}(\widehat{\boldsymbol{\beta}}^{\text{MLE}})\mathfrak{X}_{(-i)}\right)^{-1} \mathfrak{X}^\top_{(-i)} W_{(-i)}(\widehat{\boldsymbol{\beta}}^{\text{MLE}})^{1/2}\, \widetilde{\boldsymbol{Y}}_{(-i)},
\end{aligned}$$

where all lower indices $_{(-i)}$ indicate that we drop the corresponding row or/and column from the matrices and vectors, and where $\widetilde{\boldsymbol{Y}}$ has been defined in the previous subsection. This allows us to compare $\widehat{\boldsymbol{\beta}}^{\text{MLE}}$ and $\widehat{\boldsymbol{\beta}}^{(1)}_{(-i)}$ to analyze the influence of observation Y_i.

To reformulate this approximation, we come back to the hat matrix $H = H(\widehat{\boldsymbol{\beta}}^{\text{MLE}}) = (h_{i,j})_{1\le i,j\le n}$ defined in (5.66). It fulfills

$$W(\widehat{\boldsymbol{\beta}}^{\text{MLE}})^{1/2}\mathfrak{X}\widehat{\boldsymbol{\beta}}^{\text{MLE}} = H\widetilde{\boldsymbol{Y}} = \left(\sum_{j=1}^n h_{1,j}\widetilde{Y}_j, \ldots, \sum_{j=1}^n h_{n,j}\widetilde{Y}_j\right)^\top \in \mathbb{R}^n.$$

Thus, for predicting Y_i we can consider the linear predictor (for the chosen link g)

$$\widehat{\eta}_i = g(\widehat{\mu}_i) = \langle \widehat{\boldsymbol{\beta}}^{\text{MLE}}, \boldsymbol{x}_i\rangle = (\mathfrak{X}\widehat{\boldsymbol{\beta}}^{\text{MLE}})_i = W_{i,i}(\widehat{\boldsymbol{\beta}}^{\text{MLE}})^{-1/2} \sum_{j=1}^n h_{i,j}\widetilde{Y}_j.$$

A computation of the linear predictor of Y_i using the leave-one-out approximation $\widehat{\boldsymbol{\beta}}^{(1)}_{(-i)}$ gives

$$\widehat{\eta}_i^{(-i,1)} = \langle \widehat{\boldsymbol{\beta}}^{(1)}_{(-i)}, \boldsymbol{x}_i \rangle = \frac{1}{1-h_{i,i}} \widehat{\eta}_i - W_{i,i}(\widehat{\boldsymbol{\beta}}^{\text{MLE}})^{-1/2} \frac{h_{i,i}}{1-h_{i,i}} \widetilde{Y}_i.$$

This allows one to efficiently calculate a leave-one-out prediction using the hat matrix H. This also motivates to study the *generalized cross-validation* (GCV) loss which is an approximation to leave-one-out cross-validation, see Sect. 4.2.2,

$$\begin{aligned} \widehat{\mathfrak{D}}^{\text{GCV}} &= \frac{1}{n} \sum_{i=1}^{n} \frac{v_i}{\varphi} \, \mathfrak{d}\left(Y_i, g^{-1}(\widehat{\eta}_i^{(-i,1)})\right) \qquad (5.67) \\ &= \frac{2}{n} \sum_{i=1}^{n} \frac{v_i}{\varphi} \Big[Y_i h(Y_i) - \kappa(h(Y_i)) - Y_i h\left(g^{-1}(\widehat{\eta}_i^{(-i,1)})\right) + \kappa\left(h\left(g^{-1}(\widehat{\eta}_i^{(-i,1)})\right)\right) \Big]. \end{aligned}$$

Example 5.30 (Generalized Cross-Validation Loss in the Gaussian Case) We study the generalized cross-validation loss $\widehat{\mathfrak{D}}^{\text{GCV}}$ in the homoskedastic Gaussian case $v_i/\varphi \equiv 1/\sigma^2$ with cumulant function $\kappa(\theta) = \theta^2/2$ and canonical link $g(\mu) = h(\mu) = \mu$. The generalized cross-validation loss in the Gaussian case is given by

$$\widehat{\mathfrak{D}}^{\text{GCV}} = \frac{1}{n} \sum_{i=1}^{n} \frac{1}{\sigma^2} \left(Y_i - \widehat{\eta}_i^{(-i,1)} \right)^2,$$

with (linear) leave-one-out predictor

$$\widehat{\eta}_i^{(-i,1)} = \langle \widehat{\boldsymbol{\beta}}^{(1)}_{(-i)}, \boldsymbol{x}_i \rangle = \sum_{j=1, j\neq i}^{n} \frac{h_{i,j}}{1-h_{i,i}} Y_j = \frac{1}{1-h_{i,i}} \widehat{\eta}_i - \frac{h_{i,i}}{1-h_{i,i}} Y_i.$$

This gives us generalized cross-validation loss in the Gaussian case

$$\widehat{\mathfrak{D}}^{\text{GCV}} = \frac{1}{n} \sum_{i=1}^{n} \frac{1}{\sigma^2} \left(\frac{Y_i - \widehat{\eta}_i}{1-h_{i,i}} \right)^2,$$

with $\boldsymbol{\beta}$ independent hat matrix

$$H = \mathfrak{X} \left(\mathfrak{X}^\top \mathfrak{X} \right)^{-1} \mathfrak{X}^\top.$$

The generalized cross-validation loss is used, for instance, for generalized additive model (GAM) fitting where an efficient and fast cross-validation method is required to select regularization parameters. Generalized cross-validation has been introduced by Craven–Wahba [84] but these authors replaced $h_{i,i}$ by $\sum_{j=1}^{n} h_{j,j}/n$. It holds that $\sum_{j=1}^{n} h_{j,j} = \text{trace}(H) = q + 1$, thus, using this approximation we receive

$$\widehat{\mathfrak{D}}^{\text{GCV}} \approx \frac{1}{n}\sum_{i=1}^{n} \frac{1}{\sigma^2}\left(\frac{Y_i - \widehat{\eta}_i}{1-\sum_{j=1}^{n} h_{j,j}/n}\right)^2 = \frac{n}{(n-(q+1))^2}\sum_{i=1}^{n}\frac{(Y_i - \widehat{\eta}_i)^2}{\sigma^2}$$
$$= \frac{n}{n-(q+1)}\frac{\widehat{\varphi}^{\text{P}}}{\sigma^2},$$

with $\widehat{\varphi}^{\text{P}}$ being Pearson's dispersion estimate in the Gaussian model, see (5.30). ∎

We give a numerical example based on the gamma GLM for the claim sizes studied in Sect. 5.3.7.

Example 5.31 (Leave-One-Out Cross-Validation) The aim of this example is to compare the generalized cross-validation loss $\widehat{\mathfrak{D}}^{\text{GCV}}$ to the leave-one-out cross-validation loss $\widehat{\mathfrak{D}}^{\text{loo}}$, see (4.34), the former being an approximation to the latter. We do this for the gamma claim size model studied in Sect. 5.3.7. In this example it is feasible to exactly calculate the leave-one-out cross-validation loss because we have only 656 claims.

The results are presented in Table 5.16. Firstly, the different cross-validation losses confirm that the model slightly (in-sample) over-fits to the data, which is not a surprise when estimating 7 regression parameters based on 656 observations. Secondly, the cross-validation losses provide similar numbers with leave-one-out being slightly bigger than tenfold cross-validation, here. Thirdly, the generalized cross-validation loss $\widehat{\mathfrak{D}}^{\text{GCV}}$ manages to approximate the leave-one-out cross-validation loss $\widehat{\mathfrak{D}}^{\text{loo}}$ very well in this example.

Table 5.17 gives the corresponding results for model Poisson GLM1 of Sect. 5.2.4. Firstly, in this example with 610'206 observations it is not feasible to calculate the leave-one-out cross-validation loss (for computational reasons). Therefore, we rely on the generalized cross-validation loss as an approximation. From the results of Table 5.17 it seems that this approximation (rather) underestimates the loss (compared to tenfold cross-validation). Indeed, this is an observation that we have made also in other examples. ∎

Table 5.16 Comparison of different cross-validation losses for model Gamma GLM2

	Gamma GLM2
In-sample loss $\mathfrak{D}(\mathcal{L}, \widehat{\mu}_{\mathcal{L}}^{\text{MLE}})$	1.719
Tenfold CV loss $\widehat{\mathfrak{D}}^{\text{CV}}$	1.747
Leave-one-out CV loss $\widehat{\mathfrak{D}}^{\text{loo}}$	1.756
Generalized CV loss $\widehat{\mathfrak{D}}^{\text{GCV}}$	1.758

Table 5.17 Comparison of different cross-validation losses for model Poisson GLM1

	Poisson GLM1
In-sample loss $\mathfrak{D}(\mathcal{L}, \widehat{\mu}_{\mathcal{L}}^{\text{MLE}})$	24.101
Tenfold CV loss $\widehat{\mathfrak{D}}^{\text{CV}}$	24.121
Leave-one-out CV loss $\widehat{\mathfrak{D}}^{\text{loo}}$	N/A
Generalized CV loss $\widehat{\mathfrak{D}}^{\text{GCV}}$	24.105

5.7 Generalized Linear Models with Categorical Responses

The reader will have noticed that the discussion of GLMs in this chapter has been focusing on the single-parameter linear EDF case (5.1). In many actuarial applications we also want to study examples of the vector-valued parameter EF (2.2). We briefly discuss the categorical case since this case is frequently used.

5.7.1 *Logistic Categorical Generalized Linear Model*

We recall the EF representation of the categorical distribution studied in Sect. 2.1.4. We choose as ν the counting measure on the finite set $\mathcal{Y} = \{1, \ldots, k+1\}$. A random variable Y taking values in $\mathcal{Y}$ is called categorical, and the levels $y \in \mathcal{Y}$ can either be ordinal or nominal. This motivates dummy coding of the categorical random variable Y providing

$$T(Y) = (\mathbb{1}_{\{Y=1\}}, \ldots, \mathbb{1}_{\{Y=k\}})^\top \in \{0, 1\}^k, \tag{5.68}$$

thus, $k + 1$ has been chosen as reference level. For the canonical parameter $\boldsymbol{\theta} = (\theta_1, \ldots, \theta_k)^\top \in \boldsymbol{\Theta} = \mathbb{R}^k$ we have cumulant function and mean functional, respectively,

$$\kappa(\boldsymbol{\theta}) = \log\left(1 + \sum_{j=1}^k e^{\theta_j}\right), \qquad \boldsymbol{p} = \mathbb{E}_{\boldsymbol{\theta}}[T(Y)] = \nabla_{\boldsymbol{\theta}}\kappa(\boldsymbol{\theta}) = \frac{e^{\boldsymbol{\theta}}}{1 + \sum_{j=1}^k e^{\theta_j}}.$$

With these choices we receive the EF representation of the categorical distribution (set $\theta_{k+1} = 0$)

$$dF(y; \boldsymbol{\theta}) = \exp\left\{\boldsymbol{\theta}^\top T(y) - \log\left(1 + \sum_{j=1}^k e^{\theta_j}\right)\right\} d\nu(y) = \prod_{l=1}^{k+1}\left(\frac{e^{\theta_l}}{\sum_{j=1}^{k+1} e^{\theta_j}}\right)^{\mathbb{1}_{\{y=l\}}} d\nu(y).$$

The covariance matrix of $T(Y)$ is given by

$$\Sigma(\boldsymbol{\theta}) = \text{Var}_{\boldsymbol{\theta}}(T(Y)) = \nabla_{\boldsymbol{\theta}}^2\kappa(\boldsymbol{\theta}) = \text{diag}\,(\boldsymbol{p}) - \boldsymbol{p}\boldsymbol{p}^\top \in \mathbb{R}^{k\times k}.$$

Assume that we have feature information $\boldsymbol{x} \in \mathcal{X} \subset \{1\} \times \mathbb{R}^q$ for response variable Y. This allows us to lift this categorical model to a GLM. The *logistic GLM* assumes for $\boldsymbol{p} = (p_1, \ldots, p_k)^\top \in (0, 1)^k$ a regression function, $1 \leq l \leq k$,

$$\boldsymbol{x} \mapsto p_l = p_l(\boldsymbol{x}) = \mathbb{P}_{\boldsymbol{\beta}}[Y = l] = \frac{\exp\langle \boldsymbol{\beta}_l, \boldsymbol{x}\rangle}{1 + \sum_{j=1}^k \exp\langle \boldsymbol{\beta}_j, \boldsymbol{x}\rangle}, \tag{5.69}$$

for regression parameter $\boldsymbol{\beta} = (\boldsymbol{\beta}_1^\top, \ldots, \boldsymbol{\beta}_k^\top)^\top \in \mathbb{R}^{k(q+1)}$. Equivalently, we can rewrite these regression probabilities relative to the reference level, that is, we consider linear predictors for $1 \leq l \leq k$

$$\eta_l(\boldsymbol{x}) = \log\left(\frac{\mathbb{P}_{\boldsymbol{\beta}}[Y = l]}{\mathbb{P}_{\boldsymbol{\beta}}[Y = k+1]}\right) = \langle \boldsymbol{\beta}_l, \boldsymbol{x}\rangle. \tag{5.70}$$

Note that this naturally gives us the canonical link h which we have already derived in Sect. 2.1.4. Define the matrix for feature $\boldsymbol{x} \in \mathcal{X} \subset \{1\} \times \mathbb{R}^q$

$$\boldsymbol{X} = \begin{pmatrix} \boldsymbol{x}^\top & 0 & 0 & \cdots & 0 \\ 0 & \boldsymbol{x}^\top & 0 & \cdots & 0 \\ 0 & 0 & \boldsymbol{x}^\top & \cdots & 0 \\ \vdots & \vdots & \vdots & \ddots & \vdots \\ 0 & 0 & 0 & \cdots & \boldsymbol{x}^\top \end{pmatrix} \in \mathbb{R}^{k \times k(q+1)}. \tag{5.71}$$

This gives linear predictor and canonical parameter, respectively, under the canonical link h

$$\boldsymbol{\theta} = h(\boldsymbol{p}(\boldsymbol{x})) = \boldsymbol{\eta}(\boldsymbol{x}) = \boldsymbol{X}\boldsymbol{\beta} = \big(\langle \boldsymbol{\beta}_1, \boldsymbol{x}\rangle, \ldots, \langle \boldsymbol{\beta}_k, \boldsymbol{x}\rangle\big)^\top \in \boldsymbol{\Theta} = \mathbb{R}^k. \tag{5.72}$$

5.7.2 Maximum Likelihood Estimation in Categorical Models

Assume we have n independent observations Y_i following the logistic categorical GLM (5.69) with features $\boldsymbol{x}_i \in \mathbb{R}^{q+1}$ and $\boldsymbol{X}_i \in \mathbb{R}^{k \times k(q+1)}$, respectively, for $1 \leq i \leq n$. The joint log-likelihood function is given by, we use (5.72),

$$\boldsymbol{\beta} \mapsto \ell_{\boldsymbol{Y}}(\boldsymbol{\beta}) = \sum_{i=1}^n (\boldsymbol{X}_i\boldsymbol{\beta})^\top T(Y_i) - \kappa(\boldsymbol{X}_i\boldsymbol{\beta}).$$

This provides us with score equations

$$s(\boldsymbol{\beta}, \boldsymbol{Y}) = \nabla_{\boldsymbol{\beta}} \ell_{\boldsymbol{Y}}(\boldsymbol{\beta}) = \sum_{i=1}^n \boldsymbol{X}_i^\top \left[T(Y_i) - \nabla_{\boldsymbol{\theta}} \kappa(\boldsymbol{X}_i\boldsymbol{\beta})\right] = \sum_{i=1}^n \boldsymbol{X}_i^\top \left[T(Y_i) - \boldsymbol{p}(\boldsymbol{x}_i)\right] = 0,$$

with logistic regression function (5.69) for $\boldsymbol{p}(\boldsymbol{x})$. For the score equations with canonical link we also refer to the second case in Proposition 5.1. Next, we calculate Fisher's information matrix, we also refer to (3.16),

$$\mathcal{I}_n(\boldsymbol{\beta}) = -\mathbb{E}_{\boldsymbol{\beta}}\left[\nabla^2_{\boldsymbol{\beta}}\ell_{\boldsymbol{Y}}(\boldsymbol{\beta})\right] = \sum_{i=1}^n \boldsymbol{X}_i^\top \Sigma_i(\boldsymbol{\beta})\boldsymbol{X}_i,$$

with covariance matrix of $T(Y_i)$

$$\Sigma_i(\boldsymbol{\beta}) = \nabla^2_{\boldsymbol{\theta}}\kappa(\boldsymbol{X}_i\boldsymbol{\beta}) = \operatorname{diag}(\boldsymbol{p}(\boldsymbol{x}_i)) - \boldsymbol{p}(\boldsymbol{x}_i)\boldsymbol{p}(\boldsymbol{x}_i)^\top.$$

We rewrite the score in a similar way as in Sect. 5.1.4. This requires for general link $g(\boldsymbol{p}) = \boldsymbol{\eta}$ and inverse link $\boldsymbol{p} = g^{-1}(\boldsymbol{\eta})$, respectively, the following block diagonal matrix

$$\begin{aligned} W(\boldsymbol{\beta}) &= \operatorname{diag}\left(\left(\nabla_{\boldsymbol{\eta}} g^{-1}(\boldsymbol{\eta})\Big|_{\boldsymbol{\eta}=\boldsymbol{X}_i\boldsymbol{\beta}}\right)\Sigma_i(\boldsymbol{\beta})^{-1}\left(\nabla_{\boldsymbol{\eta}} g^{-1}(\boldsymbol{\eta})\Big|_{\boldsymbol{\eta}=\boldsymbol{X}_i\boldsymbol{\beta}}\right)^\top\right)_{1\le i\le n} \\ &= \operatorname{diag}\left(\left(\nabla_{\boldsymbol{p}} g(\boldsymbol{p})\Big|_{\boldsymbol{p}=g^{-1}(\boldsymbol{X}_i\boldsymbol{\beta})}\right)^\top \Sigma_i(\boldsymbol{\beta})\left(\nabla_{\boldsymbol{p}} g(\boldsymbol{p})\Big|_{\boldsymbol{p}=g^{-1}(\boldsymbol{X}_i\boldsymbol{\beta})}\right)\right)^{-1}_{1\le i\le n}, \end{aligned} \tag{5.73}$$

and the working residuals

$$\boldsymbol{R}(\boldsymbol{Y},\boldsymbol{\beta}) = \left(\left(\nabla_{\boldsymbol{p}} g(\boldsymbol{p})\Big|_{\boldsymbol{p}=g^{-1}(\boldsymbol{X}_i\boldsymbol{\beta})}\right)^\top (T(Y_i) - \boldsymbol{p}(\boldsymbol{x}_i))\right)_{1\le i\le n}. \tag{5.74}$$

Because we work with the canonical link $g = h$ and $g^{-1} = \nabla_{\boldsymbol{\theta}}\kappa$, we can use the simplified block diagonal matrix

$$W(\boldsymbol{\beta}) = \operatorname{diag}(\Sigma_1(\boldsymbol{\beta}), \ldots, \Sigma_n(\boldsymbol{\beta})) \in \mathbb{R}^{kn\times kn},$$

and the working residuals

$$\boldsymbol{R}(\boldsymbol{Y},\boldsymbol{\beta}) = \left(\Sigma_i(\boldsymbol{\beta})^{-1}(T(Y_i) - \boldsymbol{p}(\boldsymbol{x}_i))\right)_{1\le i\le n} \in \mathbb{R}^{kn}.$$

Finally, we define the design matrix

$$\mathfrak{X} = \begin{pmatrix} \boldsymbol{X}_1 \\ \boldsymbol{X}_2 \\ \vdots \\ \boldsymbol{X}_n \end{pmatrix} \in \mathbb{R}^{kn\times k(q+1)}.$$

Putting everything together we receive the score equations

$$s(\boldsymbol{\beta}, \boldsymbol{Y}) = \nabla_{\boldsymbol{\beta}} \ell_{\boldsymbol{Y}}(\boldsymbol{\beta}) = \mathfrak{X}^\top W(\boldsymbol{\beta}) \boldsymbol{R}(\boldsymbol{Y}, \boldsymbol{\beta}) = 0. \tag{5.75}$$

This is now exactly in the same form as in Proposition 5.1. Fisher's scoring method/IRLS algorithm then allows us to recursively calculate the MLE of $\boldsymbol{\beta} \in \mathbb{R}^{k(q+1)}$ by

$$\widehat{\boldsymbol{\beta}}^{(t)} \mapsto \widehat{\boldsymbol{\beta}}^{(t+1)} = \left(\mathfrak{X}^\top W(\widehat{\boldsymbol{\beta}}^{(t)})\mathfrak{X}\right)^{-1} \mathfrak{X}^\top W(\widehat{\boldsymbol{\beta}}^{(t)}) \left(\mathfrak{X}\widehat{\boldsymbol{\beta}}^{(t)} + \boldsymbol{R}(\boldsymbol{Y}, \widehat{\boldsymbol{\beta}}^{(t)})\right).$$

We have asymptotic normality of the MLE (under suitable regularity conditions)

$$\widehat{\boldsymbol{\beta}}_n^{\mathrm{MLE}} \overset{(\mathrm{d})}{\approx} \mathcal{N}(\boldsymbol{\beta}, \mathcal{I}_n(\boldsymbol{\beta})^{-1}),$$

for large sample sizes n. This allows us to apply the Wald test (5.32) for backward parameter elimination. Moreover, in-sample and out-of-sample losses can be analyzed with unit deviances coming from the categorical cross-entropy loss function (4.19).

Remarks 5.32 The above derivations have been done for the categorical distribution under the canonical link choice. However, these considerations hold true for more general links g within the vector-valued parameter EF. That is, the block diagonal matrix $W(\boldsymbol{\beta})$ in (5.73) and the working residuals $\boldsymbol{R}(\boldsymbol{Y}, \boldsymbol{\beta})$ in (5.74) provide score equations (5.75) for general vector-valued parameter EF examples, and where we replace the categorical probability $\boldsymbol{p}$ by the mean $\mu = \mathbb{E}_{\boldsymbol{\beta}}[T(Y)]$.

5.8 Further Topics of Regression Modeling

There are several special topics and tools in regression modeling that we have not discussed, yet. Some of them will be considered in selected chapters below, and some points are mentioned here, without going into detail.

5.8.1 Longitudinal Data and Random Effects

The GLMs studied above have been considering cross-sectional data, meaning that we have fixed one time period t and studied this time period in an isolated fashion. Time-dependent extensions are called longitudinal or panel data. Consider a time series of data $(Y_{i,t}, \boldsymbol{x}_{i,t})$ for policies $1 \le i \le n$ and time points $t \ge 1$. For the prediction of response variable $Y_{i,t}$ we may then regress on the individual past history of policy i, given by the data

$$\mathcal{D}_{i,t} = \left\{Y_{i,1}, \ldots, Y_{i,t-1}, \boldsymbol{x}_{i,1}, \ldots, \boldsymbol{x}_{i,t}\right\}.$$

In particular, we may explore the distribution of $Y_{i,t}$, conditionally given $\mathcal{D}_{i,t}$,

$$Y_{i,t}|_{\mathcal{D}_{i,t}} \sim F(\cdot|\mathcal{D}_{i,t}; \theta),$$

for canonical parameter $\theta \in \boldsymbol{\Theta}$ and $F(\cdot|\mathcal{D}_{i,t}; \theta)$ being a member of the EDF. For a GLM we choose a link function g and make the assumption

$$g\left(\mathbb{E}_{\boldsymbol{\beta}}[Y_{i,t}|\mathcal{D}_{i,t}]\right) = \langle \boldsymbol{\beta}, \boldsymbol{z}_{i,t}\rangle, \tag{5.76}$$

where $\boldsymbol{z}_{i,t} \in \mathbb{R}^{q+1}$ is a $(q+1)$-dimensional and $\sigma(\mathcal{D}_{i,t})$-measurable feature vector, and regression parameter $\boldsymbol{\beta} \in \mathbb{R}^{q+1}$ describes the common systematic effects across all policies $1 \le i \le n$. This gives a generalized auto-regressive model, and if we have the Markov property

$$F(\cdot|\mathcal{D}_{i,t}; \theta) \stackrel{\text{(d)}}{=} F(\cdot|Y_{i,t-1}, \boldsymbol{x}_{i,t}; \theta) \qquad \text{for all } t \ge 2 \text{ and } \theta \in \boldsymbol{\Theta},$$

we obtain a generalized auto-regressive model of order 1. These longitudinal models allow one to model experience rating, for instance, in car insurance where the past claims history directly influences the future insurance prices, we refer to Remark 5.15 on bonus-malus systems (BMS).

The next level of complexity is obtained by extending regression structure (5.76) by policy i specific random effects $\boldsymbol{B}_i$ such that we may postulate

$$g\left(\mathbb{E}_{\boldsymbol{\beta}}[Y_{i,t}|\mathcal{D}_{i,t}, \boldsymbol{B}_i]\right) = \langle \boldsymbol{\beta}, \boldsymbol{z}_{i,t}\rangle + \langle \boldsymbol{B}_i, \boldsymbol{w}_{i,t}\rangle, \tag{5.77}$$

with $\sigma(\mathcal{D}_{i,t})$-measurable feature vector $\boldsymbol{w}_{i,t}$. Regression parameter $\boldsymbol{\beta}$ then describes the fixed systematic effects that are common over the entire portfolio $1 \le i \le n$ and $\boldsymbol{B}_i$ describes the policy dependent random effects (assumed to be normalized $\mathbb{E}[\boldsymbol{B}_i] = 0$). Typically one assumes that $\boldsymbol{B}_1, \ldots, \boldsymbol{B}_n$ are centered and i.i.d. Such effects are called static random effects because they are not time-dependent, and they may also be interpreted in a Bayesian sense.

Finally, extending these static random effects to dynamic random effects $\boldsymbol{B}_{i,t}$, $t \ge 1$, leads to so-called state-space models, the linear state-space model being the most popular example and being fitted using the Kalman filter [207].

5.8.2 *Regression Models Beyond the GLM Framework*

There are several ways in which the GLM framework can be modified.

Siblings of Generalized Linear Regression Functions

The most common modification of GLMs concerns the regression structure, namely, that the scalar product in the linear predictor

$$\boldsymbol{x} \mapsto g(\mu) = \eta = \langle \boldsymbol{\beta}, \boldsymbol{x} \rangle,$$

is replaced by another regression function. A popular alternative is the framework of generalized additive models (GAMs). GAMs go back to Hastie–Tibshirani [181, 182] and the standard reference is Wood [384]. GAMs consider the regression functions

$$\boldsymbol{x} \mapsto g(\mu) = \eta = \beta_0 + \sum_j \beta_j s_j(x_j), \tag{5.78}$$

where $s_j : \mathbb{R} \to \mathbb{R}$ are natural cubic splines. Natural cubic splines s_j are obtained by concatenating cubic functions in so-called nodes. A GAM can have as many nodes in each cubic spline s_j as there are different levels $x_{i,j}$ in the data $1 \le i \le n$. In general, this leads to very flexible regression models, and to control in-sample over-fitting regularization is applied, for regularization we also refer to Sect. 6.2. Regularization requires setting a tuning parameter, and an efficient determination of this tuning parameter uses generalized cross-validation, see Sect. 5.6. Nevertheless, fitting GAMs can be very computational, already for portfolios with 1 million policies and involving 20 feature components the calibration can be very slow. Moreover, regression function (5.78) does not (directly) allow for a data driven method of finding interactions between feature components. For these reasons, we do not further study GAMs in this monograph.

A modification in the regression function that is able to consider interactions between feature components is the framework of classification and regression trees (CARTs). CARTs have been introduced by Breiman et al. [54] in 1984, and they are still used in its original form today. Regression trees aim to partition the feature space $\mathcal{X}$ into a finite number of disjoint subsets $\mathcal{X}_t$, $1 \le t \le T$, such that all policies $(Y_i, \boldsymbol{x}_i)$ in the same subset $\boldsymbol{x}_i \in \mathcal{X}_t$ satisfy a certain homogeneity property w.r.t. the regression task (and the chosen loss function). The CART regression function is then defined by

$$\boldsymbol{x} \mapsto \mu(\boldsymbol{x}) = \sum_{t=1}^{T} \widehat{\mu}_t \, \mathbb{1}_{\{\boldsymbol{x} \in \mathcal{X}_t\}},$$

where $\widehat{\mu}_t$ is the homogeneous mean estimator on $\mathcal{X}_t$. These CARTs are popular building blocks for ensemble methods where different regression functions are combined, we mention random forests and boosting algorithms that mainly rely on CARTs. Random forests have been introduced by Breiman [52], and boosting has been popularized by Valiant [362], Kearns–Valiant [209, 210], Schapire [328],

Freund [139] and Freund–Schapire [140]. Today boosting belongs to the most powerful predictive regression methods, we mention the XGBoost algorithm of Chen–Guestrin [71] that has won many competitions. We will not further study CARTs and boosting in these notes because these methods also have some drawbacks. For instance, resulting regression functions are not continuous nor do they easily allow to extrapolate data beyond the (observed) feature space, e.g., if we have a time component. Moreover, they are more difficult in the use of unstructured data such as text data. For more on CARTs and boosting in actuarial science we refer to Denuit et al. [100] and Ferrario–Hämmerli [125].

Other Distributional Models

The theory above has been relying on the EDF, but, of course, we could also study any other family of distribution functions. A clear drawback of the EDF is that it only considers light-tailed distribution functions, i.e., distribution functions for which the moment generating function exists around the origin. If the data is more heavy-tailed, one may need to transform this data and then use the EDF on the transformed data (with the drawback that one loses the balance property) or one chooses another family of distribution functions. Transformations have already been discussed in Remarks 2.11 and Sect. 5.3.9. Another two families of distributions that have been studied in the actuarial literature are the generalized beta of the second kind (GB2) distribution, see Venter [369], Frees et al. [137] and Chan et al. [66], and inhomogeneous phase type (IHP) distributions, see Albrecher et al. [8] and Bladt [37]. The GB2 family is a 4-parameter family, and it nests several examples such as the gamma, the Weibull, the Pareto and the Lomax distributions, see Table B1 in Chan et al. [66]. The density of the GB2 distribution is for $y > 0$ given by

$$f(y; a, b, \alpha_1, \alpha_2) = \frac{\frac{|a|}{b}\left(\frac{y}{b}\right)^{a\alpha_1 - 1}}{B(\alpha_1, \alpha_2)\left(1 + \left(\frac{y}{b}\right)^a\right)^{\alpha_1 + \alpha_2}} \tag{5.79}$$

$$= \frac{\frac{|a|}{y}}{B(\alpha_1, \alpha_2)} \left(\frac{\left(\frac{y}{b}\right)^a}{1 + \left(\frac{y}{b}\right)^a}\right)^{\alpha_1} \left(\frac{1}{1 + \left(\frac{y}{b}\right)^a}\right)^{\alpha_2},$$

with scale parameter $b > 0$, shape parameters $a \in \mathbb{R}$ and $\alpha_1, \alpha_2 > 0$, and beta function

$$B(\alpha_1, \alpha_2) = \frac{\Gamma(\alpha_1)\Gamma(\alpha_2)}{\Gamma(\alpha_1 + \alpha_2)}.$$

Consider a modified logistic transformation of variable $y \mapsto z = (y/b)^a/(1 + (y/b)^a) \in (0, 1)$. This gives us the beta density

$$f(z; \alpha_1, \alpha_2) = \frac{z^{\alpha_1 - 1}(1 - z)^{\alpha_2 - 1}}{B(\alpha_1, \alpha_2)}.$$

Thus, the GB2 distribution can be obtained by a transformation of the beta distribution. The latter provides that a GB2 distributed random variable Y can be simulated from $Y \stackrel{(d)}{=} b(Z/(1-Z))^{1/a}$ with $Z \sim \text{Beta}(\alpha_1, \alpha_2)$.

A GB2 distributed random variable Y has first moment

$$\mathbb{E}_{a,b,\alpha_1,\alpha_2}[Y] \;=\; \frac{B(\alpha_1+1/a, \alpha_2-1/a)}{B(\alpha_1,\alpha_2)}\, b,$$

for $-\alpha_1 a < 1 < \alpha_2 a$. Observe that for $a > 0$ we have that the survival function of Y is regularly varying with tail index $\alpha_2 a > 0$. Thus, we can model Pareto-like tails with the GB2 family; for regular variation we refer to (1.3).

As proposed in Frees et al. [137], one can introduce a regression structure for $b > 0$ by choosing a log-link and setting

$$\log\left(\mathbb{E}_{a,b,\alpha_1,\alpha_2}[Y]\right) = \log\left(\frac{B(\alpha_1+1/a, \alpha_2-1/a)}{B(\alpha_1,\alpha_2)}\right) + \langle \boldsymbol{\beta}, \boldsymbol{x}\rangle.$$

MLE of $\boldsymbol{\beta}$ may pose some challenge because it depends on nuisance parameters a, α_1, α_2. In a recent paper Li et al. [251], there is a proposal to extend this GB2 regression to a composite regression model; composite models are discussed in Sect. 6.4.4, below. This closes this short section, and for more examples we refer to the literature.

5.8.3 *Quantile Regression*

Pinball Loss Function

The GLMs introduced above aim at estimating the means $\mu(\boldsymbol{x}) = \mathbb{E}_{\theta(\boldsymbol{x})}[Y]$ of random variables Y being explained by features $\boldsymbol{x}$. Since mean estimation can be rather sensitive in situations where we have large claims, the more robust quantile regression has attracted some attention, recently. Quantile regression has been introduced by Koenker–Bassett [220]. The idea is that instead of estimating the mean μ of a random variable Y, we rather try to estimate its τ-quantile for given $\tau \in (0, 1)$. The τ-quantile is given by the generalized inverse $F^{-1}(\tau)$ of the distribution function F of Y, that is,

$$F^{-1}(\tau) = \inf\{y \in \mathbb{R};\; F(y) \geq \tau\}. \tag{5.80}$$

Consider the *pinball loss function* for $y \in \mathfrak{C}$ (convex closure of the support of Y) and actions $a \in \mathbb{A} = \mathbb{R}$

$$(y, a) \mapsto L_\tau(y, a) = (y - a)\left(\tau - \mathbb{1}_{\{y-a<0\}}\right) \;\geq\; 0. \tag{5.81}$$

This provides us with the expected loss for $Y \sim F$ and action $a \in \mathbb{A}$

$$\begin{aligned}\mathbb{E}_F\left[L_\tau(Y,a)\right] &= \mathbb{E}_F\left[(Y-a)\left(\tau - \mathbb{1}_{\{Y<a\}}\right)\right]\\ &= (\tau-1)\mathbb{E}_F\left[(Y-a)\mathbb{1}_{\{Y<a\}}\right] + \tau\mathbb{E}_F\left[(Y-a)\mathbb{1}_{\{Y\geq a\}}\right]\\ &= (\tau-1)\int_{-\infty}^{a}(y-a)dF(y) + \tau\int_{a}^{\infty}(y-a)dF(y).\end{aligned}$$

The aim is to find an optimal action $\widehat{a}(F)$ that minimizes this expected loss, see (4.24),

$$\widehat{a}(F) \;\in\; \mathfrak{A}(F) \;=\; \underset{a\in\mathbb{A}}{\arg\min}\, \mathbb{E}_F\left[L_\tau(Y,a)\right].$$

Note that for the time being we do not know whether the solution to this minimization problem is a singleton. For this reason, we state the solution (subject to existence) as a set-valued functional $\mathfrak{A}$, see (4.25).

We calculate the score equation of the expected loss using the Leibniz rule

$$\begin{aligned}\frac{\partial}{\partial a}\mathbb{E}_F\left[L_\tau(Y,a)\right] &= -(\tau-1)\int_{-\infty}^{a}dF(y) - \tau\int_{a}^{\infty}dF(y)\\ &= -(\tau-1)F(a) - \tau\left(1-F(a)\right) \;=\; F(a)-\tau \;\overset{!}{=}\; 0.\end{aligned}$$

Assume the distribution F is continuous. This implies $F(F^{-1}(\tau)) = \tau$, and we have

$$F^{-1}(\tau) \;\in\; \mathfrak{A}(F) \;=\; \underset{a\in\mathbb{A}}{\arg\min}\, \mathbb{E}_F\left[L_\tau(Y,a)\right].$$

In fact, using the pinball loss, we have just seen that the τ-quantile is elicitable within the class of continuous distributions, see Definition 4.18.

For a more general result we need a more general definition of a (set-valued) τ-quantile

$$Q_\tau(F) = \left\{y \in \mathbb{R};\; \lim_{z\uparrow y} F(z) \leq \tau \leq F(y)\right\}. \tag{5.82}$$

This defines a closed interval and its lower endpoint corresponds to the generalized inverse $F^{-1}(\tau)$ given in (5.80). In complete analogy to Theorem 4.19 on the elicitability of the mean functional, we have the following statement for the τ-quantile; this result goes back to Thomson [351] and Saerens [326].

Theorem 5.33 (Gneiting [162, Theorem 9], Without Proof) *Let $\mathcal{F}$ be the class of distribution functions on an interval $\mathfrak{C} \subseteq \mathbb{R}$ and choose quantile level $\tau \in (0,1)$.*

- *The τ-quantile* (5.82) *is elicitable relative to $\mathcal{F}$.*

- *Assume the loss function* $L : \mathfrak{C} \times \mathbb{A} \to \mathbb{R}_+$ *satisfies (L0)-(L2) on page 92 for interval* $\mathfrak{C} = \mathbb{A} \subseteq \mathbb{R}$. *L is consistent for the* τ*-quantile* (5.82) *relative to the class* $\mathcal{F}$ *of compactly supported distributions on* $\mathfrak{C}$ *if and only if L is of the form*

$$L(y, a) = (G(y) - G(a)) \left(\tau - \mathbb{1}_{\{y-a<0\}}\right),$$

for a non-decreasing function G on $\mathfrak{C}$.
- *If G is strictly increasing on* $\mathfrak{C}$ *and if* $\mathbb{E}_F[G(Y)]$ *exists and is finite for all* $F \in \mathcal{F}$, *then the above loss function L is strictly consistent for the* τ*-quantile* (5.82) *relative to the class* $\mathcal{F}$.

Theorem 5.33 characterizes the strictly consistent loss functions for quantile estimation, the pinball loss being the special case $G(y) = y$.

Quantile Regression

The idea behind quantile regression is that we build a regression model for the τ-quantile. Assume we have a datum $(Y, \boldsymbol{x})$ whose conditional τ-quantile, given $\boldsymbol{x} \in \{1\} \times \mathbb{R}^q$, can be described by the regression function

$$\boldsymbol{x} \mapsto g\left(F_{Y|\boldsymbol{x}}^{-1}(\tau)\right) = \langle \boldsymbol{\beta}_\tau, \boldsymbol{x} \rangle,$$

for a strictly monotone and smooth link function $g : \mathfrak{C} \to \mathbb{R}$, and for a regression parameter $\boldsymbol{\beta}_\tau \in \mathbb{R}^{q+1}$. The aim now is to estimate this regression parameter from independent data $(Y_i, \boldsymbol{x}_i)$, $1 \le i \le n$. The pinball loss L_τ, given in (5.81), provides us with the following optimization problem

$$\widehat{\boldsymbol{\beta}}_\tau = \underset{\boldsymbol{\beta} \in \mathbb{R}^{q+1}}{\arg\min} \sum_{i=1}^{n} L_\tau \left(Y_i, g^{-1}\langle \boldsymbol{\beta}, \boldsymbol{x}_i \rangle\right).$$

This then allows us to estimate the corresponding τ-quantile as a function of the feature information $\boldsymbol{x}$. For $\tau = 1/2$ we estimate the median by

$$\widehat{F}_{Y|\boldsymbol{x}}^{-1}(1/2) = g^{-1}\left\langle \widehat{\boldsymbol{\beta}}_{1/2}, \boldsymbol{x} \right\rangle.$$

We conclude from this short section that we can regress any quantity $a(F)$ that is elicitable, i.e., for which a loss function exists that is strictly consistent for $a(F)$ on $F \in \mathcal{F}$. For more on quantile regression we refer to the monograph of Uribe–Guillén [361], and an interesting paper is Dimitriades et al. [106]. We will study quantile regression within deep networks in Chap. 11.2, below.

Chapter 6
Bayesian Methods, Regularization and Expectation-Maximization

The previous chapter has been focusing on MLE of regression parameters within GLMs. Alternatively, we could address the parameter estimation problem within a Bayesian setting. The purpose of this chapter is to discuss the Bayesian estimation approach. This leads us to the notion of regularization within GLMs. Bayesian methods are also used in the Expectation-Maximization (EM) algorithm for MLE in the case of incomplete data. For literature on Bayesian theory we recommend Gelman et al. [157], Congdon [79], Robert [319], Bühlmann–Gisler [58] and Gilks et al. [158]. A nice historical (non-mathematical) review of Bayesian methods is presented in McGrayne [266]. Regularization is discussed in the book of Hastie et al. [184], and a good reference for the EM algorithm is McLachlan–Krishnan [267].

6.1 Bayesian Parameter Estimation

The Bayesian estimator has been introduced in Definition 3.6. Assume that the observation $\boldsymbol{Y}$ has independent components Y_i that can be described by a GLM with link function g and regression parameter $\boldsymbol{\beta} \in \mathbb{R}^{q+1}$, i.e., the random variables Y_i have densities

$$Y_i \stackrel{\text{ind.}}{\sim} f(y; \boldsymbol{\beta}, \boldsymbol{x}_i, v_i/\varphi) = \exp\left\{ \frac{y(h \circ g^{-1})\langle \boldsymbol{\beta}, \boldsymbol{x}_i\rangle - (\kappa \circ h \circ g^{-1})\langle \boldsymbol{\beta}, \boldsymbol{x}_i\rangle}{\varphi/v_i} + a(y; v_i/\varphi) \right\},$$

with canonical link $h = (\kappa')^{-1}$. In a Bayesian approach one models the regression parameter $\boldsymbol{\beta}$ with a prior distribution[1] $\pi(\boldsymbol{\beta})$ on the parameter space $\mathbb{R}^{q+1}$, and the independence assumption between the components of $\boldsymbol{Y}$ needs to be understood

[1] Often, in Bayesian arguing, distribution and density is used in an interchangeable (and not fully precise) way, and it is left to the reader to give the right meaning to π.

M. V. Wüthrich, M. Merz, *Statistical Foundations of Actuarial Learning and its Applications*, Springer Actuarial, https://doi.org/10.1007/978-3-031-12409-9_6

conditionally, given the regression parameter $\boldsymbol{\beta}$. In other words, all observations Y_i share the same regression parameter $\boldsymbol{\beta}$, which itself is modeled by a prior distribution π.

The joint density of $\boldsymbol{Y}$ and $\boldsymbol{\beta}$ is given by

$$p(\boldsymbol{y}, \boldsymbol{\beta}) = \left(\prod_{i=1}^{n} f(y_i; \boldsymbol{\beta}, \boldsymbol{x}_i, v_i/\varphi)\right) \pi(\boldsymbol{\beta}) = \exp\left\{\ell_{\boldsymbol{Y}=\boldsymbol{y}}(\boldsymbol{\beta}) + \log \pi(\boldsymbol{\beta})\right\}. \tag{6.1}$$

For the given observation $\boldsymbol{Y}$, this allows us to calculate the posterior density of $\boldsymbol{\beta}$ using Bayes' rule

$$\pi(\boldsymbol{\beta}|\boldsymbol{Y}) = \frac{p(\boldsymbol{Y}, \boldsymbol{\beta})}{\int p(\boldsymbol{Y}, \widetilde{\boldsymbol{\beta}}) d\widetilde{\boldsymbol{\beta}}} \propto \left(\prod_{i=1}^{n} f(Y_i; \boldsymbol{\beta}, \boldsymbol{x}_i, v_i/\varphi)\right) \pi(\boldsymbol{\beta}), \tag{6.2}$$

where the proportionality sign $\propto$ indicates that we have dropped the terms that do not depend on $\boldsymbol{\beta}$. Thus, the functional form in $\boldsymbol{\beta}$ of the posterior density $\pi(\boldsymbol{\beta}|\boldsymbol{Y})$ is fully determined by the joint density $p(\boldsymbol{Y}, \boldsymbol{\beta})$, and the remaining term is a normalization to obtain a proper probability distribution. In many situations, the knowledge of the functional form of the posterior density in $\boldsymbol{\beta}$ is sufficient to perform Bayesian parameter estimation, at least, numerically. We will give some references, below.

The Bayesian estimator for $\boldsymbol{\beta}$ is given by the posterior mean (supposed it exists)

$$\widehat{\boldsymbol{\beta}}^{\text{Bayes}} = \mathbb{E}_\pi\left[\boldsymbol{\beta}|\,\boldsymbol{Y}\right] = \int \boldsymbol{\beta}\, \pi(\boldsymbol{\beta}|\boldsymbol{Y}) d\nu(\boldsymbol{\beta}).$$

If we want to calculate the expectation of a new random variable Y_{n+1} that is conditionally, given $\boldsymbol{\beta}$, independent of $\boldsymbol{Y}$ and follows the same GLM as $\boldsymbol{Y}$, we can directly calculate, using the tower property and conditional independence,[2]

$$\begin{aligned} \mathbb{E}_\pi\left[Y_{n+1}|\,\boldsymbol{Y}\right] = \mathbb{E}_\pi\left[\mathbb{E}\left[Y_{n+1}|\,\boldsymbol{\beta}, \boldsymbol{Y}\right]|\,\boldsymbol{Y}\right] &= \mathbb{E}_\pi\left[\mathbb{E}\left[Y_{n+1}|\,\boldsymbol{\beta}\right]|\,\boldsymbol{Y}\right] \\ = \mathbb{E}_\pi\left[g^{-1}\langle\boldsymbol{\beta}, \boldsymbol{x}_{n+1}\rangle\middle|\,\boldsymbol{Y}\right] &= \int g^{-1}\langle\boldsymbol{\beta}, \boldsymbol{x}_{n+1}\rangle\, \pi(\boldsymbol{\beta}|\boldsymbol{Y}) d\nu(\boldsymbol{\beta}), \end{aligned}$$

supposed that this first moment exists and that $\boldsymbol{x}_{n+1}$ is the feature of Y_{n+1}. We see that it all boils down to have sufficiently explicit knowledge about the posterior density $\pi(\boldsymbol{\beta}|\boldsymbol{Y})$ given in (6.2).

Remark 6.1 (Conditional MSEP) Based on the assumption that the posterior distribution $\pi(\boldsymbol{\beta}|\boldsymbol{Y})$ can be determined, we can analyze the GL. In a Bayesian setup one

[2] Note that we identify probabilities $\mathbb{P}_{\boldsymbol{\beta}}[\cdot] = \mathbb{P}[\cdot|\boldsymbol{\beta}]$ for given $\boldsymbol{\beta}$.

usually does not calculate the MSEP as described in Theorem 4.1, but one rather studies the conditional MSEP, conditioned exactly on the collected information $\boldsymbol{Y}$. That is,

$$\begin{aligned}
\mathbb{E}_\pi\left[(Y_{n+1} - \mathbb{E}_\pi\left[Y_{n+1}|\,\boldsymbol{Y}\right])^2 \middle|\, \boldsymbol{Y}\right] &= \mathrm{Var}_\pi\left(Y_{n+1}|\,\boldsymbol{Y}\right) \\
&= \mathrm{Var}_\pi\left(\mathbb{E}\left[Y_{n+1}|\,\boldsymbol{\beta}, \boldsymbol{Y}\right]|\,\boldsymbol{Y}\right) + \mathbb{E}_\pi\left[\mathrm{Var}\left(Y_{n+1}|\,\boldsymbol{\beta}, \boldsymbol{Y}\right)|\,\boldsymbol{Y}\right] \\
&= \mathrm{Var}_\pi\left(g^{-1}\langle\boldsymbol{\beta}, \boldsymbol{x}_{n+1}\rangle \middle|\, \boldsymbol{Y}\right) + \frac{\varphi}{v_{n+1}}\mathbb{E}_\pi\left[(\kappa'' \circ h \circ g^{-1})\langle\boldsymbol{\beta}, \boldsymbol{x}_{n+1}\rangle \middle|\, \boldsymbol{Y}\right] \\
&= \mathrm{Var}_\pi\left(g^{-1}\langle\boldsymbol{\beta}, \boldsymbol{x}_{n+1}\rangle \middle|\, \boldsymbol{Y}\right) + \frac{\varphi}{v_{n+1}}\mathbb{E}_\pi\left[V(g^{-1}\langle\boldsymbol{\beta}, \boldsymbol{x}_{n+1}\rangle) \middle|\, \boldsymbol{Y}\right],
\end{aligned}$$

where we need to assume existence of second moments. Similar to Theorem 4.1, the first term is the estimation variance (in a Bayesian setting) and the second term is the average process variance (using the EDF variance function $\mu \mapsto V(\mu)$).

The remaining difficulty is the calculation of the posterior expectation of functions of $\boldsymbol{\beta}$, based on posterior density (6.2). In very well-designed experiments the posterior density $\pi(\boldsymbol{\beta}|\boldsymbol{Y})$ can be determined explicitly, for instance, in the homogeneous EDF case with so-called conjugate priors, see Chapter 2 in Bühlmann–Gisler [58]. But in most cases, there is no closed from solution for the posterior distribution. Major progress in Bayesian modeling has been made with the emergence of computational methods like the Markov chain Monte Carlo (MCMC) method, Gibbs sampling, the Metropolis–Hastings (MH) algorithm [185, 274], sequential Monte Carlo (SMC) sampling, non-linear particle filters, and the Hamilton Monte Carlo (HMC) algorithm. These methods help us to empirically approximate the posterior density $\pi(\boldsymbol{\beta}|\boldsymbol{Y})$ in different modeling setups. These methods have in common that the explicit knowledge of the normalizing constant in (6.2) is not necessary, but it suffices to know the functional form in $\boldsymbol{\beta}$ of the posterior density $\pi(\boldsymbol{\beta}|\boldsymbol{Y})$.

For a detailed description of MCMC methods in general, which includes Gibbs sampling and MH algorithms, we refer to Gilks et al. [158], Green [169, 170], Johansen et al. [199]; SMC sampling and non-linear particle filters are explained in Del Moral et al. [92, 93], Johansen–Evers [199], Doucet–Johansen [111], Creal [85] and Wüthrich [389]; the HMC algorithm is described in Neal [281]. We do not present these algorithms here, but for the description of the most popular algorithms we refer to Section 4.4 in Wüthrich–Buser [392]. The reason for not presenting these algorithms here is that they still face the curse of dimensionality, which makes it difficult to use Bayesian methods for high-dimensional data sets in large models; we provide another short discussion in Sect. 11.6.3, below.

6.2 Regularization

6.2.1 *Maximal a Posterior Estimator*

In the previous section we have proposed to approximate the posterior density $\pi(\boldsymbol{\beta}|\boldsymbol{Y})$ of the regression parameter $\boldsymbol{\beta}$, given $\boldsymbol{Y}$, using MCMC methods. The posterior log-likelihood in the Bayesian GLM is given by, see (6.2),

$$\begin{aligned}\log \pi(\boldsymbol{\beta}|\boldsymbol{Y}) &\propto \ell_{\boldsymbol{Y}}(\boldsymbol{\beta}) + \log \pi(\boldsymbol{\beta}) \\ &\propto \sum_{i=1}^{n} \frac{Y_i (h \circ g^{-1})\langle \boldsymbol{\beta}, \boldsymbol{x}_i \rangle - (\kappa \circ h \circ g^{-1})\langle \boldsymbol{\beta}, \boldsymbol{x}_i \rangle}{\varphi / v_i} + \log \pi(\boldsymbol{\beta}).\end{aligned}$$

Compared to the classical log-likelihood function $\ell_{\boldsymbol{Y}}(\boldsymbol{\beta})$ for MLE, there is an additional log-density term $\log \pi(\boldsymbol{\beta})$ that comes from the prior distribution of $\boldsymbol{\beta}$. Thus, the posterior log-likelihood is a balanced version of the log-likelihood $\ell_{\boldsymbol{Y}}(\boldsymbol{\beta})$ of the data $\boldsymbol{Y}$ and the prior log-density $\log \pi(\boldsymbol{\beta})$ of the regression parameter $\boldsymbol{\beta}$. We interpret this as *regularization* because the prior π smooths extremes in the log-likelihood of the observation $\boldsymbol{Y}$. This gives rise to estimate the regression parameter $\boldsymbol{\beta}$ by the so-called maximal a posterior (MAP) estimator

$$\widehat{\boldsymbol{\beta}}^{\text{MAP}} = \underset{\boldsymbol{\beta} \in \mathbb{R}^{q+1}}{\arg\max} \; \log \pi(\boldsymbol{\beta}|\boldsymbol{Y}) = \underset{\boldsymbol{\beta} \in \mathbb{R}^{q+1}}{\arg\max} \; \ell_{\boldsymbol{Y}}(\boldsymbol{\beta}) + \log \pi(\boldsymbol{\beta}). \tag{6.3}$$

This π-regularized (MAP) parameter estimation has gained much popularity because it is a useful tool to prevent the model from over-fitting under suitable prior choices. Moreover, under specific choices, it allows for parameter selection. This is especially useful in high-dimensional problems; for a reference we refer to Hastie et al. [184].

Popular choices for π are prior densities coming from L^p-norms for some $p \geq 1$, that is, $\pi(\boldsymbol{\beta}) \propto \exp\{-\lambda \|\boldsymbol{\beta}\|_p^p\}$ for $\lambda > 0$. Optimization problem (6.3) then becomes

$$\widehat{\boldsymbol{\beta}}^{\text{MAP}} = \underset{\boldsymbol{\beta} \in \mathbb{R}^{q+1}}{\arg\max} \; \ell_{\boldsymbol{Y}}(\boldsymbol{\beta}) - \lambda \|\boldsymbol{\beta}\|_p^p,$$

for a fixed *regularization parameter* $\lambda > 0$ (also called tuning parameter). In practical applications we should exclude the intercept parameter $\beta_0 \in \mathbb{R}$ from regularization: if we work with the canonical link within the GLM framework we have the balance property which implies unbiasedness, see Corollary 5.7. This property gets lost if β_0 is included in the regularization term. For this reason, we set $\boldsymbol{\beta}_- = (\beta_1, \ldots, \beta_q)^\top \in \mathbb{R}^q$ and we let regularization only act on these components

$$\widehat{\boldsymbol{\beta}}^{\text{MAP}} = \widehat{\boldsymbol{\beta}}^{\text{MAP}}(\lambda) = \underset{\boldsymbol{\beta} \in \mathbb{R}^{q+1}}{\arg\max}\ \frac{1}{n}\ell_Y(\boldsymbol{\beta}) - \lambda \|\boldsymbol{\beta}_-\|_p^p, \tag{6.4}$$

we also scale with the sample size n to make the units of the tuning parameter λ independent of the sample size n.

Remarks 6.2

- The regularization term $\lambda\|\boldsymbol{\beta}_-\|_p^p$ keeps the components of the regression parameter $\boldsymbol{\beta}_-$ close to zero, thus, it prevents from over-fitting by letting parameters only take moderate values. The magnitudes of the parameter values are controlled by the regularization parameter $\lambda > 0$ which acts as a hyper-parameter. Optimal hyper-parameters are determined by cross-validation.
- In (6.4) all components of $\boldsymbol{\beta}_-$ are treated equally. This may not be appropriate if the feature components of $\boldsymbol{x}$ live on different scales. This problem of different scales can be solved by either scaling the components of $\boldsymbol{x}$ to a unit scale, or by introducing a diagonal importance matrix $T = \text{diag}(t_1, \ldots, t_q)$ with $t_j > 0$ that describes the scales of the components of $\boldsymbol{x}$. This allows us to regularize $\|T^{-1}\boldsymbol{\beta}_-\|_p^p$ instead of $\|\boldsymbol{\beta}_-\|_p^p$. Thus, in this latter case we replace (6.4) by the weighted version

$$\widehat{\boldsymbol{\beta}}^{\text{MAP}} = \underset{\boldsymbol{\beta}}{\arg\max}\ \frac{1}{n}\ell_Y(\boldsymbol{\beta}) - \lambda \sum_{j=1}^{q} t_j^{-p}|\beta_j|^p.$$

- Often, the features have a natural group structure $\boldsymbol{x} = (x_0, \boldsymbol{x}_1, \ldots, \boldsymbol{x}_K)$, for instance, $\boldsymbol{x}_k \in \{0, 1\}^{q_k}$ may represent dummy coding of a categorical feature component with $q_k + 1$ levels. In that case regularization should equally act on all components of $\boldsymbol{\beta}_k \in \mathbb{R}^{q_k}$ (that correspond to $\boldsymbol{x}_k$) because these components describe the same systematic effect. Yuan–Lin [398] proposed for this problem grouped penalties of the form

$$\widehat{\boldsymbol{\beta}}^{\text{MAP}} = \underset{\boldsymbol{\beta}}{\arg\max}\ \frac{1}{n}\ell_Y(\boldsymbol{\beta}) - \lambda \sum_{k=1}^{K} \|\boldsymbol{\beta}_k\|_2. \tag{6.5}$$

  This proposal leads to sparsity, i.e., for large regularization parameters λ the entire $\boldsymbol{\beta}_k$ may be shrunk (exactly) to zero; this is discussed in Sect. 6.2.5, below. We also refer to Section 4.3 in Hastie et al. [184], and Devriendt et al. [104] proposed this approach in the actuarial literature.
- There are more versions of regularization, e.g., in the fused LASSO approach we ensure that the first differences $\beta_j - \beta_{j-1}$ remain small.

Our motivation for considering regularization has been inspired by Bayesian theory, but we can also come from a completely different angle, namely, we can consider a constraint optimization problem with a given budget constraint $c > 0$. That is, we can consider

$$\underset{\boldsymbol{\beta} \in \mathbb{R}^{q+1}}{\arg\max} \frac{1}{n} \ell_{\boldsymbol{Y}}(\boldsymbol{\beta}) \qquad \text{subject to } \|\boldsymbol{\beta}_-\|_p^p \le c. \tag{6.6}$$

This optimization problem can be tackled by the method of Karush, Kuhn and Tucker (KKT) [208, 228]. Optimization problem (6.4) corresponds by Lagrangian duality to the constraint optimization problem (6.6). For every c for which the budget constraint in (6.6) is binding $\|\boldsymbol{\beta}_-\|_p^p = c$, there is a corresponding regularization parameter $\lambda = \lambda(c)$, and, conversely, the solution of (6.4) solves (6.6) with $c = \|\widehat{\boldsymbol{\beta}}_-^{\text{MAP}}(\lambda)\|_p^p$.

6.2.2 *Ridge vs. LASSO Regularization*

We compare the two special cases of $p = 1, 2$ in this section, and in the subsequent Sects. 6.2.3 and 6.2.4 we discuss how these two cases can be solved numerically.

Ridge Regularization $p = 2$ For $p = 2$, the prior distribution π in (6.4) is a centered Gaussian distribution. This L^2-regularization is called *ridge regularization* or Tikhonov regularization [353], and we have

$$\widehat{\boldsymbol{\beta}}^{\text{ridge}} = \widehat{\boldsymbol{\beta}}^{\text{ridge}}(\lambda) = \underset{\boldsymbol{\beta} \in \mathbb{R}^{q+1}}{\arg\max} \frac{1}{n} \ell_{\boldsymbol{Y}}(\boldsymbol{\beta}) - \lambda \sum_{j=1}^{q} \beta_j^2. \tag{6.7}$$

LASSO Regularization $p = 1$ For $p = 1$, the prior distribution π in (6.4) is a Laplace distribution. This L^1-regularization is called *LASSO regularization* (least absolute shrinkage and selection operator), see Tibshirani [352], and we have

$$\widehat{\boldsymbol{\beta}}^{\text{LASSO}} = \widehat{\boldsymbol{\beta}}^{\text{LASSO}}(\lambda) = \underset{\boldsymbol{\beta} \in \mathbb{R}^{q+1}}{\arg\max} \frac{1}{n} \ell_{\boldsymbol{Y}}(\boldsymbol{\beta}) - \lambda \sum_{j=1}^{q} |\beta_j|. \tag{6.8}$$

LASSO regularization has the advantage that it shrinks (unimportant) regression components to exactly zero, i.e., LASSO regularization can be used for parameter elimination and model reduction. This is discussed in the next paragraphs.

Ridge vs. LASSO Regularization Ridge ($p = 2$) and LASSO ($p = 1$) regularization behave rather differently. This can be understood best by using the budget constraint (6.6) interpretation which gives us a nice geometric illustration. The crucial part is that the side constraint gives us either a budget constraint $\|\boldsymbol{\beta}_-\|_2^2 = \sum_{j=1}^q \beta_j^2 \le c$ (squared Euclidean norm) or $\|\boldsymbol{\beta}_-\|_1 = \sum_{j=1}^q |\beta_j| \le c$ (Manhattan norm). In Fig. 6.1 we illustrate these two cases, the left-hand side shows the Euclidean ball in blue color (in two dimensions) and the right-hand side shows the corresponding Manhattan square in blue color; this figure is similar to Figure 2.2 in Hastie et al. [184].

The (unconstraint) MLE $\widehat{\boldsymbol{\beta}}^{\rm MLE}$ is illustrated by the red dot in Fig. 6.1. If the red dot would lie within the blue area, the budget constraint would not be binding. In Fig. 6.1 the red dot (MLE) does not lie within the blue budget constraint, and we need to compromise in the optimality of the MLE. Assume that the log-likelihood $\boldsymbol{\beta} \mapsto \ell_{\boldsymbol{Y}}(\boldsymbol{\beta})$ is a concave function in $\boldsymbol{\beta}$, then we receive convex level sets $\{\boldsymbol{\beta};\, \ell_{\boldsymbol{Y}}(\boldsymbol{\beta}) \ge \gamma_0\}$ around the MLE $\widehat{\boldsymbol{\beta}}^{\rm MLE}$. The critical constant γ_0 for which this level set is tangential to the blue budget constraint exactly gives us the solution to (6.6); this solution corresponds to the yellow dots in Fig. 6.1. The crucial difference between ridge and LASSO regularization is that in the latter case the yellow dot will eventually be in the corner of the Manhattan square if we shrink the budget constraint c to zero. Or in other words, some of the components of $\boldsymbol{\beta}$ are set exactly equal to zero for small c or large λ, respectively; in Fig. 6.1 (rhs) this happens to the first component of $\widehat{\boldsymbol{\beta}}^{\rm LASSO}$ (under the given budget constraint c). In

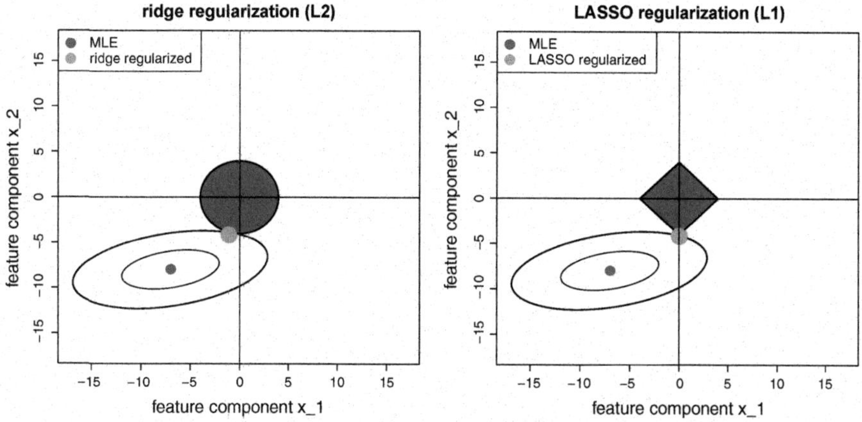

Fig. 6.1 Illustration of optimization problem (6.6) under a budget constraint (lhs) for $p = 2$ (Euclidean norm) and (rhs) $p = 1$ (Manhattan norm)

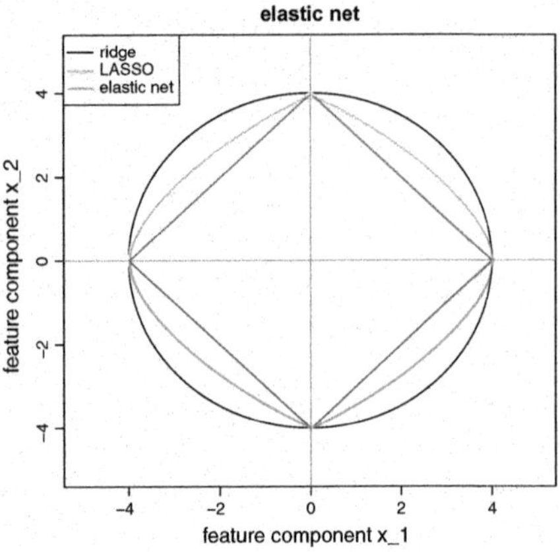

Fig. 6.2 Elastic net regularization

ridge regularization this is not the case, except for special situations concerning the position of the red MLE. Thus, ridge regression makes components of parameter estimates generally smaller, whereas LASSO shrinks some of these components exactly to zero (this also explains the name LASSO).

Remark 6.3 (Elastic Net) LASSO regularization faces difficulties with collinearity in feature components. In particular, if we have a group of highly correlated feature components, LASSO fails to do a grouped selection, but it selects one component and ignores the other ones. On the other hand, ridge regularization can deal with this issue. For this reason, Zou–Hastie [409] proposed the *elastic net regularization*, which uses a combined regularization term

$$\widehat{\boldsymbol{\beta}}^{\text{elastic net}} = \underset{\boldsymbol{\beta}\in\mathbb{R}^{q+1}}{\arg\max}\ \frac{1}{n}\ell_{\boldsymbol{Y}}(\boldsymbol{\beta}) - \lambda\left[(1-\alpha)\|\boldsymbol{\beta}\|_2^2 + \alpha\|\boldsymbol{\beta}\|_1\right],$$

for some $\alpha \in (0,1)$. The L^1-term gives sparsity and the quadratic term removes the limitation on the number of selected variables, providing a grouped selection. In Fig. 6.2 we compare the elastic net regularization (orange color) to ridge and LASSO regularization (black and blue color). Ridge regularization provides a smooth strictly convex boundary (black), whereas LASSO provides a boundary that is non-differentiable in the corners (blue). The elastic net is still non-differentiable in the corners, this is needed for variable selection, and at the same time it is strictly convex between the corners which is needed for grouping.

6.2.3 Ridge Regression

In this section we consider ridge regression ($p = 2$) in more detail and we provide an example. The ridge estimator $\widehat{\boldsymbol{\beta}}^{\text{ridge}}$ in (6.7) is found by solving the score equations

$$\widetilde{s}(\boldsymbol{\beta}, \boldsymbol{Y}) = \nabla_{\boldsymbol{\beta}} \left(\ell_{\boldsymbol{Y}}(\boldsymbol{\beta}) - n\lambda \|\boldsymbol{\beta}_-\|_2^2 \right) = \mathfrak{X}^\top W(\boldsymbol{\beta}) \boldsymbol{R}(\boldsymbol{Y}, \boldsymbol{\beta}) - 2n\lambda \boldsymbol{\beta}_- = 0, \tag{6.9}$$

note that we exclude the intercept β_0 from regularization (we use a slight abuse of notation, here), and we also refer to Proposition 5.1. The negative expected Hessian of this optimization problem is given by

$$\mathcal{J}(\boldsymbol{\beta}) = -\mathbb{E}_{\boldsymbol{\beta}} \left[\nabla_{\boldsymbol{\beta}}^2 \left(\ell_{\boldsymbol{Y}}(\boldsymbol{\beta}) - n\lambda \|\boldsymbol{\beta}_-\|_2^2 \right) \right] = \mathcal{I}(\boldsymbol{\beta}) + 2n\lambda \text{diag}(0, 1, \ldots, 1) \in \mathbb{R}^{(q+1)\times(q+1)},$$

where $\mathcal{I}(\boldsymbol{\beta}) = \mathfrak{X}^\top W(\boldsymbol{\beta}) \mathfrak{X}$ is Fisher's information matrix of the unconstraint MLE problem. This provides us with Fisher's scoring updates for $t \geq 0$, see (5.13),

$$\widehat{\boldsymbol{\beta}}^{(t)} \mapsto \widehat{\boldsymbol{\beta}}^{(t+1)} = \widehat{\boldsymbol{\beta}}^{(t)} + \mathcal{J}(\widehat{\boldsymbol{\beta}}^{(t)})^{-1} \widetilde{s}(\widehat{\boldsymbol{\beta}}^{(t)}, \boldsymbol{Y}). \tag{6.10}$$

Lemma 6.4 *Fisher's scoring update* (6.10) *can be rewritten as follows*

$$\widehat{\boldsymbol{\beta}}^{(t)} \mapsto \widehat{\boldsymbol{\beta}}^{(t+1)} = \mathcal{J}(\widehat{\boldsymbol{\beta}}^{(t)})^{-1} \mathfrak{X}^\top W(\widehat{\boldsymbol{\beta}}^{(t)}) \left(\mathfrak{X} \widehat{\boldsymbol{\beta}}^{(t)} + \boldsymbol{R}(\boldsymbol{Y}, \widehat{\boldsymbol{\beta}}^{(t)}) \right).$$

Proof A straightforward calculation shows

$$\begin{aligned}
\widehat{\boldsymbol{\beta}}^{(t+1)} &= \widehat{\boldsymbol{\beta}}^{(t)} + \mathcal{J}(\widehat{\boldsymbol{\beta}}^{(t)})^{-1} \widetilde{s}(\widehat{\boldsymbol{\beta}}^{(t)}, \boldsymbol{Y}) \\
&= \mathcal{J}(\widehat{\boldsymbol{\beta}}^{(t)})^{-1} \left(\mathcal{J}(\widehat{\boldsymbol{\beta}}^{(t)}) \widehat{\boldsymbol{\beta}}^{(t)} + \mathfrak{X}^\top W(\widehat{\boldsymbol{\beta}}^{(t)}) \boldsymbol{R}(\boldsymbol{Y}, \widehat{\boldsymbol{\beta}}^{(t)}) - 2n\lambda \widehat{\boldsymbol{\beta}}_-^{(t)} \right) \\
&= \mathcal{J}(\widehat{\boldsymbol{\beta}}^{(t)})^{-1} \left(\mathcal{I}(\widehat{\boldsymbol{\beta}}^{(t)}) \widehat{\boldsymbol{\beta}}^{(t)} + \mathfrak{X}^\top W(\widehat{\boldsymbol{\beta}}^{(t)}) \boldsymbol{R}(\boldsymbol{Y}, \widehat{\boldsymbol{\beta}}^{(t)}) \right) \\
&= \mathcal{J}(\widehat{\boldsymbol{\beta}}^{(t)})^{-1} \mathfrak{X}^\top W(\widehat{\boldsymbol{\beta}}^{(t)}) \left(\mathfrak{X} \widehat{\boldsymbol{\beta}}^{(t)} + \boldsymbol{R}(\boldsymbol{Y}, \widehat{\boldsymbol{\beta}}^{(t)}) \right).
\end{aligned}$$

This proves the claim. □

Lemma 6.4 allows us to fit a ridge regularized GLM. To determine an optimal regularization parameter $\lambda \geq 0$ one uses cross-validation, in particular, generalized cross-validation is used to receive an efficient cross-validation method, see (5.67).

Example 6.5 (Ridge Regression) We revisit the gamma claim size example of Sect. 5.3.7, and we choose model Gamma GLM1, see Listing 5.11. This example does not consider any categorical features, but only continuous ones. We directly

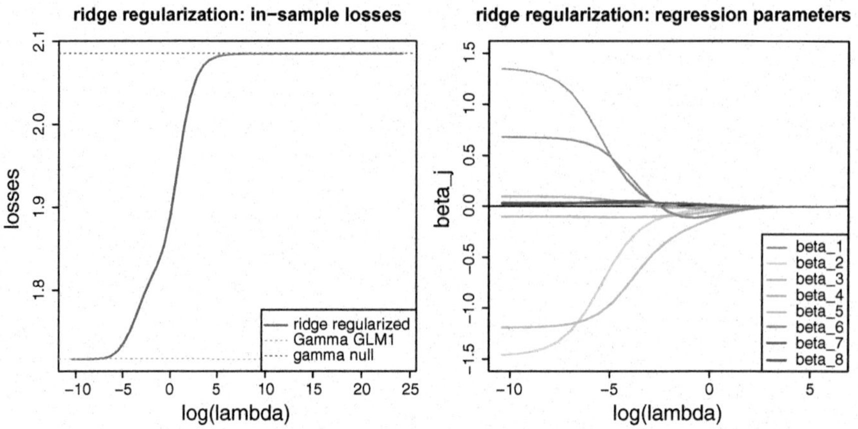

Fig. 6.3 Ridge regularized MLEs in model Gamma GLM1: (lhs) in-sample deviance losses as a function of the regularization parameter $\lambda > 0$, (rhs) resulting $\widehat{\beta}_j^{\text{ridge}}(\lambda)$ for $1 \leq j \leq q = 8$

apply Fisher's scoring updates (6.10).[3] For this analysis we center and normalize (to unit variance) the columns of the design matrix (except for the initial column of $\mathfrak{X}$ encoding the intercept).

Figure 6.3 (lhs) shows the resulting in-sample deviance losses as a function of $\lambda > 0$. Regularization parameter λ allows us to continuously connect the in-sample deviance losses of the null model (2.085) and model Gamma GLM1 (1.717), see Table 5.13. Figure 6.3 (rhs) shows the regression parameter estimates $\widehat{\beta}_j^{\text{ridge}}(\lambda)$, $1 \leq j \leq q = 8$, as a function of $\lambda > 0$. Overall they decrease because the budget constraint gets more tight for increasing λ, however, the individual parameters do not need to be monotone, since one parameter may (better) compensate a decrease of another (through correlations in feature components).

Finally, we need to choose the optimal regularization parameter $\lambda > 0$. This is done by cross-validation. We exploit the generalized cross-validation loss, see (5.67), and the hat matrix in this ridge regularized case is given by

$$H_\lambda = W(\widehat{\boldsymbol{\beta}}^{\text{ridge}})^{1/2} \mathfrak{X} \, \mathcal{J}(\widehat{\boldsymbol{\beta}}^{\text{ridge}})^{-1} \, \mathfrak{X}^\top W(\widehat{\boldsymbol{\beta}}^{\text{ridge}})^{1/2}.$$

In contrast to (5.66), this hat matrix H_λ is not a projection but we would need to work in an augmented model to receive the projection property (accounting for the regularization part).

Figure 6.4 plots the generalized cross-validation loss as a function of $\lambda > 0$. We observe the minimum in parameter $\lambda = e^{-9.4}$. The resulting generalized cross-validation loss is 1.76742. This is bigger than the one received in model Gamma

[3] The R command `glmnet` [142] allows for regularized MLE, however, the current version does not include the gamma distribution. Therefore, we have implemented our own routine.

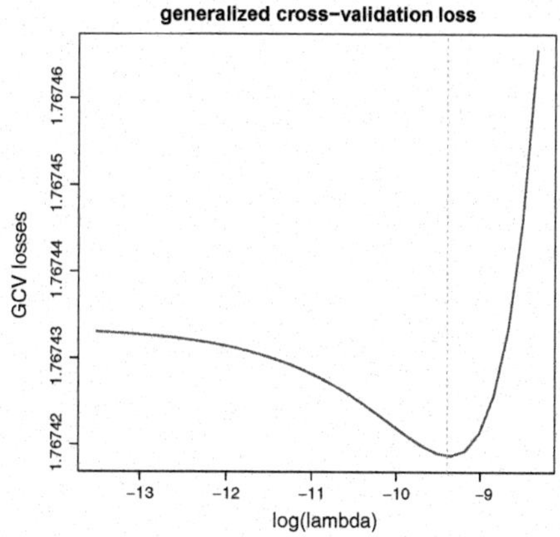

Fig. 6.4 Generalized cross-validation loss $\widehat{\mathfrak{D}}^{\mathrm{GCV}}(\lambda)$ as a function of $\lambda > 0$

GLM2, see Table 5.16, thus, we still prefer model Gamma GLM2 over the optimally ridge regularized model GLM1. Note that for model Gamma GLM2 we did variable selection, whereas ridge regression just generally shrinks regression parameters. For more interpretation we refer to Example 6.8, below, which considers LASSO regularization. ■

6.2.4 LASSO Regularization

In this section we consider LASSO regularization ($p = 1$). This is more challenging than ridge regularization because of the non-differentiability of the budget constraint, see Fig. 6.1 (rhs). This section follows Chapters 2 and 5 of Hastie et al. [184] and Parikh–Boyd [292].

Gaussian Case

We start with the homoskedastic Gaussian model having unit variance $\sigma^2 = 1$. In a first step, the regression model only involves one feature component $q = 1$. Thus, we aim at solving LASSO optimization

$$\widehat{\boldsymbol{\beta}}^{\mathrm{LASSO}} = \underset{\boldsymbol{\beta}\in\mathbb{R}^2}{\arg\max} \; -\frac{1}{2n}\sum_{i=1}^{n}(Y_i - \beta_0 - \beta_1 x_i)^2 - \lambda|\beta_1|.$$

We standardize the observations and features $(Y_i, x_i)_{1\le i\le n}$ such that we have $\sum_{i=1}^n Y_i = 0$, $\sum_{i=1}^n x_i = 0$ and $n^{-1}\sum_{i=1}^n x_i^2 = 1$. This implies that we can omit the intercept parameter β_0, as the optimal intercept satisfies for this standardized data (and any $\beta_1 \in \mathbb{R}$)

$$\widehat{\beta}_0 = \frac{1}{n}\sum_{i=1}^{n} Y_i - \beta_1 x_i = 0. \tag{6.11}$$

Thus, w.l.o.g., we assume to work with standardized data in this section, this gives us the optimization problem (we drop the lower index in β_1 because we only have one component)

$$\widehat{\beta}^{\text{LASSO}} = \widehat{\beta}^{\text{LASSO}}(\lambda) = \underset{\beta \in \mathbb{R}}{\arg\max} \; -\frac{1}{2n}\sum_{i=1}^{n}(Y_i - \beta x_i)^2 - \lambda|\beta|. \tag{6.12}$$

The difficulty is that the regularization term is not differentiable in zero. Since this term is convex we can express its derivative in terms of a sub-gradient $\mathfrak{s}$. This provides score

$$\frac{\partial}{\partial \beta}\left(-\frac{1}{2n}\sum_{i=1}^{n}(Y_i - \beta x_i)^2 - \lambda|\beta|\right) = \frac{1}{n}\sum_{i=1}^{n}(Y_i - \beta x_i)\,x_i - \lambda\mathfrak{s} = \frac{1}{n}\langle \boldsymbol{Y}, \boldsymbol{x}\rangle - \beta - \lambda\mathfrak{s},$$

where we use standardization $n^{-1}\sum_{i=1}^{n} x_i^2 = 1$ in the second step, $\langle \boldsymbol{Y}, \boldsymbol{x}\rangle$ is the scalar product of $\boldsymbol{Y}, \boldsymbol{x} = (x_1, \ldots, x_n)^\top \in \mathbb{R}^n$, and where we consider the sub-gradient

$$\mathfrak{s} = \mathfrak{s}(\beta) = \begin{cases} +1 & \text{if } \beta > 0, \\ -1 & \text{if } \beta < 0, \\ \in [-1, 1] & \text{otherwise.} \end{cases}$$

Henceforth, we receive the score equation for $\beta \neq 0$

$$n^{-1}\langle \boldsymbol{Y}, \boldsymbol{x}\rangle - \beta - \lambda\mathfrak{s} = n^{-1}\langle \boldsymbol{Y}, \boldsymbol{x}\rangle - \beta - \text{sign}(\beta)\lambda \overset{!}{=} 0.$$

This score equation has a proper solution $\widehat{\beta} > 0$ if $n^{-1}\langle \boldsymbol{Y}, \boldsymbol{x}\rangle > \lambda$, and it has a proper solution $\widehat{\beta} < 0$ if $n^{-1}\langle \boldsymbol{Y}, \boldsymbol{x}\rangle < -\lambda$. In any other case we have a boundary solution $\widehat{\beta} = 0$ for our maximization problem (6.12).

This solution can be written in terms of the following *soft-thresholding operator* for $\lambda \geq 0$

$$\widehat{\beta}^{\text{LASSO}} = \mathcal{S}_\lambda\left(n^{-1}\langle \boldsymbol{Y}, \boldsymbol{x}\rangle\right) \qquad \text{with} \qquad \mathcal{S}_\lambda(x) = \text{sign}(x)(|x| - \lambda)_+. \tag{6.13}$$

This soft-thresholding operator is illustrated in Fig. 6.5 for $\lambda = 4$.

This approach can be generalized to multiple feature components $\boldsymbol{x} \in \mathbb{R}^q$. We standardize the observations and features $\sum_{i=1}^{n} Y_i = 0$, $\sum_{i=1}^{n} x_{i,j} = 0$ and

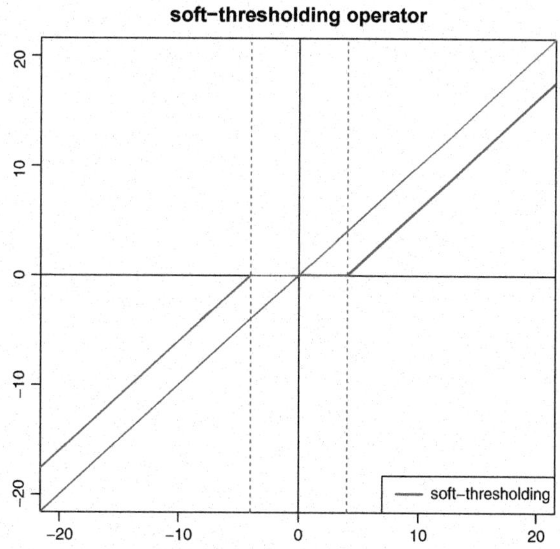

Fig. 6.5 Soft-thresholding operator $x \mapsto \mathcal{S}_\lambda(x)$ for $\lambda = 4$ (red dotted lines)

$n^{-1}\sum_{i=1}^n x_{i,j}^2 = 1$ for all $1 \le j \le q$. This allows us again to drop the intercept term and to directly consider

$$\widehat{\boldsymbol{\beta}}^{\mathrm{LASSO}} = \widehat{\boldsymbol{\beta}}^{\mathrm{LASSO}}(\lambda) = \underset{\boldsymbol{\beta}\in\mathbb{R}^q}{\arg\max} \; -\frac{1}{2n}\sum_{i=1}^n \left(Y_i - \sum_{j=1}^q \beta_j x_{i,j}\right)^2 - \lambda\|\boldsymbol{\beta}\|_1.$$

Since this is a concave (quadratic) maximization problem with a separable (convex) penalty term, we can apply a *cycle coordinate descent method* that iterates a cyclic coordinate-wise maximization until convergence. Thus, if we want to maximize in the t-th iteration the j-th coordinate of the regression parameter we consider recursively

$$\widehat{\beta}_j^{(t)} = \underset{\beta_j\in\mathbb{R}}{\arg\max} \; -\frac{1}{2n}\sum_{i=1}^n \left(Y_i - \sum_{l=1}^{j-1}\beta_l^{(t)}x_{i,l} - \sum_{l=j+1}^{q}\beta_l^{(t-1)}x_{i,l} - \beta_j x_{i,j}\right)^2 - \lambda|\beta_j|.$$

Using the soft-thresholding operator (6.13) we find the optimal solution

$$\widehat{\beta}_j^{(t)} = \mathcal{S}_\lambda\left(n^{-1}\left\langle \boldsymbol{Y} - \sum_{l=1}^{j-1}\beta_l^{(t)}\boldsymbol{x}_l - \sum_{l=j+1}^{q}\beta_l^{(t-1)}\boldsymbol{x}_l,\; \boldsymbol{x}_j\right\rangle\right),$$

with vectors $\boldsymbol{x}_l = (x_{1,l},\ldots,x_{n,l})^\top \in \mathbb{R}^n$ for $1 \le l \le q$. Iteration until convergence provides the LASSO regularized estimator $\widehat{\boldsymbol{\beta}}^{\mathrm{LASSO}}(\lambda)$ for given regularization parameter $\lambda > 0$.

Typically, we want to explore $\widehat{\boldsymbol{\beta}}^{\text{LASSO}}(\lambda)$ for multiple λ's. For this, one runs a *pathwise cyclic coordinate descent method*. We start with a large value for λ, namely, we define

$$\lambda^{\max} = \max_{1 \le j \le q} n^{-1} \left| \langle \boldsymbol{Y}, \boldsymbol{x}_j \rangle \right|.$$

For $\lambda \ge \lambda^{\max}$, we have $\widehat{\boldsymbol{\beta}}^{\text{LASSO}}(\lambda) = 0$, i.e., we have the null model. Pathwise cycle coordinate descent starts with this solution for $\lambda_0 = \lambda^{\max}$. In a next step, one slightly decreases λ_0 and runs the cyclic coordinate descent algorithm until convergence for this slightly smaller $\lambda_1 < \lambda_0$, and with starting value $\widehat{\boldsymbol{\beta}}^{\text{LASSO}}(\lambda_0)$. This is then iterated for $\lambda_{t+1} < \lambda_t$, $t \ge 0$, which provides a sequence of LASSO regularized estimators $\widehat{\boldsymbol{\beta}}^{\text{LASSO}}(\lambda_t)$ along the path $(\lambda_t)_{t \ge 0}$.

For further remarks we refer to Section 2.6 in Hastie et al. [184]. This concerns statements about uniqueness for general design matrices, also in the set-up where $q > n$, i.e., where we have more parameters than observations. Moreover, references to convergence results are given in Section 2.7 of Hastie et al. [184]. This closes the Gaussian case.

Gradient Descent Algorithm for LASSO Regularization

In Sect. 7.2.3 we will discuss gradient descent methods for network fitting. In this section we provide preliminary considerations on gradient descent methods because these are also useful to fit LASSO regularized parameters within GLMs (different from Gaussian GLMs). Remark that we do a sign switch in what follows, and we aim at minimizing an objective function g.

Choose a convex and differentiable function $g : \mathbb{R}^{q+1} \to \mathbb{R}$. Assuming that the global minimum of g is achieved, a necessary and sufficient condition for the optimality of $\boldsymbol{\beta}^* \in \mathbb{R}^{q+1}$ in this convex setting is $\nabla_{\boldsymbol{\beta}} g(\boldsymbol{\beta})|_{\boldsymbol{\beta}=\boldsymbol{\beta}^*} = 0$. *Gradient descent algorithms* find this optimal point by iterating for $t \ge 0$

$$\boldsymbol{\beta}^{(t)} \mapsto \boldsymbol{\beta}^{(t+1)} = \boldsymbol{\beta}^{(t)} - \varrho_{t+1} \nabla_{\boldsymbol{\beta}} g(\boldsymbol{\beta}^{(t)}), \tag{6.14}$$

for tempered *learning rates* $\varrho_{t+1} > 0$. This algorithm is motivated by a first order Taylor expansion that determines the direction of the maximal local decrease of the objective function g supposed we are in position $\boldsymbol{\beta}$, i.e.,

$$g(\widetilde{\boldsymbol{\beta}}) = g(\boldsymbol{\beta}) + \nabla_{\boldsymbol{\beta}} g(\boldsymbol{\beta})^\top \left(\widetilde{\boldsymbol{\beta}} - \boldsymbol{\beta}\right) + o\left(\|\widetilde{\boldsymbol{\beta}} - \boldsymbol{\beta}\|_2\right) \qquad \text{as } \|\widetilde{\boldsymbol{\beta}} - \boldsymbol{\beta}\|_2 \to 0.$$

The gradient descent algorithm (6.14) leads to the (unconstraint) minimum of the objective function g at convergence. A budget constraint like (6.6) leads to a convex constraint $\boldsymbol{\beta} \in \mathcal{C} \subset \mathbb{R}^{q+1}$. Consideration of such a convex constraint requires that we reformulate the gradient descent algorithm (6.14). The gradient descent step (6.14) can also be found, for given learning rate ϱ_{t+1}, by solving the following

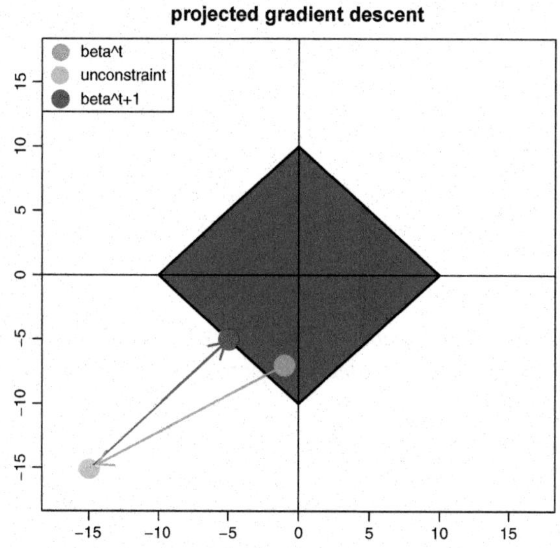

Fig. 6.6 Projected gradient descent step, first, mapping $\boldsymbol{\beta}^{(t)}$ to the unconstraint solution $\boldsymbol{\beta}^{(t)} - \varrho_{t+1}\nabla_{\boldsymbol{\beta}} g(\boldsymbol{\beta}^{(t)})$ of (6.15) and, second, projecting this unconstraint solution back to the convex set $\mathcal{C}$ giving $\boldsymbol{\beta}^{(t+1)}$; see also Figure 5.5 in Hastie et al. [184]

linearized problem for g with the Euclidean square distance penalty term (ridge regularization) for too big gradient descent steps

$$\underset{\boldsymbol{\beta}\in\mathbb{R}^{q+1}}{\arg\min}\left\{ g(\boldsymbol{\beta}^{(t)}) + \nabla_{\boldsymbol{\beta}} g(\boldsymbol{\beta}^{(t)})^\top \left(\boldsymbol{\beta} - \boldsymbol{\beta}^{(t)}\right) + \frac{1}{2\varrho_{t+1}}\|\boldsymbol{\beta} - \boldsymbol{\beta}^{(t)}\|_2^2 \right\}. \tag{6.15}$$

The solution to this optimization problem exactly gives the gradient descent step (6.14). This is now adapted to a constraint gradient descent update for convex constraint $\mathcal{C}$:

$$\boldsymbol{\beta}^{(t+1)} = \underset{\boldsymbol{\beta}\in\mathcal{C}}{\arg\min}\left\{ g(\boldsymbol{\beta}^{(t)}) + \nabla_{\boldsymbol{\beta}} g(\boldsymbol{\beta}^{(t)})^\top \left(\boldsymbol{\beta} - \boldsymbol{\beta}^{(t)}\right) + \frac{1}{2\varrho_{t+1}}\|\boldsymbol{\beta} - \boldsymbol{\beta}^{(t)}\|_2^2 \right\}. \tag{6.16}$$

The solution to this constraint convex optimization problem is obtained by, first, taking an unconstraint gradient descent step $\boldsymbol{\beta}^{(t)} \mapsto \boldsymbol{\beta}^{(t)} - \varrho_{t+1}\nabla_{\boldsymbol{\beta}} g(\boldsymbol{\beta}^{(t)})$, and, second, if this step is not within the convex set $\mathcal{C}$, it is projected back to $\mathcal{C}$; this is illustrated in Fig. 6.6, and it is called *projected gradient descent step* (justification is given in Lemma 6.6 below). Thus, the only difficulty in applying this projected gradient descent step is to find an efficient method of projecting the unconstraint solution (6.14)–(6.15) back to the convex constraint set $\mathcal{C}$.

Assume that the convex constraint set $\mathcal{C}$ is expressed by a convex function h (not necessarily being differentiable). To solve (6.16) and to motivate the projected gradient descent step, we use the *proximal gradient method* discussed in Section 5.3.3 of Hastie et al. [184]. The proximal gradient method helps us to do the projection in the projected gradient descent step. We introduce the *generalized projection operator*, for $z \in \mathbb{R}^{q+1}$

$$\mathrm{prox}_h(z) = \underset{\beta \in \mathbb{R}^{q+1}}{\arg\min} \left\{ \frac{1}{2} \|z - \beta\|_2^2 + h(\beta) \right\}. \tag{6.17}$$

This generalized projection operator should be interpreted as a square minimization problem $\|z - \beta\|_2^2/2$ on a convex set $\mathcal{C}$ being expressed by its dual Lagrangian formulation described by the regularization term $h(\beta)$. The following lemma shows that the generalized projection operator solves the Lagrangian form of (6.16).

Lemma 6.6 *Assume the convex constraint $\mathcal{C}$ is expressed by the convex function h. The generalized projection operator solves*

$$\begin{aligned} \beta^{(t+1)} &= \mathrm{prox}_{\varrho_{t+1}h}\left(\beta^{(t)} - \varrho_{t+1}\nabla_\beta g(\beta^{(t)})\right) \qquad (6.18)\\ &= \underset{\beta \in \mathbb{R}^{q+1}}{\arg\min} \left\{ g(\beta^{(t)}) + \nabla_\beta g(\beta^{(t)})^\top \left(\beta - \beta^{(t)}\right) + \frac{1}{2\varrho_{t+1}} \|\beta - \beta^{(t)}\|_2^2 + h(\beta) \right\}. \end{aligned}$$

Proof of Lemma 6.6 It suffices to consider the following calculation

$$\begin{aligned} &\frac{1}{2}\left\|\beta^{(t)} - \varrho_{t+1}\nabla_\beta g(\beta^{(t)}) - \beta\right\|_2^2 + \varrho_{t+1}h(\beta)\\ &= \frac{1}{2}\varrho_{t+1}^2 \left\|\nabla_\beta g(\beta^{(t)})\right\|_2^2 - \varrho_{t+1}\left\langle \nabla_\beta g(\beta^{(t)}), \beta^{(t)} - \beta \right\rangle + \frac{1}{2}\left\|\beta^{(t)} - \beta\right\|_2^2 + \varrho_{t+1}h(\beta)\\ &= \frac{1}{2}\varrho_{t+1}^2 \left\|\nabla_\beta g(\beta^{(t)})\right\|_2^2 + \varrho_{t+1}\left(\nabla_\beta g(\beta^{(t)})^\top \left(\beta - \beta^{(t)}\right) + \frac{1}{2\varrho_{t+1}}\left\|\beta^{(t)} - \beta\right\|_2^2 + h(\beta)\right). \end{aligned}$$

This is exactly the right objective function (in the round brackets) if we ignore all terms that are independent of β. This proves the lemma. □

Thus, to solve the constraint optimization problem (6.16) we bring it into its dual Lagrangian form (6.18). Then we apply the generalized projection operator to the unconstraint solution to find the constraint solution, see Lemma 6.6. This approach will be successful if we can explicitly compute the generalized projection operator $\mathrm{prox}_h(\cdot)$.

Lemma 6.7 *The generalized projection operator* (6.17) *satisfies for LASSO constraint $h(\beta) = \lambda\|\beta_-\|_1$*

$$\mathrm{prox}_h(z) = \mathcal{S}_\lambda^{\mathrm{LASSO}}(z) \overset{\mathrm{def.}}{=} \left(z_0, \mathrm{sign}(z_1)(|z_1| - \lambda)_+, \ldots, \mathrm{sign}(z_q)(|z_q| - \lambda)_+\right)^\top,$$

for $z \in \mathbb{R}^{q+1}$.

Proof of Lemma 6.7 We need to solve for function $\boldsymbol{\beta} \mapsto h(\boldsymbol{\beta}) = \lambda \|\boldsymbol{\beta}_-\|_1$

$$\mathrm{prox}_{\lambda\|(\cdot)_-\|_1}(z) = \underset{\boldsymbol{\beta}\in\mathbb{R}^{q+1}}{\arg\min}\left\{\frac{1}{2}\|z-\boldsymbol{\beta}\|_2^2 + \lambda\|\boldsymbol{\beta}_-\|_1\right\} = \underset{\boldsymbol{\beta}\in\mathbb{R}^{q+1}}{\arg\min}\left\{\frac{1}{2}\sum_{j=0}^{q}(z_j-\beta_j)^2 + \lambda\sum_{j=1}^{q}|\beta_j|\right\}.$$

This decouples into $q+1$ independent optimization problems. The first one is solved by $\beta_0 = z_0$ and the remaining ones are solved by the soft-thresholding operator (6.13). This finishes the proof. □

We conclude that the constraint optimization problem (6.16) for the (convex) LASSO constraint $\mathcal{C} = \{\boldsymbol{\beta};\ \|\boldsymbol{\beta}_-\|_1 \le c\}$ is brought into its dual Lagrangian form (6.18) of Lemma 6.6 with $h(\boldsymbol{\beta}) = \lambda\|\boldsymbol{\beta}_-\|_1$ for suitable $\lambda = \lambda(c)$. The LASSO regularized parameter estimation is then solved by first performing an unconstraint gradient descent step $\boldsymbol{\beta}^{(t)} \mapsto \boldsymbol{\beta}^{(t)} - \varrho_{t+1}\nabla_{\boldsymbol{\beta}} g(\boldsymbol{\beta}^{(t)})$, and this updated parameter is projected back to $\mathcal{C}$ using the generalized projection operator of Lemma 6.7 with $h(\boldsymbol{\beta}) = \varrho_{t+1}\lambda\|\boldsymbol{\beta}_-\|_1$.

Proximal gradient descent algorithm for LASSO

---

1. Make the gradient descent step for a suitable learning rate $\varrho_{t+1} > 0$

$$\boldsymbol{\beta}^{(t)} \mapsto \widetilde{\boldsymbol{\beta}}^{(t+1)} = \boldsymbol{\beta}^{(t)} - \varrho_{t+1}\nabla_{\boldsymbol{\beta}} g(\boldsymbol{\beta}^{(t)}).$$

2. Perform soft-thresholding of the gradient descent solution

$$\widetilde{\boldsymbol{\beta}}^{(t+1)} \mapsto \boldsymbol{\beta}^{(t+1)} = \mathcal{S}^{\mathrm{LASSO}}_{\varrho_{t+1}\lambda}\left(\widetilde{\boldsymbol{\beta}}^{(t+1)}\right),$$

where the latter soft-thresholding function is defined in Lemma 6.7.
3. Iterate these two steps until a stopping criterion is met.

If the gradient $\nabla_{\boldsymbol{\beta}} g(\cdot)$ is Lipschitz continuous with Lipschitz constant $L > 0$, the proximal gradient descent algorithm will converge at rate $O(1/t)$ for a fixed step size $0 < \varrho = \varrho_{t+1} \le L$, see Section 4.2 in Parikh–Boyd [292].

Example 6.8 (LASSO Regression) We revisit Example 6.5 which considers claim size modeling using model Gamma GLM1. In order to apply the proximal gradient descent algorithm for LASSO regularization we need to calculate the gradient of the negative log-likelihood. In the gamma case with log-link, it is given by, see Example 5.5,

$$\begin{aligned}
-\nabla_{\boldsymbol{\beta}}\ell_{\boldsymbol{Y}}(\boldsymbol{\beta}) &= -\mathfrak{X}^\top W(\boldsymbol{\beta})\boldsymbol{R}(\boldsymbol{Y},\boldsymbol{\beta}) \\
&= -\mathfrak{X}^\top \mathrm{diag}\left(\frac{n_1}{\varphi},\ldots,\frac{n_m}{\varphi}\right)\left(\frac{Y_1}{\mu_1}-1,\ldots,\frac{Y_m}{\mu_m}-1\right)^\top,
\end{aligned}$$

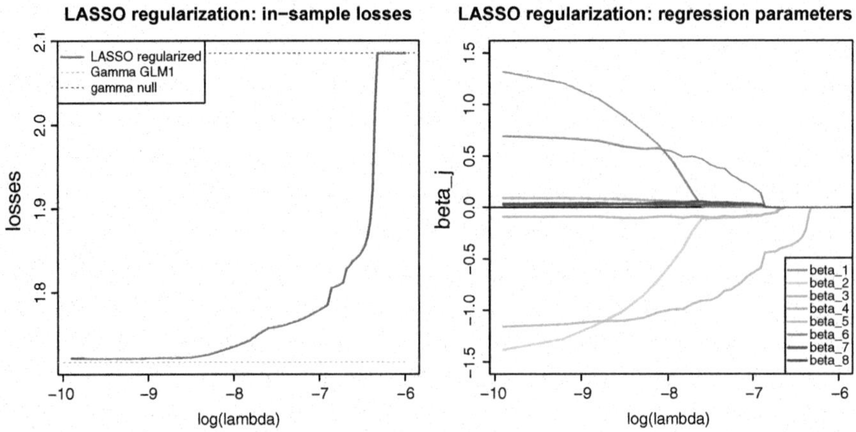

Fig. 6.7 LASSO regularized MLEs in model Gamma GLM1: (lhs) in-sample losses as a function of the regularization parameter $\lambda > 0$, (rhs) resulting $\widehat{\beta}_j^{\text{LASSO}}(\lambda)$ for $1 \le j \le q$

where $m \in \mathbb{N}$ is the number of policies with claims, and $\mu_i = \mu_i(\boldsymbol{\beta}) = \exp\langle \boldsymbol{\beta}, \boldsymbol{x}_i \rangle$. We set $\varphi = 1$ as this constant can be integrated into the learning rates ϱ_{t+1}.

We have implemented the proximal gradient descent algorithm ourselves using an equidistant grid for the regularization parameter $\lambda > 0$, a fixed learning rate $\varrho_{t+1} = 0.05$ and normalized features. Since this has been done rather brute force, the results presented in Fig. 6.7 look a bit wiggly. These results should be compared to Fig. 6.3. We see that, in contrast to ridge regularization, less important regression parameters are shrunk exactly to zero in LASSO regularization. We give the order in which the parameters are shrunk to zero: β_1 (`OwnerAge`), β_4 (`RiskClass`), β_6 (`VehAge`^{2}), β_8 (`BonusClass`), β_7 (`GenderMale`), β_2 (`OwnerAge`^{2}), β_3 (`AreaGLM`) and β_5 (`VehAge`). In view of Listing 5.11 this order seems a bit surprising. The reason for this surprising order is that we have grouped features here, and, obviously, these should be considered jointly. In particular, we first drop `OwnerAge` because this can also be partially explained by `OwnerAge`^{2}, therefore, we should not treat these two variables individually. Having this weakness in mind supports the conclusions drawn from the Wald tests in Listing 5.11, and we come back to this in Example 6.10, below.

■

Oracle Property

An interesting question is whether the chosen regularization fulfills the so-called oracle property. For simplicity, we assume to work in the normalized Gaussian case that allows us to exclude the intercept β_0, see (6.11). Thus, we work with a regression parameter $\boldsymbol{\beta} \in \mathbb{R}^q$. Assume that there is a true data model that can be described by the (true) regression parameter $\boldsymbol{\beta}^* \in \mathbb{R}^q$. Denote by $\mathcal{A}^* = \{j \in \{1, \ldots, q\}; \beta_j^* \neq 0\}$ the set of feature components of $\boldsymbol{x} \in \mathbb{R}^q$ that determine the true regression function, and we assume $|\mathcal{A}^*| < q$. Denote by $\widehat{\boldsymbol{\beta}}_n(\lambda)$ the parameter estimate that has been received by the regularized MAP estimation for a given regularization parameter $\lambda \geq 0$ and based on i.i.d. data of sample size n. We say that $(\widehat{\boldsymbol{\beta}}_n(\lambda_n))_{n\in\mathbb{N}}$ fulfills the *oracle property* if there exists a sequence $(\lambda_n)_{n\in\mathbb{N}}$ of regularization parameters $\lambda_n \geq 0$ such that

$$\lim_{n\to\infty} \mathbb{P}[\widehat{\mathcal{A}}_n = \mathcal{A}^*] = 1, \tag{6.19}$$

$$\sqrt{n}\left(\widehat{\boldsymbol{\beta}}_{n,\mathcal{A}^*}(\lambda_n) - \boldsymbol{\beta}^*_{\mathcal{A}^*}\right) \Rightarrow \mathcal{N}\left(0, \mathcal{I}_{\mathcal{A}^*}^{-1}\right) \qquad \text{as } n \to \infty, \tag{6.20}$$

where $\widehat{\mathcal{A}}_n = \{j \in \{1, \ldots, q\}; (\widehat{\boldsymbol{\beta}}_n(\lambda_n))_j \neq 0\}$, $\boldsymbol{\beta}_{\mathcal{A}}$ only considers the components in $\mathcal{A} \subset \{1, \ldots, q\}$, and $\mathcal{I}_{\mathcal{A}^*}$ is Fisher's information matrix on the true feature components. The first oracle property (6.19) tells us that asymptotically we choose the right feature components, and the second oracle property (6.20) tells us that we have asymptotic normality and, in particular, consistency on the right feature components.

Zou [408] states that LASSO regularization, in general, does not satisfy the oracle property. LASSO regularization can perform variable selection, however, as Zou [408] argues, there are situations where consistency is violated and, therefore, the oracle property cannot hold in general. Zou [408] therefore proposes an adaptive LASSO regularization method. Alternatively, Fan–Li [124] introduced smoothly clipped absolute deviation (SCAD) regularization which is a non-convex regularization that possesses the oracle property. SCAD regularization of $\boldsymbol{\beta}$ is obtained by penalizing

$$J_\lambda(\boldsymbol{\beta}) = \sum_{j=1}^{q} \lambda|\beta_j|\mathbb{1}_{\{|\beta_j|\leq\lambda\}} - \frac{|\beta_j|^2 - 2a\lambda|\beta_j| + \lambda^2}{2(a-1)}\mathbb{1}_{\{\lambda<|\beta_j|\leq a\lambda\}} + \frac{(a+1)\lambda^2}{2}\mathbb{1}_{\{|\beta_j|>a\lambda\}},$$

for a hyperparameter $a > 2$. This function is continuous and differentiable except in $\beta_j = 0$ with partial derivatives for $\beta > 0$

$$\lambda\left(\mathbb{1}_{\{\beta\leq\lambda\}} + \frac{(a\lambda - \beta)_+}{\lambda(a-1)}\mathbb{1}_{\{\beta>\lambda\}}\right).$$

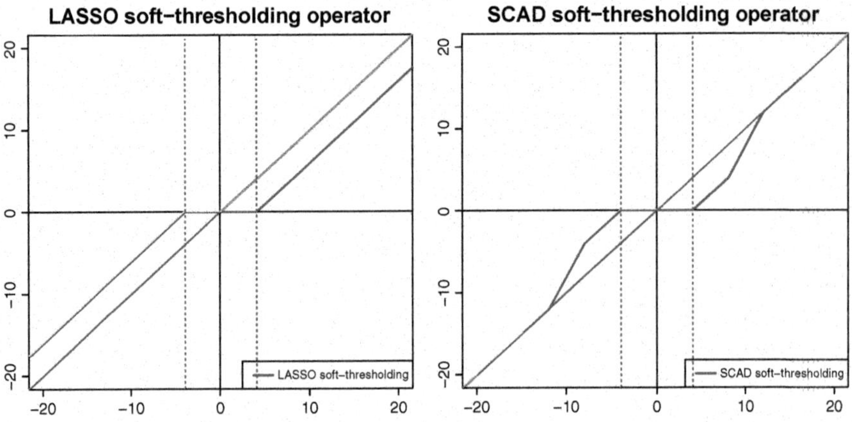

Fig. 6.8 (lhs) LASSO soft-thresholding operator $x \mapsto \mathcal{S}_\lambda(x)$ for $\lambda = 4$ (red dotted lines), (rhs) SCAD thresholding operator $x \mapsto \mathcal{S}_\lambda^{\mathrm{SCAD}}(x)$ for $\lambda = 4$ and $a = 3$

Thus, we have a constant LASSO-like slope $\lambda > 0$ for $0 < \beta \le \lambda$, shrinking some components exactly to zero. For $\beta > a\lambda$ the slope is 0, removing regularization, and it is concatenated between the two scenarios. The thresholding operator for SCAD regularization is given by, see Fan–Li [124],

$$\mathcal{S}_\lambda^{\mathrm{SCAD}}(x) = \begin{cases} \mathrm{sign}(x)(|x| - \lambda)_+ & \text{for } |x| \le 2\lambda, \\ \frac{(a-1)x - \mathrm{sign}(x)a\lambda}{a-2} & \text{for } 2\lambda < |x| \le a\lambda, \\ x & \text{for } |x| > a\lambda. \end{cases}$$

Figure 6.8 compares the two thresholding operators of LASSO and SCAD.

Alternatively, we propose to do variable selection with LASSO regularization in a first step. Since the resulting LASSO regularized estimator may not be consistent, one should explore a second regression step where one uses an un-penalized regression model on the LASSO selected components, we also refer to Lee et al. [237].

6.2.5 *Group LASSO Regularization*

In Example 6.8 we have seen that if there are natural groups within the feature components they should be treated simultaneously. Assume we have a group structure $\boldsymbol{x} = (x_0, \boldsymbol{x}_1, \ldots, \boldsymbol{x}_K)$ with groups $\boldsymbol{x}_k \in \mathbb{R}^{q_k}$ that should be treated simultaneously. This motivates the grouped penalties proposed by Yuan–Lin [398], see (6.5),

$$\widehat{\boldsymbol{\beta}}^{\text{group}} = \widehat{\boldsymbol{\beta}}^{\text{group}}(\lambda) = \underset{\boldsymbol{\beta}=(\beta_0,\boldsymbol{\beta}_1,\ldots,\boldsymbol{\beta}_K)}{\arg\max} \frac{1}{n}\ell_{\boldsymbol{Y}}(\boldsymbol{\beta}) - \lambda \sum_{k=1}^{K} \|\boldsymbol{\beta}_k\|_2, \tag{6.21}$$

where we assume a group structure in the linear predictor providing

$$\boldsymbol{x} \mapsto \eta(\boldsymbol{x}) = \langle \boldsymbol{\beta}, \boldsymbol{x} \rangle = \beta_0 + \sum_{k=1}^{K} \langle \boldsymbol{\beta}_k, \boldsymbol{x}_k \rangle.$$

LASSO regularization is a special case of this grouped regularization, namely, if all groups $1 \le k \le K$ only contain one single component, i.e., $K = q$, we have $\widehat{\boldsymbol{\beta}}^{\text{group}} = \widehat{\boldsymbol{\beta}}^{\text{LASSO}}$.

The side constraint in (6.21) is convex, and the optimization problem (6.21) can again be solved by the proximal gradient descent algorithm. That is, in view of Lemma 6.6, the only difficulty is the calculation of the generalized projection operator for regularization term $h(\boldsymbol{\beta}) = \lambda \sum_{k=1}^{K} \|\boldsymbol{\beta}_k\|_2$. We therefore need to solve for $z = (z_0, z_1, \ldots, z_K)$, $z_k \in \mathbb{R}^{q_k}$,

$$\begin{aligned}
\text{prox}_h(z) &= \underset{\boldsymbol{\beta}=(\beta_0,\boldsymbol{\beta}_1,\ldots,\boldsymbol{\beta}_K)}{\arg\min} \left\{ \frac{1}{2} \|z - \boldsymbol{\beta}\|_2^2 + \lambda \sum_{k=1}^{K} \|\boldsymbol{\beta}_k\|_2 \right\} \\
&= \left(z_0, \left(\underset{\boldsymbol{\beta}_k \in \mathbb{R}^{q_k}}{\arg\min} \left\{ \frac{1}{2} \left\| z_k - \boldsymbol{\beta}_k \right\|_2^2 + \lambda \|\boldsymbol{\beta}_k\|_2 \right\} \right)_{1 \le k \le K} \right).
\end{aligned}$$

The latter highlights that the problem decouples into K independent problems. Thus, we need to solve for all $1 \le k \le K$ the optimization problems

$$\underset{\boldsymbol{\beta}_k \in \mathbb{R}^{q_k}}{\arg\min} \left\{ \frac{1}{2} \left\| z_k - \boldsymbol{\beta}_k \right\|_2^2 + \lambda \|\boldsymbol{\beta}_k\|_2 \right\}.$$

Lemma 6.9 *The group LASSO generalized soft-thresholding operator satisfies for $z_k \in \mathbb{R}^{q_k}$*

$$\mathcal{S}_\lambda^{q_k}(z_k) = \underset{\boldsymbol{\beta}_k \in \mathbb{R}^{q_k}}{\arg\min} \left\{ \frac{1}{2} \left\| z_k - \boldsymbol{\beta}_k \right\|_2^2 + \lambda \|\boldsymbol{\beta}_k\|_2 \right\} = z_k \left(1 - \frac{\lambda}{\|z_k\|_2}\right)_+ \in \mathbb{R}^{q_k},$$

and for the generalized projection operator for $h(\boldsymbol{\beta}) = \lambda \sum_{k=1}^K \|\boldsymbol{\beta}_k\|_2$ we have

$$\operatorname{prox}_h(z) = \mathcal{S}_\lambda^{\text{group}}(z) \stackrel{\text{def.}}{=} \left(z_0, \mathcal{S}_\lambda^{q_1}(z_1), \ldots, \mathcal{S}_\lambda^{q_K}(z_K)\right),$$

for $z = (z_0, z_1, \ldots, z_K)$ with $z_k \in \mathbb{R}^{q_k}$.

Proof We prove this lemma. In a first step we have

$$\underset{\boldsymbol{\beta}_k}{\arg\min} \left\{ \frac{1}{2} \left\| z_k - \boldsymbol{\beta}_k \right\|_2^2 + \lambda \|\boldsymbol{\beta}_k\|_2 \right\} = \underset{\boldsymbol{\beta}_k = \varrho z_k/\|z_k\|_2,\ \varrho \geq 0}{\arg\min} \left\{ \frac{1}{2} \|z_k\|_2^2 \left(1 - \frac{\varrho}{\|z_k\|_2}\right)^2 + \lambda \varrho \right\},$$

this follows because the square distance $\left\| z_k - \boldsymbol{\beta}_k \right\|_2^2 = \|z_k\|_2^2 - 2\langle z_k, \boldsymbol{\beta}_k \rangle + \left\| \boldsymbol{\beta}_k \right\|_2^2$ is minimized if z_k and $\boldsymbol{\beta}_k$ point into the same direction. Thus, there remains the minimization of the objective function in $\varrho \geq 0$. The first derivative is given by

$$\frac{\partial}{\partial \varrho} \left(\frac{1}{2} \|z_k\|_2^2 \left(1 - \frac{\varrho}{\|z_k\|_2}\right)^2 + \lambda \varrho \right) = - \|z_k\|_2 \left(1 - \frac{\varrho}{\|z_k\|_2}\right) + \lambda = \lambda - \|z_k\|_2 + \varrho.$$

If $\|z_k\|_2 > \lambda$ we have $\varrho = \|z_k\|_2 - \lambda > 0$, and otherwise we need to set $\varrho = 0$. This implies

$$\mathcal{S}_\lambda^{q_k}(z_k) = (\|z_k\|_2 - \lambda)_+ \, z_k / \|z_k\|_2.$$

This completes the proof. □

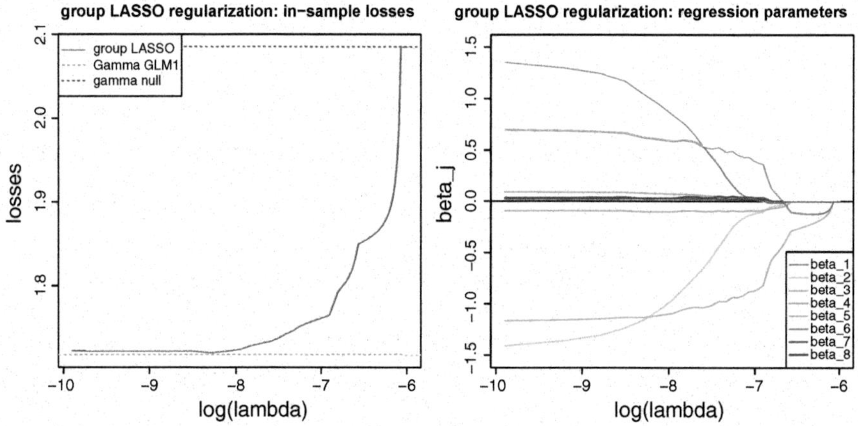

Fig. 6.9 Group LASSO regularized MLEs in model Gamma GLM1: (lhs) in-sample losses as a function of the regularization parameter $\lambda > 0$, (rhs) resulting $\widehat{\beta}_j^{\text{group}}(\lambda)$ for $1 \le j \le q$

Proximal gradient descent algorithm for group LASSO

---

1. Make the gradient descent step for a suitable learning rate $\varrho_{t+1} > 0$

$$\boldsymbol{\beta}^{(t)} \mapsto \widetilde{\boldsymbol{\beta}}^{(t+1)} = \boldsymbol{\beta}^{(t)} - \varrho_{t+1} \nabla_{\boldsymbol{\beta}} g(\boldsymbol{\beta}^{(t)}).$$

2. Perform soft-thresholding of the gradient descent solution

$$\widetilde{\boldsymbol{\beta}}^{(t+1)} \mapsto \boldsymbol{\beta}^{(t+1)} = \mathcal{S}^{\text{group}}_{\varrho_{t+1}\lambda}\left(\widetilde{\boldsymbol{\beta}}^{(t+1)}\right),$$

where the latter soft-thresholding function is defined in Lemma 6.9.
3. Iterate these two steps until a stopping criterion is met.

Example 6.10 (Group LASSO Regression) We revisit Example 6.8 which considers claim size modeling using model Gamma GLM1. This time we group the variables `OwnerAge` and `OwnerAge`^{2} (β_1, β_2) as well as `VehAge` and `VehAge`^{2} (β_5, β_6). The results are shown in Fig. 6.9.

The order in which the parameters are regularized to zero is: β_4 (`RiskClass`), β_8 (`BonusClass`), β_7 (`GenderMale`), (β_1, β_2) (`OwnerAge`, `OwnerAge`^{2}), β_3 (`AreaGLM`) and (β_5, β_6) (`VehAge`, `VehAge`^{2}). This order now reflects more the variable importance as received from the Wald statistics of Listing 5.11, and it shows that grouped features should be regularized jointly in order to determine their importance. ∎

6.3 Expectation-Maximization Algorithm

6.3.1 Mixture Distributions

In many applied problems there does not exist a simple off-the-shelf distribution that is suitable to model the whole range of observations. We think of claim size modeling which may range from small to very large claims; the main body of the data may look like, say, gamma distributed, but the tail of the data being regularly varying. Another related problem is that claims may come from different insurance policy modules. For instance, in property insurance, one can insure water damage, fire, glass and theft claims on the same insurance policy, and feature information about the claim type may not always be available. In such cases, it looks attractive to choose a mixture or a composition of different distributions. In this section we focus on mixtures.

Choose a fixed integer K bigger than 1 and define the $(K-1)$-unit simplex excluding the edges by

$$\Delta_K = \left\{ \boldsymbol{p} \in (0,1)^K;\ \sum_{k=1}^K p_k = 1 \right\}. \tag{6.22}$$

Δ_K defines the family of categorical distributions with K levels (all levels having a strictly positive probability). These distributions belong to the vector-valued parameter EF which we have met in Sects. 2.1.4 and 5.7.

The idea behind mixture distributions is to mix K different distributions with a mixture probability $\boldsymbol{p} \in \Delta_K$. For instance, we can mix K different EDF densities f_k by considering

$$Y \sim \sum_{k=1}^K p_k f_k(y; \theta_k, v/\varphi_k) = \sum_{k=1}^K p_k \exp\left\{ \frac{y\theta_k - \kappa_k(\theta_k)}{\varphi_k/v} + a_k(y; v/\varphi_k) \right\}, \tag{6.23}$$

with cumulant functions $\theta_k \in \boldsymbol{\Theta}_k \mapsto \kappa_k(\theta_k)$, exposure $v > 0$ and dispersion parameters $\varphi_k > 0$, for $1 \le k \le K$.

At the first sight, this does not look very spectacular and parameter estimation seems straightforward. If we consider the log-likelihood of n independent random variables $\boldsymbol{Y} = (Y_1, \ldots, Y_n)^\top$ following mixture density (6.23) we receive log-likelihood function

$$(\boldsymbol{\theta}, \boldsymbol{p}) \mapsto \ell_{\boldsymbol{Y}}(\boldsymbol{\theta}, \boldsymbol{p}) = \sum_{i=1}^n \ell_{Y_i}(\boldsymbol{\theta}, \boldsymbol{p}) = \sum_{i=1}^n \log\left(\sum_{k=1}^K p_k f_k(Y_i; \theta_k, v_i/\varphi_k) \right), \tag{6.24}$$

for canonical parameter $\boldsymbol{\theta} = (\theta_1, \ldots, \theta_K)^\top \in \boldsymbol{\Theta} = \boldsymbol{\Theta}_1 \times \cdots \times \boldsymbol{\Theta}_K$ and mixture probability $\boldsymbol{p} \in \Delta_K$. Unfortunately, MLE of $(\boldsymbol{\theta}, \boldsymbol{p})$ in (6.24) is not that simple. Note, the summation over $1 \le k \le K$ is inside of the logarithmic function, and the use of the Newton–Raphson algorithm may be cumbersome. The Expectation-Maximization (EM) algorithm presented in Sect. 6.3.3, below, makes parameter estimation feasible. In a nutshell, the EM algorithm leads to a sequence of parameter estimates for $(\boldsymbol{\theta}, \boldsymbol{p})$ that monotonically increases the log-likelihood in each iteration of the algorithm. Thus, we can receive an approximation to the MLE of $(\boldsymbol{\theta}, \boldsymbol{p})$.

Nevertheless, model fitting may still be difficult for the following reasons. Firstly, the log-likelihood function of a mixture distribution does not need to be bounded, we highlight this in Example 6.13, below. In that case, MLE is not a well-defined problem. Secondly, even in very simple situations, the log-likelihood function (6.24) can have multiple local maximums. This usually happens if the data is clustered and the clusters are well separated. In that case of multiple local maximums, convergence of the EM algorithm does not guarantee that we have found the global maximum. Thirdly, convergence of the log-likelihood function through the EM algorithm does not guarantee that also the sequence of parameter estimates of $(\boldsymbol{\theta}, \boldsymbol{p})$ converges. The latter needs additional examination and regularity conditions.

Figure 6.10 (lhs) shows a density of a mixture distribution mixing $K = 3$ gamma densities with shape parameters $\alpha_k = 1, 20, 40$ (orange, green and blue) and mixture probability $\boldsymbol{p} = (0.7, 0.1, 0.2)^\top$; the mixture components are already multiplied with $\boldsymbol{p}$. The resulting mixture density in red color is continuous. Figure 6.10 (rhs) replaces the blue gamma component of the plot on the left-hand side by a Pareto component (in blue). As a result we observe that the resulting mixture density in red is no longer continuous. This example is often used in practice, however, the discontinuity may be a serious issue in applications and one may use a Lomax (Pareto Type II) component instead, we refer to Sect. 2.2.5.

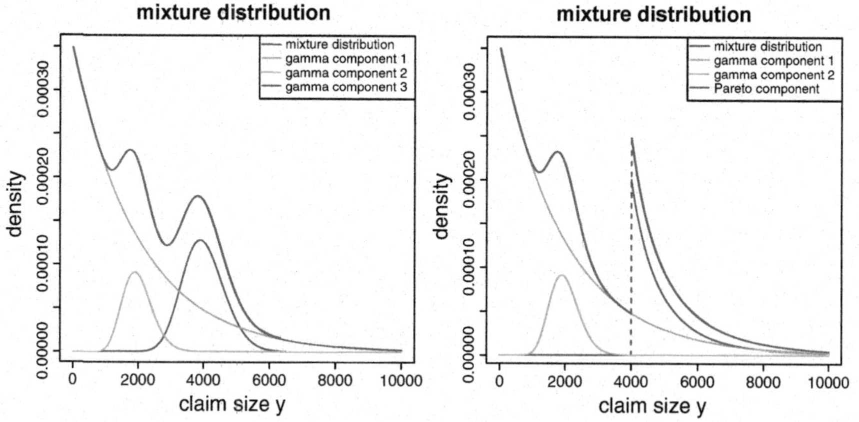

Fig. 6.10 (lhs) Mixture distribution mixing three gamma densities, and (rhs) mixture distributions mixing two gamma components and a Pareto component with mixture probabilities $\boldsymbol{p} = (0.7, 0.1, 0.2)^\top$ for orange, green and blue components (the density components are already multiplied with $\boldsymbol{p}$)

6.3.2 Incomplete and Complete Log-Likelihoods

A mixture distribution can be defined (brute force) by just defining a mixture density as in (6.23). Alternatively, we could define a mixture distribution in a more constructive way. In the following we discuss this constructive derivation which will allow us to efficiently fit mixture distributions to data $\boldsymbol{Y}$. For our outline we focus on (6.23), but all results presented below hold true in much more generality.

Choose a categorical random variable Z with $K \geq 2$ levels having probabilities $\mathbb{P}[Z = k] = p_k > 0$ for $1 \leq k \leq K$, that is, with $\boldsymbol{p} \in \Delta_K$. The main idea is to sample in a first step level $Z = k \in \{1, \ldots, K\}$, and in a second step $Y|_{\{Z=k\}} \sim f_k(y; \theta_k, v/\varphi_k)$, based on the selected level $Z = k$. The random tuple (Y, Z) has joint density

$$(Y, Z) \sim f_{\boldsymbol{\theta},\boldsymbol{p}}(y, k) = p_k f_k(y; \theta_k, v/\varphi_k),$$

and the marginal density of Y is exactly given by (6.23). In this interpretation we have a hierarchical model (Y, Z). If only Y is available for parameter estimation, then we are in the situation of *incomplete information* because information about the first hierarchy Z is missing. If both Y and Z are available we say that we have *complete information*.

For the subsequent derivations we use a different coding of the categorical random variable Z, namely, Z can be represented in the following one-hot encoding version

$$\mathbf{Z} = (Z_1, \ldots, Z_K)^\top = (\mathbb{1}_{\{Z=1\}}, \ldots, \mathbb{1}_{\{Z=K\}})^\top, \tag{6.25}$$

these are the K corners of the $(K-1)$-unit simplex Δ_K. One-hot encoding differs from dummy coding (5.21). One-hot encoding does not lead to a full rank design matrix because there is a redundancy, that is, we can drop one component of $\mathbf{Z}$ and still have the same information. One-hot encoding $\mathbf{Z}$ of Z allows us to extend the *incomplete (data) log-likelihood* $\ell_Y(\boldsymbol{\theta}, \boldsymbol{p})$, see (6.23)–(6.24), under complete information $(Y, \mathbf{Z})$ as follows

$$\begin{aligned}
\ell_{(Y,\mathbf{Z})}(\boldsymbol{\theta}, \boldsymbol{p}) &= \log\left(\prod_{k=1}^K \left(p_k f_k(Y; \theta_k, v/\varphi_k)\right)^{Z_k}\right) \\
&= \log\left(\prod_{k=1}^K \left(p_k \exp\left\{\frac{Y\theta_k - \kappa_k(\theta_k)}{\varphi_k/v} + a_k(Y; v/\varphi_k)\right\}\right)^{Z_k}\right) \qquad (6.26) \\
&= \sum_{k=1}^K Z_k \left(\log(p_k) + \frac{Y\theta_k - \kappa_k(\theta_k)}{\varphi_k/v} + a_k(Y; v/\varphi_k)\right).
\end{aligned}$$

$\ell_{(Y,Z)}(\boldsymbol{\theta}, \boldsymbol{p})$ is called *complete (data) log-likelihood*. As a consequence of this last expression we observe that under complete information $(Y_i, \boldsymbol{Z}_i)_{1 \le i \le n}$, the MLE of $\boldsymbol{\theta}$ and $\boldsymbol{p}$ can be determined completely analogously to above. Namely, θ_k is estimated from all observations Y_i for which $\boldsymbol{Z}_i$ belongs to level k, and the level indicators $(\boldsymbol{Z}_i)_{1 \le i \le n}$ are used to estimate the mixture probability $\boldsymbol{p}$. Thus, the objective function nicely decouples under complete information into independent parts for θ_k and $\boldsymbol{p}$ estimation. There remains the question of how to fit this model under incomplete information $\boldsymbol{Y}$. The next section will discuss this problem.

6.3.3 Expectation-Maximization Algorithm for Mixtures

The EM algorithm is a general purpose tool for parameter estimation under incomplete information. The EM algorithm has been introduced within the EF by Sundberg [348, 349]. Sundberg's developments have been based on the vector-valued parameter EF with statistics $S(\boldsymbol{Y}) \in \mathbb{R}^k$, see (3.17), and he solved the estimation problem under the assumption that $S(\boldsymbol{Y})$ is not fully known. These results have been generalized to MLE under incomplete data in the celebrated work of Dempster et al. [96] and Wu [385]. The monograph of McLachlan–Krishnan [267] gives the theory behind the EM algorithm, and it also provides a historical review in Section 1.8. In actuarial science the EM algorithm is increasingly used to solve various kinds of problems of incomplete data. Mixture models of Erlang distributions are considered in Lee–Lin [240], Yin–Lin [396] and Fung et al. [146, 147]; general Erlang mixtures are universal approximators to positive distributions (in the weak convergence sense), and regularized Erlang mixtures and mixtures of experts models are determined using the EM algorithm to receive approximations to the true underlying model. Miljkovic–Grün [278], Parodi [295] and Fung et al. [148] consider the EM algorithm for mixtures of general distributions, in particular, mixtures of small and large claims distributions. Verbelen et al. [371], Blostein–Miljkovic [40], Grün–Miljkovic [173] and Fung et al. [147] use the EM algorithm for censored and/or truncated observations, and dispersion modeling is performed with the EM algorithm in Tzougas–Karlis [359]. (Inhomogeneous) phase-type and matrix Mittag–Leffler distributions are fitted with the EM algorithm in Asmussen et al. [14], Albrecher et al. [8] and Bladt [37], and the EM algorithm is used to fit mixture density networks (MDNs) in Delong et al. [95]. Parameter uncertainty is investigated in O'Hagan et al. [289] using the bootstrap method. The present section is mainly based on McLachlan–Krishnan [267].

As mentioned above, the EM algorithm is a general purpose tool for parameter estimation under incomplete data, and we describe the variant of the EM algorithm which is useful for our mixture distribution setup given in (6.26). We give a justification for its functioning below. The EM algorithm is an iterative algorithm that performs a Bayesian expectation step (E-step) to infer the latent variable $\boldsymbol{Z}$, given the model parameters and Y. Next, it performs a maximization step (M-step) for MLE of the parameters given the observation Y and the estimated latent variable $\widehat{\boldsymbol{Z}}$. More specifically, the E-step and the M-step look as follows.

- **E-step.** Calculate the posterior probability of the event that a given observation Y has been generated from the k-th component of the mixture distribution. Bayes' rule allows us to infer this posterior probability (for given $\boldsymbol{\theta}$ and $\boldsymbol{p}$) from (6.26)

$$\mathbb{P}_{\boldsymbol{\theta},\boldsymbol{p}}[Z_k = 1|Y] = \frac{p_k f_k(Y; \theta_k, v/\varphi_k)}{\sum_{l=1}^{K} p_l f_l(Y; \theta_l, v/\varphi_l)}.$$

The posterior (Bayesian) estimate for Z_k after having observed Y is given by

$$\widehat{Z}_k(\boldsymbol{\theta}, \boldsymbol{p}|Y) \stackrel{\text{def.}}{=} \mathbb{E}_{\boldsymbol{\theta},\boldsymbol{p}}[Z_k|Y] = \mathbb{P}_{\boldsymbol{\theta},\boldsymbol{p}}[Z_k = 1|Y] \qquad \text{for } 1 \leq k \leq K. \tag{6.27}$$

This posterior mean $\widehat{\boldsymbol{Z}} = \widehat{\boldsymbol{Z}}(\boldsymbol{\theta}, \boldsymbol{p}|Y) = (\widehat{Z}_1(\boldsymbol{\theta}, \boldsymbol{p}|Y), \ldots, \widehat{Z}_K(\boldsymbol{\theta}, \boldsymbol{p}|Y))^\top \in \Delta_K$ is used as an estimate for the (unobserved) latent variable $\boldsymbol{Z}$; note that this posterior mean depends on the unknown parameters $(\boldsymbol{\theta}, \boldsymbol{p})$.
- **M-step.** Based on Y and $\widehat{\boldsymbol{Z}}$ the parameters $\boldsymbol{\theta}$ and $\boldsymbol{p}$ are estimated with MLE.

Alternation of these two steps provide the following recursive algorithm. We assume to have independent responses $(Y_i, \boldsymbol{Z}_i)$, $1 \leq i \leq n$, following the mixture distribution (6.26), where, for simplicity, we assume that only the volumes $v_i > 0$ are dependent on i.

EM algorithm for mixture distributions

---

(0) Choose an initial parameter $(\widehat{\boldsymbol{\theta}}^{(0)}, \widehat{\boldsymbol{p}}^{(0)}) \in \boldsymbol{\Theta} \times \Delta_K$.
(1) Repeat for $t \geq 1$ until a stopping criterion is met:

- **E-step.** Given parameter $(\widehat{\boldsymbol{\theta}}^{(t-1)}, \widehat{\boldsymbol{p}}^{(t-1)}) \in \boldsymbol{\Theta} \times \Delta_K$ estimate the latent variables $\boldsymbol{Z}_i$, $1 \leq i \leq n$, by their conditional expectations, see (6.27),

$$\widehat{\boldsymbol{Z}}_i^{(t)} = \widehat{\boldsymbol{Z}}\left(\widehat{\boldsymbol{\theta}}^{(t-1)}, \widehat{\boldsymbol{p}}^{(t-1)}\middle| Y_i\right) = \mathbb{E}_{\widehat{\boldsymbol{\theta}}^{(t-1)}, \widehat{\boldsymbol{p}}^{(t-1)}}[\boldsymbol{Z}_i|Y_i] \in \Delta_K. \tag{6.28}$$

- **M-step.** Calculate the MLE $(\widehat{\boldsymbol{\theta}}^{(t)}, \widehat{\boldsymbol{p}}^{(t)}) \in \boldsymbol{\Theta} \times \Delta_K$ based on (complete) observations $((Y_1, \widehat{\boldsymbol{Z}}_1^{(t)}), \ldots, (Y_n, \widehat{\boldsymbol{Z}}_n^{(t)}))$, i.e., solve the score equations, see (6.26),

$$\nabla_{\boldsymbol{\theta}} \left(\sum_{i=1}^{n} \sum_{k=1}^{K} \widehat{Z}_{i,k}^{(t)} \frac{Y_i \theta_k - \kappa_k(\theta_k)}{\varphi_k / v_i} \right) = 0, \tag{6.29}$$

$$\nabla_{\boldsymbol{p}_-} \left(\sum_{i=1}^{n} \sum_{k=1}^{K} \widehat{Z}_{i,k}^{(t)} \log(p_k) \right) = 0, \tag{6.30}$$

where $\boldsymbol{p}_- = (p_1, \ldots, p_{K-1})^\top$ and setting $p_K = 1 - \sum_{k=1}^{K-1} p_k \in (0, 1)$.

Remarks 6.11

- The E-step uses Bayes' rule. This motivates to consider the EM algorithm in this Bayesian chapter; alternatively, it also fits to the MLE chapters.
- We have formulated the M-step in (6.29)–(6.30) in a general way because the canonical parameter $\boldsymbol{\theta}$ and the mixture probability $\boldsymbol{p}$ could be modeled by GLMs, and, henceforth, they may be feature $\boldsymbol{x}_i$ dependent. Moreover, (6.29) is formulated for a mixture of single-parameter EDF distributions, but, of course, this holds in much more generality.
- Equations (6.29)–(6.30) are the score equations received from (6.26). There is a subtle point here, namely, $Z_k \in \{0, 1\}$ in (6.26) are observations, whereas $\widehat{Z}_{i,k}^{(t)} \in (0, 1)$ in (6.29)–(6.30) are their estimates. Thus, in the EM algorithm the unknown latent variables are replaced by their estimates which, in our setup, results in two different types of variables with disjoint ranges. This may matter in software implementations, for instance, a categorical GLM may ask for a categorical random variable $Z \in \{1, \ldots, K\}$ (of `factor` type), whereas $\widehat{\boldsymbol{Z}}$ is in the interior of the unit simplex Δ_K.
- For mixture distributions one can replace the latent variables $\boldsymbol{Z}_i$ by their conditionally expected values $\widehat{\boldsymbol{Z}}_i$, see (6.29)–(6.30). In general, this does not hold true in EM algorithm applications: in our case we benefit from the fact that Z_k influences the complete log-likelihood *linearly*, see (6.26). In the general (non-linear) case of the EM algorithm application, different from mixture distribution problems, one needs to calculate the conditional expectation of the log-likelihood function.

- If we calculate the scores element-wise we receive

$$\frac{\partial}{\partial \theta_k} \sum_{i=1}^n \frac{Y_i \theta_k - \kappa_k(\theta_k)}{\varphi_k / (v_i \widehat{Z}_{i,k}^{(t)})} = 0,$$

$$\frac{\partial}{\partial p_k} \sum_{i=1}^n \left(\widehat{Z}_{i,k}^{(t)} \log(p_k) + \widehat{Z}_{i,K}^{(t)} \log(p_K) \right) = 0,$$

recall normalization $p_K = 1 - \sum_{k=1}^{K-1} p_k \in (0,1)$.

From the first score equation we see that we receive the classical MLE/GLM framework, and all tools introduced above for parameter estimation can directly be used. The only part that changes are the weights $v_i \mapsto v_i \widehat{Z}_{i,k}^{(t)}$. In the homogeneous case, i.e., in the null model we have MLE after the t-th iteration of the EM algorithm

$$\widehat{\theta}_k^{(t)} = h_k \left(\frac{\sum_{i=1}^n v_i \widehat{Z}_{i,k}^{(t)} Y_i}{\sum_{i=1}^n v_i \widehat{Z}_{i,k}^{(t)}} \right),$$

where h_k is the canonical link that corresponds to cumulant function κ_k.

If we choose the null model for the mixture probabilities we receive MLEs

$$\widehat{p}_k^{(t)} = \frac{1}{n} \sum_{i=1}^n \widehat{Z}_{i,k}^{(t)} \qquad \text{for } 1 \le k \le K. \tag{6.31}$$

In Sect. 6.3.4, below, we will present an example that uses the null model for the mixture probabilities $\boldsymbol{p}$, and we present an other example that uses a logistic categorical GLM for these mixture probabilities.

Justification of the EM Algorithm So far, we have neither given any argument why the EM algorithm is reasonable for parameter estimation nor have we said anything about convergence. The purpose of this paragraph is to justify the above EM algorithm. We aim at solving the incomplete log-likelihood maximization problem, see (6.24),

$$(\widehat{\boldsymbol{\theta}}^{\mathrm{MLE}}, \widehat{\boldsymbol{p}}^{\mathrm{MLE}}) = \underset{(\boldsymbol{\theta},\boldsymbol{p})}{\arg\max}\, \ell_{\boldsymbol{Y}}(\boldsymbol{\theta},\boldsymbol{p}) = \underset{(\boldsymbol{\theta},\boldsymbol{p})}{\arg\max} \sum_{i=1}^n \log\left(\sum_{k=1}^K p_k f_k(Y_i;\theta_k, v_i/\varphi_k) \right),$$

subject to existence and uniqueness. We introduce some notation. Let $f(y, z; \boldsymbol{\theta}, \boldsymbol{p}) = \exp\{\ell_{(y,z)}(\boldsymbol{\theta}, \boldsymbol{p})\}$ be the joint density of $(Y, \boldsymbol{Z})$ and let $f(y; \boldsymbol{\theta}, \boldsymbol{p}) =$

$\exp\{\ell_y(\boldsymbol{\theta}, \boldsymbol{p})\}$ be the marginal density of Y. This allows us to rewrite the incomplete log-likelihood as follows for any value of $\boldsymbol{z}$

$$\ell_Y(\boldsymbol{\theta}, \boldsymbol{p}) = \log f(Y; \boldsymbol{\theta}, \boldsymbol{p}) = \log\left(\frac{f(Y, z; \boldsymbol{\theta}, \boldsymbol{p})}{f(z|Y; \boldsymbol{\theta}, \boldsymbol{p})}\right),$$

thus, we bring in the complete log-likelihood by using Bayes' rule. Choose an arbitrary categorical distribution $\pi \in \Delta_K$ with K levels. We have using the previous step

$$\begin{aligned}
\ell_Y(\boldsymbol{\theta}, \boldsymbol{p}) = \log f(Y; \boldsymbol{\theta}, \boldsymbol{p}) &= \sum_z \pi(z) \log f(Y; \boldsymbol{\theta}, \boldsymbol{p}) \\
&= \sum_z \pi(z) \log\left(\frac{f(Y, z; \boldsymbol{\theta}, \boldsymbol{p})/\pi(z)}{f(z|Y; \boldsymbol{\theta}, \boldsymbol{p})/\pi(z)}\right) \\
&= \sum_z \pi(z) \log\left(\frac{f(Y, z; \boldsymbol{\theta}, \boldsymbol{p})}{\pi(z)}\right) + \sum_z \pi(z) \log\left(\frac{\pi(z)}{f(z|Y; \boldsymbol{\theta}, \boldsymbol{p})}\right) \\
&= \sum_z \pi(z) \log\left(\frac{f(Y, z; \boldsymbol{\theta}, \boldsymbol{p})}{\pi(z)}\right) + D_{\mathrm{KL}}\left(\pi || f(\cdot|Y; \boldsymbol{\theta}, \boldsymbol{p})\right) \qquad (6.32) \\
&\geq \sum_z \pi(z) \log\left(\frac{f(Y, z; \boldsymbol{\theta}, \boldsymbol{p})}{\pi(z)}\right),
\end{aligned}$$

the inequality follows because the KL divergence is always non-negative, see Lemma 2.21. This provides us with a lower bound for the incomplete log-likelihood $\ell_Y(\boldsymbol{\theta}, \boldsymbol{p})$ for any categorical distribution $\pi \in \Delta_K$ and any $(\boldsymbol{\theta}, \boldsymbol{p}) \in \boldsymbol{\Theta} \times \Delta_K$:

$$\begin{aligned}
\ell_Y(\boldsymbol{\theta}, \boldsymbol{p}) &\geq \sum_z \pi(z) \log\left(\frac{f(Y, z; \boldsymbol{\theta}, \boldsymbol{p})}{\pi(z)}\right) \qquad (6.33) \\
&= \mathbb{E}_{\boldsymbol{Z}\sim\pi}\left[\ell_{(Y,\boldsymbol{Z})}(\boldsymbol{\theta}, \boldsymbol{p})\middle| Y\right] - \sum_z \pi(z) \log(\pi(z)) \overset{\text{def.}}{=} \mathcal{Q}(\boldsymbol{\theta}, \boldsymbol{p}; \pi).
\end{aligned}$$

Thus, we have a lower bound $\mathcal{Q}(\boldsymbol{\theta}, \boldsymbol{p}; \pi)$ on the incomplete log-likelihood $\ell_Y(\boldsymbol{\theta}, \boldsymbol{p})$. This lower bound is based on the conditionally expected complete log-likelihood $\ell_{(Y,\boldsymbol{Z})}(\boldsymbol{\theta}, \boldsymbol{p})$, given Y, and under an arbitrary choice π for $\boldsymbol{Z}$. The difference between this arbitrary π and the true conditional posterior distribution is given by the KL divergence $D_{\mathrm{KL}}(\pi || f(\cdot|Y; \boldsymbol{\theta}, \boldsymbol{p}))$, see (6.32).

The general idea of the EM algorithm is to make this lower bound $\mathcal{Q}(\boldsymbol{\theta}, \boldsymbol{p}; \pi)$ as large as possible in $\boldsymbol{\theta}$, $\boldsymbol{p}$ and π by iterating the following two alternating steps for $t \geq 1$:

$$\widehat{\pi}^{(t)} = \underset{\pi}{\arg\max}\ \mathcal{Q}\left(\widehat{\boldsymbol{\theta}}^{(t-1)}, \widehat{\boldsymbol{p}}^{(t-1)}; \pi\right), \tag{6.34}$$

$$(\widehat{\boldsymbol{\theta}}^{(t)}, \widehat{\boldsymbol{p}}^{(t)}) = \underset{\boldsymbol{\theta}, \boldsymbol{p}}{\arg\max}\ \mathcal{Q}\left(\boldsymbol{\theta}, \boldsymbol{p}; \widehat{\pi}^{(t)}\right). \tag{6.35}$$

The first step (6.34) can be solved explicitly and it results in the E-step. Namely, from (6.32) we see that maximizing $\mathcal{Q}(\widehat{\boldsymbol{\theta}}^{(t-1)}, \widehat{\boldsymbol{p}}^{(t-1)}; \pi)$ in π is equivalent to minimizing the KL divergence $D_{\mathrm{KL}}(\pi || f(\cdot|Y; \widehat{\boldsymbol{\theta}}^{(t-1)}, \widehat{\boldsymbol{p}}^{(t-1)}))$ in π because the left-hand side of (6.32) is independent of π. Thus, we have to solve

$$\widehat{\pi}^{(t)} = \underset{\pi}{\arg\max}\ \mathcal{Q}\left(\widehat{\boldsymbol{\theta}}^{(t-1)}, \widehat{\boldsymbol{p}}^{(t-1)}; \pi\right) = \underset{\pi}{\arg\min}\ D_{\mathrm{KL}}\left(\pi \,\middle\|\, f(\cdot|Y; \widehat{\boldsymbol{\theta}}^{(t-1)}, \widehat{\boldsymbol{p}}^{(t-1)})\right).$$

This optimization is solved by choosing the density $\widehat{\pi}^{(t)} = f(\cdot|Y; \widehat{\boldsymbol{\theta}}^{(t-1)}, \widehat{\boldsymbol{p}}^{(t-1)})$, see Lemma 2.21, and this gives us exactly (6.28) if we calculate the corresponding conditional expectation of the latent variable $\boldsymbol{Z}$. Moreover, importantly, this step provides us with an identity in (6.33):

$$\ell_Y(\widehat{\boldsymbol{\theta}}^{(t-1)}, \widehat{\boldsymbol{p}}^{(t-1)}) = \mathcal{Q}\left(\widehat{\boldsymbol{\theta}}^{(t-1)}, \widehat{\boldsymbol{p}}^{(t-1)}; \widehat{\pi}^{(t)}\right). \tag{6.36}$$

The second step (6.35) then increases the right-hand side of (6.36). This second step is equivalent to

$$(\widehat{\boldsymbol{\theta}}^{(t)}, \widehat{\boldsymbol{p}}^{(t)}) = \underset{\boldsymbol{\theta}, \boldsymbol{p}}{\arg\max}\ \mathcal{Q}\left(\boldsymbol{\theta}, \boldsymbol{p}; \widehat{\pi}^{(t)}\right) = \underset{\boldsymbol{\theta}, \boldsymbol{p}}{\arg\max}\ \mathbb{E}_{\boldsymbol{Z} \sim \widehat{\pi}^{(t)}}\left[\ell_{(Y,\boldsymbol{Z})}(\boldsymbol{\theta}, \boldsymbol{p}) \middle| Y\right], \tag{6.37}$$

and this maximization is solved by the solution of the score equations (6.29)–(6.30) of the M-step. In this step we explicitly use the linearity in $\boldsymbol{Z}$ of the log-likelihood $\ell_{(Y,\boldsymbol{Z})}$, which allows us to calculate the objective function in (6.37) explicitly resulting in replacing $\boldsymbol{Z}$ by $\widehat{\boldsymbol{Z}}^{(t)}$. For other incomplete data problems, where we do not have this linearity, this step will be more complicated.

Summarizing, alternating optimizations (6.34) and (6.35) gives us a sequence of parameters $(\widehat{\boldsymbol{\theta}}^{(t)}, \widehat{\boldsymbol{p}}^{(t)})_{t \geq 0}$ with monotonically increasing incomplete log-likelihoods

$$\ldots \leq \ell_Y(\widehat{\boldsymbol{\theta}}^{(t-1)}, \widehat{\boldsymbol{p}}^{(t-1)}) \leq \ell_Y(\widehat{\boldsymbol{\theta}}^{(t)}, \widehat{\boldsymbol{p}}^{(t)}) \leq \ell_Y(\widehat{\boldsymbol{\theta}}^{(t+1)}, \widehat{\boldsymbol{p}}^{(t+1)}) \leq \ldots. \tag{6.38}$$

Therefore, the EM algorithm converges supposed that the incomplete log-likelihood $\ell_Y(\boldsymbol{\theta}, \boldsymbol{p})$ is a bounded function.

Remarks 6.12

- In general, the log-likelihood function $(\boldsymbol{\theta}, \boldsymbol{p}) \mapsto \ell_Y(\boldsymbol{\theta}, \boldsymbol{p})$ does not need to be bounded. In that case the EM algorithm may not converge (unless it converges to a local maximum). An illustrative example is given in Example 6.13, below, which shows what can go wrong in MLE of mixture distributions.
- Even if the log-likelihood function $(\boldsymbol{\theta}, \boldsymbol{p}) \mapsto \ell_Y(\boldsymbol{\theta}, \boldsymbol{p})$ is bounded, one may not expect a unique solution to the parameter estimation problem with the EM algorithm. Firstly, a monotonically increasing sequence (6.38) only guarantees that we have convergence of that sequence. But the sequence may not converge to the global maximum and different starting points of the algorithm need to be explored. Secondly, convergence of sequence (6.38) does not necessarily imply that the parameters $(\widehat{\boldsymbol{\theta}}^{(t)}, \widehat{\boldsymbol{p}}^{(t)})$ converge for $t \to \infty$. On the one hand, we may have an identifiability issue because the components f_k of the mixture distribution may be exchangeable, and secondly one needs stronger conditions to ensure that not only the log-likelihoods converge but also their arguments (parameters) $(\widehat{\boldsymbol{\theta}}^{(t)}, \widehat{\boldsymbol{p}}^{(t)})$. This is the point studied in Wu [385].
- Even in very simple examples of mixture distributions we can have multiple local maximums. In this case the role of the starting point plays a crucial role. It is advantageous that in the starting configuration every component k shares roughly the same number of observations for the initial estimates $(\widehat{\boldsymbol{\theta}}^{(0)}, \widehat{\boldsymbol{p}}^{(0)})$ and $\widehat{\boldsymbol{Z}}^{(1)}$, otherwise one may start in a so-called spurious configuration where only a few observations almost fully determine a component k of the mixture distribution. This may result in similar singularities as in Example 6.13, below. Therefore, there are three common ways to determine a starting configuration of the EM algorithm, see Miljkovic–Grün [278]: (a) Euclidean distance-based initialization: cluster centers are selected at random, and all observations are allocated to these centers according to the shortest Euclidean distance; (b) K-means clustering allocation; or (c) completely random allocation to K bins. Using one of these three options, f_k and $\boldsymbol{p}$ are initialized.
- We have formulated the EM algorithm in the homogeneous situation. However, we can easily expand it to GLMs by, for instance, assuming that the canonical parameters θ_k are modeled by linear predictors $\langle \boldsymbol{\beta}_k, \boldsymbol{x} \rangle$ and/or likewise for the mixture probabilities $\boldsymbol{p}$. The E-step will not change in this setup. For the M-step, we will solve a different maximization problem, however, this maximization problem respects monotonicity (6.38), and therefore a modified version of the above EM algorithm applies. We emphasize that the crucial point is monotonicity (6.38) that makes the EM algorithm a valid procedure.

6.3.4 Lab: Mixture Distribution Applications

In this section we are going to present different mixture distribution examples that use the EM algorithm for parameter estimation. On the one hand this illustrates the functioning of the EM algorithm, and on the other hand it also highlights pitfalls that need to be avoided.

Example 6.13 (Gaussian Mixture) We directly fit a mixture model to the observation $\boldsymbol{Y} = (Y_1, \ldots, Y_n)^\top$. Assume that the log-likelihood of $\boldsymbol{Y}$ is given by a mixture of two Gaussian distributions

$$\ell_{\boldsymbol{Y}}(\boldsymbol{\theta}, \boldsymbol{\sigma}, \boldsymbol{p}) = \sum_{i=1}^{n} \log\left(\sum_{k=1}^{2} p_k \frac{1}{\sqrt{2\pi}\sigma_k} \exp\left\{-\frac{1}{2\sigma_k^2}(Y_i - \theta_k)^2\right\}\right),$$

with $\boldsymbol{p} \in \Delta_2$, mean vector $\boldsymbol{\theta} = (\theta_1, \theta_2)^\top \in \mathbb{R}^2$ and standard deviations $\boldsymbol{\sigma} = (\sigma_1, \sigma_2)^\top \in \mathbb{R}_+^2$. Choose estimate $\widehat{\theta}_1 = Y_1$, then we have

$$\lim_{\sigma_1 \to 0} \frac{1}{\sqrt{2\pi}\sigma_1} \exp\left\{-\frac{1}{2\sigma_1^2}(Y_1 - \widehat{\theta}_1)^2\right\} = \lim_{\sigma_1 \to 0} \frac{1}{\sqrt{2\pi}\sigma_1} = \infty.$$

For any $i \neq 1$ we have $Y_i \neq \widehat{\theta}_1$ (note that the Gaussian distribution is absolutely continuous and observations are distinct, a.s.). Henceforth for $i \neq 1$

$$\lim_{\sigma_1 \to 0} \frac{1}{\sqrt{2\pi}\sigma_1} \exp\left\{-\frac{1}{2\sigma_1^2}(Y_i - \widehat{\theta}_1)^2\right\} = \lim_{\sigma_1 \to 0} \frac{1}{\sqrt{2\pi}} \exp\left\{-\frac{1}{2\sigma_1^2}(Y_i - \widehat{\theta}_1)^2 - \log \sigma_1\right\} = 0.$$

If we choose any $\widehat{\theta}_2 \in \mathbb{R}$, $\boldsymbol{p} \in \Delta_2$ and $\sigma_2 > 0$, we receive for $\widehat{\theta}_1 = Y_1$

$$\begin{aligned}\lim_{\sigma_1 \to 0} \ell_{\boldsymbol{Y}}(\widehat{\boldsymbol{\theta}}, \boldsymbol{\sigma}, \boldsymbol{p}) = \lim_{\sigma_1 \to 0} & \log\left(\sum_{k=1}^{2} p_k \frac{1}{\sqrt{2\pi}\sigma_k} \exp\left\{-\frac{1}{2\sigma_k^2}(Y_1 - \widehat{\theta}_k)^2\right\}\right) \\ & + \sum_{i=2}^{n} \log\left(\frac{p_2}{\sqrt{2\pi}\sigma_2}\right) - \frac{1}{2\sigma_2^2}(Y_i - \widehat{\theta}_2)^2 = \infty.\end{aligned}$$

Thus, we can make the log-likelihood of this mixture Gaussian model arbitrarily large by fitting a degenerate Gaussian model to one observation in one mixture component, and letting the remaining observations be described by the other mixture component. This shows that the MLE problem may not be well-posed for mixture distributions because the log-likelihood can be unbounded.

If the data has well separated clusters, the log-likelihood of a mixture Gaussian distribution will have multiple local maximums. One can construct for any given number $B \in \mathbb{N}$ a data set $\boldsymbol{Y}$ such that the number of local maximums exceeds this number B, see Theorem 3 in Améndola et al. [11]. ■

Example 6.14 (Gamma Claim Size Modeling) In this example we consider claim size modeling of the French MTPL example given in Chap. 13.1. In view of Fig. 13.15 this seems quite difficult because we have three modes and heavy-tailedness. We choose a mixture of 5 distribution functions, we choose four gamma distributions and the Lomax distribution

$$Y \sim \sum_{k=1}^{4} \left(p_k \frac{\beta_k^{\alpha_k}}{\Gamma(\alpha_k)} y^{\alpha_k - 1} \exp\{-\beta_k y\} \right) + p_5 \frac{\beta_5}{M} \left(\frac{y+M}{M} \right)^{-(\beta_5+1)}, \tag{6.39}$$

with shape parameters α_k and scale parameters β_k, $1 \le k \le 4$, for the gamma densities; scale parameter M and tail parameter β_5 for the Lomax density; and with mixture probability $\boldsymbol{p} \in \Delta_5$. The idea behind this choice is that three gamma distributions take care of the three modes of the empirical density, see Fig. 13.15, the fourth gamma distribution models the remaining claims in the body of the distribution, and the Lomax distribution takes care of the regularly varying tail of the data. For the gamma distribution, we refer to Sect. 2.1.3, and for the Lomax distribution, we refer to Sect. 2.2.5.

We choose the null model for both the mixture probabilities $\boldsymbol{p} \in \Delta_5$ and the densities f_k, $1 \le k \le 5$. This model can directly be fitted with the EM algorithm as presented above, in particular, we can estimate the mixture probabilities by (6.31). The remaining shape, scale and tail parameters are directly estimated by MLE. To initialize the EM algorithm we use the interpretation of the components as explained above. We partition the entire data into $K = 5$ bins according to their claim sizes Y_i being in $(0, 300]$, $(300, 1'000]$, $(1'000, 1'200]$, $(1'200, 5'000]$ or $(5'000, \infty)$. The first three intervals will initialize the three modes of the empirical density, see Fig. 13.15 (lhs). This will correspond to the categorical variable taking values $Z = 1, 2, 3$; the fourth interval will correspond to $Z = 4$ and it will model the main body of the claims; and the last interval will correspond to $Z = 5$, modeling the Lomax tail of the claims. These choices provide the initialization given in Table 6.1 with upper indices $^{(0)}$. We remark that we choose a fixed threshold of $M = 2'000$ for the Lomax distribution, this choice will be further discussed below.

Based on these choices we run the EM algorithm for mixture distributions. We observe convergence after roughly 80 iterations, and the resulting parameters after 100 iterations are presented in Table 6.1. We observe rather large shape parameters $\widehat{\alpha}_k^{(100)}$ for the first three components $k = 1, 2, 3$. This indicates that these three components model the three modes of the empirical density and these three modes collect almost $\widehat{p}_1^{(100)} + \widehat{p}_2^{(100)} + \widehat{p}_3^{(100)} \approx 50\%$ of all claims. The remaining claims are modeled by the gamma density $k = 4$ having mean 1'304 and by the Lomax distribution having tail parameter $\widehat{\beta}_5^{(100)} = 1.416$, thus, this tail has finite first moment $M/(\widehat{\beta}_5^{(100)} - 1) = 4'812$ and infinite second moment.

Table 6.1 Parameter choices in the mixture model (6.39)

	$k = 1$	$k = 2$	$k = 3$	$k = 4$	$k = 5$
$\widehat{p}_k^{(0)}$	0.13	0.18	0.25	0.39	0.05
$\widehat{\alpha}_k^{(0)}$	2.43	11.24	1'299.44	5.63	–
$\widehat{\beta}_k^{(0)}$	0.019	0.018	1.141	0.003	0.517
$\widehat{\mu}_k^{(0)} = \widehat{\alpha}_k^{(0)}/\widehat{\beta}_k^{(0)}$	125	623	1'138	1'763	–
$\widehat{p}_k^{(100)}$	0.04	0.03	0.42	0.25	0.26
$\widehat{\alpha}_k^{(100)}$	93.05	650.94	1'040.37	1.34	–
$\widehat{\beta}_k^{(100)}$	1.207	1.108	0.888	0.001	1.416
$\widehat{\mu}_k^{(100)} = \widehat{\alpha}_k^{(100)}/\widehat{\beta}_k^{(100)}$	77	588	1'172	1'304	–

Figure 6.11 shows the resulting estimated mixture distribution. It gives the individual mixture components (top-lhs), the resulting mixture density (top-rhs), the QQ plot (bottom-lhs) and the log-log plot (bottom-rhs). Overall we find a rather good fit; maybe the first mode is a bit too spiky. However, this plot may also be misleading because the empirical density plot relies on kernel smoothing having a given bandwidth. Thus, the true observations may be more spiky than the plot indicates. The third mode suggests that there are two different values in the observations around 1'100, this is also visible in the QQ plot. Nevertheless, the overall result seems satisfactory. These results (based on 13 estimated parameters) are also summarized in Table 6.2.

We mention a couple of limitations of these results. Firstly, the log-likelihood of this mixture model is unbounded, similarly to Example 6.13 we can precisely fit one degenerate gamma mixture component to an individual observation Y_i which results in an infinite log-likelihood value. Thus, the found solution corresponds to a local maximum of the log-likelihood function and we should not state AIC values in Table 6.2, see also Remarks 4.28. Secondly, it is crucial to initialize three components to the three modes, if we randomly allocate all claims to 5 bins as initial configuration, the EM algorithm only finds mode $Z = 3$ but not necessarily the first two modes, at least, in our specifically chosen random initialization this was the case. In fact, the likelihood value of our latter solution was worse than in the first calibration which shows that we ended up in a worse local maximum.

We may be tempted to also estimate the Lomax threshold M with MLE. In Fig. 6.12 we plot the maximal log-likelihood as a function of M (if we start the EM algorithm always in the same configuration given in Table 6.1). From this figure a threshold of $M = 1'600$ seems optimal. Choosing this threshold of $M = 1'600$ leads to a slightly bigger log-likelihood of -199'304 and a slightly smaller tail parameter of $\widehat{\beta}_5^{(100)} = 1.318$. However, overall the model is very similar to the one with $M = 2'000$. In general, we do *not* recommend to estimate M with MLE, but this should be treated as a hyper-parameter selected by the modeler. The reason for this recommendation is that this threshold is crucial in deciding for large claims modeling and its estimation from data is, typically, not very robust; we also refer to Remarks 6.15, below.

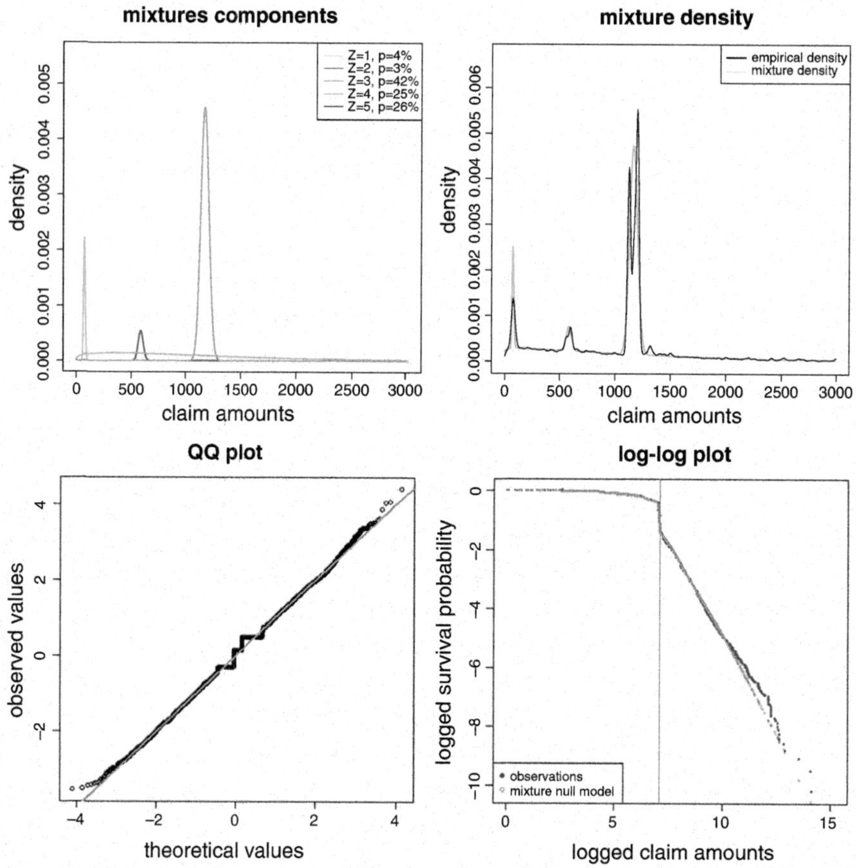

Fig. 6.11 Mixture null model: (top-lhs) individual estimated gamma components $f_k(\cdot; \widehat{\alpha}_k^{(100)}, \widehat{\beta}_k^{(100)})$, $1 \leq k \leq K$, and Lomax component $f_5(\cdot; \widehat{\beta}_5^{(100)})$, (top-rhs) estimated mixture density $\sum_{k=1}^{4} \widehat{p}_k^{(100)} f_k(\cdot; \widehat{\alpha}_k^{(100)}, \widehat{\beta}_k^{(100)}) + \widehat{p}_5^{(100)} f_5(\cdot; \widehat{\beta}_5^{(100)})$, (bottom-lhs) QQ plot of the estimated model, (bottom-rhs) log-log plot of the estimated model

Table 6.2 Mixture models for French MTPL claim size modeling

	# Param.	$\ell_Y(\widehat{\boldsymbol{\theta}}, \widehat{\boldsymbol{p}})$	AIC	$\widehat{\mu} = \mathbb{E}_{\widehat{\boldsymbol{\theta}}, \widehat{\boldsymbol{p}}}[Y]$
Empirical				2'266
Null model ($M = 2000$)	13	−199'306	398'637	2'381
Logistic GLM ($M = 2000$)	193	−198'404	397'193	2'176

In a next step we enhance the mixture modeling by including feature information $\boldsymbol{x}_i$ to explain the responses Y_i. In view of Fig. 13.17 we have decided to only model the mixture probabilities $\boldsymbol{p} = \boldsymbol{p}(\boldsymbol{x})$ feature dependent because feature information seems to mainly influence the heights of the peaks. We do not consider features `VehPower` and `VehGas` because these features do not seem to contribute, and

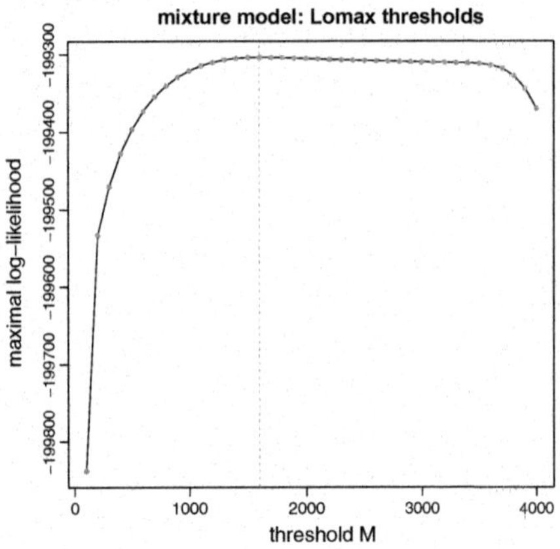

Fig. 6.12 Choice of Lomax threshold M

we do not consider `Density` because of the high co-linearity with `Area`, see Fig. 13.12 (rhs). Thus, we are left with the features `Area`, `VehAge`, `DrivAge`, `BonusMalus`, `VehBrand` and `Region`. Pre-processing of these features is done as in Listing 5.1, except that we keep `Area` categorical. Using these features $\boldsymbol{x} \in \mathcal{X} \subset \{1\} \times \mathbb{R}^q$ we choose a logistic categorical GLM for the mixture probabilities

$$\boldsymbol{x} \mapsto (p_1(\boldsymbol{x}), \ldots, p_{K-1}(\boldsymbol{x}))^\top = \frac{\exp\{\boldsymbol{X}\boldsymbol{\gamma}\}}{1 + \sum_{l=1}^4 \exp\langle \boldsymbol{\gamma}_l, \boldsymbol{x}\rangle}, \tag{6.40}$$

that is, we choose $K = 5$ as reference level, feature matrix $\boldsymbol{X} \in \mathbb{R}^{(K-1)\times(K-1)(q+1)}$ is defined in (5.71), and with regression parameter $\boldsymbol{\gamma} = (\boldsymbol{\gamma}_1^\top, \ldots, \boldsymbol{\gamma}_{K-1}^\top)^\top \in \mathbb{R}^{(K-1)(q+1)}$; this regression parameter $\boldsymbol{\gamma}$ should not be confused with the shape parameters $\beta_1, \ldots, \beta_4$ of the gamma components and the tail parameter β_5 of the Lomax component, see (6.39). Note that the notation in this section slightly differs from Sect. 5.7 on the logistic categorical GLM. In this section we consider mixture probabilities $\boldsymbol{p}(\boldsymbol{x}) \in \Delta_{K=5}$ (which corresponds to one-hot encoding), whereas in Sect. 5.7 we model $(p_1(\boldsymbol{x}), \ldots, p_{K-1}(\boldsymbol{x}))^\top$ with a categorical GLM (which corresponds to dummy coding), and normalization provides us with $p_K(\boldsymbol{x}) = 1 - \sum_{l=1}^{K-1} p_l(\boldsymbol{x}) \in (0, 1)$.

This logistic categorical GLM requires that we replace in the M-step the probability estimation (6.31) by Fisher's scoring method for GLMs as outlined in Sect. 5.7.2, but there is a small difference to that section. In the working residuals (5.74) we use dummy coding $T(Z) \in \{0, 1\}^{K-1}$ of a categorical variable Z, this now needs to be replaced by the estimated vector $(\widehat{Z}_1(\boldsymbol{\theta}, \boldsymbol{p}|Y), \ldots, \widehat{Z}_{K-1}(\boldsymbol{\theta}, \boldsymbol{p}|Y))^\top \in (0, 1)^{K-1}$ which is used as an estimate for the latent variable $T(Z)$. Apart from that everything is done as described in Sect. 5.7.2; in R this can be done with the procedure `multinom` from the package `nnet` [368]. We start the EM algorithm exactly in the final configuration of the

Table 6.3 Parameter choices in the mixture models: upper part null model, lower part GLM for estimated mixture probabilities $\widehat{p}(x_i)$

	$k=1$	$k=2$	$k=3$	$k=4$	$k=5$
Null: $\widehat{p}_k^{(100)}$	0.04	0.03	0.42	0.25	0.26
Null: $\widehat{\alpha}_k^{(100)}$	93.05	650.94	1'040.37	1.34	–
Null: $\widehat{\beta}_k^{(100)}$	1.207	1.108	0.888	0.001	1.416
Null: $\widehat{\mu}_k^{(100)} = \widehat{\alpha}_k^{(100)}/\widehat{\beta}_k^{(100)}$	77	588	1'172	1'304	–
GLM: average mixture probabilities	0.04	0.03	0.42	0.25	0.26
GLM: $\widehat{\alpha}_k^{(100)}$	94.03	597.20	1'043.38	1.28	–
GLM: $\widehat{\beta}_k^{(100)}$	1.223	1.019	0.891	0.001	1.365
GLM: $\widehat{\mu}_k^{(100)} = \widehat{\alpha}_k^{(100)}/\widehat{\beta}_k^{(100)}$	77	586	1'172	1'268	–

estimated mixture null model, and we run this algorithm for 20 iterations (which provides convergences).

The resulting parameters are given in the lower part of Table 6.3. We observe that the resulting parameters remain essentially the same, the second mode $Z = 2$ is a bit less spiky, and the tail parameter is slightly smaller. The summary of this model is given on the last line of Table 6.2. Regression modeling adds another $4 \cdot 45 = 180$ parameters to the model because we have $q = 45$ feature components in $\boldsymbol{x}$ (different from the intercept component). In view of AIC we give preference to the logistic mixture probability case (though AIC has to be interpreted with care, here, because we do not consider the MLE but rather a local maximum).

Figure 6.13 plots the individual estimated mixture probabilities $\boldsymbol{x}_i \mapsto \widehat{\boldsymbol{p}}(\boldsymbol{x}_i) \in \Delta_5$ over the insurance policies $1 \le i \le n$; these plots are inspired by the thesis of Frei [138]. The upper plots consider these probabilities against the estimated claim sizes $\widehat{\mu}(\boldsymbol{x}_i) = \sum_{k=1}^{5} \widehat{p}_k(\boldsymbol{x}_i)\widehat{\mu}_k$ and the lower plots against the ranks of $\widehat{\mu}(\boldsymbol{x}_i)$, the latter gives a different scaling on the x-axis because of the heavy-tailedness of the claims. The plots on the left-hand side show all individual policies $1 \le i \le n$, and the plots on the right-hand side show a quadratic spline fit to these observations. Not surprisingly, we observe that the claim size estimate $\widehat{\mu}(\boldsymbol{x}_i)$ is mainly driven by the large claims probability $\widehat{p}_5(\boldsymbol{x}_i)$ describing the Lomax contribution.

In Fig. 6.14 we compare the QQ plots of the mixture null model and the one where we model the mixture probabilities with the logistic categorical GLM. We see that the latter (more complex) model clearly outperforms the more simple one, in fact, this QQ plot looks quite convincing for the French MTPL claim size data. Finally, we perform a Wald test (5.32). We simultaneously treat all parameters that belong to the same feature variable (similar to the ANOVA analysis); for instance, for the 22 `Regions` the corresponding part of the regression parameter $\boldsymbol{\gamma}$ contains $4 \cdot 21 = 84$ components. The resulting p-values of dropping such components are all close to 0 which says that we should not eliminate one of the feature variables. This closes the example. ∎

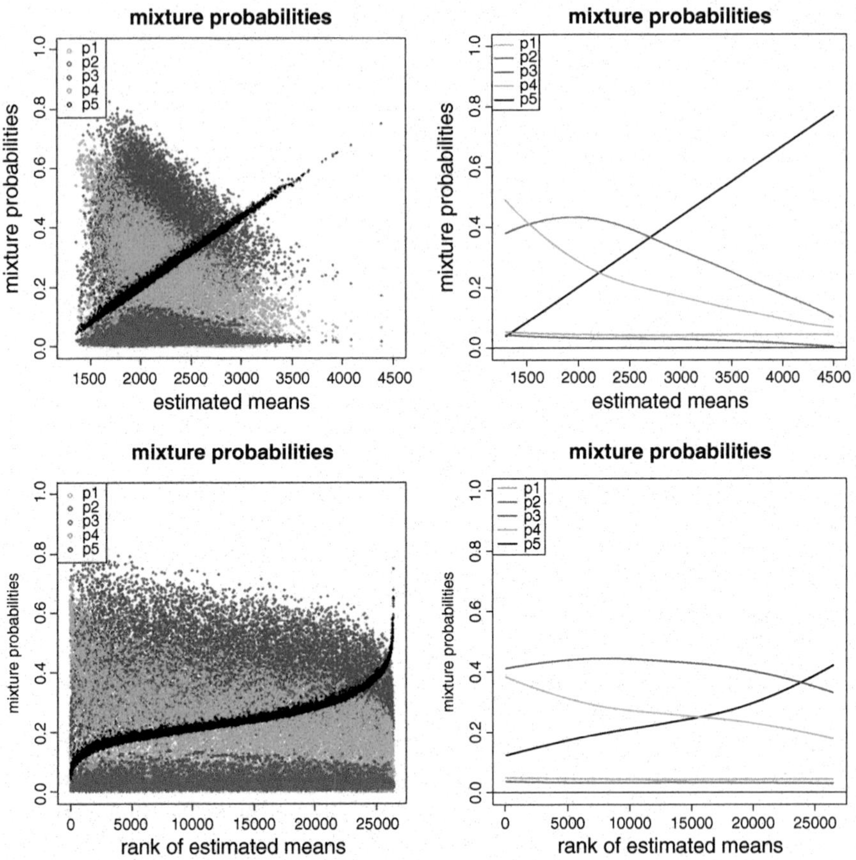

Fig. 6.13 Mixture probabilities $x_i \mapsto \widehat{p}(x_i)$ on individual policies $1 \le i \le n$: (top) against the estimated means $\widehat{\mu}(x_i)$ and (bottom) against the ranks of the estimated means $\widehat{\mu}(x_i)$; (lhs) over policies $1 \le i \le n$ and (rhs) quadratic spline fit

Remarks 6.15

- In Example 6.14 we have chosen a mixture distribution with four gamma components and one Lomax component. The reason for choosing the Lomax component has been two-fold. Firstly, we need a regularly varying tail to model the heavy-tailed property of the data. Secondly, we have preferred the Lomax distribution over the Pareto distribution because this provides us with a continuous density in (6.39). The results in Example 6.14 have been satisfactory. In most practical approaches, however, this approach will not work, even when fixing the threshold M of the Lomax component. Often, the nature of the data is such that the chosen gamma mixture distribution is not able to fully explain the small data in the body of the distribution, and in that situation the Lomax tail will assist in fitting the small claims. The typical result is that the Lomax part

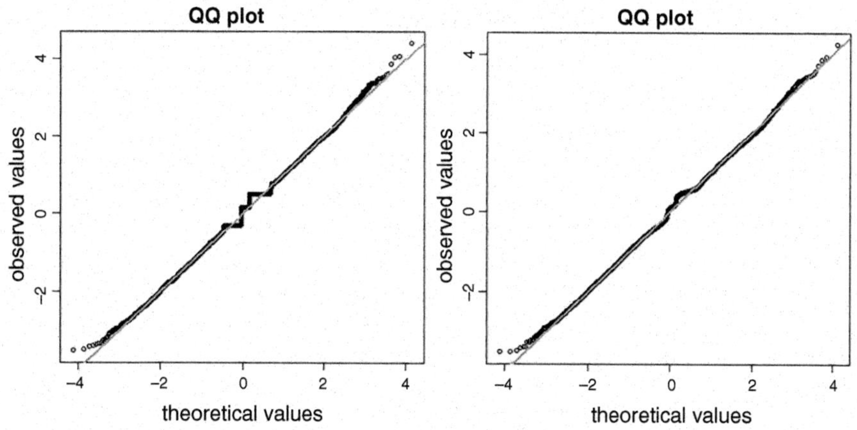

Fig. 6.14 QQ plots of the mixture models: (lhs) null model and (rhs) logistic categorical GLM for mixture probabilities

then pays more attention to small claims (through the log-likelihood function of numerous small claims) and the fitting of the tail turns out to be poor (because a few large claims do not sufficiently contribute to the log-likelihood). There are two ways to solve this dilemma. Either one works with composite distributions, see (6.56) below, and one drops the continuity property of the density; this is the approach taken in Fung et al. [148]. Or one fits the Lomax distribution solely to large observations in a first step, and then fixes the parameters of the Lomax distribution during the second step when fitting the full model to all data, this is the approach taken in Frei [138]. Both of these two approaches have been providing good results on real insurance data.

- There is an asymptotic theory for the optimal selection of the number of mixture components, we refer to Khalili–Chen [214] and Khalili [213]. Fung et al. [148] combine this asymptotic theory of mixture component selection with feature selection within these mixture components using LASSO and SCAD regularization.
- In Example 6.14 we have only modeled the mixture probabilities feature dependent, but not the parameters of the gamma mixture components. Introducing regressions for the gamma mixture components needs some care in fitting. For policy independent shape parameters $\alpha_1, \ldots, \alpha_4$, we can estimate the regression functions for the means of the mixture components without explicitly specifying α_k because these shape parameters cancel in the score equations. However, these shape parameters will be needed in the E-step, which requires also MLE of α_k. For more discussion on shape parameter estimation we refer to Sect. 5.3.7 (GLM with constant shape parameter) and Sect. 5.5.4 (double GLM).

6.4 Truncated and Censored Data

6.4.1 *Lower-Truncation and Right-Censoring*

A common problem in insurance is that we often have truncated or censored observations. Truncation naturally occurs if we sell insurance products that have a deductible $d > 0$ because in that case only the insurance claim $(Y - d)_+$ is compensated, and claims below the deductible d are usually not reported to the insurance company. This case is called *lower-truncation*, because claims below the deductible are not observed. If we lower-truncate an original claim $Y \sim f(\cdot; \theta)$ with lower-truncation point $\tau \in \mathbb{R}$ we obtain the density

$$f_{(\tau,\infty)}(y; \theta) = \frac{f(y; \theta)\mathbb{1}_{\{y>\tau\}}}{1 - F(\tau, \theta)}, \tag{6.41}$$

if $F(\cdot; \theta)$ is the distribution function corresponding to the density $f(\cdot; \theta)$. The lower-truncated density $f_{(\tau,\infty)}(y; \theta)$ only considers claims that fall into the interval (τ, ∞). Obviously, we can define upper-truncation completely analogously by considering an interval $(-\infty, \tau]$ instead. Figure 6.15 (lhs) gives an example of a lower-truncated density, and Fig. 6.15 (rhs) gives an example of a lower- and upper-truncated density.

Censoring occurs by selling insurance products with a maximal cover $M > 0$ because in that case only the insurance claim $Y \wedge M = \min\{Y, M\}$ is compensated, and the exact claim size above the maximal cover M may not be available. This case is called *right-censoring* because the exact claim amount above M is not known. Right-censoring of an original claim $Y \sim F(\cdot; \theta)$ with censoring point $M \in \mathbb{R}$

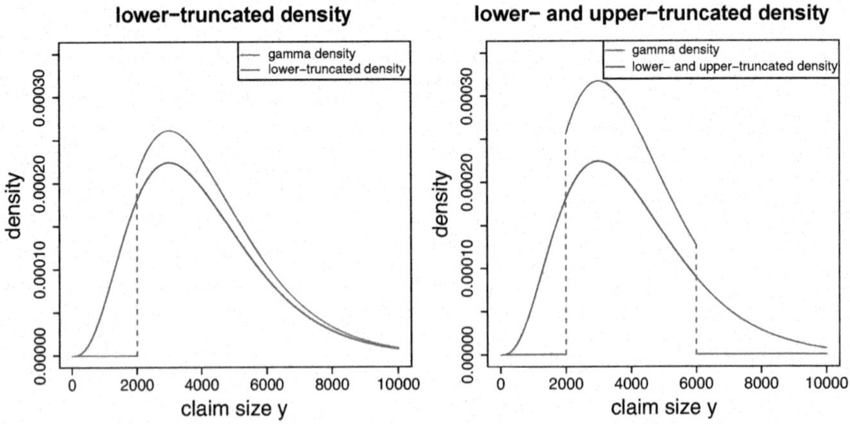

Fig. 6.15 (lhs) Lower-truncated gamma density with $\tau = 2'000$, and (rhs) lower- and upper-truncated gamma density with truncation points $2'000$ and $6'000$

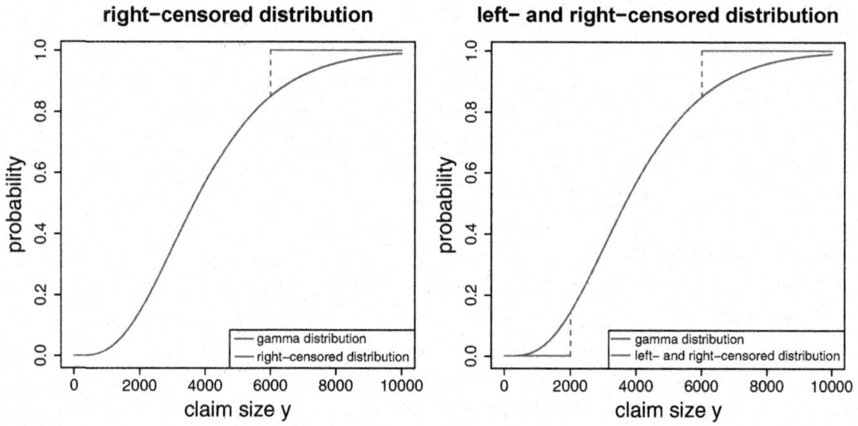

Fig. 6.16 (lhs) Right-censored gamma distribution with $M = 6'000$, and (rhs) left- and right-censored gamma distribution with censoring points $2'000$ and $6'000$

gives the distribution

$$F_{Y \wedge M}(y; \theta) = F(y; \theta)\mathbb{1}_{\{y<M\}} + \mathbb{1}_{\{y \geq M\}},$$

that is, we have a point mass in the censoring point M. We can define left-censoring analogously by considering the claim $Y \vee M = \max\{Y, M\}$. Figure 6.16 (lhs) shows a right-censored gamma distribution with censoring point $M = 6'000$, and Fig. 6.16 (rhs) shows a left- and right-censored example with censoring points $2'000$ and $6'000$.

Often in re-insurance, deductibles (also called retention levels) and maximal covers are combined, for instance, an excess-of-loss (XL) insurance cover of size $u > 0$ above the retention level $d > 0$ covers the claim

$$(Y - d)_+ \wedge u = (Y - d)\mathbb{1}_{\{d \leq Y < d+u\}} + u\mathbb{1}_{\{Y \geq d+u\}} = (Y - d)_+ - (Y - (d + u))_+.$$

Obviously, truncation and censoring pose some challenges in regression modeling because at the same time we need to consider the density $f(\cdot; \theta)$ and the distribution function $F(\cdot; \theta)$ to estimate a parameter θ. Both cases can be understood as missing data problems, with censoring providing the number of claims but not necessarily the exact claim size, and with truncation leaving also the number of claims unknown. These two cases are studied in Fung et al. [147] within the mixture of experts models using a variant of the EM algorithm. We use their techniques within the EDF framework for right-censored or lower-truncated data. This is done in the next sections.

6.4.2 Parameter Estimation Under Right-Censoring

Assume we have a fixed censoring point $M > 0$ that applies to independent observations Y_i following EDF densities $f(\cdot; \theta_i, v_i/\varphi)$; for simplicity we assume to work with an absolutely continuous EDF in this section. The (incomplete) log-likelihood function of canonical parameters $\boldsymbol{\theta} = (\theta_i)_{1\le i\le n}$ for observations $\boldsymbol{Y} \wedge M$ is given by

$$\ell_{\boldsymbol{Y}\wedge M}(\boldsymbol{\theta}) = \sum_{i:\, Y_i<M} \log f(Y_i; \theta_i, v_i/\varphi) + \sum_{i:\, Y_i\wedge M=M} \log\left(1 - F(M; \theta_i, v_i/\varphi)\right). \tag{6.42}$$

We interpret this as an incomplete data problem because the claim sizes Y_i above the censoring point M are not known. The complete log-likelihood is given by

$$\ell_{\boldsymbol{Y}}(\boldsymbol{\theta}) = \sum_{i=1}^{n} \log f(Y_i; \theta_i, v_i/\varphi).$$

Similarly to (6.32) we calculate a lower bound to the incomplete log-likelihood. We focus on one component of $\boldsymbol{Y}$ and drop the lower index i in Y_i for this consideration. Firstly, if $Y \wedge M < M$ we are in the situation of full claim size information and, obviously, we have log-likelihood in that case $Y < M$

$$\ell_{Y\wedge M}(\theta) = \ell_Y(\theta) = \frac{Y\theta - \kappa(\theta)}{\varphi/v} + a(Y; v/\varphi). \tag{6.43}$$

In the second case $Y \wedge M = M$ we do not have precise claim size information. In that case we have conditional density of claim $Y|_{\{Y\wedge M=M\}} = Y|_{\{Y\ge M\}}$ above M

$$f(z|Y \ge M; \theta, v/\varphi) = \frac{f(z; \theta, v/\varphi)\mathbb{1}_{\{z\ge M\}}}{1 - F(M; \theta, v/\varphi)} = \frac{f(z; \theta, v/\varphi)\mathbb{1}_{\{z\ge M\}}}{\exp\{\ell_{Y\wedge M}(\theta)\}}, \tag{6.44}$$

the latter follows because $Y\wedge M = M$ has the corresponding point mass in censoring point M (we work with an absolutely continuous EDF here). Choose an arbitrary density π having the same support as $Y|_{\{Y\ge M\}}$, and consider a random variable $Z \sim \pi$. Using (6.44) and the EDF structure on the last line, we have for $Y \ge M$

$$\begin{aligned}
\ell_{Y\wedge M}(\theta) &= \int \pi(z)\, \ell_{Y\wedge M}(\theta)\, d\nu(z) \\
&= \int \pi(z) \log\left(\frac{f(z; \theta, v/\varphi)/\pi(z)}{f(z|Y \ge M; \theta, v/\varphi)/\pi(z)}\right) d\nu(z) \\
&= \int \pi(z) \log\left(\frac{f(z; \theta, v/\varphi)}{\pi(z)}\right) d\nu(z) + D_{\mathrm{KL}}\left(\pi || f(\cdot|Y \ge M; \theta, v/\varphi)\right)
\end{aligned}$$

$$\begin{aligned} &\geq \int \pi(z) \log\left(\frac{f(z;\theta,v/\varphi)}{\pi(z)}\right) d\nu(z) \\ &= \frac{\mathbb{E}_\pi\left[Z\right]\theta - \kappa(\theta)}{\varphi/v} + \mathbb{E}_\pi\left[a(Z;v/\varphi)\right] - \mathbb{E}_\pi\left[\log \pi(Z)\right] \stackrel{\text{def.}}{=} \mathcal{Q}(\theta;\pi). \end{aligned}$$

This allows us to explore the E-step and the M-step similarly to (6.34) and (6.35).

The **E-step** in the case $Y \geq M$ for given canonical parameter estimate $\widehat{\theta}^{(t-1)}$ reads as

$$\begin{aligned} \widehat{\pi}^{(t)} = \underset{\pi}{\arg\max}\ \mathcal{Q}\left(\widehat{\theta}^{(t-1)};\pi\right) &= \underset{\pi}{\arg\min}\ D_{\mathrm{KL}}\left(\pi \left\| f(\cdot|Y\geq M;\widehat{\theta}^{(t-1)},v/\varphi)\right.\right) \\ = f(\cdot|Y\geq M;\widehat{\theta}^{(t-1)},v/\varphi). \end{aligned}$$

This allows us to calculate the estimation of the claim size above M, i.e., under $\widehat{\pi}^{(t)}$

$$\widehat{Y}^{(t)} = \mathbb{E}_{\widehat{\pi}^{(t)}}\left[Z\right] = \int z\, f(z|Y\geq M;\widehat{\theta}^{(t-1)},v/\varphi)\, d\nu(z). \tag{6.45}$$

Note that this is an estimate of the censored claim $Y|_{\{Y\geq M\}}$. This completes the E-step.

The **M-step** considers in the EDF case for censored claim sizes $Y \geq M$

$$\begin{aligned} \widehat{\theta}^{(t)} = \underset{\theta}{\arg\max}\ \mathcal{Q}\left(\theta;\widehat{\pi}^{(t)}\right) &= \underset{\theta}{\arg\max}\ \frac{\mathbb{E}_{\widehat{\pi}^{(t)}}\left[Z\right]\theta - \kappa(\theta)}{\varphi/v} \\ = \underset{\theta}{\arg\max}\ \ell_{\widehat{Y}^{(t)}}(\theta), \end{aligned} \tag{6.46}$$

the latter uses that the normalizing term $a(\cdot;v/\varphi)$ is not relevant for the MLE of θ. That is, (6.46) describes the regular MLE step under the observation $\widehat{Y}^{(t)}$ in the case of a censored observation $Y \geq M$; and if $Y < M$ we simply use the log-likelihood (6.43).

EM algorithm for right-censored data within the EDF

---

(0) Choose an initial parameter $\widehat{\boldsymbol{\theta}}^{(0)} = (\widehat{\theta}_i^{(0)})_{1\leq i\leq n}$.
(1) Repeat for $t \geq 1$:

- **E-step.** Given parameter $\widehat{\boldsymbol{\theta}}^{(t-1)} = (\widehat{\theta}_i^{(t-1)})_{1\leq i\leq n}$, estimate for the right-censored claims $Y_i \geq M$ their sizes by, see (6.45),

$$\widehat{Y}_i^{(t)} = \int z\, f\left(z\left|Y_i\geq M;\widehat{\theta}_i^{(t-1)},v_i/\varphi\right.\right) d\nu(z).$$

This provides us with an estimated observation

$$\widehat{\boldsymbol{Y}}^{(t)} = \left(Y_i \mathbb{1}_{\{Y_i < M\}} + \widehat{Y}_i^{(t)} \mathbb{1}_{\{Y_i \geq M\}} \right)_{1 \leq i \leq n}^{\top}.$$

- **M-step.** Calculate the MLE $\widehat{\boldsymbol{\theta}}^{(t)} = (\widehat{\theta}_i^{(t)})_{1 \leq i \leq n}$ based on observation $\widehat{\boldsymbol{Y}}^{(t)}$, i.e., solve

$$\widehat{\boldsymbol{\theta}}^{(t)} = \underset{\boldsymbol{\theta}}{\arg\max}\ \ell_{\widehat{\boldsymbol{Y}}^{(t)}}(\boldsymbol{\theta}).$$

Note that the above EM algorithm uses that the log-likelihood $\ell_{\boldsymbol{Y}}(\boldsymbol{\theta})$ of the EDF is linear in the observations that interact with parameter $\boldsymbol{\theta}$. We revisit the gamma claim size example of Sect. 5.3.7.

Example 6.16 (Right-Censored Gamma Claim Sizes) We revisit the gamma claim size GLM introduced in Sect. 5.3.7. The claim sizes are illustrated in Fig. 13.22. In total we have $n = 656$ observations Y_i, and they range from 16 SEK to 211'254 SEK. We right-censor this data at $M = 50'000$, this results in 545 uncensored observations and 111 censored observations equal to M. Thus, for the 17% largest claims we assume to not have any knowledge about the exact claim sizes. We use the EM algorithm for right-censored data to fit a GLM to this problem.

In order to calculate the E-step we need to evaluate the conditional expectation (6.45) under the gamma model

$$\begin{aligned} \widehat{Y}^{(t)} &= \int z\, f(z|Y \geq M; \widehat{\theta}^{(t-1)}, v/\varphi)\, d\nu(z) \qquad (6.47)\\ &= \int_M^\infty z \frac{\frac{\beta^\alpha}{\Gamma(\alpha)} z^{\alpha-1} \exp\{-\beta z\}}{1 - \mathcal{G}(\alpha, \beta M)} dz \;=\; \frac{\alpha}{\beta} \frac{1 - \mathcal{G}(\alpha+1, \beta M)}{1 - \mathcal{G}(\alpha, \beta M)}, \end{aligned}$$

with shape parameter $\alpha = v/\varphi$, scale parameter $\beta = -\widehat{\theta}^{(t-1)} v/\varphi$, see (5.45), and scaled incomplete gamma function

$$\mathcal{G}(\alpha, y) = \frac{1}{\Gamma(\alpha)} \int_0^y z^{\alpha-1} \exp\{-z\}\, dz \in (0, 1) \qquad \text{for } y \in (0, \infty). \qquad (6.48)$$

Thus, we receive a simple formula that allows us to efficiently calculate the E-step, and the M-step is exactly the gamma GLM explained in Sect. 5.3.7 for the (estimated) data $\widehat{\boldsymbol{Y}}^{(t)}$.

For the modeling we choose exactly the features as used for model Gamma GLM2, this gives $q + 1 = 7$ regression parameter components and additionally we set for the dispersion parameter $\widehat{\varphi}^{\text{MLE}} = 1.427$, this is the MLE in model Gamma

Table 6.4 Comparison of the complete log-likelihood and the incomplete log-likelihood (right-censoring $M = 50'000$) results

	# Param.	Log-likelihood $\ell_Y(\widehat{\theta}^{\text{MLE}}, \widehat{\varphi}^{\text{MLE}})$	Dispersion est. $\widehat{\varphi}^{\text{MLE}}$	Average amount	Rel. change
Gamma GLM2 (complete data)	$7+1$	$-7'129$	1.427	25'130	
Crude GLM2 (right-censored)	$7+1$	$-7'158$		18'068	-28%
EM est. GLM2 (right-censored)	$7+1$	$-7'132$		26'687	$+6\%$

GLM2. This dispersion parameter we keep fixed in all our models studied in this example. In a first step we simply fit a gamma GLM to the right-censored data $Y_i \wedge M$. We call this model 'crude GLM2', and it underestimates the empirical claim sizes by 28% because it ignores the fact of having right-censored data.

To initialize the EM algorithm for right-censored data we use the model crude GLM2. We then iterate the algorithm for 15 steps which provides convergence. The results are presented in Table 6.4. We observe that the resulting log-likelihood of the model fitted on the censored data and evaluated on the complete data ℓ_Y (which is available here) is almost the same as for model Gamma GLM2, which has been estimated on the complete data. Moreover, this right-censored EM algorithm fitted model slightly over-estimates the average claim sizes.

Figure 6.17 shows the estimated means $\widehat{\mu}_i$ on an individual claims level. The x-axis always gives the estimates from the complete log-likelihood model Gamma GLM2. The y-axis on the left-hand side shows the estimates from the crude GLM and the right-hand side the estimates from the EM algorithm fitted counterpart (fitted on the right-censored data). We observe that the crude model underestimates the claims (being below the diagonal), and the largest estimate lies below $M = 50'000$

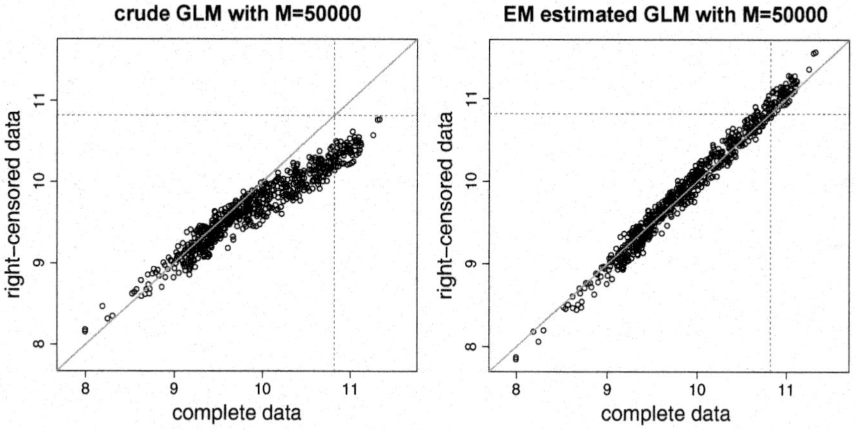

Fig. 6.17 Comparison of the estimated means $\widehat{\mu}_i$ in model Gamma GLM2 against (lhs) the crude GLM and (rhs) the EM fitted right-censored model; both axis are on the log-scale, the dotted lines shows the censoring point $\log(M)$

in our example (horizontal dotted line). The EM algorithm fitted model, considering the fact that we have right-censored data, corrects for the censoring, and the resulting estimates resemble the ones from the complete log-likelihood model quite well. In fact, we probably slightly over-estimate under right-censoring, here. Note that all these considerations have been done under an identical dispersion parameter estimate $\widehat{\varphi}^{\mathrm{MLE}}$. For the complete log-likelihood case, this is not really needed for mean estimation because it cancels in the score equations for mean estimation. However, a reasonable dispersion parameter estimate is crucial for the incomplete case as it enters $\widehat{Y}^{(t)}$ in the E-step, see (6.47), thus, the caveat here is that we need a reasonable dispersion estimate from the right-censored data (which we did not discuss, here, and which requires further research). ■

6.4.3 Parameter Estimation Under Lower-Truncation

Compared to censoring we have less information under truncation because not only the claim sizes below the lower-truncation point are unknown, but we also do not know how many claims there are below that truncation point τ. Assume we work with responses belonging to the EDF. The incomplete log-likelihood is given by

$$\ell_{\boldsymbol{Y}>\tau}(\boldsymbol{\theta}) = \sum_{i=1}^{n} \log f(Y_i;\theta_i, v_i/\varphi) - \log\left(1 - F(\tau;\theta_i, v_i/\varphi)\right),$$

assuming that $\boldsymbol{Y} = (Y_i)_{1\le i\le n} > \tau$ collects all claims above the truncation point $Y_i > \tau$, see (6.41). We proceed as in Fung et al. [147] to construct a complete log-likelihood; there are different ways to do so, but this proposal is convenient for parameter estimation. Firstly, we equip each observed claim $Y_i > \tau$ with an independent count random variable $K_i \sim p(\cdot;\theta_i, v_i/\varphi)$ that determines the number of claims below the truncation point that correspond to claim i above the truncation point. Secondly, we assume that these claims are given by independent observations $Z_{i,1}, \ldots, Z_{i,K_i} \le \tau$, a.s., with a distribution obtained from an un-truncated version of Y_i, i.e., we consider the upper-truncated version of $f(\cdot;\theta_i, v_i/\varphi)$ for $Z_{i,j}$. This gives us the complete log-likelihood

$$\ell_{(\boldsymbol{Y},\boldsymbol{K},\boldsymbol{Z})}(\boldsymbol{\theta}) = \sum_{i=1}^{n} \left(\log\left(\frac{f(Y_i;\theta_i, v_i/\varphi)}{1 - F(\tau;\theta_i, v_i/\varphi)}\right) \right. \qquad (6.49)$$

$$\left. + \log p(K_i;\theta_i, v_i/\varphi) + \sum_{j=1}^{K_i} \log\left(\frac{f(Z_{i,j};\theta_i, v_i/\varphi)}{F(\tau;\theta_i, v_i/\varphi)}\right)\right),$$

with $\boldsymbol{K} = (K_i)_{1 \le i \le n}$, and $\boldsymbol{Z}$ collects all (latent) claims $Z_{i,j} \le \tau$, an empty sum is set equal to zero. Next, we assume that K_i is following the geometric distribution

$$\mathbb{P}_{\theta_i}[K_i = k] \;=\; p(k;\theta_i, v_i/\varphi) \;=\; F(\tau;\theta_i, v_i/\varphi)^k \left(1 - F(\tau;\theta_i, v_i/\varphi)\right). \tag{6.50}$$

As emphasized in Fung et al. [147], this complete log-likelihood is an artificial construct that supports parameter estimation of lower-truncated data. It does *not* claim that the true un-truncated data follows this model (6.49) but it provides a distributional extension below the truncation point $\tau > 0$ that is convenient for parameter estimation. Namely, inserting this geometric distribution assumption into (6.49) gives us complete log-likelihood

$$\ell_{(\boldsymbol{Y},\boldsymbol{K},\boldsymbol{Z})}(\boldsymbol{\theta}) = \sum_{i=1}^{n} \left(\log f(Y_i;\theta_i, v_i/\varphi) + \sum_{j=1}^{K_i} \log f(Z_{i,j};\theta_i, v_i/\varphi) \right). \tag{6.51}$$

Within the EDF this allows us to do the same EM algorithm considerations as above; note that this expression no longer involves the distribution function. We consider one observation $Y_i > \tau$ and we drop the lower index i. This gives us complete observation $(Y, K, Z = (Z_j)_{1 \le j \le K})$ and conditional density

$$f(k, z|y;\theta, v/\varphi) = \frac{f(y,k,z;\theta,v/\varphi)}{f_{(\tau,\infty)}(y;\theta,v/\varphi)} = \frac{f(y,k,z;\theta,v/\varphi)}{\exp\{\ell_{Y=y>\tau}(\theta)\}},$$

where $\ell_{Y>\tau}(\theta)$ is the log-likelihood of the lower-truncated datum $Y > \tau$. Choose an arbitrary density π modeling the random vector (K, Z) below the truncation point τ. This gives us for the random vector $(K, Z) \sim \pi$

$$\begin{aligned}
\ell_{Y>\tau}(\theta) &= \int \pi(k,z)\,\ell_{Y>\tau}(\theta)\, d\nu(k,z) \\
&= \int \pi(k,z) \log\left(\frac{f(Y,k,z;\theta,v/\varphi)/\pi(k,z)}{f(k,z|Y;\theta,v/\varphi)/\pi(k,z)} \right) d\nu(k,z) \\
&= \int \pi(k,z) \log\left(\frac{f(Y,k,z;\theta,v/\varphi)}{\pi(k,z)} \right) d\nu(k,z) + D_{\mathrm{KL}}\left(\pi \,||\, f(\cdot|Y;\theta,v/\varphi)\right) \\
&\ge \int \pi(k,z) \log\left(\frac{f(Y,k,z;\theta,v/\varphi)}{\pi(k,z)} \right) d\nu(k,z) \\
&= \mathbb{E}_\pi\left[\ell_{(Y,K,Z)}(\theta) \middle| Y \right] - \mathbb{E}_\pi\left[\log \pi(K,Z)\right] \\
&= \log f(Y;\theta,v/\varphi) + \mathbb{E}_\pi\left[\sum_{j=1}^{K} \log f(Z_j;\theta,v/\varphi) \right] - \mathbb{E}_\pi\left[\log \pi(K,Z)\right] \\
&\stackrel{\text{def.}}{=} \mathcal{Q}(\theta;\pi),
\end{aligned}$$

where the second last identity uses that the log-likelihood (6.51) has a simple form under the geometric distribution chosen for K; this is exactly the step where we benefit from this specific choice of the probability extension below the truncation point. There is a subtle point here. Namely, $\ell_{Y>\tau}(\theta)$ is the log-likelihood of the lower-truncated datum $Y > \tau$, whereas $\log f(Y;\theta, v/\varphi)$ is the log-likelihood not using any lower-truncation.

The **E-step** for given canonical parameter estimate $\widehat{\theta}^{(t-1)}$ reads as

$$\begin{aligned}
\widehat{\pi}^{(t)} &= \underset{\pi}{\arg\max}\ \mathcal{Q}\left(\widehat{\theta}^{(t-1)};\pi\right) = \underset{\pi}{\arg\min}\ D_{\mathrm{KL}}\left(\pi \,\middle\|\, f(\cdot|Y;\widehat{\theta}^{(t-1)}, v/\varphi)\right) \\
&= f\left(\cdot \,\middle|\, Y;\widehat{\theta}^{(t-1)}, v/\varphi\right) \\
&= p\left(\cdot;\widehat{\theta}^{(t-1)}, v/\varphi\right) \prod_{j=1}^{\cdot} \frac{f(\cdot_j;\widehat{\theta}^{(t-1)}, v/\varphi)}{F(\tau;\widehat{\theta}^{(t-1)}, v/\varphi)}.
\end{aligned}$$

The latter describes a compound distribution for $\sum_{j=1}^{K} Z_j$ with a geometric count random variable K and independent i.i.d. random variables $Z_1, Z_2, \ldots,$ having upper-truncated densities $f_{(-\infty,\tau]}(\cdot;\widehat{\theta}^{(t-1)}, v/\varphi)$. This allows us to calculate the expected compound claim below the truncation point

$$\begin{aligned}
\widehat{Y}^{(t)}_{\leq\tau} = \mathbb{E}_{\widehat{\pi}^{(t)}}\left[\sum_{j=1}^{K} Z_j\right] &= \mathbb{E}_{\widehat{\pi}^{(t)}}[K]\ \mathbb{E}_{\widehat{\pi}^{(t)}}[Z_1] \\
&= \frac{F(\tau;\widehat{\theta}^{(t-1)}, v/\varphi)}{1-F(\tau;\widehat{\theta}^{(t-1)}, v/\varphi)} \int z\, f_{(-\infty,\tau]}(z;\widehat{\theta}^{(t-1)}, v/\varphi)\, d\nu(z).
\end{aligned}$$

This completes the E-step.

The **M-step** considers within the EDF

$$\begin{aligned}
\widehat{\theta}^{(t)} &= \underset{\theta}{\arg\max}\ \mathcal{Q}\left(\theta;\widehat{\pi}^{(t)}\right) \\
&= \underset{\theta}{\arg\max}\ \frac{\left(Y + \mathbb{E}_{\widehat{\pi}^{(t)}}\left[\sum_{j=1}^{K} Z_j\right]\right)\theta - (1+\mathbb{E}_{\widehat{\pi}^{(t)}}[K])\kappa(\theta)}{\varphi/v} \\
&= \underset{\theta}{\arg\max}\ \frac{v(1+\mathbb{E}_{\widehat{\pi}^{(t)}}[K])}{\varphi}\left[\left(\frac{Y+\widehat{Y}^{(t)}_{\leq\tau}}{1+\mathbb{E}_{\widehat{\pi}^{(t)}}[K]}\right)\theta - \kappa(\theta)\right].
\end{aligned}$$

That is, the M-step applies the classical MLE step, we only need to change weights and observations

$$v \mapsto v^{(t)} = v\left(1 + \mathbb{E}_{\widehat{\pi}^{(t)}}[K]\right) = \frac{v}{1 - F(\tau; \widehat{\theta}^{(t-1)}, v/\varphi)},$$

$$Y \mapsto \widehat{Y}^{(t)} = \frac{Y + \widehat{Y}^{(t)}_{\leq \tau}}{1 + \mathbb{E}_{\widehat{\pi}^{(t)}}[K]} = \frac{Y + \mathbb{E}_{\widehat{\pi}^{(t)}}[K]\,\mathbb{E}_{\widehat{\pi}^{(t)}}[Z_1]}{1 + \mathbb{E}_{\widehat{\pi}^{(t)}}[K]}.$$

Note that this uses the specific structure of the EDF, in particular, we benefit from linearity here which allows for closed-form solutions.

EM algorithm for lower-truncated data within the EDF

---

(0) Choose an initial parameter $\widehat{\boldsymbol{\theta}}^{(0)} = (\widehat{\theta}_i^{(0)})_{1 \leq i \leq n}$.
(1) Repeat for $t \geq 1$:

- **E-step.** Given parameter $\widehat{\boldsymbol{\theta}}^{(t-1)} = (\widehat{\theta}_i^{(t-1)})_{1 \leq i \leq n}$, estimate the number of claims $\boldsymbol{K}$ and the corresponding claim sizes $Z_{i,j}$ by

$$\widehat{K}_i^{(t)} = \frac{F(\tau; \widehat{\theta}_i^{(t-1)}, v_i/\varphi)}{1 - F(\tau; \widehat{\theta}_i^{(t-1)}, v_i/\varphi)},$$

$$\widehat{Z}_{i,1}^{(t)} = \int z\, f_{(-\infty,\tau]}(z; \widehat{\theta}_i^{(t-1)}, v_i/\varphi)\, d\nu(z). \tag{6.52}$$

  This provides us with estimated weights and observations for $1 \leq i \leq n$

$$v_i^{(t)} = v_i\left(1 + \widehat{K}_i^{(t)}\right) \qquad \text{and} \qquad \widehat{Y}_i^{(t)} = \frac{Y_i + \widehat{K}_i^{(t)} \widehat{Z}_{i,1}^{(t)}}{1 + \widehat{K}_i^{(t)}}.$$

- **M-step.** Calculate the MLE $\widehat{\boldsymbol{\theta}}^{(t)} = (\widehat{\theta}_i^{(t)})_{1 \leq i \leq n}$ based on observations $\widehat{\boldsymbol{Y}}^{(t)} = (\widehat{Y}_i^{(t)})^\top_{1 \leq i \leq n}$ and weights $\widehat{\boldsymbol{v}}^{(t)} = (\widehat{v}_i^{(t)})^\top_{1 \leq i \leq n}$, i.e., solve

$$\widehat{\boldsymbol{\theta}}^{(t)} = \underset{\boldsymbol{\theta}}{\arg\max}\; \ell_{\widehat{\boldsymbol{Y}}^{(t)}}(\boldsymbol{\theta}; \widehat{\boldsymbol{v}}^{(t)}/\varphi) = \underset{\boldsymbol{\theta}}{\arg\max} \sum_{i=1}^n \log f(\widehat{Y}_i^{(t)}; \theta_i, \widehat{v}_i^{(t)}/\varphi).$$

Remarks 6.17 Essentially, the above algorithm uses that the MLE in the EDF is based on a sufficient statistics of the observations, and in our case this sufficient statistics is $\widehat{Y}_i^{(t)}$.

Example 6.18 (Lower-Truncated Claim Sizes) We revisit the gamma claim size GLM introduced in Sect. 5.3.7, see also Example 6.16 on right-censored claims. We choose as lower-truncation point $\tau = 1'000$, i.e., we get rid of the very small claims that mainly generate administrative expenses at a rather small claim compensation. We have 70 claims below this truncation point, and there remain $n = 586$ claims above the truncation point that can be used for model fitting in the lower-truncated case. We use the EM algorithm for lower-truncated data to fit a GLM to this problem.

In order to calculate the E-step we need to evaluate the conditional expectation (6.52) under the gamma model for truncation probability

$$F(\tau; \widehat{\theta}^{(t-1)}, v/\varphi) = \int_0^\tau \frac{\beta^\alpha}{\Gamma(\alpha)} z^{\alpha-1} \exp\{-\beta z\}\, dz = \mathcal{G}(\alpha, \beta\tau),$$

with shape parameter $\alpha = v/\varphi$ and scale parameter $\beta = -\widehat{\theta}^{(t-1)} v/\varphi$. In complete analogy to (6.47) we have

$$\widehat{Z}_1^{(t)} = \int z\, f_{(\infty,\tau]}(z; \widehat{\theta}^{(t-1)}, v/\varphi)\, d\nu(z) = \frac{\alpha}{\beta} \frac{\mathcal{G}(\alpha+1, \beta\tau)}{\mathcal{G}(\alpha, \beta\tau)}.$$

For the modeling we choose again the features as used for model Gamma GLM2, this gives $q+1 = 7$ regression parameter components and additionally we set for the dispersion parameter $\widehat{\varphi}^{\text{MLE}} = 1.427$. This dispersion parameter we keep fixed in all the models studied in this example. In a first step we simply fit a gamma GLM to the lower-truncated data $Y_i > \tau$. We call this model 'crude GLM2', and it overestimates the true claim sizes because it ignores the fact of having lower-truncated data.

To initialize the EM algorithm for lower-truncated data we use the model crude GLM2. We then iterate the algorithm for 10 steps which provides convergence. The results are presented in Table 6.5. We observe that the resulting log-likelihood fitted on the lower-truncated data and evaluated on the complete data ℓ_Y (which is available here) is the same as for model Gamma GLM2 which has been estimated on the complete data. Moreover, this lower-truncated EM algorithm fitted model slightly under-estimates the average claim sizes.

Figure 6.18 shows the estimated means $\widehat{\mu}_i$ on an individual claims level. The x-axis always gives the estimates from the complete log-likelihood model Gamma GLM2. The y-axis on the left-hand side shows the estimates from the crude GLM and the right-hand side the estimates from the EM algorithm fitted counterpart (fitted on the lower-truncated data). We observe that the crude model overestimates

Table 6.5 Comparison of the complete log-likelihood and the incomplete log-likelihood (lower-truncation $\tau = 1'000$) results

	# Param.	Log-likelihood $\ell_Y(\widehat{\theta}^{\text{MLE}}, \widehat{\varphi}^{\text{MLE}})$	Dispersion est. $\widehat{\varphi}^{\text{MLE}}$	Average amount	Rel. change
Gamma GLM2 (complete data)	7 + 1	−7'129	1.427	25'130	
Crude GLM2 (lower-truncated)	7 + 1	−7'133		26'879	+7%
EM est. GLM2 (lower-truncated)	7 + 1	−7'129		24'900	−1%

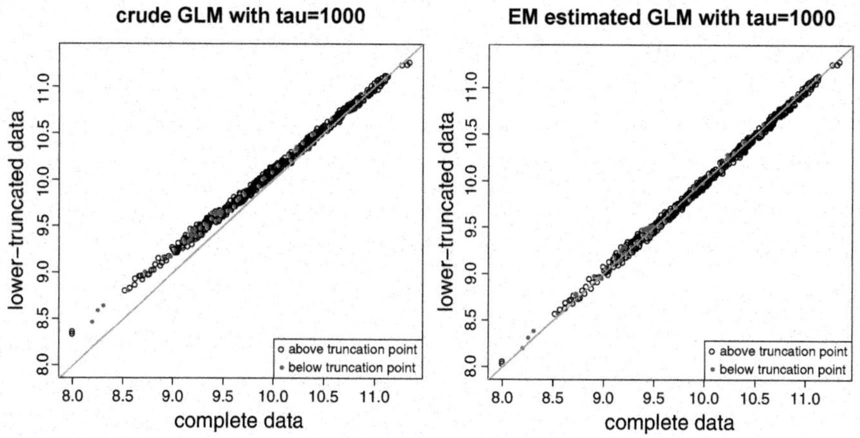

Fig. 6.18 Comparison of the estimated means $\widehat{\mu}_i$ in model Gamma GLM2 against (lhs) the crude GLM and (rhs) the EM fitted lower-truncated model; both axis are on the log-scale

the claims (being above the orange diagonal), in particular, this applies to claims with lower expected claim amounts. The EM algorithm fitted model, considering the fact that we have lower-truncated data, corrects for the truncation, and the resulting estimates almost completely coincide with the ones from the complete log-likelihood model. Again we remark that we use an identical dispersion parameter estimate $\widehat{\varphi}^{\text{MLE}}$, and it is an open problem to select a reasonable value from lower-truncated data. ■

Example 6.19 (Zero-Truncated Claim Counts and the Hurdle Poisson Model) In Sect. 5.3.6, we have been studying the ZIP model that has assigned an additional probability weight to the event $\{N = 0\}$ of having zero claims. This model can be understood as a hierarchical model with a latent variable Z indicating whether we have an excess zero claim or not, see (5.41). In that situation we have a mixture distribution of a Poisson distribution and a degenerate distribution. Fitting in Example 5.25 has been done brute force by using a general purpose optimizer, but we could also use the EM algorithm for mixture distributions.

An alternative way of modeling excess zeros is the hurdle approach which combines a lower-truncated count distribution with a point mass in zero. For the Poisson case this reads as, see (5.42),

$$f_{\text{hurdle Poisson}}(k; \lambda, v, \pi_0) = \begin{cases} \pi_0 & \text{for } k = 0, \\ (1-\pi_0)\frac{e^{-v\lambda}\frac{(v\lambda)^k}{k!}}{1-e^{-v\lambda}} & \text{for } k \in \mathbb{N}, \end{cases} \tag{6.53}$$

for $\pi_0 \in (0, 1)$ and $\lambda, v > 0$. If we ignore any observation $\{N = 0\}$ we obtain a lower-truncated Poisson model, also called zero-truncated Poisson (ZTP) model. This ZTP model can be fitted with the EM algorithm for lower-truncated data. In the following we only consider insurance policies i with $N_i > 0$. The log-likelihood of the ZTP model $N > 0$ is given by (we consider one single component only and drop the lower index in the notation)

$$\theta \mapsto \ell_{N>0}(\theta) = N\theta - ve^{\theta} - \log(N!) + N\log(v) - \log(1 - e^{-ve^{\theta}}), \tag{6.54}$$

with exposure $v > 0$ and canonical parameter $\theta \in \boldsymbol{\Theta} = \mathbb{R}$ such that $\lambda = \exp\{\theta\}$. The ZTP model provides for the random variable K the following geometric distribution (for the number of claims below the truncation point), see (6.50),

$$\mathbb{P}_{\theta}[K = k] \;=\; \mathbb{P}_{\theta}[N = 0]^k \, \mathbb{P}_{\theta}[N > 0] \;=\; e^{-kve^{\theta}} \left(1 - e^{-ve^{\theta}}\right).$$

In view of (6.51), this gives us complete log-likelihood (note that $Z_j = 0$ for all j)

$$\begin{aligned}
\ell_{(N,K,Z)}(\theta) &= N\theta - ve^{\theta} - \log(N!) + N\log(v) + \sum_{j=1}^{K} \left(Z_j\theta - ve^{\theta} - \log(Z_j!) + Z_j \log(v)\right) \\
&= N\theta - (1+K)\, ve^{\theta} - \log(N!) + N\log(v).
\end{aligned}$$

We can now directly apply a simplified version of the EM algorithm for lower-truncated data. For the E-step we have, given parameter $\widehat{\theta}^{(t-1)}$,

$$\widehat{K}^{(t)} = \frac{\mathbb{P}_{\widehat{\theta}^{(t-1)}}[N=0]}{1 - \mathbb{P}_{\widehat{\theta}^{(t-1)}}[N=0]} = \frac{e^{-ve^{\widehat{\theta}^{(t-1)}}}}{1 - e^{-ve^{\widehat{\theta}^{(t-1)}}}} \qquad \text{and} \qquad \widehat{Z}_1^{(t)} = 0.$$

This provides us with the estimated weights and observations (set $Y = N/v$)

$$v^{(t)} = v\left(1 + \widehat{K}^{(t)}\right) = \frac{v}{1 - e^{-ve^{\widehat{\theta}^{(t-1)}}}} \qquad \text{and} \qquad \widehat{Y}^{(t)} = \frac{Y}{1 + \widehat{K}^{(t)}} = \frac{N}{v^{(t)}}. \tag{6.55}$$

Thus, the EM algorithm iterates Poisson MLEs, and the E-Step modifies the weights $v^{(t)}$ in each step of the loop correspondingly. We remark that the ZTP model has an EF representation which allows one to directly estimate the corresponding parameters without using the EM algorithm, see Remark 6.20, below.

We revisit the French MTPL claim frequency data, and, in particular, we use model Poisson GLM3 as a benchmark, we refer to Tables 5.5 and 5.10. The feature engineering is done exactly as in model Poisson GLM3. We then select only the insurance policies from the learning data $\mathcal{L}$ that have suffered at least one claim, i.e., $N_i > 0$. These are $m = 22'434$ out of $n = 610'206$ insurance policies. Thus, we only consider $m/n = 3.68\%$ of all insurance policies, and we fit the lower-truncated log-likelihood (ZTP model) to this data

$$\ell_{N>0}(\boldsymbol{\beta}) = \sum_{i=1}^{m} N_i\theta_i - v_i e^{\theta_i} - \log(N_i!) + N_i \log(v_i) - \log(1 - e^{-v_i e^{\theta_i}}),$$

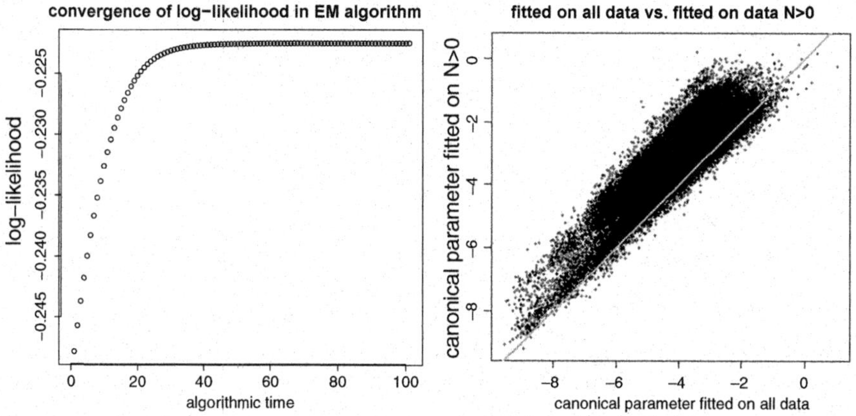

Fig. 6.19 (lhs) Convergence of the EM algorithm for the lower-truncated data in the Poisson hurdle case; (rhs) canonical parameters of the Poisson GLMs fitted on all data $\mathcal{L}$ vs. fitted only on policies with $N_i > 0$

Table 6.6 Run times, number of parameters, AICs, in-sample and out-of-sample deviance losses (units are in 10^{-2}) and in-sample average frequency of the Poisson null model and the Poisson, negative-binomial, ZIP and hurdle Poisson GLMs

	Run time	# Param.	AIC	In-sample loss on $\mathcal{L}$	Out-of-sample loss on $\mathcal{T}$	Aver. freq.
Poisson null	–	1	199'506	25.213	25.445	7.36%
Poisson GLM3	15 s	50	192'716	24.084	24.102	7.36%
NB GLM3 $\widehat{\alpha}_{\mathrm{NB}}^{\mathrm{MLE}} = 1.810$	85 s	51	192'113	20.722	20.674	7.38%
ZIP GLM3 (null π_0)	270 s	51	192'393	–	–	7.37%
Hurdle Poisson GLM3	300 s	100	**191'851**	–	–	7.39%

where $1 \leq i \leq m$ runs over all insurance policies with at least one claim and where the canonical parameter θ_i is given by the linear predictor $\theta_i = \langle \boldsymbol{\beta}, \boldsymbol{x}_i \rangle$. We fit this model using the EM algorithm for lower-truncated data. In each loop this requires that the offset $o_i^{(t)} = \log(v_i^{(t)})$ is adjusted according to (6.55); for the discussion of offsets we refer to Sect. 5.2.3. Convergence of the EM algorithm is achieved after roughly 75 iterations, see Fig. 6.19 (lhs).

In our first analysis we do not consider the Poisson hurdle model, but we simply consider model Poisson GLM3. However, this Poisson model with regression parameter $\boldsymbol{\beta}$ is fitted only on the data $N_i > 0$ (exactly using the results of the EM algorithm for lower-truncated data $N_i > 0$). The resulting predictive model is presented in Table 6.7. We observe that model Poisson GLM3 that is only fitted on the data $N_i > 0$ is clearly not competitive, i.e., we cannot simply extrapolate this estimated model to $\{N_i = 0\}$. This extrapolation results in a Poisson GLM that has a much too large average frequency of 15.11%, see last column of Table 6.7; this bias can clearly be seen in Fig. 6.19 (rhs) where we compare the two fits. From this we conclude that either the Poisson model assumption in general does not

Table 6.7 Number of parameters, in-sample and out-of-sample deviance losses on all data (units are in 10^{-2}), out-of-sample lower-truncated log-likelihood $\ell_{N>0}$ and in-sample average frequency of the Poisson null model and model Poisson GLM3 fitted on all data $\mathcal{L}$ and fitted on the data $N_i > 0$ only

	#	In-sample	Out-of-sample		Aver.
	Param.	Loss on $\mathcal{L}$	Loss on $\mathcal{T}$	$\ell_{N>0}$	freq.
Poisson null	1	25.213	25.445	–	7.36%
Poisson GLM3 fitted on all data	50	24.084	24.102	−0.2278	7.36%
Poisson GLM3 fitted on $N_i > 0$	50	28.064	28.211	−0.2195	15.11%

match the data, or that we have excess zeros (which do not influence the estimation procedure if we only consider the policies with at least one claim). Let us compare the lower-truncated log-likelihood $\ell_{N>0}$ out-of-sample only on the policies with at least one claim (ZTP model). We observe that the EM fitted model provides a better description of the data, as we have a bigger log-likelihood than the model fitted on all data $\mathcal{L}$ (i.e. −0.2195 vs. −0.2278 for the ZTP log-likelihood). Thus, the lower-truncated fitting procedure finds a better model on $\{N_i > 0\}$ when only fitted on these lower-truncated claim counts.

This analysis concludes that we need to fit the full hurdle Poisson model (6.53). That is, we cannot simply extrapolate the model fitted on the ZTP log-likelihood $\ell_{N>0}$ because, typically, $\pi_0(\boldsymbol{x}_i) \neq \exp\{-v_i e^{\langle \boldsymbol{\beta}, \boldsymbol{x}_i\rangle}\}$, the latter coming from the Poisson GLM with regression parameter $\boldsymbol{\beta}$. We model the zero claim probability $\pi_0(\boldsymbol{x}_i)$ by the logistic Bernoulli GLM indicating whether we have claims or not. We set up the logistic GLM for $p(\boldsymbol{x}_i) = 1 - \pi_0(\boldsymbol{x}_i)$ of describing the indicator $Y_i = \mathbb{1}_{\{N_i>0\}}$ of having claims. The difficulty compared to the Poisson model is that we cannot easily integrate the time exposure v_i as a pro rata temporis variable like in the Poisson case. We therefore make the following considerations. The canonical link in the logistic Bernoulli GLM is the logit function $p \mapsto \text{logit}(p) = \log(p/(1 - p)) = \log(p) - \log(1 - p)$ for $p \in (0, 1)$. Typically, in our application, $p \ll 1$ is fairly small because claims are rare events. This implies $\log(p/(1 - p)) \approx \log(p)$, i.e., the logit link behaves similarly to the log-link for small default probabilities p. This motivates to integrate the logged exposures $\log v_i$ as offsets into the logistic probabilities. That is, we make the following model assumption

$$(\boldsymbol{x}, v) \mapsto \text{logit}(p(\boldsymbol{x}_i, v_i)) = \log(v_i) + \langle \widetilde{\boldsymbol{\beta}}, \boldsymbol{x}_i\rangle,$$

with offset $o_i = \log(v_i)$ and regression parameter $\widetilde{\boldsymbol{\beta}} \in \mathbb{R}^{q+1}$. We fit this model using the R command `glm` using `family=binomial()`. The results then allow us to define the estimated hurdle Poisson model by, recall $p(\boldsymbol{x}_i, v_i) = 1 - \pi_0(\boldsymbol{x}_i, v_i)$,

$$f_{\text{hurdle Poisson}}(k; \boldsymbol{x}_i, v_i) = \begin{cases} 1 - p(\boldsymbol{x}_i, v_i) = \left(1 + \exp\{\log(v_i) + \langle \widetilde{\boldsymbol{\beta}}, \boldsymbol{x}_i\rangle\}\right)^{-1} & \text{for } k = 0, \\ \frac{p(\boldsymbol{x}_i, v_i)}{1 - e^{-\mu(\boldsymbol{x}_i, v_i)}} e^{-\mu(\boldsymbol{x}_i, v_i)} \frac{\mu(\boldsymbol{x}_i, v_i)^k}{k!} & \text{for } k \in \mathbb{N}, \end{cases}$$

Table 6.8 Contingency table of the observed numbers of policies against predicted numbers of policies with given claim counts `ClaimNb` (in-sample)

	Numbers of claims `ClaimNb`					
	0	1	2	3	4	5
Observed number of policies	587'772	21'198	1'174	57	4	1
Poisson predicted number of policies	587'325	22'064	779	34	3	0.3
NB predicted number of policies	587'902	20'982	1'200	100	15	4
ZIP predicted number of policies	587'829	21'094	1'191	79	9	4
Hurdle Poisson predicted number of policies	587'772	21'119	1'233	76	6	1

where $\widetilde{\boldsymbol{\beta}} \in \mathbb{R}^{q+1}$ is the regression parameter from the logistic Bernoulli GLM, and where $\mu(\boldsymbol{x}_i, v_i) = v_i \exp\langle \boldsymbol{\beta}, \boldsymbol{x}_i \rangle$ is the Poisson GLM estimated with the EM algorithm on the lower-truncated data $N_i > 0$ (ZTP model). The results are presented in Table 6.6.

Table 6.6 compares the hurdle Poisson model to the approaches studied in Table 5.10. Firstly, fitting the hurdle Poisson model is more time intensive, the EM algorithm takes some time and we need to fit the Bernoulli logistic GLM which is of a similar complexity as fitting model Poisson GLM3. The results in terms of AIC look convincing. The hurdle Poisson model provides an excellent model for the indicator of having a claim (here it outperforms model ZIP GLM3). It also tries to optimally fit a ZTP model to all insurance policies having at least one claim. This can also be seen from Table 6.8 which determines the expected number of policies that suffer the different numbers of claims.

We close this example by concluding that the hurdle Poisson model provides the best description, at the price of using more parameters. The ZIP model could be lifted to a similar level, however, we consider fitting the hurdle approach to be more convenient, see also Remark 6.20, below. In particular, feature engineering seems simpler in the hurdle approach because the different effects are clearly separated, whereas in the ZIP approach it is more difficult to suitably model the excess zeros, see also Listing 5.10. This closes this example. ■

Remark 6.20 In (6.54) we have been considering the ZTP model for different exposures $v > 0$. If we set these exposures to $v = 1$, we obtain the ZTP log-likelihood

$$\ell_{N>0}(\theta) = N\theta - \left(e^{\theta} + \log(1 - e^{-e^{\theta}})\right) - \log(N!).$$

Note that this describes a single-parameter linear EF with cumulant function

$$\kappa(\theta) = e^{\theta} + \log(1 - e^{-e^{\theta}}),$$

for canonical parameter in the effective domain $\theta \in \boldsymbol{\Theta} = \mathbb{R}$. The mean of this EF model is given by

$$\mu = \mathbb{E}_\theta[N] = \kappa'(\theta) = \frac{e^\theta}{1 - e^{-e^\theta}} = \frac{\lambda}{1 - e^{-\lambda}},$$

where we set $\lambda = e^\theta$. The variance is given by

$$\mathrm{Var}_\theta(N) \; = \; \kappa''(\theta) \; = \; \mu \left(\frac{e^\lambda - (1+\lambda)}{e^\lambda - 1} \right) \; = \; \mu \left(1 - \mu e^{-\lambda}\right) \; > \; 0.$$

Note that the term in brackets is positive but less than one. The latter implies that the ZTP model has under-dispersion. Alternatively to the EM algorithm, we can also directly fit a GLM to this ZTP model. The only difficulty is that we need to appropriately integrate the time exposures. The original Poisson model suggests that if we choose the canonical parameter being equal to the linear predictor, we should integrate the logged exposures as offsets into the linear predictors. Along these lines, if we choose the canonical link $h = (\kappa')^{-1}$ of the ZTP model, we receive that the canonical parameter θ is equal to the linear predictor $\langle \boldsymbol{\beta}, \boldsymbol{x} \rangle$, and we can directly integrate the logged exposures as offsets into the canonical parameters, see (5.25). This then allows us to directly fit this ZTP model with exposures using Fisher's scoring method. In this case of a concave log-likelihood function, the result will be identical to the solution of the EM algorithm found in Example 6.19, and, in fact, this direct approach is more straightforward and more time-efficient. Similar considerations can be done for other hurdle models.

6.4.4 Composite Models

In Sect. 6.3.1 we have promoted to mix distributions in cases where the data cannot be modeled by a single EDF distribution. Alternatively, one can also consider to compose densities which leads to so-called *composite models* (also called splicing models). This idea has been introduced to the actuarial literature by Cooray–Ananda [81] and Scollnik [332]. Assume we have two absolutely continuous densities $f^{(i)}(\cdot; \theta_i)$ with corresponding distribution functions $F^{(i)}(\cdot; \theta_i)$, $i = 1, 2$. These two densities can easily be composed at a splicing value τ and with weight $p \in (0, 1)$ by considering the following composite density

$$f(y; p, \theta_1, \theta_2) \; = \; p \, \frac{f^{(1)}(y; \theta_1) \mathbb{1}_{\{y \le \tau\}}}{F^{(1)}(\tau; \theta_1)} + (1-p) \, \frac{f^{(2)}(y; \theta_2) \mathbb{1}_{\{y > \tau\}}}{1 - F^{(2)}(\tau; \theta_2)}, \tag{6.56}$$

supposed that both denominators are non-zero. In this notation we treat splicing value τ as a hyper-parameter that is chosen by the modeler, and is not estimated from data. In view of (6.41) we can rewrite this in terms for lower- and upper-truncated densities

$$f(y; p, \theta_1, \theta_2) = p\, f^{(1)}_{(-\infty,\tau]}(y; \theta_1) + (1-p)\, f^{(2)}_{(\tau,\infty)}(y; \theta_2).$$

In this notation, we see that a composite model can also be interpreted as a mixture model with mixture probability $p \in (0, 1)$ and mixing densities $f^{(1)}_{(-\infty,\tau]}$ and $f^{(2)}_{(\tau,\infty)}$ having disjoint supports $(\infty, \tau]$ and (τ, ∞), respectively.

These disjoint supports allow for simpler MLE, i.e., we do not need to rely on the 'EM algorithm for mixture distributions' to fit this model. The log-likelihood of $Y \sim f(y; p, \theta_1, \theta_2)$ is given by

$$\ell_Y(p, \theta_1, \theta_2) = \left(\log(p) + \log f^{(1)}_{(-\infty,\tau]}(Y; \theta_1)\right) \mathbb{1}_{\{Y \le \tau\}} + \left(\log(1-p) + \log f^{(2)}_{(\tau,\infty)}(Y; \theta_2)\right) \mathbb{1}_{\{Y > \tau\}}.$$

This shows that the log-likelihood nicely decouples in the composite case and all parameters can directly be estimated with MLE: parameter θ_1 uses all observations smaller or equal to τ, parameter θ_2 uses all observations bigger than τ, and p is estimated by the proportions of claims below and above the splicing point τ. This holds for a null model as well as for a GLM approach for θ_1, θ_2 and p.

Nevertheless, the EM algorithm may still be used for parameter estimation, namely, truncation may ask for the 'EM algorithm for truncated data'. Alternatively, we could also use the 'EM algorithm for censored data' to estimate the truncated densities, because we have knowledge of the number of claims above and below the splicing point τ, thus, we could right- or left-censor these claims. The latter may lead to more stability in the estimation procedure since we use more information in parameter estimation, i.e., the two truncated densities will not be independent because they simultaneously consider all claim counts (but not identical claim sizes due to censoring).

For composite models one sometimes requires more regularity in the densities, we may, e.g., require continuity in the density in the splicing point which provides mixture probability

$$p = \frac{f^{(2)}(\tau; \theta_2) F^{(1)}(\tau; \theta_1)}{f^{(1)}(\tau; \theta_1)(1 - F^{(2)}(\tau; \theta_2)) + f^{(2)}(\tau; \theta_2) F^{(1)}(\tau; \theta_1)}.$$

This reduces the number of parameters to be estimated but complicates the score equations. If we require a differential condition in τ we receive requirement

$$p = \frac{f^{(2)}_y(\tau; \theta_2) F^{(1)}(\tau; \theta_1)}{f^{(1)}_y(\tau; \theta_1)(1 - F^{(2)}(\tau; \theta_2)) + f^{(2)}_y(\tau; \theta_2) F^{(1)}(\tau; \theta_1)},$$

where $f_y^{(i)}(y;\theta_i)$ denotes the first derivative w.r.t. y. Together with the continuity this provides requirement for having differentiability in τ

$$\frac{f^{(2)}(\tau;\theta_2)}{f^{(1)}(\tau;\theta_1)} = \frac{f_y^{(2)}(\tau;\theta_2)}{f_y^{(1)}(\tau;\theta_1)}.$$

Again this reduces the degrees of freedom in parameter estimation but complicates the score equations. We refrain from giving an example and close this section; we will consider a deep composite regression model in Sect. 11.3.2, below, where we replace the fixed splicing point by a quantile for a fixed quantile level.

Chapter 7
Deep Learning

In the sequel, we introduce deep learning models. In this chapter these deep learning models will be based on fully-connected feed-forward neural networks. We present these networks as an extension of GLMs. These networks perform feature engineering themselves. We discuss how networks achieve this, and we explain how networks are used for predictive modeling. There is a vastly growing literature on deep learning with networks, the classical reference is the book of Goodfellow et al. [166], but also the numerous tutorials around the open-source deep learning libraries TensorFlow [2], Keras [77] or PyTorch [296] give an excellent overview of the state-of-the-art in this field.

7.1 Deep Learning and Representation Learning

In Chap. 5 on GLMs, we have been modeling the mean structure of the responses Y, given features $\boldsymbol{x}$, by the following regression function, see (5.6),

$$\boldsymbol{x} \mapsto \mu(\boldsymbol{x}) = \mathbb{E}_{\theta(\boldsymbol{x})}[Y] = g^{-1}\langle \boldsymbol{\beta}, \boldsymbol{x} \rangle. \tag{7.1}$$

The crucial assumption has been that the regression function (7.1) provides a reasonable functional description of the expected value $\mathbb{E}_{\theta(\boldsymbol{x})}[Y]$ of datum $(Y, \boldsymbol{x})$. As described in Sect. 5.2.2, this typically requires *manual feature engineering* of $\boldsymbol{x}$, bringing feature information into the right structural form.

In contrast to manual feature engineering, deep learning aims at performing an *automated feature engineering* within the statistical model by massaging information through different transformations. Deep learning uses a finite sequence of functions $(z^{(m)})_{1 \le m \le d}$, called *layers*,

$$z^{(m)} : \{1\} \times \mathbb{R}^{q_{m-1}} \to \{1\} \times \mathbb{R}^{q_m},$$

M. V. Wüthrich, M. Merz, *Statistical Foundations of Actuarial Learning and its Applications*, Springer Actuarial, https://doi.org/10.1007/978-3-031-12409-9_7

of (fixed) dimensions $q_m \in \mathbb{N}$, $1 \le m \le d$, and initialization $q_0 = q$ being the dimension of the (raw) feature information $\boldsymbol{x} \in \mathcal{X} \subset \{1\} \times \mathbb{R}^q$. Each of these layers presents a new *representation of the features*, that is, after layer m we have a q_m-dimensional representation of the raw feature $\boldsymbol{x} \in \mathcal{X}$

$$z^{(m:1)}(\boldsymbol{x}) \stackrel{\text{def.}}{=} \left(z^{(m)} \circ \cdots \circ z^{(1)}\right)(\boldsymbol{x}) \in \{1\} \times \mathbb{R}^{q_m}. \tag{7.2}$$

Note that the first component is always identically equal to 1. For this reason we call the representation $z^{(m:1)}(\boldsymbol{x}) \in \{1\} \times \mathbb{R}^{q_m}$ of $\boldsymbol{x}$ to be q_m-dimensional.

Deep learning now assumes that we have $d \in \mathbb{N}$ appropriate transformations (layers) $z^{(m)}$, $1 \le m \le d$, such that $z^{(d:1)}(\boldsymbol{x})$ provides a suitable q_d-dimensional representation of the raw feature $\boldsymbol{x} \in \mathcal{X}$, that then enters a GLM

$$\mu(\boldsymbol{x}) = \mathbb{E}_{\theta(\boldsymbol{x})}[Y] = g^{-1}\langle \boldsymbol{\beta}, z^{(d:1)}(\boldsymbol{x})\rangle, \tag{7.3}$$

with link function $g : \mathcal{M} \to \mathbb{R}$ and regression parameter $\boldsymbol{\beta} \in \mathbb{R}^{q_d+1}$. This regression architecture is called a *feed-forward network* of *depth* $d \in \mathbb{N}$ because information $\boldsymbol{x}$ is processed in a directed acyclic (feed-forward) path through the d layers $z^{(1)}, \ldots, z^{(d)}$ before entering the final GLM.

Each layer $z^{(m)}$ involves parameters. Successful deep learning simultaneously fits these parameters as well as the regression parameter $\boldsymbol{\beta}$ to the available learning data $\mathcal{L}$ so that we obtain an optimal predictive model on the test data $\mathcal{T}$. That is, the learned model should optimally generalize to unseen data, we refer to Chap. 4 on predictive modeling. Thus, the process of optimal representation learning is also part of the model fitting procedure. In contrast to GLMs, the resulting log-likelihood functions are non-concave in their parameters because, typically, each layer involves non-linear transformations. This makes model fitting a challenge. State-of-the-art model fitting in deep learning uses variants of the gradient descent algorithm which we have already met in Sect. 6.2.4.

Remark 7.1 Representation learning $\boldsymbol{x} \mapsto z^{(d:1)}(\boldsymbol{x})$ is closely related to Mercer's kernel [272]. If we have a portfolio with features $\boldsymbol{x}_1, \ldots, \boldsymbol{x}_n$, we obtain a Mercer's kernel by considering the matrix

$$\mathbf{K} = \left(K(\boldsymbol{x}_i, \boldsymbol{x}_j)\right)_{1\le i,j\le n} = \left(\left\langle z^{(d:1)}(\boldsymbol{x}_i), z^{(d:1)}(\boldsymbol{x}_j)\right\rangle\right)_{1\le i,j\le n} \in \mathbb{R}^{n\times n}. \tag{7.4}$$

In many regression problems it can be shown that one can equivalently work with the design matrix $\mathfrak{Z} = (z^{(d:1)}(\boldsymbol{x}_1), \ldots, z^{(d:1)}(\boldsymbol{x}_n))^\top \in \mathbb{R}^{n\times(q_d+1)}$ or with

Mercer's kernel $\mathbf{K} \in \mathbb{R}^{n \times n}$. Mercer's kernel does not require the full knowledge of the learned representations $z^{(d:1)}(\boldsymbol{x}_i)$, but it suffices to know the discrepancies between $z^{(d:1)}(\boldsymbol{x}_i)$ and $z^{(d:1)}(\boldsymbol{x}_j)$ measured by the scalar products $K(\boldsymbol{x}_i, \boldsymbol{x}_j)$. This is also closely related to the cosine similarity in word embeddings, see (10.11). This approach then results in replacing the search for an optimal representation learning by a search of the optimal Mercer's kernel for the given data; this is called the kernel trick in machine learning.

7.2 Generic Feed-Forward Neural Networks

Feed-forward neural (FN) *networks* use special layers $z^{(m)}$ in (7.2)–(7.3), whose components are called *neurons*. This is discussed and studied in detail in this section.

7.2.1 *Construction of Feed-Forward Neural Networks*

FN networks are regression functions of type (7.3) where each neuron $z_j^{(m)}$, $1 \leq j \leq q_m$, of the layers $z^{(m)} = (1, z_1^{(m)}, \ldots, z_{q_m}^{(m)})^\top$, $1 \leq m \leq d$, has the structure of a GLM; the first component $z_0^{(m)} = 1$ always plays the role of the intercept and does not need any modeling.

A first important choice is the *activation function* $\phi : \mathbb{R} \to \mathbb{R}$ which plays the role of the inverse link function g^{-1}. To perform non-linear representation learning, this activation function should be non-linear, too. The most popular choices of activation functions are listed in Table 7.1.

The first three examples in Table 7.1 are smooth functions with simple derivatives, see the last column of Table 7.1. Having simple derivatives is an advantage in gradient descent algorithms for model fitting. The derivative of the ReLU activation function for $x \neq 0$ is given by the step function activation, and in 0 one typically considers a sub-gradient. We briefly comment on these activation functions.

Table 7.1 Popular choices of non-linear activation functions and their derivatives; the last two examples are not strictly monotone

	Activation function	Derivative
Sigmoid (logistic) activation	$\phi(x) = (1 + e^{-x})^{-1}$	$\phi' = \phi(1 - \phi)$
Hyperbolic tangent activation	$\phi(x) = \tanh(x)$	$\phi' = 1 - \phi^2$
Exponential activation	$\phi(x) = \exp(x)$	$\phi' = \phi$
Step function activation	$\phi(x) = \mathbb{1}_{\{x \geq 0\}}$	
Rectified linear unit (ReLU) activation	$\phi(x) = x\mathbb{1}_{\{x \geq 0\}}$	

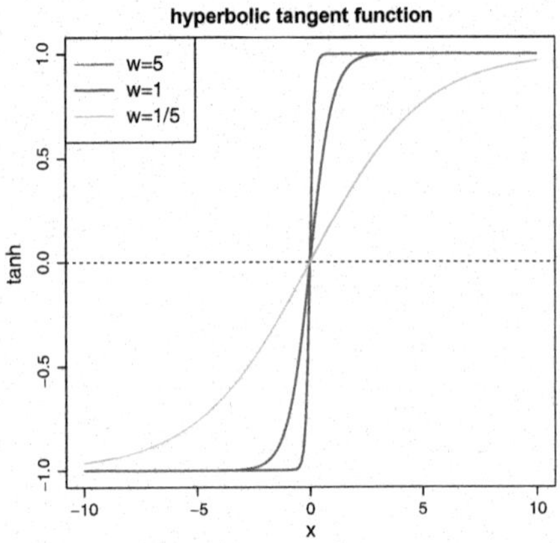

Fig. 7.1 Hyperbolic tangent activation function $x \mapsto \tanh(wx) \in (-1, 1)$ for (fixed) weights $w \in \{1/5, 1, 5\}$ and $x \in (-10, 10)$

- We are mainly going to use the *hyperbolic tangent activation* function

$$x \mapsto \tanh(x) = \frac{e^x - e^{-x}}{e^x + e^{-x}} = 2\left(1 + e^{-2x}\right)^{-1} - 1 \in (-1, 1).$$

  Figure 7.1 illustrates the hyperbolic tangent activation function.

  The hyperbolic tangent activation function is anti-symmetric w.r.t. the origin with range $(-1, 1)$. This anti-symmetry and boundedness is an advantage in fitting deep FN network architectures. For this reason we usually prefer the hyperbolic tangent over other activation functions.
- The *sigmoid activation* function corresponds to the logistic function that was used in the Bernoulli and the categorical EFs, see Sects. 2.1.2 and 5.7. The sigmoid activation function can be obtained from the hyperbolic tangent activation function by setting $\phi(x) = (\tanh(x/2) + 1)/2$.
- The *step function activation* is not really used in applications. However, it allows for nice interpretations, and it links FN networks to the theory of regression and classification trees (CARTs); see Breiman et al. [54] for CARTs.
- The *exponential activation* function is a nice differentiable choice whenever the range should be one-sided bounded.
- The *ReLU activation* function is also called hinge function or ramp function. This is the preferred choice in the machine learning community. However, typically, we will not use it because in our experience it is less robust in fitting compared to the hyperbolic tangent activation function. This may be for two reasons, firstly, the ReLU activation is unbounded, and secondly, it is identically equal to zero for $x < 0$, which implies that there is no sensitivity in negative choices of x.

A *FN layer* with activation function ϕ is a mapping

$$z^{(m)} : \{1\} \times \mathbb{R}^{q_{m-1}} \to \{1\} \times \mathbb{R}^{q_m} \tag{7.5}$$
$$z \mapsto z^{(m)}(z) = \left(1, z_1^{(m)}(z), \ldots, z_{q_m}^{(m)}(z)\right)^\top,$$

having neurons for $1 \le j \le q_m$

$$z_j^{(m)}(z) = \phi\langle w_j^{(m)}, z\rangle = \phi\left(\sum_{l=0}^{q_{m-1}} w_{l,j}^{(m)} z_l\right), \tag{7.6}$$

with given *network weights* $w_j^{(m)} = (w_{l,j}^{(m)})_{0 \le l \le q_{m-1}} \in \mathbb{R}^{q_{m-1}+1}$.

Interpretation Every neuron $z \mapsto z_j^{(m)}(z)$ describes a GLM regression function with link function ϕ^{-1} and regression parameter $w_j^{(m)} \in \mathbb{R}^{q_{m-1}+1}$ for features $z \in \{1\} \times \mathbb{R}^{q_{m-1}}$. These GLM regression functions can be interpreted as data compression, i.e., in each neuron the q_{m-1}-dimensional feature z is projected to a real number $\langle w_j^{(m)}, z\rangle \in \mathbb{R}$ which is then (non-linearly) activated by ϕ. Since this leads to a substantial loss of information, we perform this procedure of data compression q_m times in FN layer $z^{(m)}$, so that each neuron in $(z_j^{(m)}(z))_{1 \le j \le q_m}$ represents a different projection of input z. Choosing suitable weights $w_j^{(m)}$ will allow us to extract the crucial feature information from z to receive good explanatory variables for the regression task at hand.

A FN network of depth $d \in \mathbb{N}$ is obtained by composing d FN layers $z^{(1)}, \ldots, z^{(d)}$ to receive the mapping

$$z^{(d:1)} : \{1\} \times \mathbb{R}^{q_0 = q} \to \{1\} \times \mathbb{R}^{q_d} \tag{7.7}$$
$$x \mapsto z^{(d:1)}(x) = \left(z^{(d)} \circ \cdots \circ z^{(1)}\right)(x).$$

Choosing a strictly monotone and smooth link function g and a regression parameter $\beta \in \mathbb{R}^{q_d+1}$ we receive the FN network regression function

$$x \in \mathcal{X} \mapsto \mu(x) = g^{-1}\langle \beta, z^{(d:1)}(x)\rangle. \tag{7.8}$$

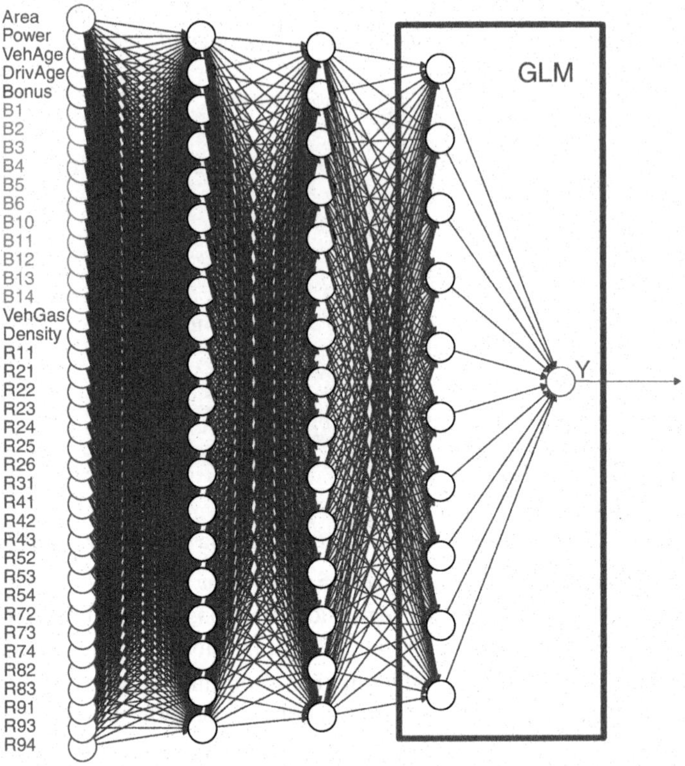

Fig. 7.2 FN network of depth $d = 3$, with number of neurons $(q_1, q_2, q_3) = (20, 15, 10)$ and input dimension $q_0 = 40$. This gives us a network parameter $\boldsymbol{\vartheta} \in \mathbb{R}^r$ of dimension $r = 1'306$

This FN network regression function (7.8) has a *network parameter* $\boldsymbol{\vartheta} = (\boldsymbol{w}_1^{(1)}, \ldots, \boldsymbol{w}_{q_d}^{(d)}, \boldsymbol{\beta})^\top \in \mathbb{R}^r$ of dimension

$$r = \sum_{m=1}^{d} q_m(q_{m-1} + 1) + (q_d + 1).$$

In Fig. 7.2 we illustrate a FN network of depth $d = 3$, FN layers of dimensions $(q_1, q_2, q_3) = (20, 15, 10)$ and input dimension $q_0 = 40$.[1] This gives us a network parameter $\boldsymbol{\vartheta} \in \mathbb{R}^r$ of dimension $r = 1'306$. On the left-hand side we have the raw features $\boldsymbol{x} \in \mathcal{X} \subset \{1\} \times \mathbb{R}^{q_0}$, these are processed through the three FN layers, where the black circles illustrate the neurons $z_j^{(m)}$. The third FN layer $\boldsymbol{z}^{(3)}$ has dimension

[1] Figures 7.2 and 7.9 are similar to Figure 1 in [122], and all FN network plots have been created with modified versions of the plot functions of the R package `neuralnet` [144].

$q_3 = 10$ providing the learned representation $z^{(3:1)}(x) \in \{1\} \times \mathbb{R}^{q_3}$ of x. This is used in the final GLM step (7.8) in the green box of Fig. 7.2.

Remarks 7.2

- One distinguishes between FN networks of depth $d = 1$, called *shallow networks*, and FN networks of depth $d > 1$, called *deep networks*. In this sense, deep learning means that we learn suitable feature representations through multiple FN layers $d > 1$. We come back to this in Sect. 7.2.2, below. Remark that some people would only call a network deep if $d \gg 1$, here $d > 1$ will be chosen for the definition of deep (which is also a precise definition).
- There are two ways of receiving a GLM. If we have a (trivial) FN network of depth $d = 0$, this naturally corresponds to a GLM, see Fig. 7.2. In that case, one works with the original features $x \in \mathcal{X}$ in (7.8). The second way of receiving a GLM is given by choosing the identity function as activation function $\phi(x) = x$. This implies that $x \mapsto z^{(d:1)}(x) = Ax$ is a linear function for some matrix $A \in \mathbb{R}^{(q_d+1)\times(q+1)}$ and, henceforth, we receive a GLM.
- Under the above interpretation of the representation learning structure (7.7), we may also give a different intuition for the FN layers. Typically, we expect that the first FN layers decompose feature information x into bits and pieces, which are then recomposed in a suitable way for the prediction task. In this sense, we typically choose a larger dimension for the early FN layers otherwise we may lose too much information already from the very beginning.
- The neural network introduced in (7.7) is called FN network because the signals propagate from one layer to the next (directed acyclic graph). If the network has loops it is called a *recurrent neural* (RN) network. RN networks have been applied very successfully in image and speech recognition, for instance, long short-term memory (LSTM) networks are very useful for time-series analysis. We study RN networks in Chap. 8, below. A third type of neural networks are *convolutional neural* (CN) networks which are very successfully applied to image recognition because they are capable to detect similar structures at different places in images, i.e., CN networks learn local representations. We will discuss CN network architectures in Chap. 9, below.
- The generic FN network architecture (7.8) can be complemented by drop-out layers, normalization layers, skip connections, embedding layers, etc. Such layers are special purpose layers, for instance, taking care of over-fitting. We introduce and discuss these below.
- The regression function (7.8) has a one-dimensional output for regression modeling. Of course, categorical classification can be done completely analogously by choosing a link function g suitable for classification, see Sect. 5.7. A similar approach also works if, for instance, we want to model simultaneously the mean and the dispersion of the data with a two-dimensional output function g^{-1}.

7.2.2 Universality Theorems

The use of FN networks for representation learning is motivated by the so-called *universality theorems* which say that any compactly supported continuous (regression) function can be approximated arbitrarily well by a suitably large FN network. As such, we can understand the FN network framework as an approximation tool which, of course, is useful far beyond statistical modeling. In Chapter 12 we give some proofs of selected universality statements to illustrate the flavor of such results. In particular, Cybenko [86], Hornik et al. [192], Hornik [191], Leshno et al. [247], Park–Sandberg [293, 294], Petrushev [302] and Isenbeck–Rüschendorf [198] have shown (under mild conditions on the activation function) that shallow FN networks can approximate any compactly supported continuous function arbitrarily well (in supremum norm or in L^2-norm), if we allow for an arbitrary number of neurons $q_1 \in \mathbb{N}$ in the single FN layer. Roughly speaking, such a result for shallow FN networks holds true if and only if the chosen activation function is non-polynomial, see Leshno et al. [247]. Such results are proved either by algebraic methods of Stone–Weierstrass type or by Wiener–Tauberian denseness type arguments. Moreover, approximation results are studied in Barron [25, 26], Yukich et al. [399], Makavoz [262], Pinkus [303] and Döhler–Rüschendorf [108].

The above stated universality theorems say that shallow FN networks are sufficient from an approximation point of view. Nevertheless, we will mainly use deep (multiple layers) FN networks, below. These have better convergence properties to given function classes because they more easily promote interactions in feature components compared to shallow ones. Such questions have been studied, e.g., by Elbrächter et al. [120], Kidger–Lyons [215], Lu et al. [260] or Cheridito et al. [75]. For instance, Elbrächter et al. [120] compare finite-depth wide networks to finite-width deep networks (under the choice of the ReLU activation function), and they conclude that for many function classes deep networks lead to exponential approximation rates, whereas shallow networks only provide polynomial approximation rates at the same number of network parameters. This motivates to consider sufficiently deep FN networks for representation learning because these typically have a better approximation capacity compared to shallow ones.

We motivate this by two simple examples. For this motivation we use the step function activation $\phi(x) = \mathbb{1}_{\{x \geq 0\}} \in \{0, 1\}$. If we have the step function activation, each neuron partitions $\mathbb{R}^{q_{m-1}}$ along a hyperplane, i.e.,

$$z \mapsto z_j^{(m)}(z) = \phi\langle \boldsymbol{w}_j^{(m)}, z\rangle = \mathbb{1}_{\left\{\sum_{l=1}^{q_{m-1}} w_{l,j}^{(m)} z_l \geq -w_{0,j}^{(m)}\right\}} \in \{0, 1\}. \tag{7.9}$$

For a shallow FN network we can study the question of the maximal complexity of the resulting partition of the feature space $\mathcal{X} \subset \{1\} \times \mathbb{R}^{q_0}$ when considering q_1

neurons (7.9) in the single FN layer $z^{(1)}$. Zaslavsky [400] proved that q_1 hyperplanes can partition the Euclidean space $\mathbb{R}^{q_0}$ in at most

$$\sum_{j=0}^{\min\{q_0,q_1\}} \binom{q_1}{j} \qquad \text{disjoint sets.} \tag{7.10}$$

This number (7.10) can be seen as a maximal upper complexity bound for shallow FN networks with step function activation. It grows exponentially for $q_1 \leq q_0$, and it slows down to a polynomial growth for $q_1 > q_0$. Thus, the complexity of shallow FN networks grows comparably slow as the width q_1 of the network exceeds q_0, and therefore we often need a huge network to receive a good approximation.

This result (7.10) should be contrasted to Theorem 4 in Montúfar et al. [280] who give a lower bound on the complexity of regression functions of deep FN networks (under the ReLU activation function). Assume $q_m \geq q_0$ for all $1 \leq m \leq d$. The maximal complexity is bounded below by

$$\left(\prod_{m=1}^{d-1} \left\lfloor \frac{q_m}{q_0} \right\rfloor^{q_0}\right) \sum_{j=0}^{q_0} \binom{q_d}{j} \qquad \text{disjoint linear regions.} \tag{7.11}$$

If we choose as an example a FN network with fixed width $q_m = 4$ for all $m \geq 1$ and an input of dimension $q_0 = 2$, we receive from (7.11) a lower bound of

$$4^{d-1}\left(\binom{4}{0} + \binom{4}{1} + \binom{4}{2}\right) = \frac{11}{4}\exp\{d\log(4)\}.$$

Thus, we have an exponential growth in depth $d \to \infty$. This contrasts the polynomial complexity growth (7.10) of shallow FN networks.

Example 7.3 (Shallow vs. Deep Networks: Partitions) We give a second more explicit example that compares shallow and deep FN networks. Choose $q_0 = 2$ and assume we want to describe a regression function

$$\mu : \mathbb{R}^2 \to \mathbb{R}, \qquad \boldsymbol{x} \mapsto \mu(\boldsymbol{x}).$$

If we think of a tool box of basis functions to build regression function μ we may want to choose indicator functions $\boldsymbol{x} \mapsto \chi_A(\boldsymbol{x}) \in \{0, 1\}$ for arbitrary rectangles $A = [x_1^-, x_1^+) \times [x_2^-, x_2^+) \subset \mathbb{R}^2$. We show that we can easily construct such indicator functions $\chi_A(\boldsymbol{x})$ for given rectangles $A \subset \mathbb{R}^2$ with FN networks of depth $d = 2$, but not with shallow FN networks.

For illustrative purposes, we fix a square $A = [-1/2, 1/2) \times [-1/2, 1/2) \subset \mathbb{R}^2$, and we want to construct $\chi_A(\boldsymbol{x})$ with a network of depth $d = 2$. This indicator function χ_A is illustrated in Fig. 7.3.

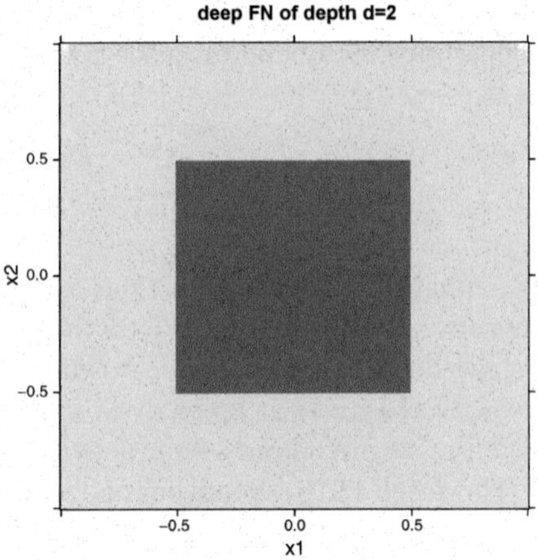

Fig. 7.3 Indicator function $\chi_A(\boldsymbol{x})$ for square $A = [-1/2, 1/2) \times [-1/2, 1/2) \subset \mathbb{R}^2$

We choose the step function activation for ϕ and a first FN layer with $q_1 = 4$ neurons

$$\begin{aligned}\boldsymbol{x} \mapsto \boldsymbol{z}^{(1)}(\boldsymbol{x}) &= \left(1, z_1^{(1)}(\boldsymbol{x}), \ldots, z_4^{(1)}(\boldsymbol{x})\right)^\top \\ &= \left(1, \mathbb{1}_{\{x_1 \geq -1/2\}}, \mathbb{1}_{\{x_2 \geq -1/2\}}, \mathbb{1}_{\{x_1 \geq 1/2\}}, \mathbb{1}_{\{x_2 \geq 1/2\}}\right)^\top \in \{1\} \times \{0, 1\}^4.\end{aligned}$$

This FN layer has a network parameter, see also (7.9),

$$\left(\boldsymbol{w}_1^{(1)}, \ldots, \boldsymbol{w}_4^{(1)}\right) = \left(\begin{pmatrix}1/2\\1\\0\end{pmatrix}, \begin{pmatrix}1/2\\0\\1\end{pmatrix}, \begin{pmatrix}-1/2\\1\\0\end{pmatrix}, \begin{pmatrix}-1/2\\0\\1\end{pmatrix}\right), \tag{7.12}$$

having dimension $q_1(q_0 + 1) = 12$. For the second FN layer with $q_2 = 4$ neurons we choose the step function activation and

$$\begin{aligned}\boldsymbol{z} \mapsto \boldsymbol{z}^{(2)}(\boldsymbol{z}) &= \left(1, z_1^{(2)}(\boldsymbol{z}), \ldots, z_4^{(2)}(\boldsymbol{z})\right)^\top \\ &= \left(1, \mathbb{1}_{\{z_1+z_2 \geq 3/2\}}, \mathbb{1}_{\{z_2+z_3 \geq 3/2\}}, \mathbb{1}_{\{z_1+z_4 \geq 3/2\}}, \mathbb{1}_{\{z_3+z_4 \geq 3/2\}}\right)^\top.\end{aligned}$$

This FN layer has a network parameter

$$\left(\boldsymbol{w}_1^{(2)}, \ldots, \boldsymbol{w}_4^{(2)}\right) = \left(\begin{pmatrix}-3/2\\1\\1\\0\\0\end{pmatrix}, \begin{pmatrix}-3/2\\0\\1\\1\\0\end{pmatrix}, \begin{pmatrix}-3/2\\1\\0\\0\\1\end{pmatrix}, \begin{pmatrix}-3/2\\0\\0\\1\\1\end{pmatrix}\right),$$

having dimension $q_2(q_1+1)=20$. For the output layer we choose the identity link $g(x)=x$, and the regression parameter $\boldsymbol{\beta}=(0,1,-1,-1,1)^\top \in \mathbb{R}^5$. As a result, we obtain

$$\chi_A(\boldsymbol{x}) = \left\langle \boldsymbol{\beta}, \boldsymbol{z}^{(2:1)}(\boldsymbol{x}) \right\rangle. \tag{7.13}$$

That is, this network of depth $d=2$, number of neurons $(q_1,q_2)=(4,4)$, step function activation and identity link can perfectly replicate the indicator function for the square $A=[-1/2,1/2)\times[-1/2,1/2)$, see Fig. 7.3. This network has $r=37$ parameters.

We now consider a shallow FN network with q_1 neurons. The resulting regression function with identity link is given by

$$\begin{aligned}\boldsymbol{x} \mapsto \left\langle \boldsymbol{\beta}, \boldsymbol{z}^{(1:1)}(\boldsymbol{x}) \right\rangle &= \left\langle \boldsymbol{\beta}, (1, z_1^{(1)}(\boldsymbol{x}), \ldots, z_{q_1}^{(1)}(\boldsymbol{x}))^\top \right\rangle \\ &= \left\langle \boldsymbol{\beta}, \left(1, \mathbb{1}_{\{\langle \boldsymbol{w}_1^{(1)}, \boldsymbol{x}\rangle \ge 0\}}, \ldots, \mathbb{1}_{\{\langle \boldsymbol{w}_{q_1}^{(1)}, \boldsymbol{x}\rangle \ge 0\}}\right)^\top \right\rangle,\end{aligned}$$

where we have used the step function activation $\phi(x)=\mathbb{1}_{\{x\ge 0\}}$. As in (7.9), each of these neurons leads to a partition of the space $\mathbb{R}^2$ with a straight line. Importantly these straight lines go *across* the *entire feature space*, and, therefore, we cannot exactly construct the indicator function of Fig. 7.3 with a shallow FN network. This can nicely be seen in Fig. 7.4 (lhs), where we consider a shallow FN network with $q_1=4$ neurons, weights (7.12), and $\boldsymbol{\beta}=(0,1/2,1/2,-1/2,-1/2)^\top$.

However, from the universality theorems we know that shallow FN networks can approximate any compactly supported (continuous) function arbitrarily well for sufficiently large q_1. In this example we can introduce additional neurons and let the resulting hyperplanes rotate around the origin. In Fig. 7.4 (middle, rhs) we show this for $q_1=8$ and $q_1=64$ neurons. We observe that this allows us to approximate a circle, see Fig. 7.4 (rhs), and having circles of different sizes at

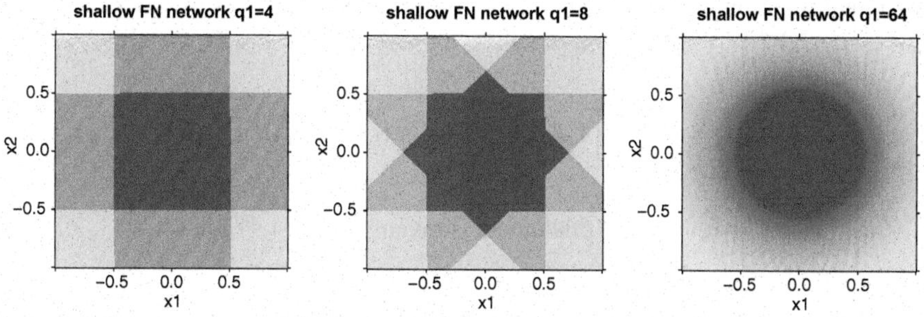

Fig. 7.4 Shallow FN networks with $q_1=4$ (lhs), $q_1=8$ (middle) and $q_1=64$ (rhs)

different locations will allow us to approximate the square A considered above. However, of course, this is a much less efficient way compared to the deep FN network (7.13).

Intuitively speaking, shallow FN networks act like additions where we add more and more separating hyperplanes for $q_1 \to \infty$ (*superposition of basis functions*). In contrast to that, going deep allows us to not only use additions but to also use multiplications (*composition of basis functions*). This is the reason, why we can easily construct the indicator function χ_A in the deep case (where we multiply zero's along the boundary of A), but not in the shallow case. ■

7.2.3 Gradient Descent Methods

We describe gradient descent methods in this section. These are used to fit FN networks. Gradient descent algorithms have already been used in Sect. 6.2.4 for fitting LASSO regularized regression models. We will give the full methodological part here, without relying on Sect. 6.2.4.

Plain Vanilla Gradient Descent Algorithm

Assume we have independent instances $(Y_i, \boldsymbol{x}_i)$, $1 \le i \le n$, that follow the same member of the EDF. We choose a regression function

$$\boldsymbol{x}_i \mapsto \mu(\boldsymbol{x}_i) = \mu_{\boldsymbol{\vartheta}}(\boldsymbol{x}_i) = \mathbb{E}_{\theta(\boldsymbol{x}_i)}[Y_i] = g^{-1}\left\langle \boldsymbol{\beta}, \boldsymbol{z}^{(d:1)}(\boldsymbol{x}_i)\right\rangle,$$

for a strictly monotone and smooth link function g, and a FN network $\boldsymbol{z}^{(d:1)}$ with network parameter $\boldsymbol{\vartheta} \in \mathbb{R}^r$. We assume that the chosen activation function ϕ is differentiable. We highlight in the notation that the mean functional $\mu_{\boldsymbol{\vartheta}}(\cdot)$ depends on the network parameter $\boldsymbol{\vartheta}$. The canonical parameter of the response Y_i is given by $\theta(\boldsymbol{x}_i) = h(\mu_{\boldsymbol{\vartheta}}(\boldsymbol{x}_i)) \in \boldsymbol{\Theta}$, where $h = (\kappa')^{-1}$ is the canonical link and κ the cumulant function of the chosen member of the EDF. This gives us (under constant dispersion φ) the log-likelihood function, for given data $\boldsymbol{Y} = (Y_1, \ldots, Y_n)^\top$,

$$\boldsymbol{\vartheta} \mapsto \ell_{\boldsymbol{Y}}(\boldsymbol{\vartheta}) = \sum_{i=1}^{n} \frac{v_i}{\varphi}\Big[Y_i h(\mu_{\boldsymbol{\vartheta}}(\boldsymbol{x}_i)) - \kappa\left(h(\mu_{\boldsymbol{\vartheta}}(\boldsymbol{x}_i))\right)\Big] + a(Y_i; v_i/\varphi).$$

The deviance loss function in this model is given by, see (4.9) and (4.8),

$$\mathfrak{D}(\boldsymbol{Y}, \boldsymbol{\vartheta}) = \frac{2}{n}\sum_{i=1}^{n} \frac{v_i}{\varphi}\Big(Y_i h\left(Y_i\right) - \kappa\left(h\left(Y_i\right)\right) - Y_i h\left(\mu_{\boldsymbol{\vartheta}}(\boldsymbol{x}_i)\right) + \kappa\left(h\left(\mu_{\boldsymbol{\vartheta}}(\boldsymbol{x}_i)\right)\right)\Big) \ge 0. \tag{7.14}$$

The MLE of ϑ is found by either maximizing the log-likelihood function or by minimizing the deviance loss function in ϑ. This problem cannot be solved in general because of complexity. Typically, the deviance loss function is non-convex in ϑ and it may have many local minimums. This is one of the reasons, why we are less ambitious here, and why we just try to find a network parameter $\widehat{\vartheta}$ which provides a "small" deviance loss $\mathfrak{D}(\boldsymbol{Y}, \widehat{\vartheta})$ for the given data $\boldsymbol{Y}$. We discuss this further, below, in fact, this is a crucial point in FN network fitting that is related to *in-sample over-fitting* and, therefore, this point will require a broader discussion.

For the moment, we just try to find a network parameter $\widehat{\vartheta}$ that provides a small deviance loss $\mathfrak{D}(\boldsymbol{Y}, \widehat{\vartheta})$ for the given data $\boldsymbol{Y}$. Gradient descent algorithms suggest that we try to step-wise locally improve our current position by changing the network parameter into the direction of the maximal local decrease of the deviance loss function. By assumption, our deviance loss function is differentiable in ϑ. This allows us to consider the following first order Taylor expansion in ϑ

$$\mathfrak{D}(\boldsymbol{Y}, \widetilde{\vartheta}) = \mathfrak{D}(\boldsymbol{Y}, \vartheta) + \nabla_{\vartheta}\mathfrak{D}(\boldsymbol{Y}, \vartheta)^{\top}\left(\widetilde{\vartheta} - \vartheta\right) + o\left(\|\widetilde{\vartheta} - \vartheta\|_2\right) \qquad \text{as } \|\widetilde{\vartheta} - \vartheta\|_2 \to 0.$$

This shows that the locally optimal change $\vartheta \mapsto \widetilde{\vartheta}$ points into the opposite direction of the gradient of the deviance loss function. This motivates the following gradient descent step.

Assume that at algorithmic time $t \in \mathbb{N}$ we have a network parameter $\vartheta^{(t)} \in \mathbb{R}^r$. Choose a suitable *learning rate* $\varrho_{t+1} > 0$, and consider the gradient descent update

$$\vartheta^{(t)} \mapsto \vartheta^{(t+1)} = \vartheta^{(t)} - \varrho_{t+1}\nabla_{\vartheta}\mathfrak{D}(\boldsymbol{Y}, \vartheta^{(t)}). \tag{7.15}$$

This gradient descent update gives us the new (smaller) deviance loss at algorithmic time $t + 1$

$$\mathfrak{D}(\boldsymbol{Y}, \vartheta^{(t+1)}) = \mathfrak{D}(\boldsymbol{Y}, \vartheta^{(t)}) - \varrho_{t+1}\left\|\nabla_{\vartheta}\mathfrak{D}(\boldsymbol{Y}, \vartheta^{(t)})\right\|_2^2 + o\left(\varrho_{t+1}\right) \qquad \text{for } \varrho_{t+1} \downarrow 0.$$

Under suitably tempered learning rates $(\varrho_t)_{t\geq 1}$, this algorithm will converge to a local minimum of the deviance loss function as $t \to \infty$ (supposed that we do not get trapped in a saddlepoint).

Remarks 7.4 We give a couple of (preliminary) remarks on the gradient descent algorithm (7.15), more explanation, further derivations, and variants of the gradient descent algorithm will be discussed below.

- In the applications we will *early stop* the gradient descent algorithm before reaching a local minimum (to prevent from over-fitting). This is going to be discussed in the next paragraphs.
- Fine-tuning the learning rate $(\varrho_t)_t$ is important, in particular, there is a trade-off between smaller and bigger learning rates: they need to be sufficiently small so that the first order Taylor expansion is still a valid approximation, and they should be sufficiently big otherwise the convergence of the algorithm will be very slow because it needs many iterations.
- The gradient descent algorithm is a first order algorithm, and one is tempted to study higher order approximations, e.g., leading to the Newton–Raphson algorithm. Unfortunately, higher order derivatives are computationally not feasible if the size n of the data $\boldsymbol{Y} = (Y_1, \ldots, Y_n)^\top$ and the dimension r of the network parameter $\boldsymbol{\vartheta}$ are large. In fact, even the calculation of the first order derivatives may be challenging and, therefore, stochastic gradient descent methods are considered below. Nevertheless, it is beneficial to have a notion of a second order term. Momentum-based methods originate from approximating the second order terms, these will be studied in (7.19)–(7.20), below.
- The gradient descent step (7.15) solves an unconstraint local optimization. Similarly to (6.15)–(6.16) we could change the gradient descent algorithm to a constraint optimization problem, e.g., involving a LASSO constraint that can be solved with the generalized projection operator (6.17).

Gradient Calculation via Back-Propagation

Fast gradient descent algorithms essentially rely on fast gradient calculations of the deviance loss function. Under the EDF setup we have gradient w.r.t. $\boldsymbol{\vartheta}$

$$\nabla_{\boldsymbol{\vartheta}} \mathfrak{D}(\boldsymbol{Y}, \boldsymbol{\vartheta}) = \frac{2}{n} \sum_{i=1}^{n} \frac{v_i}{\varphi} \Big(\mu_{\boldsymbol{\vartheta}}(\boldsymbol{x}_i) - Y_i \Big) h'\left(\mu_{\boldsymbol{\vartheta}}(\boldsymbol{x}_i)\right) \nabla_{\boldsymbol{\vartheta}} \mu_{\boldsymbol{\vartheta}}(\boldsymbol{x}_i) \tag{7.16}$$

$$= \frac{2}{n} \sum_{i=1}^{n} \frac{v_i}{\varphi} \frac{\mu_{\boldsymbol{\vartheta}}(\boldsymbol{x}_i) - Y_i}{V\left(\mu_{\boldsymbol{\vartheta}}(\boldsymbol{x}_i)\right)} \frac{1}{g'(\mu_{\boldsymbol{\vartheta}}(\boldsymbol{x}_i))} \nabla_{\boldsymbol{\vartheta}} \left\langle \boldsymbol{\beta}, \boldsymbol{z}^{(d:1)}(\boldsymbol{x}_i) \right\rangle,$$

where the last step uses the variance function $V(\cdot)$ of the chosen EDF, we also refer to (5.9). The main difficulty is the calculation of the gradient

$$\nabla_{\boldsymbol{\vartheta}} \left\langle \boldsymbol{\beta}, \boldsymbol{z}^{(d:1)}(\boldsymbol{x}) \right\rangle = \nabla_{\boldsymbol{\vartheta}} \left\langle \boldsymbol{\beta}, \left(\boldsymbol{z}^{(d)} \circ \cdots \circ \boldsymbol{z}^{(1)}\right)(\boldsymbol{x}) \right\rangle,$$

w.r.t. the network parameter $\boldsymbol{\vartheta} = (\boldsymbol{w}_1^{(1)}, \ldots, \boldsymbol{w}_{q_d}^{(d)}, \boldsymbol{\beta})^\top \in \mathbb{R}^r$, and where each FN layer $\boldsymbol{z}^{(m)}$ involves the weights $W^{(m)} = (\boldsymbol{w}_1^{(m)}, \ldots, \boldsymbol{w}_{q_m}^{(m)}) \in \mathbb{R}^{(q_{m-1}+1) \times q_m}$. The workhorse for these gradient calculations is the back-propagation method of Rumelhart et al. [324]. Basically, the back-propagation method is a clever reparametrization of the problem so that the gradients can be calculated more easily. We therefore modify the weight matrices $\mathcal{W}^{(m)}$ by dropping the first row containing the intercept parameters $w_{0,j}^{(m)}$, $1 \le j \le q_m$. Define for $1 \le m \le d+1$

$$\mathcal{W}_{(-0)}^{(m)} = \left(w_{j_{m-1},j_m}^{(m)}\right)_{1 \le j_{m-1} \le q_{m-1};\ 1 \le j_m \le q_m} \in \mathbb{R}^{q_{m-1} \times q_m},$$

where $w_{j_{m-1},j_m}^{(m)}$ denotes component j_{m-1} of $\boldsymbol{w}_{j_m}^{(m)}$, and where we set $q_{d+1} = 1$ (output dimension) and $w_{j_d,1}^{(d+1)} = \beta_{j_d}$ for $0 \le j_d \le q_d$.

Proposition 7.5 (Back-Propagation for the Hyperbolic Tangent Activation) *Choose a FN network of depth $d \in \mathbb{N}$ and with hyperbolic tangent activation function $\phi(x) = \tanh(x)$.*

- *Define recursively*
  - *initialize $q_{d+1} = 1$ and $\boldsymbol{\delta}^{(d+1)}(\boldsymbol{x}) = \boldsymbol{1} \in \mathbb{R}^{q_{d+1}}$;*
  - *iterate for $d \ge m \ge 1$*

$$\boldsymbol{\delta}^{(m)}(\boldsymbol{x}) = diag\left(1 - \left(z_{j_m}^{(m:1)}(\boldsymbol{x})\right)^2\right)_{1 \le j_m \le q_m} \mathcal{W}_{(-0)}^{(m+1)} \boldsymbol{\delta}^{(m+1)}(\boldsymbol{x}) \in \mathbb{R}^{q_m}.$$

- *We obtain for $0 \le m \le d$*

$$\left(\frac{\partial \langle \boldsymbol{\beta}, \boldsymbol{z}^{(d:1)}(\boldsymbol{x})\rangle}{\partial w_{j_m,j_{m+1}}^{(m+1)}}\right)_{0 \le j_m \le q_m;\ 1 \le j_{m+1} \le q_{m+1}} = \boldsymbol{z}^{(m:1)}(\boldsymbol{x})\, \boldsymbol{\delta}^{(m+1)}(\boldsymbol{x})^\top \in \mathbb{R}^{(q_m+1) \times q_{m+1}},$$

where $\boldsymbol{z}^{(0:1)}(\boldsymbol{x}) = \boldsymbol{x} \in \mathbb{R}^{q_0+1}$ and $\boldsymbol{w}_1^{(d+1)} = \boldsymbol{\beta} \in \mathbb{R}^{q_d+1}$.

Proof of Proposition 7.5 Choose $1 \le m \le d$ and define for the neurons $1 \le j_m \le q_m$ the variables

$$\zeta_{j_m}^{(m)}(\boldsymbol{x}) = \left\langle \boldsymbol{w}_{j_m}^{(m)}, \boldsymbol{z}^{(m-1:1)}(\boldsymbol{x}) \right\rangle.$$

The learned representation in the m-th FN layer is obtained by activating these variables

$$\boldsymbol{z}^{(m:1)}(\boldsymbol{x}) = \left(1, \phi\left(\zeta_1^{(m)}(\boldsymbol{x})\right), \ldots, \phi\left(\zeta_{q_m}^{(m)}(\boldsymbol{x})\right)\right)^\top \in \mathbb{R}^{q_m+1}.$$

For the output we define

$$\zeta_1^{(d+1)}(\boldsymbol{x}) = \langle \boldsymbol{\beta}, \boldsymbol{z}^{(d:1)}(\boldsymbol{x})\rangle.$$

The main idea is to calculate the derivatives of $\langle \boldsymbol{\beta}, \boldsymbol{z}^{(d:1)}(\boldsymbol{x})\rangle$ w.r.t. these new variables $\zeta_j^{(m)}(\boldsymbol{x})$.

Initialization for $m = d+1$ This provides for $m = d+1$ and $1 \le j_{d+1} \le q_{d+1} = 1$

$$\frac{\partial \langle \boldsymbol{\beta}, \boldsymbol{z}^{(d:1)}(\boldsymbol{x})\rangle}{\partial \zeta_1^{(d+1)}(\boldsymbol{x})} = 1 = \delta_1^{(d+1)}(\boldsymbol{x}).$$

Recursion for $m < d+1$ Next, we calculate the derivatives w.r.t. $\zeta_{j_d}^{(d)}(\boldsymbol{x})$, for $m = d$ and $1 \le j_d \le q_d$. They are given by (note $q_{d+1} = 1$)

$$\begin{aligned}
\frac{\partial \langle \boldsymbol{\beta}, \boldsymbol{z}^{(d:1)}(\boldsymbol{x})\rangle}{\partial \zeta_{j_d}^{(d)}(\boldsymbol{x})} &= \frac{\partial \langle \boldsymbol{\beta}, \boldsymbol{z}^{(d:1)}(\boldsymbol{x})\rangle}{\partial \zeta_1^{(d+1)}(\boldsymbol{x})} \frac{\partial \zeta_1^{(d+1)}(\boldsymbol{x})}{\partial \zeta_{j_d}^{(d)}(\boldsymbol{x})} \\
&= \delta_1^{(d+1)}(\boldsymbol{x})\, \beta_{j_d}\, \phi'(\zeta_{j_d}^{(d)}(\boldsymbol{x})) \qquad (7.17)\\
&= \delta_1^{(d+1)}(\boldsymbol{x})\, w_{j_d,1}^{(d+1)} \left(1 - (z_{j_d}^{(d:1)}(\boldsymbol{x}))^2\right) \;=\; \delta_{j_d}^{(d)}(\boldsymbol{x}),
\end{aligned}$$

where we have used $w_{j_d,1}^{(d+1)} = \beta_{j_d}$ and for the hyperbolic tangent activation function $\phi' = 1 - \phi^2$. Continuing recursively for $d > m \ge 1$ and $1 \le j_m \le q_m$ we obtain

$$\begin{aligned}
\frac{\partial \langle \boldsymbol{\beta}, \boldsymbol{z}^{(d:1)}(\boldsymbol{x})\rangle}{\partial \zeta_{j_m}^{(m)}(\boldsymbol{x})} &= \sum_{j_{m+1}=1}^{q_{m+1}} \frac{\partial \langle \boldsymbol{\beta}, \boldsymbol{z}^{(d:1)}(\boldsymbol{x})\rangle}{\partial \zeta_{j_{m+1}}^{(m+1)}(\boldsymbol{x})} \frac{\partial \zeta_{j_{m+1}}^{(m+1)}(\boldsymbol{x})}{\partial \zeta_{j_m}^{(m)}(\boldsymbol{x})} \\
&= \sum_{j_{m+1}=1}^{q_{m+1}} \delta_{j_{m+1}}^{(m+1)}(\boldsymbol{x})\, w_{j_m,j_{m+1}}^{(m+1)} \left(1 - (z_{j_m}^{(m:1)}(\boldsymbol{x}))^2\right) \;=\; \delta_{j_m}^{(m)}(\boldsymbol{x}).
\end{aligned}$$

Thus, the vectors $\boldsymbol{\delta}^{(m)}(\boldsymbol{x}) = (\delta_1^{(m)}(\boldsymbol{x}), \ldots, \delta_{q_m}^{(m)}(\boldsymbol{x}))^\top$ are calculated recursively in $d \ge m \ge 1$ with initialization $\boldsymbol{\delta}^{(d+1)}(\boldsymbol{x}) = \mathbf{1}$ and the recursion

$$\boldsymbol{\delta}^{(m)}(\boldsymbol{x}) \;=\; \operatorname{diag}\left(1 - (z_{j_m}^{(m:1)}(\boldsymbol{x}))^2\right)_{1 \le j_m \le q_m} \mathcal{W}_{(-0)}^{(m+1)}\, \boldsymbol{\delta}^{(m+1)}(\boldsymbol{x}) \;\in\; \mathbb{R}^{q_m}.$$

Finally, we need to show how these derivatives are related to the original derivatives in the gradient descent method. We have for $0 \le j_d \le q_d$ and $j_{d+1} = 1$

$$\frac{\partial \langle \boldsymbol{\beta}, \boldsymbol{z}^{(d:1)}(\boldsymbol{x})\rangle}{\partial \beta_{j_d}} \;=\; \frac{\partial \langle \boldsymbol{\beta}, \boldsymbol{z}^{(d:1)}(\boldsymbol{x})\rangle}{\partial \zeta_1^{(d+1)}(\boldsymbol{x})} \frac{\partial \zeta_1^{(d+1)}(\boldsymbol{x})}{\partial \beta_{j_d}} \;=\; \delta_{j_{d+1}}^{(d+1)}(\boldsymbol{x})\, z_{j_d}^{(d:1)}(\boldsymbol{x}).$$

For $1 \le m < d$, and $0 \le j_m \le q_m$ and $1 \le j_{m+1} \le q_{m+1}$ we have

$$\frac{\partial \langle \boldsymbol{\beta}, \boldsymbol{z}^{(d:1)}(\boldsymbol{x})\rangle}{\partial w^{(m+1)}_{j_m, j_{m+1}}} = \frac{\partial \langle \boldsymbol{\beta}, \boldsymbol{z}^{(d:1)}(\boldsymbol{x})\rangle}{\partial \zeta^{(m+1)}_{j_{m+1}}(\boldsymbol{x})} \frac{\partial \zeta^{(m+1)}_{j_{m+1}}(\boldsymbol{x})}{\partial w^{(m+1)}_{j_m, j_{m+1}}} = \delta^{(m+1)}_{j_{m+1}}(\boldsymbol{x})\, z^{(m:1)}_{j_m}(\boldsymbol{x}).$$

For $m = 0$, and $0 \le l \le q_0$ and $1 \le j_1 \le q_1$ we have

$$\frac{\partial \langle \boldsymbol{\beta}, \boldsymbol{z}^{(d:1)}(\boldsymbol{x})\rangle}{\partial w^{(1)}_{l, j_1}} = \frac{\partial \langle \boldsymbol{\beta}, \boldsymbol{z}^{(d:1)}(\boldsymbol{x})\rangle}{\partial \zeta^{(1)}_{j_1}(\boldsymbol{x})} \frac{\partial \zeta^{(1)}_{j_1}(\boldsymbol{x})}{\partial w^{(1)}_{l, j_1}} = \delta^{(1)}_{j_1}(\boldsymbol{x})\, x_l.$$

This completes the proof of Proposition 7.5. □

Remark 7.6 Proposition 7.5 gives the back-propagation method for the hyperbolic tangent activation function which has derivative $\phi' = 1 - \phi^2$. This becomes visible in the definition of $\boldsymbol{\delta}^{(m)}(\boldsymbol{x})$ where we consider the diagonal matrix

$$\operatorname{diag}\left(1 - \left(z^{(m:1)}_{j_m}(\boldsymbol{x})\right)^2\right)_{1 \le j_m \le q_m}.$$

For a general differentiable activation function ϕ this needs to be replaced by, see (7.17),

$$\operatorname{diag}\left(\phi'\left\langle \boldsymbol{w}^{(m)}_{j_m}, \boldsymbol{z}^{(m-1:1)}(\boldsymbol{x})\right\rangle\right)_{1 \le j_m \le q_m}.$$

In the case of the sigmoid activation function this gives us, see also Table 7.1,

$$\operatorname{diag}\left(z^{(m:1)}_{j_m}(\boldsymbol{x})\left(1 - z^{(m:1)}_{j_m}(\boldsymbol{x})\right)\right)_{1 \le j_m \le q_m}.$$

Plain vanilla gradient descent algorithm for FN networks

---

1. Choose an initial network parameter $\boldsymbol{\vartheta}^{(0)} \in \mathbb{R}^r$.
2. Iterate for $t \ge 0$ until a stopping criterion is met:
   (a) Calculate the gradient $\nabla_{\boldsymbol{\vartheta}} \mathfrak{D}(\boldsymbol{Y}, \boldsymbol{\vartheta})$ in network parameter $\boldsymbol{\vartheta} = \boldsymbol{\vartheta}^{(t)}$ using (7.16) and the back-propagation method of Proposition 7.5 (for the hyperbolic tangent activation function).
   (b) Make the gradient descent step for a suitable learning rate $\varrho_{t+1} > 0$

$$\boldsymbol{\vartheta}^{(t)} \mapsto \boldsymbol{\vartheta}^{(t+1)} = \boldsymbol{\vartheta}^{(t)} - \varrho_{t+1} \nabla_{\boldsymbol{\vartheta}} \mathfrak{D}(\boldsymbol{Y}, \boldsymbol{\vartheta}^{(t)}).$$

Remark 7.7 The initialization $\boldsymbol{\vartheta}^{(0)} \in \mathbb{R}^r$ of the gradient descent algorithm needs some care. A FN network has many symmetries, for instance, we can permute neurons within a FN layer and we receive the same predictive model. For this reason, the initial network weights $\mathcal{W}^{(m)} = (\boldsymbol{w}_1^{(m)}, \ldots, \boldsymbol{w}_{q_m}^{(m)}) \in \mathbb{R}^{(q_{m-1}+1)\times q_m}$, $1 \le m \le d$, should not be chosen with identical components because this will result in a saddlepoint of the corresponding objective function, and gradient descent will not work. For this reason, these weights are initialized randomly either using a uniform or a Gaussian distribution. The former is related to the `glorot_uniform` initializer in `keras`,[2] see (16) in Glorot–Bengio [160]. This initializer scales the support of the uniform distribution with the sizes of the FN layers that are connected by the corresponding weights $\boldsymbol{w}_j^{(m)}$.

For the output parameter we usually set as initial value $\boldsymbol{\beta}^{(0)} = (\widehat{\beta}_0^{(0)}, 0, \ldots, 0)^\top \in \mathbb{R}^{q_d+1}$, where $\widehat{\beta}_0^{(0)}$ is the MLE in the corresponding null model (not considering any features) and transformed to the chosen link g. This choice implies that the gradient descent algorithm starts in the null model, and any decrease in deviance loss can be seen as an improved in-sample loss of using the FN network regression structure over the null model.

Stochastic Gradient Descent

The gradient in (7.16) has two parts. We have a vector

$$\boldsymbol{v}(\boldsymbol{Y}) = \left(\frac{v_i}{\varphi}\left(\mu_{\boldsymbol{\vartheta}}(\boldsymbol{x}_i) - Y_i\right)\frac{1}{V\left(\mu_{\boldsymbol{\vartheta}}(\boldsymbol{x}_i)\right)}\frac{1}{g'(\mu_{\boldsymbol{\vartheta}}(\boldsymbol{x}_i))}\right)^\top_{1\le i\le n} \in \mathbb{R}^n,$$

and we have a matrix

$$\mathbf{M} = \left(\nabla_{\boldsymbol{\vartheta}}\left\langle \boldsymbol{\beta}, \boldsymbol{z}^{(d:1)}(\boldsymbol{x}_1)\right\rangle, \ldots, \nabla_{\boldsymbol{\vartheta}}\left\langle \boldsymbol{\beta}, \boldsymbol{z}^{(d:1)}(\boldsymbol{x}_n)\right\rangle\right) \in \mathbb{R}^{r\times n}.$$

The gradient of the deviance loss function is obtained by the matrix multiplication

$$\nabla_{\boldsymbol{\vartheta}}\mathfrak{D}(\boldsymbol{Y}, \boldsymbol{\vartheta}) = \frac{2}{n}\,\mathbf{M}\,\boldsymbol{v}(\boldsymbol{Y}).$$

Matrix multiplication can be very slow in numerical implementations if the sample size n is large. For this reason, one typically uses the *stochastic gradient descent* (SGD) method that does not consider the entire data $\boldsymbol{Y} = (Y_1, \ldots, Y_n)^\top$ simultaneously.

---

[2] For our examples we use the R library `keras` [77] which is an API to TensorFlow [2].

For the SGD method one chooses a fixed *batch size* $b \in \mathbb{N}$, and one randomly partitions the entire data $\boldsymbol{Y}$ into *(mini-)batches* $\boldsymbol{Y}_1, \ldots, \boldsymbol{Y}_{\lfloor n/b \rfloor}$ of approximately the same size b (up to cardinality). Each gradient descent update

$$\boldsymbol{\vartheta}^{(t)} \mapsto \boldsymbol{\vartheta}^{(t+1)} = \boldsymbol{\vartheta}^{(t)} - \varrho_{t+1} \nabla_{\boldsymbol{\vartheta}} \mathfrak{D}(\boldsymbol{Y}_s, \boldsymbol{\vartheta}^{(t)}),$$

is then only based on the observations $\boldsymbol{Y}_s$ in the corresponding batch $1 \le s \le \lfloor n/b \rfloor$. Typically, one sequentially visits all batches, and screening each batch once is called an *epoch*. Thus, if we run the SGD algorithm over K epochs on batches of size $b \le n$, then we perform $K \lfloor n/b \rfloor$ gradient descent steps.

Choosing batches of size b reduces the complexity of the matrix multiplication from n to b, and, henceforth, leads to much faster run times in one gradient descent step. On the other hand, batches should have a minimal size so that the gradient descent updates are not too erratic, i.e., if the batches are too small, the randomness in the data may point too often into a (completely) wrong direction for the optimal gradient descent step. For this reason, optimal batch sizes should be chosen carefully. For instance, if we study a low frequency claims count problem, say, with an expected frequency of $\lambda = 10\%$, we can determine confidence bounds for parameter estimation. This will provide an estimate of a minimal batch size b for a reliable parameter estimate.

To have a few erratic steps in SGD, however, can also be beneficial, as long as there are not too many of those. Sometimes, the algorithm gets trapped in saddlepoints or in flat areas of the objective function (vanishing gradient problem). If this is the case, an erratic step may be beneficial because it may perturb the algorithm out of its bottleneck. In fact, often SGD has a better performance than the plain vanilla gradient descent algorithm that is based on the entire data $\boldsymbol{Y}$ because of these noisy contributions.

Momentum-Based Gradient Descent Methods

The gradient descent method only considers a first order Taylor expansion and one is tempted to consider higher order terms to improve the approximation. For instance, Newton's method uses a second order Taylor term by updating

$$\boldsymbol{\vartheta}^{(t)} \mapsto \boldsymbol{\vartheta}^{(t+1)} = \boldsymbol{\vartheta}^{(t)} - \left(\nabla^2_{\boldsymbol{\vartheta}} \mathfrak{D}(\boldsymbol{Y}, \boldsymbol{\vartheta}^{(t)})\right)^{-1} \nabla_{\boldsymbol{\vartheta}} \mathfrak{D}(\boldsymbol{Y}, \boldsymbol{\vartheta}^{(t)}). \tag{7.18}$$

In many practical applications this calculation is not feasible as the Hessian $\nabla^2_{\boldsymbol{\vartheta}} \mathfrak{D}(\boldsymbol{Y}, \boldsymbol{\vartheta}^{(t)})$ cannot be calculated in a reasonable amount of time. Another (simple) way of considering the changes in the gradients is the *momentum-based gradient descent method* of Rumelhart et al. [324]. This is inspired by mechanics in physics and it is achieved by considering the gradients over several iterations of the algorithm (with exponentially decaying weights). Choose a momentum coefficient $\nu \in [0, 1)$ and define the initial speed $\mathbf{v}^{(0)} = 0 \in \mathbb{R}^r$.

Replace the gradient descent update (7.15) by

$$\mathbf{v}^{(t)} \mapsto \mathbf{v}^{(t+1)} = \nu \mathbf{v}^{(t)} - \varrho_{t+1} \nabla_{\vartheta} \mathfrak{D}(\boldsymbol{Y}, \boldsymbol{\vartheta}^{(t)}), \tag{7.19}$$

$$\boldsymbol{\vartheta}^{(t)} \mapsto \boldsymbol{\vartheta}^{(t+1)} = \boldsymbol{\vartheta}^{(t)} + \mathbf{v}^{(t+1)}. \tag{7.20}$$

For $\nu = 0$ we have the plain vanilla gradient descent method, for $\nu > 0$ we also memorize the previous gradients (with exponentially decaying weights). Typically this leads to better convergence properties.

Nesterov [284] has noticed that for convex functions the gradient descent updates may have a zig-zag behavior. Therefore, he proposed the so-called Nesterov-accelerated version

$$\begin{aligned} \mathbf{v}^{(t)} \mapsto \mathbf{v}^{(t+1)} &= \nu \mathbf{v}^{(t)} - \varrho_{t+1} \nabla_{\vartheta} \mathfrak{D}(\boldsymbol{Y}, \boldsymbol{\vartheta}^{(t)} + \nu \mathbf{v}^{(t)}), \\ \boldsymbol{\vartheta}^{(t)} \mapsto \boldsymbol{\vartheta}^{(t+1)} &= \boldsymbol{\vartheta}^{(t)} + \mathbf{v}^{(t+1)}. \end{aligned} \tag{7.21}$$

Thus, the calculation of the momentum $\mathbf{v}^{(t+1)}$ uses a look-ahead $\boldsymbol{\vartheta}^{(t)} + \nu \mathbf{v}^{(t)}$ in the gradient calculation (anticipating part of the next step). This provides for the update (7.21) the following equivalent versions, under reparametrization $\widetilde{\boldsymbol{\vartheta}}^{(t)} = \boldsymbol{\vartheta}^{(t)} + \nu \mathbf{v}^{(t)}$,

$$\begin{aligned} \boldsymbol{\vartheta}^{(t+1)} &= \boldsymbol{\vartheta}^{(t)} + \left(\nu \mathbf{v}^{(t)} - \varrho_{t+1} \nabla_{\vartheta} \mathfrak{D}(\boldsymbol{Y}, \boldsymbol{\vartheta}^{(t)} + \nu \mathbf{v}^{(t)})\right) \\ &= \boldsymbol{\vartheta}^{(t)} + \left(\nu \mathbf{v}^{(t)} - \varrho_{t+1} \nabla_{\vartheta} \mathfrak{D}(\boldsymbol{Y}, \widetilde{\boldsymbol{\vartheta}}^{(t)})\right) \\ &= \widetilde{\boldsymbol{\vartheta}}^{(t)} + \left(\nu \mathbf{v}^{(t+1)} - \varrho_{t+1} \nabla_{\vartheta} \mathfrak{D}(\boldsymbol{Y}, \widetilde{\boldsymbol{\vartheta}}^{(t)})\right) - \nu \mathbf{v}^{(t+1)}. \end{aligned} \tag{7.22}$$

For the Nesterov accelerated update we can also study, we use the last line of (7.22),

$$\begin{aligned} \mathbf{v}^{(t)} \mapsto \mathbf{v}^{(t+1)} &= \nu \mathbf{v}^{(t)} - \varrho_{t+1} \nabla_{\vartheta} \mathfrak{D}(\boldsymbol{Y}, \widetilde{\boldsymbol{\vartheta}}^{(t)}), \\ \widetilde{\boldsymbol{\vartheta}}^{(t)} \mapsto \widetilde{\boldsymbol{\vartheta}}^{(t+1)} &= \widetilde{\boldsymbol{\vartheta}}^{(t)} + \left(\nu \mathbf{v}^{(t+1)} - \varrho_{t+1} \nabla_{\vartheta} \mathfrak{D}(\boldsymbol{Y}, \widetilde{\boldsymbol{\vartheta}}^{(t)})\right). \end{aligned} \tag{7.23}$$

Compared to (7.19)–(7.20), we just shift the index by 1 in the momentum $\mathbf{v}^{(t)}$ in the round brackets of (7.23). The typical way how the Nesterov-acceleration is formulated is, yet, another equivalent formulation, namely, only in terms of $\boldsymbol{\vartheta}^{(t)}$ and $\widetilde{\boldsymbol{\vartheta}}^{(t)}$. From the second line of (7.22) and (7.21) we have the updates

$$\begin{aligned} \boldsymbol{\vartheta}^{(t+1)} &= \widetilde{\boldsymbol{\vartheta}}^{(t)} - \varrho_{t+1} \nabla_{\vartheta} \mathfrak{D}(\boldsymbol{Y}, \widetilde{\boldsymbol{\vartheta}}^{(t)}), \\ \widetilde{\boldsymbol{\vartheta}}^{(t+1)} &= \boldsymbol{\vartheta}^{(t+1)} + \nu \left(\boldsymbol{\vartheta}^{(t+1)} - \boldsymbol{\vartheta}^{(t)}\right). \end{aligned} \tag{7.24}$$

Typically, one chooses the momentum coefficient ν in (7.24) time-dependent by setting $\nu_t = t/(t+3)$.

In our applications we will use the R interface to the `keras` library [77]. This library has a couple of standard momentum-based gradient descent methods implemented which use pre-defined learning rates and momentum coefficients. In our analysis we are mainly relying on the variants `rmsprop` and the Nesterov-accelerated version of `adam`, called `nadam`. Therefore, we briefly describe these three variants, and for more information we refer to Sections 8.3 and 8.5 in Goodfellow et al. [166].

Predefined Gradient Descent Methods

- `rmsprop` stands for 'root mean square propagation', and its origin can be found in a lecture of Hinton et al. [187]. Denote by $\odot$ the Hadamard product that computes the component-wise products of two matrices. Choose a weight $\alpha \in (0,1)$ and calculate the accumulated squared gradients, set $\mathbf{r}^{(0)} = 0 \in \mathbb{R}^r$,

$$\mathbf{r}^{(t)} \mapsto \mathbf{r}^{(t+1)} = \alpha \mathbf{r}^{(t)} + (1-\alpha)\left(\nabla_{\boldsymbol{\vartheta}} \mathfrak{D}(\boldsymbol{Y}, \boldsymbol{\vartheta}^{(t)}) \odot \nabla_{\boldsymbol{\vartheta}} \mathfrak{D}(\boldsymbol{Y}, \boldsymbol{\vartheta}^{(t)})\right) \in \mathbb{R}^r.$$

The sequence $(\mathbf{r}^{(t)})_{t\geq 1}$ memorizes the (squared) magnitudes of the components of the gradients $\nabla_{\boldsymbol{\vartheta}} \mathfrak{D}(\boldsymbol{Y}, \boldsymbol{\vartheta}^{(t)})$, $t \geq 1$. This is done individually for each component because we may have directional differences in magnitudes (and momentum). In contrast to (7.19), $\mathbf{r}^{(t)}$ does not model the speed, but rather an inverse weight. This then motivates the gradient descent update

$$\boldsymbol{\vartheta}^{(t)} \mapsto \boldsymbol{\vartheta}^{(t+1)} = \boldsymbol{\vartheta}^{(t)} - \frac{\varrho}{\sqrt{\varepsilon + \mathbf{r}^{(t+1)}}} \odot \nabla_{\boldsymbol{\vartheta}} \mathfrak{D}(\boldsymbol{Y}, \boldsymbol{\vartheta}^{(t)}),$$

where the square-root is taken component-wise, for a global decay rate $\varrho > 0$, and for a small positive constant $\varepsilon > 0$ to ensure that everything is well-defined.

- `adam` stands for 'adaptive moment' estimation, and it has been proposed by Kingma–Ba [216]. The momentum is determined by the first two moments in `adam`, namely, we set $\mathbf{v}^{(0)} = \mathbf{r}^{(0)} = 0 \in \mathbb{R}^r$ and we consider

$$\mathbf{v}^{(t)} \mapsto \mathbf{v}^{(t+1)} = \nu \mathbf{v}^{(t)} + (1-\nu)\nabla_{\boldsymbol{\vartheta}} \mathfrak{D}(\boldsymbol{Y}, \boldsymbol{\vartheta}^{(t)}), \tag{7.25}$$

$$\mathbf{r}^{(t)} \mapsto \mathbf{r}^{(t+1)} = \alpha \mathbf{r}^{(t)} + (1-\alpha)\left(\nabla_{\boldsymbol{\vartheta}} \mathfrak{D}(\boldsymbol{Y}, \boldsymbol{\vartheta}^{(t)}) \odot \nabla_{\boldsymbol{\vartheta}} \mathfrak{D}(\boldsymbol{Y}, \boldsymbol{\vartheta}^{(t)})\right), \tag{7.26}$$

for given weights $\nu, \alpha \in (0,1)$. Similar to Bayesian credibility theory, $\mathbf{v}^{(t)}$ and $\mathbf{r}^{(t)}$ are biased because these two processes have been initialized in zero. Therefore, they are rescaled by $1/(1-\nu^t)$ and $1/(1-\alpha^t)$, respectively. This gives us the gradient descent update

$$\boldsymbol{\vartheta}^{(t)} \mapsto \boldsymbol{\vartheta}^{(t+1)} = \boldsymbol{\vartheta}^{(t)} - \frac{\varrho}{\varepsilon + \sqrt{\frac{\mathbf{r}^{(t+1)}}{1-\alpha^t}}} \odot \frac{\mathbf{v}^{(t+1)}}{1-\nu^t},$$

where the square-root is taken component-wise, for a global decay rate $\varrho > 0$, and for a small positive constant $\varepsilon > 0$ to ensure that everything is well-defined.

- `nadam` is the Nesterov-accelerated [284] version of `adam`. Similarly as when going from (7.19)–(7.20) to (7.23), the acceleration is obtained by a shift of 1 in the velocity parameter, thus, consider the Nesterov-accelerated `adam` update

$$\boldsymbol{\vartheta}^{(t)} \mapsto \boldsymbol{\vartheta}^{(t+1)} = \boldsymbol{\vartheta}^{(t)} - \frac{\varrho}{\varepsilon + \sqrt{\frac{\mathbf{r}^{(t+1)}}{1-\alpha^t}}} \odot \frac{\nu \mathbf{v}^{(t+1)} + (1-\nu)\nabla_{\boldsymbol{\vartheta}} \mathfrak{D}(\boldsymbol{Y}, \boldsymbol{\vartheta}^{(t)})}{1-\nu^t},$$

using (7.25) and (7.26).

Maximum Likelihood Estimation and Over-fitting

As explained above, we model the mean of the datum $(Y, \boldsymbol{x})$ by a deep FN network

$$\boldsymbol{x} \mapsto \mu(\boldsymbol{x}) = \mu_{\boldsymbol{\vartheta}}(\boldsymbol{x}) = \mathbb{E}_{\theta(\boldsymbol{x})}[Y] = g^{-1}\left\langle \boldsymbol{\beta}, \boldsymbol{z}^{(d:1)}(\boldsymbol{x}) \right\rangle,$$

for a network parameter $\boldsymbol{\vartheta} \in \mathbb{R}^r$. MLE of this network parameter requires solving for given data $\boldsymbol{Y}$

$$\widehat{\boldsymbol{\vartheta}}^{\text{MLE}} = \underset{\boldsymbol{\vartheta}}{\arg\min}\, \mathfrak{D}(\boldsymbol{Y}, \boldsymbol{\vartheta}).$$

In Fig. 7.5 we give a schematic figure of a loss surface $\boldsymbol{\vartheta} \mapsto \mathfrak{D}(\boldsymbol{Y}, \boldsymbol{\vartheta})$ for a (low-dimensional) example $\boldsymbol{\vartheta} \in \mathbb{R}^2$. The two plots show the same loss surface from two different angles. This loss surface has three (local) minimums (red color), and the smallest one (global minimum) gives the MLE $\widehat{\boldsymbol{\vartheta}}^{\text{MLE}}$.

In general, this global minimum cannot be found for more complex network architectures because the loss surface typically has a complicated structure for high-dimensional parameter spaces. Is this a problem in FN network fitting? Not really! We are going to explain why. The universality theorems in Sect. 7.2.2 state that more complex FN networks have an excellent approximation capacity. If we translate this to our statistical modeling problem it means that the observations $\boldsymbol{Y}$ can be approximated arbitrarily well by sufficiently complex FN networks. In particular, for a given complex network architecture, the MLE $\widehat{\boldsymbol{\vartheta}}^{\text{MLE}}$ will provide the optimal fit of this architecture to the data $\boldsymbol{Y}$, and, as a result, this network does not only reflect the systematic effects in the data but also the noisy part. This behavior is called *(in-sample) over-fitting* to the learning data $\mathcal{L}$. It implies that such statistical models typically have a poor generalization to unseen (out-of-sample) test data $\mathcal{T}$; this is illustrated by the red color in Fig. 7.6. For this reason, in general, we are not interested in finding the MLE $\widehat{\boldsymbol{\vartheta}}^{\text{MLE}}$ of $\boldsymbol{\vartheta}$ in FN network regression modeling, but we would like to find a parameter estimate $\widehat{\boldsymbol{\vartheta}}$ that (only) extracts the systematic effects from the learning data $\mathcal{L}$. This is illustrated by the different colors in Figs. 7.5

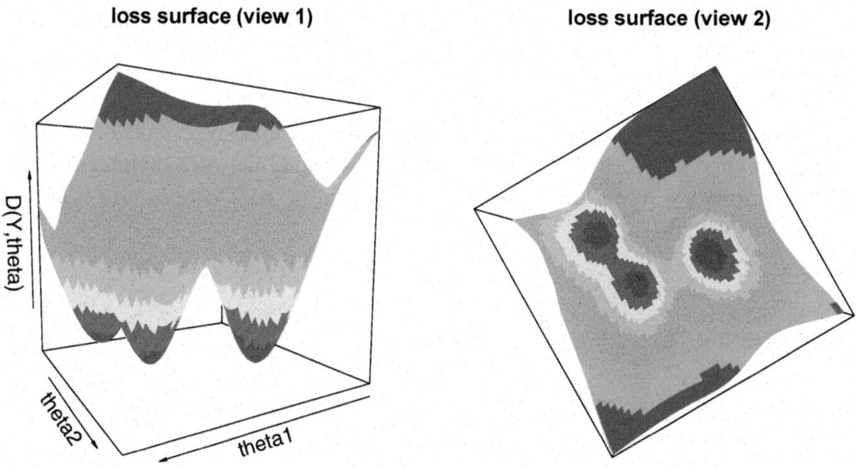

Fig. 7.5 Schematic figure of a loss surface $\vartheta \mapsto \mathfrak{D}(\boldsymbol{Y}, \vartheta)$ from two different angles for a two-dimensional parameter $\vartheta \in \mathbb{R}^2$

Fig. 7.6 Schematic figure of in-sample over-fitting (red), under-fitting (blue) and extracting systematic effects (green)

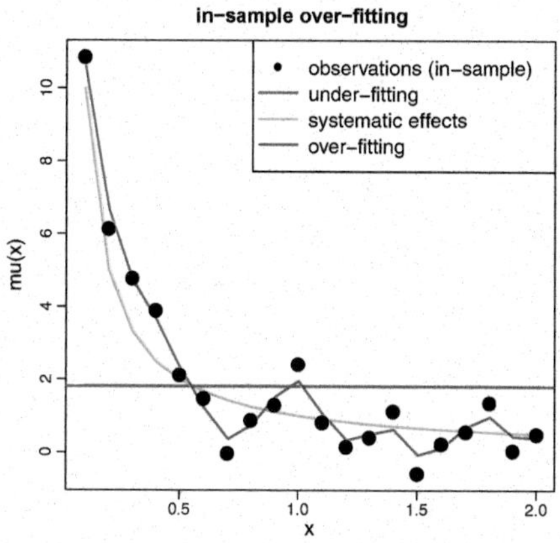

and 7.6, where we assume: (a) red color provides models with a poor generalization power due to over-fitting, (b) blue color provides models with a poor generalization power, too, because these parametrizations do not explain the systematic effects in the data at all (called under-fitting), and (c) green color gives good parametrizations that explain the systematic effects in the data and generalize well to unseen data. Thus, the aim is to find parametrizations that are in the green area of Fig. 7.5. This green area emphasizes that we lose the notion of uniqueness because there are infinitely many models in the green area that have a comparable generalization

power. Next we explain how we can exploit the gradient descent algorithm to make it useful for finding parametrizations in the green area.

Remark 7.8 The loss surface considerations in Fig. 7.5 are based on a fixed network architecture. Recent research promotes the so-called Graph HyperNetwork (GHN) that is a (hyper-)network which tries to find the optimal network architecture and its parametrization by an additional network, we refer to Zhang et al. [402] and Knyazev et al. [219].

Regularization Through Early Stopping

As stated above, if we run the gradient descent algorithm with properly tempered learning rates it will converge to a local minimum of the loss function, which means that the resulting FN network over-fits to the learning data. For this reason we need to *early stop* the gradient descent algorithm beforehand. Coming back to Fig. 7.5, typically, we start the gradient descent algorithm somewhere in the blue area of the loss surface (supposed that the red area is a sparse set on the loss surface). Visually speaking, the gradient descent algorithm then walks down the valley (green, yellow and red area) by exploiting locally optimal steps. Since at the early stage of the algorithm the systematic effects play a dominant role over the noisy part, the gradient descent algorithm learns these systematic effects at this first stage (blue area in Fig. 7.5). When the algorithm arrives at the green area the noisy part in the data starts to increasingly influence the model calibration (gradient descent steps), and, henceforth, at this stage the algorithm should be stopped, and the learned parameter should be selected for predictive modeling. This early stopping is an implicit way of regularization, because it implies that we stop the parameter fitting before the parameters start to learn very individual features of the (noisy) data (and take extreme values).

This early stopping point is determined by doing an out-of-sample analysis. This requires the learning data $\mathcal{L}$ to be further split into *training data* $\mathcal{U}$ and *validation data* $\mathcal{V}$. The training data $\mathcal{U}$ is used for gradient descent parameter learning, and the validation data $\mathcal{V}$ is used for tracking the over-fitting by an instantaneous (out-of-sample) validation analysis. This partition is illustrated in Fig. 7.7, which also highlights that the validation data $\mathcal{V}$ is disjoint from the test data $\mathcal{T}$, the latter only being used in the final step for comparing different statistical models (e.g., a GLM vs. a FN network). That is, model comparison is done in a proper out-of-sample manner on $\mathcal{T}$, and each of these models is only fit on $\mathcal{U}$ and $\mathcal{V}$. Thus, for FN network fitting with early stopping we need a reasonable amount of data that can be split into 3 sufficiently large data sets so that each is suitable for its purpose.

For early stopping we partition the learning data $\mathcal{L}$ into training data $\mathcal{U}$ and validation data $\mathcal{V}$. The plain vanilla gradient descent algorithm can then be changed as follows.

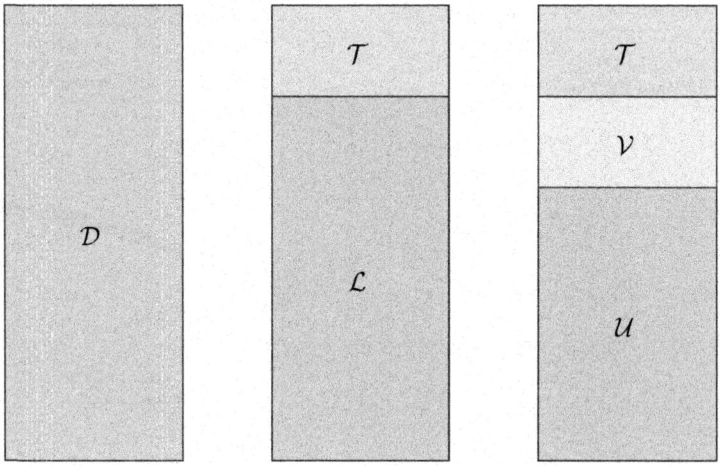

Fig. 7.7 Partition of entire data $\mathcal{D}$ (lhs) into learning data $\mathcal{L}$ and test data $\mathcal{T}$ (middle), and into training data $\mathcal{U}$, validation data $\mathcal{V}$ and test data $\mathcal{T}$ (rhs)

Plain vanilla gradient descent algorithm with early stopping

---

1. Choose an initial network parameter $\boldsymbol{\vartheta}^{(0)} \in \mathbb{R}^r$.
2. Iterate for $t \geq 0$ until the early stopping criterion is met:

   (a) Calculate the gradient $\nabla_{\boldsymbol{\vartheta}} \mathfrak{D}(\mathcal{U}, \boldsymbol{\vartheta})$ in network parameter $\boldsymbol{\vartheta} = \boldsymbol{\vartheta}^{(t)}$ on the training data $\mathcal{U}$ using (7.16) and the back-propagation method of Proposition 7.5 (for the hyperbolic tangent activation function).

   (b) Make the gradient descent step for a suitable learning rate $\varrho_{t+1} > 0$

   $$\boldsymbol{\vartheta}^{(t)} \mapsto \boldsymbol{\vartheta}^{(t+1)} = \boldsymbol{\vartheta}^{(t)} - \varrho_{t+1} \nabla_{\boldsymbol{\vartheta}} \mathfrak{D}(\mathcal{U}, \boldsymbol{\vartheta}^{(t)}).$$

   (c) Calculate the validation loss $\mathfrak{D}(\mathcal{V}, \boldsymbol{\vartheta}^{(t)})$ on the validation data $\mathcal{V}$.

   (d) Stop the algorithm if the validation loss increases, i.e., if

   $$\mathfrak{D}(\mathcal{V}, \boldsymbol{\vartheta}^{(t)}) > \mathfrak{D}(\mathcal{V}, \boldsymbol{\vartheta}^{(t-1)}), \tag{7.27}$$

   and return the learned parameter (estimate) $\widehat{\boldsymbol{\vartheta}} = \boldsymbol{\vartheta}^{(t-1)}$.

In applications we use the SGD algorithm that can also have erratic steps because not all random (mini-)batches are necessarily typical representations of the data. In such cases we should use more sophisticated stopping criteria than (7.27), for instance, early stop if the validation loss increases five times in a row.

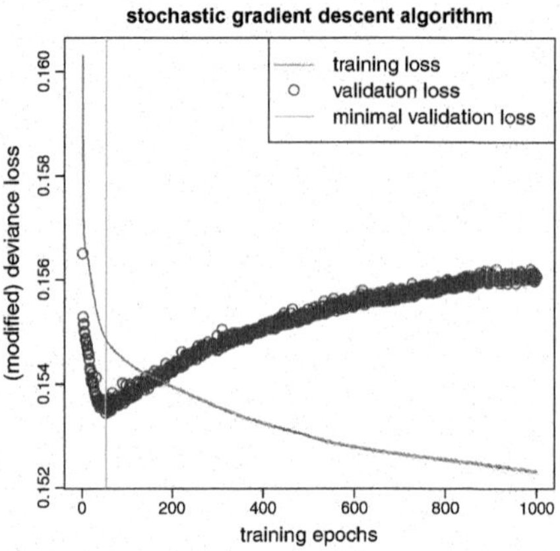

Fig. 7.8 Training loss $\mathfrak{D}(\mathcal{U}, \boldsymbol{\vartheta}^{(t)})$ vs. validation loss $\mathfrak{D}(\mathcal{V}, \boldsymbol{\vartheta}^{(t)})$ over different iterations $t \geq 0$ of the SGD algorithm

Figure 7.8 provides an example of the application of the SGD algorithm on training data $\mathcal{U}$ and validation data $\mathcal{V}$. The training loss is in blue color and the validation loss in green color. We observe that the validation loss has its minimum after 52 epochs (orange vertical line), and hence the fitting algorithm should be stopped at this point. We give a couple of remarks concerning Fig. 7.8:

- The learning data $\mathcal{L}$ exactly corresponds to the claims frequency data of Sect. 5.2.4, see also Table 5.2. We take 10% as validation data which gives $|\mathcal{U}| = 549'185$ and $|\mathcal{V}| = 61'021$. For the SGD algorithm we use batches of size $10'000$ which implies that one epoch corresponds to $\lfloor 549'185/10'000 \rfloor = 54$ gradient descent steps. For batches of size $10'000$ we expect an approximate estimation precision on an average frequency of $\bar{\lambda} = 7.36\%$ in the Poisson model of

$$\left[\bar{\lambda} - 2\sqrt{\frac{\bar{\lambda}}{10'000\bar{v}}},\ \bar{\lambda} + 2\sqrt{\frac{\bar{\lambda}}{10'000\bar{v}}}\right] = [6.62\%, 8.11\%],$$

  with an average exposure $\bar{v} = 0.5283$ on our learning data, we also refer to Example 3.22.
- The FN network architecture used in Fig. 7.8 is the one shown in Fig. 7.2 using one-hot encoding for categorical variables, see Sect. 7.3.1, below, and the responses are modeled by a Poisson distribution.
- The training loss $\mathfrak{D}(\mathcal{U}, \boldsymbol{\vartheta}^{(t)})$, blue curve in Fig. 7.8, is a bit wiggly which comes from the fact that we use a SGD where not every batch leads to the optimal decrease in loss. Remark that the loss figures in the graph correspond to average losses over an entire epoch, i.e., in our case an average over 54 SGD steps. Also remark that the y-scale does not show the Poisson deviance loss: we use the loss figures provided by `keras` [77] and these figures drop all terms of the deviance loss that are not relevant for parameter estimation.

We close this section with remarks.

Remarks 7.9

- We perform early stopping because otherwise a complex FN network would in-sample over-fit to the learning data. At this stage, one could be tempted to choose a smaller network to prevent from over-fitting. In general, this is not a sensible thing to do because the network needs sufficient flexibility to be able to be fitted to the data. That is, we need some redundancy in the model to be able to successfully apply the SGD algorithm, otherwise the algorithm may get trapped in saddlepoints or bottlenecks. Thus, the chosen network architecture should be above the bound of a necessary minimal complexity, and different architectures above this bound will provide similar accuracy (without a clear winner).
- The chosen network will contain certain elements of randomness, and different runs of the SGD algorithm will provide different solutions. Firstly, the initialization $\boldsymbol{\vartheta}^{(0)} \in \mathbb{R}^r$ of the algorithm is chosen at random, and since we early stop the algorithm and because we do not have a unique optimal point, the chosen solution will depend on this random initialization. Secondly, the split between training and validation data is done at random, and thirdly the partitioning of the training data into mini-batches is done at random. All these random elements make the early stopped SGD solution non-unique.
- Early stopping implies that the chosen network parameter estimate $\widehat{\boldsymbol{\vartheta}}$ does not correspond to a solution of the score equations and, henceforth, asymptotic results about MLEs do not apply, see Theorem 3.28.

7.3 Feed-Forward Neural Network Examples

7.3.1 Feature Pre-processing

Similarly to GLMs, we also need to pre-process the feature components in FN network regression modeling. The former Sect. 5.2.2 for GLMs has been called 'feature engineering' because we need to bring the feature components into an appropriate functional form w.r.t. the given regression task. The present section is called 'feature pre-processing' because we do not need to engineer the features for FN networks. We only need to bring them into a suitable (tabular) form to enter the network, and the network will then do an automated feature engineering through representation learning.

Categorical Feature Components: One-Hot Encoding

The categorical features have been treated by dummy coding within GLMs. Dummy coding provides full rank design matrices. For FN network regression modeling the

Table 7.2 One-hot encoding example mapping the $K = 11$ levels (colors) to the unit vectors of the 11-dimensional Euclidean space $\mathbb{R}^{11}$ showing the resulting encoding vectors $\boldsymbol{x}_j^\top$ as row vectors

a_1 = white	1	0	0	0	0	0	0	0	0	0	0
a_2 = yellow	0	1	0	0	0	0	0	0	0	0	0
a_3 = orange	0	0	1	0	0	0	0	0	0	0	0
a_4 = red	0	0	0	1	0	0	0	0	0	0	0
a_5 = magenta	0	0	0	0	1	0	0	0	0	0	0
a_6 = violet	0	0	0	0	0	1	0	0	0	0	0
a_7 = blue	0	0	0	0	0	0	1	0	0	0	0
a_8 = cyan	0	0	0	0	0	0	0	1	0	0	0
a_9 = green	0	0	0	0	0	0	0	0	1	0	0
a_{10} = beige	0	0	0	0	0	0	0	0	0	1	0
a_{11} = brown	0	0	0	0	0	0	0	0	0	0	1

full rank property is not important because, anyway, we neither have a single (local) minimum in the objective function, nor do we want to calculate the MLE of the network parameter. Typically, in FN network regression modeling one uses one-hot encoding for the categorical variables that encodes every level by a unit vector. Assume the raw feature component $\widetilde{x}_j$ is a categorical variable taking K different levels $\{a_1, \ldots, a_K\}$. One-hot encoding is obtained by the embedding map

$$\widetilde{x}_j \mapsto \boldsymbol{x}_j = \left(\mathbb{1}_{\{\widetilde{x}_j = a_1\}}, \ldots, \mathbb{1}_{\{\widetilde{x}_j = a_K\}}\right)^\top \in \{0, 1\}^K. \tag{7.28}$$

An explicit example is given in Table 7.2 which should be compared to Table 5.1.

Continuous Feature Components

The continuous feature components do not need any pre-processing but they can directly enter the FN network which will take care of representation learning. However, an efficient use of gradient descent methods typically requires that all feature components live on a similar scale and that they are roughly uniformly spread across their domains. This makes gradient descent steps more efficient in exploiting the relevant directions.

One possibility is to use the MinMaxScaler. Let x_j^- and x_j^+ be the minimal and maximal possible feature values of the continuous feature component x_j, i.e., $x_j \in [x_j^-, x_j^+]$. We transform this continuous feature component to unit scale for all data $1 \le i \le n$ by

$$x_{i,j} \mapsto x_{i,j}^{\mathrm{MM}} = 2\,\frac{x_{i,j} - x_j^-}{x_j^+ - x_j^-} - 1 \in [-1, 1]. \tag{7.29}$$

The resulting feature values $(x_{i,j}^{\mathrm{MM}})_{1 \le i \le n}$ should roughly be uniformly spread across the interval $[-1, 1]$. If this is not the case, for instance, because we have outliers in the feature values, we may first transform them non-linearly to get more uniformly spread values. For example, we consider the `Density` of the car frequency example on the log scale.

An alternative to the MinMaxScaler is to consider normalization with the empirical mean $\bar{x}_j$ and the empirical standard deviation $\hat{\sigma}_j$ over all data $x_{i,j}$. That is,

$$x_{i,j} \mapsto x_{i,j}^{\mathrm{sd}} = \frac{x_{i,j} - \bar{x}_j}{\hat{\sigma}_j}. \tag{7.30}$$

It depends on the application whether the MinMaxScaler or normalization with the empirical mean and standard deviation works better. Important in applications is that we use exactly the same values for the normalization of training data $\mathcal{U}$, validation data $\mathcal{V}$ and test data $\mathcal{T}$, to make the same network applicable to all these data sets. For notational convenience we will drop the upper index in $x_{i,j}^{\mathrm{MM}}$ or $x_{i,j}^{\mathrm{sd}}$, respectively, and we throughout assume that all feature components are appropriately pre-processed.

7.3.2 *Lab: Poisson FN Network for Car Insurance Frequencies*

We present a first FN network example applied to the French MTPL claim frequency data studied in Sect. 5.2.4. We assume that the claim counts N_i are independent and Poisson distributed with claim count density (5.26), where we replace the GLM regression function $\boldsymbol{x} \mapsto \exp\langle \boldsymbol{\beta}, \boldsymbol{x} \rangle$ by a FN network regression function

$$\boldsymbol{x} \in \mathcal{X} \mapsto \mu(\boldsymbol{x}) = \exp\langle \boldsymbol{\beta}, \boldsymbol{z}^{(d:1)}(\boldsymbol{x}) \rangle.$$

We use a FN network of depth $d = 3$ having number of neurons $(q_1, q_2, q_3) = (20, 15, 10)$ and using the hyperbolic tangent activation function. We pre-process the categorical variables `VehBrand` and `Region` by one-hot encoding providing input dimensions 11 and 22, respectively. The binary variable `VehGas` is encoded as 0–1. Because of scarcity of data we right-censor the continuous variables `VehAge` at 20, `DrivAge` at 90 and `BonusMalus` at 150, and we transform `Density` to the log scale. We then apply to each of these (modified) continuous variables `Area`, `VehPower`, `VehAge`, `DrivAge`, `BonusMalus` and log(`Density`) a MinMaxScaler. This provides us with an input dimension $q_0 = 11 + 22 + 1 + 6 = 40$. The resulting FN network is illustrated in Fig. 7.2, with the one-hot encoded variables `VehBrand` in orange color and `Region` in magenta color. It has a network parameter $\boldsymbol{\vartheta} \in \mathbb{R}^r$ of dimension $r = 1'306$.

This network is implemented in R using the library `keras` [77]. The code is provided in Listing 7.1 and the resulting network architecture is summarized in Listing 7.2. This network is now fitted to the data. We use a batch size of 10'000, we use the `nadam` version of SGD, we use 10% of the learning data $\mathcal{L}$ as validation data $\mathcal{V}$ and the remaining 90% as training data $\mathcal{U}$. We then run the corresponding

Listing 7.1 FN network of depth $d = 3$ using the R library `keras` [77]

```
library(keras)
#
Design  = layer_input(shape = c(40), dtype = 'float32', name = 'Design')
Vol     = layer_input(shape = c(1), dtype = 'float32', name = 'Vol')
#
Network = Design %>%
       layer_dense(units=20, activation='tanh', name='FNLayer1') %>%
       layer_dense(units=15, activation='tanh', name='FNLayer2') %>%
       layer_dense(units=10, activation='tanh', name='FNLayer3') %>%
       layer_dense(units=1, activation='exponential', name='Network',
             weights=list(array(0, dim=c(10,1)), array(log(lambda0), dim=c(1))))
#
Response = list(Network, Vol) %>% layer_multiply(name='Multiply')
#
model = keras_model(inputs = c(Design, Vol), outputs = c(Response))
#
summary(model)
```

Listing 7.2 FN network illustrated in Fig. 7.2

```
Layer (type)             Output Shape    Param # Connected to
=================================================================
Design (InputLayer)      (None, 40)      0
_________________________________________________________________
FNLayer1 (Dense)         (None, 20)      820     Design[0][0]
_________________________________________________________________
FNLayer2 (Dense)         (None, 15)      315     FNLayer1[0][0]
_________________________________________________________________
FNLayer3 (Dense)         (None, 10)      160     FNLayer2[0][0]
_________________________________________________________________
Network (Dense)          (None, 1)       11      FNLayer3[0][0]
_________________________________________________________________
Vol (InputLayer)         (None, 1)       0
_________________________________________________________________
Multiply (Multiply)      (None, 1)       0       Network[0][0]
                                                 Vol[0][0]
=================================================================
Total params: 1,306
Trainable params: 1,306
Non-trainable params: 0
```

Listing 7.3 Fitting a FN network using the R library `keras` [77]

```
path0 <- "path_for_callback"
CBs   <- callback_model_checkpoint(path0, monitor = "val_loss", verbose = 0,
                                   save_best_only = TRUE, save_weights_only = TRUE)
#
model %>% compile(loss = 'poisson', optimizer = 'nadam')
fit <- model %>% fit(list(Xlearn, Vlearn),  Ylearn, validation_split=0.1,
                        batch_size=10000, epochs=1000, verbose=0, callbacks=CBs)
#
load_model_weights_hdf5(model, path0)
```

Table 7.3 Run times, number of parameters, in-sample and out-of-sample deviance losses (units are in 10^{-2}) and in-sample average frequency of the Poisson null model, model Poisson GLM3 of Table 5.5 and the FN network model (with one-hot encoding of the categorical variables)

	Run time	# param.	In-sample loss on $\mathcal{L}$	Out-of-sample loss on $\mathcal{T}$	Aver. freq.
Poisson null	–	1	25.213	25.445	7.36%
Poisson GLM3	15 s	50	24.084	24.102	7.36%
One-hot FN $(q_1, q_2, q_3) = (20, 15, 10)$	51 s	1'306	23.757	23.885	6.96%

SGD algorithm and we retrieve the network with the lowest validation loss using a `callback`. This is illustrated in Listing 7.3. The fitting performance on the training and validation data is illustrated in Fig. 7.8, and we retrieve the network calibration after the 52th epoch because it has the lowest validation loss. The results are presented in Table 7.3.

From the results of Table 7.3 we conclude that the FN network outperforms model Poisson GLM3 (out-of-sample) since it has a (clearly) lower out-of-sample deviance loss on the test data $\mathcal{T}$. This may indicate that there is an interaction between the feature components that has not been captured in the GLM. The run time of 51s corresponds to the run time until the minimal validation loss is reached, of course, in practice we need to continue beyond this minimal validation loss to ensure that we have really found the minimum. Finally, and importantly, we observe that this early stopped FN network calibration does not meet the balance property because the resulting average frequency of this fitted model of 6.96% is below the empirical frequency of 7.36%. This is a major deficiency of this FN network fitting approach, and this is going to be discussed further in Sect. 7.4.2, below.

We can perform a detailed analysis about different batch sizes, variants of SGD methods, run times, etc. We briefly summarize our findings; this summary is also based on the findings in Ferrario et al. [127]. We have fitted this model on batches of sizes 2'000, 5'000, 10'000 and 20'000, and it seems that a batch size around 5'000 has the best performance, both concerning out-of-sample performance and run time to reach the minimal validation loss. Comparing the different optimizers `rmsprop`, `adam` and `nadam`, a clear preference can be given to `nadam`: the resulting prediction accuracy is similar in all three optimizers (they all reach the green area in Fig. 7.5), but `nadam` reaches this optimal point in half of the time compared to `rmsprop` and `adam`.

We conclude by highlighting that different initial points $\boldsymbol{\vartheta}^{(0)}$ of the SGD algorithm will give different network calibrations, and differences can be considerable. This is discussed in Sect. 7.4.4, below. Moreover, we could explore different network architectures, more simple ones, more complex ones, different activation functions, etc. The results of these different architectures will not be essentially different from our results, as long as the networks are above a minimal complexity bound. This closes our first example on FN networks and this example is the benchmark for refined versions that are presented in the subsequent sections.

7.4 Special Features in Networks

7.4.1 Special Purpose Layers

So far, our networks consist of stacked FN layers, and information is passed in a directed acyclic feed-forward path from one to the next FN layer. In this section we discuss special purpose layers that perform a specific task in a FN network. These include *embedding layers*, *drop-out layers* and *normalization layers*. These modules should be seen as add-ons to the FN layers. Besides these add-ons, there are also *recurrent layers* and *convolutional layers*. These two types of layers are going to be discussed in own chapters, below, because their importance goes beyond just being add-ons to the FN layers.

Embedding Layers for Categorical Feature Components

The categorical feature components have been treated either by dummy coding or by one-hot encoding, and this has resulted in numerous network parameters in the first FN layer, see Fig. 7.2. Natural language processing (NLP) treats categorical feature components differently, namely, it *embeds* categorical feature components (or words in NLP) into a Euclidean space $\mathbb{R}^b$ of a *small* dimension b. This small dimension b is a hyper-parameter that has to be selected by the modeler, and which, typically, is selected much smaller than the total number of levels of the categorical feature. This embedding technique is quite common in NLP, see Bengio et al. [27–29], but it goes beyond NLP applications, see Guo–Berkhahn [176], and it has been introduced to the actuarial community by Richman [312, 313] and the tutorial of Schelldorfer–Wüthrich [329].

We assume the same set-up as in dummy coding (5.21) and in one-hot encoding (7.28), namely, that we have a raw categorical feature component $\widetilde{x}_j$ taking K different levels $\{a_1, \ldots, a_K\}$. In one-hot encoding these K levels are mapped to the K unit vectors of the Euclidean space $\mathbb{R}^K$, and consequently all levels have the same mutual Euclidean distance. This does not seem to be the best way of comparing the different levels because in our regression analysis we would like to identify the levels that are more similar w.r.t. the regression task and, thus, these should cluster. For an *embedding layer* one chooses a Euclidean space $\mathbb{R}^b$ of a dimension $b < K$, typically being (much) smaller than K. One then considers the *embedding map*

$$\boldsymbol{e} : \{a_1, \ldots, a_K\} \to \mathbb{R}^b, \qquad a_k \mapsto \boldsymbol{e}(a_k) \stackrel{\text{def.}}{=} \boldsymbol{e}^{(k)}. \tag{7.31}$$

That is, every level a_k receives a vector representation $\boldsymbol{e}^{(k)} \in \mathbb{R}^b$ which is lower dimensional than its one-hot encoding counterpart in $\mathbb{R}^K$. Proximity of the representations $\boldsymbol{e}^{(k)}$ and $\boldsymbol{e}^{(k')}$ in $\mathbb{R}^b$, i.e., of two levels a_k and $a_{k'}$, should be related to similarity w.r.t. the regression task at hand. Such an embedding involves K

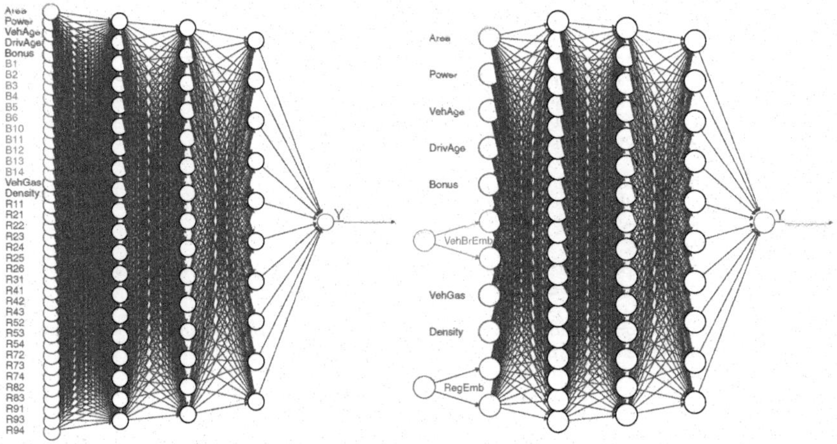

Fig. 7.9 (lhs) One-hot encoding with $q_0 = 40$, and (rhs) embedding layers for `VehBrand` and `Region` with embedding dimension $b = 2$ and $q_0 = 11$; the remaining network architecture is identical with $(q_1, q_2, q_3) = (20, 15, 10)$ for depth $d = 3$

vectors $\boldsymbol{e}^{(k)} \in \mathbb{R}^b$ of dimension b, thus, it involves Kb parameters, called *embedding weights*.

In network modeling, these embedding weights $\boldsymbol{e}^{(1)}, \ldots, \boldsymbol{e}^{(K)}$ can also be learned during gradient descent training. Basically, it just means that for the categorical variables we add an additional embedding layer before the first FN layer $z^{(1)}$, i.e., we increase the depth of the network by 1 for the categorical feature components (by a layer that is not fully connected). This is illustrated in Fig. 7.9 (rhs) for the French MTPL insurance example of Sect. 7.3.2. The graph on the left-hand side shows the network if we apply one-hot encoding to the categorical variables `VehBrand` and `Region`; this results in a network parameter of dimension $r = 1'306$. The graph on the right-hand side first embeds `VehBrand` and `Region` into two 2-dimensional spaces, illustrated by the orange and magenta circles. These embeddings are concatenated with the remaining feature components, which then provides a new dimension $q_0 = 7 + 2 + 2 = 11$ in that example. This results in a network parameter of dimension $r = 726 + 22 + 44 = 792$, where $22 + 44 = 66$ stands for the 2-dimensional embedding weights of the 11 `VehBrands` and the 22 French `Regions`, see Listing 7.5.

Example 7.10 (Embedding Layers for Categorical Features) We revisit the example of Sect. 7.3.2, but we replace one-hot encoding of the categorical variables by embedding layers of dimension $b = 2$. The corresponding R code is given in Listing 7.4 and the resulting model is illustrated in Listing 7.5 and Fig. 7.9 (rhs).

Apart from replacing one-hot encoding by embedding layers, we use exactly the same FN network architecture as in Sect. 7.3.2 and we apply the same fitting strategy in terms of batch sizes, optimizer and early stopping strategy. The results are presented in Table 7.4.

Listing 7.4 FN network of depth $d = 3$ using embedding layers

```
Design   = layer_input(shape = c(7), dtype = 'float32', name = 'Design')
VehBrand = layer_input(shape = c(1), dtype = 'int32', name = 'VehBrand')
Region   = layer_input(shape = c(1), dtype = 'int32', name = 'Region')
Vol      = layer_input(shape = c(1), dtype = 'float32', name = 'Vol')
#
BrandEmb = VehBrand %>%
 layer_embedding(input_dim=11,output_dim=2,input_length=1,name='BrandEmb') %>%
 layer_flatten(name='Brand_flat')
RegionEmb = Region %>%
 layer_embedding(input_dim=22,output_dim=2,input_length=1,name='RegionEmb') %>%
 layer_flatten(name='Region_flat')
#
Network = list(Design,BrandEmb,RegionEmb) %>% layer_concatenate(name='concate') %>%
 layer_dense(units=20, activation='tanh', name='FNLayer1') %>%
 layer_dense(units=15, activation='tanh', name='FNLayer2') %>%
 layer_dense(units=10, activation='tanh', name='FNLayer3') %>%
 layer_dense(units=1, activation='exponential', name='Network',
           weights=list(array(0, dim=c(10,1)), array(log(lambda0), dim=c(1))))
#
Response = list(Network, Vol) %>% layer_multiply(name='Multiply')
#
model = keras_model(inputs = c(Design, VehBrand, Region, Vol),
                    outputs = c(Response))
```

Table 7.4 Run times, number of parameters, in-sample and out-of-sample deviance losses (units are in 10^{-2}) and in-sample average frequency of the Poisson null model, model Poisson GLM3 of Table 5.5 and the FN network models (with one-hot encoding and embedding layers of dimension $b = 2$, respectively)

	Run time	# param.	In-sample loss on $\mathcal{L}$	Out-of-sample loss on $\mathcal{T}$	Aver. freq.
Poisson null	–	1	25.213	25.445	7.36%
Poisson GLM3	15 s	50	24.084	24.102	7.36%
One-hot FN $(q_1, q_2, q_3) = (20, 15, 10)$	51 s	1'306	23.757	23.885	6.96%
Embed FN $(q_1, q_2, q_3) = (20, 15, 10)$	120 s	792	23.694	23.820	7.24%

A first remark is that the model calibration takes longer using embedding layers compared to one-hot encoding. The main reason for this is that having an embedding layer increases the depth of the network by one layer, as can be seen from Fig. 7.9. Therefore, the back-propagation takes more time, and the convergence is slower requiring more gradient descent steps. We have less over-fitting as can be seen from Fig. 7.10. The final fitted model has a slightly better out-of-sample performance compared to the one-hot encoding one. However, this slight improvement in the performance should not be overstated because, as explained in Remarks 7.9, there are a couple of elements of randomness involved in SGD fitting, and choosing a different seed may change the results. We remark that the balance property is not fulfilled because the average frequency of the fitted model does not meet the empirical frequency, see the last column of Table 7.4; we come back to this in Sect. 7.4.2, below.

Listing 7.5 Summary of FN network of Fig. 7.9 (rhs) using embedding layers of dimension $b = 2$

```
Layer (type)                 Output Shape       Param #   Connected to
==============================================================================
VehBrand (InputLayer)         (None, 1)          0
______________________________________________________________________________
Region (InputLayer)           (None, 1)          0
______________________________________________________________________________
BrandEmb (Embedding)          (None, 1, 2)       22        VehBrand[0][0]
______________________________________________________________________________
RegionEmb (Embedding)         (None, 1, 2)       44        Region[0][0]
______________________________________________________________________________
Design (InputLayer)           (None, 7)          0
______________________________________________________________________________
Brand_flat (Flatten)          (None, 2)          0         BrandEmb[0][0]
______________________________________________________________________________
Region_flat (Flatten)         (None, 2)          0         RegionEmb[0][0]
______________________________________________________________________________
concate (Concatenate)         (None, 11)         0         Design[0][0]
                                                           Brand_flat[0][0]
                                                           Region_flat[0][0]
______________________________________________________________________________
FNLayer1 (Dense)              (None, 20)         240       concate[0][0]
______________________________________________________________________________
FNLayer2 (Dense)              (None, 15)         315       FNLayer1[0][0]
______________________________________________________________________________
FNLayer3 (Dense)              (None, 10)         160       FNLayer2[0][0]
______________________________________________________________________________
Network (Dense)               (None, 1)          11        FNLayer3[0][0]
______________________________________________________________________________
Vol (InputLayer)              (None, 1)          0
______________________________________________________________________________
Multiply (Multiply)           (None, 1)          0         Network[0][0]
                                                           Vol[0][0]
==============================================================================
Total params: 792
Trainable params: 792
Non-trainable params: 0
```

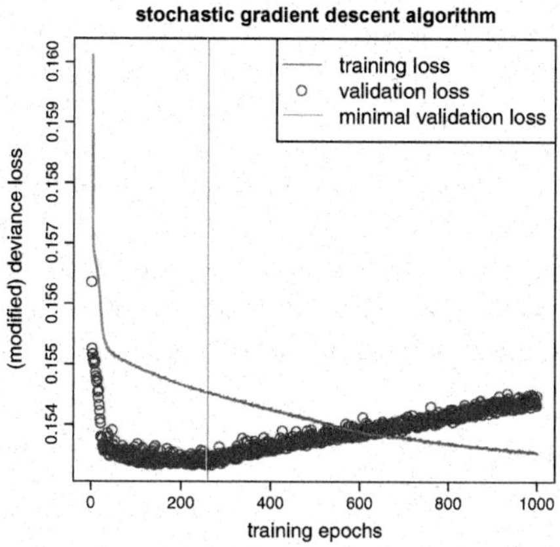

Fig. 7.10 Training loss $\mathfrak{D}(\mathcal{U}, \boldsymbol{\vartheta}^{(t)})$ vs. validation loss $\mathfrak{D}(\mathcal{V}, \boldsymbol{\vartheta}^{(t)})$ over different iterations $t \geq 0$ of the SGD algorithm in the deep FN network with embedding layers for categorical variables

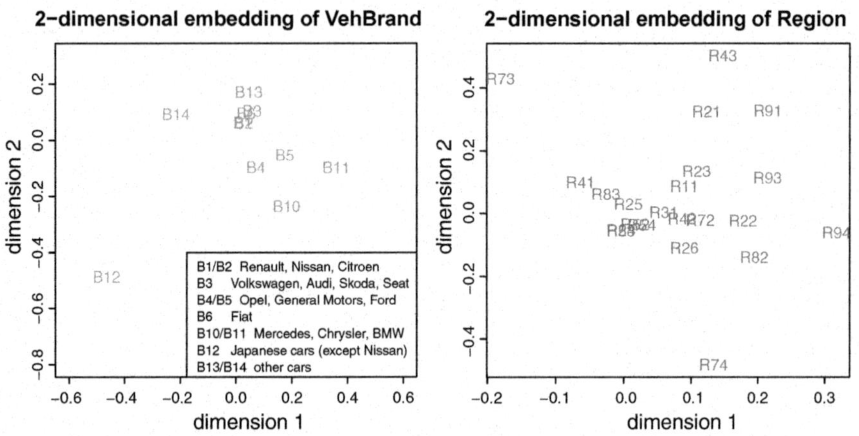

Fig. 7.11 Embedding weights $\boldsymbol{e}^{\texttt{VehBrand}} \in \mathbb{R}^2$ and $\boldsymbol{e}^{\texttt{Region}} \in \mathbb{R}^2$ of the categorical variables `VehBrand` and `Region` for embedding dimension $b = 2$

A major advantage of using embedding layers for the categorical variables is that we receive a continuous representation of nominal variables, where proximity can be interpreted as similarity for the regression task at hand. This is nicely illustrated in Fig. 7.11 which shows the resulting 2-dimensional embeddings $\boldsymbol{e}^{\texttt{VehBrand}} \in \mathbb{R}^2$ and $\boldsymbol{e}^{\texttt{Region}} \in \mathbb{R}^2$ of the categorical variables `VehBrand` and `Region`. The `Region` embedding $\boldsymbol{e}^{\texttt{Region}} \in \mathbb{R}^2$ shows surprising similarities with the French map, for instance, Paris region R11 is adjacent to R23, R22, R21, R26, R24 (which is also the case in the French map), the Isle of Corsica R94 and the South of France R93, R91 and R73 are well separated from other regions, etc. Similar observations can be made for the embedding of `VehBrand`, Japanese cars B12 are far apart from the other cars, cars B1, B2, B3 and B6 (Renault, Nissan, Citroen, Volkswagen, Audi, Skoda, Seat and Fiat) cluster, etc. ■

Drop-Out Layers and Regularization

Above, over-fitting to the learning data has been taken care of by early stopping. In view of Sect. 6.2 one could also use regularization. This can easily be obtained by replacing (7.14), for instance, by the following L^p-regularized counterpart

$$\boldsymbol{\vartheta} \mapsto \frac{2}{n}\sum_{i=1}^{n}\frac{v_i}{\varphi}\Big(Y_i h\left(Y_i\right) - \kappa\left(h\left(Y_i\right)\right) - Y_i h\left(\mu_{\boldsymbol{\vartheta}}(\boldsymbol{x}_i)\right) + \kappa\left(h\left(\mu_{\boldsymbol{\vartheta}}(\boldsymbol{x}_i)\right)\right)\Big) + \lambda \left\|\boldsymbol{\vartheta}_-\right\|_p^p,$$

for some $p \geq 1$, regularization parameter $\lambda > 0$ and where the reduced network parameter $\boldsymbol{\vartheta}_- \in \mathbb{R}^{r-1}$ excludes the intercept parameter β_0 of the output layer, we also refer to (6.4) in the context of GLMs. For grouped penalty terms we refer to (6.21). The difficulty with this approach is the tuning of the regularization parameter(s) λ: run time is one issue, suitable grouping is another issue, and non-uniqueness of the optimal network a further one that can substantially distort the selection of reasonable regularization parameters.

A more popular method to prevent from over-fitting individual neurons in a FN layer to a certain task are so-called *drop-out layers*. A drop-out layer is an additional layer between FN layers that removes at random during gradient descent training neurons from the network, i.e., in each gradient descent step, any of the earmarked neurons is offset independently from the others with a fixed probability $\delta \in (0, 1)$. This random removal will imply that the composite of the remaining neurons needs to be sufficiently well balanced to take over the role of the dropped-out neurons. Therefore, a single neuron cannot be over-trained to a certain task because it needs to be able play several different roles. Drop-out has been introduced by Srivastava et al. [345] and Wager et al. [373].

Listing 7.6 FN network of depth $d = 3$ using a drop-out layer, ridge regularization and a normalization layer

```
Network = list(Design,BrandEmb,RegionEmb) %>%
   layer_concatenate(name='concate') %>%
   layer_dense(units=20, activation='tanh', name='FNLayer1') %>%
   layer_dropout (rate = 0.01) %>%
   layer_dense(units=15, kernel_regularizer=regularizer_l2(0.0001),
                         activation='tanh', name='FNLayer2') %>%
   layer_batch_normalization() %>%
   layer_dense(units=10, activation='tanh', name='FNLayer3') %>%
   layer_dense(units=1, activation='exponential', name='Network',
             weights=list(array(0, dim=c(10,1)), array(log(lambda0), dim=c(1))))
```

Listing 7.6 gives an example, where we add a drop-out layer with a drop-out probability of $\delta = 0.01$ after the first FN layer, and in the second FN layer we apply ridge regularization to the weights $(w^{(2)}_{1,1}, \ldots, w^{(2)}_{q_1,q_2})$, i.e., excluding the intercepts $w^{(2)}_{0,j}$, $1 \le j \le q_2$. Both the drop-out layer and regularization are only used during the gradient descent fitting, and these network features are disabled during the prediction.

Drop-out is closely related to ridge regularization as the following linear Gaussian regression example shows; this consideration is taken from Section 18.6 of Efron–Hastie [117]. Assume we have a linear regression problem with square loss function

$$\mathfrak{D}(\boldsymbol{Y}, \boldsymbol{\beta}) = \frac{1}{2}\sum_{i=1}^{n} \left(Y_i - \langle \boldsymbol{\beta}, \boldsymbol{x}_i\rangle\right)^2 .$$

We assume in this Gaussian case that the observations and the features are standardized, see Sect. 6.2.4. This means that $\sum_{i=1}^{n} Y_i = 0$, $\sum_{i=1}^{n} x_{i,j} = 0$ and

$n^{-1}\sum_{i=1}^{n} x_{i,j}^2 = 1$, for all $1 \le j \le q$. This standardization implies that we can omit the intercept parameter β_0 because its MLE is equal to 0.

We introduce i.i.d. drop-out random variables $I_{i,j}$ for $1 \le i \le n$ and $1 \le j \le q$ with $(1-\delta)I_{i,j}$ being Bernoulli distributed with probability $1-\delta \in (0,1)$. This scaling implies $\mathbb{E}[I_{i,j}] = 1$. Using these Bernoulli random variables we modify the above square loss function to

$$\mathfrak{D}_I(\boldsymbol{Y}, \boldsymbol{\beta}) = \frac{1}{2}\sum_{i=1}^{n}\left(Y_i - \sum_{j=1}^{q} \beta_j I_{i,j} x_{i,j}\right)^2,$$

i.e., every individual component $x_{i,j}$ can drop out independently of the others. Gaussian MLE requires to set the gradient of $\mathfrak{D}_I(\boldsymbol{Y}, \boldsymbol{\beta})$ w.r.t. $\boldsymbol{\beta} \in \mathbb{R}^q$ equal to zero. The average score equation is given by (we average over the drop-out random variables $I_{i,j}$)

$$\begin{aligned}\mathbb{E}_\delta\left[\nabla_{\boldsymbol{\beta}}\mathfrak{D}_I(\boldsymbol{Y}, \boldsymbol{\beta})\middle|\, \boldsymbol{Y}\right] &= -\mathfrak{X}^\top \boldsymbol{Y} + \mathfrak{X}^\top\mathfrak{X}\boldsymbol{\beta} + \frac{\delta}{1-\delta}\operatorname{diag}\left(\sum_{i=1}^{n} x_{i,1}^2, \ldots, \sum_{i=1}^{n} x_{i,q}^2\right)\boldsymbol{\beta} \\ &= -\mathfrak{X}^\top \boldsymbol{Y} + \mathfrak{X}^\top\mathfrak{X}\boldsymbol{\beta} + \frac{\delta n}{1-\delta}\boldsymbol{\beta} \overset{!}{=} 0,\end{aligned}$$

where we have used the normalization of the columns of the design matrix $\mathfrak{X} \in \mathbb{R}^{n\times q}$ (we drop the intercept column). This is ridge regression in the linear Gaussian case with a regularization parameter $\lambda = \delta/(2(1-\delta)) > 0$ for $\delta \in (0,1)$, see (6.9).

Normalization Layers

In (7.29) and (7.30) we have discussed that the continuous feature components should be pre-processed so that all components live on the same scale, otherwise the gradient descent fitting may not be efficient. A similar phenomenon may occur with the learned representations $\boldsymbol{z}^{(m:1)}(\boldsymbol{x}_i)$ in the FN layers $1 \le m \le d$. In particular, this is the case if we choose an unbounded activation function ϕ. For this reason, it can be advantageous to rescale the components $z_j^{(m:1)}(\boldsymbol{x}_i)$, $1 \le j \le q_m$, in a given FN layer back to the same scale. To achieve this, a normalization step (7.30) is applied to every neuron $z_j^{(m:1)}(\boldsymbol{x}_i)$ over the given cases i in the considered (mini-)batch. This involves two more parameters (for the empirical mean and the empirical standard deviation) in each neuron of the corresponding FN layer. Note, however, that all these operations are of a linear nature. Therefore, they do not affect the predictive model (i.e., these operations cancel in the scalar products in (7.6)), but they may improve the performance of the gradient descent algorithm.

The code in Listing 7.6 uses a normalization layer on line 6. In our applications, it has not been necessary to use these normalization layers, as it has not led to better run times in SGD algorithms; note that our networks are not very deep and they use the symmetric and bounded hyperbolic tangent activation function.

7.4.2 *The Balance Property in Neural Networks*

We have seen in Table 7.4 that our FN network outperforms the GLM for claim frequency prediction in terms of a lower out-of-sample loss. We interpret this as follows. Feature engineering has not been done in the most optimal way for the GLM because the FN network finds modeling structure that is not present in the selected GLM. As a consequence, the FN network provides a better generalization to unseen data, i.e., we can better predict new data on a granular level with the FN network. However, having a more precise model on an individual policy level does not necessarily imply that the model also performs better on a global portfolio level. In our example we see that we may have smaller errors on an individual policy level, but these smaller errors do not aggregate to a more precise model in the average portfolio frequency. In our case, we have a misspecification of the average portfolio frequency, see the last column of Table 7.4. This is a major deficiency in insurance pricing because it may result in a misspecification of the overall price level, and this requires a correction. We call this correction *bias regularization*.

Simple Bias Regularization

The straightforward correction is to adjust the intercept parameter $\beta_0 \in \mathbb{R}$ accordingly. That is, compare the empirical mean

$$\bar{\mu} = \frac{\sum_{i=1}^n v_i Y_i}{\sum_{i=1}^n v_i},$$

to the model average of the fitted FN network

$$\widehat{\mu} = \frac{\sum_{i=1}^n v_i \mu_{\widehat{\boldsymbol{\vartheta}}}(\boldsymbol{x}_i)}{\sum_{i=1}^n v_i},$$

where $\widehat{\boldsymbol{\vartheta}} = (\widehat{\boldsymbol{w}}_1^{(1)}, \ldots, \widehat{\boldsymbol{w}}_{q_d}^{(d)}, \widehat{\boldsymbol{\beta}})^\top \in \mathbb{R}^r$ is the learned network parameter from the (early stopped) SGD algorithm. The output of this fitted model reads as

$$\boldsymbol{x}_i \mapsto \mu_{\widehat{\boldsymbol{\vartheta}}}(\boldsymbol{x}_i) = g^{-1}\left\langle \widehat{\boldsymbol{\beta}}, \widehat{\boldsymbol{z}}^{(d:1)}(\boldsymbol{x}_i)\right\rangle = g^{-1}\left(\widehat{\beta}_0 + \sum_{j=1}^{q_d} \widehat{\beta}_j \widehat{z}_j^{(d:1)}(\boldsymbol{x}_i)\right),$$

where the hat in $\widehat{z}^{(d:1)}$ indicates that we use the estimated weights $\widehat{\boldsymbol{w}}_l^{(m)}$, $1 \le l \le q_m$, $1 \le m \le d$, in the FN layers. The balance property can be rectified by replacing $\widehat{\beta}_0$ by the solution $\widehat{\widehat{\beta}}_0$ of the following identity

$$\sum_{i=1}^{n} v_i Y_i \stackrel{!}{=} \sum_{i=1}^{n} v_i g^{-1}\left(\widehat{\widehat{\beta}}_0 + \sum_{j=1}^{q_d} \widehat{\beta}_j \widehat{z}_j^{(d:1)}(\boldsymbol{x}_i)\right).$$

Since g^{-1} is continuous and strictly monotone, there is a unique solution to this requirement supposed that the range of g^{-1} covers the support of the Y_i's. If we work with the log-link $g(\cdot) = \log(\cdot)$, this can easily be solved and we obtain

$$\widehat{\widehat{\beta}}_0 = \widehat{\beta}_0 + \log\left(\frac{\bar{\mu}}{\bar{\widehat{\mu}}}\right).$$

Sophisticated Bias Regularization Under the Canonical Link Choice

If we work with the canonical link $g = h = (\kappa')^{-1}$, we can do better because the MLE of such a GLM automatically provides the balance property, see Corollary 5.7. Choose the SGD learned network parameter $\widehat{\boldsymbol{\vartheta}} = (\widehat{\boldsymbol{w}}_1^{(1)}, \ldots, \widehat{\boldsymbol{w}}_{q_d}^{(d)}, \widehat{\boldsymbol{\beta}})^\top \in \mathbb{R}^r$. Denote by $\widehat{z}^{(d:1)}$ the fitted network architecture that is based on the estimated weights $\widehat{\boldsymbol{w}}_1^{(1)}, \ldots, \widehat{\boldsymbol{w}}_{q_d}^{(d)}$. This allows us to study the learned representations of the raw features $\boldsymbol{x}_1, \ldots, \boldsymbol{x}_n$ in the last FN layer. We denote these learned representations by

$$\widehat{z}_1 = \widehat{z}^{(d:1)}(\boldsymbol{x}_1), \;\ldots,\; \widehat{z}_n = \widehat{z}^{(d:1)}(\boldsymbol{x}_n) \in \{1\} \times \mathbb{R}^{q_d}. \tag{7.32}$$

These learned representations can be used as new features to explain the response $\boldsymbol{Y}$. We define the feature engineered design matrix by

$$\widehat{\mathfrak{X}} = (\widehat{z}_1, \ldots, \widehat{z}_n)^\top \in \mathbb{R}^{n \times (q_d+1)}.$$

Based on this new design matrix $\widehat{\mathfrak{X}}$ we can run a classical GLM receiving a unique MLE $\widehat{\boldsymbol{\beta}}^{\text{MLE}} \in \mathbb{R}^{q_d+1}$ supposed that this design matrix has a full rank $q_d + 1 \le n$, see Proposition 5.1. Since we work with the canonical link, this re-calibrated FN network will automatically satisfy the balance property, and the resulting regression function reads as

$$\boldsymbol{x} \mapsto \widehat{\mu}(\boldsymbol{x}) = h^{-1}\left\langle \widehat{\boldsymbol{\beta}}^{\text{MLE}}, \widehat{z}^{(d:1)}(\boldsymbol{x})\right\rangle. \tag{7.33}$$

This is the proposal of Wüthrich [390]. We give some remarks.

Remarks 7.11

- This additional MLE step for the output parameter $\boldsymbol{\beta} \in \mathbb{R}^{q_d+1}$ may lead to over-fitting. In that case one might choose a lower dimensional last FN layer. Alternatively, one might explore a more early stopping rule in SGD.
- Wüthrich [390] also explores other bias correction methods like regularization using shrinkage. In combination with regression trees one can achieve averages on pre-defined sub-portfolios. We will not further explore these other approaches because they are less robust and more difficult in the applications.

Example 7.12 (Balance Property in Networks) We apply this additional MLE step to the two FN networks of Table 7.4. Note that in these two examples we consider a Poisson model using the canonical link for g, thus, the resulting adjusted network (7.33) will automatically satisfy the balance property, see Corollary 5.7.

Listing 7.7 Balance property adjustment (7.33)

```
1  glm.formula <- function(nn){
2     string <- "yy ~ X1"
3     if (nn>1){for (ll in 2:nn){ string <- paste(string, "+X",ll, sep="")}}
4     string
5      }
6  #
7  zz <- keras_model(inputs=model$input,
8                    outputs=get_layer(model, 'FNLayer3')$output)
9  xx.learn <- data.frame(zz %>% predict(list(Xlearn, Vlearn)))
10 q3 <- ncol(xx.learn)
11 xx.learn$yy <- Ylearn
12 xx.learn$Exposure <- learn$Exposure
13 #
14 glm1 <- glm(as.formula(glm.formula(q3)),
15                         data=xx.learn, offset=log(Exposure), family=poisson())
16
17 #
18 w1 <- get_weights(model)
19 w1[[7]]  <- array(glm1$coefficients[2:(q3+1)], dim=c(q3,1))
20 w1[[8]]  <- array(glm1$coefficients[1], dim=c(1))
21 set_weights(model, w1)
```

In Listing 7.7 we illustrate the necessary code that has to be added to Listings 7.1–7.3. On lines 7–8 of Listing 7.7 we retrieve the learned representations (7.32) which are used as the new features in the Poisson GLM on lines 13–14. The resulting MLE $\widehat{\boldsymbol{\beta}}^{\text{MLE}} \in \mathbb{R}^{q_d+1}$ is imputed to the network parameter $\widehat{\boldsymbol{\vartheta}}$ on lines 17–20. Table 7.5 shows the performance of the resulting bias regularized FN networks.

Firstly, we observe from the last column of Table 7.5 that, indeed, the bias regularization step (7.33) provides the balance property. In general, in-sample losses (have to) decrease because $\widehat{\boldsymbol{\beta}}^{\text{MLE}}$ is (in-sample) more optimal than the early stopped SGD solution $\widehat{\boldsymbol{\beta}}$. Out-of-sample this leads to a small improvement in the one-

Table 7.5 Run times, number of parameters, in-sample and out-of-sample deviance losses (units are in 10^{-2}) and in-sample average frequency of the Poisson null model, model Poisson GLM3 of Table 5.5 and the FN network models (with one-hot encoding and embedding layers of dimension $b = 2$, respectively), and their bias regularized counterparts

	Run time	# param.	In-sample loss on $\mathcal{L}$	Out-of-sample loss on $\mathcal{T}$	Aver. freq.
Poisson null	–	1	25.213	25.445	7.36%
Poisson GLM3	15 s	50	24.084	24.102	7.36%
One-hot FN $(q_1, q_2, q_3) = (20, 15, 10)$	51 s	1'306	23.757	23.885	6.96%
Embed FN $(q_1, q_2, q_3) = (20, 15, 10)$	120 s	792	23.694	23.820	7.24%
One-hot FN bias regularized	+4 s	1'306	23.742	23.878	7.36%
Embed FN bias regularized	+4 s	792	23.690	23.824	7.36%

hot encoded variant and a small worsening in the embedding variant, i.e., the latter slightly over-fits in this additional MLE step. However, these differences are comparably small so that we do not further worry about the over-fitting, here. This closes this example. ■

Auto-Calibration for Bias Regularization

We present another approach of correcting for the potential failure of the balance property. This method does not depend on a particular type of regression model, i.e., it can be applied to any regression model. This proposal goes back to Denuit et al. [97], and it is based on the notion of *auto-calibration* introduced by Patton [297] and Krüger–Ziegel [227]. We first describe auto-calibration and its implications.

Definition 7.13 The random variable Z is an auto-calibrated forecast of random variable Y if $\mathbb{E}[Y|Z] = Z$, a.s.

If the response Y is described by the features $\boldsymbol{X} = \boldsymbol{x}$, we consider the conditional mean of Y, given $\boldsymbol{X}$,

$$\mu(\boldsymbol{X}) = \mathbb{E}\left[Y|\boldsymbol{X}\right].$$

This conditional mean $\mu(\boldsymbol{X})$ is an auto-calibrated forecast for the response Y. Use the tower property and note that $\sigma(\mu(\boldsymbol{X})) \subset \sigma(\boldsymbol{X})$ to receive, a.s.,

$$\mathbb{E}\left[Y|\, \mu(\boldsymbol{X})\right] = \mathbb{E}\left[\mathbb{E}\left[Y|\, \boldsymbol{X}\right]|\, \mu(\boldsymbol{X})\right] = \mathbb{E}\left[\mu(\boldsymbol{X})|\, \mu(\boldsymbol{X})\right] = \mu(\boldsymbol{X}).$$

For the further understanding of auto-calibration and forecast dominance, we introduce the concept of *convex order*; forecast dominance has been introduced in Definition 4.20.

Definition 7.14 (Convex Order) A random variable Z_1 is bigger in convex order than a random variable Z_2, write $Z_1 \succeq_{cx} Z_2$, if $\mathbb{E}[\Psi(Z_1)] \geq \mathbb{E}[\Psi(Z_2)]$, for all convex functions Ψ for which the expectations exist.

By Strassen's theorem [346], $Z_1 \succeq_{cx} Z_2$ if and only if there exist random variables Z_1' and Z_2' with $Z_1 \overset{(d)}{=} Z_1'$ and $Z_2 \overset{(d)}{=} Z_2'$ and $\mathbb{E}[Z_1'|Z_2'] = Z_2'$, a.s. In particular, the convex order $Z_1 \succeq_{cx} Z_2$ implies that $\mathrm{Var}(Z_1) \geq \mathrm{Var}(Z_2)$ and $\mathbb{E}[Z_1] = \mathbb{E}[Z_2]$. The latter follows from Strassen's theorem and the tower property, and the former follows from the latter and the convex order by using the explicit choice $\Psi(x) = x^2$. Thus, the random variable Z_1 is more volatile than Z_2, both having the same mean. The following theorem shows that this additional volatility is a favorable property in terms of forecast dominance under auto-calibration.

Theorem 7.15 (Krüger–Ziegel [227, Theorem 3.1], Without Proof) *Assume that $\widehat{\mu}_1$ and $\widehat{\mu}_2$ are auto-calibrated forecasts for the random variable Y. Predictor $\widehat{\mu}_1$ forecast dominates $\widehat{\mu}_2$ if and only if $\widehat{\mu}_1 \succeq_{cx} \widehat{\mu}_2$.*

Recall that forecast dominance of $\widehat{\mu}_1$ over $\widehat{\mu}_2$ was defined as follows, see Definition 4.20,

$$\mathbb{E}\left[D_\psi(Y, \widehat{\mu}_1)\right] \leq \mathbb{E}\left[D_\psi(Y, \widehat{\mu}_2)\right],$$

for all Bregman divergences D_ψ. Strassen's theorem tells us that $\widehat{\mu}_1$ is more volatile than $\widehat{\mu}_2$ (both being auto-calibrated and unbiased for $\mathbb{E}[Y]$) and this additional volatility implies that the former auto-calibrated predictor can better follow Y. This provides the superior forecast dominance of $\widehat{\mu}_1$ over $\widehat{\mu}_2$. This relation is most easily understood by the following example. Consider $(Y, \boldsymbol{X})$ as above. Assume that the feature $\widetilde{\boldsymbol{X}}$ is a sub-variable of the feature $\boldsymbol{X}$ by dropping some of the components of $\boldsymbol{X}$. Naturally, we have $\sigma(\widetilde{\boldsymbol{X}}) \subset \sigma(\boldsymbol{X})$, and both sets of information provide auto-calibrated forecasts

$$\mu(\boldsymbol{X}) = \mathbb{E}\left[Y|\boldsymbol{X}\right] \qquad \text{and} \qquad \mu(\widetilde{\boldsymbol{X}}) = \mathbb{E}\left[Y\left|\widetilde{\boldsymbol{X}}\right.\right].$$

The tower property and Jensen's inequality give for any convex function Ψ (subject to existence)

$$\begin{aligned}\mathbb{E}\left[\Psi(\mu(\boldsymbol{X}))\right] &= \mathbb{E}\left[\Psi\left(\mathbb{E}\left[Y|\boldsymbol{X}\right]\right)\right] = \mathbb{E}\left[\mathbb{E}\left[\Psi\left(\mathbb{E}\left[Y|\boldsymbol{X}\right]\right)\left|\widetilde{\boldsymbol{X}}\right.\right]\right] \\ &\geq \mathbb{E}\left[\Psi\left(\mathbb{E}\left[\mathbb{E}\left[Y|\boldsymbol{X}\right]\left|\widetilde{\boldsymbol{X}}\right.\right]\right)\right] = \mathbb{E}\left[\Psi\left(\mathbb{E}\left[Y\left|\widetilde{\boldsymbol{X}}\right.\right]\right)\right] = \mathbb{E}\left[\Psi\left(\mu(\widetilde{\boldsymbol{X}})\right)\right].\end{aligned}$$

Thus, we have $\mu(\boldsymbol{X}) \succeq_{cx} \mu(\widetilde{\boldsymbol{X}})$ which implies forecast dominance of $\mu(\boldsymbol{X})$ over $\mu(\widetilde{\boldsymbol{X}})$. This makes perfect sense in view of $\sigma(\widetilde{\boldsymbol{X}}) \subset \sigma(\boldsymbol{X})$. Basically, this describes the construction of a $\mathbb{F}$-martingale using an integrable random variable Y and a filtration $\mathbb{F}$ on the underlying probability space $(\Omega, \mathcal{A}, \mathbb{P})$. This martingale sequence provides forecast dominance with increasing information sets described by the filtration $\mathbb{F}$.

We now turn our attention to the balance property and the unbiasedness of predictors, this follows Denuit et al. [97]. Assume we have any predictor $\widehat{\mu}(\boldsymbol{x})$ of Y, for instance, this can be any FN network predictor $\mu_{\widehat{\vartheta}}(\boldsymbol{x})$ coming from an early stopped SGD algorithm. We define its *balance-corrected* version by

$$\widehat{\mu}_{\mathrm{BC}}(\boldsymbol{x}) = \mathbb{E}\left[Y \,|\widehat{\mu}(\boldsymbol{x})\right]. \tag{7.34}$$

Proposition 7.16 (Wüthrich [391, Proposition 4.6], Without Proof) *The balance-corrected predictor $\widehat{\mu}_{BC}(\boldsymbol{X})$ is an auto-calibrated forecast for Y.*

Remarks 7.17 (Expected Deviance Generalization Loss) We return to the decomposition of the expected deviance GL given in Theorem 4.7, but we add the features $\boldsymbol{X} = \boldsymbol{x}$, now. The expected deviance GL of a predictor $\widehat{\mu}(\boldsymbol{X})$ under the unit deviance $\mathfrak{d}$ then reads as

$$\begin{aligned}\mathbb{E}_\theta\left[\mathfrak{d}\left(Y, \widehat{\mu}(\boldsymbol{X})\right)\right] = {} & \mathbb{E}_\theta\left[\mathfrak{d}\left(Y, \mu\right)\right] \\ & + 2\Big(\mu h(\mu) - \kappa(h(\mu)) - \mathbb{E}_\theta\left[Y h\left(\widehat{\mu}(\boldsymbol{X})\right)\right] + \mathbb{E}_\theta\left[\kappa\left(h\left(\widehat{\mu}(\boldsymbol{X})\right)\right)\right]\Big),\end{aligned}$$

where $\mu = \mathbb{E}_\theta[Y]$ is the unconditional mean of Y (averaging also over the feature distribution of $\boldsymbol{X}$). Note that this formula differs from (4.13) because Y and $h(\widehat{\mu}(\boldsymbol{X}))$ are no longer independent if we include the features $\boldsymbol{X}$. The term $\mathbb{E}_\theta[\mathfrak{d}\left(Y, \mu\right)]$ is called the *entropy* which is driven by the stochastic nature of the random variable Y. This is the irreducible risk if no feature information is available.

In statistical modeling one considers different decompositions of the expected deviance GL, we refer to Fissler et al. [129]. Namely, introducing the features $\boldsymbol{X}$ we can reduce the expected deviance GL compared to the unconditional mean μ in terms of forecast dominance. This allows us to decouple as follows for the prediction $\mu(\boldsymbol{X}) = \mathbb{E}_\theta[Y|\boldsymbol{X}]$

$$\begin{aligned}\mathbb{E}_\theta\left[\mathfrak{d}\left(Y, \widehat{\mu}(\boldsymbol{X})\right)\right] &= \mathbb{E}_\theta\left[\mathfrak{d}\left(Y, \mu\right)\right] - \Big(\mathbb{E}_\theta\left[\mathfrak{d}\left(Y, \mu\right)\right] - \mathbb{E}_\theta\left[\mathfrak{d}\left(Y, \mu(\boldsymbol{X})\right)\right]\Big) \\ &\quad + \Big(\mathbb{E}_\theta\left[\mathfrak{d}\left(Y, \widehat{\mu}(\boldsymbol{X})\right)\right] - \mathbb{E}_\theta\left[\mathfrak{d}\left(Y, \mu(\boldsymbol{X})\right)\right]\Big).\end{aligned}$$

This expresses the expected deviance GL of the predictor $\widehat{\mu}(\boldsymbol{X})$ as the entropy (first term), the *conditional resolution* (second term) and the *conditional calibration* (third term). The conditional resolution describes the information gain in terms of forecast dominance knowing the feature $\boldsymbol{X}$, and the conditional calibration describes how well we estimate $\mu(\boldsymbol{X})$. The conditional resolution is positive because $\mu(\boldsymbol{X}) \succeq_{\rm cx} \mu$ and the unit deviance $\mathfrak{d}(Y, \cdot)$ is a convex function, see Lemma 2.22. The conditional calibration is also positive, this can be seen by considering the deviance GL, conditional on $\boldsymbol{X}$.

We can reformulate this expected deviance GL in terms of the auto-calibration property

$$\begin{aligned} \mathbb{E}_\theta\left[\mathfrak{d}\left(Y, \widehat{\mu}(\boldsymbol{X})\right)\right] &= \mathbb{E}_\theta\left[\mathfrak{d}\left(Y, \mu\right)\right] - \Big(\mathbb{E}_\theta\left[\mathfrak{d}\left(Y, \mu\right)\right] - \mathbb{E}_\theta\left[\mathfrak{d}\left(Y, \widehat{\mu}_{\rm BC}(\boldsymbol{X})\right)\right]\Big) \\ &\quad + \Big(\mathbb{E}_\theta\left[\mathfrak{d}\left(Y, \widehat{\mu}(\boldsymbol{X})\right)\right] - \mathbb{E}_\theta\left[\mathfrak{d}\left(Y, \widehat{\mu}_{\rm BC}(\boldsymbol{X})\right)\right]\Big). \end{aligned}$$

The first term is the entropy, the second term is called the *auto-resolution* and the third term describes the *auto-calibration*. If we have an auto-calibrated forecast $\widehat{\mu}(\boldsymbol{X})$ then the last term vanishes because it is equal to its balance-corrected version $\widehat{\mu}_{\rm BC}(\boldsymbol{X})$. Again these two latter terms are positive, for the auto-calibration this can be seen by considering the deviance GL, conditioned on $\widehat{\mu}(\boldsymbol{X})$.

To rectify the balance property we directly focus on (7.34), and we *estimate* this conditional expectation. That is, the balance correction can be achieved by an additional regression step directly estimating the balance-corrected version $\widehat{\mu}_{\rm BC}(\boldsymbol{x})$ in (7.34). This additional regression step differs from (7.33) as it does not use the learned representations $\widehat{\boldsymbol{z}}^{(d:1)}(\boldsymbol{x})$ in the last FN layer (7.32), but it uses the learned representations in the output layer. That is, consider the learned features

$$\widehat{\boldsymbol{z}}_1^\star = (1, \mu_{\widehat{\vartheta}}(\boldsymbol{x}_1))^\top, \; \ldots, \; \widehat{\boldsymbol{z}}_n^\star = (1, \mu_{\widehat{\vartheta}}(\boldsymbol{x}_n))^\top \; \in \{1\} \times \mathbb{R},$$

and perform an additional linear regression step for the response $\boldsymbol{Y}$ using the design matrix

$$\widehat{\mathfrak{X}} = \left(\widehat{\boldsymbol{z}}_1^\star, \ldots, \widehat{\boldsymbol{z}}_n^\star\right)^\top \; \in \mathbb{R}^{n\times 2}.$$

This additional linear regression step gives us an estimate

$$\widehat{\boldsymbol{\beta}} = \left(\widehat{\mathfrak{X}}^\top V \widehat{\mathfrak{X}}\right)^{-1} \widehat{\mathfrak{X}}^\top V \boldsymbol{Y} \; \in \mathbb{R}^2, \tag{7.35}$$

with diagonal weight matrix $V = \mathrm{diag}(v_i)_{1\le i\le n}$. The balance property is then restored by estimating the balance-corrected means $\widehat{\mu}_{\rm BC}(\boldsymbol{x}_i)$ by

$$\widehat{\widehat{\mu}}_{\rm BC}(\boldsymbol{x}_i) = \widehat{\beta}_0 + \widehat{\beta}_1 \mu_{\widehat{\vartheta}}(\boldsymbol{x}_i), \tag{7.36}$$

for $1 \le i \le n$. Note that this can be done for any regression model since we do not rely on the network architecture in this step.

Remarks 7.18

- Balance correction (7.36) may lead to some conflict in range if the dual (mean) parameter space $\mathcal{M}$ is (one-sided) bounded. Moreover, it does not consider the deviance loss of the response $\boldsymbol{Y}$, but it rather underlies a Gaussian model by using the weighted square loss function for finding (the Gaussian MLE) $\widehat{\boldsymbol{\beta}} \in \mathbb{R}^2$. Alternatively, we could consider the canonical link h that belongs to the chosen EDF. This then allows us to study the regression problem on the canonical scale by setting for the learned representations

$$\widehat{z}_1^\theta = \left(1, h(\mu_{\widehat{\vartheta}}(\boldsymbol{x}_1))\right)^\top, \ \ldots, \ \widehat{z}_n^\theta = \left(1, h(\mu_{\widehat{\vartheta}}(\boldsymbol{x}_n))\right)^\top \in \{1\} \times \boldsymbol{\Theta}. \tag{7.37}$$

The latter motivates the consideration of a GLM under the chosen EDF

$$\boldsymbol{x}_i \mapsto h\left(\widehat{\mu}_{\mathrm{BC}}(\boldsymbol{x}_i)\right) = \langle \boldsymbol{\beta}, \widehat{z}_i^\theta \rangle = \beta_0 + \beta_1 h(\mu_{\widehat{\vartheta}}(\boldsymbol{x}_i)), \tag{7.38}$$

for regression parameter $\boldsymbol{\beta} \in \mathbb{R}^2$. The choice of the canonical link and the inclusion of an intercept will provide the balance property when estimating $\boldsymbol{\beta}$ with MLE, see Corollary 5.7. If the mean estimates $\mu_{\widehat{\vartheta}}(\boldsymbol{x}_i)$ involve the canonical link h, (7.38) reads as

$$\boldsymbol{x}_i \mapsto h\left(\widehat{\mu}_{\mathrm{BC}}(\boldsymbol{x}_i)\right) = \langle \boldsymbol{\beta}, \widehat{z}_i^\theta \rangle = \beta_0 + \beta_1 \left\langle \widehat{\boldsymbol{\beta}}, \widehat{\boldsymbol{z}}^{(d:1)}(\boldsymbol{x}_i) \right\rangle,$$

the latter scalar product is the output activation received from the FN network. From this we see that the estimated balance-corrected calibration on the canonical scale will give us a non-optimal (in-sample) estimation step compared to (7.33), if we work with the canonical link h.
- Denuit et al. [97] give a proposal to break down the global balance to a local version using a suitable kernel function, this will be further discussed in the next Example 7.19.

Example 7.19 (Auto-calibration in Networks) We apply this additional auto-calibration step (7.34) to the FN network with embedding layers that does not satisfy the balance property, i.e., having an average frequency of 7.24% < 7.36%, see Tables 7.4 and 7.5. We start by analyzing the auto-calibration property (7.34) of this network predictor $v\mu_{\widehat{\vartheta}}(\boldsymbol{x})$ by studying an empirical version of

$$z \mapsto v\widehat{\mu}_{\mathrm{BC}}(\boldsymbol{x}) = \mathbb{E}\left[vY \,\middle|\, v\mu_{\widehat{\vartheta}}(\boldsymbol{x}) = z\right]. \tag{7.39}$$

This empirical version is obtained from the R library `locfit` [254] that allows us to consider a local polynomial regression fit of degree `deg=2`, and we use a nearest neighbor fraction of `alpha=0.05`, the code is provided in Listing 7.8. We use the exposure v scaled version in (7.39) since the balance property should hold on that scale, see Corollary 5.7. The claim counts are given by $N = vY$, and the exposure

v is integrated as an offset into the FN network regression function, see line 20 of Listing 7.4.

Listing 7.8 Empirical auto-calibration using the R library `locfit` [254]

```
z      <- learn$pred
mu.BC <- predict(locfit(learn$N ~ learn$pred, alpha=0.05, deg=2), newdata=z)
```

Figure 7.12 (lhs) shows the empirical auto-calibration of (7.39) using the R code of Listing 7.8. If the auto-calibration would hold exactly, then the black dots should lie on the red diagonal line. We observe a very good match, which indicates that the auto-calibration property holds quite accurately for our network predictor $(v, \boldsymbol{x}) \mapsto v\mu_{\widehat{\vartheta}}(\boldsymbol{x})$. For very small expectations $\mathbb{E}_{\theta(\boldsymbol{x})}[N]$ we slightly underestimate, and for bigger expectations we slightly overestimate. The blue line shows the empirical density of the predictors $v_i\mu_{\widehat{\vartheta}}(\boldsymbol{x}_i)$, $1 \leq i \leq n$, highlighting heavy-tailedness and that the underestimation in the right tail will not substantially contribute to the balance property as these are only very few insurance policies.

We explore the Gaussian balance correction (7.35) considering a linear regression model with weighted square loss function. We receive the estimate $\widehat{\boldsymbol{\beta}} = (9 \cdot 10^{-4}, 1.005)^\top$, thus, $\mu_{\widehat{\vartheta}}(\boldsymbol{x})$ only gets very gently distorted, see (7.36). The results of this balance-corrected version $\widehat{\widehat{\mu}}_{\text{BC}}(\boldsymbol{x})$ are given on line 'embed FN Gauss balance-corrected' in Table 7.6. We observe that this approach is rather competitive leading to a slightly better model (out-of-sample). Figure 7.12 (rhs) shows the resulting (empirical) auto-calibration plot which is still not fully in line with Proposition 7.16; this empirical plot may be distorted by the exposures, by the fact that it is an

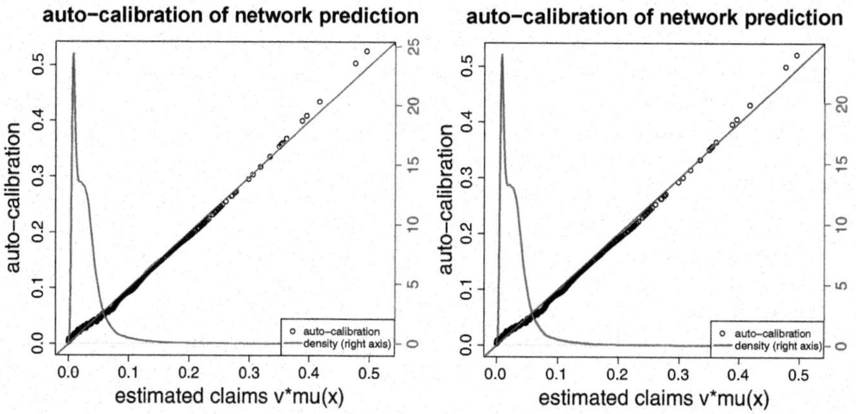

Fig. 7.12 (lhs) Empirical auto-calibration (7.39), the blue line shows the empirical density of the predictors $v_i\mu_{\widehat{\vartheta}}(\boldsymbol{x}_i)$, $1 \leq i \leq n$; (rhs) balance-corrected version using the weighted Gaussian correction (7.35)

Table 7.6 Run times, number of parameters, in-sample and out-of-sample deviance losses (units are in 10^{-2}) and in-sample average frequency of the Poisson null model, model Poisson GLM3 of Table 5.5, the FN network model (with embedding layers of dimension $b = 2$), and their bias regularized and balance-corrected counterparts, the local correction uses a GAM with 2.6 degrees of freedom in the cubic spline part

	Run time	# param.	In-sample loss on $\mathcal{L}$	Out-of-sample loss on $\mathcal{T}$	Aver. freq.
Poisson null	–	1	25.213	25.445	7.36%
Poisson GLM3	15 s	50	24.084	24.102	7.36%
Embed FN $(q_1, q_2, q_3) = (20, 15, 10)$	120 s	792	23.694	23.820	7.24%
Embed FN bias regularized	+4 s	792	23.690	23.824	7.36%
Embed FN Gauss balance-corrected	–	792 + 2	23.692	23.819	7.36%
Embed FN locally balance-corrected	–	792 + 3.6	23.692	23.818	7.36%

empirical plot fitted with `locfit`, and by fact that a linear Gaussian correction estimate may not be fully suitable.

Denuit et al. [97] propose a local balance correction that is very much in the spirit of the local polynomial regression fit with `locfit`. However, when using `locfit` we did not pay any attention to the balance property. Therefore, we proceed slightly differently, here. In formula (7.37) we give the network predictors on the canonical scale. This equips us with the data $(Y_i, v_i, \widehat{z}_i^\theta)_{1\le i\le n}$. To perform a local balance correction we fit a generalized additive model (GAM) to this data, using the canonical link, the Poisson deviance loss function, the observations Y_i, the exposures v_i and the feature information $\widehat{z}_i^\theta$; for GAMs we refer to Hastie–Tibshirani [181, 182], Wood [384] and Chapter 3 in Wüthrich–Buser [392], in particular, we proceed as in Example 3.4 of the latter reference.

The GAM regression fit on the canonical scale is illustrated in Fig. 7.13 (lhs). We essentially receive a straight line which says that the auto-calibration property is already well satisfied by the FN network predictor $\mu_{\widehat{\vartheta}}$. In fact, it is not completely a straight line, but GCV provides an optimal model with 2.6 effective degrees of freedom in the natural cubic spline part. This local (GAM) balance correction leads to another small model improvement (out-of-sample), see last line of Table 7.6.

Conclusion The balance property adjustment and the bias regularization are crucial in ensuring that the predictive model is on the right (price) level. We have presented three sophisticated methods of balance property adjustments: the additional GLM step under the canonical link choice (7.33), the model-free global Gaussian correction (7.35)–(7.36), and the local balance correction using a GAM under the canonical link choice. In our example, the results of the three different approaches are rather similar. In the sequel, we use the additional GLM step solution (7.33), the reason being that under this approach we can rely on one single regression model that directly predicts the claims. The other two approaches need two steps to get the predictions, which requires the storage of two models. ■

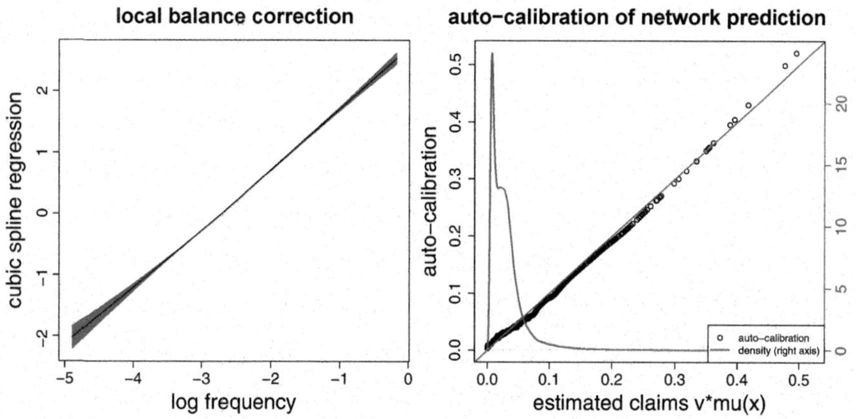

Fig. 7.13 (lhs) GAM fit on the canonical scale having 2.6 effective degrees of freedom (red shows the estimated confidence bounds); (rhs) balance-corrected version using the local GAM correction

7.4.3 Boosting Regression Models with Network Features

From Table 7.5 we conclude that the FN networks find systematic structure in the data that is not present in model Poisson GLM3, thus, the feature engineering for the GLM can be improved. Unfortunately, FN networks neither directly build on GLMs nor do they highlight the weaknesses of GLMs. In this section we discuss a proposal presented in Wüthrich–Merz [394] and Schelldorfer–Wüthrich [329] of combining two regression approaches. We are going to boost a GLM with FN network features. Typically, boosting is applied within the framework of regression trees. It goes back to the work of Valiant [362], Kearns–Valiant [209, 210], Schapire [328], Freund [139] and Freund–Schapire [140]. The idea behind boosting is to analyze the residuals of a given regression model with a second regression model to see whether this second regression model can still find systematic effects in the residuals which have not been discovered by the first one.

We start from the GLM studied in Chap. 5, and we boost this GLM with a FN network. Assume that both regression models act on the same feature space $\mathcal{X} \subset \{1\} \times \mathbb{R}^{q_0}$. The GLM provides a regression function for link function g and GLM parameter $\boldsymbol{\beta}^{\mathrm{GLM}} \in \mathbb{R}^{q_0+1}$

$$\boldsymbol{x} \mapsto \mu^{\mathrm{GLM}}(\boldsymbol{x}) = g^{-1}\left\langle \boldsymbol{\beta}^{\mathrm{GLM}}, \boldsymbol{x} \right\rangle.$$

Recall that this GLM can be interpreted as a FN network of depth 0, see Remarks 7.2. Next, we choose a FN network of depth $d \geq 1$ with the same link function g as the GLM

$$\boldsymbol{x} \mapsto \mu^{\mathrm{FN}}(\boldsymbol{x}) = g^{-1}\left\langle \boldsymbol{\beta}^{\mathrm{FN}}, \boldsymbol{z}^{(d:1)}(\boldsymbol{x})\right\rangle,$$

having a network parameter $\boldsymbol{\vartheta} = (\boldsymbol{w}_1^{(1)}, \ldots, \boldsymbol{w}_{q_d}^{(d)}, \boldsymbol{\beta}^{\mathrm{FN}})^\top \in \mathbb{R}^r$. In particular, we have the FN output parameter $\boldsymbol{\beta}^{\mathrm{FN}} \in \mathbb{R}^{q_d+1}$, we refer to Fig. 7.2.

We blend these two regression models by combining their regression functions

$$\boldsymbol{x} \mapsto \mu(\boldsymbol{x}) = g^{-1}\left\{\left\langle \boldsymbol{\beta}^{\mathrm{GLM}}, \boldsymbol{x}\right\rangle + \left\langle \boldsymbol{\beta}^{\mathrm{FN}}, \boldsymbol{z}^{(d:1)}(\boldsymbol{x})\right\rangle\right\}, \tag{7.40}$$

with parameter $\Phi = (\boldsymbol{\beta}^{\mathrm{GLM}}, \boldsymbol{\vartheta})^\top = (\boldsymbol{\beta}^{\mathrm{GLM}}, \boldsymbol{w}_1^{(1)}, \ldots, \boldsymbol{w}_{q_d}^{(d)}, \boldsymbol{\beta}^{\mathrm{FN}})^\top \in \mathbb{R}^{q_0+1+r}$.

An example is provided in Fig. 7.14. It shows the FN network using embedding layers for the categorical variables, see also Fig. 7.9 (rhs), and we add a GLM (in green color) that directly links the input $\boldsymbol{x}$ to the response variable. In machine learning this green connection is called a *skip connection* because it skips the FN layers.

Remarks 7.20

- Skip connections are a popular tool in network modeling, and they can be applied to any FN layers, i.e., a skip connection can, for instance, be added to skip the first FN layer. There are two benefits from skip connections. Firstly, they allow for more modeling flexibility, in (7.40) we directly combine a linear function

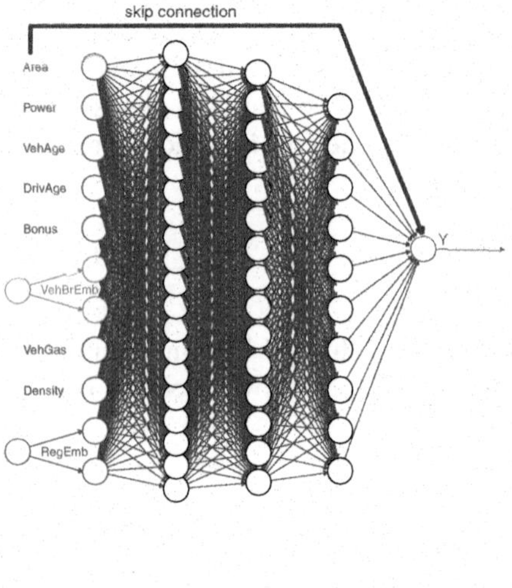

Fig. 7.14 Illustration of the combined regression function (7.40) using a GLM (in a skip connection) and a FN network

(coming from the GLM) with a non-linear one (coming form the FN network). This has the flavor of a Taylor expansion to combine terms of different orders. Secondly, skip connections can also be beneficial for gradient descent fitting because the inputs have a more direct link to the outputs, and the network only builds the functional form around the function in the skip connection.

- There are numerous variants of (7.40). A straightforward one is to choose a weight $\alpha \in (0, 1)$ and consider the regression function

$$\boldsymbol{x} \mapsto \mu(\boldsymbol{x}) = g^{-1}\left\{\alpha\left\langle\boldsymbol{\beta}^{\mathrm{GLM}}, \boldsymbol{x}\right\rangle + (1-\alpha)\left\langle\boldsymbol{\beta}^{\mathrm{FN}}, \boldsymbol{z}^{(d:1)}(\boldsymbol{x})\right\rangle\right\}. \tag{7.41}$$

  The weight α can be interpreted as the credibility assigned to the GLM.
- Regression function (7.40) considers two intercepts β_0^{GLM} and β_0^{FN}. If we do not consider the credibility version (7.41), one of the two intercepts is redundant.
- This approach also allows us to learn systematic effects across different insurance portfolios. If we have three insurance portfolios living on the same feature space and if $\chi \in \{1, 2, 3\}$ indicates which insurance portfolio we consider, we can modify the regression function (7.40) to

$$(\boldsymbol{x}, \chi) \mapsto \mu(\boldsymbol{x}, \chi) = g^{-1}\left\{\sum_{j=1}^{3}\left\langle\boldsymbol{\beta}_j^{\mathrm{GLM}}, \boldsymbol{x}\right\rangle \mathbb{1}_{\{\chi=j\}} + \left\langle\boldsymbol{\beta}^{\mathrm{FN}}, \boldsymbol{z}^{(d:1)}(\boldsymbol{x}, \chi)\right\rangle\right\}.$$

  The indicator $\mathbb{1}_{\{\chi=j\}}$ chooses the GLM that belongs to the corresponding insurance portfolio $\chi \in \{1, 2, 3\}$ with the (individual) GLM parameter $\boldsymbol{\beta}_\chi^{\mathrm{GLM}}$. The FN network term makes them related, i.e., the GLMs of the different insurance portfolios interact (jointly learn) via the FN network module. This is the approach used in Gabrielli et al. [149] to improve the chain-ladder reserving method by learning across different claims reserving triangles.

The regression function (7.40) gives the structural form of the combined regression model, but there is a second important ingredient proposed by Wüthrich–Merz [394]. Namely, the gradient descent algorithm (7.15) for model fitting can be started in an initial network parameter $\Phi^{(0)} \in \mathbb{R}^{q_0+1+r}$ that corresponds to the MLE of the GLM. Denote by $\widehat{\boldsymbol{\beta}}^{\mathrm{GLM}}$ the MLE of the GLM part, only.

Choose the initial value of the gradient descent algorithm for the fitting of the combined regression model (7.40)

$$\Phi^{(0)} = \left(\widehat{\boldsymbol{\beta}}^{\mathrm{GLM}}, \boldsymbol{w}_1^{(1)}, \ldots, \boldsymbol{w}_{q_d}^{(d)}, \boldsymbol{\beta}^{\mathrm{FN}} \equiv 0\right)^\top \in \mathbb{R}^{q_0+1+r}, \tag{7.42}$$

that is, initially, no signals traverse the FN network part because we set $\boldsymbol{\beta}^{\mathrm{FN}} \equiv 0$.

Remarks 7.21

- Using the initialization (7.42), the gradient descent algorithm starts exactly in the optimal GLM. The algorithm then tries to improve this GLM w.r.t. the given loss function using the additional FN network features. If the loss substantially reduces during the gradient descent training, the GLM misses systematic structure and it can be improved, otherwise the GLM is already good (enough).
- We can declare the MLE $\widehat{\boldsymbol{\beta}}^{\text{GLM}}$ to be *non-trainable*. In that case the original GLM always remains in the combined regression model and it acts as an offset. If we declare the MLE $\widehat{\boldsymbol{\beta}}^{\text{GLM}}$ to be non-trainable, we could choose a trainable credibility weight $\alpha \in (0, 1)$, see (7.41), which gradually reduces the influence of the GLM (if necessary).

Implementation of the general combined regression model (7.40) can be a bit cumbersome, see Listing 4 in Gabrielli et al. [149], but things can substantially be simplified by declaring the GLM part in (7.40) as being non-trainable, i.e., estimating $\boldsymbol{\beta}^{\text{GLM}}$ by $\widehat{\boldsymbol{\beta}}^{\text{GLM}}$ in the GLM, and then freeze this parameter. In view of (7.40) this simply means that we add an offset $o_i = \langle \widehat{\boldsymbol{\beta}}^{\text{GLM}}, \boldsymbol{x}_i \rangle$ to the FN network that is treated as a prior difference between the different data points, we refer to Sect. 5.2.3.

Example 7.22 (Combined GLM and FN Network) We revisit the French MTPL claim frequency GLM of Sect. 5.3.4, and we boost model Poisson GLM3 with FN network features. For the FN architecture we use the structure depicted in Fig. 7.14, i.e., a FN network of depth $d = 3$ having $(q_1, q_2, q_3) = (20, 15, 10)$ neurons, and using embedding layers of dimension $b = 2$ for the categorical feature components. Moreover, we declare the GLM part to be non-trainable which allows us to use the GLM as an offset in the FN network. Moreover, we apply bias regularization (7.33) to receive the balance property.

The results are presented in Table 7.7. A first observation is that using model Poisson GLM3 as an offset reduces the run time of gradient descent fitting because we start the algorithm already in a reasonable model. Secondly, as expected, the

Table 7.7 Run times, number of parameters, in-sample and out-of-sample deviance losses (units are in 10^{-2}) and in-sample average frequency of the Poisson null model, model Poisson GLM3 of Table 5.5, the FN network model (with embedding layers of dimension $b = 2$), and the combined regression model GLM3+FN, see (7.40)

	Run time	# param.	In-sample loss on $\mathcal{L}$	Out-of-sample loss on $\mathcal{T}$	Aver. freq.
Poisson null	–	1	25.213	25.445	7.36%
Poisson GLM3	15 s	50	24.084	24.102	7.36%
Embed FN $(q_1, q_2, q_3) = (20, 15, 10)$	120 s	792	23.694	23.820	7.24%
Embed FN bias regularized	+4 s	792	23.690	23.824	7.36%
Combined GLM+FN (20, 15, 10)	+53 s	$50 + 792$	23.772	23.834	7.24%
Combined GLM+FN bias regularized	+4 s	$50 + 792$	23.765	23.830	7.36%

FN features decrease the loss of model Poisson GLM3, this indicates that there are systematic effects that are not captured by the GLM. The final combined and regularized model has roughly the same out-of-sample loss as the corresponding FN network, showing that this approach can be beneficial in run times, and the predictive power is similar to a pure FN network. ■

Example 7.23 (Improving Model Poisson GLM3) In this example we would like to explore the deficiencies of model Poisson GLM3 by boosting it with FN network features. We do this in a systematic way by only considering two (continuous) features components at a time in the FN network. That is, we consider the combined approach (7.40) with initialization (7.42), but as feature information for the network part, we only consider two components at a time. For instance, we start with the features $(1, \texttt{Area}, \texttt{VehPower}) \in \{1\} \times \mathbb{R}^2$ for the network part, and the remaining feature information is ignored in this step. This way we can test whether the marginal modeling of `Area` and `VehPower` is suitable in model Poisson GLM3, and whether a pairwise interaction in these two components is missing. We train this FN network starting from model Poisson GLM3 (and keeping this GLM part frozen). The decrease in the out-of-sample loss during the gradient descent training is shown in Fig. 7.15 (top-left). We observe that the loss remains rather constant over 100 training epochs. This tells us that the pair (`Area`, `VehPower`) is appropriately considered in model Poisson GLM3.

Figure 7.15 gives all pairwise plots of the continuous feature components `Area`, `VehPower`, `VehAge`, `DrivAge`, `BonusMalus`, `Density`, the scale on the y-axis is identical in all plots. We observe that only the plots including the variable `BonusMalus` provide a bigger decrease in loss (in blue color in the colored version). This indicates that mainly this feature component is not modeled optimally in model Poisson GLM3, because boosting with a FN network finds systematic structure here that improves the loss of model Poisson GLM3. In model Poisson GLM3, the variable `BonusMalus` has been modeled log-linearly with an interaction term with `DrivAge` and $(\texttt{DrivAge})^2$, see (5.35). Table 7.8 shows the result if we add a FN network feature (7.40) for the pair (`DrivAge`, `BonusMalus`) to model Poisson GLM3. Indeed, we see that the resulting combined GLM-FN network model has the same GL as the full FN network approach. Thus, we conclude that model Poisson GLM3 performs fairly well and only the modeling of the pair (`DrivAge`, `BonusMalus`) should be improved. ■

7.4.4 Network Ensemble Learning

Ensemble learning is a popular way of expressing that one takes an average over different predictors. There are many established methods that belong to the family of ensemble learning, e.g., there is **b**oostrap **agg**regat**ing** (called *bagging*) introduced by Breiman [51], there are random forests, and there is boosting. Random forests

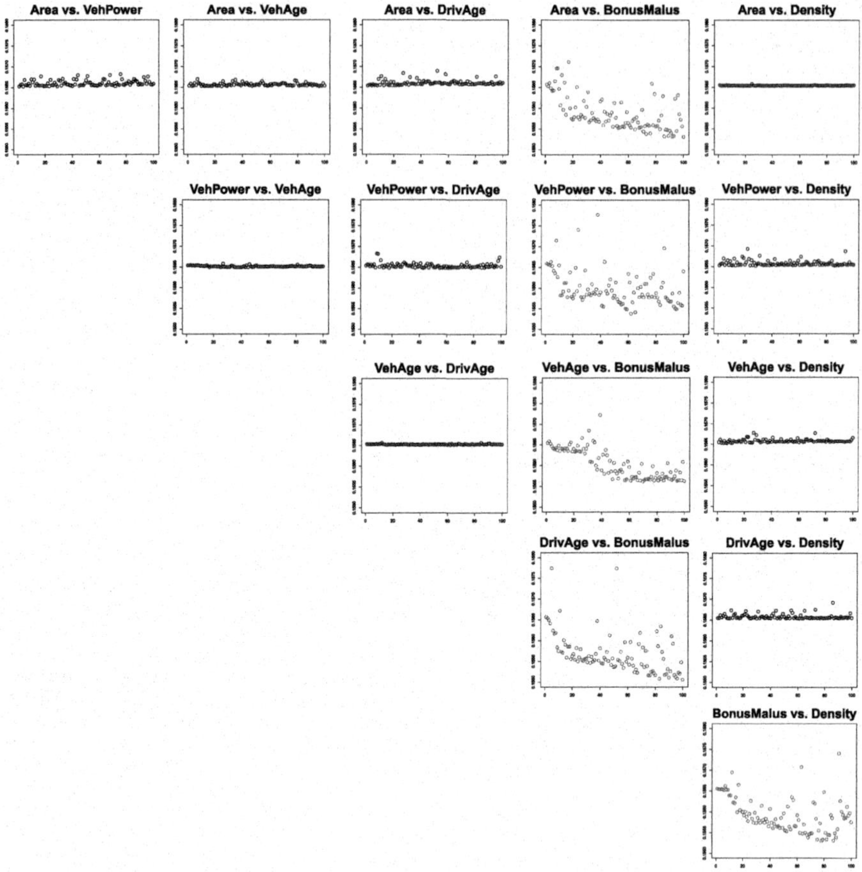

Fig. 7.15 Exploring all pairwise interactions: out-of-sample losses over 100 gradient descent epochs for all pairs of the continuous feature components `Area`, `VehPower`, `VehAge`, `DrivAge`, `BonusMalus`, `Density` (the scale on the y-axis is identical in all plots)

and boosting are mainly based on classification and regression trees (CARTs) and they belong to the most powerful machine learning methods for tabular data. These methods combine a family of predictors to a more powerful predictor. The present section is inspired by the bagging method of Breiman [51], and we perform **n**etwork **agg**regat**ing** (called *nagging*).

Stochastic Gradient Descent Fitting of Networks

We have described that network calibration involves several elements of randomness. This in combination with early stopping leads to the non-uniqueness of reasonably good networks for prediction and pricing. We have discussed this based

Table 7.8 Run times, number of parameters, in-sample and out-of-sample deviance losses (units are in 10^{-2}) and in-sample average frequency of the Poisson null model, model Poisson GLM3 of Table 5.5, model Poisson GLM3 with additional FN features for (`DrivAge`, `BonusMalus`), the FN network model (with embedding layers of dimension $b = 2$), and the combined regression model GLM3+FN, see (7.40)

	Run time	# param.	In-sample loss on $\mathcal{L}$	Out-of-sample loss on $\mathcal{T}$	Aver. freq.
Poisson null	–	1	25.213	25.445	7.36%
Poisson GLM3	15 s	50	24.084	24.102	7.36%
GLM3 +FN(`DrivAge`, `BonusMalus`)	–	$50 + 792$	23.804	23.805	7.36%
Embed FN bias regularized	124 s	792	23.690	23.824	7.36%
Combined GLM+FN bias regularized	72 s	$50 + 792$	23.765	23.830	7.36%

on Fig. 7.5, namely, for a given network architecture we have a continuum of comparably good models (w.r.t. the chosen objective function) that lie in the green area of Fig. 7.5. One SGD calibration picks one specific model from this green area, we also refer to Remarks 7.9. Of course, this is very unsatisfactory in insurance pricing because it implies that the selection of a price for an insurance policy has a substantial element of subjectivity (that cannot be explained to the customer). Naturally, we would like to combine models in the green area of Fig. 7.5, for instance, by performing some sort of integration over the models in the green area. Intuitively, this should lead to a very powerful predictive model because it diversifies the weaknesses of each individual model. This is exactly what we discuss in this section. Before doing so, we would first like to understand the different single calibrations of a given network architecture.

We consider the MTPL data of Example 7.12. We model this data with a Poisson FN network using embedding layers for the categorical features and using bias regularization (7.33) to guarantee the balance property to hold. For the FN network architecture we choose depth $d = 3$ with $(q_1, q_2, q_3) = (20, 15, 10)$ FN neurons; this setup gives us the results on the last line of Table 7.5. We now repeat this procedure $M = 1'600$ times, using exactly the same FN network architecture, the same early stopping strategy, the same SGD method and the same batch size. We only change the seeds of the starting point $\boldsymbol{\vartheta}^{(0)} \in \mathbb{R}^r$ of the SGD algorithm, the partitioning of the learning data $\mathcal{L}$ into training data $\mathcal{U}$ and validation data $\mathcal{V}$, see Fig. 7.7, and the partitioning of the training data into the (mini-)batches.

The resulting $1'600$ in-sample and out-of-sample deviance losses are presented in Fig. 7.16. We observe a considerable variation in these figures. The in-sample losses vary between 23.616 and 23.815 (mean 23.728), and the corresponding out-of-sample loss between 23.766 and 23.899 (mean 23.819), units are in 10^{-2}; note that all network calibrations are bias regularized. The in-sample loss is an average over $n = 610'206$ (individual) unit deviance losses, and the out-of-sample an average over $T = 67'801$ unit deviance losses, see also Definition 4.24. Therefore, we expect an even much bigger variation on individual insurance policies. We are going to analyze this in more detail in this section.

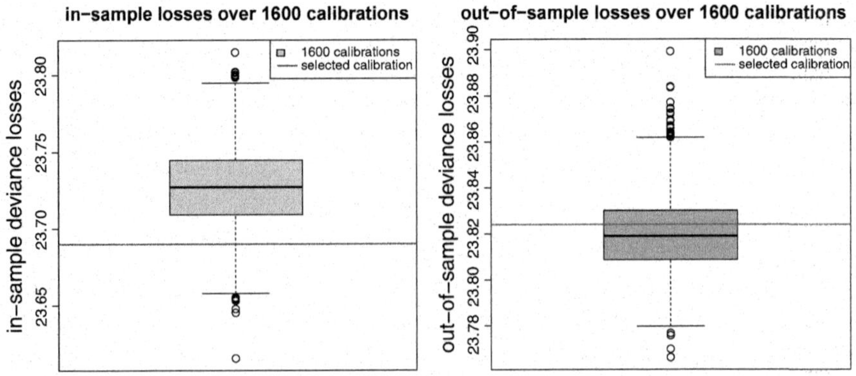

Fig. 7.16 Boxplots over $1'600$ network calibrations only differing in the seeds for the SGD algorithm and the partitioning of the learning data: (lhs) in-sample losses on $\mathcal{L}$ and (rhs) out-of-sample losses on $\mathcal{T}$, the horizontal lines show the calibration chosen in Table 7.5; units are in 10^{-2}

Before doing so, we would like to understand whether there is some dependence between the in-sample and the out-of-sample losses over the $M = 1'600$ runs of the SGD algorithm with different seeds. In Fig. 7.17 we provide a scatter plot of the out-of-sample losses vs. the in-sample losses. This plot is complemented by a cubic spline regression (in orange color). From this plot we conclude that the models with very small in-sample losses tend to over-fit, and the models with large in-sample losses tend to under-fit (always using the same early stopping rule). In view of these results we conclude that the chosen early stopping rule is sensible because on average it tends to provide the model with the smallest out-of-sample loss on $\mathcal{T}$. Recall that we do not use $\mathcal{T}$ during the SGD fitting, but only the learning data $\mathcal{L}$ that is split into the training data $\mathcal{U}$ and the validation data $\mathcal{V}$ for exercising the early stopping, see Fig. 7.7.

Fig. 7.17 Scatter plot of out-of-sample losses vs. in-sample losses for different seeds, the orange line gives a fitted cubic spline, and the cyan lines show the empirical means; units are in 10^{-2}

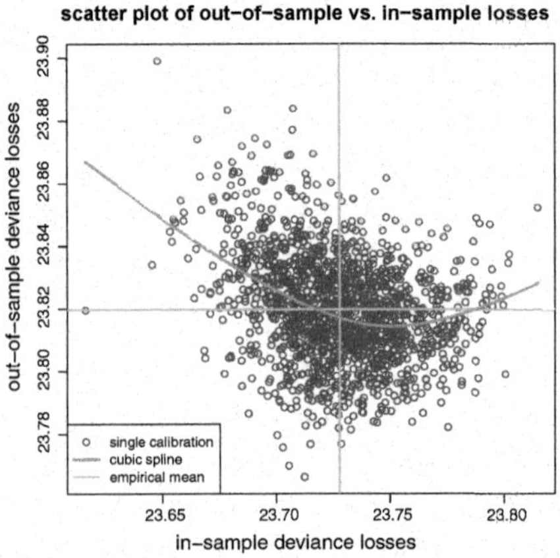

Next, we study the estimated prices on the test data (out-of-sample)

$$\mathcal{T} = \left\{ (Y_t^\dagger = N_t^\dagger / v_t^\dagger, \boldsymbol{x}_t^\dagger, v_t^\dagger) : t = 1, \ldots, T = 67'801 \right\}.$$

For each run of the SGD algorithm we receive a different (early stopped) network parameter estimate $\widehat{\boldsymbol{\vartheta}}^m \in \mathbb{R}^r$, $1 \le m \le M = 1'600$. Using these parameter estimates we receive the estimated network regression functions, for $1 \le m \le M$,

$$\boldsymbol{x} \mapsto \widehat{\mu}^m(\boldsymbol{x}) = \mu_{\widehat{\boldsymbol{\vartheta}}^m}(\boldsymbol{x}),$$

using the FN network of Listing 7.4 with network parameter $\widehat{\boldsymbol{\vartheta}}^m$. Thus, for the out-of-sample policies $1 \le t \le T$ we receive the expected frequencies

$$\boldsymbol{x}_t^\dagger \mapsto \widehat{\mu}_t^m = \widehat{\mu}^m\left(\boldsymbol{x}_t^\dagger\right) = \mu_{\widehat{\boldsymbol{\vartheta}}^m}\left(\boldsymbol{x}_t^\dagger\right).$$

Since we choose the seeds of the SGD runs *at random* we may (and will) assume that we have independence between the prices $(\widehat{\mu}_t^m)_{t \in \mathcal{T}}$ of the different runs $1 \le m \le M$ of the SGD algorithm. This allows us to estimate the average price and the coefficient of variation of these prices of a fixed insurance policy t over the different SGD runs

$$\bar{\mu}_t^{(1:M)} = \frac{1}{M} \sum_{m=1}^{M} \widehat{\mu}_t^m \quad \text{and} \quad \mathrm{Vco}_t = \frac{1}{\bar{\mu}_t^{(1:M)}} \sqrt{\frac{1}{M-1} \sum_{m=1}^{M} \left(\widehat{\mu}_t^m - \bar{\mu}_t^{(1:M)} \right)^2}. \tag{7.43}$$

These (out-of-sample) coefficients of variation are illustrated in Fig. 7.18. We observe a considerable variation on some policies. The average coefficient of variation is roughly 10% (orange horizontal line, lhs). The maximal coefficient of variation is about 40%, thus, for this policy the individual prices $\widehat{\mu}_t^m$ of the different SGD runs $1 \le m \le M$ fluctuate considerably around $\bar{\mu}_t^{(1:M)}$. This now explains why we choose $M = 1'600$ SGD runs, namely, the averaging in (7.43) reduces the coefficient of variation on this policy to $40\%/\sqrt{M} = 40\%/40 = 1\%$, note that we have independence between the different SGD runs. Thus, by averaging we receive an acceptable influence of the variation of the individual SGD fittings.

Listing 7.9 shows the 10 policies (out-of-sample) with the largest coefficients of variations Vco_t. These polices have in common that they belong to the lowest `BonusMalus` level, the drivers are very young, the cars are comparably old and they have a bigger vehicle power. From a practical point of view we should doubt these policies, since the information provided may not be correct. New drivers (at the age of 18) typically enter a bonus-malus scheme at level 100, and only after several accident-free years these drivers can reach a bonus-malus level of 50. Thus, policies as in Listing 7.9 should not exist, and our pricing framework has difficulties to (correctly) handle them. In practice, this needs further investigation because, obviously, there is a data issue, here.

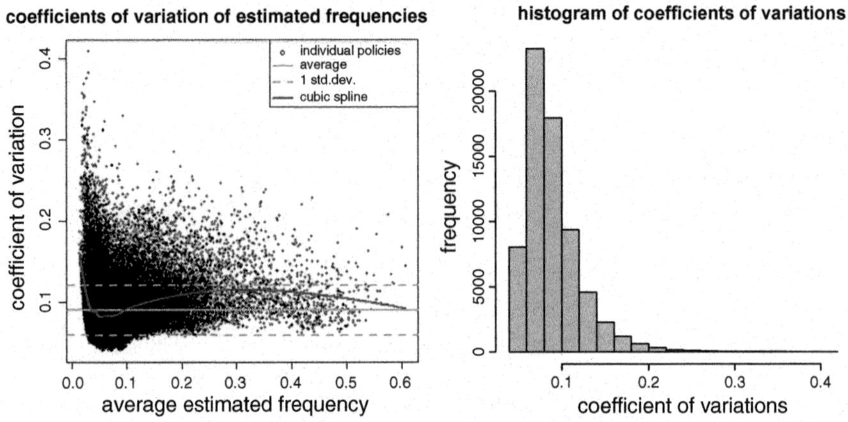

Fig. 7.18 Out-of-sample coefficients of variations Vco_t on an individual policy level $1 \leq t \leq T$ over the $1'600$ calibrations (lhs) scatter plot against the average estimated frequencies $\bar{\mu}_t^{(1:M)}$ and (rhs) resulting histogram

Listing 7.9 The 10 policies (out-of-sample) with the largest coefficients of variation

```
Area VehPower VehAge DrivAge BonusMalus VehBrand  VehGas Region       vco
   D        8     16      18         50      B11 Regular    R53 0.4089006
   D        9     17      20         50      B11 Regular    R24 0.3827665
   C        8     11      18         50       B5 Regular    R24 0.3762306
   C        9     18      18         50       B5 Regular    R24 0.3697370
   C        7     17      18         50       B1 Regular    R24 0.3579979
   C        9     19      19         50       B5 Regular    R24 0.3554879
   C        6     15      20         50       B1 Regular    R93 0.3528679
   C        7     14      19         50       B1 Regular    R53 0.3518279
   A       11     20      50         50      B13 Regular    R74 0.3442184
   D        5     14      18         50       B3  Diesel    R24 0.3403783
```

Nagging Predictor

The previously observed variations of the prices motivate to average over the different models (network calibrations). This brings us to bagging introduced by Breiman [51]. Bagging is based on averaging/aggregating over several 'independent' predictions; this is done in three steps. In a first step, a model is fitted to the data $\mathcal{L}$. In a second step, independent bootstrap samples $\mathcal{L}^{*(m)}$ are generated from this fitted model; the independence has to be understood in a conditional sense, namely, the different bootstrap samples $\mathcal{L}^{*(m)}$ are independent in m, given the data $\mathcal{L}$. In the third step, for every bootstrap sample $\mathcal{L}^{*(m)}$ one estimates a model $\widehat{\mu}^m$, and averaging (7.43) provides the bagging predictor. Bagging is mainly a *variance reduction* technique. Note that if the fitted model of the first step has a bias, then likely the bootstrap samples $\mathcal{L}^{*(m)}$ are biased, and so is the bagging predictor. Therefore, bagging does not help to reduce a potential bias. All these results have to be understood conditionally on the data $\mathcal{L}$. If this data is atypical for the problem, so will the bootstrap samples be.

We can perform a similar analysis for the fitted networks, but we do not need to bootstrap, here, because the various elements of randomness in SGD fitting allow us to generate independent predictors $\widehat{\mu}^m$, conditional on the data $\mathcal{L}$. Averaging (7.43) over these predictors then provides us with the **n**etwork **agg**regat**ing** (nagging) predictor $\bar{\mu}^{(1:M)}$; we also refer to Dietterich [105] and Richman–Wüthrich [315] for this aggregation. Thus, we replace the bootstrap step by the different runs of the SGD algorithm. Both options provide independent predictors $\widehat{\mu}^m$, conditional on the data $\mathcal{L}$. However, there is a fundamental difference between bagging and nagging. Bagging generates new (bootstrap) samples $\mathcal{L}^{*(m)}$ and, thus, bagging also involves randomness coming from sampling the new observations. Nagging always acts on the same sample $\mathcal{L}$, and it only refits the model multiple times. Therefore, the latter will typically introduce less variation. Of course, bagging and nagging can be combined, and then the full expected GL can be estimated, we come back to this in Sect. 11.4, below. We do not sample new observations, here, because we would like to understand the variations implied by the SGD algorithm with early stopping on the given (fixed) data.

In Fig. 7.18 we have seen that we need nagging over 1′600 network calibrations so that the maximal coefficient of variation on an individual policy level is below 1% in our MTPL example. In this section we would like to understand the minimal out-of-sample loss that can be achieved by nagging on the (entire) test data set, and we would like to analyze its rate of convergence.

For this we define the sequence of nagging predictors

$$\bar{\mu}^{(1:M)}(\boldsymbol{x}) = \frac{1}{M}\sum_{m=1}^{M} \widehat{\mu}^m(\boldsymbol{x}) \qquad \text{for } M \geq 1. \tag{7.44}$$

This allows us to study the out-of-sample losses on $\mathcal{T}$ in the Poisson model for $M \geq 1$

$$\mathfrak{D}(\mathcal{T}, \bar{\mu}^{(1:M)}) = \frac{2}{T}\sum_{t=1}^{T} v_t^\dagger \left(\bar{\mu}^{(1:M)}(\boldsymbol{x}_t^\dagger) - Y_t^\dagger - Y_t^\dagger \log\left(\frac{\bar{\mu}^{(1:M)}(\boldsymbol{x}_t^\dagger)}{Y_t^\dagger} \right)\right).$$

Remark 7.24 From Remarks 7.17 we know that the expected deviance GL of the estimated model is lower bounded by the expected deviance GL of the true data generating model; the difference is the conditional calibration. Within the family of Tweedie's CP models Richman–Wüthrich [315] proved that, indeed, aggregating decreases monotonically the expected deviance GL of the estimated model (Proposition 2 of [315]), convergence is established (Proposition 3 of [315]), and the speed of convergence is provided using asymptotic normality (Proposition 4 of [315]). For the Gaussian square loss results we refer to Breiman [51] and Bühlmann–Yu [60].

We revisit Proposition 2 of Richman–Wüthrich [315] which has also been proved in Proposition 3.1 of Denuit–Trufin [103]. We only consider a single case in the next proposition and we drop the feature information $\boldsymbol{x}$ (because we can condition on $\boldsymbol{X} = \boldsymbol{x}$).

Proposition 7.25 *Choose a response $Y \sim f(\cdot; \theta, v/\varphi)$ belonging to Tweedie's CP model having a power variance cumulant function $\kappa = \kappa_p$ with power variance parameter $p \in [1, 2]$, see* (2.17)*. Assume $\widehat{\mu}$ is an estimator for the mean parameter $\mu = \kappa_p'(\theta) > 0$ satisfying $\epsilon < \widehat{\mu} \le p/(p-1)\mu$, a.s., for some $\epsilon \in (0, p/(p-1)\mu)$. Choose i.i.d. copies $\widehat{\mu}^m$, $m \ge 1$, of $\widehat{\mu}$ being all independent of Y. We have for all $M \ge 1$*

$$\mathbb{E}_\theta\left[\mathfrak{d}\left(Y, \widehat{\mu}^1\right)\right] \ge \mathbb{E}_\theta\left[\mathfrak{d}\left(Y, \bar{\mu}^{(1:M)}\right)\right] \ge \mathbb{E}_\theta\left[\mathfrak{d}\left(Y, \bar{\mu}^{(1:M+1)}\right)\right] \ge \mathbb{E}_\theta\left[\mathfrak{d}(Y, \mu)\right].$$

Proof of Proposition 7.25 The lower bound on the right-hand side immediately follows from Theorem 4.19. For an estimate $\widehat{\mu} > 0$ we define the function, we also refer to (4.18) and we set for the canonical link $h_p = (\kappa_p')^{-1}$,

$$\widehat{\mu} \mapsto \psi_p(\widehat{\mu}) = \mu h_p\left(\widehat{\mu}\right) - \kappa_p\left(h_p\left(\widehat{\mu}\right)\right) = \begin{cases} \mu \log(\widehat{\mu}) - \widehat{\mu} & \text{for } p = 1, \\ \mu \frac{\widehat{\mu}^{1-p}}{1-p} - \frac{\widehat{\mu}^{2-p}}{2-p} & \text{for } p \in (1, 2), \\ -\mu/\widehat{\mu} - \log(\widehat{\mu}) & \text{for } p = 2. \end{cases}$$

This is the part of the log-likelihood (and deviance loss) that depends on the canonical parameter $\widehat{\theta} = h_p(\widehat{\mu})$, and replacing the observation Y by μ. Calculating the second derivative w.r.t. $\widehat{\mu}$ provides for $p \in [1, 2]$

$$\frac{\partial^2}{\partial \widehat{\mu}^2}\psi_p(\widehat{\mu}) = -p\mu\widehat{\mu}^{-p-1} - (1-p)\widehat{\mu}^{-p} = \widehat{\mu}^{-(1+p)}\left[-p\mu - (1-p)\widehat{\mu}\right] \le 0,$$

the last inequality uses that the square bracket is non-positive, a.s., under our assumptions on $\widehat{\mu}$. Thus, ψ_p is concave on the interval $(0, p/(p-1)\mu)$. We now focus on the inequalities for $M \ge 1$. Consider the decomposition of the nagging predictor for $M + 1$

$$\bar{\mu}^{(1:M+1)} = \frac{1}{M+1}\sum_{j=1}^{M+1} \bar{\mu}^{(-j)}, \qquad \text{where} \qquad \bar{\mu}^{(-j)} = \frac{1}{M}\sum_{m=1}^{M+1} \widehat{\mu}^m \mathbb{1}_{\{m \neq j\}}.$$

The predictors $\bar{\mu}^{(-j)}$, $j \geq 1$, are copies of $\bar{\mu}^{(1:M)}$, though not independent ones. Using the function ψ_p, the second term on the right-hand side has the same structure as the estimation risk function (4.14),

$$
\begin{aligned}
&\mathbb{E}_\theta\left[\mathfrak{d}(Y, \bar{\mu}^{(1:M)})\right] \\
&= \mathbb{E}_\theta\left[\mathfrak{d}(Y, \bar{\mu}^{(1:M+1)})\right] + 2\,\mathbb{E}_\theta\left[Y h_p\left(\bar{\mu}^{(1:M+1)}\right) - \kappa_p\left(h_p\left(\bar{\mu}^{(1:M+1)}\right)\right)\right] \\
&\qquad - 2\,\mathbb{E}_\theta\left[Y h_p\left(\bar{\mu}^{(1:M)}\right) - \kappa_p\left(h_p\left(\bar{\mu}^{(1:M)}\right)\right)\right] \\
&= \mathbb{E}_\theta\left[\mathfrak{d}(Y, \bar{\mu}^{(1:M+1)})\right] + 2\left(\mathbb{E}\left[\psi_p\left(\bar{\mu}^{(1:M+1)}\right)\right] - \mathbb{E}\left[\psi_p\left(\bar{\mu}^{(1:M)}\right)\right]\right) \\
&= \mathbb{E}_\theta\left[\mathfrak{d}(Y, \bar{\mu}^{(1:M+1)})\right] + 2\left(\mathbb{E}\left[\psi_p\left(\frac{1}{M+1}\sum_{j=1}^{M+1} \bar{\mu}^{(-j)}\right)\right] - \mathbb{E}\left[\psi_p\left(\bar{\mu}^{(1:M)}\right)\right]\right) \\
&\geq \mathbb{E}_\theta\left[\mathfrak{d}(Y, \bar{\mu}^{(1:M+1)})\right] + 2\left(\mathbb{E}\left[\frac{1}{M+1}\sum_{j=1}^{M+1} \psi_p\left(\bar{\mu}^{(-j)}\right)\right] - \mathbb{E}\left[\psi_p\left(\bar{\mu}^{(1:M)}\right)\right]\right) \\
&= \mathbb{E}_\theta\left[\mathfrak{d}(Y, \bar{\mu}^{(1:M+1)})\right],
\end{aligned}
$$

the second last step applies Jensen's inequality to the concave function ψ_p, and the last step follows from the fact that $\bar{\mu}^{(-j)}$, $j \geq 1$, are copies of $\bar{\mu}^{(1:M)}$. □

Remarks 7.26

- Proposition 7.25 says that aggregation works, i.e., aggregating i.i.d. predictors leads to monotonically decreasing expected deviance GLs. In fact, if $\widehat{\mu} \leq 2\mu$, a.s., we receive Tweedie's forecast dominance by aggregating, restricted to the power variance parameters $p \in [1, 2]$, see Definition 4.22.
- The i.i.d. assumption can be relaxed, indeed, it is sufficient that every $\bar{\mu}^{(-j)}$ in the above proof has the same distribution as $\bar{\mu}^{(1:M)}$. This does not require independence between the predictors $\widehat{\mu}^m$, $m \geq 1$, but exchangeability is sufficient.
- We need the condition $\epsilon < \widehat{\mu} \leq p/(p-1)\mu$, a.s., to ensure the monotonicity within Tweedie's CP models. For the Poisson model $p = 1$ we can drop the upper bound, and we only need the lower bound to ensure the existence of the expected deviance GL. For $p \in (1, 2]$ the upper bound is increasingly binding, in the gamma case $p = 2$ requiring $\widehat{\mu} \leq 2\mu$, a.s.
- Note that we do not require unbiasedness of $\widehat{\mu}$ for μ in Proposition 7.25. Thus, at this stage, aggregating is a variance reduction technique.

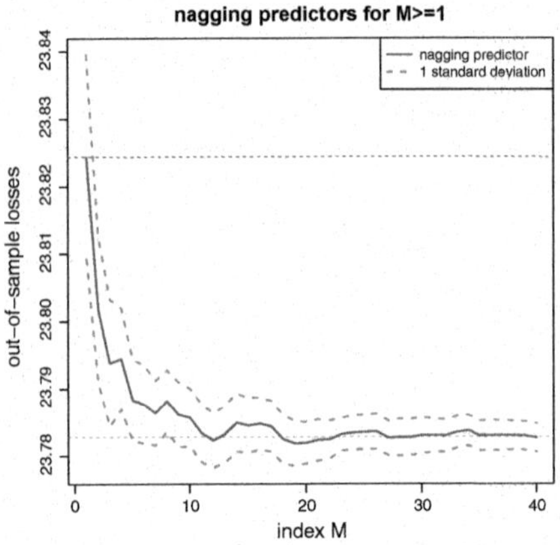

Fig. 7.19 Out-of-sample losses $\mathfrak{D}(\mathcal{T}, \bar{\mu}^{(1:M)})$ of the nagging predictors $(\bar{\mu}^{(1:M)}(\boldsymbol{x}_t^\dagger))_{1\le t\le T}$ for $1 \le M \le 40$; losses are in 10^{-2}

- If additionally we have unbiasedness of $\widehat{\mu}$ for μ and a uniformly integrable upper bound on $\bar{\mu}^{(1:M)}$, we can use Lebesgue's dominated convergence theorem and the law of large numbers to prove

$$\lim_{M\to\infty} \mathbb{E}_\theta \left[\mathfrak{d}\left(Y, \bar{\mu}^{(1:M)}\right)\right] = \mathbb{E}_\theta \left[\lim_{M\to\infty} \mathfrak{d}\left(Y, \bar{\mu}^{(1:M)}\right)\right] = \mathbb{E}_\theta \left[\mathfrak{d}(Y, \mu)\right]. \tag{7.45}$$

The uniformly integrable upper bound is only needed in the Poisson case $p = 1$, because the other cases are covered by $\epsilon < \widehat{\mu} \le p/(p-1)\mu$, a.s. Moreover, asymptotic normality can be established, we refer to Proposition 4 in Richman–Wüthrich [315].

We come back to our MTPL Poisson claim frequency example and its 1′600 network calibrations illustrated in Fig. 7.17. Figure 7.19 provides the out-of-sample portfolio losses $\mathfrak{D}(\mathcal{T}, \bar{\mu}^{(1:M)})$ of the resulting nagging predictors $(\bar{\mu}^{(1:M)}(\boldsymbol{x}_t^\dagger))_{1\le t\le T}$ for $1 \le M \le 40$ in red color, and the corresponding 1 standard deviation confidence bounds in orange color. The blue horizontal dotted line shows the case $M = 1$ which exactly refers to the (first) bias regularized FN network $\widehat{\mu}^{m=1}$ with embedding layers given in Table 7.5. Indeed, averaging over multiple networks improves the predictive model and the out-of-sample loss decreases over the first $2 \le M \le 10$ nagging steps. After the first 10 steps the picture starts to stabilize which indicates that for this size of portfolio (and this type of problem) we need to average over roughly 10–20 FN networks to receive optimal predictive models on the portfolio level. For $M \to \infty$ the out-of-sample loss converges to the green horizontal dotted line in Fig. 7.19 of $23.783 \cdot 10^{-2}$. These numbers are also reported on the last line of Table 7.9.

Figure 7.20 provides the empirical auto-calibration property (7.39) of the nagging predictor $\bar{\mu}^{(1:1600)}$; this is obtained completely analogously to Fig. 7.12.

Table 7.9 Run times, number of parameters, in-sample and out-of-sample deviance losses (units are in 10^{-2}) and in-sample average frequency of the Poisson null model, model Poisson GLM3 of Table 5.5, the FN network models (with embedding layers of dimension $b = 2$), and the nagging predictor for $M = 1'600$

	Run time	# param.	In-sample loss on $\mathcal{L}$	Out-of-sample loss on $\mathcal{T}$	Aver. freq.
Poisson null	–	1	25.213	25.445	7.36%
Poisson GLM3	15 s	50	24.084	24.102	7.36%
Embed FN bias regularized $\widehat{\mu}^{m=1}$	+4 s	792	23.690	23.824	7.36%
Average over $1'600$ SGDs (Fig. 7.16)	–	792	23.728	23.819	7.36%
Nagging FN $\bar{\mu}^{(1:M)}$, $M = 1'600$	∞	'792'	23.691	23.783	7.36%

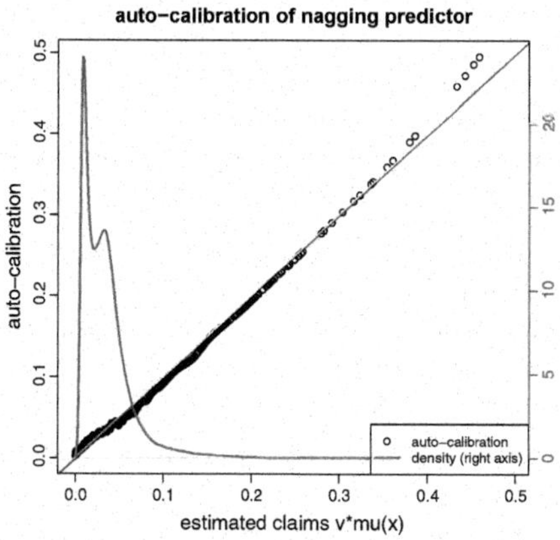

Fig. 7.20 Empirical auto-calibration (7.39) of the Poisson nagging predictor, the blue line shows the empirical density of $v_i \bar{\mu}^{(1:1600)}(\boldsymbol{x}_i)$, $1 \le i \le n$

The nagging predictors are (already) bias regularized, and Fig. 7.20 supports that the auto-calibration property holds rather accurately.

At this stage, we have fully arrived at Breiman's [53] two modeling cultures dilemma, see also Sect. 1.1. We have started from a parametric data model, and in order to boost its predictive performance we have combined such models in an algorithmic way. Working with many blended networks is not really practical, therefore, in such situations, a meta model can be fitted to the resulting nagging predictor.

Meta Model

Since working with $M = 1'600$ different FN networks is not practical, we fit a meta model to the nagging predictors $\bar{\mu}^{(1:M)}(\cdot)$. This can easily be done by just selecting an additional FN network and fit this additional network to the working data

$$\mathcal{D}^* = \left\{ \left(\bar{\mu}^{(1:M)}(\boldsymbol{x}_i), \boldsymbol{x}_i, v_i \right) : i = 1, \ldots, n \right\} \cup \left\{ \left(\bar{\mu}^{(1:M)}(\boldsymbol{x}_t^\dagger), \boldsymbol{x}_t^\dagger, v_t^\dagger \right) : t = 1, \ldots, T \right\}.$$

Table 7.10 Run times, number of parameters, in-sample and out-of-sample deviance losses (units are in 10^{-2}) and in-sample average frequency of the Poisson null model, model Poisson GLM3 of Table 5.5, the FN network model (with embedding layers of dimension $b = 2$), the nagging predictor, and the meta network model

	Run time	# param.	In-sample loss on $\mathcal{L}$	Out-of-sample loss on $\mathcal{T}$	Aver. freq.
Poisson null	–	1	25.213	25.445	7.36%
Poisson GLM3	15 s	50	24.084	24.102	7.36%
Embed FN bias regularized $\widehat{\mu}^{m=1}$	+4 s	792	23.690	23.824	7.36%
Nagging FN $\bar{\mu}^{(1:M)}$	∞	'792'	23.691	23.783	7.36%
Meta FN network $\widehat{\mu}^{\text{meta}}$	–	792	23.714	23.777	7.36%

For this calibration step we can consider all data, since we would like to fit a regression model as accurately as possible to the entire regression surface formed by all nagging predictors from the learning and the test data sets $\mathcal{L}$ and $\mathcal{T}$. Moreover, this step should not over-fit since this regression surface of nagging predictors does not include any noise, but it is on the level of expected values. As network architecture we choose again the same FN network of depth $d = 3$. The only change to the fitting procedure above is replacing the Poisson deviance loss by the square loss function, since we do not work with the Poisson responses N_i but rather with their mean estimates $\bar{\mu}^{(1:M)}(\boldsymbol{x}_i)$ and $\bar{\mu}^{(1:M)}(\boldsymbol{x}_t^\dagger)$ in this fitting step. Since the resulting meta network model may still have a bias we apply the bias regularization step of Listing 7.7 to the Poisson observations with the Poisson deviance loss on the learning data $\mathcal{L}$ (only). The results are presented in Table 7.10.

From these results we observe that in our case the meta network performs similarly well to the nagging predictor, and it seems to be a very reasonable choice.

Finally, in Fig. 7.21 (lhs) we analyze the resulting frequencies on an individual policy level on the test data set $\mathcal{T}$. We plot the estimated frequencies $\widehat{\mu}^{m=1}(\boldsymbol{x}_t^\dagger)$ of the first FN network (this corresponds to 'embed FN bias regularized' in Table 7.10 with an out-of-sample loss of 23.824) against the nagging predictor $\bar{\mu}^{(1:M)}(\boldsymbol{x}_t^\dagger)$ which averages over $M = 1'600$ networks. From Fig. 7.21 (lhs) we conclude that there are quite some differences between these two predictors, this exactly reflects the variations obtained in Fig. 7.18 (lhs). The nagging predictor removes this variation by averaging. Figure 7.21 (rhs) compares the nagging predictor $\bar{\mu}^{(1:M)}(\boldsymbol{x}_t^\dagger)$ to the one of the meta model $\widehat{\mu}^{\text{meta}}(\boldsymbol{x}_t^\dagger)$. This scatter plot shows that the predictors lie almost perfectly on the diagonal line which suggests that the meta model can be used as a substitute for the nagging predictor. This completes this claim frequency modeling example.

Remark 7.27 The meta model concept can also be useful in other situations. For instance, we can fit a gradient boosting regression model to the observations. Typically, this is much faster than calculating a nagging predictor (because it directly focuses on the weaknesses of the existing model). If the gradient boosting model is based on regression trees, it has the disadvantage that the resulting regression

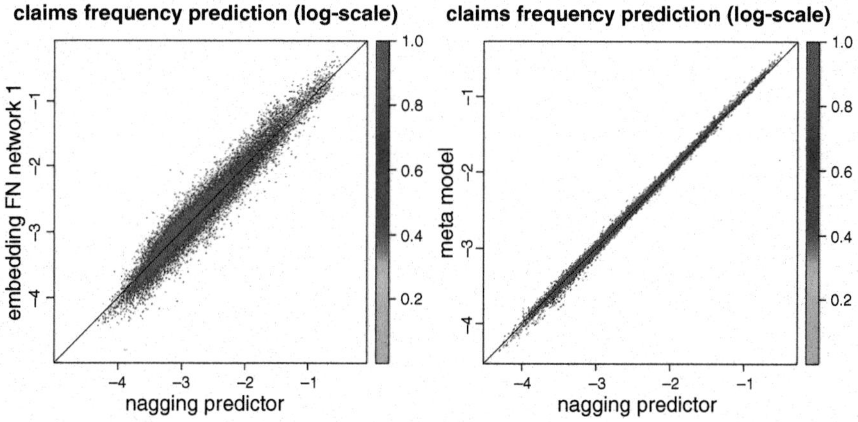

Fig. 7.21 Scatter plot of the out-of-sample predictions $\widehat{\mu}^{m=1}(\boldsymbol{x}_t^\dagger)$, $\bar{\mu}^{(1:M)}(\boldsymbol{x}_t^\dagger)$ and $\widehat{\mu}^{\text{meta}}(\boldsymbol{x}_t^\dagger)$ over all polices $1 \leq t \leq T$ on the test data set $\mathcal{T}$: (lhs) $\widehat{\mu}^{m=1}(\boldsymbol{x}_t^\dagger)$ vs. $\bar{\mu}^{(1:M)}(\boldsymbol{x}_t^\dagger)$ and (rhs) $\widehat{\mu}^{\text{meta}}(\boldsymbol{x}_t^\dagger)$ vs. $\bar{\mu}^{(1:M)}(\boldsymbol{x}_t^\dagger)$; the color scale shows the exposures $v_t^\dagger \in (0, 1]$

function is not continuous, and a non-constant extrapolation might be an issue. In a second step we can fit a meta FN network model to the former regression model, lifting the boosting model to a smooth network that allows for a non-constant extrapolation.

Example 7.28 (Gamma Claim Size Modeling) We revisit the gamma claim size example of Sect. 5.3.7. The data comprises Swedish motorcycle claim amounts. We have seen that this claim size data is not heavy-tailed, thus, a gamma distribution may be a reasonable choice for this data. For the modeling of this data we use the same normalization is in (5.45), this parametrization does not require the explicit knowledge of the (constant) shape parameter of the gamma distribution for mean estimation.

The difficulty with this data is that only 656 insurance policies suffer a claim, and likely a single FN network will not lead to stable results in this example. As FN network architecture we again choose a network of depth $d = 3$ and with $(q_1, q_2, q_3) = (20, 15, 10)$ neurons. Since the input layer has dimension $q_0 = 1 + 6 = 7$ we receive a network parameter of dimension $r = 626$. As loss function we choose the gamma deviance loss, see Table 4.1. Moreover, we choose the `nadam` optimizer, a batch size of 300, a training-validation split of 8:2, and we retrieve the network calibration with the lowest validation loss with a callback.

Figure 7.22 shows the results of 1′000 different SGD runs (only differing in the initial seeds and the splits of the training-validation sets as well as the batches). We see a considerable variation between the different SGD runs, both in in-sample deviance losses but also in the average estimated claims. Note that we did not bias-regularize the resulting networks (we work with the log-link here which is not the canonical one). This is why we receive fluctuating portfolio averages in Fig. 7.22

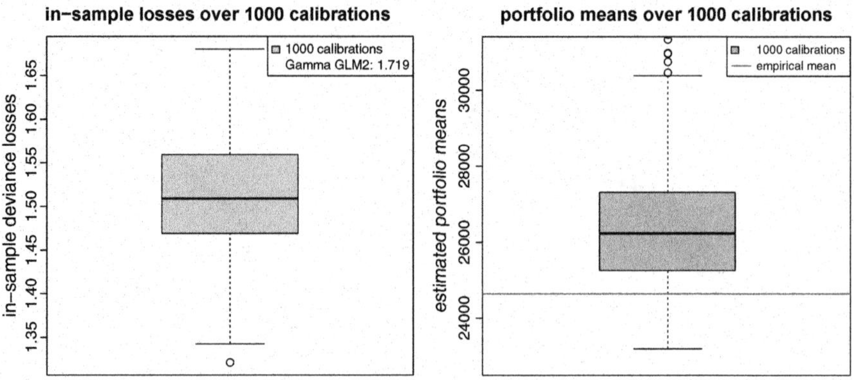

Fig. 7.22 Boxplots over 1′000 network calibrations only differing in the seeds for the SGD algorithm and the partitioning of the learning-validation data: (lhs) in-sample losses on the (entire) data $\mathcal{L}$ and (rhs) average estimated claims

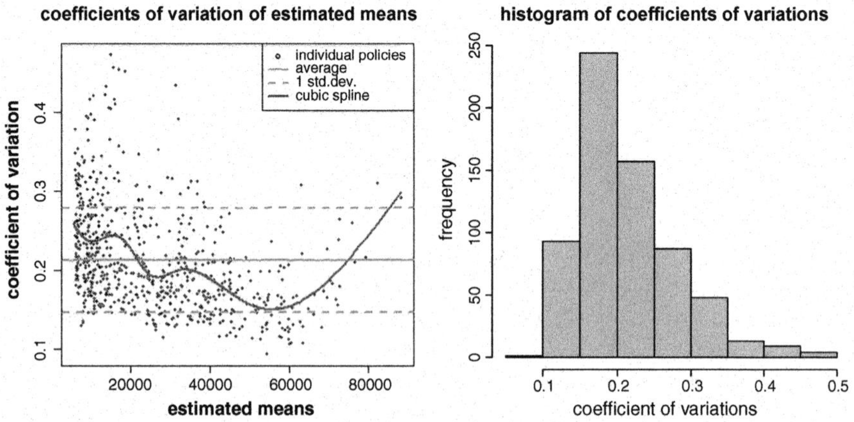

Fig. 7.23 Coefficients of variations Vco_i on an individual claim level $1 \leq i \leq n$ over the 1′000 calibrations (lhs) scatter plot against the nagging predictor $\bar{\mu}^{(1:M)}(\boldsymbol{x}_i)$ and (rhs) histogram

(rhs), the red line illustrates the empirical mean. Obviously, these FN networks are (on average) positively biased, and they will need a bias correction for the final prediction.

Figure 7.23 analyzes the variations on an individual claim level by studying the in-sample version of the coefficient of variation given in (7.43). We see that these coefficients of variation are bigger than in the claim frequency example, see Fig. 7.18. Thus, to receive stable results the nagging predictors $\bar{\mu}^{(1:M)}(\boldsymbol{x}_i)$ have to be calculated over many networks. Figure 7.24 confirms that aggregating reduces (in-sample) losses also in this case. From this figure we also see that the convergence is slower compared to the MTPL frequency example of Fig. 7.19, of course, because we have a much smaller claims portfolio.

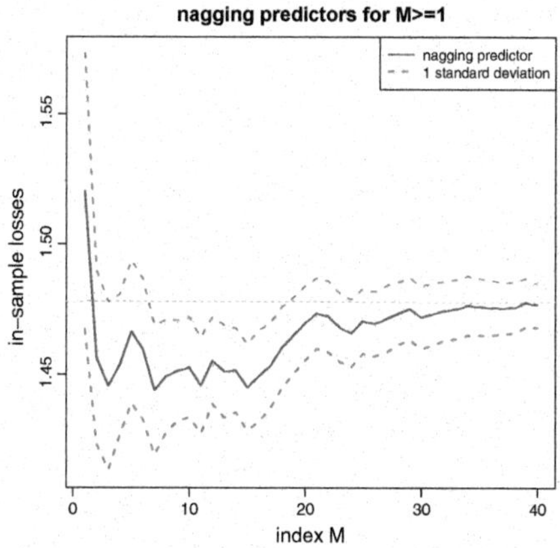

Fig. 7.24 In-sample losses $\mathfrak{D}(\mathcal{L}, \bar{\mu}^{(1:M)})$ of the nagging predictors $(\bar{\mu}^{(1:M)}(\boldsymbol{x}_i))_{1\le i\le n}$ for $1 \le M \le 40$ on the motorcycle claim size data

Table 7.11 Number of parameters, Pearson's dispersion estimate, MLE dispersion estimate, in-sample losses and in-sample average claim amounts of the null model (gamma intercept model), the gamma GLMs and the network nagging predictor; for the GLMs we refer to Table 5.13

	#	Dispersion		In-sample	Average
	param.	$\widehat{\varphi}^{\text{P}}$	$\widehat{\varphi}^{\text{MLE}}$	loss on $\mathcal{L}$	amount
Gamma null	$1+1$	2.057	1.690	2.085	24'641
Gamma GLM1	$9+1$	1.537	1.426	1.717	25'105
Gamma GLM2	$7+1$	1.544	1.427	1.719	25'130
Gamma FN network nagging	$626+1$	–	–	1.478	26'387
Gamma FN network nagging (bias reg)	$626+1$	*1.050*	1.240	1.465	24'641

Table 7.11 presents the results if we take the nagging predictor over $1'000$ different networks. The first observation is that we receive a much smaller in-sample loss compared to the GLMs, thus, there seems to be much room for improvements in the GLMs. Secondly, the nagging predictor has a substantial bias. For this reason we shift the intercept parameter in the output layer so that the portfolio average of the nagging predictor is equal to the empirical mean, see the last column of Table 7.11.

A main difficulty in this model is the estimation of the dispersion parameter $\varphi > 0$ and the shape parameter $\alpha = 1/\varphi$ of the gamma distribution, respectively. Pearson's dispersion estimate does not work because we do not know the degrees of freedom of the nagging predictor, see also (5.49). In Table 7.11 we calculate Pearson's dispersion estimate by simply dividing by the number of observations; this should be understood as a lower bound; this number is highlighted in italic. Alternatively, we can calculate the MLE, however, this may be rather different from Pearson's estimate, as indicated in Table 7.11. Figure 7.25 (lhs) shows the resulting QQ plot of the nagging predictor if we use the MLE $\widehat{\varphi}^{\text{MLE}} = 1.240$, and the right-hand side shows the same plot for $\widehat{\varphi} = 1.050$. From these plots it seems that we should rather go for a smaller dispersion parameter, the MLE being probably too much dominated by the small claims. This observation should also be understood as a red flag, as it tells us that the chosen gamma model is not fully suitable. This may

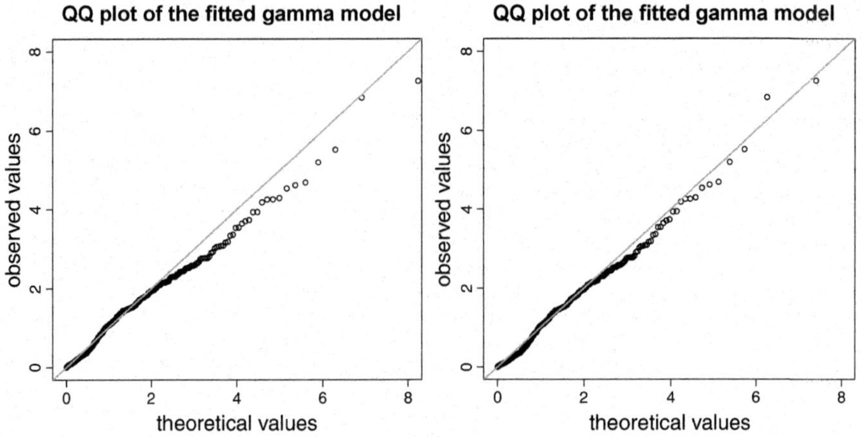

Fig. 7.25 QQ plots of the nagging predictors against the gamma density with (lhs) $\widehat{\varphi}^{\text{MLE}} = 1.240$ and (rhs) $\widehat{\varphi} = 1.050$

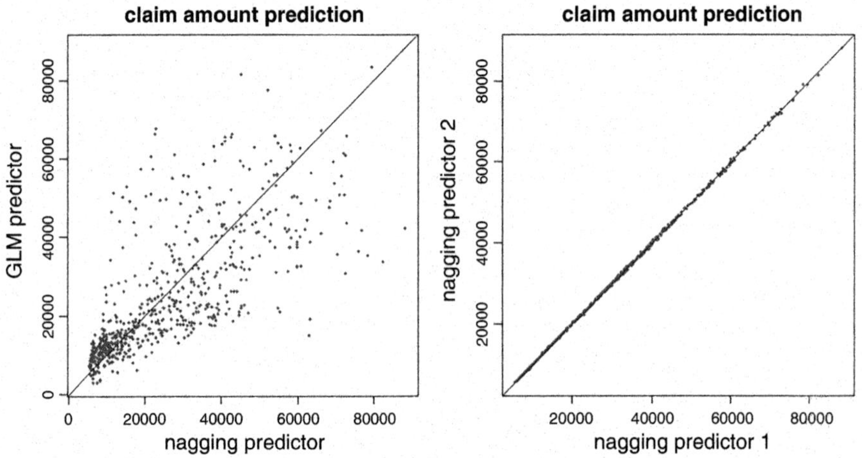

Fig. 7.26 (lhs) Scatter plot of model Gamma GLM2 predictors against the nagging predictors $\bar{\mu}^{(1:M)}(\boldsymbol{x}_i)$ over all instances $1 \leq i \leq n$, (rhs) scatter plot of two (independent) nagging predictors

be for various reasons: (1) the dispersion is not constant and should be modeled policy dependent, (2) the features are not sufficient to explain the observations, or (3) the gamma distribution is not suitable and should be replaced by another distribution.

In Fig. 7.26 (lhs) we compare the predictions received from model Gamma GLM2 against the nagging predictors $\bar{\mu}^{(1:M)}(\boldsymbol{x}_i)$ over all instances $1 \leq i \leq n$. The scatter plot spreads quite wildly around the diagonal which seriously questions at least one of the two models. To ensure that this variability between the two models is not caused by the (complex) FN network architecture, we verify the nagging

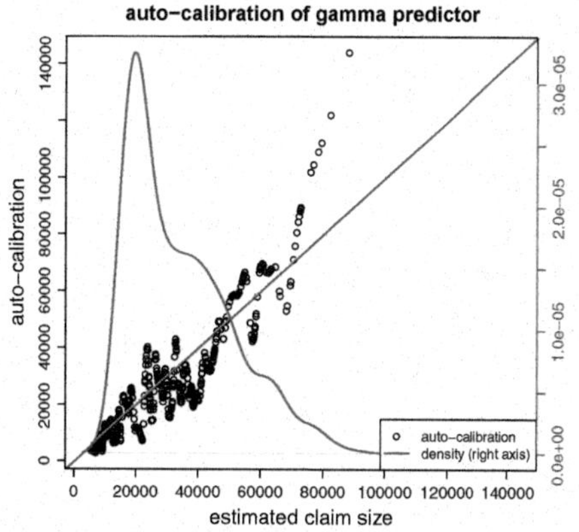

Fig. 7.27 Empirical auto-calibration (7.39) of the Gamma FN network nagging predictor of Table 7.11, the blue line shows the empirical density of $\bar{\mu}^{(1:M)}(\boldsymbol{x}_i)$, $1 \le i \le n$

predictor $\bar{\mu}^{(1:M)}$, $M = 1'000$, by computing a second independent one. Indeed, Fig. 7.26 shows that these two independent nagging predictors come to the same conclusion on the individual instance level. Thus, the network finds/uses systematic effects that are not present in model Gamma GLM2. If we perform a pairwise interaction analysis for boosting the GLM as in Example 7.23, we find that we should add interactions to the GLM between (`VehAge`, `RiskClass`), (`VehAge`, `BonusClass`), (`OwnerAge`, `Area`), and (`OwnerAge`, `VehAge`); recall that model Gamma GLM2 neither includes `BonusClass` nor `Gender` as supported by a `drop1` backward elimination analysis from model Gamma GLM1. However, it turns out, here, that we should have `BonusClass` in the model by letting it interact with `VehAge`.

Finally, Fig. 7.27 shows the empirical auto-calibration behavior (7.39) of the Gamma FN network nagging predictor of Table 7.11. The resulting black dots are rather volatile which shows that we do not (fully) have the auto-calibration property, here, but it also expresses that we fit a model on only 656 claims. The prediction of these claims is highlighted by the blue empirical density given by $\bar{\mu}^{(1:M)}(\boldsymbol{x}_i)$, $1 \le i \le n$. On the positive side, the auto-calibration plot shows that we neither systematically under- nor over-estimate because the black dots fluctuate around the diagonal red line, only the upper tail seems to under-estimate the true claim size. ■

Ensembling over Selected Networks vs. All Networks

Zhou et al. [406] ask the question whether ensembling over 'selected' networks is better than ensembling over all networks. In their proposal they introduce a weighted averaging scheme over the different network predictors $\widehat{\mu}^m$, $1 \le m \le M$. We perform a slightly different analysis here. We are re-using the $M = 1'600$ SGD calibrations of the Poisson FN network illustrated in Fig. 7.17. We order these SGD calibrations w.r.t. their in-sample losses $\mathfrak{D}(\mathcal{L}, \widehat{\mu}^m)$, $1 \le m \le M$, and partition this ordered sample into three equally sized sets: the first one containing the smallest

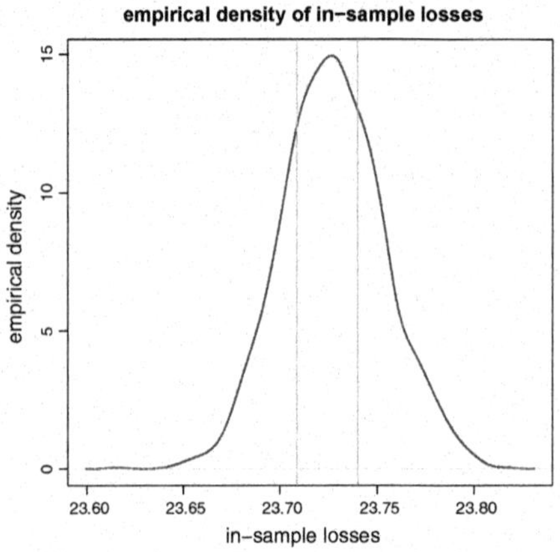

Fig. 7.28 Empirical density of the in-sample losses $\mathfrak{D}(\mathcal{L}, \widehat{\mu}^m)$, $1 \leq m \leq M$, of Fig. 7.17

in-sample losses, the second one the middle sized in-sample losses, and the third one the largest in-sample losses. Figure 7.28 shows the empirical density of these in-sample losses, and the vertical lines give the partition into the three sets, we call the resulting (disjoint) index sets $\mathcal{I}^{\text{small}}, \mathcal{I}^{\text{middle}}, \mathcal{I}^{\text{large}} \subset \{1, \ldots, M\}$. Remark that this partition is done fully *in-sample*, based on the learning data $\mathcal{L}$, only.

We then consider the nagging predictors on each of these index sets separately, i.e.,

$$\bar{\mu}^{\text{small}}(\boldsymbol{x}) = \frac{1}{|\mathcal{I}^{\text{small}}|} \sum_{m \in \mathcal{I}^{\text{small}}} \widehat{\mu}^m(\boldsymbol{x}),$$

$$\bar{\mu}^{\text{middle}}(\boldsymbol{x}) = \frac{1}{|\mathcal{I}^{\text{middle}}|} \sum_{m \in \mathcal{I}^{\text{middle}}} \widehat{\mu}^m(\boldsymbol{x}), \tag{7.46}$$

$$\bar{\mu}^{\text{large}}(\boldsymbol{x}) = \frac{1}{|\mathcal{I}^{\text{large}}|} \sum_{m \in \mathcal{I}^{\text{large}}} \widehat{\mu}^m(\boldsymbol{x}).$$

If we believe into the orange cubic spline in Fig. 7.17, the middle nagging predictor $\bar{\mu}^{\text{middle}}$ should out-perform the other two nagging predictors. Indeed, this is the case, here. We receive the out-of-sample losses (in 10^{-2}) on the three subsets

$$\mathfrak{D}(\mathcal{T}, \bar{\mu}^{\text{small}}) = 23.784, \qquad \mathfrak{D}(\mathcal{T}, \bar{\mu}^{\text{middle}}) = 23.272, \qquad \mathfrak{D}(\mathcal{T}, \bar{\mu}^{\text{large}}) = 23.782. \tag{7.47}$$

This approach boosts by far any other approach considered, see Table 7.10; note that this analysis relies on a fully proper in-sample and out-of-sample testing strategy. Moreover, this also supports our early stopping strategy because, obviously, the optimal networks are centered around our early stopping rule. How does this result match Proposition 7.25 saying that the nagging predictor has a monotonically

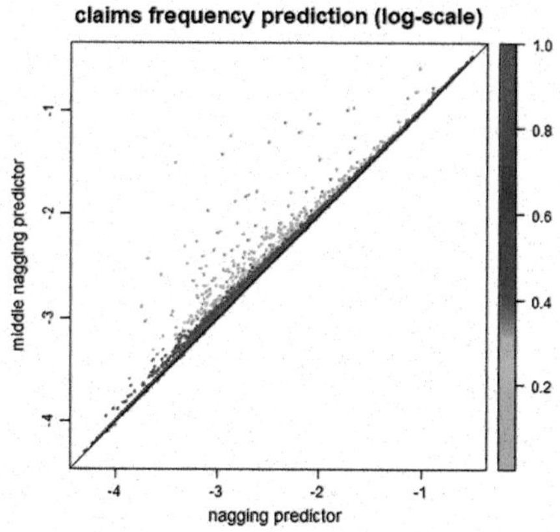

Fig. 7.29 Scatter plot of the nagging predictors $\bar{\mu}^{\text{middle}}(\boldsymbol{x}_t^\dagger)$ and $\bar{\mu}^{(1:M)}(\boldsymbol{x}_t^\dagger)$ over all out-of-sample polices $1 \leq t \leq T$; the color scale shows the sizes of the exposures $v_t^\dagger \in (0, 1]$

decreasing deviance loss. For the convergence (7.45) we need unbiasedness, and (7.47) indicates that averaging over all M network calibrations results in biases on an *individual* policy level; on the aggregate portfolio level, we have applied the bias regularization step (7.33), but this does not act on an individual policy level. The latter would require a local balance correction similar to the GAM approach presented in Example 7.19.

Figure 7.29 is truly striking! It compares the nagging predictors $\bar{\mu}^{(1:M)}(\boldsymbol{x}_t^\dagger)$ to the ones $\bar{\mu}^{\text{middle}}(\boldsymbol{x}_t^\dagger)$ only using the calibrations $m \in \mathcal{I}^{\text{middle}}$, i.e., only using the calibrations with middle sized in-sample losses. The different colors show the exposures $v_t^\dagger \in (0, 1]$. We observe that only portfolios with short exposures do not lie on the diagonal line. Thus, there seems to be an issue with insurance policies with short exposures. Recall that we model the Poisson claim counts N_i using the assumption, see (5.27),

$$N_i \sim \text{Poi}(v_i \mu(\boldsymbol{x}_i)). \tag{7.48}$$

That is, the expected claim count $\mathbb{E}_{\theta_i}[N_i] = v_i \mu(\boldsymbol{x}_i)$ is assumed to scale proportionally in the exposure $v_i > 0$. Figure 7.29 raises some doubts whether this is really the case, or at least SGD fitting has some difficulties to assess the expected frequencies $\mu(\boldsymbol{x}_i)$ on the policies i with short exposures $v_i > 0$. We discuss this further in the next subsection. Table 7.12 gives a summary of our results.

Analysis of Over-dispersion

With all the excitement of Fig. 7.29, the above models do not fit the observations since the over-dispersion is too large, see the last column of Table 7.12. This has motivated the study of the negative binomial model in Sect. 5.3.5, the ZIP model in Sect. 5.3.6, and the hurdle Poisson model in Example 6.19. These models have led to an improvement in terms of AIC, see Table 6.6. We could go down the same

Table 7.12 Number of parameters, in-sample and out-of-sample deviance losses (units are in 10^{-2}), in-sample average frequency and (over-)dispersion of the Poisson null model, model Poisson GLM3 of Table 5.5, the FN network model (with embedding layers of dimension $b = 2$), the nagging predictor, the meta network model, and the middle nagging predictor

	# param.	In-sample loss on $\mathcal{L}$	Out-of-sample loss on $\mathcal{T}$	Aver. freq.	Disp. $\widehat{\varphi}^P$
Poisson null	1	25.213	25.445	7.36%	1.7160
Poisson GLM3	50	24.084	24.102	7.36%	1.6644
Embed FN bias regularized $\widehat{\mu}^{m=1}$	792	23.690	23.824	7.36%	1.6812
Nagging FN $\bar{\mu}^{(1:M)}$	'792'	23.691	23.783	7.36%	1.6592
Meta FN network $\widehat{\mu}^{\text{meta}}$	792	23.714	23.777	7.36%	1.6737
Middle nagging FN $\bar{\mu}^{\text{middle}}$	'792'	23.698	23.272	7.36%	1.6618

route here by substituting the Poisson model. We refrain from doing so, as we want to further analyze the Poisson model. Suppose we calculate an AIC value for the Poisson FN network using 792 as the number of parameters involved. In that case, we receive a value of 191′790, thus, clearly lower than the one of the negative binomial GLM, and also slightly lower than the one of the hurdle Poisson model, see Table 6.6. Remark that AIC values within FN networks are not supported by any theory as we neither use the MLE nor do we have a reasonable evaluation of the number of parameters involved in networks. Thus, such a value may serve at best as a rough rule of thumb.

This lower AIC value suggests that we should try to improve the modeling of the systematic effects by better regression functions. In particular, there may be more explanatory variables involved that have predictive power. If these explanatory variables are latent, we can rely on the negative binomial model, as it can be interpreted as a mixture model averaging over latent variables. In view of Fig. 7.29, the exposures v_i seem to have a predictive power different from proportional scaling, see (7.48); we also mention some peculiarities of the exposures on page 556. This motivates to change the FN network regression model such that the exposures are considered non-proportionally. We choose a FN network that directly models the mean of the claim counts

$$(\boldsymbol{x}, v) \in \mathcal{X} \times (0, 1] \mapsto \mu(\boldsymbol{x}, v) = \exp\left\langle \boldsymbol{\beta}, \boldsymbol{z}^{(d:1)}(\boldsymbol{x}, v)\right\rangle > 0, \tag{7.49}$$

modeling the mean $\mathbb{E}_{\vartheta}[N] = \mu(\boldsymbol{x}, v)$ of the Poisson datum $(N, \boldsymbol{x}, v)$. The expected frequency is then given by $\mathbb{E}_{\vartheta}[Y] = \mathbb{E}_{\vartheta}[N/v] = \mu(\boldsymbol{x}, v)/v$.

Remark 7.29 At this stage we clearly have to distinguish between statistical modeling and actuarial modeling. In statistical modeling it makes perfect sense to choose the regression function (7.49), since including the exposure in a non-proportional way may increase the predictive power of the model, at least this is what our data suggests.

From an actuarial point of view this approach should clearly be doubted. The typical exposure of car insurance policies is one calendar year, i.e., $v = 1$, if the renewals of insurance policies are accounted correctly. Shorter exposures may have a specific (non-predictable) reason, for example, the policyholder or the insurance company may terminate an insurance contract after a claim. Thus, if this is possible, the exposure is a random variable, too, and it clearly has a predictive power for claims prediction; in that case we lose the properties of the Poisson count process (having independent and stationary increments).

As a consequence, we should include the exposure proportionally from an actuarial modeling point of view. Nevertheless we do the modeling exercise based on the regression function (7.49), here. This will indicate the predictive power of the exposure, which may be thought of a proxy for another (non-available) explanatory variable. Moreover, if (7.49) allows for a good Poisson regression model, we have a simple way of bootstrapping from our data (conditionally on given exposures v).

We would also like to emphasize that if one feature component dominates all others in terms of the predictive power, then likely there is a leakage of information through this component, and this needs a more careful analysis.

We implement the FN network regression model (7.49) using again a network architecture of depth $d = 3$ with $(q_1, q_2, q_3) = (20, 15, 10)$ neurons. We use embedding layers for the two categorical variables `VehBrand` and `Region`, and we have 8 continuous/binary feature components. This is one more compared to Fig. 7.9 (rhs) because we also model the exposure v_i as a continuous input to the network. As a result, the dimension r of the network parameter $\boldsymbol{\vartheta} \in \mathbb{R}^r$ increases from 792 to 812 (because we have $q_1 = 20$ neurons in the first FN layer). We calculate the nagging predictor $\bar{\mu}^{(1:M)}$ of this network averaging over $M = 500$ individual (early stopped) FN network calibrations, the results are presented in Table 7.13.

Table 7.13 Number of parameters, in-sample and out-of-sample deviance losses (units are in 10^{-2}), in-sample average frequency and (over-)dispersion of the Poisson null model, model Poisson GLM3 of Table 5.5, the FN network models (with embedding layers of dimension $b = 2$), the nagging predictors, and the middle nagging predictors excluding and including exposures v_i as continuous network inputs

	# param.	In-sample loss on $\mathcal{L}$	Out-of-sample loss on $\mathcal{T}$	Aver. freq.	Disp. $\widehat{\varphi}^P$
Poisson null	1	25.213	25.445	7.36%	1.7160
Poisson GLM3	50	24.084	24.102	7.36%	1.6644
Embed FN $\widehat{\mu}^{m=1}$	792	23.690	23.824	7.36%	1.6812
Nagging FN $\bar{\mu}^{(1:M)}$	'792'	23.691	23.783	7.36%	1.6592
Middle nagging FN $\bar{\mu}^{\text{middle}}$	'792'	23.698	23.272	7.36%	1.6618
Exposure v: FN $\widehat{\mu}^{m=1}$	812	23.358	23.496	7.36%	1.0650
Exposure v: nagging FN $\bar{\mu}^{(1:M)}$	'812'	23.299	23.382	7.36%	1.0416
Exposure v: middle nagging FN $\bar{\mu}^{\text{middle}}$	'812'	23.303	23.299	7.36%	1.0427

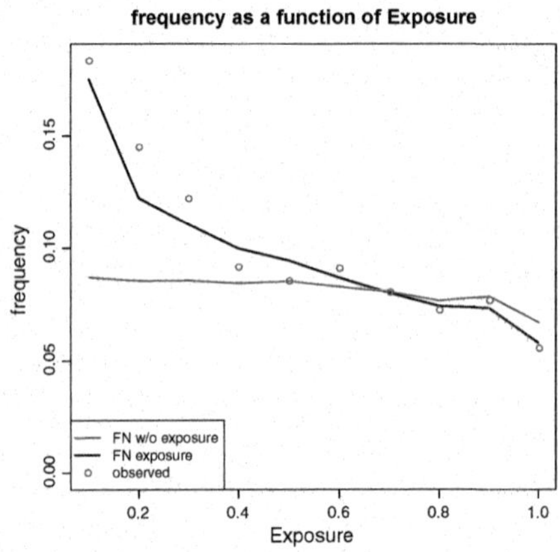

Fig. 7.30 Average frequency as a function of the exposure $v \in (0, 1]$: nagging predictors considering the exposures proportionally (blue), the model including exposures non-proportionally through the FN network (black) and observed (red)

We observe a major improvement when including the exposure v as an input to the network, i.e., by including the exposure non-proportionally into the mean estimate. This is true in-sample (we use early stopping here), and in terms of Pearson's dispersion estimate; we set $r = 812$ for the number of parameters in Pearson's dispersion estimate (5.30) which may be too big because we do not perform proper MLE, here. In particular, we receive a dispersion estimate close to one which, now, is in support of modeling the claim counts by Poisson random variables (using this regression function). That is, this regression function explains the systematic effects so that we no longer observe much over-dispersion in the data relative to the chosen model. However, we would like to remind of Remark 7.29 which needs a careful consideration for the use of this regression model in insurance practice.

This is also supported by Fig. 7.30 which studies the average frequency as a function of the exposure $v \in (0, 1]$. The red observed average frequency has a clear decreasing slope which can be modeled by running the exposure v through the FN network (black), but not by including it proportionally (blue). From an actuarial modeling point of view this plot clearly questions the quality of the data, because there seem to be effects in the exposures that certainly require more investigation. Unfortunately, we cannot do this here because we do not have additional insight into this data set. This closes the example.

7.4.5 *Identifiability in Feed-Forward Neural Networks*

In the previous section we have studied ensembles of FN networks. One may also aim at directly comparing these networks to each other in terms of the fitted network parameters $\widehat{\vartheta}^j$ over the different calibrations $1 \le j \le M$ (of the same FN network architecture). Such a comparison may, e.g., be useful if one wants to choose a prior parameter distribution π for $\boldsymbol{\vartheta}$ in a Bayesian setting. Comparing the different network calibrations $\widehat{\boldsymbol{\vartheta}}^j$, $1 \leq j \leq M$, of an architecture needs some care because networks have many symmetries that make the parameters non-identifiable. We can, for instance, permute the neurons in a FN layer $z^{(m)}$, with the corresponding permutation of the weights that connect this layer to the previous layer $z^{(m-1)}$ and to the succeeding layer $z^{(m+1)}$. The resulting predictive model under this permutation is the same as the original one. For this reason we need to introduce some order in a FN network to make the parameters identifiable.

Rüger–Ossen [323] have introduced the notion of a fundamental domain for the network parameter $\boldsymbol{\vartheta}$, and we briefly review this idea. We start with an explicit example. Assume that the activation function fulfills the anti-symmetry property $-\phi(x) = \phi(-x)$ for all $x \in \mathbb{R}$, this is the case for the hyperbolic tangent. This implies several symmetries in the FN network parametrization. E.g., if we consider the output of a shallow FN network $d = 1$ with link function g, we can do a sign switch in a fixed neuron $1 \leq k \leq q_1$

$$
\begin{aligned}
g(\mu(\boldsymbol{x})) = \beta_0 + \sum_{j=1}^{q_1} \beta_j z_j^{(1:1)}(\boldsymbol{x}) &= \beta_0 + \sum_{j=1}^{q_1} \beta_j\, \phi\left\langle \boldsymbol{w}_j^{(1)}, \boldsymbol{x}\right\rangle \\
&= \beta_0 + \sum_{j \neq k} \beta_j\, \phi\left\langle \boldsymbol{w}_j^{(1)}, \boldsymbol{x}\right\rangle + (-\beta_k)\, \phi\left\langle -\boldsymbol{w}_k^{(1)}, \boldsymbol{x}\right\rangle. \qquad (7.50)
\end{aligned}
$$

From this we see that the following two network parameters (we switch signs in all the parameters that belong to index k)

$$
\begin{aligned}
\boldsymbol{\vartheta} &= (\boldsymbol{w}_1^{(1)}, \ldots, \boldsymbol{w}_k^{(1)}, \ldots, \boldsymbol{w}_{q_1}^{(1)}, \beta_0, \ldots, \beta_k, \ldots, \beta_{q_1})^\top \qquad \text{and} \\
\widetilde{\boldsymbol{\vartheta}} &= (\boldsymbol{w}_1^{(1)}, \ldots, -\boldsymbol{w}_k^{(1)}, \ldots, \boldsymbol{w}_{q_1}^{(1)}, \beta_0, \ldots, -\beta_k, \ldots, \beta_{q_1})^\top
\end{aligned}
$$

give the same FN network predictions. Beside these sign switches, we can also permute the enumeration of the neurons in a given FN layer, giving the same predictions. We discuss Theorem 2 of Rüger–Ossen [323] to solve this identifiability issue. First, we consider the network weights from the input $\boldsymbol{x}$ to the first FN layer $z^{(1)}(\boldsymbol{x})$. Apply the sign switch operation (7.50) to the neurons in the first FN layer so that all the resulting intercepts $w_{0,1}^{(1)}, \ldots, w_{0,q_1}^{(1)}$ are positive while not changing the regression function $\boldsymbol{x} \mapsto g(\mu(\boldsymbol{x}))$. Next, apply a permutation to the indices $1 \leq j \leq q_1$ so that we receive ordered intercepts

$$
w_{0,1}^{(1)} > \ldots > w_{0,q_1}^{(1)} > 0,
$$

with an unchanged regression function $\boldsymbol{x} \mapsto g(\mu(\boldsymbol{x}))$. To make these transformations well-defined we need to assume that all intercepts are non-zero and mutually different (which we assume for the time-being).

Then, we move recursively through the FN layers $2 \leq m \leq d$ applying the sign switch operations and the permutations so that the regression function $\boldsymbol{x} \mapsto g(\mu(\boldsymbol{x}))$ remains unchanged and such that for all $1 \leq m \leq d$

$$w_{0,1}^{(m)} > \ldots > w_{0,q_m}^{(m)} > 0.$$

This provides us with a unique representation of every network parameter $\boldsymbol{\vartheta} \in \mathbb{R}^r$ in the *fundamental domain*

$$\left\{ \boldsymbol{\vartheta} \in \mathbb{R}^r;\ w_{0,1}^{(m)} > \ldots > w_{0,q_m}^{(m)} > 0 \text{ for all } 1 \leq m \leq d \right\} \subset \mathbb{R}^r, \tag{7.51}$$

supposed that all intercepts are different from zero and mutually different in the same FN layers. As stated in Section 2.2 of Rüger–Ossen [323], there may still exist different parameters in this fundamental domain that provide the same predictive model, but these are of zero Lebesgue measure. The same applies to the intercepts $w_{0,j}^{(m)}$ being zero or having equal intercepts for different neurons. Basically, this means that we are fine if we work with absolutely continuous prior distributions on the fundamental domain when we want to work within a Bayesian setup.

7.5 Auto-encoders

Auto-encoders are tools that aim at reducing the dimension of high-dimensional data such that the reconstruction error of the original data is small, i.e., such that the loss of information by the dimension reduction is minimized. The most popular auto-encoder is the principal components analysis (PCA) which we are going to present here. The PCA is a linear dimension reduction technique. Bottleneck neural (BN) networks can be viewed as a non-linear extension of the PCA. This is going to be discussed in Sect. 7.5.5, below. Dimension reduction techniques belong to the family of unsupervised learning methods because they do not consider a response variable, but they aim at finding common structure in the features. Unsupervised learning methods can roughly be categorized into three classes: dimension reduction techniques (studied in this section), clustering methods and visualization methods. For a discussion of clustering and visualization methods we refer to the tutorial of Rentzmann–Wüthrich [310].

7.5.1 Standardization of the Data Matrix

Assume we have q-dimensional data points $\boldsymbol{y}_i \in \mathbb{R}^q$, $1 \le i \le n$. This provides us with a data matrix

$$\boldsymbol{Y} = (\boldsymbol{y}_1, \ldots, \boldsymbol{y}_n)^\top = \begin{pmatrix} y_{1,1} & \cdots & y_{1,q} \\ \vdots & \ddots & \vdots \\ y_{n,1} & \cdots & y_{n,q} \end{pmatrix} \in \mathbb{R}^{n \times q}.$$

We assume that each of the q columns of $\boldsymbol{Y}$ measures a quantity in a given unit. The first column may, for instance, describe the age of a car driver in years, the second column his body weight in kilograms, etc. That is, each column $1 \le j \le q$ of $\boldsymbol{Y}$ describes a specific quantity, and each row $\boldsymbol{y}_i^\top$ of $\boldsymbol{Y}$ describes these quantities for a given instance $1 \le i \le n$. Since often the analysis should not depend on the units of the columns of $\boldsymbol{Y}$, one centers the columns with the empirical means $\bar{y}_j = \sum_{i=1}^n y_{i,j}/n$, and one normalizes them with the empirical standard deviations $\widehat{\sigma}_j = (\sum_{i=1}^n (y_{i,j} - \bar{y}_j)^2/n)^{1/2}$, $1 \le j \le q$. This gives the normalized data matrix

$$\begin{pmatrix} \frac{y_{1,1}-\bar{y}_1}{\widehat{\sigma}_1} & \cdots & \frac{y_{1,q}-\bar{y}_q}{\widehat{\sigma}_q} \\ \vdots & \ddots & \vdots \\ \frac{y_{n,1}-\bar{y}_1}{\widehat{\sigma}_1} & \cdots & \frac{y_{n,q}-\bar{y}_q}{\widehat{\sigma}_q} \end{pmatrix} \in \mathbb{R}^{n \times q}. \tag{7.52}$$

We typically center the data matrix $\boldsymbol{Y}$, providing $\sum_{i=1}^n y_{i,j} = 0$ for all $1 \le j \le q$, normalization w.r.t. the standard deviation can be done, but is not always necessary. Centering implies that we can interpret $\boldsymbol{Y}$ as a q-dimensional empirical distribution with each component (column) being centered. The covariance matrix of this (centered) empirical distribution is calculated as

$$\widehat{\Sigma} = \frac{1}{n} \left(\sum_{i=1}^n y_{i,j} y_{i,k} \right)_{1 \le j,k \le q} = \frac{1}{n} \boldsymbol{Y}^\top \boldsymbol{Y} \in \mathbb{R}^{q \times q}. \tag{7.53}$$

This is a covariance matrix, and if the columns of $\boldsymbol{Y}$ are normalized with the empirical standard deviations $\widehat{\sigma}_j$, $1 \le j \le q$, this is a correlation matrix.

7.5.2 Introduction to Auto-encoders

An auto-encoder encodes a high-dimensional vector $\boldsymbol{y} \in \mathbb{R}^q$ to a low-dimensional representation so that the dimension reduction leads to a minimal loss of information. A function $L(\cdot,\cdot) : \mathbb{R}^q \times \mathbb{R}^q \to \mathbb{R}_+$ is called *dissimilarity function* if $L(\boldsymbol{y}, \boldsymbol{y}') = 0$ if and only if $\boldsymbol{y} = \boldsymbol{y}'$.

An auto-encoder is a pair (Φ, Ψ) of mappings, for given dimensions $p < q$,

$$\Phi : \mathbb{R}^q \to \mathbb{R}^p \qquad \text{and} \qquad \Psi : \mathbb{R}^p \to \mathbb{R}^q, \tag{7.54}$$

such that their composition $\Psi \circ \Phi$ has a small reconstruction error w.r.t. the chosen dissimilarity function $L(\cdot, \cdot)$, that is,

$$\boldsymbol{y} \mapsto L\left(\boldsymbol{y}, \Psi \circ \Phi(\boldsymbol{y})\right) \quad \text{is small for all cases } \boldsymbol{y} \text{ of interest.} \tag{7.55}$$

Note that we want (7.55) for selected cases $\boldsymbol{y}$, and if they are within a p-dimensional manifold the auto-encoding will be successful. The first mapping $\Phi : \mathbb{R}^q \to \mathbb{R}^p$ is called encoder, and the second mapping $\Psi : \mathbb{R}^p \to \mathbb{R}^q$ is called decoder. The object $\Phi(\boldsymbol{y}) \in \mathbb{R}^p$ is a p-dimensional encoding (representation) of $\boldsymbol{y} \in \mathbb{R}^q$ which contains maximal information of $\boldsymbol{y}$ up to the reconstruction error (7.55).

7.5.3 Principal Components Analysis

PCA gives us a linear auto-encoder (7.54). If the data matrix $\boldsymbol{Y} \in \mathbb{R}^{n \times q}$ has rank q, there exist q linearly independent rows of $\boldsymbol{Y}$ that span $\mathbb{R}^q$. PCA determines a different, very specific basis of $\mathbb{R}^q$. It looks for an orthonormal basis $\boldsymbol{v}_1, \ldots, \boldsymbol{v}_q \in \mathbb{R}^q$ such that $\boldsymbol{v}_1$ explains the direction of the biggest variability in $\boldsymbol{Y}$, $\boldsymbol{v}_2$ the direction of the second biggest variability in $\boldsymbol{Y}$ orthogonal to $\boldsymbol{v}_1$, and so forth. Variability is understood in the sense of maximal empirical variance under the assumption that the columns of $\boldsymbol{Y}$ are centered, see (7.52)–(7.53). Such an orthonormal basis can be found by determining q linearly independent eigenvectors of the symmetric and positive definite matrix

$$A = n\widehat{\Sigma} = \boldsymbol{Y}^\top \boldsymbol{Y} \in \mathbb{R}^{q \times q}.$$

For this we can solve recursively the following convex Lagrange problems. The first basis vector $\boldsymbol{v}_1 \in \mathbb{R}^q$ is determined by the solution of[3]

$$\boldsymbol{v}_1 \;=\; \underset{\|\boldsymbol{w}\|_2 = 1}{\arg\max}\, \|\boldsymbol{Y}\boldsymbol{w}\|_2^2 \;=\; \underset{\boldsymbol{w}^\top \boldsymbol{w} = 1}{\arg\max} \left(\boldsymbol{w}^\top \boldsymbol{Y}^\top \boldsymbol{Y} \boldsymbol{w}\right), \tag{7.56}$$

and the j-th basis vector $\boldsymbol{v}_j \in \mathbb{R}^q, 2 \le j \le q$, is received recursively by the solution of

$$\boldsymbol{v}_j \;=\; \underset{\|\boldsymbol{w}\|_2 = 1}{\arg\max}\, \|\boldsymbol{Y}\boldsymbol{w}\|_2^2 \qquad \text{subject to } \langle \boldsymbol{v}_k, \boldsymbol{w} \rangle = 0 \text{ for all } 1 \le k \le j-1. \tag{7.57}$$

[3] If the q eigenvalues of A are distinct, the solution to (7.56) and (7.57) is unique up to the sign, otherwise this requires more care.

Singular value decomposition (SVD) gives an alternative way of computing this orthonormal basis, we refer to Section 14.5.1 in Hastie et al. [183]. The algorithm of Golub–Van Loan [165] gives an efficient way of performing a SVD. There exist orthogonal matrices $U \in \mathbb{R}^{n\times q}$ and $V \in \mathbb{R}^{q\times q}$ (with $U^\top U = V^\top V = \mathbb{1}_q$), and a diagonal matrix $\Lambda = \text{diag}(\lambda_1, \ldots, \lambda_q) \in \mathbb{R}^{q\times q}$ with singular values $\lambda_1 \geq \ldots \geq \lambda_q > 0$ such that we have the SVD

$$\boldsymbol{Y} = U\Lambda V^\top. \tag{7.58}$$

The matrix U is called left-singular matrix of $\boldsymbol{Y}$, and the matrix V is called right-singular matrix of $\boldsymbol{Y}$. Observe by using the SVD (7.58)

$$V^\top A V = V^\top \boldsymbol{Y}^\top \boldsymbol{Y} V = V^\top V \Lambda U^\top U \Lambda V^\top V = \Lambda^2 = \text{diag}(\lambda_1^2, \ldots, \lambda_q^2).$$

That is, the squared singular values $(\lambda_j^2)_{1\leq j\leq q}$ are the eigenvalues of matrix A, and the column vectors of the right-singular matrix $V = (\boldsymbol{v}_1, \ldots, \boldsymbol{v}_q)$ (eigenvectors of A) give an orthonormal basis $\boldsymbol{v}_1, \ldots, \boldsymbol{v}_q$. This motivates to define the q principal components of $\boldsymbol{Y}$ by the column vectors of

$$\begin{aligned} \boldsymbol{Y}V &= U\Lambda = U\text{diag}(\lambda_1, \ldots, \lambda_q) \\ &= \left(\lambda_1 \boldsymbol{u}_1, \ldots, \lambda_q \boldsymbol{u}_q\right) \in \mathbb{R}^{n\times q}. \end{aligned} \tag{7.59}$$

E.g., the first principal component of the instances $1 \leq i \leq n$ is given by $\boldsymbol{Y}\boldsymbol{v}_1 = \lambda_1 \boldsymbol{u}_1 \in \mathbb{R}^n$. Considering the first $p \leq q$ principal components gives the rank p matrix

$$\boldsymbol{Y}_p = U\text{diag}(\lambda_1, \ldots, \lambda_p, 0, \ldots, 0)V^\top \in \mathbb{R}^{n\times q}. \tag{7.60}$$

The Eckart–Young–Mirsky theorem [114, 279][4] proves that this rank p matrix $\boldsymbol{Y}_p$ minimizes the Frobenius norm relative to $\boldsymbol{Y}$ among all rank p matrices, that is,

$$\boldsymbol{Y}_p \in \underset{B\in\mathbb{R}^{n\times q}}{\arg\min} \|\boldsymbol{Y} - B\|_F \qquad \text{subject to rank}(B) \leq p, \tag{7.61}$$

where the Frobenius norm is given by $\|C\|_F^2 = \sum_{i,j} c_{i,j}^2$ for a matrix $C = (c_{i,j})_{i,j}$.

The orthonormal basis $\boldsymbol{v}_1, \ldots, \boldsymbol{v}_q \in \mathbb{R}^q$ gives the (linear) encoder (projection)

$$\Phi : \mathbb{R}^q \to \mathbb{R}^p, \qquad \boldsymbol{y} \mapsto \Phi(\boldsymbol{y}) = \left(\boldsymbol{y}^\top \boldsymbol{v}_1, \ldots, \boldsymbol{y}^\top \boldsymbol{v}_p\right)^\top = (\boldsymbol{v}_1, \ldots, \boldsymbol{v}_p)^\top \boldsymbol{y}.$$

[4] In fact, (7.61) holds for both the Frobenius norm and the spectral norm.

These gives the first p principal components in (7.59) if we insert the transposed data matrix $\boldsymbol{Y}^\top = (\boldsymbol{y}_1, \ldots, \boldsymbol{y}_n) \in \mathbb{R}^{q\times n}$ for $\boldsymbol{y} \in \mathbb{R}^q$. The (linear) decoder Ψ is given by

$$\Psi : \mathbb{R}^p \to \mathbb{R}^q, \qquad \boldsymbol{z} \mapsto \Psi(\boldsymbol{z}) = (\boldsymbol{v}_1, \ldots, \boldsymbol{v}_p)\boldsymbol{z}.$$

The following is understood column-wise for the transposed data matrix $\boldsymbol{Y}^\top$,

$$\begin{aligned}\Psi \circ \Phi(\boldsymbol{Y}^\top) &= \Psi\left((\boldsymbol{v}_1, \ldots, \boldsymbol{v}_p)^\top \boldsymbol{Y}^\top\right)\\ &= \left(\boldsymbol{Y}(\boldsymbol{v}_1, \ldots, \boldsymbol{v}_p)(\boldsymbol{v}_1, \ldots, \boldsymbol{v}_p)^\top\right)^\top\\ &= \left(\boldsymbol{Y}(\boldsymbol{v}_1, \ldots, \boldsymbol{v}_p, 0, \ldots, 0)(\boldsymbol{v}_1, \ldots, \boldsymbol{v}_p, \boldsymbol{v}_{p+1}, \ldots, \boldsymbol{v}_q)^\top\right)^\top\\ &= \left(U\mathrm{diag}(\lambda_1, \ldots, \lambda_p, 0, \ldots, 0)V^\top\right)^\top \;=\; \boldsymbol{Y}_p^\top.\end{aligned}$$

Thus, $\Psi \circ \Phi(\boldsymbol{Y}^\top)$ minimizes the Frobenius reconstruction error (7.61) on the data matrix $\boldsymbol{Y}^\top$ among all linear maps of rank p. In view of (7.55) we can express the squared Frobenius reconstruction error as

$$\|\boldsymbol{Y} - \boldsymbol{Y}_p\|_F^2 = \sum_{i=1}^n \left\|\boldsymbol{y}_i - \Psi \circ \Phi(\boldsymbol{y}_i)\right\|_2^2 = \sum_{i=1}^n L\left(\boldsymbol{y}_i, \Psi \circ \Phi(\boldsymbol{y}_i)\right), \tag{7.62}$$

thus, we choose the squared Euclidean distance as the dissimilarity measure, here, that we minimize simultaneously on all cases $\boldsymbol{y}_i$, $1 \le i \le n$.

Remark 7.30 The PCA gives a linear approximation to the data matrix $\boldsymbol{Y}$ by minimizing (7.61) and (7.62) for given rank p. This may not be appropriate if the non-linear terms are dominant. Figure 7.31 (lhs) gives a situation where the PCA works well; this data has been generated by i.i.d. multivariate Gaussian random vectors $\boldsymbol{y}_i \sim \mathcal{N}(\boldsymbol{0}, \Sigma)$. Figure 7.31 (middle) gives a non-linear example where the PCA does not work well, the data matrix $\boldsymbol{Y} \in \mathbb{R}^{n\times 2}$ is a column-centered matrix that builds a circle around the origin.

Another nice example where the PCA fails is Fig. 7.31 (rhs). This figure is inspired by Shlens [337] and Ruckstuhl [321]. It shows a situation where the level sets are non-convex, and the principal components point into a completely wrong direction to explain the structure of the data.

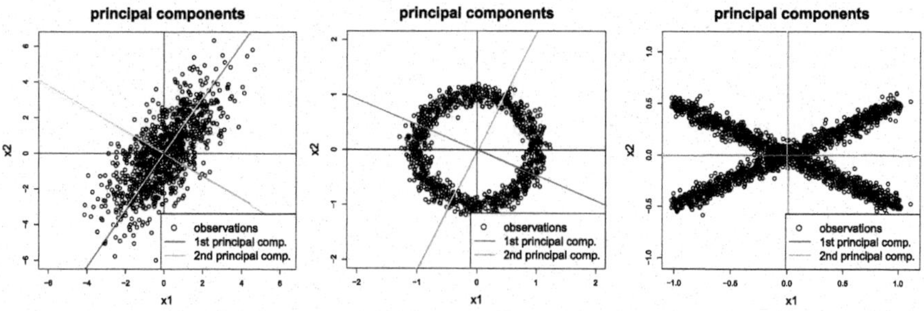

Fig. 7.31 Two-dimensional PCAs in different situations of the data matrix $\boldsymbol{Y} \in \mathbb{R}^{n\times 2}$

7.5.4 Lab: Lee–Carter Mortality Model

We use the SVD to fit the most popular stochastic mortality model, the Lee–Carter (LC) model [238], to (raw) mortality data. The raw mortality data considers for each calendar year t and each age x the number of people $D_{x,t}$ who died (in that year t at age x) divided by the corresponding population exposure $e_{x,t}$. In practice this requires some care. Due to migration, often, the exposures $e_{x,t}$ are non-observable figures and need to be estimated. Moreover, also the death counts $D_{x,t}$ in year t at age x can be defined differently, age cohorts are usually defined by the year of birth. We denote the (observed) raw mortality rates by $M_{x,t} = D_{x,t}/e_{x,t}$. The subsequent derivations consider the raw log-mortality rates $\log(M_{x,t})$, for this reason we assume that $M_{x,t} > 0$ for all calendar years t and ages x. The goal is to model these raw log-mortality rates (for each country, region, risk group and gender separately).

The LC model defines the force of mortality as

$$\log(\mu_{x,t}) = a_x + b_x k_t, \tag{7.63}$$

where $\log(\mu_{x,t})$ is the (deterministic) log-mortality rate in calendar year t for a person aged x (for a fixed country, region and gender). The individual terms in (7.63) have the following meaning: a_x is the average force of mortality at age x, b_x is the rate of change of the force of mortality broken down to the different ages x, and k_t is the time index describing the change of the force of mortality in calendar year t.

Strictly speaking, we do not have a stochastic model, here, that can explain the observations $M_{x,t}$, but we try to fit a deterministic mortality surface $(\mu_{x,t})_{x,t}$ to these noisy observations $(M_{x,t})_{x,t}$. For this we use the PCA and the Frobenius norm as the measure of dissimilarity (on the log-scale).

In a first step, we center the raw log-mortality rates for all ages x, i.e., over the calendar years $t \in \mathcal{T}$ under consideration. We define the centered raw log-mortality rates $y_{x,t}$ and the estimate $\widehat{a}_x$ of the average force of mortality at age x as follows

$$Y_{x,t} \;=\; \log(M_{x,t}) - \widehat{a}_x \;=\; \log(M_{x,t}) - \frac{1}{|\mathcal{T}|}\sum_{s\in\mathcal{T}} \log(M_{x,s}), \tag{7.64}$$

where the last identity defines the estimate $\widehat{a}_x$. Strictly speaking we have a slight difference to the centering in Sect. 7.5.1 because we center the rows and not the columns of the data matrix, here, but the role of rows and columns is exchangeable in the PCA. The optimal (parameter) values $(\widehat{b}_x)_x$ and $(\widehat{k}_t)_t$ are determined as follows, see (7.63),

$$\underset{(b_x)_x,(k_t)_t}{\arg\min} \sum_{x,t} \left(Y_{x,t} - b_x k_t\right)^2,$$

where the sum runs over the years $t \in \mathcal{T}$ and the ages $x_0 \leq x \leq x_1$, with x_0 and x_1 being the lower and upper age boundaries. This can be rewritten as an optimization problem (7.61)–(7.62). Consider the data matrix $\boldsymbol{Y} = (Y_{x,t})_{x_0 \leq x \leq x_1; t \in \mathcal{T}} \in \mathbb{R}^{n \times q}$, and set $n = x_1 - x_0 + 1$ and $q = |\mathcal{T}|$. Assume $\boldsymbol{Y}$ has rank q. This allows us to consider

$$\boldsymbol{Y}_1 \in \underset{B \in \mathbb{R}^{n \times q}}{\arg\min} \|\boldsymbol{Y} - B\|_{\mathrm{F}} \qquad \text{subject to } \mathrm{rank}(B) \leq 1.$$

A solution to this problem is given, see (7.60),

$$\boldsymbol{Y}_1 = U \mathrm{diag}(\lambda_1, 0, \ldots, 0) V^\top = (\lambda_1 \boldsymbol{u}_1)\, \boldsymbol{v}_1^\top = (\boldsymbol{Y} \boldsymbol{v}_1)\, \boldsymbol{v}_1^\top \in \mathbb{R}^{n \times q},$$

with left-singular matrix $U = (\boldsymbol{u}_1, \ldots, \boldsymbol{u}_q) \in \mathbb{R}^{n \times q}$ and right-singular matrix $V = (\boldsymbol{v}_1, \ldots, \boldsymbol{v}_q) \in \mathbb{R}^{q \times q}$ of $\boldsymbol{Y}$. This implies that the first principal component $\lambda_1 \boldsymbol{u}_1 = \boldsymbol{Y} \boldsymbol{v}_1 \in \mathbb{R}^n$ gives an estimate for $(b_x)_{x_0 \leq x \leq x_1}$, and the first column vector $\boldsymbol{v}_1 \in \mathbb{R}^q$ of V gives an estimate for the time index $(k_t)_{t \in \mathcal{T}}$. For parameter identifiability we normalize

$$\sum_{x=x_0}^{x_1} \widehat{b}_x = 1 \qquad \text{and} \qquad \sum_{t \in \mathcal{T}} \widehat{k}_t = 0, \tag{7.65}$$

the latter being consistent with the centering of the rows of $\boldsymbol{Y}$ with $\widehat{a}_x$ in (7.64).

We fit the LC model to the Swiss mortality data of females and males separately. The raw log-mortality rates $\log(M_{x,t})$ for the years $t \in \mathcal{T} = \{1950, \ldots, 2016\}$ and the ages $0 \leq x \leq 99$ are illustrated in Fig. 7.32; both plots use the same color scale. This mortality data has been obtained from the Human Mortality Database (HMD) [195]. In general, we observe a diagonal structure that indicates mortality improvements over time.

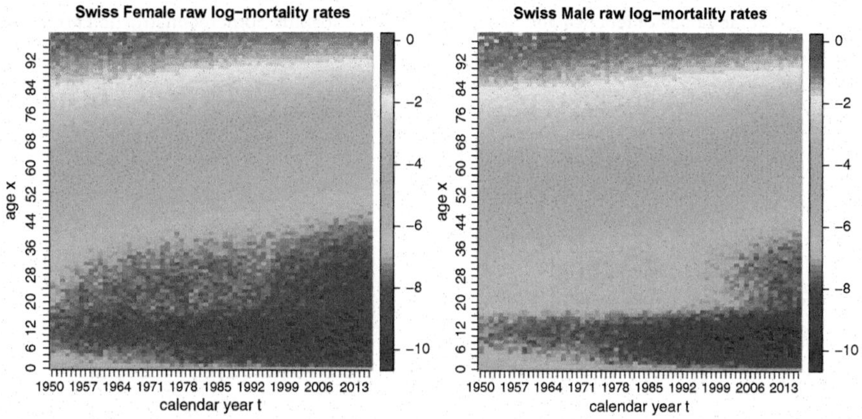

Fig. 7.32 Raw log-mortality rates $\log(M_{x,t})$ for the calendar years $1950 \leq t \leq 2016$ and the ages $x_0 = 0 \leq x \leq x_1 = 99$ of Swiss females (lhs) and Swiss males (rhs); both plots use the same color scale

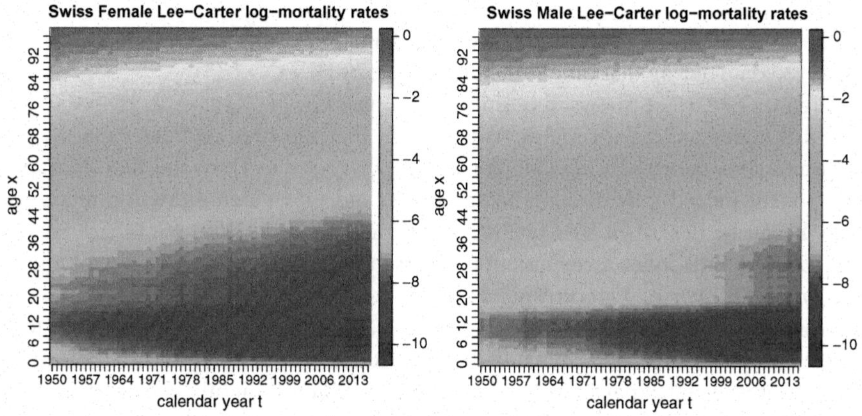

Fig. 7.33 LC fitted log-mortality rates $\log(\widehat{\mu}_{x,t})$ for the calendar years $1950 \leq t \leq 2016$ and the ages $x_0 = 0 \leq x \leq x_1 = 99$ of Swiss females (lhs) and Swiss males (rhs); the plots use the same color scale as Fig. 7.32

Define the fitted log-mortality surface

$$\log(\widehat{\mu}_{x,t}) = \widehat{a}_x + \widehat{b}_x \widehat{k}_t \qquad \text{for } x_0 \leq x \leq x_1 \text{ and } t \in \mathcal{T}.$$

Figure 7.33 shows the LC fitted log-mortality surface $(\log(\widehat{\mu}_{x,t}))_{0 \leq x \leq 99; t \in \mathcal{T}}$ separately for Swiss females and Swiss males, the color scale is the same as in Fig. 7.32. The plots show a huge similarity between the raw log-mortality data and the LC fitted log-mortality surface which clearly supports the LC model for the Swiss data. In general, the LC surface is a smoothed version of the raw log-mortality surface. The main difference in our LC fit concerns the male population for ages

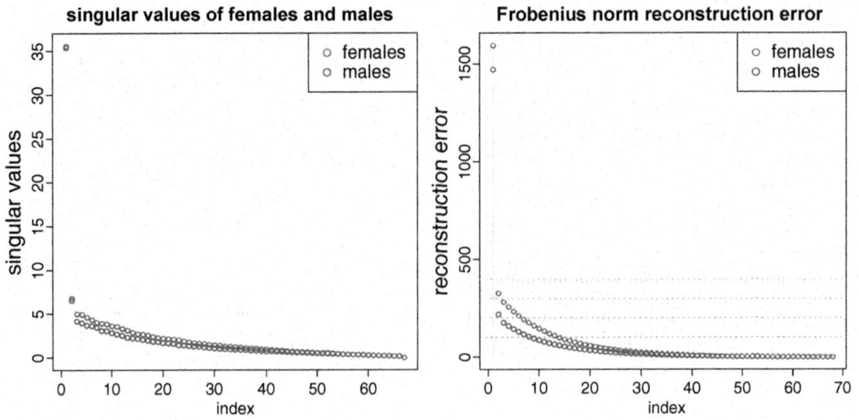

Fig. 7.34 (lhs) Singular values λ_j, $1 \leq j \leq |\mathcal{T}|$, of the SVD of the data matrix $\boldsymbol{Y} \in \mathbb{R}^{n \times |\mathcal{T}|}$, and (rhs) the reconstruction errors $\|\boldsymbol{Y} - \boldsymbol{Y}_p\|_F^2$ for $0 \leq p \leq |\mathcal{T}|$

$20 \leq x \leq 40$ from 1980 to 2000, one explanation of the special pattern in the observed data during that time is the emergence of HIV.

Figure 7.34 (lhs) shows the singular values $\lambda_1 \geq \ldots \geq \lambda_{|\mathcal{T}|} > 0$ for Swiss females and Swiss males. We observe that the first singular value λ_1 by far dominates the remaining singular values λ_j, $j \geq 2$. Thus, the first principal component indeed may already be sufficient, and the centered raw log-mortality data $\boldsymbol{Y}$ can be described by a matrix $\boldsymbol{Y}_1$ of rank $p = 1$. Figure 7.34 (rhs) gives the squared Frobenius reconstruction errors of the approximations $\boldsymbol{Y}_p$ of ranks $0 \leq p \leq |\mathcal{T}|$, where $\boldsymbol{Y}_0$ corresponds to the zero matrix where we do not use any approximation, but use just the average observed log-mortality rate. We observe that the first singular value leads by far to the biggest decrease in the reconstruction error, and the subsequent expansions λ_j, $j \geq 2$, improve it only slightly in each step. This supports the use of the LC model using a rank $p = 1$ approximation to the centered raw log-mortality rates $\boldsymbol{Y}$. The higher rank PCA within mortality modeling has been studied in Renshaw–Haberman (RH) [308], and the RH(p) mortality model considers the rank p approximation $\boldsymbol{Y}_p$ to the raw log-mortality rates $\boldsymbol{Y}$ given by

$$\log(\mu_{x,t}) = a_x + \langle \boldsymbol{b}_x, \boldsymbol{k}_t \rangle,$$

for $\boldsymbol{b}_x, \boldsymbol{k}_t \in \mathbb{R}^p$.

We have (only) fitted a mortality surface to the raw log-mortality rates on the rectangle $\{x_0, \ldots, x_1\} \times \mathcal{T}$. This does not allow us to forecast mortality into the future. Forecasting requires a two step procedure, which, after this first estimation step, extrapolates the time index (time-series) $(\widehat{k}_t)_{t \in \mathcal{T}}$ beyond the latest observation point in $\mathcal{T}$. The simplest (meaningful) model for this second (extrapolation) step is a random walk with drift for the time index process $(\widehat{k}_t)_{t \geq 0}$. Figure 7.35 shows the estimated two-dimensional process $(\widehat{\boldsymbol{k}}_t)_{t \in \mathcal{T}}$, i.e., for $p = 2$, on the rectangle

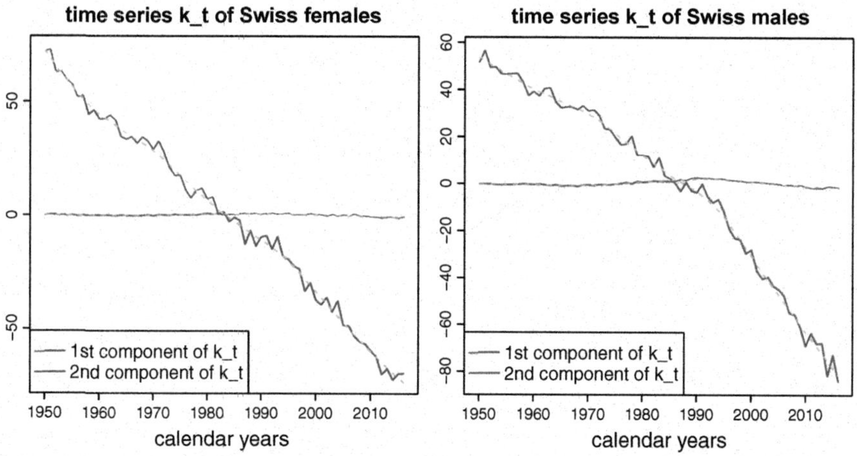

Fig. 7.35 Estimated two-dimensional processes $(\widehat{\boldsymbol{k}}_t)_{t\in\mathcal{T}}$ for Swiss females (lhs) and Swiss males (rhs); these are normalized such that they are centered and such that the components of $\widehat{\boldsymbol{b}}_x$ add up to 1

$\{x_0, \ldots, x_1\} \times \mathcal{T}$ which needs to be extrapolated to predict within the RH ($p = 2$) mortality model. We refrain from doing this step, but extrapolation will be studied in Sect. 8.4, below.

7.5.5 Bottleneck Neural Network

BN networks have become popular in studying non-linear generalizations of PCA, we refer to Kramer [225] and Hinton–Salakhutdinov [186]. The BN network architecture is such that (1) the input dimension q_0 is equal to the output dimension q_{d+1} of a FN network, and (2) in between there is a FN layer $1 \le m \le d$ that has a very low dimension $q_m \ll q_0$, called the bottleneck. Figure 7.36 (lhs) shows such a BN network of depth $d = 3$ and neurons

$$(q_0, q_1, q_2, q_3, q_4) = (20, 7, 2, 7, 20).$$

The input and output neurons have blue color, and the bottleneck of dimension $q_2 = 2$ is shown in red color in Fig. 7.36 (lhs).

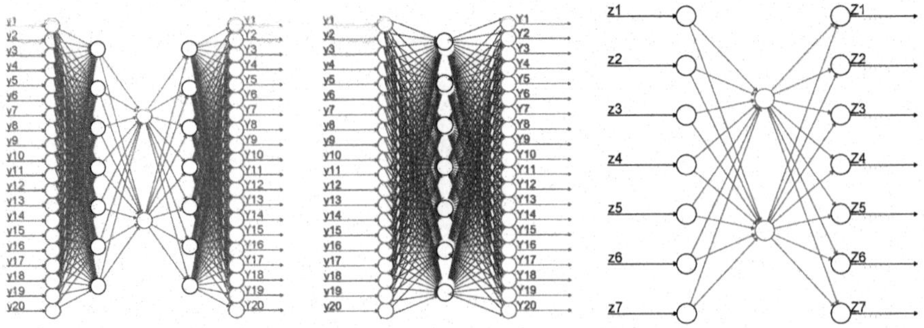

Fig. 7.36 (lhs) BN network of depth $d = 3$ with $(q_0, q_1, q_2, q_3, q_4) = (20, 7, 2, 7, 20)$, (middle and rhs) shallow BN networks with a bottleneck of dimensions 7 and 2, respectively

The motivation is as follows. Assume we have a given dissimilarity function $L(\cdot, \cdot) : \mathbb{R}^q \times \mathbb{R}^q \to \mathbb{R}_+$ that measures the reconstruction error of an auto-encoder $\Psi \circ \Phi(\boldsymbol{y}) \in \mathbb{R}^q$ relative to the original input $\boldsymbol{y} \in \mathbb{R}^q$, see (7.55). We try to find a BN network with input and output dimensions $q_0 = q_{d+1} = q$ (we drop the intercepts in the entire construction) and a bottleneck in layer m having a low dimension q_m, such that the BN network provides a small reconstruction error. Choose a FN network

$$\boldsymbol{y} \in \mathbb{R}^q \;\mapsto\; \Psi \circ \Phi(\boldsymbol{y}) = \boldsymbol{z}^{(d+1:1)}(\boldsymbol{y}) = \left(\boldsymbol{z}^{(d+1)} \circ \boldsymbol{z}^{(d)} \circ \cdots \circ \boldsymbol{z}^{(1)}\right)(\boldsymbol{y}) \in \mathbb{R}^q,$$

with FN layers for $1 \le m \le d$ (excluding intercepts)

$$\boldsymbol{z}^{(m)} : \mathbb{R}^{q_{m-1}} \to \mathbb{R}^{q_m}, \qquad \boldsymbol{z} \mapsto \boldsymbol{z}^{(m)}(\boldsymbol{z}) = \left(\phi\langle \boldsymbol{w}_1^{(m)}, \boldsymbol{z}\rangle, \ldots, \phi\langle \boldsymbol{w}_{q_m}^{(m)}, \boldsymbol{z}\rangle\right)^\top,$$

and having network weights $\boldsymbol{w}_j^{(m)} \in \mathbb{R}^{q_{m-1}}$, $1 \le j \le q_m$. For the output we choose the identity function as activation function

$$\boldsymbol{z}^{(d+1)} : \mathbb{R}^{q_d} \to \mathbb{R}^{q_{d+1}}, \qquad \boldsymbol{z} \mapsto \boldsymbol{z}^{(d+1)}(\boldsymbol{z}) = \left(\langle \boldsymbol{w}_1^{(d+1)}, \boldsymbol{z}\rangle, \ldots, \langle \boldsymbol{w}_{q_{d+1}}^{(d+1)}, \boldsymbol{z}\rangle\right)^\top,$$

and having network weights $\boldsymbol{w}_j^{(d+1)} \in \mathbb{R}^{q_d}$, $1 \le j \le q_{d+1}$. The resulting network parameter $\boldsymbol{\vartheta}$ is now fitted to the data matrix $\boldsymbol{Y} = (\boldsymbol{y}_1, \ldots, \boldsymbol{y}_n)^\top \in \mathbb{R}^{n \times q}$ such that the reconstruction error is minimized over all instances

$$\widehat{\boldsymbol{\vartheta}} \;=\; \underset{\boldsymbol{\vartheta} \in \mathbb{R}^r}{\arg\min} \sum_{i=1}^n L\left(\boldsymbol{y}_i, \Psi \circ \Phi(\boldsymbol{y}_i)\right) \;=\; \underset{\boldsymbol{\vartheta} \in \mathbb{R}^r}{\arg\min} \sum_{i=1}^n L\left(\boldsymbol{y}_i, \boldsymbol{z}^{(d+1:1)}(\boldsymbol{y}_i)\right).$$

We use this fitted network parameter $\widehat{\boldsymbol{\vartheta}}$ and denote the resulting FN layers by $\widehat{\boldsymbol{z}}^{(m)}$ for $1 \le m \le d+1$.

This allows us to define the BN encoder, set $q = q_0$ and $p = q_m$,

$$\Phi : \mathbb{R}^{q_0} \to \mathbb{R}^{q_m}, \qquad y \mapsto \Phi(y) = \hat{z}^{(m:1)}(y) = \left(\hat{z}^{(m)} \circ \cdots \circ \hat{z}^{(1)}\right)(y), \tag{7.66}$$

and the BN decoder is given by, set $q_m = p$ and $q_{d+1} = q$,

$$\Psi : \mathbb{R}^{q_m} \to \mathbb{R}^{q_{d+1}}, \quad z \mapsto \Psi(z) = \hat{z}^{(d+1:m+1)}(z) = \left(\hat{z}^{(d+1)} \circ \cdots \circ \hat{z}^{(m+1)}\right)(z).$$

The BN encoder (7.66) gives us a q_m-dimensional representation of the data. A linear rank p representation $\boldsymbol{Y}_p$ of $\boldsymbol{Y}$, see (7.61), can be found by a BN network architecture that has a minimal FN layer width of dimension $p = \min_{1\le j\le d} q_j$, and with the identity activation function $\phi(x) = x$. Such a BN network is a linear map of maximal rank p. Using the Euclidean square distance as dissimilarity measure provides us an optimal network parameter $\widehat{\vartheta}$ for this linear map such that we receive $\boldsymbol{Y}_p^\top = \hat{z}^{(d+1:1)}(\boldsymbol{Y}^\top)$. There is one point to be considered, here, why the bottleneck activations $\Phi(\boldsymbol{y}) = \hat{z}^{(m:1)}(\boldsymbol{y}) \in \mathbb{R}^p$ in the linear activation case are not directly comparable to the principal components $(\boldsymbol{y}^\top \boldsymbol{v}_1, \ldots, \boldsymbol{y}^\top \boldsymbol{v}_p)^\top$ of the PCA. Namely, the PCA uses an orthonormal basis $\boldsymbol{v}_1, \ldots, \boldsymbol{v}_p$ whereas the linear BN network case uses any p-dimensional basis, i.e., to directly bring these two representations in line we still need a coordinate transformation of the bottleneck activations.

Hinton–Salakhutdinov [186] noticed that the gradient descent fitting of a BN network needs some care, otherwise we may find a local minimum of the loss function that has a poor reconstruction performance. In order to implement a more sophisticated way of SGD fitting we require that the depth d of the network is an odd number and that the network architecture is symmetric around the central FN layer $(d + 1)/2$. This is the case in Fig. 7.36 (lhs). Fitting of this network of depth $d = 3$ is now done in three steps:

1. The symmetry around the central FN layer $m = 2$ allows us to collapse this central layer by merging layers 1 and 3 (because $q_1 = q_3$). Merging these two layers provides us a shallow BN network with neurons $(q_0, q_1 = q_3, q_{d+1} = q_0) = (20, 7, 20)$. This shallow BN network is shown in Fig. 7.36 (middle). In a first step we fit this simpler network to the data $\boldsymbol{Y}$. This gives us the preliminary estimates for the network weights $\boldsymbol{w}_1^{(1)}, \ldots, \boldsymbol{w}_{q_1}^{(1)}$ and $\boldsymbol{w}_1^{(4)}, \ldots, \boldsymbol{w}_{q_4}^{(4)}$ of the full BN network. From this fitted shallow BN network we receive the learned representations $z_i = z^{(1)}(\boldsymbol{y}_i) \in \mathbb{R}^{q_1}$, $1 \le i \le n$, in the central layer using the preliminary estimates of the network weights.
2. In the second step we use the learned representations $z_i \in \mathbb{R}^{q_1}$, $1 \le i \le n$, to fit the inner part of the original network (using a suitable dissimilarity function). This inner part is a shallow network with neurons $(q_1, q_2, q_3 = q_1) = (7, 2, 7)$,

see Fig. 7.36 (rhs). This second step gives us the preliminary estimates for the network weights $\boldsymbol{w}_1^{(2)}, \ldots, \boldsymbol{w}_{q_2}^{(2)}$ and $\boldsymbol{w}_1^{(3)}, \ldots, \boldsymbol{w}_{q_3}^{(3)}$ of the full BN network.

3. In the final step we fit the full BN network on the data $\boldsymbol{Y}$ and use the preliminary estimates of the weights (of the previous two steps) as initialization of the gradient descent algorithm.

Example 7.31 (BN Network Mortality Model) We apply this BN network approach to modify the LC model of Sect. 7.5.4. Hainaut [178] considered such a BN network application. For computational reasons, Hainaut [178] proposed a calibration strategy different from Hinton–Salakhutdinov [186]. We use this latter calibration strategy as it has turned out to work well in our setting.

As BN network architecture we choose a FN network of depth $d = 3$. The input and output dimensions are equal to $q_0 = q_4 = 67$, this exactly corresponds to the number of available calendar years $1950 \le t \le 2016$, see Fig. 7.32. Then, we select a symmetric architecture around the central FN layer $m = 2$ with $q_1 = q_3 = 20$ neurons. That is, in a first step, the 67 calendar years are compressed to a 20-dimensional representation. For the bottleneck we then explore different numbers of neurons $q_2 = p \in \{1, \ldots, 20\}$. These BN networks are implemented and fitted in R with the library `keras` [77]. We have fitted these models separately to the Swiss female and male populations. The raw log-mortality rates are illustrated in Fig. 7.32, and for comparability with the LC approach we have centered these log-mortality rates according to (7.64), and we use the squared Euclidean distance as the objective function.

Figure 7.37 compares the squared Frobenius reconstruction errors of the linear LC approximations $\boldsymbol{Y}_p$ to their non-linear BN network counterparts with bottlenecks $q_2 = p$. We observe that the BN figures are clearly smaller saying that a non-linear auto-encoding provides a better reconstruction, this is true, in particular, for $2 \le q_2 < 20$. For $q_2 \ge 20$ the learning with the BN networks seems saturated, note that the outer layers have $q_1 = q_3 = 20$ neurons which limits the learning at the bottleneck for bigger q_2. In view of Fig. 7.37 there seems to be a kink at $q_2 = 4$,

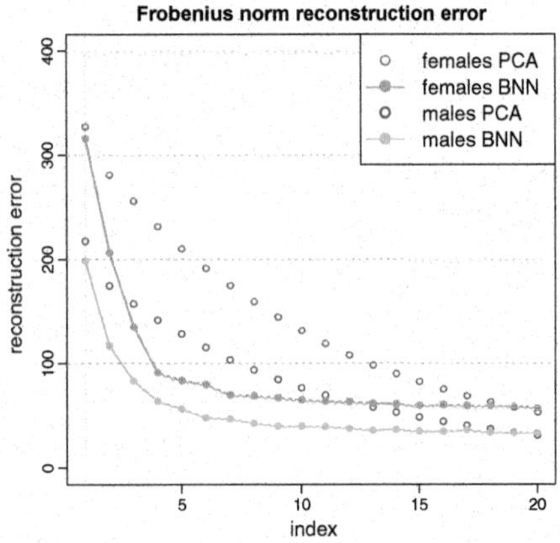

Fig. 7.37 Frobenius reconstruction errors $\|\boldsymbol{Y} - \boldsymbol{Y}_p\|_F^2$ for $1 \le p = q_2 \le 20$ in the linear LC approach and the non-linear BN approach

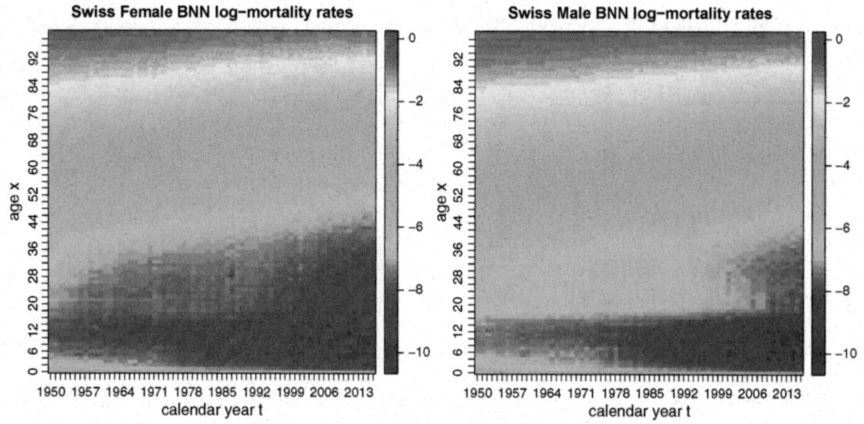

Fig. 7.38 BN network $(q_1, q_2, q_3) = (20, 2, 20)$ fitted log-mortality rates $\log(\widehat{\mu}_{x,t})$ for the calendar years $1950 \le t \le 2016$ and the ages $x_0 = 0 \le x \le x_1 = 99$ of Swiss females (left) and Swiss males (right); the plots use the same color scale as Fig. 7.32

and an "elbow" criterion says that this is the critical bottleneck size that should not be exceeded.

The resulting estimated log-mortality surfaces for the bottleneck $q_2 = 2$ are illustrated in Fig. 7.38. These strongly resemble the raw log-mortality rates in Fig. 7.32, in particular, for the male population we get a better fit for ages $20 \le x \le 40$ from 1980 to 2000 compared to the LC model. In a further analysis we should check whether this BN network does not over-fit to the data. We could, e.g., explore drop-outs during calibration or smaller FN (compression) layers $q_1 = q_3$.

Finally, we analyze the resulting activations at the bottleneck by considering the BN encoder (7.66). Note that we assume $\boldsymbol{y} \in \mathbb{R}^q$ in (7.66) with $q = |\mathcal{T}|$ being the rank of the data matrix $\boldsymbol{Y} \in \mathbb{R}^{n \times q}$. Thus, the encoder takes a fixed age $0 \le x \le 99$ and encodes the corresponding time-series observation $\boldsymbol{y}_x \in \mathbb{R}^{|\mathcal{T}|}$ by the bottleneck activations. This parametrization has been inspired by the PCA which typically considers a data matrix that has more rows than columns. This results in at most $q = \text{rank}(\boldsymbol{Y})$ singular values, supposed $n \ge q$. However, we can easily exchange the role of rows and columns, e.g., by transposing all matrices involved. For mortality forecasting it is advantageous to exchange these roles because we would like to extrapolate a time-series beyond $\mathcal{T}$. For this reason we set for the input dimension $q_0 = q = 100$, which provides us with $|\mathcal{T}|$ observations $\boldsymbol{y}_t \in \mathbb{R}^{100}$. We then fit the BN encoder (7.66) to receive the bottleneck activations

$$\boldsymbol{Y} = (\boldsymbol{y}_t)_{t \in \mathcal{T}} \mapsto \Phi(\boldsymbol{Y}) = (\Phi(\boldsymbol{y}_t))_{t \in \mathcal{T}} \in \mathbb{R}^{q_2 \times |\mathcal{T}|}.$$

Figure 7.39 shows these figures for a bottleneck $q_2 = 2$. We observe that these bottleneck time-series $(\Phi(\boldsymbol{y}_t))_{t \in \mathcal{T}}$ are much more difficult to understand than the LC/RH ones given in Fig. 7.35. Firstly, we see that we have quite some dependence

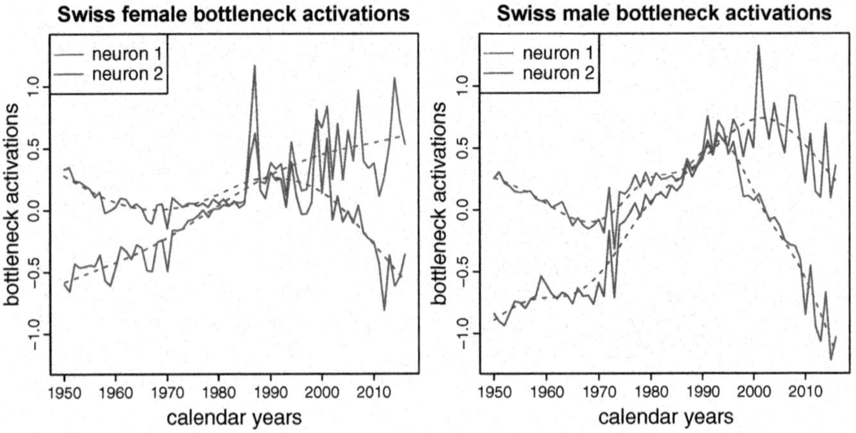

Fig. 7.39 BN network $(q_1, q_2, q_3) = (20, 2, 20)$: bottleneck activations showing $\Phi(\boldsymbol{y}_t) \in \mathbb{R}^2$ for $t \in \mathcal{T}$

between the components of the time-series. Secondly, in contrast to the LC/RH case of Fig. 7.35, there is not one component that dominates. Note that this dominance has been obtained by scaling the components of $(\boldsymbol{b}_x)_x$ to add up to 1 (which, of course, reflects the magnitudes of the singular values). In the non-linear case, these scales are hidden in the decoder which is more difficult to extract. Thirdly, the extrapolation may not work if the time-series has a trend and if we use the hyperbolic tangent activation function that has a bounded range. In general, a trend extrapolation has to be considered very carefully with FN networks with non-linear activation functions, and often there is no good solution to this problem within the FN network framework. We conclude that this approach improves in-sample mortality surface modeling, but it leaves open the question about forecasting the future mortality rates because an extrapolation seems more difficult. ∎

Remark 7.32 The concept of BN networks has also been considered in the actuarial literature to encode geographic information, see Blier-Wong et al. [39]. Since geographic information has a natural spatial component, these authors propose to use a convolutional neural network to encode the spatial information before processing the learned features through a BN network. The proposed decoder may have different forms, either it tries to reconstruct the whole (spatial) neighborhood of a given location or it only tries to reconstruct the site of a given location.

7.6 Model-Agnostic Tools

We collect some model-agnostic tools in this section that help us to better understand and analyze the networks, their calibrations and predictions. Model-agnostic tools are techniques that are not specific to a certain model type and can be used for any regression model. Most methods presented here are nicely presented in the tutorial of Lorentzen–Mayer [258]. There are several ways of getting a better understanding of a regression model. First, we can analyze variable importance which tries to answer similar questions to the GLM variable selection tools of Sect. 5.3 on model validation. However, in general, we cannot rely on any asymptotic likelihood theory for such an analysis. Second, we can try to understand the predictive model. For a GLM with the log-link function this is quite simple because the systematic effects are of a multiplicative nature. For networks this is much more complicated because we allow for much more general regression functions. We can either try to understand these functions on a global portfolio level (by averaging the effects over many insurance policies) or we can try to understand these functions locally for individual insurance policies. The latter refers to local sensitivities around a chosen feature value $\boldsymbol{x} \in \mathcal{X}$, and the former to global model-agnostics.

7.6.1 Variable Permutation Importance

For GLMs we have studied the LRT and the Wald test that have been assisting us in reducing the GLM by the feature components that do not contribute sufficiently to the regression task at hand, see Sects. 5.3.2 and 5.3.3. These variable reduction techniques rely on an asymptotic likelihood theory. Here, we need to proceed differently, and we just aim at ranking the variables by their importance, similarly to a `drop1` analysis, see Listing 5.6.

For a given FN network regression model

$$\boldsymbol{x} \in \mathcal{X} \mapsto \mu(\boldsymbol{x}) = g^{-1}\langle \boldsymbol{\beta}, \boldsymbol{z}^{(d:1)}(\boldsymbol{x})\rangle,$$

we randomize one component of $\boldsymbol{x} = (x_1, \ldots, x_q)^\top$ at a time, and we study the resulting change in the objective function. More precisely, for given (learning) data $\mathcal{L}$, with features $\boldsymbol{x}_1, \ldots, \boldsymbol{x}_n$, we select one feature component $1 \le j \le q$ and permute $(x_{i,j})_{1\le i\le n}$ randomly across the entire portfolio $1 \le i \le n$. We denote by $\mathcal{L}^{(j)}$ the resulting data with the j-th component being permuted. We then compare the resulting deviance loss $\mathfrak{D}(\mathcal{L}^{(j)}, \mu)$ to the one $\mathfrak{D}(\mathcal{L}, \mu)$ on the original data $\mathcal{L}$ using the same regression model μ. We call this approach variable permutation importance (VPI). Note that such a permutation does not only act on the marginal effects, but it also distorts the interaction effects of the different feature components.

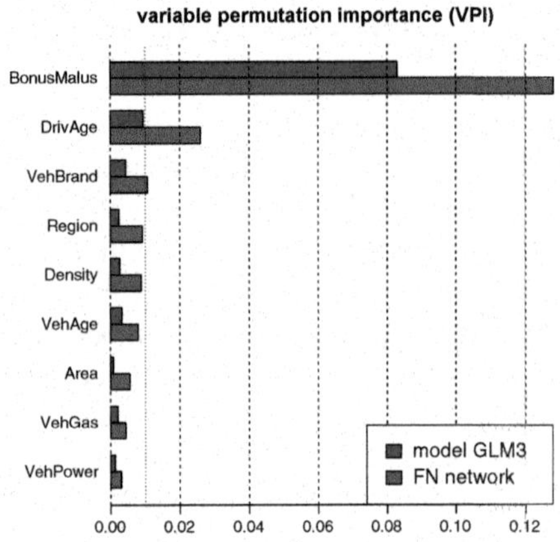

Fig. 7.40 VPI measured by the relative change $\text{vpi}^{(j)}$, $1 \le j \le q$, of model Poisson GLM3 of Table 5.5 and the FN network regression model $\widehat{\mu}^{m=1}$ of Table 7.9

We calculate the VPI on the MTPL claim frequency data of model Poisson GLM3 of Table 5.5 and the FN network regression model $\widehat{\mu}^{m=1}$ of Table 7.9; we use this example throughout this section on model-agnostic tools. Figure 7.40 shows the relative increases

$$\text{vpi}^{(j)} = \frac{\mathfrak{D}(\mathcal{L}^{(j)}, \mu) - \mathfrak{D}(\mathcal{L}, \mu)}{\mathfrak{D}(\mathcal{L}, \mu)},$$

of the deviance losses by permuting one feature component $1 \le j \le q$ at a time.

Obviously, the `BonusMalus` level followed by `DrivAge` and `VehBrand` are the most important variables according to this VPI method. This is in alignment for both models. Thereafter, there are smaller disagreements between the two models. These disagreements may (also) be caused by a non-optimal feature pre-processing in the GLM where, for instance, we have to add the interaction effects manually, see (5.35). Overall, these VPI results are in line with the findings of the classical methods on GLMs, see for instance the `drop1` table in Listing 5.6.

One point that is worth mentioning (and which makes the VPI results not fully reliable) is the use of feature components that are highly correlated. In our case, `Density` and `Area` are highly correlated, see Fig. 13.12. Therefore, it may not make sense to randomly permute one component while keeping the other one unchanged. This issue will also arise in other methods described below.

Remark 7.33 (Global Surrogate Model) There are other machine learning methods that offer different measures of variable importance. For instance, (binary split) classification and regression trees (CARTs) offer popular methods for measuring variable importance; for binary split CARTs we refer to Breiman et al. [54] and Denuit et al. [100]. These CARTs select individual feature components for partitioning the feature space $\mathcal{X}$, and variable importance is measured by analyzing the contribution of each feature component to the total decrease of the objective function. Binary split CARTs have the advantage that this can be done in an additive way.

More complex regression models like FN networks can then be analyzed by using a binary split regression tree as a global surrogate model. That is, we can fit a CART to the network regression function (as a surrogate model) and then analyze variable importance in this surrogate regression tree model using the tools of regression trees. We will not give an explicit example here because we have not formally introduced regression trees in this manuscript, but this concept is fairly straightforward and well-understood.

7.6.2 *Partial Dependence Plots*

There are several graphical tools that study the individual behavior in the feature components. Some of these tools select individual insurance policies and others study global portfolio properties. They have in common that they are based on marginal considerations, i.e., some sort of projection.

Individual Conditional Expectation

Individual conditional expectation (ICE) selects individual insurance policies $(Y_i, \boldsymbol{x}_i, v_i)$ and varies the feature components of $\boldsymbol{x}_i$ over their entire domain; we refer to Goldstein et al. [164]. Similarly to the VPI of Sect. 7.6.1, ICE does not respect collinearity in feature components, but it is rather an isolated view of individual components.

In Fig. 7.41 we provide the ICE plots of model Poisson GLM3 of Table 5.5 and the FN network regression model $\widehat{\mu}^{m=1}$ of Table 7.9 of 100 randomly selected insurance policies $\boldsymbol{x}_i$. For these randomly selected insurance policies we let the variable `DrivAge` vary over its domain $\{18, \ldots, 90\}$. Each color corresponds to one insurance policy i, and the colors in the two plots coincide. In the GLM we observe that the lines are roughly parallel which reflects that we have an additive regression structure on the canonical scale (note that these plots are on the canonical parameter scale). The lines are not perfectly parallel because we allow for an interaction between `DrivAge` and `BonusMalus` in model Poisson GLM3, see (5.35). The plot of the FN network is more difficult to interpret. Overall the levels (colors) coincide in the two plots, but in the FN network plot the lines are not increasing for ages approaching 18, the reason for this is that we have interactions with other feature components that are important. In particular, for ages close to 18 we cannot have a `BonusMalus` level of 50% and, therefore, the FN network cannot be trained on this part of the feature space. Nevertheless, the ICE plot allows for such feature configurations (by just extrapolating the FN network regression function beyond the set of available insurance policies). This difficulty is confirmed

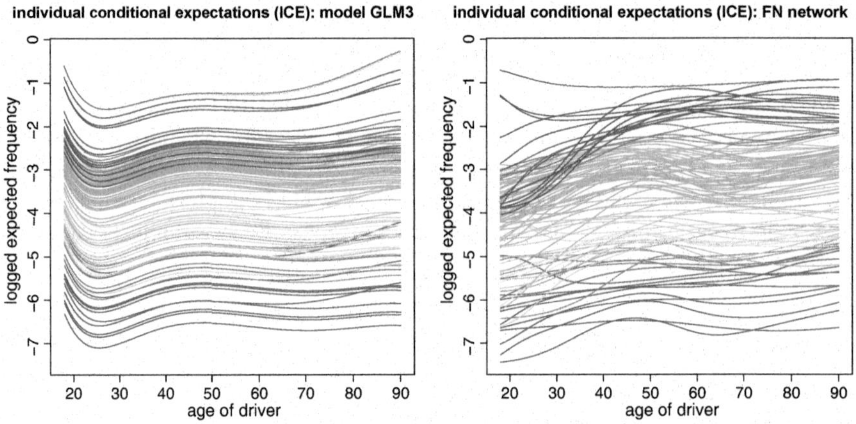

Fig. 7.41 ICE plots of 100 randomly selected insurance policies $\boldsymbol{x}_i$ of (lhs) model Poisson GLM3 and (rhs) FN network $\widehat{\mu}^{m=1}$ letting the variable `DrivAge` vary over its domain; the y-axis is on the canonical parameter scale

by exploiting the same plot only on insurance policies that have a `BonusMalus` level of at least 100%. In that case the lines for small ages are non-decreasing when approaching the age of 18, thus, providing a more reasonable interpretation. We conclude that if we have strong dependence and/or interactions between the feature components this method may not provide any reasonable interpretations.

Partial Dependence Plot

Partial dependence plots (PDPs) have been introduced by Friedman [141], see also Zhao–Hastie [405]. PDPs are closely related to the do-operator in causal inference in statistics; we refer to Pearl [298] and Pearl et al. [299] for the do-operator. A PDP and the do-operator, respectively, are obtained by breaking the dependence structure between different feature components. Namely, we decompose the feature $\boldsymbol{x} = (x_j, \boldsymbol{x}_{\setminus j})$ into two parts with $\boldsymbol{x}_{\setminus j}$ denoting all feature components except of component x_j; we will use a slight abuse of notation because the components need to be permuted correspondingly in the following regression function $\boldsymbol{x} \to \mu(\boldsymbol{x}) = \mu(x_j, \boldsymbol{x}_{\setminus j})$. Since, typically, there is dependence between x_j and $\boldsymbol{x}_{\setminus j}$ one can infer $\boldsymbol{x}_{\setminus j}$ from x_j, and vice versa. A PDP breaks this inference potential so that the sensitivity can be studied purely in x_j. In particular, the partial dependence profile is obtained by

$$x_j \mapsto \bar{\mu}^j(x_j) = \int \mu(x_j, \boldsymbol{x}_{\setminus j}) \, dp(\boldsymbol{x}_{\setminus j}), \tag{7.67}$$

where $p(\boldsymbol{x}_{\setminus j})$ is the marginal (portfolio) distribution of the feature components $\boldsymbol{x}_{\setminus j}$. Observe that this differs from the conditional expectation which reads as

$$x_j \mapsto \mu(x_j) = \mathbb{E}_p\left[\mu(x_j, \boldsymbol{x}_{\setminus j}) \middle| x_j\right] = \int \mu(x_j, \boldsymbol{x}_{\setminus j})\, dp(\boldsymbol{x}_{\setminus j}|x_j),$$

the latter allowing for inferring $\boldsymbol{x}_{\setminus j}$ from x_j through the conditional probability $dp(\boldsymbol{x}_{\setminus j}|x_j)$.

Remark 7.34 (Discrimination-Free Insurance Pricing) Recent actuarial literature discusses discrimination-free insurance pricing which aims at developing a pricing framework that is free of discrimination w.r.t. so-called protected characteristics such as gender and ethnicity; we refer to Guillén [174], Chen et al. [69, 70], Lindholm et al. [253] and Frees–Huang [136] for discussions on discrimination in insurance. In general, part of the problem also lies in the fact that one can often infer the protected characteristics from the non-protected feature information. This is called indirect discrimination or proxy discrimination. The proposal of Lindholm et al. [253] for achieving discrimination-free prices exactly follows the construction (7.67), by breaking the link, which infers the protected characteristics from the non-protected ones.

The partial dependence profile on our portfolio $\mathcal{L}$ with given features $\boldsymbol{x}_1, \ldots, \boldsymbol{x}_n$ is now obtained by just using the portfolio distribution as an empirical distribution for p in (7.67). That is, for a selected component x_j of $\boldsymbol{x}$, we consider the partial dependence profile

$$x_j \mapsto \bar{\mu}^j(x_j) = \frac{1}{n}\sum_{i=1}^n \mu(x_j, \boldsymbol{x}_{i,\setminus j}) = \frac{1}{n}\sum_{i=1}^n \mu\left(x_{i,0}, x_{i,1}, \ldots, x_{i,j-1}, x_j, x_{i,j+1}, \ldots, x_{i,q}\right),$$

thus, we average the ICE plots over $\boldsymbol{x}_{i,\setminus j}$ of our portfolio $1 \le i \le n$.

Figure 7.42 (lhs, middle) give the PDPs of the variables `BonusMalus` and `DrivAge` of model Poisson GLM3 and the FN network $\widehat{\mu}^{m=1}$. Overall they

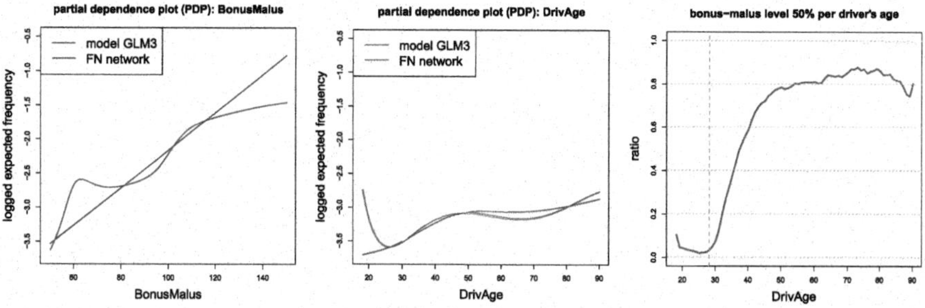

Fig. 7.42 PDPs of (lhs) `BonusMalus` level and (middle) `DrivAge`; the y-axis is on the canonical parameter scale; (rhs) ratio of policies with a bonus-malus level of 50% per driver's age

look reasonable. However, we are again facing the difficulty that these partial dependence profiles consider feature configurations that should not appear in our portfolio. Roughly 57% of all insurance policies have a bonus-malus level of 50%, which means that these driver's did not suffer any claims in the past couple of years. Obviously a driver of age 18 cannot be on this bonus-malus level, simply because she/he is not in a state where she/he can have multiple years of driving experience without an accident. However, the PDP does not respect this fact, and just extrapolates the regression function into that part of the feature space. Therefore, the PDP at driver's age 18 is based on 57% of the insurance policies being on a bonus-malus level of 50% because this corresponds to the empirical portfolio distribution $p(\boldsymbol{x}_{\setminus j})$ excluding the driver's age x_j = `DrivAge` information. Figure 7.42 (rhs) shows the ratio of insurance policies that have a bonus-malus level of 50%. We observe that this ratio is roughly zero up to age 28 (orange vertical dotted line), which indicates that a driver needs 10 successive accident-free years to reach the lowest bonus-malus level (starting from 100%). We consider it to be data error that this ratio below age 28 is not identically equal to zero. We conclude that these PDPs need to be interpreted very carefully because the insurance portfolio is not uniformly distributed across the feature space. In some parts of the feature space the regression function $\boldsymbol{x} \mapsto \mu(\boldsymbol{x})$ may not even be well-defined because certain combinations of feature values $\boldsymbol{x}$ may not exist (e.g., a driver of age 18 on bonus-malus level 50% or a boy at a girl's college).

Accumulated Local Effects Profile

PDPs have the problem that they do not respect the dependencies between the feature components, as explained in the previous paragraphs. The accumulated local effects (ALE) profile tries to take account for these dependencies by only studying a local feature perturbation, we refer to Apley–Zhu [13]. We present a smooth (gradient-based) version of ALE because our regression functions are differentiable. Consider the local effect in the individual feature $\boldsymbol{x}$ w.r.t. the component x_j by studying the partial derivative

$$\mu_j(\boldsymbol{x}) = \frac{\partial \mu(\boldsymbol{x})}{\partial x_j}. \tag{7.68}$$

The average local effect of component j is received by

$$x_j \mapsto \Delta_j(x_j; \mu) = \int \mu_j(x_j, \boldsymbol{x}_{\setminus j}) dp(\boldsymbol{x}_{\setminus j} | x_j). \tag{7.69}$$

ALE integrate the average local effects $\Delta_j(\cdot)$ over their domain, and the ALE profile is defined by

$$x_j \mapsto \int_{x_{j0}}^{x_j} \Delta_j(z_j; \mu) dz_j = \int_{x_{j0}}^{x_j} \int \mu_j(z_j, \boldsymbol{x}_{\setminus j}) dp(\boldsymbol{x}_{\setminus j} | z_j) dz_j, \tag{7.70}$$

where x_{j0} is a given initialization point. The difference between PDPs and ALE is that the latter correctly considers the dependence structure between x_j and $\boldsymbol{x}_{\setminus j}$, see (7.69).

Listing 7.10 Local effects through the gradients of FN networks in `keras` [77]

```
1  Input  = layer_input(shape = c(11), dtype = 'float32', name = 'Design')
2  #
3  Output = Input %>%
4            layer_dense(units=20, activation='tanh', name='FNLayer1') %>%
5            layer_dense(units=15, activation='tanh', name='FNLayer2') %>%
6            layer_dense(units=10, activation='tanh', name='FNLayer3') %>%
7            layer_dense(units=1, activation='linear', name='Network')
8  #
9  model = keras_model(inputs = c(Input), outputs = c(Output))
10 #
11 grad = Output %>%
12         layer_lambda(function(x) k_gradients(model$outputs, model$inputs))
13 model.grad = keras_model(inputs = c(Input), outputs = c(grad))
14 theta.grad <- data.frame(model.grad %>% predict(XX))
```

Example We come back to our MTPL claim frequency FN network example. The local effects (7.68) can directly be calculated in the R library `keras` [77] for a FN network, see Listing 7.10. In order to do so we need to drop the embedding layers, compared to Listing 7.4, and directly work on the learned embeddings. This gives an input layer of dimension $q = 7 + 2 + 2 = 11$ because we have two categorical features that have been embedded into 2-dimensional Euclidean spaces $\mathbb{R}^2$. Then, we can formally calculate the gradient of the FN network w.r.t. its inputs which is done on lines 11–13 of Listing 7.10. Remark that we work on the canonical scale because we use the linear activation function on line 7 of the listing.

There remain the averaging (7.69) and the integration (7.70) which can be done empirically

$$x_j \mapsto \Delta_j(x_j; \mu) = \frac{1}{|\mathcal{E}(x_j)|} \sum_{i \in \mathcal{E}(x_j)} \mu_j(\boldsymbol{x}_i), \tag{7.71}$$

where $\mathcal{E}(x_j)$ denotes the indices i of all cases $\boldsymbol{x}_i$, $1 \leq i \leq n$, with $x_{i,j} = x_j$, assuming of having discrete feature data observations. Note that this empirical averaging respects the dependence within $\boldsymbol{x}$. The (uncentered) ALE profile is then obtained by aggregating these local effects, that is,

$$x_j \mapsto \widetilde{\mu}_j(x_j) = \int_{x_{j0}}^{x_j} \Delta_j(z_j; \mu) dz_j,$$

where this integration is typically understood in a discrete sense because the observed feature components $x_{i,j}$ are discrete. Often, this uncentered ALE profile is still translated (centered) by the portfolio average.

Remarks 7.35

- We have only introduced ALE for continuous feature variables. For nominal categorical feature components it is not immediately clear how to reasonably integrate the average local effects $\Delta_j(x_j; \mu)$, and one typically directly analyzes these average local effects.
- For GLMs the ALEs are rather simple if we work on the canonical scale and under the canonical link, since

$$\theta_j(\boldsymbol{x}) = \frac{\partial \theta(\boldsymbol{x})}{\partial x_j} = \beta_j \equiv \Delta_j(x_j; \theta).$$

  In the case of model Poisson GLM3 presented in Sect. 5.3.4 the situation is more delicate as we model the interactions in the GLM as follows, see (5.34) and (5.35),

$$\begin{aligned}&(\texttt{DrivAge}, \texttt{BonusMalus})\\ &\quad\mapsto \beta_l\, \texttt{DrivAge} + \beta_{l+1}\log(\texttt{DrivAge}) + \sum_{j=2}^{4} \beta_{l+j}(\texttt{DrivAge})^j\\ &\qquad+\beta_{l+5}\texttt{BonusMalus} + \beta_{l+6}\,\texttt{BonusMalus} \cdot \texttt{DrivAge}\\ &\qquad+\beta_{l+7}\texttt{BonusMalus} \cdot (\texttt{DrivAge})^2.\end{aligned}$$

  In that case, though we work with a GLM, the resulting local effects are different if we calculate the derivatives w.r.t. DrivAge and BonusMalus, respectively, because we explicitly (manually) include non-linear effects into the GLM.

Figure 7.43 shows the ALE profiles of the variables BonusMalus and DrivAge. The shapes of these profiles can directly be compared to the PDPs of Fig. 7.42 (the scale on the y-axis should be ignored because this will depend on the applied centering, however, we hold on to the canonical scale). The main difference between these two plots can be observed for the variable DrivAge at low ages. Namely, the ALE profiles have a different shape at low ages respecting the dependencies in the feature components by only considering real local feature configurations.

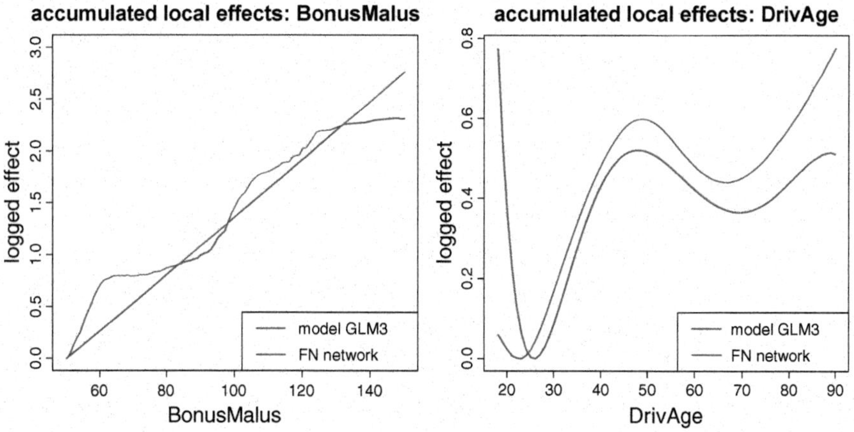

Fig. 7.43 ALE profiles of (lhs) `BonusMalus` level and (rhs) `DrivAge`; the y-axis is on the log-scale

7.6.3 *Interaction Strength*

Next we are going to discuss pairwise interaction strength. Friedman–Popescu [143] made the following proposal. Roughly speaking, there is an interaction between the two feature components x_j and x_k of $\boldsymbol{x}$ in the regression function $\boldsymbol{x} \mapsto \mu(\boldsymbol{x})$ if

$$\mu_{j,k}(\boldsymbol{x}) = \frac{\partial^2 \mu(\boldsymbol{x})}{\partial x_j \partial x_k} \neq 0. \tag{7.72}$$

This means that the magnitude of a change of the regression function $\mu(\boldsymbol{x})$ in x_j depends on the current value of x_k. If there is no such interaction, we can additively decompose the regression function $\mu(\boldsymbol{x})$ into two independent terms. This then reads as $\mu(\boldsymbol{x}) = \mu_{\setminus j}(\boldsymbol{x}_{\setminus j}) + \mu_{\setminus k}(\boldsymbol{x}_{\setminus k})$. This motivation is now applied to the PDP profiles given in (7.67). We define the centered versions $x_j \mapsto \breve{\mu}^j(x_j)$ and $x_k \mapsto \breve{\mu}^k(x_k)$ of the PDP profiles by centering the PDP profiles $x_j \mapsto \bar{\mu}^j(x_j)$ and $x_k \mapsto \bar{\mu}^k(x_k)$ over the portfolio values $\boldsymbol{x}_i$, $1 \leq i \leq n$. Next, we consider an analogous two-dimensional version for (x_j, x_k). Let $(x_j, x_k) \mapsto \breve{\mu}^{j,k}(x_j, x_k)$ be the centered version of a two-dimensional PDP profile $(x_j, x_k) \mapsto \bar{\mu}^{j,k}(x_j, x_k)$.

Friedman's H-statistics measures the pairwise interaction strength between the components x_j and x_k, and it is defined by

$$H_{j,k}^2 = \frac{\sum_{i=1}^n \left(\breve{\mu}^{j,k}(x_{i,j}, x_{i,k}) - \breve{\mu}^j(x_{i,j}) - \breve{\mu}^k(x_{i,k})\right)^2}{\sum_{i=1}^n \breve{\mu}^{j,k}(x_{i,j}, x_{i,k})^2}, \tag{7.73}$$

we refer to formula (44) in Friedman–Popescu [143]. While $H_{j,k}^2$ measures the proportion of the joint interaction effect, as we normalize by the variability of the joint effect $\sum_{i=1}^{n} \breve{\mu}^{j,k}(x_{i,j}, x_{i,k})^2$, sometimes also the absolute measure is considered by taking the square root of the enumerator in (7.73). Of course, this can be extended to interactions of three components, etc., we refer to Friedman–Popescu [143].

We do not give an example here, because calculating Friedman's H-statistics can be computationally demanding if one has many feature components with many levels in FN network modeling.

7.6.4 Local Model-Agnostic Methods

The above methods like the PDP and the ALE profile have been analyzing the global behavior of the regression functions. We briefly mention some tools that describe the local sensitivity and explanation of regression results.

Probably the most popular method is the locally interpretable model-agnostic explanation (LIME) introduced by Ribeiro et al. [311]. This analyzes locally the expected response of a given feature $\boldsymbol{x}$ by perturbing $\boldsymbol{x}$. In a nutshell, the idea is to select an environment $\mathcal{E}(\boldsymbol{x}) \subset \mathcal{X}$ of a chosen feature $\boldsymbol{x}$ and to study the regression function $\boldsymbol{x}' \mapsto \mu(\boldsymbol{x}')$ in this environment $\boldsymbol{x}' \in \mathcal{E}(\boldsymbol{x})$. This is done by fitting a (much) simpler surrogate model to μ on this environment $\mathcal{E}(\boldsymbol{x})$. If the environment is small, often a linear regression model is chosen. This then allows one to interpret the regression function $\mu(\cdot)$ locally using the simpler surrogate model, and if we have a high-dimensional feature space, this linear regression is complemented with LASSO regularization to only select the most important feature components.

The second method considered in the literature is the Shapley additive explanation (SHAP). The SHAP is based on Shapley values [335] which is a method of allocating rewards to players in cooperative games, where a team of individual players jointly contributes to a potential success. Shapley values solve this allocation problem under the requirements of additivity and fairness. This concept can be translated to analyzing how individual feature components of $\boldsymbol{x}$ contribute to the total prediction $\mu(\boldsymbol{x})$ of a given case. Shapley values allow one to do such a contribution analysis in the aforementioned additive and fair way, see Lundberg–Lee [261]. The calculation of SHAP values is combinatorially demanding and therefore several approximations have been proposed, many of them having their own caveats, we refer to Aas et al. [1]. We will not further consider these but refer to the relevant literature.

7.6.5 Marginal Attribution by Conditioning on Quantiles

The above model-agnostic tools have mainly been studying the sensitivities of the expected response $\mu(\boldsymbol{x})$ in the feature components of $\boldsymbol{x}$. This becomes apparent from considering the partial derivatives (7.68) to calculate the local effects. Alternatively, we could try to understand how the feature components of $\boldsymbol{x}$ contribute to a given response $\mu(\boldsymbol{x})$, see Ancona et al. [12]; this section follows Merz et al. [273]. The marginal attribution on an input component j of the response $\mu(\boldsymbol{x})$ can be studied by the directional derivative

$$x_j \mapsto x_j \mu_j(\boldsymbol{x}) = x_j \frac{\partial \mu(\boldsymbol{x})}{\partial x_j}. \tag{7.74}$$

This was first proposed to the data science community by Shrikumar et al. [340]. Basically, it means that we replace the partial derivative $\mu_j(\boldsymbol{x})$ by the directional derivative along the vector $x_j \boldsymbol{e}_j = (0, \ldots, 0, x_j, 0, \ldots, 0)^\top \in \mathbb{R}^{q+1}$

$$\begin{aligned}\lim_{\epsilon \to 0} & \frac{\mu(\boldsymbol{x} + \epsilon x_j \boldsymbol{e}_j) - \mu(\boldsymbol{x})}{\epsilon} \\ &= \lim_{\epsilon \to 0} \frac{\mu\left((1, x_1, \ldots, x_{j-1}, (1+\epsilon)x_j, x_{j+1}, \ldots, x_q)^\top\right) - \mu(\boldsymbol{x})}{\epsilon} = x_j \mu_j(\boldsymbol{x}),\end{aligned}$$

where $\boldsymbol{e}_j$ is the $(j+1)$-st basis vector in $\mathbb{R}^{q+1}$ (index $j = 0$ corresponds to the intercept component $x_0 = 1$).

We start by recalling the sensitivity analysis of Hong [189] and Tsanakas–Millossovich [355] in the context of risk measurement. Assume the features have a portfolio distribution $\boldsymbol{X} \sim p$. This describes the random selection of an insurance policy $\boldsymbol{X} = \boldsymbol{x}$ from the portfolio described by p. The average price over the entire portfolio is then given by

$$\mu = \mathbb{E}_p[\mu(\boldsymbol{X})] = \int \mu(\boldsymbol{x}) dp(\boldsymbol{x}).$$

We implicitly interpret $\mu(\boldsymbol{X}) = \mathbb{E}[Y|\boldsymbol{X}]$ as the price of the response Y, here, though we do not need the response distribution in this section. Assume $\mu(\boldsymbol{X})$ has a continuous distribution function $F_{\mu(\boldsymbol{X})}$; and we drop the intercept component $X_0 = x_0 = 1$ from these considerations (but we still keep it in the regression model). This implies that $U_{\mu(\boldsymbol{X})} = F_{\mu(\boldsymbol{X})}(\mu(\boldsymbol{X}))$ is uniformly distributed on $[0, 1]$. Choosing a density ζ on $[0, 1]$ gives us a probability distortion $\zeta(U_{\mu(\boldsymbol{X})})$ as we have the normalization

$$\mathbb{E}_p\left[\zeta(U_{\mu(\boldsymbol{X})})\right] = \int_0^1 \zeta(u) du = 1.$$

This allows us to define a distorted portfolio price in the sense of a Radon–Nikodým derivative, namely, we set for the distorted portfolio price

$$\varrho(\mu(\boldsymbol{X}); \zeta) = \mathbb{E}_p\left[\mu(\boldsymbol{X})\zeta(U_{\mu(\boldsymbol{X})})\right].$$

This functional $\varrho(\mu(\boldsymbol{X}); \zeta)$ is a so-called distortion risk measure. Our goal is to study the sensitivities of this distortion risk measure in the components of $\boldsymbol{X}$. Assume existence of the following directional derivatives for all $1 \leq j \leq q$

$$S_j(\mu; \zeta) = \frac{\partial}{\partial \epsilon} \varrho\left(\mu\left((1, X_1, \ldots, X_{j-1}, (1+\epsilon)X_j, X_{j+1}, \ldots X_q)^\top\right); \zeta\right)\bigg|_{\epsilon=0}.$$

$S_j(\mu; \zeta)$ can be used to describe the sensitivities of the regression function $\boldsymbol{X} \mapsto \mu(\boldsymbol{X})$ in the feature components X_j. Under different sets of assumptions, Hong [189] and Tsanakas–Millossovich [355] have proved the following identity

$$S_j(\mu; \zeta) = \mathbb{E}_p\left[X_j \mu_j(\boldsymbol{X}) \zeta(U_{\mu(\boldsymbol{X})})\right],$$

the right-hand side exactly uses the marginal attribution (7.74). There remains the freedom of the choice of the density ζ on $[0, 1]$, which allows us to study the sensitivities of different distortion risk measures. For the uniform distribution $\zeta \equiv 1$ on $[0, 1]$ we simply have the average (best-estimate) price and its average marginal attributions

$$\varrho(\mu(\boldsymbol{X}); \zeta \equiv 1) = \mathbb{E}_p[\mu(\boldsymbol{X})] = \mu \qquad \text{and} \qquad S_j(\mu; \zeta \equiv 1) = \mathbb{E}_p[X_j \mu_j(\boldsymbol{X})].$$

If we want to consider a quantile risk measure, called value-at-risk (VaR), we choose a Dirac measure for the density ζ. That is, choose a point measure of mass 1 in $\alpha \in (0, 1)$, i.e., the density ζ is concentrated in the single point α. In that case, the event $\{F_{\mu(\boldsymbol{X})}(\mu(\boldsymbol{X})) = U_{\mu(\boldsymbol{X})} = \alpha\}$ receives probability one, and therefore we have the α-quantile

$$\varrho(\mu(\boldsymbol{X}); \alpha) = F^{-1}_{\mu(\boldsymbol{X})}(\alpha),$$

and the corresponding sensitivities for $1 \leq j \leq q$

$$S_j(\mu; \alpha) \; = \; \mathbb{E}_p\left[X_j \mu_j(\boldsymbol{X}) \,\middle|\, \mu(\boldsymbol{X}) = F^{-1}_{\mu(\boldsymbol{X})}(\alpha)\right]. \tag{7.75}$$

Remarks 7.36

- In the introduction to this section we have assumed that $\mu(\boldsymbol{X})$ has a continuous distribution function. This emphasizes that this sensitivity analysis is most suitable for continuous feature components. Categorical and discrete feature components can be embedded into a Euclidean space, e.g., using embedding layers, and then they can be treated as continuous variables.
- Sensitivities (7.75) respect the local portfolio structure as they are calculated w.r.t. p.
- In applications, we will work with the empirical portfolio distribution for p provided by $(\boldsymbol{x}_i)_{1 \leq i \leq n}$. This gives an empirical approximation to (7.75) and, in particular, it will require a choice of a bandwidth for the evaluation of the

conditional probability, conditioned on the event $\{\mu(\boldsymbol{X}) = F^{-1}_{\mu(\boldsymbol{X})}(\alpha)\}$. This is done with a local smoother similarly to Listing 7.8.

In analogy to Merz et al. [273] we give a different interpretation to the sensitivities (7.75), which allows us to further expand this formula. We have 1st order Taylor expansion

$$\mu(\boldsymbol{x}+\boldsymbol{\epsilon}) = \mu(\boldsymbol{x}) + (\nabla_{\boldsymbol{x}}\mu(\boldsymbol{x}))^\top \boldsymbol{\epsilon} + o(\|\boldsymbol{\epsilon}\|_2) \qquad \text{for } \|\boldsymbol{\epsilon}\|_2 \to 0.$$

Obviously, this is a local approximation in $\boldsymbol{x}$. Setting $\boldsymbol{\epsilon} = -\boldsymbol{x}$, we get (a possibly crude) approximation

$$\mu(\boldsymbol{0}) \;\approx\; \mu(\boldsymbol{x}) - (\nabla_{\boldsymbol{x}}\mu(\boldsymbol{x}))^\top \boldsymbol{x}.$$

By bringing the gradient term to the other side, using (7.75) and conditionally averaging, we receive the 1st order marginal attributions

$$F^{-1}_{\mu(\boldsymbol{X})}(\alpha) = \mathbb{E}_p\left[\mu(\boldsymbol{X}) \middle| \mu(\boldsymbol{X}) = F^{-1}_{\mu(\boldsymbol{X})}(\alpha)\right] \;\approx\; \mu(\boldsymbol{0}) + \sum_{j=1}^{q} S_j(\mu;\alpha). \tag{7.76}$$

Thus, the sensitivities $S_j(\mu;\alpha)$ provide a 1st order description of the quantiles $F^{-1}_{\mu(\boldsymbol{X})}(\alpha)$ of $\mu(\boldsymbol{X})$. We call this approach marginal attribution by conditioning on quantiles (MACQ) because it shows how the components X_j of $\boldsymbol{X}$ contribute to a given quantile level.

Example 7.37 (MACQ for Linear Regression) The simplest case is the linear regression case because the 1st order marginal attributions (7.76) are exact in this case. Consider a linear regression function with regression parameter $\boldsymbol{\beta} \in \mathbb{R}^{q+1}$

$$\boldsymbol{x} \;\mapsto\; \mu(\boldsymbol{x}) = \langle \boldsymbol{\beta}, \boldsymbol{x}\rangle = \beta_0 + \sum_{j=1}^{q} \beta_j x_j.$$

The 1st order marginal attributions for fixed $\alpha \in (0, 1)$ are given by

$$\begin{aligned} F^{-1}_{\mu(\boldsymbol{X})}(\alpha) &= \mu(\boldsymbol{0}) + \sum_{j=1}^{q} S_j(\mu;\alpha) \\ &= \beta_0 + \sum_{j=1}^{q} \beta_j \mathbb{E}_p\left[X_j \middle| \mu(\boldsymbol{X}) = F^{-1}_{\mu(\boldsymbol{X})}(\alpha)\right]. \end{aligned} \tag{7.77}$$

That is, we replace the feature components X_j by their expected contributions on a given quantile level $F^{-1}_{\mu(\boldsymbol{X})}(\alpha)$ in (7.77). We compare this explanation to the ALE profile (7.70). Set initial value $x_{j_0} = 0$, the ALE profile for the linear regression model is given by

$$x_j \mapsto \int_0^{x_j} \Delta_j(z_j)dz_j = \beta_j x_j.$$

This is the sensitivity of the linear regression function in component x_j, whereas (7.77) describes the contribution of each feature component to an expected response level $\mu(\boldsymbol{x})$, in particular, $\mathbb{E}_p[X_j|\mu(\boldsymbol{X}) = F^{-1}_{\mu(\boldsymbol{X})}(\alpha)]$ describes the average feature value in component j on a given quantile level. ■

A natural next step is to expand the 1st order attributions to 2nd orders. This allows us to consider the interaction terms. Consider the 2nd order Taylor expansion

$$\mu(\boldsymbol{x}+\boldsymbol{\epsilon}) = \mu(\boldsymbol{x}) + (\nabla_{\boldsymbol{x}}\mu(\boldsymbol{x}))^\top \boldsymbol{\epsilon} + \frac{1}{2}\boldsymbol{\epsilon}^\top \nabla^2_{\boldsymbol{x}}\mu(\boldsymbol{x})\boldsymbol{\epsilon} + o(\|\boldsymbol{\epsilon}\|_2^2) \qquad \text{for } \|\boldsymbol{\epsilon}\|_2 \to 0.$$

Similar to (7.76), setting $\boldsymbol{\epsilon} = -\boldsymbol{x}$, this gives us the 2nd order marginal attributions

$$\begin{aligned} F^{-1}_{\mu(\boldsymbol{X})}(\alpha) &\approx \mu(\boldsymbol{0}) + \sum_{j=1}^q S_j(\mu;\alpha) - \frac{1}{2}\sum_{j,k=1}^q T_{j,k}(\mu;\alpha) \qquad (7.78)\\ &= \mu(\boldsymbol{0}) + \sum_{j=1}^q \left(S_j(\mu;\alpha) - \frac{1}{2}T_{j,j}(\mu;\alpha)\right) - \sum_{1\le j<k\le q} T_{j,k}(\mu;\alpha), \end{aligned}$$

where for $1 \le j,k \le q$ we define $\mu_{j,k}(\boldsymbol{x}) = \partial_{x_j}\partial_{x_k}\mu(\boldsymbol{x})$, see (7.72), and

$$T_{j,k}(\mu;\alpha) = \mathbb{E}_p\left[X_jX_k\mu_{j,k}(\boldsymbol{X})\,\middle|\,\mu(\boldsymbol{X}) = F^{-1}_{\mu(\boldsymbol{X})}(\alpha)\right]. \qquad (7.79)$$

Remarks 7.38

- The first line of (7.78) separates the 1st order attributions from the 2nd order attributions, the second line splits w.r.t. the individual component j attributions and the interaction attributions $j \neq k$.
- The 1st order attributions (7.75) have been motivated by considering the directional derivatives of the VaR distortion risk measure. Unfortunately, the 2nd order consideration has no simple equivalent motivation, as the 2nd order directional derivatives are much more involved, even in the linear case, we refer to Property 1 in Gourieroux et al. [167].

- Interestingly, we can precisely evaluate the accuracy of approximation (7.78) by analyzing for a given regression function $\mu(\cdot)$

$$\sup_{\alpha\in(0,1)} \left| F^{-1}_{\mu(\boldsymbol{X})}(\alpha) - \mu\left(\boldsymbol{0}\right) - \sum_{j=1}^{q} S_j(\mu;\alpha) + \frac{1}{2}\sum_{j,k=1}^{q} T_{j,k}(\mu;\alpha)\right|. \tag{7.80}$$

  Intuitively, in order to have a uniform good approximation, the origin $\boldsymbol{0}$ should be somehow centered in the feature distribution $\boldsymbol{X} \sim p$. This will be studied next.

Above we have implicitly assumed that $\boldsymbol{0}$ is a suitable reference point that makes the approximation error (7.80) small. For FN network fitting we typically normalize the features either using the MinMaxScaler (7.29) or we center and normalize the components of $(\boldsymbol{x}_i)_{1\le i\le n}$ according to (7.30). That is, the reference point is chosen such that the gradient descent fitting works efficiently. However, this may not be an optimal reference point for studying the 2nd order attributions. Therefore, we analyze this question in more detail, and the following reparametrization can still be done after model fitting.

If we choose an arbitrary translation $\boldsymbol{a} \in \mathbb{R}^q$, we can set $\boldsymbol{\epsilon} = \boldsymbol{a} - \boldsymbol{x}$ in the above 2nd order Taylor expansion to receive another 2nd order marginal attribution representation

$$\begin{aligned} F^{-1}_{\mu(\boldsymbol{X})}(\alpha) \approx \mu\left(\boldsymbol{a}\right) - \mathbb{E}_p\left[(\boldsymbol{a}-\boldsymbol{X})^\top \nabla_{\boldsymbol{x}}\mu(\boldsymbol{X}) \,\middle|\, \mu(\boldsymbol{X}) = F^{-1}_{\mu(\boldsymbol{X})}(\alpha)\right] \\ -\frac{1}{2}\mathbb{E}_p\left[(\boldsymbol{a}-\boldsymbol{X})^\top \nabla^2_{\boldsymbol{x}}\mu(\boldsymbol{X})(\boldsymbol{a}-\boldsymbol{X}) \,\middle|\, \mu(\boldsymbol{X}) = F^{-1}_{\mu(\boldsymbol{X})}(\alpha)\right]. \end{aligned} \tag{7.81}$$

Essentially, this means that we shift the feature distribution p to considering the shifted random vectors $\boldsymbol{X}^{\boldsymbol{a}} = \boldsymbol{X} - \boldsymbol{a}$ and while setting $\mu^{\boldsymbol{a}}(\cdot) = \mu(\boldsymbol{a} + \cdot)$, thus, this simply says that we pre-process the features differently. In view of approximation (7.81) we can now select a reference point $\boldsymbol{a} \in \mathbb{R}^q$ that makes the 2nd order marginal attributions as precise as possible. Define the events $\mathcal{A}_l = \{\mu(\boldsymbol{X}) = F^{-1}_{\mu(\boldsymbol{X})}(\alpha_l)\}$ for a discrete quantile grid $0 < \alpha_1 < \ldots < \alpha_L < 1$. We define the objective function

$$\begin{aligned} \boldsymbol{a} \mapsto G(\boldsymbol{a};\mu) = \sum_{l=1}^{L}\Bigg(F^{-1}_{\mu(\boldsymbol{X})}(\alpha_l) - \mu\left(\boldsymbol{a}\right) + \mathbb{E}_p\left[(\boldsymbol{a}-\boldsymbol{X})^\top \nabla_{\boldsymbol{x}}\mu(\boldsymbol{X}) \,\middle|\, \mathcal{A}_l\right] \\ + \frac{1}{2}\,\mathbb{E}_p\left[(\boldsymbol{a}-\boldsymbol{X})^\top \nabla^2_{\boldsymbol{x}}\mu(\boldsymbol{X})(\boldsymbol{a}-\boldsymbol{X})^\top \,\middle|\, \mathcal{A}_l\right]\Bigg)^2. \end{aligned} \tag{7.82}$$

Making this objective function $G(\boldsymbol{a};\mu)$ small in $\boldsymbol{a}$ will provide us with a good reference point for the selected quantile levels $(\alpha_l)_{1\le l\le L}$; this is exactly the MACQ proposal of Merz et al. [273]. A local minimum can be found by applying a gradient descent algorithm

$$\boldsymbol{a}^{(t)} \mapsto \boldsymbol{a}^{(t+1)} = \boldsymbol{a}^{(t)} - \delta_{t+1} \nabla_{\boldsymbol{a}} G(\boldsymbol{a}^{(t)}; \mu),$$

for tempered learning rates $\delta_{t+1} > 0$. The gradient of G w.r.t. $\boldsymbol{a}$ is given by

$$\begin{aligned}\nabla_{\boldsymbol{a}} G(\boldsymbol{a}; \mu) = 2 \sum_{l=1}^{L} \Bigg(& F^{-1}_{\mu(\boldsymbol{X})}(\alpha_l) - \mu(\boldsymbol{a}) + \mathbb{E}_p \left[(\boldsymbol{a} - \boldsymbol{X})^\top \nabla_{\boldsymbol{x}} \mu(\boldsymbol{X}) \middle| \mathcal{A}_l \right] \\ & + \frac{1}{2} \mathbb{E}_p \left[(\boldsymbol{a} - \boldsymbol{X})^\top \nabla^2_{\boldsymbol{x}} \mu(\boldsymbol{X}) (\boldsymbol{a} - \boldsymbol{X})^\top \middle| \mathcal{A}_l \right] \Bigg) \\ & \times \Bigg(- \nabla_{\boldsymbol{a}} \mu(\boldsymbol{a}) + \mathbb{E}_p \left[\nabla_{\boldsymbol{x}} \mu(\boldsymbol{X}) \middle| \mathcal{A}_l \right] \\ & - \mathbb{E}_p \left[\boldsymbol{X}^\top \nabla^2_{\boldsymbol{x}} \mu(\boldsymbol{X}) \middle| \mathcal{A}_l \right] + \frac{1}{2} \boldsymbol{a}^\top \mathbb{E}_p \left[\nabla^2_{\boldsymbol{x}} \mu(\boldsymbol{X}) \middle| \mathcal{A}_l \right] \Bigg).\end{aligned}$$

All subsequent considerations and interpretations are done w.r.t. an optimal reference point $\boldsymbol{a} \in \mathbb{R}^q$ by minimizing the objective function (7.82) on the chosen quantile grid. Mathematically speaking, this optimal choice is w.l.o.g. because the origin $\mathbf{0}$ of the coordinate system of the feature space $\mathcal{X}$ is arbitrary, and any other origin can be chosen by a translation, see formula (7.81) and the subsequent discussion. For interpretations, however, the choice of the reference point $\boldsymbol{a}$ matters because the directional derivative $X_j \mu_j(\boldsymbol{X})$ can be small either because X_j is small or because $\mu_j(\boldsymbol{X})$ is small. Having a small X_j means that this feature value is close to the chosen reference point.

Example 7.39 (MACQ Analysis) We revisit the MTPL claim frequency example using the FN network regression model of depth $d = 3$ having $(q_1, q_2, q_3) = (20, 15, 10)$ neurons. Importantly, we use the hyperbolic tangent as the activation function in the FN layers which provides smoothness of the regression function. Figure 7.40 shows the VPI plot of this fitted model. Obviously, the variable `BonusMalus` plays the most important role in this predictive model. Remark that the VPI plot does not properly respect the dependence structure in the features as it independently permutes each feature component at a time. The aim in this example is to determine variable importance by doing the MACQ analysis (7.78).

Figure 7.44 (lhs) shows the empirical density of the fitted canonical parameter $\theta(\boldsymbol{x}_i)$, $1 \le i \le n$; all plots in this example refer to the canonical scale. We then minimize the objective function (7.82) which provides us with an optimal reference point $\boldsymbol{a} \in \mathbb{R}^q$; we choose equidistant quantile grid $1\% < 2\% < \ldots < 99\%$ and all conditional expectations in $\nabla_{\boldsymbol{a}} G(\boldsymbol{a}; \mu)$ are empirically approximated by a local smoother similar to Listing 7.8. Figure 7.44 (rhs) gives the resulting marginal attributions w.r.t. this reference point. The orange line shows the 1st order marginal

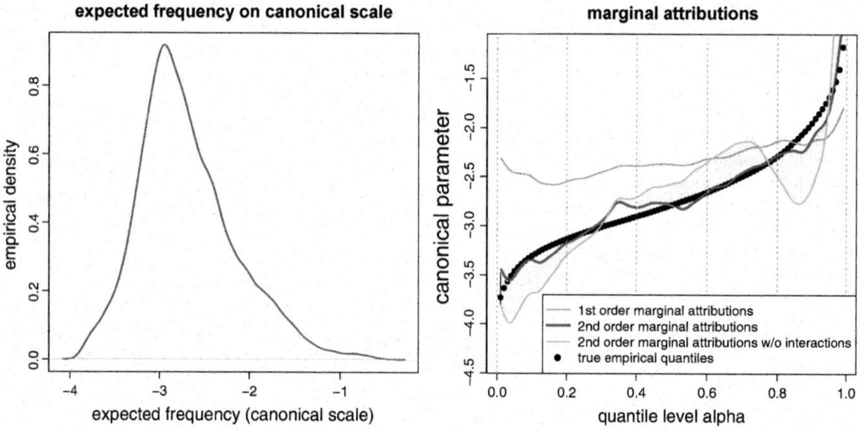

Fig. 7.44 (lhs) Empirical density of the fitted canonical parameter $\theta(\boldsymbol{x}_i)$, $1 \leq i \leq n$, (rhs) 1st and 2nd order marginal attributions

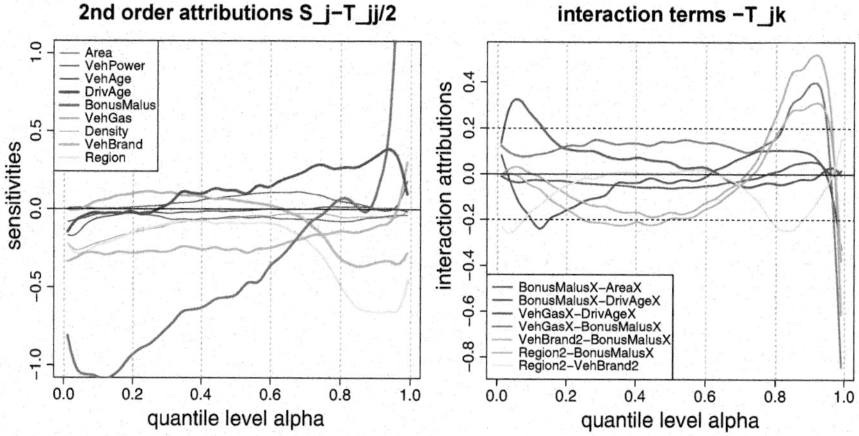

Fig. 7.45 (lhs) Second order marginal attributions $S_j(\mu; \alpha) - \frac{1}{2}T_{j,j}(\mu; \alpha)$ excluding interaction terms, and (rhs) interaction terms $-\frac{1}{2}T_{j,k}(\mu; \alpha)$, $j \neq k$

attributions (7.76), and the red line the 2nd order marginal attributions (7.78). The cyan line drops the interaction terms $T_{j,k}(\mu; \alpha)$, $j \neq k$, from the 2nd order marginal attributions. From the shaded cyan area we see the importance of the interaction terms. We note that the 2nd order marginal attributions (red line) match the true empirical quantiles (black dots) quite well for the chosen reference point $\boldsymbol{a}$.

Figure 7.45 gives the 2nd order marginal attributions $S_j(\mu; \alpha) - \frac{1}{2}T_{j,j}(\mu; \alpha)$ of the individual components $1 \leq j \leq q$ on the left-hand side, and the interaction terms $-\frac{1}{2}T_{j,k}(\mu; \alpha)$, $j \neq k$ on the right-hand side. We identify the following components as being important `BonusMalus`, `DrivAge`, `VehGas`, `VehBrand` and `Region`; these components show a behavior substantially different from being equal to 0, i.e.,

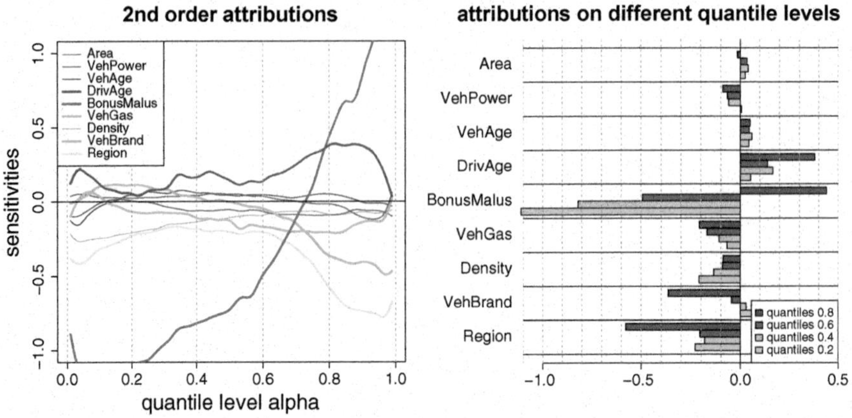

Fig. 7.46 (lhs) Second order marginal attributions $S_j(\mu;\alpha) - \frac{1}{2}\sum_{k=1}^{q} T_{j,k}(\mu;\alpha)$ including interaction terms, and (rhs) slices at the quantile levels $\alpha \in \{20\%, 40\%, 60\%, 80\%\}$

these components differentiate from the reference point $\boldsymbol{a}$. These components also have major interactions that contribute to the quantiles above the level 80%.

If we allocate the interaction terms to the corresponding components $1 \le j \le q$ we receive the second order marginal attributions $S_j(\mu;\alpha) - \frac{1}{2}\sum_{k=1}^{q} T_{j,k}(\mu;\alpha)$. These are illustrated in Fig. 7.46 (lhs) and the quantile slices at the levels $\alpha \in \{20\%, 40\%, 60\%, 80\%\}$ are given in Fig. 7.46 (rhs). These graphs illustrate variable importance on different quantile levels (and respecting the dependence within the features). In particular, we identify the main variables that distinguish the given quantile levels from the reference level $\theta(\boldsymbol{a})$, i.e., Fig. 7.46 (rhs) should be understood as the relative differences to the chosen reference level. Once more we see that `BonusMalus` is the main driver, but also other variables contribute to the differentiation of the high quantile levels.

Figure 7.47 shows the individual attributions $x_{i,j}\mu_j(\boldsymbol{x}_i)$ of 1'000 randomly selected cases $\boldsymbol{x}_i$ for the feature components j = `BonusMalus`, `DrivAge`, `VehGas`, `VehBrand`; the colors illustrate the corresponding feature values $x_{i,j}$ of the individual car drivers i, and the black solid line corresponds to $S_j(\mu;\alpha) - \frac{1}{2}T_{j,j}(\mu;\alpha)$ excluding the interaction terms (the black dotted line is one empirical standard deviation around the black solid line). Focusing on the variable `BonusMalus` we observe that the lower quantiles are almost completely dominated by insurance policies on the lowest bonus-malus level. The bonus-malus levels 70–80 provide little sensitivity (are concentrated around the zero line) because the reference point $\boldsymbol{a}$ reflects these bonus-malus levels, and, finally, the large quantiles are dominated by high bonus-malus levels (red dots).

The plot of the variable `DrivAge` is interpreted similarly. The reference point $\boldsymbol{a}$ is close to the young drivers, therefore, young drivers are concentrated around the zero line. At the low quantile levels, higher ages contribute positively to the low expected frequencies, whereas these ages have an unfavorable impact at higher

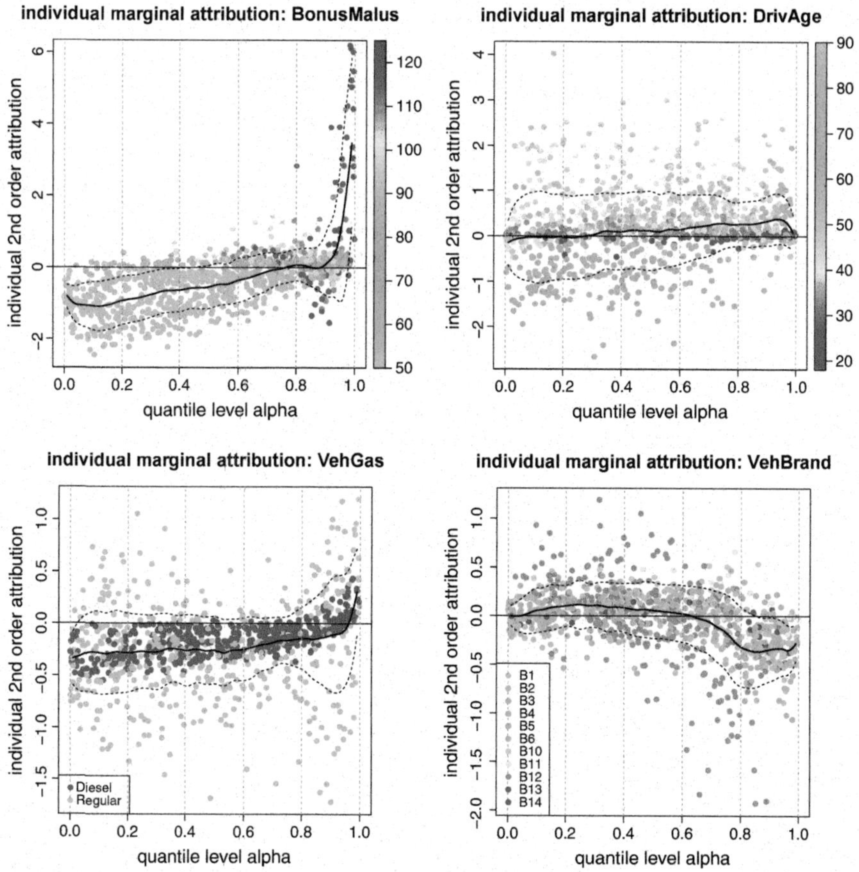

Fig. 7.47 Individual attributions $x_{i,j}\mu_j(\boldsymbol{x}_i)$ of 1'000 randomly selected cases $\boldsymbol{x}_i$ for j = `BonusMalus`, `DrivAge`, `VehGas`, `VehBrand`; the plots have different y-scales

quantile levels (this should be considered in combination with their bonus-malus levels). We also observe a few outliers in this plot, for instance, we can identify a driver of age 20 at a quantile level of 20%. Further inspection of this driver raises some doubts whether this data is correct since this driver is at a bonus-malus level of 68% (which should technically not be possible) and she/he has an exposure of 2 days. Surely, this insurance policy would need further investigation.

The plot of `VehGas` shows that the chosen reference level $\theta(\boldsymbol{a})$ is closer to `Diesel` fuel cars as the red dots fluctuate less around the zero line; in different runs of the gradient descent algorithm (with different seeds) this order has been changing (as it depends on the reference point $\boldsymbol{a}$). We skip a detailed analysis of the variable `VehBrand`. ■

7.7 Lab: Analysis of the Fitted Networks

In the previous section we have studied some model-agnostic tools that can be used for any (differentiable) regression model. In this section we give some network specific plots. For simplicity we choose one specific example, namely, the FN network $\widehat{\mu} \stackrel{\text{def.}}{=} \widehat{\mu}^{m=1}$ of Table 7.9. We start by analyzing the learned representations in the different FN layers, this links to our introduction in Sect. 7.1.

For any FN layer $1 \leq m \leq d$ we can study the learned representations $z^{(m:1)}(\boldsymbol{x})$. For Fig. 7.48 we select at random 1'000 insurance policies $\boldsymbol{x}_i$, and the dots show the activations of these insurance policies in neurons $j = 4$ (x-axis) and $j = 9$ (y-axis) in the corresponding FN layers. These neuron activations are in the interval $(-1, 1)$ because we work with the hyperbolic tangent activation function for ϕ. The color scale shows the resulting estimated frequencies $\widehat{\mu}(\boldsymbol{x}_i)$ of the selected policies. We observe that the layers are increasingly (in the depth of the network) separating the low frequency policies (light blue-green colors) from the high frequency policies (red color). This is a quite typical picture that we obtain here, though, this sparsity in the 3rd FN layer is not the case for every neuron $1 \leq j \leq q_d$.

In higher dimensional FN architectures it will be difficult to analyze the learned representations on each individual neuron, but at least one can try to understand the main effects learned. For this, on the one hand, we can focus on the important feature components, see, e.g., Sect. 7.6.1, and, on the other hand, we can try to study the main effects learned using a PCA in each FN layer, see Sect. 7.5.3. Figure 7.49 shows the singular values $\lambda_1 \geq \lambda_2 \geq \ldots \geq \lambda_{q_m} > 0$ in each of the three FN layers $1 \leq m \leq d = 3$; we center the neuron activations to mean zero before applying the SVD. These plots support the previously made statement that the layers are increasingly separating the high frequency from the low frequency policies. An elbow criterion tells us that in the first FN layer we have 8 important principal components (out of 20), in the second FN layer 3 (out of 15) and in the third FN layer 1 (out of 10). This is also reflected in Fig. 7.48 where we see more and more

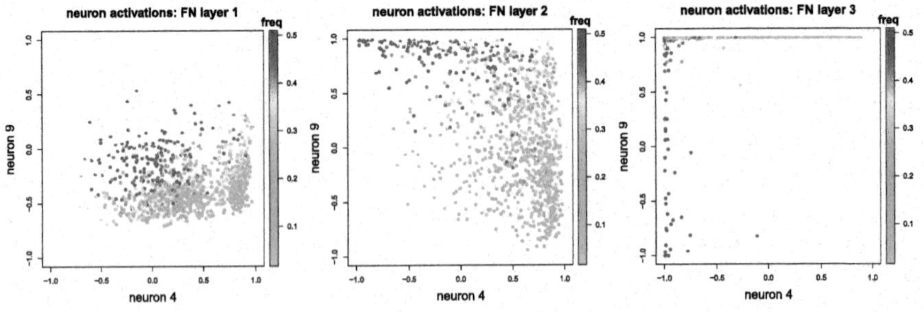

Fig. 7.48 Observed activations in the three FN layers $m = 1, 2, 3$ (left-middle-right) in the corresponding neurons $j = 4, 9$, the color key shows the estimated frequencies $\widehat{\mu}(\boldsymbol{x}_i)$

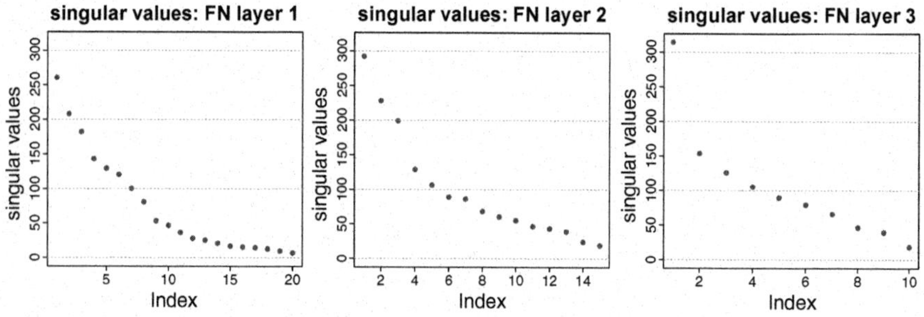

Fig. 7.49 Singular values $\lambda_1 \geq \lambda_2 \geq \ldots \geq \lambda_{q_m} > 0$ in the FN layers $1 \leq m \leq d = 3$

concentration in the neuron activations. It is important to notice that the chosen FN network calibration $\widehat{\mu}$ does not involve any drop-out layers during the gradient descent fitting, see Sect. 7.4.1. Drop-out layers prevent individual neurons to over-train to a specific task. Consequently, we will receive a network calibration that is more equally balanced across all neurons under drop-outs, because if one neuron drops out, the composite of the remaining neurons needs to be able to take over the task of the dropped out neuron. This leads to less sparsity and to singular values that are more similarly sized.

In Fig. 7.50 we analyze the first two principal components in each FN layer, thus, these are the two principal components that correspond to the two biggest singular values (λ_1, λ_2) in each of the three FN layers. The first row shows the input variables (`BonusMalus`, `DrivAge`) $\in [50, 125] \times [18, 90]$ of the 1'000 randomly selected policies $\boldsymbol{x}_i$; these are the two most important feature components according to the VPI analysis. All three columns show the same data, however, in different color scales: (lhs) gives the color scale $\widehat{\mu}$, (middle) gives the color scale `BonusMalus`, and (rhs) gives the color scale `DrivAge`. These color scales also apply to the other rows. The 2nd row shows the first two principal components in the 1st FN layer, the 3rd row in the 2nd FN layer, and the last row in the third FN layer. Focusing on the first column we observe that the layers cluster the high and the low frequency policies in the 1st principal component more and more across the FN layers. Not surprisingly this leads to a quite clear-cut separation w.r.t. the bonus-malus level which can be verified from the second column of Fig. 7.50. For the driver's age variable this sharp separation gets lost across the layers, see third column of Fig. 7.50, which indicates that the variable `DrivAge` does not influence the frequency monotonically and it interacts with the variable `BonusMalus`.

Figure 7.51 shows the second order marginal attributions (7.78) for the different inputs. The graph on the left-hand side shows the plot w.r.t. the original inputs $\boldsymbol{x}_i$, the graph in the middle w.r.t. the learned representations $\boldsymbol{z}^{(1:1)}(\boldsymbol{x}_i) \in \mathbb{R}^{q_1}$ in the first FN layer, and on the right-hand side w.r.t. the learned representations $\boldsymbol{z}^{(2:1)}(\boldsymbol{x}_i) \in \mathbb{R}^{q_2}$ in the second FN layer. We interpret these plots as follows: the FN network disentangles the different effects through the FN layers by making

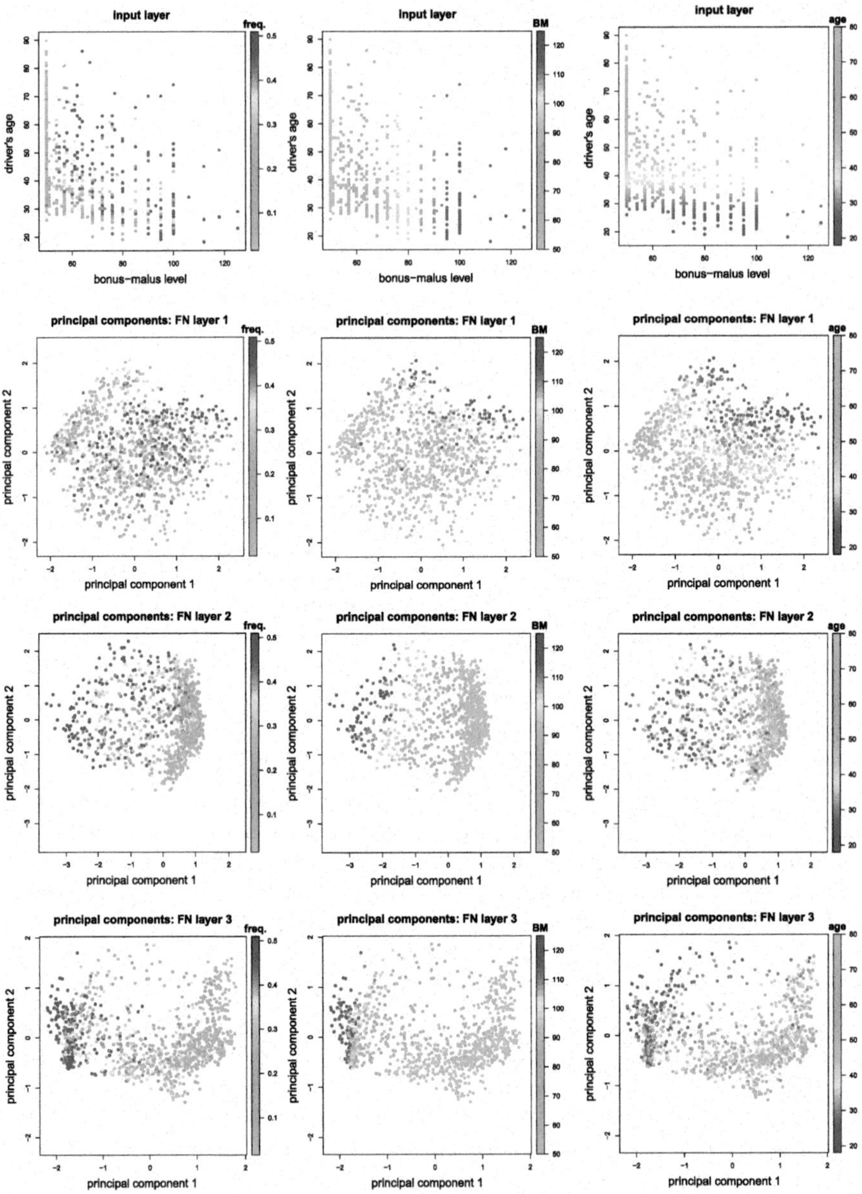

Fig. 7.50 (First row) Input variables (BonusMalus, DrivAge), (Second–fourth row) first two principal components in FN layers $m = 1, 2, 3$; (lhs) gives the color scale of estimated frequency $\widehat{\mu}$, (middle) gives the color scale BonusMalus, and (rhs) gives the color scale DrivAge

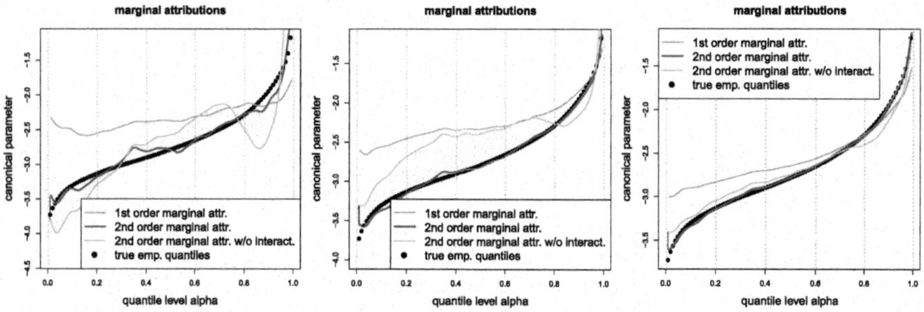

Fig. 7.51 Second order marginal attributions: (lhs) w.r.t. the input layer $\boldsymbol{x} \in \mathbb{R}^{q_0}$, (middle) w.r.t. the first FN layer $\boldsymbol{z}^{(1:1)}(\boldsymbol{x}) \in \mathbb{R}^{q_1}$, and (rhs) w.r.t. the second FN layer $\boldsymbol{z}^{(2:1)}(\boldsymbol{x}) \in \mathbb{R}^{q_2}$

the plots more smooth and making the interactions between the neurons smaller. Note that the learned representations $\boldsymbol{z}^{(3:1)}(\boldsymbol{x}_i) \in \mathbb{R}^{q_3}$ in the last FN layer go into a classical GLM for the output layer, which does not have any interactions in the canonical predictor (because it is additive on the canonical scale), thus, being of the same type as the linear regression of Example 7.37. In the Poisson model with the log-link function, the interactions can only be of a multiplicative type in GLMs. Therefore, the network feature-engineers the input $\boldsymbol{x}_i$ (in an automated way) such that the learned representation $\boldsymbol{z}^{(d:1)}(\boldsymbol{x}_i)$ in the last FN layer is exactly in this GLM structure. This is verified by the small interaction part in Fig. 7.51 (rhs). This closes this part on model-agnostic tools.

Chapter 8
Recurrent Neural Networks

Chapter 7 has discussed fully-connected *feed-forward neural* (FN) networks. Feed-forward means that information is passed in a directed acyclic path from the input layer to the output layer. A natural extension is to allow these networks to have cycles. In that case, we call the architecture a *recurrent neural* (RN) network. A RN network architecture is particularly useful for time-series modeling. The discussion on time-series data also links to Sect. 5.8.1 on longitudinal and panel data. RN networks have been introduced in the 1980s, and the two most popular RN network architectures are the long short-term memory (LSTM) architecture proposed by Hochreiter–Schmidhuber [188] and the gated recurrent unit (GRU) architecture introduced by Cho et al. [76]. These two architectures will be described in detail in this chapter.

8.1 Motivation for Recurrent Neural Networks

We start from a deep FN network providing the regression function, see (7.2)–(7.3),

$$\boldsymbol{x} \;\mapsto\; \mu(\boldsymbol{x}) = g^{-1}\langle \boldsymbol{\beta}, \boldsymbol{z}^{(d:1)}(\boldsymbol{x})\rangle, \tag{8.1}$$

with a composition $\boldsymbol{z}^{(d:1)}$ of d FN layers $\boldsymbol{z}^{(m)}$, $1 \le m \le d$, link function g and with output parameter $\boldsymbol{\beta} \in \mathbb{R}^{q_d+1}$. In principle, we could directly use this FN network architecture for time-series forecasting. We explain here why this is not the best option to deal with time-series data.

Assume we want to predict a random variable Y_{T+1} at time $T \ge 0$ based on the time-series information $\boldsymbol{x}_0, \boldsymbol{x}_1, \ldots, \boldsymbol{x}_T$. This information is assumed to be available at time T for predicting the response Y_{T+1}. The past response information Y_t, $1 \le$

M. V. Wüthrich, M. Merz, *Statistical Foundations of Actuarial Learning and its Applications*, Springer Actuarial, https://doi.org/10.1007/978-3-031-12409-9_8

$t \le T$, is typically included in $\boldsymbol{x}_t$.[1] Using the above FN network architecture we could directly try to predict Y_{T+1}, based on this past information. Therefore, we define the feature information $\boldsymbol{x}_{0:T} = (\boldsymbol{x}_0, \ldots, \boldsymbol{x}_T)$ and we aim at designing a FN network (8.1) for modeling

$$\boldsymbol{x}_{0:T} \mapsto \mu_T(\boldsymbol{x}_{0:T}) = \mathbb{E}[Y_{T+1}|\boldsymbol{x}_{0:T}] = \mathbb{E}[Y_{T+1}|\boldsymbol{x}_0, \ldots, \boldsymbol{x}_T].$$

In principle we could work with such an approach, however, it has a couple of severe drawbacks. Obviously, the length of the feature vector $\boldsymbol{x}_{0:T}$ depends on time T, that is, it will grow with every time step. Therefore, the regression function (network architecture) $\boldsymbol{x}_{0:T} \mapsto \mu_T(\boldsymbol{x}_{0:T})$ is time-dependent. Consequently, with this approach we have to fit a network for every T. This deficiency can be circumvented if we assume a Markov property that does not require of carrying forward the whole past history. Assume that it is sufficient to consider a history of a certain length. Choose $\tau \ge 0$ fixed, then, for $T \ge \tau$, we can set for the feature information $\boldsymbol{x}_{T-\tau:T} = (\boldsymbol{x}_{T-\tau}, \ldots, \boldsymbol{x}_T)$, which has a fixed length $\tau + 1 \ge 1$, now. In this situation we could try to design a FN network

$$\boldsymbol{x}_{T-\tau:T} \mapsto \mu(\boldsymbol{x}_{T-\tau:T}) = \mathbb{E}[Y_{T+1}|\boldsymbol{x}_{T-\tau:T}] = \mathbb{E}[Y_{T+1}|\boldsymbol{x}_{T-\tau}, \ldots, \boldsymbol{x}_T].$$

This network regression function can be chosen independent of T since the relevant history $\boldsymbol{x}_{T-\tau:T}$ always has the same length $\tau+1$. The time variable T could be used as a feature component in $\boldsymbol{x}_{T-\tau:T}$. The disadvantage of this approach is that such a FN network architecture does not respect the temporal causality. Observe that we feed the past history into the first FN layer

$$\boldsymbol{x}_{T-\tau:T} \mapsto \boldsymbol{z}^{(1)}(\boldsymbol{x}_{T-\tau:T}) \in \{1\} \times \mathbb{R}^{q_1}.$$

This operation typically does not respect any topology in the time index of $\boldsymbol{x}_{T-\tau+1:T}$. Thus, the FN network does not recognize that the feature $\boldsymbol{x}_{t-1}$ has been experienced just before the next feature $\boldsymbol{x}_t$. For this reason we are looking for a network architecture that can handle the time-series information in a temporal causal way.

[1] More mathematically speaking, we assume to have a filtration $(\mathcal{A}_t)_{t\ge0}$ on the probability space $(\Omega, \mathcal{A}, \mathbb{P})$. The basic assumption then is that both sequences $(\boldsymbol{x}_t)_t$ and $(Y_t)_t$ are $(\mathcal{A}_t)_t$-adapted, and we aim at predicting Y_{T+1}, based on the information $\mathcal{A}_T$. In the above case this information $\mathcal{A}_T$ is generated by $\boldsymbol{x}_0, \boldsymbol{x}_1, \ldots, \boldsymbol{x}_T$, where $\boldsymbol{x}_t$ typically includes the observation Y_t. We could also shift the time index in $\boldsymbol{x}_t$ by one time unit, and in that case we would assume that $(\boldsymbol{x}_t)_t$ is previsible w.r.t. the filtration $(\mathcal{A}_t)_t$. We do not consider this shift in time index as it only makes the notation unnecessarily more complicated, but the results remain the same by including the information correspondingly into the features.

8.2 Plain-Vanilla Recurrent Neural Network

8.2.1 Recurrent Neural Network Layer

We explain the basic idea of RN networks in a shallow network architecture, and deep network architectures will be discussed in Sect. 8.2.2, below. We start from the time-series input variable $\boldsymbol{x}_{0:T} = (\boldsymbol{x}_0, \ldots, \boldsymbol{x}_T)$, all components having the same structure $\boldsymbol{x}_t \in \mathcal{X} \subset \{1\} \times \mathbb{R}^{q_0}$, $0 \leq t \leq T$. The aim is to design a network architecture that allows us to predict the random variable Y_{T+1}, based on this time-series information $\boldsymbol{x}_{0:T}$.

The main idea is to feed one component $\boldsymbol{x}_t$ of the time-series $\boldsymbol{x}_{0:T}$ at a time into the network, and at the same time we use the output $\boldsymbol{z}_{t-1}$ of the previous loop as an input for the next loop. This variable $\boldsymbol{z}_{t-1}$ carries forward a memory of the past variables $\boldsymbol{x}_{0:t-1}$. We explain this with a single RN layer having $q_1 \in \mathbb{N}$ neurons. A RN layer is given (recursively) by a mapping, $t \geq 1$,

$$\begin{aligned} \boldsymbol{z}^{(1)} : \{1\} \times \mathbb{R}^{q_0} \times \mathbb{R}^{q_1} &\to \mathbb{R}^{q_1}, \\ (\boldsymbol{x}_t, \boldsymbol{z}_{t-1}) &\mapsto \boldsymbol{z}_t = \boldsymbol{z}^{(1)}(\boldsymbol{x}_t, \boldsymbol{z}_{t-1}), \end{aligned} \tag{8.2}$$

where the RN layer $\boldsymbol{z}^{(1)}$ has the same structure as the FN layer given in (7.5), but based on feature input $(\boldsymbol{x}_t, \boldsymbol{z}_{t-1}) \in \mathcal{X} \times \mathbb{R}^{q_1} \subset \{1\} \times \mathbb{R}^{q_0} \times \mathbb{R}^{q_1}$, and not including an intercept component $\{1\}$ in the output.

More formally, a *RN layer* with activation function ϕ is a mapping

$$\begin{aligned} \boldsymbol{z}^{(1)} : \{1\} \times \mathbb{R}^{q_0} \times \mathbb{R}^{q_1} &\to \mathbb{R}^{q_1} \\ (\boldsymbol{x}, \boldsymbol{z}) &\mapsto \boldsymbol{z}^{(1)}(\boldsymbol{x}, \boldsymbol{z}) = \left(z_1^{(1)}(\boldsymbol{x}, \boldsymbol{z}), \ldots, z_{q_1}^{(1)}(\boldsymbol{x}, \boldsymbol{z})\right)^\top, \end{aligned} \tag{8.3}$$

having neurons, $1 \leq j \leq q_1$,

$$z_j^{(1)}(\boldsymbol{x}, \boldsymbol{z}) = \phi\left(\left\langle \boldsymbol{w}_j^{(1)}, \boldsymbol{x} \right\rangle + \left\langle \boldsymbol{u}_j^{(1)}, \boldsymbol{z} \right\rangle\right), \tag{8.4}$$

for given network weights $\boldsymbol{w}_j^{(1)} \in \mathbb{R}^{q_0+1}$ and $\boldsymbol{u}_j^{(1)} \in \mathbb{R}^{q_1}$.

Thus, the FN layers (7.5)–(7.6) and the RN layers (8.3)–(8.4) are structurally equivalent, only the input $\boldsymbol{x} \in \mathcal{X}$ is adapted to the time-series structure $(\boldsymbol{x}_t, \boldsymbol{z}_{t-1}) \in \mathcal{X} \times \mathbb{R}^{q_1}$. Before giving more interpretation and before explaining how this single RN network structure can be extended to a deep RN network we illustrate this RN layer.

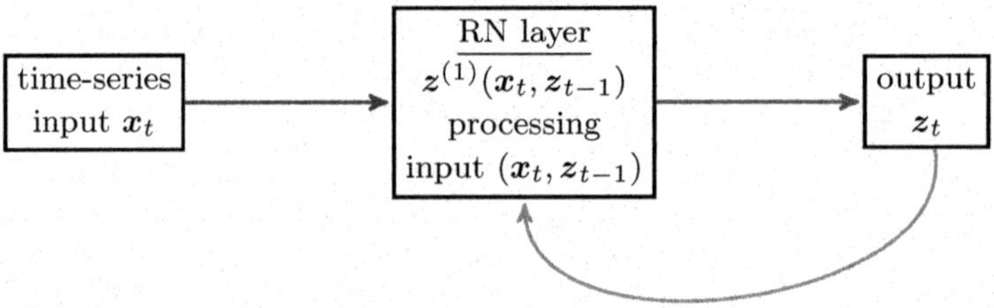

Fig. 8.1 RN layer $z^{(1)}$ processing the input $(\boldsymbol{x}_t, z_{t-1})$

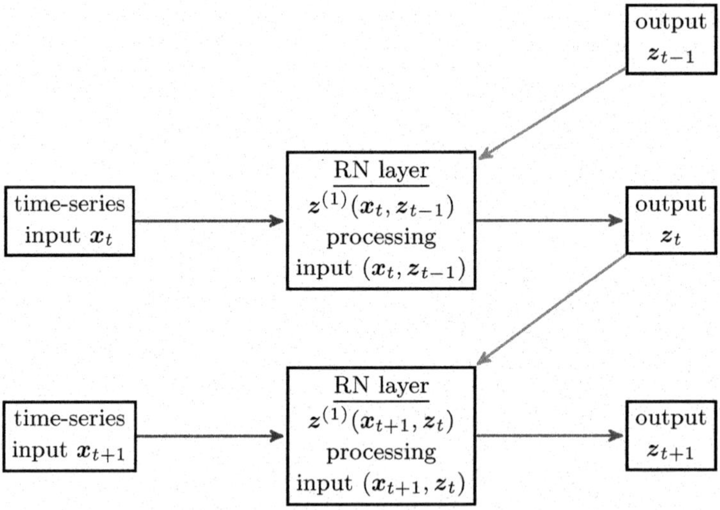

Fig. 8.2 Unfolded representation of RN layer $z^{(1)}$ processing the input $(\boldsymbol{x}_t, z_{t-1})$

Figure 8.1 shows an RN layer $z^{(1)}$ processing the input $(\boldsymbol{x}_t, z_{t-1})$, see (8.2). From this graph, the recurrent structure becomes clear since we have a loop (cycle) feeding the output z_t back into the RN layer to process the next input $(\boldsymbol{x}_{t+1}, z_t)$.
Often one depicts the RN architecture in a so-called unfolded way. This is done in Fig. 8.2. Instead of plotting the loop (cycle) as in Fig. 8.1 (orange arrow in the colored version), we unfold this loop by plotting the RN layer multiple times. Note that this RN layer in Fig. 8.2 uses always the same network weights $\boldsymbol{w}_j^{(1)}$ and $\boldsymbol{u}_j^{(1)}$, $1 \le j \le q_1$, for all t. Moreover, the use of the colors of the arrows (in the colored version) in the two figures coincides.

Remarks 8.1

- The neurons of the RN layer (8.4) have the following structure

$$z_j^{(1)}(\boldsymbol{x}, z) = \phi\left(\langle \boldsymbol{w}_j^{(1)}, \boldsymbol{x}\rangle + \langle \boldsymbol{u}_j^{(1)}, z\rangle\right) = \phi\left(w_{0,j}^{(1)} + \sum_{l=1}^{q_0} w_{l,j}^{(1)} x_l + \sum_{l=1}^{q_1} u_{l,j}^{(1)} z_l\right).$$

The network weights $W^{(1)} = (\boldsymbol{w}_j^{(1)})_{1 \le j \le q_1} \in \mathbb{R}^{(q_0+1) \times q_1}$ include an intercept component $w_{0,j}^{(1)}$ and the network weights $U^{(1)} = (\boldsymbol{u}_j^{(1)})_{1 \le j \le q_1} \in \mathbb{R}^{q_1 \times q_1}$ do not include an intercept component, otherwise we would have a redundancy.

- The RN network architecture generates a new process $(z_t)_t$. This process encodes the part of the past history $(\boldsymbol{x}_{0:t})_t$ which is relevant for forecasting the next step. Thus, $(z_t)_t$ can be interpreted as a (latent) *memory process*, or as the process of learned (relevant) time-series representation giving us $z_t = z_t(\boldsymbol{x}_{0:t})$.
- The same activation function ϕ and the same network weights $(\boldsymbol{w}_j^{(1)})_{1 \le j \le q_1}$ and $(\boldsymbol{u}_j^{(1)})_{1 \le j \le q_1}$ are shared across all time periods $t \ge 0$. This means that we assume a stationary (stochastic) process.
- The upper index $^{(1)}$ indicates the fact that this is the first (and single) RN layer in this example. In this sense, Figs. 8.1 and 8.2 show a shallow RN network. In the next section we are going to discuss deep RN networks, and below we are also going to discuss how the output is modeled, i.e., how the response Y_{T+1} is predicted based on the pre-processed features $(z_t)_{0 \le t \le T} \in \mathbb{R}^{q_1 \times (T+1)}$.

8.2.2 *Deep Recurrent Neural Network Architectures*

There are many different ways of extending a shallow RN network to a deep RN network. Assume we want to model a RN network of depth $d \ge 2$. A first (obvious) way of receiving a deep RN network architecture is

$$z_t^{[1]} = z^{(1)}\left(\boldsymbol{x}_t, z_{t-1}^{[1]}\right) \in \mathbb{R}^{q_1}, \tag{8.5}$$

$$z_t^{[m]} = z^{(m)}\left(z_t^{[m-1]}, z_{t-1}^{[m]}\right) \in \mathbb{R}^{q_m} \qquad \text{for } 2 \le m \le d, \tag{8.6}$$

where all RN layers $z^{(m)}$, $1 \le m \le d$, are of type (8.3)–(8.4), and additionally we include an intercept component in the RN layers $z^{(m)}$, $2 \le m \le d$. We add the upper indices (in square brackets $[\cdot]$) to the time-series $(z_t^{[m]})_t$ to indicate which RN layer outputs these learned representations (memory processes). In fact, we could also write $z_t^{[m:1]}$ instead of $z_t^{[m]}$, because in $z_t^{[m:1]}$ the feature input $\boldsymbol{x}_{0:t}$ has been processed through m RN layers $z^{(1)}, \ldots, z^{(m)}$. For simplicity, we just use the notation $z_t^{[m]} = z_t^{[m]}(\boldsymbol{x}_{0:t})$.

We are going to use the following abbreviation for a RN layer $m \geq 1$

$$z_t^{[m]} = z^{(m)}\left(z_t^{[m-1]}, z_{t-1}^{[m]}\right) = \phi\left(\left\langle W^{(m)}, z_t^{[m-1]}\right\rangle + \left\langle U^{(m)}, z_{t-1}^{[m]}\right\rangle\right) \in \mathbb{R}^{q_m}, \tag{8.7}$$

where the weights $W^{(m)} = (\boldsymbol{w}_1^{(m)}, \ldots, \boldsymbol{w}_{q_m}^{(m)}) \in \mathbb{R}^{(q_{m-1}+1)\times q_m}$ include the intercept components, and the weights $U^{(m)} = (\boldsymbol{u}_1^{(m)}, \ldots, \boldsymbol{u}_{q_m}^{(m)}) \in \mathbb{R}^{q_m\times q_m}$ do not include any intercept components. The scalar product is understood column-wise in the weight matrices $W^{(m)}$ and $U^{(m)}$, and the activation ϕ is understood component-wise. Moreover, we initialize for the input $z_t^{[0]} = \boldsymbol{x}_t$.

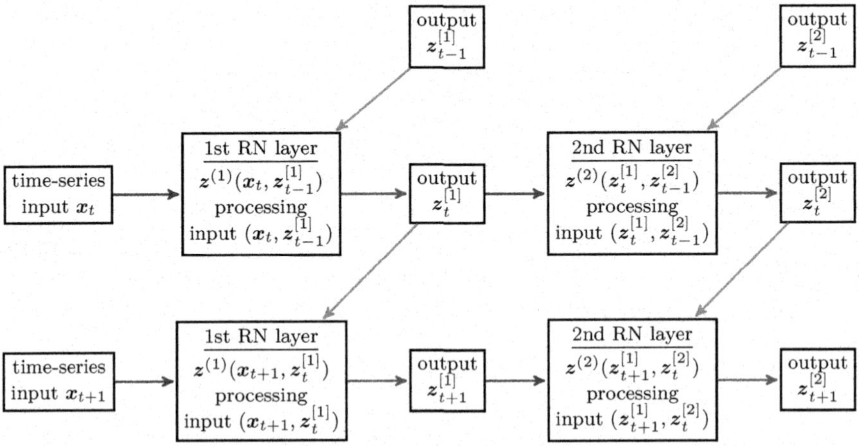

Fig. 8.3 Unfolded representation of a RN network architecture of depth $d = 2$

Figure 8.3 shows the RN network architecture of depth $d = 2$ defined in (8.5)–(8.6). The dimension of the input $z_t^{[0]} = \boldsymbol{x}_t \in \mathcal{X} \subseteq \{1\} \times \mathbb{R}^{q_0}$ is $q_0 + 1$, the first RN layer has q_1 neurons and the second RN layer q_2 neurons. From this graph it becomes clear how a RN network architecture of any depth $d \in \mathbb{N}$ can be constructed (recursively).

Remark 8.2 There are many alternative ways in building deep RN networks. E.g., we can add a loop that connects the output of the second RN layer back to the first one

$$z_t^{[1]} = z^{(1)}\left(\boldsymbol{x}_t, z_{t-1}^{[1]}, z_{t-1}^{[2]}\right),$$

$$z_t^{[2]} = z^{(2)}\left(z_t^{[1]}, z_{t-1}^{[2]}\right),$$

or we can add a skip connection from the input variable $\boldsymbol{x}_t$ to the second RN layer

$$z_t^{[1]} = z^{(1)}\left(\boldsymbol{x}_t, z_{t-1}^{[1]}\right),$$

$$z_t^{[2]} = z^{(2)}\left(\boldsymbol{x}_t, z_t^{[1]}, z_{t-1}^{[2]}\right).$$

We refrain from explicitly studying such RN network variants any further.

8.2.3 Designing the Network Output

There remains to explain how to predict the response variable Y_{T+1} based on the pre-processed features (memory processes) $z_T^{[1]}, \ldots, z_T^{[d]}$, outputted by the RN network of depth $d \geq 1$. Typically, only the final output of the last RN layer $z_T^{[d]} = z_T^{[d]}(\boldsymbol{x}_{0:T}) \in \mathbb{R}^{q_d}$ is considered to predict the response Y_{T+1}. We take this output and feed it into a FN network $\bar{z}^{(D:1)} : \{1\} \times \mathbb{R}^{q_d} \to \{1\} \times \mathbb{R}^{\bar{q}_D}$ of depth $D \in \mathbb{N}$ and with FN layers $\bar{z}^{(m)}$, $1 \leq m \leq D$, given by (7.5). Moreover, we choose a strictly monotone and smooth link function g.

This then provides us with the regression function, see (7.7)–(7.8),

$$\boldsymbol{x}_{0:T} \mapsto \mathbb{E}[Y_{T+1}|\boldsymbol{x}_{0:T}] = \mu(\boldsymbol{x}_{0:T}) = g^{-1}\left\langle \boldsymbol{\beta}, \bar{z}^{(D:1)}\left(z_T^{[d]}(\boldsymbol{x}_{0:T})\right)\right\rangle. \tag{8.8}$$

Thus, we first process the time-series features $\boldsymbol{x}_{0:T}$ through a RN network to receive the learned representation $z_T^{[d]}(\boldsymbol{x}_{0:T}) \in \mathbb{R}^{q_d}$ at time T. This learned representation is then used as a feature input to a FN network $\bar{z}^{(D:1)}$ that allows us to predict the response Y_{T+1}. This is illustrated in Fig. 8.4 for depth $d = 1$.

Remarks 8.3

- From the graph in Fig. 8.4 it also becomes apparent that we can consider different insurance policies $1 \leq i \leq n$ having different lengths of the corresponding histories $\boldsymbol{x}_{i,T-\tau_i:T} \in \mathbb{R}^{(q_0+1)\times(\tau_i+1)}$, $\tau_i \in \{0, \ldots, T\}$. The stationarity assumption allows us to enter the network in Fig. 8.4 at any time $T - \tau_i$. The RN network encodes this history into a learned feature $z_T^{[1]}(\boldsymbol{x}_{i,T-\tau_i:T})$ which is then decoded by the FN network $\bar{z}^{(D:1)}$ to forecast $Y_{i,T+1}$.
- If there is additional insurance policy dependent feature information $\widetilde{\boldsymbol{x}}_i$ that is not of a time-series structure, we can concatenate the feature information $(z_T^{[d]}(\boldsymbol{x}_{i,0:T}), \widetilde{\boldsymbol{x}}_i)$ which then enters the FN network (8.8).

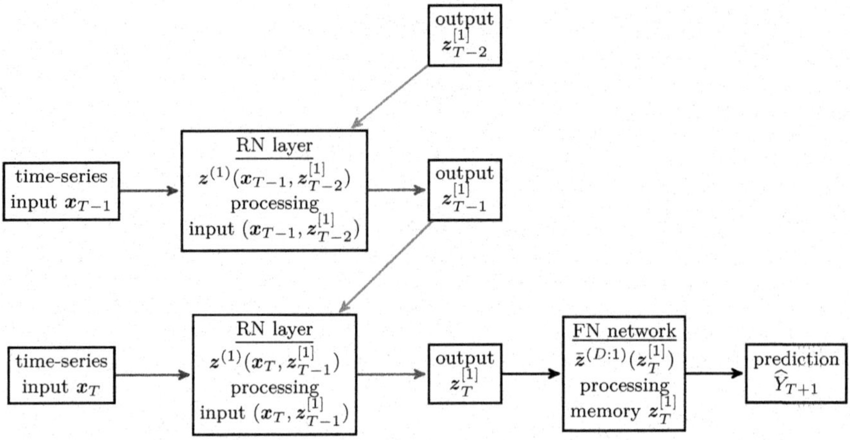

Fig. 8.4 Forecasting the response Y_{T+1} using a RN network (8.8) based on a single RN layer $d = 1$ and on a FN network of depth D

There remains to fit this network architecture having d RN layers and D FN layers to the available data. The RN layers involve the network weights $W^{(m)} \in \mathbb{R}^{(q_{m-1}+1)\times q_m}$ and $U^{(m)} \in \mathbb{R}^{q_m \times q_m}$, for $1 \le m \le d$, and the FN layers involve the network weights $(\bar{w}^{(m)}_j)_{1\le j\le \bar{q}_m} \in \mathbb{R}^{(\bar{q}_{m-1}+1)\times \bar{q}_m}$, for $1 \le m \le D$, and with $\bar{q}_0 = q_d$. Moreover, we have an output parameter $\beta \in \mathbb{R}^{\bar{q}_D+1}$. The fitting is again done by a gradient descent algorithm minimizing the corresponding objective function.

Assume we have independent (in i) data $(Y_{i,T+1}, x_{i,0:T}, v_{i,T+1})$ of the cases $1 \le i \le n$. We then assume that the responses $Y_{i,T+1}$ can be modeled by a fixed member of the EDF having unit deviance $\mathfrak{d}$. We consider the deviance loss function, see (4.9),

$$\vartheta \mapsto \mathfrak{D}(Y_{T+1}, \vartheta) = \frac{1}{n}\sum_{i=1}^{n} \frac{v_{i,T+1}}{\varphi}\, \mathfrak{d}\Big(Y_{i,T+1}, \mu_\vartheta(x_{i,0:T})\Big), \tag{8.9}$$

for the observations $Y_{T+1} = (Y_{1,T+1}, \ldots, Y_{n,T+1})^\top$, and where ϑ collects all the RN and FN network weights/parameters of the regression function (8.8). This model can now be fitted using a variant of the gradient descent algorithm. The variant uses back-propagation through time (BPTT) which is an adaption of the back-propagation method to calculate the gradient w.r.t. the network parameter ϑ.

8.2.4 *Time-Distributed Layer*

There is a special feature in RN network modeling which is called a *time-distributed layer*. Observe from Fig. 8.4 that the deviance loss function (8.9) only focuses on the final observation $Y_{i,T+1}$. However, the stationarity assumption allows us to output and study any (previous) observation $Y_{i,t+1}$, $0 \le t \le T$. A time-distributed layer considers applying the deep FN network (8.8) *simultaneously* at all time points $0 \le t \le T$; simultaneously meaning that we use the same FN network weights for all t. The latter is justified under the assumption of having stationarity.

This then provides us with the regressions

$$\boldsymbol{x}_{0:t} \mapsto \mathbb{E}[Y_{t+1}|\boldsymbol{x}_{0:t}] = \mu(\boldsymbol{x}_{0:t}) = g^{-1}\left\langle \boldsymbol{\beta}, \bar{z}^{(D:1)}\left(z_t^{[d]}(\boldsymbol{x}_{0:t})\right)\right\rangle \qquad \text{for all } t \ge 0. \tag{8.10}$$

Figure 8.5 illustrates a time-distributed output where we predict $(Y_{t+1})_t$ based on the history $(\boldsymbol{x}_{0:t})_t$, and we always apply the same FN network $\bar{z}^{(D:1)}$ to the memory $z_t^{[1]} = z_t^{[1]}(\boldsymbol{x}_{0:t})$.

A time-distributed layer changes the fitting procedure. Instead of considering the objective function (8.9) for the final observation $Y_{i,T+1}$, we now include all observations $\boldsymbol{Y} = (Y_{i,t+1})_{0 \le t \le T, 1 \le i \le n}$ into the objective function. This results in studying the deviance loss function

$$\boldsymbol{\vartheta} \mapsto \mathfrak{D}(\boldsymbol{Y}, \boldsymbol{\vartheta}) = \frac{1}{n}\sum_{i=1}^{n} \frac{1}{T+1}\sum_{t=0}^{T} \frac{v_{i,t+1}}{\varphi}\, \mathfrak{d}\Big(Y_{i,t+1}, \mu_{\boldsymbol{\vartheta}}(\boldsymbol{x}_{i,0:t})\Big). \tag{8.11}$$

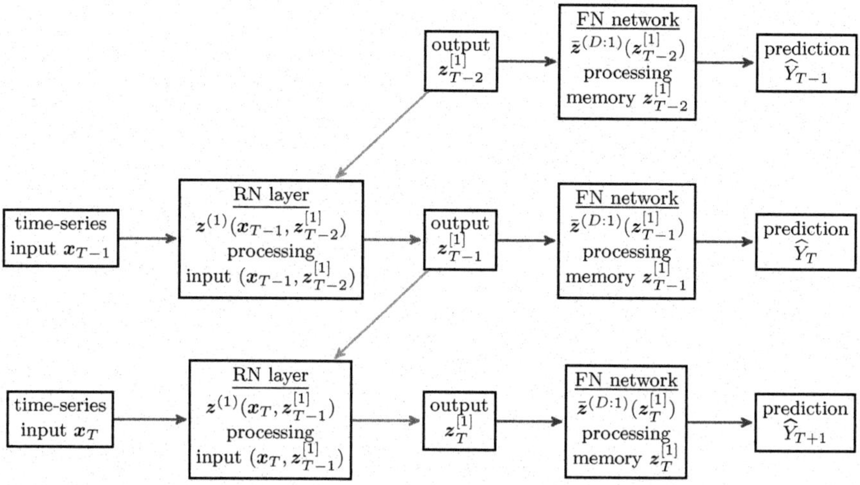

Fig. 8.5 Forecasting $(Y_{t+1})_t$ using a RN network (8.10) based on a single RN layer $d = 1$ and using a time-distributed FN layer for the outputs

Note that this can easily be adapted if the different cases $1 \leq i \leq n$ have different lengths in their histories. An example is provided in Listing 10.8, below.

8.3 Special Recurrent Neural Networks

In the plain-vanilla RN networks introduced above we have defined the memory processes $(z_t^{[m]})_{t\geq 0}$, $1 \leq m \leq d$, which encode the information history $(\boldsymbol{x}_t)_{t\geq 0}$ through different RN layers in a temporal causal way. This is naturally done through the use of a time-series structure as illustrated, e.g., in Fig. 8.5. There are more specific RN network architectures that allow the memory processes to be of a long memory or a short memory type. In this section, we present the two most popular architectures that pay a special attention to the memory storage. This is the long short-term memory (LSTM) architecture introduced by Hochreiter–Schmidhuber [188] and the gated recurrent unit (GRU) architecture proposed by Cho et al. [76].

8.3.1 Long Short-Term Memory Network

The LSTM network of Hochreiter–Schmidhuber [188] is the most commonly used RN network architecture. The LSTM network uses simultaneously three different activation functions for different purposes, the sigmoid and hyperbolic tangent activation functions, respectively,

$$\phi_\sigma(x) = \frac{1}{1+e^{-x}} \in (0,1) \qquad \text{and} \qquad \phi_{\tanh}(x) = \frac{e^x - e^{-x}}{e^x + e^{-x}} \in (-1,1),$$

and a general activation function $\phi : \mathbb{R} \to \mathbb{R}$, see also Table 7.1.

The LSTM network relies on several RN layers that are of the same structure as the plain-vanilla RN layer given in (8.7). We start by defining three different so-called *gates* that all have the RN layer structure (8.7). These three gates are used to model the memory cell of the LSTM network. Choose a layer index $m \geq 1$ and assume that $z_t^{[m-1]}$ is modeled by the previous layer $m-1$; for $m = 1$ we initialize $z_t^{[0]} = \boldsymbol{x}_t$. The three gates are then defined as follows, set $t \geq 1$:

- The *forget gate* models the loss of memory rate

$$f_t^{[m]} = f^{(m)}\left(z_t^{[m-1]}, z_{t-1}^{[m]}\right) = \phi_\sigma^f\left(\left\langle W_f^{(m)}, z_t^{[m-1]}\right\rangle + \left\langle U_f^{(m)}, z_{t-1}^{[m]}\right\rangle\right) \in (0,1)^{q_m},$$

with the network weights $W_f^{(m)} \in \mathbb{R}^{(q_{m-1}+1)\times q_m}$ and $U_f^{(m)} \in \mathbb{R}^{q_m \times q_m}$, and with the sigmoid activation function $\phi_\sigma^f = \phi_\sigma$, we also refer to (8.7).

- The *input gate* models the memory update rate

$$\boldsymbol{i}_t^{[m]} = \boldsymbol{i}^{(m)}\left(\boldsymbol{z}_t^{[m-1]}, \boldsymbol{z}_{t-1}^{[m]}\right) = \boldsymbol{\phi}_\sigma^i\left(\left\langle W_i^{(m)}, \boldsymbol{z}_t^{[m-1]}\right\rangle + \left\langle U_i^{(m)}, \boldsymbol{z}_{t-1}^{[m]}\right\rangle\right) \in (0,1)^{q_m},$$

  with the network weights $W_i^{(m)} \in \mathbb{R}^{(q_{m-1}+1)\times q_m}$ and $U_i^{(m)} \in \mathbb{R}^{q_m \times q_m}$, and with the sigmoid activation function $\phi_\sigma^i = \phi_\sigma$.
- The *output gate* models the release of memory information rate

$$\boldsymbol{o}_t^{[m]} = \boldsymbol{o}^{(m)}\left(\boldsymbol{z}_t^{[m-1]}, \boldsymbol{z}_{t-1}^{[m]}\right) = \boldsymbol{\phi}_\sigma^o\left(\left\langle W_o^{(m)}, \boldsymbol{z}_t^{[m-1]}\right\rangle + \left\langle U_o^{(m)}, \boldsymbol{z}_{t-1}^{[m]}\right\rangle\right) \in (0,1)^{q_m}, \tag{8.12}$$

  with the network weights $W_o^{(m)} \in \mathbb{R}^{(q_{m-1}+1)\times q_m}$ and $U_o^{(m)} \in \mathbb{R}^{q_m \times q_m}$, and with the sigmoid activation function $\phi_\sigma^o = \phi_\sigma$.

These gates have outputs in $(0, 1)$, and they determine the relative amount of memory that is updated and released in each step. The so-called *cell state process* $(\boldsymbol{c}_t^{[m]})_t$ is used to store the relevant memory. Given $\boldsymbol{z}_t^{[m-1]}$, $\boldsymbol{z}_{t-1}^{[m]}$ and $\boldsymbol{c}_{t-1}^{[m]}$, the updated cell state is defined by

$$\begin{aligned}\boldsymbol{c}_t^{[m]} &= \boldsymbol{c}^{(m)}\left(\boldsymbol{z}_t^{[m-1]}, \boldsymbol{z}_{t-1}^{[m]}, \boldsymbol{c}_{t-1}^{[m]}\right) \qquad (8.13)\\ &= \boldsymbol{f}_t^{[m]} \odot \boldsymbol{c}_{t-1}^{[m]} + \boldsymbol{i}_t^{[m]} \odot \boldsymbol{\phi}_{\tanh}\left(\left\langle W_c^{(m)}, \boldsymbol{z}_t^{[m-1]}\right\rangle + \left\langle U_c^{(m)}, \boldsymbol{z}_{t-1}^{[m]}\right\rangle\right) \in \mathbb{R}^{q_m},\end{aligned}$$

with the network weights $W_c^{(m)} \in \mathbb{R}^{(q_{m-1}+1)\times q_m}$ and $U_c^{(m)} \in \mathbb{R}^{q_m \times q_m}$, and $\odot$ denotes the Hadamard product. This defines how the memory (cell state) is updated and passed forward using the forget and the input gates $\boldsymbol{f}_t^{[m]}$ and $\boldsymbol{i}_t^{[m]}$, respectively.

The neuron activations $\boldsymbol{z}_t^{[m]}$ are updated, given $\boldsymbol{z}_t^{[m-1]}$, $\boldsymbol{z}_{t-1}^{[m]}$ and $\boldsymbol{c}_t^{[m]}$, by

$$\boldsymbol{z}_t^{[m]} = \boldsymbol{z}^{(m)}\left(\boldsymbol{z}_t^{[m-1]}, \boldsymbol{z}_{t-1}^{[m]}, \boldsymbol{c}_t^{[m]}\right) = \boldsymbol{o}_t^{[m]} \odot \boldsymbol{\phi}\left(\boldsymbol{c}_t^{[m]}\right) \in \mathbb{R}^{q_m}, \tag{8.14}$$

with the cell state $\boldsymbol{c}_t^{[m]}$ given in (8.13) and the output gate $\boldsymbol{o}_t^{[m]}$ defined in (8.12).

Figure 8.6[2] shows a LSTM cell (8.13)–(8.14) which includes four RN layers (8.7) for the forget gate $\boldsymbol{f}^{(m)}$, the input gate $\boldsymbol{i}^{(m)}$, the output gate $\boldsymbol{o}^{(m)}$ and in the cell state update (8.13). These RN layers are combined using the Hadamard product $\odot$ resulting in the updated cell state $\boldsymbol{c}_t^{[m]}$ and the learned representation $\boldsymbol{z}_t^{[m]}$ both being functions of the inputs $\boldsymbol{x}_{0:t}$.

[2] This figure is based on colah's blog explaining LSTMs https://colah.github.io/posts/2015-08-Understanding-LSTMs/.

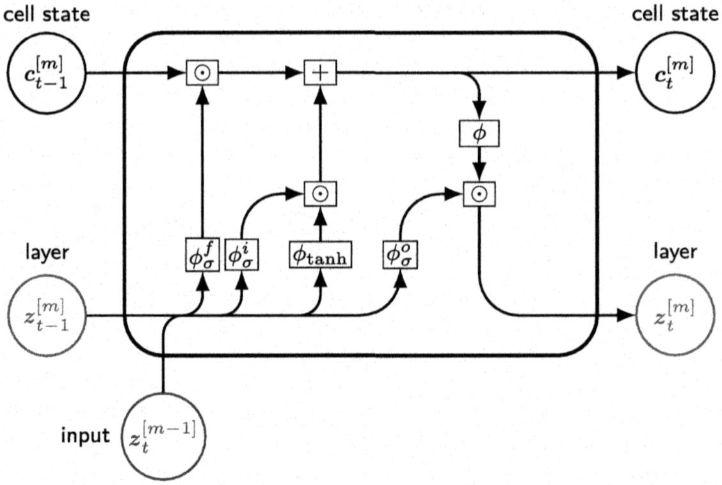

Fig. 8.6 LSTM cell $z^{(m)}$ with forget gate ϕ_σ^f, input gate ϕ_σ^i and output gate ϕ_σ^o

Below, we are going to summarize the LSTM cell update (8.13)–(8.14) as follows

$$\left(z_t^{[m-1]}, z_{t-1}^{[m]}, c_{t-1}^{[m]}\right) \mapsto \left(z_t^{[m]}, c_t^{[m]}\right) = z^{\mathrm{LSTM}(m)}\left(z_t^{[m-1]}, z_{t-1}^{[m]}, c_{t-1}^{[m]}\right). \tag{8.15}$$

The update (8.15) involves the eight network weight matrices $W_f^{(m)}$, $W_i^{(m)}$, $W_o^{(m)}$, $W_c^{(m)} \in \mathbb{R}^{(q_{m-1}+1)\times q_m}$ and $U_f^{(m)}$, $U_i^{(m)}$, $U_o^{(m)}$, $U_c^{(m)} \in \mathbb{R}^{q_m \times q_m}$. Altogether we have $4(q_{m-1} + 1 + q_m)q_m$ network parameters in each LSTM cell $1 \le m \le d$. These are learned with the gradient descent method. Moreover, we need to initialize the LSTM cell update (8.15). From the previous layer $m-1$ we have the input $z_t^{[m-1]}$ which we initialize as $z_t^{[0]} = \boldsymbol{x}_t$ for $m = 1$ and $t \ge 0$. The initial states $z_0^{[m]}$ and $c_0^{[m]}$ are usually set to zero.

8.3.2 *Gated Recurrent Unit Network*

The LSTM architecture of the previous section seems quite complex and involves many parameters. Cho et al. [76] have introduced the GRU architecture that is simpler and uses less parameters, but has similar properties. The GRU architecture uses two gates that are defined as follows for $t \ge 1$, see also (8.7):

- The *reset gate* models the memory reset rate

$$\boldsymbol{r}_t^{[m]} = \boldsymbol{r}^{(m)}\left(z_t^{[m-1]}, z_{t-1}^{[m]}\right) = \phi_\sigma^r\left(\left\langle W_r^{(m)}, z_t^{[m-1]}\right\rangle + \left\langle U_r^{(m)}, z_{t-1}^{[m]}\right\rangle\right) \in (0,1)^{q_m},$$

with the network weights $W_r^{(m)} \in \mathbb{R}^{(q_{m-1}+1)\times q_m}$ and $U_r^{(m)} \in \mathbb{R}^{q_m\times q_m}$, and with the sigmoid activation function $\phi_\sigma^r = \phi_\sigma$.
- The *update gate* models the memory update rate

$$\boldsymbol{u}_t^{[m]} = \boldsymbol{u}^{(m)}\left(z_t^{[m-1]}, z_{t-1}^{[m]}\right) = \phi_\sigma^u\left(\left\langle W_u^{(m)}, z_t^{[m-1]}\right\rangle + \left\langle U_u^{(m)}, z_{t-1}^{[m]}\right\rangle\right) \in (0,1)^{q_m},$$

with the network weights $W_u^{(m)} \in \mathbb{R}^{(q_{m-1}+1)\times q_m}$ and $U_u^{(m)} \in \mathbb{R}^{q_m\times q_m}$, and with the sigmoid activation function $\phi_\sigma^u = \phi_\sigma$.

The neuron activations $z_t^{[m]}$ are updated, given $z_t^{[m-1]}$ and $z_{t-1}^{[m]}$, by

$$\begin{aligned} z_t^{[m]} &= z^{(m)}\left(z_t^{[m-1]}, z_{t-1}^{[m]}\right) \qquad (8.16)\\ &= \boldsymbol{r}_t^{[m]} \odot z_{t-1}^{[m]} + (\boldsymbol{1} - \boldsymbol{r}_t^{[m]}) \odot \phi\left(\langle W^{(m)}, z_t^{[m-1]}\rangle + \boldsymbol{u}_t^{[m]} \odot \langle U^{(m)}, z_{t-1}^{[m]}\rangle\right) \in \mathbb{R}^{q_m}, \end{aligned}$$

with the network weights $W^{(m)} \in \mathbb{R}^{(q_{m-1}+1)\times q_m}$ and $U^{(m)} \in \mathbb{R}^{q_m\times q_m}$, and for a general activation function ϕ.

The GRU and the LSTM architectures are similar, the former using less parameters because we do not explicitly model the cell state process. For an illustration of a GRU cell we refer to Fig. 8.7. In the sequel we focus on the LSTM architecture;

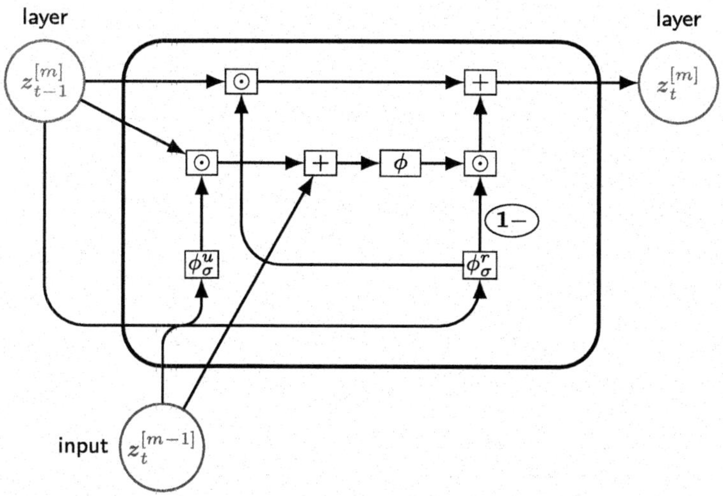

Fig. 8.7 GRU cell $z^{(m)}$ with reset gate ϕ_σ^r and update gate ϕ_σ^u

though the GRU architecture is simpler and has less parameters, it is less robust in fitting.

8.4 Lab: Mortality Forecasting with RN Networks

8.4.1 Lee–Carter Model, Revisited

The mortality data has a natural time-series structure, and for this reason mortality forecasting is an obvious problem that can be studied within RN networks. For instance, the LC mortality model (7.63) involves a stochastic process $(k_t)_t$ that needs to be extrapolated into the future. This extrapolation problem can be done in different ways. The original proposal of Lee and Carter [238] has been to analyze ARIMA time-series models, and to use standard statistical tools, Lee and Carter found that the random walk with drift gives a good stochastic description of the time index process $(k_t)_t$. Nigri et al. [286] proposed to fit a LSTM network to this stochastic process, this approach is also studied in Lindholm–Palmborg [252] where an efficient use of the mortality data for network fitting is discussed. These approaches still rely on the classical LC calibration using the SVD of Sect. 7.5.4, and the LSTM network is (only) used to extrapolate the LC time index process $(k_t)_t$.

More generally, one can design a RN network architecture that directly processes the raw mortality data $M_{x,t} = D_{x,t}/e_{x,t}$, not specifically relying on the LC structure. This has been done in Richman–Wüthrich [316] using a FN network architecture, in Perla et al. [301] using a RN network and a convolutional neural (CN) network architecture, and in Schürch–Korn [330] extending this analysis to the study of prediction uncertainty using bootstrapping. A similar CN network approach has been taken by Wang et al. [375] interpreting the raw mortality data of Fig. 7.32 as an image.

Lee–Carter Mortality Model: Random Walk with Drift Extrapolation

We revisit the LC mortality model [238] presented in Sect. 7.5.4. The LC log-mortality rate is assumed to have the following structure, see (7.63),

$$\log(\mu_{x,t}^{(p)}) = a_x^{(p)} + b_x^{(p)} k_t^{(p)},$$

for the ages $x_0 \le x \le x_1$ and for the calendar years $t \in \mathcal{T}$. We now add the upper indices $^{(p)}$ to consider different populations p. The SVD gives us the estimates $\widehat{a}_x^{(p)}$, $\widehat{k}_t^{(p)}$ and $\widehat{b}_x^{(p)}$ based on the observed centered raw log-mortality rates, see Sect. 7.5.4. The SVD is applied to each population p separately, i.e., there is no interaction between the different populations. This approach allows us to fit a separate log-mortality surface estimate $(\log(\widehat{\mu}_{x,t}^{(p)}))_{x_0 \le x \le x_1; t \in \mathcal{T}}$ to each population p. Figure 7.33

shows an example for two populations p, namely, for Swiss females and for Swiss males.

The mortality forecasting requires to extrapolate the time index processes $(\widehat{k}_t^{(p)})_{t\in\mathcal{T}}$ beyond the latest observed calendar year $t_1 = \max\{\mathcal{T}\}$. As mentioned in Lee–Carter [238] a random walk with drift provides a suitable model for modeling $(\widehat{k}_t^{(p)})_{t\geq 0}$ for many populations p, see Fig. 7.35 for the Swiss population. Assume that

$$\widehat{k}_{t+1}^{(p)} = \widehat{k}_t^{(p)} + \varepsilon_{t+1}^{(p)} \qquad t \geq 0, \tag{8.17}$$

with $\varepsilon_t^{(p)} \stackrel{\text{i.i.d.}}{\sim} \mathcal{N}(\delta_p, \sigma_p^2)$, $t \geq 1$, having drift $\delta_p \in \mathbb{R}$ and variance $\sigma_p^2 > 0$.

Model assumption (8.17) allows us to estimate the (constant) drift δ_p with MLE. For observations $(\widehat{k}_t^{(p)})_{t\in\mathcal{T}}$ we receive the log-likelihood function

$$\delta_p \mapsto \ell_{(\widehat{k}_t^{(p)})_{t\in\mathcal{T}}}(\delta_p) = \sum_{t=t_0+1}^{t_1} -\log(\sqrt{2\pi}\sigma_p) - \frac{1}{2\sigma_p^2}\left(\widehat{k}_t^{(p)} - \widehat{k}_{t-1}^{(p)} - \delta_p\right)^2,$$

with first observed calendar year $t_0 = \min\{\mathcal{T}\}$. The MLE is given by

$$\widehat{\delta}_p^{\text{MLE}} = \frac{\widehat{k}_{t_1}^{(p)} - \widehat{k}_{t_0}^{(p)}}{t_1 - t_0}. \tag{8.18}$$

This allows us to forecast the time index process for $t > t_1$ by

$$\widehat{k}_t^{(p)} = \widehat{k}_{t_1}^{(p)} + (t - t_1)\widehat{\delta}_p^{\text{MLE}}.$$

We explore this extrapolation for different Western European countries from the HMD [195]. We consider separately females and males of the countries {AUT, BE, CH, ESP, FRA, ITA, NL, POR}, thus, we choose $2 \cdot 8 = 16$ different populations p. For these countries we have observations for the ages $0 = x_0 \leq x \leq x_1 = 99$ and for the calendar years $1950 \leq t \leq 2018$.[3] For the following analysis we choose $\mathcal{T} = \{t_0 \leq t \leq t_1\} = \{1950 \leq t \leq 2003\}$, thus, we fit the models on 54 years of mortality history. This fitted models are then extrapolated to the calendar years $2004 \leq t \leq 2018$. These 15 calendar years from 2004 to 2018 allow us to perform an out-of-sample evaluation because we have the observations $M_{x,t}^{(p)} = D_{x,t}^{(p)}/e_{x,t}^{(p)}$ for these years from the HMD [195].
Figure 8.8 shows the estimated time index process $(\widehat{k}_t^{(p)})_{t\in\mathcal{T}}$ to the left of the dotted lines, and to the right of the dotted lines we have the random walk with drift extrapolation $(\widehat{k}_t^{(p)})_{t>t_1}$. The general observation is that, indeed, the random walk with drift seems to be a suitable model for $(\widehat{k}_t^{(p)})_t$. Moreover, there is a huge

[3] We exclude Germany from this consideration of (continental) Western European countries because the German mortality history is shorter due to the reunification in 1990.

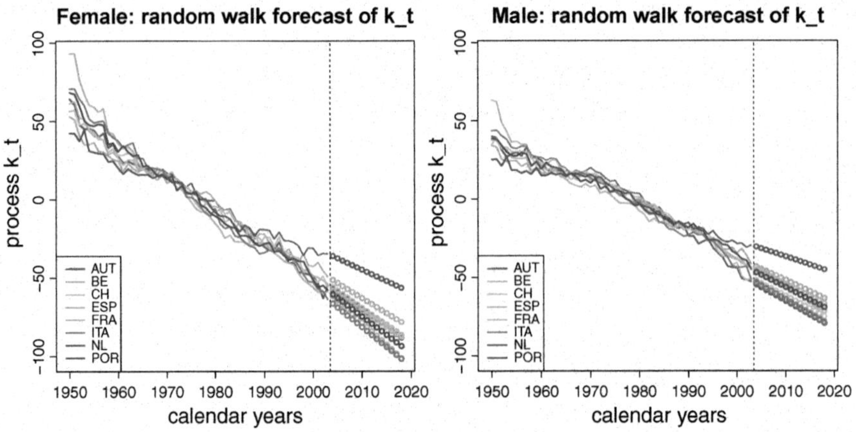

Fig. 8.8 Random walk with drift extrapolation of the time index process $(\widehat{k}_t)_t$ for different countries and genders; the y-scale is the same in both plots

similarity between the different countries, only with the Netherlands (NL) being somewhat an outlier.

Remarks 8.4

- For Fig. 8.8 we did not explore any fine-tuning, for instance, the estimation of the drift δ_p is very sensitive in the selection of the time span $\mathcal{T}$. ESP has the biggest negative drift estimate, but this is partially caused by the corresponding observations in the calendar years between 1950 and 1960, see Fig. 8.8, which may no longer be relevant for a decline in mortality in the new millennium.
- For all countries, the females have a bigger negative drift than the males (the y-scale in both plots is the same). Moreover, note that we use the normalization $\sum_{x=x_0}^{x_1} \widehat{b}_x^{(p)} = 1$ and $\sum_{t\in\mathcal{T}} \widehat{k}_t^{(p)} = 0$, see (7.65). This normalization is discussed and questioned in many publications as the extrapolation becomes dependent on these choices; see De Jong et al. [90] and the references therein, who propose different identification schemes.
- Another issue is an age coherence in forecasting, meaning that for long term forecasts the mortality rates across the different ages should not diverge, see Li et al. [250], Li–Lu [248] and Gao–Shi [153] and the references therein.
- There are many modifications and extensions of the LC model, we just mention a few of them. Brouhns et al. [56] embed the LC model into a Poisson modeling framework which provides a proper stochastic model for mortality modeling. Renshaw–Haberman [308] extend the one-factor LC model to a multifactor model, and in Renshaw–Haberman [309] a cohort effect is added. Hyndman–Ullah [197] and Hainaut–Denuit [179] explore a functional data method and a wavelet-based decomposition, respectively. The static PCA can be adopted to a dynamic PCA version, see Shang [333], and a long memory behavior in the time-series is studied in Yan et al. [395].

- The LC model is fitted to each population p separately, without exploring any common structure across the populations. There are many multi-population extensions that try to learn common structure across different populations. We mention the common age effect (CAE) model of Kleinow [218], the augmented common factor (ACF) model of Li–Lee [249] and the functional time-series models of Hyndman et al. [196] and Shang–Haberman [334]. A direct multi-population extension of the SVD matrix decomposition of the LC model is obtained by the tensor decomposition approaches of Russolillo et al. [325] and Dong et al. [110].

Lee–Carter Mortality Model: LSTM Extrapolation

Our aim here is to replace the individual random walk with drift extrapolations (8.17) by a common extrapolation across all considered populations p. For this we design a LSTM architecture. A second observation is that the increments $\varepsilon_t^{(p)} = \widehat{k}_t^{(p)} - \widehat{k}_{t-1}^{(p)}$ have an average empirical auto-correlation (for lag 1) of -0.33. This clearly questions the Gaussian i.i.d. assumption in (8.17).

We first discuss the available data and we construct the input data. We have the time-series observations $(\widehat{k}_t^{(p)})_{t\in\mathcal{T}}$, and the population index $p = (c, g)$ has two categorical labels c for country and g for gender. We are going to use two-dimensional embedding layers for these two categorical variables, see (7.31) for embedding layers. The time-series observations $(\widehat{k}_t^{(p)})_{t\in\mathcal{T}}$ will be pre-processed such that we do not simultaneously feed the entire time-series into the LSTM layer, but we divide them into shorter time-series. We will directly forecast the increments $\varepsilon_t^{(p)} = \widehat{k}_t^{(p)} - \widehat{k}_{t-1}^{(p)}$ and not the time index process $(\widehat{k}_t^{(p)})_{t\geq t_0}$; in extrapolations with drift it is easier to forecast the increments with the networks. We choose a *lookback period* of $\tau = 3$ calendar years, and we aim at predicting the response $Y_t = \varepsilon_t^{(p)}$ based on the time-series features $\boldsymbol{x}_{t-\tau:t-1} = (\varepsilon_{t-\tau}^{(p)}, \ldots, \varepsilon_{t-1}^{(p)})^\top \in \mathbb{R}^\tau$. This provides us with the following data structure for each population $p = (c, g)$:

year	country	gender	feature $\boldsymbol{x}_{t-\tau:t-1}$	Y_t
$t_0+\tau+1$	c	g	$\varepsilon_{t_0+1}^{(p)} \cdots \varepsilon_{t_0+\tau}^{(p)}$	$\varepsilon_{t_0+\tau+1}^{(p)}$
$\vdots$	$\vdots$	$\vdots$	$\vdots \quad \vdots$	$\vdots$
t	c	g	$\varepsilon_{t-\tau}^{(p)} \cdots \varepsilon_{t-1}^{(p)}$	$\varepsilon_t^{(p)}$
$\vdots$	$\vdots$	$\vdots$	$\vdots \quad \vdots$	$\vdots$
t_1	c	g	$\varepsilon_{t_1-\tau}^{(p)} \cdots \varepsilon_{t_1-1}^{(p)}$	$\varepsilon_{t_1}^{(p)}$

(8.19)

Thus, each observation $Y_t = \varepsilon_t^{(p)}$ is equipped with the feature information $(t, c, g, \boldsymbol{x}_{t-\tau:t-1})$. As discussed in Lindholm–Palmborg [252], one should highlight that there is a dependence across t, since we have a diagonal cohort structure in the features and the observations $(\boldsymbol{x}_{t-\tau:t-1}, Y_t)$. Usually, this dependence is not harmful in stochastic gradient descent fitting.

Listing 8.1 LSTM architecture example

```
1  TS      = layer_input(shape=c(lookback,1), dtype='float32', name='TS')
2  Country = layer_input(shape=c(1), dtype='int32',   name='Country')
3  Gender  = layer_input(shape=c(1), dtype='int32',   name='Gender')
4  Time    = layer_input(shape=c(1), dtype='float32', name='Time')
5  #
6  CountryEmb = Country %>%
7   layer_embedding(input_dim=8,output_dim=2,input_length=1,name='CountryEmb') %>%
8   layer_flatten(name='Country_flat')
9  #
10 GenderEmb = Gender %>%
11  layer_embedding(input_dim=2,output_dim=2,input_length=1,name='GenderEmb') %>%
12  layer_flatten(name='Gender_flat')
13 #
14 LSTM = TS %>%
15  layer_lstm(units=15,activation='tanh',recurrent_activation='sigmoid',
16                name='LSTM')
17 #
18 Output = list(LSTM,CountryEmb,GenderEmb,Time) %>% layer_concatenate() %>%
19  layer_dense(units=10, activation='tanh',   name='FNLayer') %>%
20  layer_dense(units=1, activation='linear', name='Network')
21 #
22 model = keras_model(inputs = list(TS, Country, Gender, Time),
23                     outputs = c(Output))
```

In Fig. 8.9 we plot the LSTM architecture used to forecast $\varepsilon_t^{(p)}$ for $t > t_1$, and Listing 8.1 gives the corresponding R code. We process the time-series $\boldsymbol{x}_{t-\tau:t-1}$ through a LSTM cell, see lines 14–16 of Listing 8.1. We choose a shallow LSTM network ($d = 1$) and therefore drop the upper index $m = 1$ in (8.15), but we add an upper index $^{[\text{LSTM}]}$ to highlight the output of the LSTM cell. This gives us the

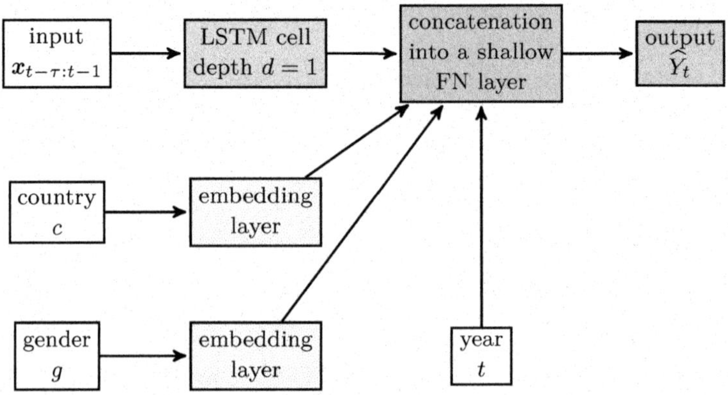

Fig. 8.9 LSTM architecture used to forecast $\varepsilon_t^{(p)}$ for $t > t_1$

LSTM cell updates for $t-\tau \le s \le t-1$

$$\left(\boldsymbol{x}_s, \boldsymbol{z}_{s-1}^{[\mathrm{LSTM}]}, \boldsymbol{c}_{s-1}\right) \mapsto \left(\boldsymbol{z}_s^{[\mathrm{LSTM}]}, \boldsymbol{c}_s\right) = \boldsymbol{z}^{\mathrm{LSTM}}\left(\boldsymbol{x}_s, \boldsymbol{z}_{s-1}^{[\mathrm{LSTM}]}, \boldsymbol{c}_{s-1}\right).$$

This LSTM recursion to process the time-series $\boldsymbol{x}_{t-\tau:t-1}$ gives us the LSTM output $\boldsymbol{z}_{t-1}^{[\mathrm{LSTM}]} \in \mathbb{R}^{q_1}$, and it involves $4(q_0+1+q_1)q_1 = 4(2+15)15 = 1'020$ network parameters for the input dimension $q_0 = 1$ and the output dimension $q_1 = 15$.

For the categorical country code c and the binary gender g we choose two-dimensional embedding layers, see (7.31),

$$c \mapsto \boldsymbol{e}^{\mathrm{C}}(c) \in \mathbb{R}^2 \qquad \text{and} \qquad g \mapsto \boldsymbol{e}^{\mathrm{G}}(g) \in \mathbb{R}^2,$$

these embedding maps give us $2(8+2) = 20$ embedding weights. Finally, we concatenate the LSTM output $\boldsymbol{z}_{t-1}^{[\mathrm{LSTM}]} \in \mathbb{R}^{15}$, the embeddings $\boldsymbol{e}^{\mathrm{C}}(c), \boldsymbol{e}^{\mathrm{G}}(g) \in \mathbb{R}^2$ and the continuous calendar year variable $t \in \mathbb{R}$ and process this vector through a shallow FN network with $q_2 = 10$ neurons, see lines 18–20 of Listing 8.1. This FN layer gives us $(q_1+2+2+1+1)q_2 = (15+2+2+1+1)10 = 210$ parameters. Together with the output parameter of dimension $q_2+1 = 11$, we receive 1'261 network parameters to be fitted, which seems quite a lot.

To fit this model we have $8 \cdot 2 = 16$ populations, and for each population we have the observations $\widehat{k}_t^{(p)}$ for the calendar years $1950 \le t \le 2003$. Considering the increments $\varepsilon_t^{(p)}$ and a lookback period of $\tau = 3$ calendar years gives us $2003 - 1950 - \tau = 50$ observations, rows in (8.19), per population p, thus, we have in total 800 observations. For the gradient descent fitting and the early stopping we choose a training to validation split of $8:2$. As loss function we choose the squared error loss function. This implicitly implies that we assume that the increments $Y_t = \varepsilon_t^{(p)}$ are Gaussian distributed, or in other words, minimizing the squared error loss function means maximizing the Gaussian log-likelihood function. We then fit this model to the data using early stopping as described in (7.27). We analyze this fitted model. Figure 8.10 provides the learned embeddings for the country codes c. These learned embeddings have some similarity with the European map.
The final step is the extrapolation $\widehat{k}_t$, $t > t_1$. These updates need to be done recursively. We initialize for $t = t_1+1$ the time-series feature

$$\boldsymbol{x}_{t_1+1-\tau:t_1} = (\varepsilon_{t_1+1-\tau}^{(p)}, \ldots, \varepsilon_{t_1}^{(p)})^\top \in \mathbb{R}^\tau. \tag{8.20}$$

Using the feature information $(t_1+1, c, g, \boldsymbol{x}_{t_1+1-\tau:t_1})$ allows us to forecast the next increment $Y_{t_1+1} = \varepsilon_{t_1+1}^{(p)}$ by $\widehat{Y}_{t_1+1}$, using the fitted LSTM architecture of Fig. 8.9. Thus, this LSTM network allows us to perform a *one-period-ahead forecast* to receive

$$\widehat{k}_{t_1+1} = \widehat{k}_{t_1} + \widehat{Y}_{t_1+1}. \tag{8.21}$$

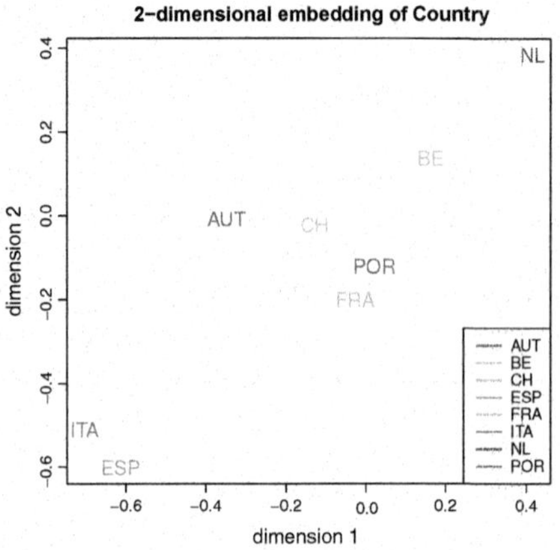

Fig. 8.10 Learned country embeddings for forecasting $(\widehat{k}_t)_t$

This update (8.21) needs to be iterated recursively. For the next period $t = t_1 + 2$ we set for the time-series feature

$$\boldsymbol{x}_{t_1+2-\tau:t_1+1} = (\varepsilon_{t_1+2-\tau}^{(p)}, \ldots, \varepsilon_{t_1}^{(p)}, \widehat{Y}_{t_1+1})^\top \in \mathbb{R}^\tau, \tag{8.22}$$

which gives us the next predictions $\widehat{Y}_{t_1+2}$ and $\widehat{k}_{t_1+2}$, etc.

In Fig. 8.11 we present the extrapolation of $(\varepsilon_t^{(p)})_t$ for Belgium females and males. The blue curve shows the observed increments $(\varepsilon_t^{(p)})_{1951 \le t \le 2003}$ and the LSTM fitted (in-sample) values $(\widehat{Y}_t)_{1954 \le t \le 2003}$ are in red color. Firstly, we observe a negative correlation (zig-zag behavior) in both the blue observations $(\varepsilon_t^{(p)})_{1951 \le t \le 2003}$ and in their red estimated means $(\widehat{Y}_t)_{1954 \le t \le 2003}$. Thus, the LSTM finds this negative correlation (and it does not propose i.i.d. residuals). Secondly, the volatility in the

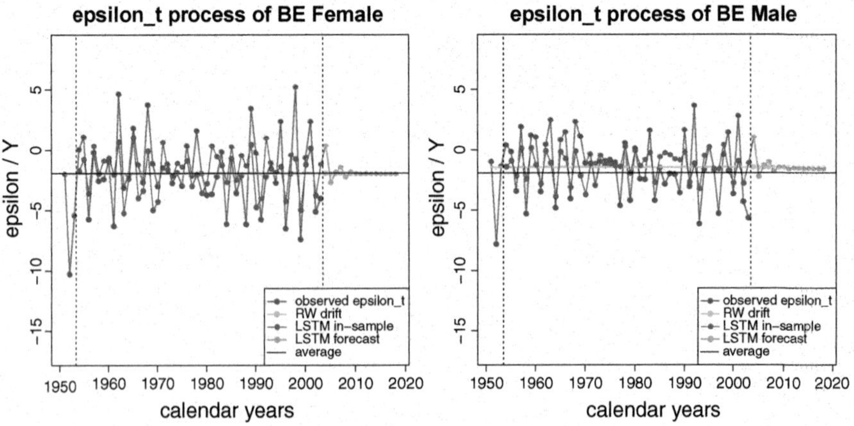

Fig. 8.11 LSTM network extrapolation $(\widehat{Y}_t)_{t>t_1}$ for Belgium (BE) females and males

red curve is smaller than in the blue curve, the former relates to expected values and the latter to observations of the random variables (which should be more volatile). The light-blue color shows the random walk with drift extrapolation (which is just a horizontal straight line at level $\widehat{\delta}_p^{\mathrm{MLE}}$, see (8.18)). The orange color shows the LSTM extrapolation using the recursive one-period-ahead updates (8.20)–(8.22), which has a zig-zag behavior that vanishes over time. This vanishing behavior is critical and is going to be discussed next.

There is one issue with this recursive one-period-ahead updating algorithm. This updating algorithm is not fully consistent in how the data is being used. The original LSTM architecture calibration is based on the feature components $\varepsilon_t^{(p)}$, see (8.20). Since these increments are not known for the later periods $t > t_1$, we replace their unknown values by the predictors, see (8.22). The subtle point here is that the predictors are on the level of expected values, and not on the level of random variables. Thus, $\widehat{Y}_t$ is typically less volatile than $\varepsilon_t^{(p)}$, but in (8.22) we pretend that we can use these predictors as a one-to-one replacement. A more consistent way would be to simulate/bootstrap $\varepsilon_t^{(p)}$ from $\mathcal{N}(\widehat{Y}_t, \sigma^2)$ so that the extrapolation receives the same volatility as the original process. For simplicity we refrain from doing so, but Fig. 8.11 indicates that this would be a necessary step because the volatility in the orange curve is going to vanish after the calendar year 2003, i.e., the zig-zag behavior vanishes, which is clearly not appropriate.

The LSTM extrapolation of $(\widehat{k}_t)_t$ is shown in Fig. 8.12. We observe quite some similarity to the random walk with drift extrapolation in Fig. 8.8, and, indeed, the random walk with drift seems to work very well (though the auto-correlation has not been specified correctly). Note that Fig. 8.8 is based on the individual extrapolations in p, whereas in Fig. 8.12 we have a common model for all populations.

Table 8.1 shows how often one model outperforms the other one (out-of-sample on calendar years $2004 \leq t \leq 2018$ and per gender). On the male populations of

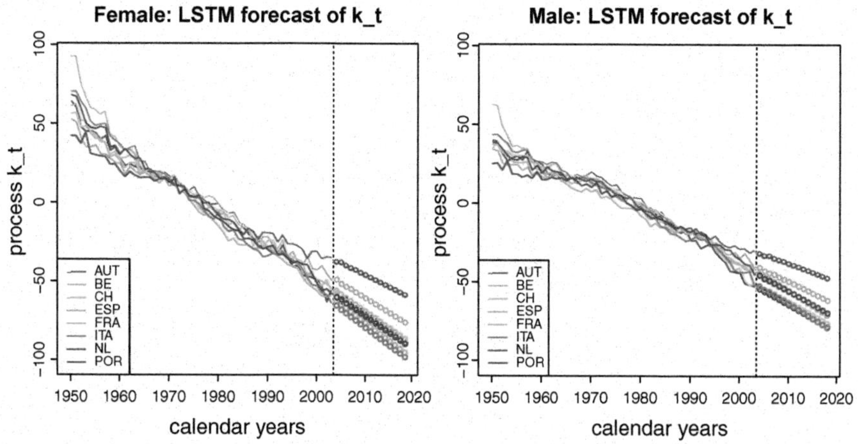

Fig. 8.12 LSTM network extrapolation of $(\widehat{k}_t)_t$ for different countries and genders

Table 8.1 Comparison of the out-of-sample mean squared error losses for the calendar years $2004 \le t \le 2018$: the numbers show how often one approach outperforms the other one on each gender

	Female	Male
Random walk with drift	5/8	4/8
LSTM architecture	3/8	4/8

the 8 European countries both models outperform the other one 4 times, whereas for the female population the random walk with drift gives 5 times the better out-of-sample prediction. Of course, this seems disappointing for the LSTM approach. This observation is quite common, namely, that the deep learning approach outperforms the classical methods on complex problems. However, on simple problems, as the one here, we should go for a classical (simpler) model like a random walk with drift or an ARIMA model.

8.4.2 *Direct LSTM Mortality Forecasting*

The previous section has been relying on the LC mortality model and only the extrapolation of the time-series $(\widehat{k}_t)_t$ has been based on a RN network architecture. In this section we aim at directly processing the raw mortality rates $M_{x,t} = D_{x,t}/e_{x,t}$ through a network, thus, we perform the representation learning directly on the raw data. We therefore use a simplified version of the network architecture proposed in Perla et al. [301].

As input to the network we use the raw mortality rates $M_{x,t}$. We choose a lookback period of $\tau = 5$ years and we define the time-series feature information to forecast the mortality in calendar year t by

$$\boldsymbol{x}_{t-\tau:t-1} = (\boldsymbol{x}_{t-\tau}, \ldots, \boldsymbol{x}_{t-1}) = \left(M_{x,s}\right)_{x_0 \le x \le x_1, t-\tau \le s \le t-1} \in \mathbb{R}^{(x_1 - x_0 + 1)\times\tau} = \mathbb{R}^{100\times 5}. \tag{8.23}$$

Thus, we directly process the raw mortality rates (simultaneously for all ages x) through the network architecture; in the corresponding R code we need to input the transposed features $\boldsymbol{x}_{t-\tau:t-1}^\top \in \mathbb{R}^{5\times 100}$, see line 1 of Listing 8.2.

We choose a shallow LSTM network ($d = 1$) and drop the upper index $m = 1$ in (8.15). This gives us the LSTM cell updates for $t - \tau \le s \le t - 1$

$$\left(\boldsymbol{x}_s, \boldsymbol{z}_{s-1}^{[\text{LSTM}]}, \boldsymbol{c}_{s-1}\right) \mapsto \left(\boldsymbol{z}_s^{[\text{LSTM}]}, \boldsymbol{c}_s\right) = \boldsymbol{z}^{\text{LSTM}}\left(\boldsymbol{x}_s, \boldsymbol{z}_{s-1}^{[\text{LSTM}]}, \boldsymbol{c}_{s-1}\right).$$

This LSTM recursion to process the time-series $\boldsymbol{x}_{t-\tau:t-1}$ gives us the LSTM output $\boldsymbol{z}_{t-1}^{[\text{LSTM}]} \in \mathbb{R}^{q_1}$, see lines 14–15 of Listing 8.2. It involves $4(q_0 + 1 + q_1)q_1 = 4(100 + 1 + 20)20 = 9'680$ network parameters for the input dimension $q_0 = 100$

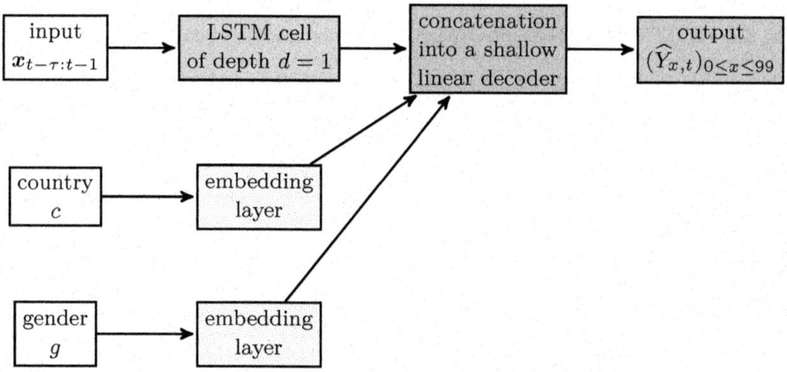

Fig. 8.13 LSTM architecture used to process the raw mortality rates $(M_{x,t})_{x,t}$

and the output dimension $q_1 = 20$. Many statisticians would probably stop at this point with this approach, as it seems highly over-parametrized. Let's see what we get.

For the categorical country code c and the binary gender g we choose two one-dimensional embeddings, see (7.31),

$$c \mapsto \boldsymbol{e}^{\mathrm{C}}(c) \in \mathbb{R} \qquad \text{and} \qquad g \mapsto \boldsymbol{e}^{\mathrm{G}}(g) \in \mathbb{R}. \tag{8.24}$$

These embeddings give us $8 + 2 = 10$ embedding weights. Figure 8.13 shows the LSTM cell in orange color and the embeddings in yellow color (in the colored version).

The LSTM output and the two embeddings are then concatenated to a learned representation $z_{t-1} = (z_{t-1}^{[\mathrm{LSTM}]}, \boldsymbol{e}^{\mathrm{C}}(c), \boldsymbol{e}^{\mathrm{G}}(g))^\top \in \mathbb{R}^{q_1\times 1\times 1} = \mathbb{R}^{22}$. The 22-dimensional learned representation z_{t-1} *encodes* the 500-dimensional input $\boldsymbol{x}_{t-\tau:t-1} \in \mathbb{R}^{100\times 5}$ and the two categorical variables c and g. The last step is to *decode* this representation $z_{t-1} \in \mathbb{R}^{22}$ to predict the log-mortality rates $(Y_{x,t})_{0\le x\le 99} = (\log M_{x,t})_{0\le x\le 99} \in \mathbb{R}^{100}$, simultaneously for all ages x. This decoding is obtained by the code on lines 17–19 of Listing 8.2; this reads as

$$z_{t-1} \mapsto \left(\beta_x^0 + \beta_x^C \boldsymbol{e}^{\mathrm{C}}(c) + \beta_x^G \boldsymbol{e}^{\mathrm{G}}(g) + \left\langle \boldsymbol{\beta}_x, z_{t-1}^{[\mathrm{LSTM}]} \right\rangle\right)_{0\le x\le 99}. \tag{8.25}$$

This decoding involves another $(1 + 22)100 = 2'300$ parameters $(\beta_x^0, \beta_x^G, \beta_x^C, \boldsymbol{\beta}_x)_{0\le x\le 99}$. Thus, altogether this LSTM network has $r = 11'990$ parameters.

Summarizing: the above architecture follows the philosophy of the auto-encoder of Sect. 7.5. A high-dimensional observation $(\boldsymbol{x}_{t-\tau:t-1}, c, g)$ is encoded to a low-dimensional bottleneck activation $z_{t-1} \in \mathbb{R}^{22}$, which is then decoded by (8.25) to give the forecast $(\widehat{Y}_{x,t})_{0\le x\le 99}$ for the log-mortality rates. It is not precisely an auto-encoder because the response is different from the input, as we forecast the log-mortality rates in the next calendar year t based on the information z_{t-1} that

Listing 8.2 LSTM architecture to directly process the raw mortality rates $(M_{x,t})_{x,t}$

```
1  TS      = layer_input(shape=c(lookback,100), dtype='float32', name='TS')
2  Country = layer_input(shape=c(1), dtype='int32',    name='Country')
3  Gender  = layer_input(shape=c(1), dtype='int32',    name='Gender')
4  Time    = layer_input(shape=c(1), dtype='float32', name='Time')
5  #
6  CountryEmb = Country %>%
7   layer_embedding(input_dim=8,output_dim=1,input_length=1,name='CountryEmb') %>%
8   layer_flatten(name='Country_flat')
9  #
10 GenderEmb = Gender %>%
11  layer_embedding(input_dim=2,output_dim=1,input_length=1,name='GenderEmb') %>%
12  layer_flatten(name='Gender_flat')
13 #
14 LSTM = TS %>%
15  layer_lstm(units=20,activation='linear',recurrent_activation='sigmoid',
16                  name='LSTM')
17 #
18 Output = list(LSTM,CountryEmb,GenderEmb) %>% layer_concatenate() %>%
19  layer_dense(units=100, activation='linear',  name='scalarproduct') %>%
20  layer_reshape(c(1,100), name = 'Output')
21 #
22 model = keras_model(inputs = list(TS, Country, Gender),
23                     outputs = c(Output))
```

is available at the end of the previous calendar year $t - 1$. In contrast to the LC mortality model, we no longer rely on the two-step approach by first fitting the parameters with a SVD, and performing a random walk with drift extrapolation. This encoder-decoder network performs both steps simultaneously.

We fit this network architecture to the available data. We have $r = 11'990$ network parameters. Based on a lookback period of $\tau = 5$ years, we have $2003 - 1950 - \tau + 1 = 49$ observations per population $p = (c, g)$. Thus, we have in total 784 observations $\left(\boldsymbol{x}_{t-\tau:t-1}, c, g, (Y_{x,t})_{0 \le x \le 99}\right)$. We fit this network using the nadam version of the gradient descent algorithm. We choose a training to validation split of $8 : 2$ and we explore 10'000 gradient descent epochs. A crucial observation is that the algorithm converges rather slowly and it does not show any signs of over-fitting, i.e., there is no strong need for the early stopping. This seems surprising because we have 11'990 parameters and only 784 observations. There are a couple of important ingredients that make this work. The features and observations themselves are high-dimensional, the low-dimensional encoding (compression) leads to a natural regularization, Moreover, this is combined with linear activation functions, see lines 15 and 19 of Listing 8.2. The gradient descent fitting has a certain inertness, and it seems that high-dimensional problems on comparably smooth high-dimensional data do not over-fit to individual components because the gradients are not very sensitive in the individual partial derivatives (in high dimensions). These high-dimensional approaches only work if we have sufficiently many populations across which we can learn, here we have 16 populations, Perla et al. [301] even use 76 populations.

Since every gradient descent fit still involves several elements of randomness, we consider the nagging predictor (7.44), averaging over 10 fitted networks, see

Table 8.2 Comparison of the out-of-sample mean squared losses for the calendar years $2004 \leq t \leq 2018$; the figures are in 10^{-4}

	LC female	LSTM female	LC male	LSTM male
Austria AUT	0.765	**0.312**	2.527	**1.169**
Belgium BE	0.371	**0.311**	2.835	**0.960**
Switzerland CH	0.654	**0.478**	1.609	**1.134**
Spain ESP	1.446	**0.514**	1.742	**0.245**
France FRA	**0.175**	1.684	**0.333**	0.363
Italy ITA	**0.179**	0.330	0.874	**0.320**
The Netherlands NL	0.426	**0.315**	1.978	**0.601**
Portugal POR	2.097	**0.464**	1.848	**1.239**

Sect. 7.4.4. The out-of-sample prediction results on the calendar years 2004 to 2018, i.e., $t > t_1 = 2004$, are presented in Table 8.2. These results verify the appropriateness of this LSTM approach. It outperforms the LC model on the female population in 6 out of 8 cases and on the male population on 7 out of 8 cases, only for the French population this LSTM approach seems to have some difficulties (compared to the LC model). Note that these are out-of-sample figures because the LSTM has only been fitted on the data prior to 2004. Moreover, we did not pre-process the raw mortality rates $M_{x,t}$, $t \leq 2003$, and the prediction is done recursively in a one-period-ahead prediction approach, we also refer to (8.22). A more detailed analysis of the results shows that the LC and the LSTM approaches have a rather similar behavior for females. For males the LSTM prediction clearly outperforms the LC model prediction, this out-performance is across different ages x and different calendar years $t \geq 2004$.

The advantage of this LSTM approach is that we can directly predict by processing the raw data. The disadvantage compared to the LC approach is that the LSTM network approach is more complex and more time-consuming. Moreover, unlike in the LC approach, we cannot (easily) assess the prediction uncertainty. In the LC approach the prediction uncertainty is obtained from assessing the uncertainty in the extrapolation and the uncertainty in the parameter estimates, e.g., using a bootstrap. The LSTM approach is not sufficiently robust (at least not on our data) to provide any reasonable uncertainty estimates.

We close this section and example by analyzing the functional form of the decoder (8.25). We observe that this decoder has much similarity with the LC model assumption (7.63)

$$\widehat{Y}_{x,t} = \beta_x^0 + \beta_x^C \boldsymbol{e}^{\mathrm{C}}(c) + \beta_x^G \boldsymbol{e}^{\mathrm{G}}(g) + \left\langle \boldsymbol{\beta}_x, \boldsymbol{z}_{t-1}^{[\mathrm{LSTM}]} \right\rangle,$$

$$\log(\mu_{x,t}^{(p)}) = a_x^{(p)} + b_x^{(p)} k_t^{(p)}.$$

The LC model considers the average force of mortality $a_x^{(p)} \in \mathbb{R}$ for each population $p = (c, g)$ and each age x; the LSTM architecture has the same term $\beta_x^0 + \beta_x^C \boldsymbol{e}^{\mathrm{C}}(c) +$

$\beta_x^G \boldsymbol{e}^{\mathrm{G}}(g)$. In the LC model, the change of force of mortality is considered by a population-dependent term $b_x^{(p)} k_t^{(p)}$, whereas the LSTM architecture has a term $\langle \boldsymbol{\beta}_x, \boldsymbol{z}_{t-1}^{[\mathrm{LSTM}]} \rangle$. This latter term is also population-dependent because the LSTM cell directly processes the raw mortality data $M_{x,t}$ coming from the different populations p. Note that this is the only time-t-dependent term in the LSTM architecture. We conclude that the main difference between these two forecast approaches is how the past mortality observations are processed. Apart from that the general structure is the same.

Chapter 9
Convolutional Neural Networks

The previous two chapters have been considering fully-connected feed-forward neural (FN) networks and recurrent neural (RN) networks. Fully-connected FN networks are the prototype of networks for deep representation learning on tabular data. This type of networks extracts *global properties* from the features $\boldsymbol{x}$. RN networks are an adaption of FN networks to time-series data. Convolutional neural (CN) networks are a third type of networks, and their specialty is to extract *local structure* from the features. Originally, they have been introduced for speech and image recognition aiming at finding similar structure in different parts of the feature $\boldsymbol{x}$. For instance, if $\boldsymbol{x}$ is a picture consisting of pixels, and if we want to classify this picture according to its contents, then we try to find similar structure (objects) in different locations of this picture. CN networks are suitable for this task as they work with filters (kernels) that have a fixed window size. These filters then screen across the picture to detect similar local structure at different locations in the picture. CN networks were introduced in the 1980s by Fukushima [145] and LeCun et al. [234, 235], and they have been celebrating great success in many applications. Our introduction to CN networks is based on the tutorial of Meier–Wüthrich [269]. For real data applications there are many pre-trained CN network libraries that can be downloaded and used for several different tasks, an example for image recognition is the AlexNet of Krizhevsky et al. [226].

9.1 Plain-Vanilla Convolutional Neural Network Layer

Structurally, the CN network architectures are similar to the FN network architectures, only they replace certain FN layers by CN layers. Therefore, we start by introducing the CN layer, and one should keep the structure of the FN layer (7.5)

M. V. Wüthrich, M. Merz, *Statistical Foundations of Actuarial Learning and its Applications*, Springer Actuarial, https://doi.org/10.1007/978-3-031-12409-9_9

in mind. In a nutshell, FN layers consider non-linearly activated inner products $\langle \boldsymbol{w}_j^{(m)}, \boldsymbol{z}\rangle$, and CN layers replace these inner products by a type of convolution $\boldsymbol{W}_j^{(m)} * \boldsymbol{z}$.

9.1.1 Input Tensors and Channels

We start from an *input tensor* $\boldsymbol{z} \in \mathbb{R}^{q^{(1)}\times\cdots\times q^{(K)}}$ that has dimension $q^{(1)}\times\cdots\times q^{(K)}$. This input tensor $\boldsymbol{z}$ is a *multi-dimensional array of order (length)* $K \in \mathbb{N}$ and with elements $z_{i_1,\ldots,i_K} \in \mathbb{R}$ for $1 \le i_k \le q^{(k)}$ and $1 \le k \le K$. The special case of order $K = 2$ is a matrix $\boldsymbol{z} \in \mathbb{R}^{q^{(1)}\times q^{(2)}}$. This matrix can illustrate a black and white image of dimension $q^{(1)} \times q^{(2)}$ with the matrix entries $z_{i_1,i_2} \in \mathbb{R}$ describing the intensities of the gray scale in the corresponding pixels (i_1, i_2). A color image typically has the three color channels Red, Green and Blue (RGB), and such a RGB image can be represented by a tensor $\boldsymbol{z} \in \mathbb{R}^{q^{(1)}\times q^{(2)}\times q^{(3)}}$ of order 3 with $q^{(1)} \times q^{(2)}$ being the dimension of the image and $q^{(3)} = 3$ describing the three color channels, i.e., $(z_{i_1,i_2,1}, z_{i_1,i_2,2}, z_{i_1,i_2,3})^\top \in \mathbb{R}^3$ describes the intensities of the colors RGB in the pixel (i_1, i_2).

Typically, the structure of black and white images and RGB images is unified by representing the black and white picture by a tensor $\boldsymbol{z} \in \mathbb{R}^{q^{(1)}\times q^{(2)}\times q^{(3)}}$ of order 3 with a single channel $q^{(3)} = 1$. This philosophy is going to be used throughout this chapter. Namely, if we consider a tensor $\boldsymbol{z} \in \mathbb{R}^{q^{(1)}\times\cdots\times q^{(K-1)}\times q^{(K)}}$ of order K, the first $K-1$ components $(i_1, \ldots, i_{K-1})$ will play the role of the *spatial components* that have a natural topology, and the last components $1 \le i_K \le q^{(K)}$ are called the *channels* reflecting, e.g., a gray scale (for $q^{(K)} = 1$) or the RGB intensities (for $q^{(K)} = 3$).

In Sect. 9.1.3, below, we will also study time-series data where we have 2nd order tensors (matrices). The first component reflects time $1 \le t \le q^{(1)}$, i.e., the spatial component is temporal for time-series data, and the second component (channels) describes the different elements $\boldsymbol{z}_t = (z_{t,1}, \ldots, z_{t,q^{(2)}})^\top \in \mathbb{R}^{q^{(2)}}$ that are measured/observed at each time point t.

9.1.2 Generic Convolutional Neural Network Layer

We start from an input tensor $\boldsymbol{z} \in \mathbb{R}^{q_{m-1}^{(1)}\times\cdots\times q_{m-1}^{(K)}}$ of order K. The first $K-1$ components of this tensor have a spatial structure and the K-th component stands for the channels. A CN layer applies (local) convolution operations to this tensor. We choose a *filter size*, also called *window size* or *kernel size*, $(f_m^{(1)}, \ldots, f_m^{(K)})^\top \in \mathbb{N}^K$ with $f_m^{(k)} \le q_{m-1}^{(k)}$, for $1 \le k \le K-1$, and $f_m^{(K)} = q_{m-1}^{(K)}$. This filter size determines the output dimension of the CN operation by

$$q_m^{(k)} \stackrel{\text{def.}}{=} q_{m-1}^{(k)} - f_m^{(k)} + 1, \tag{9.1}$$

for $1 \leq k \leq K$. Thus, the size of the image is reduced by the window size of the filter. In particular, the output dimension of the channels component $k = K$ is $q_m^{(K)} = 1$, i.e., all channels are compressed to a scalar output. The spatial components $1 \leq k \leq K - 1$ retain their spatial structure but the dimension is reduced according to (9.1).

A *CN operation* is a mapping (note that the order of the tensor is reduced from K to $K - 1$ because the channels are compressed; index j is going to be explained later)

$$z_j^{(m)} : \mathbb{R}^{q_{m-1}^{(1)} \times \cdots \times q_{m-1}^{(K)}} \to \mathbb{R}^{q_m^{(1)} \times \cdots \times q_m^{(K-1)}} \tag{9.2}$$

$$z \mapsto z_j^{(m)}(z) = \left(z_{i_1,\ldots,i_{K-1};j}^{(m)}(z) \right)_{1 \leq i_k \leq q_m^{(k)}; 1 \leq k \leq K-1},$$

taking the values for a fixed activation function $\phi : \mathbb{R} \to \mathbb{R}$

$$z_{i_1,\ldots,i_{K-1};j}^{(m)}(z) = \phi \left(w_{0,j}^{(m)} + \sum_{l_1=1}^{f_m^{(1)}} \cdots \sum_{l_K=1}^{f_m^{(K)}} w_{l_1,\ldots,l_K;j}^{(m)} \, z_{i_1+l_1-1,\ldots,i_{K-1}+l_{K-1}-1,l_K} \right), \tag{9.3}$$

for given intercept $w_{0,j}^{(m)} \in \mathbb{R}$ and *filter weights*

$$\boldsymbol{W}_j^{(m)} = \left(w_{l_1,\ldots,l_K;j}^{(m)} \right)_{1 \leq l_k \leq f_m^{(k)}; 1 \leq k \leq K} \in \mathbb{R}^{f_m^{(1)} \times \cdots \times f_m^{(K)}}; \tag{9.4}$$

the network parameter has dimension $r_m = 1 + \prod_{k=1}^{K} f_m^{(k)}$.

At first sight this CN operation looks quite complicated. Let us give some remarks that allow for a better understanding and a more compact notation. The operation in (9.3) chooses the corner $(i_1, \ldots, i_{K-1}, 1)$ as base point, and then it reads the tensor elements in the (discrete) window

$$(i_1, \ldots, i_{K-1}, 1) + \left[0 : f_m^{(1)} - 1\right] \times \cdots \times \left[0 : f_m^{(K-1)} - 1\right] \times \left[0 : f_m^{(K)} - 1\right], \tag{9.5}$$

with given filter weights $\boldsymbol{W}_j^{(m)}$. This window is then moved across the entire tensor z by changing the base point $(i_1, \ldots, i_{K-1}, 1)$ accordingly, but with fixed filter weights $\boldsymbol{W}_j^{(m)}$. This operation resembles a convolution, however, in (9.3) the indices in $z_{i_1+l_1-1,\ldots,i_{K-1}+l_{K-1}-1,l_K}$ run in reverse direction compared to a classical (mathematical) convolution. By a slight abuse of notation, nevertheless, we use the symbol of the convolution operator $*$ to abbreviate (9.2). This gives us the compact notation:

$$z_j^{(m)} : \mathbb{R}^{q_{m-1}^{(1)} \times \cdots \times q_{m-1}^{(K)}} \to \mathbb{R}^{q_m^{(1)} \times \cdots \times q_m^{(K-1)}}$$

$$z \mapsto z_j^{(m)}(z) = \phi\left(w_{0,j}^{(m)} + \boldsymbol{W}_j^{(m)} * z\right), \qquad (9.6)$$

having the activations for $1 \le i_k \le q_m^{(k)}$, $1 \le k \le K-1$,

$$\phi\left(w_{0,j}^{(m)} + \boldsymbol{W}_j^{(m)} * z\right)_{i_1,\ldots,i_{K-1}} = z_{i_1,\ldots,i_{K-1};j}^{(m)}(z),$$

where the latter is given by (9.3).

Remarks 9.1

- The beauty of this notation is that we can now see the analogy to the FN layer. Namely, (9.6) exactly plays the role of a FN neuron (7.6), but the CN operation $w_{0,j}^{(m)} + \boldsymbol{W}_j^{(m)} * z$ replaces the inner product $\langle \boldsymbol{w}_j^{(m)}, z \rangle$, and correspondingly accounting for the intercept.
- A FN neuron (7.6) can be seen as a special case of CN operation (9.6). Namely, if we have a tensor of order $K = 1$, the input tensor (vector) reads as $z \in \mathbb{R}^{q_{m-1}^{(1)}}$. That is, we do not have a spatial component, but only $q_{m-1} = q_{m-1}^{(1)}$ channels. In that case we have $\boldsymbol{W}_j^{(m)} * z = \langle \boldsymbol{W}_j^{(m)}, z \rangle$ for the filter weights $\boldsymbol{W}_j^{(m)} \in \mathbb{R}^{q_{m-1}^{(1)}}$, and where we assume that z does not include an intercept component. Thus, the CN operation boils down to a FN neuron in the case of a tensor of order 1.
- In the CN operation we take advantage of having a spatial structure in the tensor z, which is not the case in the FN operation. The CN operation takes a spatial input of dimension $\prod_{k=1}^{K} q_{m-1}^{(k)}$ and it maps this input to a spatial object of dimension $\prod_{k=1}^{K-1} q_m^{(k)}$. For this it uses $r_m = 1 + \prod_{k=1}^{K} f_m^{(k)}$ filter weights. The FN operation takes an input of dimension q_{m-1} and it maps it to a 1-dimensional neuron activation, for this it uses $1 + q_{m-1}$ parameters. If we identify the input dimensions $q_{m-1} \stackrel{!}{=} \prod_{k=1}^{K} q_{m-1}^{(k)}$ we can observe that $r_m \ll 1 + q_{m-1}$ because, typically, the filter sizes $f_m^{(k)} \ll q_{m-1}^{(k)}$, for $1 \le k \le K-1$. Thus, the CN operation uses much less parameters as the filters only act locally through the $*$-operation by translating the filter window (9.5).

This understanding now allows us to define a CN layer. Note that the mappings (9.6) have a lower index j which indicates that this is one single projection (filter extraction), called a *filter*. By choosing multiple different filters $(w_{0,j}^{(m)}, \mathbf{W}_j^{(m)})$, we can define the CN layer as follows.

Choose $q_m^{(K)} \in \mathbb{N}$ filters, each having a r_m-dimensional filter weight $(w_{0,j}^{(m)}, \mathbf{W}_j^{(m)})$, $1 \leq j \leq q_m^{(K)}$. A *CN layer* is a mapping

$$z^{(m)} : \mathbb{R}^{q_{m-1}^{(1)} \times \cdots \times q_{m-1}^{(K)}} \to \mathbb{R}^{q_m^{(1)} \times \cdots \times q_m^{(K)}} \tag{9.7}$$

$$z \mapsto z^{(m)}(z) = \left(z_1^{(m)}(z), \ldots, z_{q_m^{(K)}}^{(m)}(z) \right),$$

with filters $z_j^{(m)}(z) \in \mathbb{R}^{q_m^{(1)} \times \cdots \times q_m^{(K-1)}}$, $1 \leq j \leq q_m^{(K)}$, given by (9.6).

A CN layer (9.7) converts the $q_{m-1}^{(K)}$ input channels to $q_m^{(K)}$ output filters by preserving the spatial structure on the first $K - 1$ components of the input tensor z. More mathematically, CN layers and networks have been studied, among others, by Zhang et al. [403, 404], Mallat [263] and Wiatowski–Bölcskei [382]. These authors prove that CN networks have certain translation invariance properties and deformation stability. This exactly explains why these networks allow one to recognize similar objects at different locations in the input tensor. Basically, by translating the filter windows (9.5) across the tensor, we try to extract the local structure from the tensor that provides similar signals in different locations of that tensor. Thinking of an image where we try to recognize, say, a dog, such a dog can be located at different sites in the image, and a filter (window) that moves across that image tries to locate the dogs in the image.

A CN layer (9.7) defines one layer indexed by the upper index $^{(m)}$, and for deep representation learning we now have to compose multiple of these CN layers, but we can also compose CN layers with FN layers or RN layers. Before doing so, we need to introduce some special purpose layers and tools that are useful for CN network modeling, this is done in Sect. 9.2, below.

9.1.3 Example: Time-Series Analysis and Image Recognition

Most CN network examples are based on time-series data or images. The former has a 1-dimensional temporal component, and the latter has a 2-dimensional spatial component. Thus, these two examples are giving us tensors of orders $K = 2$ and $K = 3$, respectively. We briefly discuss such examples as specific applications of a tensors of a general order $K \geq 2$.

Time-Series Analysis with CN Networks

For a time-series analysis we often have observations $\boldsymbol{x}_t \in \mathbb{R}^{q_0}$ for the time points $0 \le t \le T$. Bringing this time-series data into a tensor form gives us

$$\boldsymbol{x} = \boldsymbol{x}_{0:T}^\top = (\boldsymbol{x}_0, \ldots, \boldsymbol{x}_T)^\top \in \mathbb{R}^{(T+1)\times q_0} = \mathbb{R}^{q_0^{(1)}\times q_0^{(2)}},$$

with $q_0^{(1)} = T + 1$ and $q_0^{(2)} = q_0$. We have met such examples in Chap. 8 on RN networks. Thus, for time-series data the input to a CN network is a tensor of order $K = 2$ with a temporal component having the dimension $T + 1$ and at each time point t we have q_0 measurements (channels) $\boldsymbol{x}_t \in \mathbb{R}^{q_0}$. A CN network tries to find similar structure at different time points in this time-series data $\boldsymbol{x}_{0:T}$. For a first CN layer $m = 1$ we therefore choose $q_1 \in \mathbb{N}$ filters and consider the mapping

$$\begin{aligned} z^{(1)} : \mathbb{R}^{(T+1)\times q_0} &\to \mathbb{R}^{(T-f_1+2)\times q_1} \\ \boldsymbol{x}_{0:T}^\top &\mapsto z^{(1)}(\boldsymbol{x}_{0:T}^\top) = \left(z_1^{(1)}(\boldsymbol{x}_{0:T}^\top), \ldots, z_{q_1}^{(1)}(\boldsymbol{x}_{0:T}^\top)\right), \end{aligned} \tag{9.8}$$

with filters $z_j^{(1)}(\boldsymbol{x}_{0:T}^\top) \in \mathbb{R}^{T-f_1+2}$, $1 \le j \le q_1$, given by (9.6) and for a fixed window size $f_1 \in \mathbb{N}$. From (9.8) we observe that the length of the time-series is reduced from $T + 1$ to $T - f_1 + 2$ accounting for the window size f_1. In financial mathematics, a structure (9.8) is often called a rolling window that moves across the time-series $\boldsymbol{x}_{0:T}$ and extracts the corresponding information.

We have introduced two different architectures to process time-series information $\boldsymbol{x}_{0:T}$, and these different architectures serve different purposes. A RN network architecture is most suitable if we try to forecast the next response of a time-series. I.e., we typically process the past observations through a recurrent structure to predict the next response, this is the motivation, e.g., behind Figs. 8.4 and 8.5. The motivation for the use of a CN network architecture is different as we try to find similar structure at different times, e.g., in a financial time-series we may be interested in finding the downturns of more than 20%. The latter is a local analysis which is explored by local filters (of a finite window size).

Image Recognition

Image recognition extends (9.8) by one order to a tensor of order $K = 3$. Typically, we have images of dimensions (pixels) $I \times J$, and having three color channels RGB. These images then read as

$$\boldsymbol{x} = (\boldsymbol{x}_1, \boldsymbol{x}_2, \boldsymbol{x}_3) \in \mathbb{R}^{I\times J\times 3} = \mathbb{R}^{q_0^{(1)}\times q_0^{(2)}\times q_0^{(3)}},$$

where $\boldsymbol{x}_1 \in \mathbb{R}^{I\times J}$ is the intensity of red, $\boldsymbol{x}_2 \in \mathbb{R}^{I\times J}$ is the intensity of green, and $\boldsymbol{x}_3 \in \mathbb{R}^{I\times J}$ is the intensity of blue.

Chose a window size of $f_1^{(1)} \times f_1^{(2)}$ and $q_1 \in \mathbb{N}$ filters to receive the CN layer

$$\begin{aligned} z^{(1)} : \mathbb{R}^{I\times J\times 3} &\to \mathbb{R}^{(I-f_1^{(1)}+1)\times(J-f_1^{(2)}+1)\times q_1} \\ (\boldsymbol{x}_1,\boldsymbol{x}_2,\boldsymbol{x}_3) &\mapsto z^{(1)}(\boldsymbol{x}_1,\boldsymbol{x}_2,\boldsymbol{x}_3) = \left(z_1^{(1)}(\boldsymbol{x}_1,\boldsymbol{x}_2,\boldsymbol{x}_3),\ldots,z_{q_1}^{(1)}(\boldsymbol{x}_1,\boldsymbol{x}_2,\boldsymbol{x}_3)\right), \end{aligned} \tag{9.9}$$

with filters $z_j^{(1)}(\boldsymbol{x}_1,\boldsymbol{x}_2,\boldsymbol{x}_3) \in \mathbb{R}^{(I-f_1^{(1)}+1)\times(J-f_1^{(2)}+1)}$, $1 \le j \le q_1$. Thus, we compress the 3 channels in each filter j, but we preserve the spatial structure of the image (by the convolution operation $*$).

For black and white pictures which only have one color channel, we preserve the spatial structure of the picture, and we modify the input tensor to a tensor of order 3 and of the form

$$\boldsymbol{x} = (\boldsymbol{x}_1) \in \mathbb{R}^{I\times J\times 1}.$$

9.2 Special Purpose Tools for Convolutional Neural Networks

9.2.1 Padding with Zeros

We have seen that the CN operation reduces the size of the output by the filter sizes, see (9.1). Thus, if we start from an image of size $100 \times 50 \times 1$, and if the filter sizes are given by $f_m^{(1)} = f_m^{(2)} = 9$, then the output will be of dimension $92 \times 42 \times q_1^{(3)}$, see (9.9). Sometimes, this reduction in dimension is impractical, and padding helps to keep the original shape. Padding a tensor z with $p_m^{(k)}$ parameters, $1 \le k \le K-1$, means that the tensor is extended in all $K-1$ spatial directions by (typically) adding zeros of that size, so that the padded tensor has dimension

$$\left(p_m^{(1)} + q_{m-1}^{(1)} + p_m^{(1)}\right) \times \cdots \times \left(p_m^{(K-1)} + q_{m-1}^{(K-1)} + p_m^{(K-1)}\right) \times q_{m-1}^{(K)}.$$

This implies that the output filters will have the dimensions

$$q_m^{(k)} = q_{m-1}^{(k)} + 2p_m^{(k)} - f_m^{(k)} + 1,$$

for $1 \le k \le K-1$. The spatial dimension of the original tensor size is preserved if $2p_m^{(k)} - f_m^{(k)} + 1 = 0$. Padding does not add any additional parameters, but it is only used to reshape the tensors.

9.2.2 Stride

Strides are used to skip part of the input tensor z in order to reduce the size of the output. This may be useful if the input tensor is a very high resolution image. Choose the stride parameters $s_m^{(k)}$, $1 \leq k \leq K-1$. We can then replace the summation in (9.3) by the following term

$$\sum_{l_1=1}^{f_m^{(1)}} \cdots \sum_{l_K=1}^{f_m^{(K)}} w_{l_1,\ldots,l_K;j}^{(m)} \, z_{s_m^{(1)}(i_1-1)+l_1,\ldots,s_m^{(K-1)}(i_{K-1}-1)+l_{K-1},l_K}.$$

This only extracts the tensor entries on a discrete grid of the tensor by translating the window by multiples of integers, see also (9.5),

$$\left(s_m^{(1)}(i_1-1),\ldots,s_m^{(K-1)}(i_{K-1}-1),1\right)+\left[1:f_m^{(1)}\right]\times\cdots\times\left[1:f_m^{(K-1)}\right]\times\left[0:f_m^{(K)}-1\right],$$

and the size of the output is reduced correspondingly. If we choose strides $s_m^{(k)} = f_m^{(k)}$, $1 \leq k \leq K-1$, we receive a partition of the spatial part of the input tensor z, this is going to be used in the max-pooling layer (9.11).

9.2.3 Dilation

Dilation is similar to stride, though, different in that it enlarges the filter sizes instead of skipping certain positions in the input tensor. Choose the dilation parameters $e_m^{(k)}$, $1 \leq k \leq K-1$. We can then replace the summation in (9.3) by the following term

$$\sum_{l_1=1}^{f_m^{(1)}} \cdots \sum_{l_K=1}^{f_m^{(K)}} w_{l_1,\ldots,l_K;j}^{(m)} \, z_{i_1+e_m^{(1)}(l_1-1),\ldots,i_{K-1}+e_m^{(K-1)}(l_{K-1}-1),l_K}.$$

This applies the filter weights to the tensor entries on discrete grids

$$(i_1,\ldots,i_{K-1},1)+e_m^{(1)}\left[0:f_m^{(1)}-1\right]\times\cdots\times e_m^{(K-1)}\left[0:f_m^{(K-1)}-1\right]\times\left[0:f_m^{(K)}-1\right],$$

where the intervals $e_m^{(k)}[0:f_m^{(k)}-1]$ run over the grids of span sizes $e_m^{(k)}$, $1 \leq k \leq K-1$. Thus, in comparably smoothing images we do not read all the pixels but only every $e_m^{(k)}$-th pixel in the window. Also this reduces the size of the output tensor.

9.2.4 Pooling Layer

As we have seen above, the dimension of the tensor is reduced by the filter size in each spatial direction if we do not apply padding with zeros. In general, deep representation learning follows the paradigm of auto-encoding by reducing a high-dimensional input to a low-dimensional representation. In CN networks this is usually (efficiently) done by so-called pooling layers. In spirit, pooling layers work similarly to CN layers (having a fixed window size), but we do not apply a convolution operation $*$, but rather a maximum operation to the window to extract the dominant tensor elements.

We choose a fixed window size $(f_m^{(1)}, \ldots, f_m^{(K-1)})^\top \in \mathbb{N}^{K-1}$ and strides $s_m^{(k)} = f_m^{(k)}$, $1 \le k \le K-1$, for the spatial components of the tensor z of order K. A *max-pooling layer* is given by

$$z^{(m)} : \mathbb{R}^{q_{m-1}^{(1)} \times \cdots \times q_{m-1}^{(K)}} \to \mathbb{R}^{q_m^{(1)} \times \cdots \times q_m^{(K)}}$$

$$z \mapsto z^{(m)}(z) = \text{MaxPool}(z), \tag{9.10}$$

with dimensions $q_m^{(K)} = q_{m-1}^{(K)}$ and for $1 \le k \le K-1$

$$q_m^{(k)} = \left\lfloor q_{m-1}^{(k)} / f_m^{(k)} \right\rfloor, \tag{9.11}$$

having the activations for $1 \le i_k \le q_m^{(k)}$, $1 \le k \le K$,

$$\text{MaxPool}(z)_{i_1,\ldots,i_K} = \max_{\substack{1 \le l_k \le f_m^{(k)}, \\ 1 \le k \le K-1}} z_{f_m^{(1)}(i_1-1)+l_1,\ldots,f_m^{(K-1)}(i_{K-1}-1)+l_{K-1},i_K}.$$

Alternatively, the floors in (9.11) could be replaced by ceilings and padding with zeros to receive the right cardinality. This extracts the maximums from the (spatial) windows

$$\left(f_m^{(1)}(i_1-1), \ldots, f_m^{(K-1)}(i_{K-1}-1), i_K\right) + \left[1 : f_m^{(1)}\right] \times \cdots \times \left[1 : f_m^{(K-1)}\right] \times [0]$$
$$= \left[f_m^{(1)}(i_1-1)+1 : f_m^{(1)} i_1\right] \times \cdots \times \left[f_m^{(K-1)}(i_{K-1}-1)+1 : f_m^{(K-1)} i_{K-1}\right] \times [i_K],$$

for each channel $1 \le i_K \le q_{m-1}^{(K)}$ individually. Thus, the max-pooling operator is chosen such that it extracts the maximum of each channel and each window, the windows providing a partition of the spatial part of the tensor. This reduces the dimension of the tensor according to (9.11), e.g., if we consider a tensor of order 3 of an RGB image of dimension $I \times J = 180 \times 50$ and apply a max-pooling layer with window sizes $f_m^{(1)} = 10$ and $f_m^{(2)} = 5$, we receive a dimension reduction

$$180 \times 50 \times 3 \mapsto 18 \times 10 \times 3.$$

Replacing the maximum operator in (9.10) by an averaging operator is sometimes also used, and this is called an *average-pooling layer*.

9.2.5 Flatten Layer

A *flatten layer* performs the transformation of rearranging a tensor to a vector, so that the output of a flatten layer can be used as an input to a FN layer. That is,

$$z^{(m)} : \mathbb{R}^{q_{m-1}^{(1)} \times \cdots \times q_{m-1}^{(K)}} \to \mathbb{R}^{q_m}$$

$$z \mapsto z^{(m)}(z) = \left(z_{1,\ldots,1}, \ldots, z_{q_{m-1}^{(1)},\ldots,q_{m-1}^{(K)}}\right)^\top, \tag{9.12}$$

with $q_m = \prod_{k=1}^K q_{m-1}^{(k)}$. We have already used flatten layers after embedding layers on lines 8 and 11 of Listing 7.4.

9.3 Convolutional Neural Network Architectures

9.3.1 Illustrative Example of a CN Network Architecture

We are now ready to patch everything together. Assume we have RGB images described by tensors $\boldsymbol{x}^{(0)} \in \mathbb{R}^{I \times J \times 3}$ of order 3 modeling the three RGB channels of images of a fixed size $I \times J$. Moreover, we have the tabular feature information $\boldsymbol{x}^{(1)} \in \mathcal{X} \subset \{1\} \times \mathbb{R}^q$ that describes further properties of the data. That is, we have an input variable $(\boldsymbol{x}^{(0)}, \boldsymbol{x}^{(1)})$, and we aim at predicting a response variable Y by a using a suitable regression function

$$(\boldsymbol{x}^{(0)}, \boldsymbol{x}^{(1)}) \mapsto \mu(\boldsymbol{x}^{(0)}, \boldsymbol{x}^{(1)}) = \mathbb{E}\left[Y \middle| \boldsymbol{x}^{(0)}, \boldsymbol{x}^{(1)}\right]. \tag{9.13}$$

We choose two convolutional layers $z^{(\mathrm{CN1})}$ and $z^{(\mathrm{CN2})}$, each followed by a max-pooling layer $z^{(\mathrm{Max1})}$ and $z^{(\mathrm{Max2})}$, respectively. Then we apply a flatten layer $z^{(\mathrm{flatten})}$ to bring the learned representation into a vector form. These layers are chosen according to (9.7), (9.10) and (9.12) with matching input and output dimensions so that the following composition is well-defined

$$z^{(5:1)} = \left(z^{(\mathrm{flatten})} \circ z^{(\mathrm{Max2})} \circ z^{(\mathrm{CN2})} \circ z^{(\mathrm{Max1})} \circ z^{(\mathrm{CN1})}\right) : \mathbb{R}^{I \times J \times 3} \to \mathbb{R}^{q_5}.$$

Listing 9.1 provides an example starting from a $I \times J \times 3 = 180 \times 50 \times 3$ input tensor $\boldsymbol{x}^{(0)}$ and receiving a $q_5 = 60$ dimensional learned representation $z^{(5:1)}(\boldsymbol{x}^{(0)}) \in \mathbb{R}^{60}$.

Listing 9.1 CN network architecture in `keras`

```
shape <- c(180,50,3)
#
model = keras_model_sequential()
model %>%
  layer_conv_2d(filters = 10, kernel_size = c(11,6), activation='tanh',
                                                    input_shape = shape) %>%
  layer_max_pooling_2d(pool_size = c(10,5)) %>%
  layer_conv_2d(filters = 5, kernel_size = c(6,4), activation='tanh') %>%
  layer_max_pooling_2d(pool_size = c(3,2)) %>%
  layer_flatten()
```

Listing 9.2 Summary of CN network architecture

```
Layer (type)                      Output Shape              Param #
=======================================================================
conv2d_1 (Conv2D)                 (None, 170, 45, 10)       1990
-----------------------------------------------------------------------
max_pooling2d_1 (MaxPooling2D) (None, 17, 9, 10)            0
-----------------------------------------------------------------------
conv2d_2 (Conv2D)                 (None, 12, 6, 5)          1205
-----------------------------------------------------------------------
max_pooling2d_2 (MaxPooling2D) (None, 4, 3, 5)              0
-----------------------------------------------------------------------
flatten_1 (Flatten)               (None, 60)                0
=======================================================================
Total params: 3,195
Trainable params: 3,195
Non-trainable params: 0
```

Listing 9.2 gives the summary of this architecture providing the dimension reduction mappings (encodings)

$$180 \times 50 \times 3 \overset{\text{CN1}}{\mapsto} 170 \times 45 \times 10 \overset{\text{Max1}}{\mapsto} 17 \times 9 \times 10 \overset{\text{CN2}}{\mapsto} 12 \times 6 \times 5 \overset{\text{Max2}}{\mapsto} 4 \times 3 \times 5 \overset{\text{flatten}}{\mapsto} 60.$$

The first CN layer ($m = 1$) involves $q_1^{(3)} r_1 = 10 \cdot (1 + 11 \cdot 6 \cdot 3) = 1'990$ filter weights $(w_{0,j}^{(1)}, \boldsymbol{W}_j^{(1)})_{1 \le j \le q_1^{(3)}}$ (including the intercepts), and the second CN layer ($m = 3$) involves $q_3^{(3)} r_3 = 5 \cdot (1+6 \cdot 4 \cdot 10) = 1'205$ filter weights $(w_{0,j}^{(3)}, \boldsymbol{W}_j^{(3)})_{1 \le j \le q_3^{(3)}}$. Altogether we have a network parameter of dimension $3'195$ to be fitted in this CN network architecture.

To perform the prediction task (9.13) we concatenate the learned representation $\boldsymbol{z}^{(5:1)}(\boldsymbol{x}^{(0)}) \in \mathbb{R}^{q_5}$ of the RGB image $\boldsymbol{x}^{(0)}$ with the tabular feature $\boldsymbol{x}^{(1)} \in \mathcal{X} \subset \{1\} \times \mathbb{R}^q$. This concatenated vector is processed through a FN network architecture $\boldsymbol{z}^{(d+5:6)}$ of depth $d \ge 1$ providing the output

$$\left(\boldsymbol{z}^{(5:1)}(\boldsymbol{x}^{(0)}), \boldsymbol{x}^{(1)}\right) \mapsto \mathbb{E}\left[Y \middle| \boldsymbol{x}^{(0)}, \boldsymbol{x}^{(1)}\right] = g^{-1}\left\langle \boldsymbol{\beta}, \boldsymbol{z}^{(d+5:6)}\left(\boldsymbol{z}^{(5:1)}(\boldsymbol{x}^{(0)}), \boldsymbol{x}^{(1)}\right)\right\rangle,$$

for given link function g. This last step can be done in complete analogy to Chap. 7, and fitting of such a network architecture uses variants of the SGD algorithm.

9.3.2 *Lab: Telematics Data*

We present a CN network example that studies time-series of telematics car driving data. Unfortunately, this data is not publicly available. Recently, telematics car driving data has gained much popularity in actuarial science, because this data provides information of car drivers that goes beyond the classical features (age of driver, year of driving test, etc.), and it provides a better discrimination of good and bad drivers as it is directly based on the driving habits and the driving styles.

The telematics data has many different aspects. Raw telematics data typically consists of high-frequency GPS location data, say, second by second, from which several different statistics such as speed, acceleration and change of direction can be calculated. Besides the GPS location data, it often contains vehicle speeds from the vehicle instrumental panel, and acceleration in all directions from an accelerometer. Thus, often, there are 3 different sources from which the speed and the acceleration can be extracted. In practice, the data quality is often an issue as these 3 different sources may give substantially different numbers, Meng et al. [271] give a broader discussion on these data quality issues. The telematics GPS data is often complemented by further information such as engine revolutions, daytime of trips, road and traffic conditions, weather conditions, traffic rule violations, etc. This raw telematics data is then pre-processed, e.g., special maneuvers are extracted (speeding, sudden acceleration, hard braking, extreme right- and left-turns), total distances are calculated, driving distances at different daytimes and weekdays are analyzed. For references analyzing such statistics for predictive modeling we refer to Ayuso et al. [17–19], Boucher et al. [42], Huang–Meng [193], Lemaire et al. [246], Paefgen et al. [291], So et al. [344], Sun et al. [347] and Verbelen et al. [370]. A different approach has been taken by Wüthrich [388] and Gao et al. [151, 154, 155], namely, these authors aggregate the telematics data of speed and acceleration to so-called speed-acceleration v-a heatmaps. These v-a heatmaps are understood as images which can be analyzed, e.g., by CN networks; such an analysis has been performed in Zhu–Wüthrich [407] for image classification and in Gao et al. [154] for claim frequency modeling. Finally, the work of Weidner et al. [377, 378] directly acts on the time-series of the telematics GPS data by performing a Fourier analysis.

In this section, we aim at allocating individual car driving trips to the right drivers by directly analyzing the time-series of the telematics data of these trips using CN networks. We therefore replicate the analysis of Gao–Wüthrich [156] on slightly different data. For our illustrative example we select 3 car drivers and we call them driver A, driver B and driver C. For each of these 3 drivers we choose individual car driving trips of 180 seconds, and we analyze their speed-acceleration-change in angle (v-a-Δ) pattern every second. Thus, for $t = 1, \ldots, T = 180$, we study the three input channels

$$\boldsymbol{x}_{s,t} = (v_{s,t}, a_{s,t}, \Delta_{s,t})^\top \in [2, 50]\text{km/h} \times [-3, 3]\text{m/s}^2 \times [0, 1/2] \subset \mathbb{R}^3,$$

where $1 \le s \le S$ labels all individual trips of the considered drivers. This data has been pre-processed by cutting-out the idling phase and the speeds above 50km/h and concatenating the remaining pieces. We perform this pre-processing since we do not want to identify the drivers because they have a special idling phase picture or because they are more likely on the highway. Acceleration has been censored at $\pm 3\text{m/s}^2$ because we cannot exclude that more extreme observations are caused by data quality issues (note that the acceleration is calculated from the GPS coordinates and if the signals are not fully precise it can lead to extreme acceleration observations). Finally, change in angle is measured in absolute values of sine per second (censored at 1/2), i.e., we do not distinguish between left and right turns. This then provides us with three time-series channels giving tensors of order 2

$$\boldsymbol{x}_s = \left((v_{s,1}, a_{s,1}, \Delta_{s,1})^\top, \ldots, (v_{s,180}, a_{s,180}, \Delta_{s,180})^\top \right)^\top \in \mathbb{R}^{180 \times 3},$$

for $1 \le s \le S$. Moreover, there is a categorical response $Y_s \in \{\text{A}, \text{B}, \text{C}\}$ indicating which driver has been driving trip s.

Figure 9.1 illustrates the first three trips $\boldsymbol{x}_s$ of $T = 180$ seconds of each of these three drivers A (top), B (middle) and C (bottom); note that the 180 seconds have been chosen at a random location within each trip. The first lines in red color show the acceleration patterns $(a_t)_{1 \le t \le T}$, the second lines in black color the change in angle patterns $(\Delta_t)_{1 \le t \le T}$, and the last lines in blue color the speed patterns $(v_t)_{1 \le t \le T}$. Table 9.1 summarizes the available data. In total we have 932 individual trips, and we randomly split these trips in the learning data $\mathcal{L}$ consisting of 744 trips and the test data $\mathcal{T}$ collecting the remaining trips. The goal is to train a classification model that correctly allocates the test data $\mathcal{T}$ to the right driver. As feature information, we use the telematics data $\boldsymbol{x}_s$ of length 180 seconds. We design a logistic categorical regression model with response set $\mathcal{Y} = \{\text{A}, \text{B}, \text{C}\}$. Hence, we obtain a vector-valued parameter EF with a response having 3 levels, see Sect. 2.1.4.

To process the telematics data $\boldsymbol{x}_s$, we design a CN network architecture having three convolutional layers $\boldsymbol{z}^{(\text{CN}j)}$, $1 \le j \le 3$, each followed by a max-pooling layer $\boldsymbol{z}^{(\text{Max}j)}$, then we apply a drop-out layer $\boldsymbol{z}^{(\text{DO})}$ and finally a fully-connected FN layer $\boldsymbol{z}^{(\text{FN})}$ providing the logistic response classification; this is the same network architecture as used in Gao–Wüthrich [156]. The code is given in Listing 9.3 and it describes the mapping

$$\boldsymbol{z}^{(8:1)} = \left(\boldsymbol{z}^{(\text{FN})} \circ \boldsymbol{z}^{(\text{DO})} \circ \boldsymbol{z}^{(\text{Max3})} \circ \boldsymbol{z}^{(\text{CN3})} \circ \boldsymbol{z}^{(\text{Max2})} \circ \boldsymbol{z}^{(\text{CN2})} \circ \boldsymbol{z}^{(\text{Max1})} \circ \boldsymbol{z}^{(\text{CN1})} \right) : \\ \mathbb{R}^{T \times 3} \to (0,1)^3.$$

The first CN and pooling layer $\boldsymbol{z}^{(\text{Max1})} \circ \boldsymbol{z}^{(\text{CN1})}$ maps the dimension 180×3 to a tensor of dimension 58×12 using 12 filters; the max-pooling uses the floor (9.11). The second CN and pooling layer $\boldsymbol{z}^{(\text{Max2})} \circ \boldsymbol{z}^{(\text{CN2})}$ maps to 18×10 using 10 filters, and the third CN and pooling layer $\boldsymbol{z}^{(\text{Max3})} \circ \boldsymbol{z}^{(\text{CN3})}$ maps to 1×8 using 8 filters. Actually, this last max-pooling layer is a global max-pooling layer extracting the maximum in each of the 8 filters. Next, we apply a drop-out layer with a drop-out

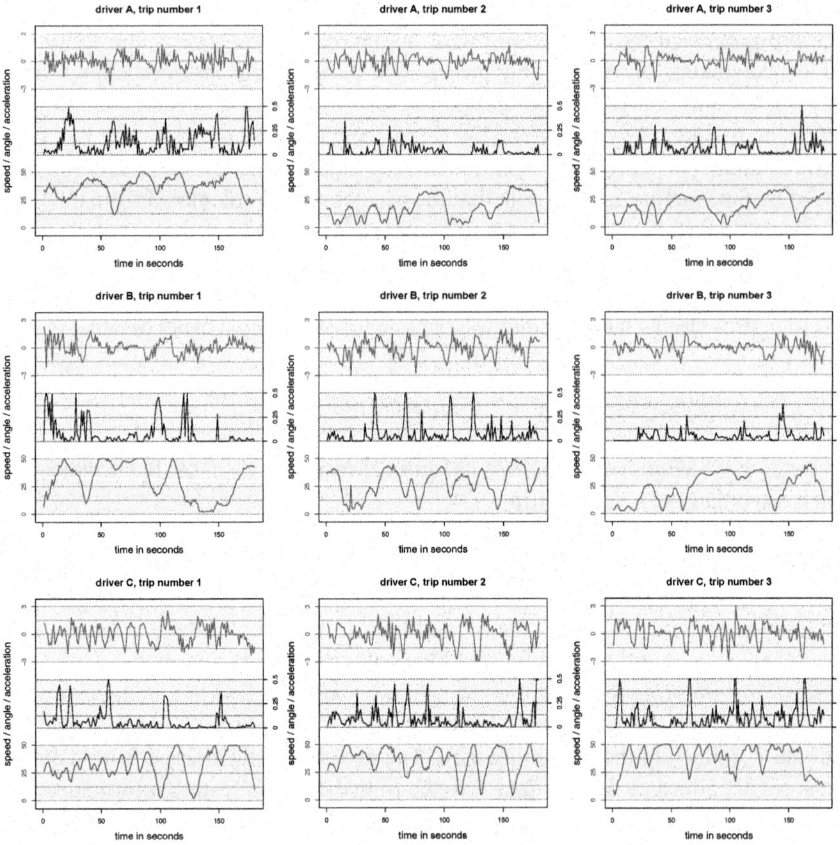

Fig. 9.1 First 3 trips of driver A (top), driver B (middle) and driver C (bottom); each trip is 180 seconds, red color shows the acceleration pattern $(a_t)_t$, black color the change in angle pattern $(\Delta_t)_t$ and blue color the speed pattern $(v_t)_t$

Table 9.1 Summary of the trips and the choice of learning and test data sets $\mathcal{L}$ and $\mathcal{T}$

	Driver A	Driver B	Driver C	Total
Number of trips S	261	385	286	932
Learning data $\mathcal{L}$	209	307	228	744
Test data $\mathcal{T}$	52	78	58	188
Average speed v_t	24.8	30.4	30.2	km/h
Average acceleration/braking $\lvert a_t \rvert$	0.56	0.61	0.74	m/s^2
Average change in angle Δ_t	0.065	0.054	0.076	|sin|/s

rate of 30% to prevent from over-fitting. Finally we apply a fully-connected FN layer that maps the 8 neurons to the 3 categorical outputs using the softmax output activation function, which provides the canonical link of the logistic categorical EF.

Listing 9.3 CN network architecture for the individual car trip allocation

```
shape <- c(180,3)
#
model = keras_model_sequential()
model %>%
layer_conv_1d(filters = 12, kernel_size = 5, activation='tanh',
                                                input_shape = shape) %>%
layer_max_pooling_1d(pool_size = 3) %>%
layer_conv_1d(filters = 10, kernel_size = 5, activation='tanh') %>%
layer_max_pooling_1d(pool_size = 3) %>%
layer_conv_1d(filters = 8, kernel_size = 5, activation='tanh') %>%
layer_global_max_pooling_1d() %>%
layer_dropout(rate = .3) %>%
layer_dense(units = 3, activation = 'softmax')
```

For a summary of the network architecture see Listing 9.4. Altogether this involves 1'237 network parameters that need to be fitted.

Listing 9.4 Summary of CN network architecture for the individual car trip allocation

```
Layer (type)                          Output Shape                  Param #
===========================================================================
conv1d_1 (Conv1D)                     (None, 176, 12)               192
---------------------------------------------------------------------------
max_pooling1d_1 (MaxPooling1D)        (None, 58, 12)                0
---------------------------------------------------------------------------
conv1d_2 (Conv1D)                     (None, 54, 10)                610
---------------------------------------------------------------------------
max_pooling1d_2 (MaxPooling1D)        (None, 18, 10)                0
---------------------------------------------------------------------------
conv1d_3 (Conv1D)                     (None, 14, 8)                 408
---------------------------------------------------------------------------
global_max_pooling1d_1 (GlobalMaxPool (None, 8)                     0
---------------------------------------------------------------------------
dropout_1 (Dropout)                   (None, 8)                     0
---------------------------------------------------------------------------
dense_1 (Dense)                       (None, 3)                     27
===========================================================================
Total params: 1,237
Trainable params: 1,237
Non-trainable params: 0
```

We choose the 744 trips of the learning data $\mathcal{L}$ to train this network to the classification task, see Table 9.1. We use the multi-class cross-entropy loss function, see (4.19), with 80% of the learning data $\mathcal{L}$ as training data $\mathcal{U}$ and the remaining 20% as validation data $\mathcal{V}$ to track over-fitting. We retrieve the network with the smallest validation loss using a callback, we refer to Listing 7.3 for a callback. Since the learning data is comparably small and to reduce randomness, we use the nagging predictor averaging over 10 different network fits (using different seeds).

Table 9.2 Out-of-sample confusion matrix

	True labels		
	Driver A	Driver B	Driver C
Predicted label A	**39**	10	2
Predicted label B	9	**66**	6
Predicted label C	4	2	**50**
% correctly allocated	75.0%	84.6%	86.2%
# of trips in test data	52	78	58

These fitted networks then provide us with a mapping

$$z^{(8:1)} : \mathbb{R}^{T\times 3} \to (0,1)^3, \qquad \boldsymbol{x} \mapsto z^{(8:1)}(\boldsymbol{x}) = \left(z_{\mathrm{A}}^{(8:1)}(\boldsymbol{x}), z_{\mathrm{B}}^{(8:1)}(\boldsymbol{x}), z_{\mathrm{C}}^{(8:1)}(\boldsymbol{x})\right)^\top,$$

and for each trip $\boldsymbol{x}_s \in \mathbb{R}^{T\times 3}$ we receive the classification

$$\widehat{Y}_s = \underset{y\in\{\mathrm{A,B,C}\}}{\arg\max}\; z_y^{(8:1)}(\boldsymbol{x}_s).$$

Table 9.2 shows the out-of-sample results on the test data $\mathcal{T}$. On average more than 80% of all trips are correctly allocated; a purely random allocation would provide a success rate of 33%. This shows that this allocation problem can be solved rather successfully and, indeed, the CN network architecture is able to learn structure in the telematics trip data $\boldsymbol{x}_s$ that allows one to discriminate car drivers. This sounds very promising. In fact, the telematics car driving data seems to be very transparent which, of course, also raises privacy issues. On the downside we should mention that from this approach we cannot really see what the network has learned and how it manages to distinguish the different trips.

There are several approaches that try to visualize what the network has learned in the different layers by extracting the filter activations in the CN layers, others try to invert the networks trying to backtrack which activations and weights mostly contribute to a certain output, we mention, e.g., DeepLIFT of Shrikumar et al. [339]. For more analysis and references we refer to Sect. 4 of the tutorial Meier–Wüthrich [269]. We do not further discuss this and close this example.

9.3.3 Lab: Mortality Surface Modeling

We revisit the mortality example of Sect. 8.4.2 where we used a LSTM architecture to process the raw mortality data for forecasting, see Fig. 8.13. We are going to do a (small) change to that architecture by simply replacing the LSTM encoder by a CN network encoder. This approach has been promoted in the literature, e.g., by Perla et al. [301], Schnürch–Korn [330] and Wang et al. [375]. A main difference between these references is whether the mortality tensor is considered as a tensor of order 2 (reflecting time-series data) or of order 3 (reflecting the mortality surface as an image). In the present example we are going to interpret the mortality tensor as a monochrome image, and this requires that we extend (8.23) by an additional channels component

$$\begin{aligned} \boldsymbol{x}_{t-\tau:t-1} &= (\boldsymbol{x}_{t-\tau}, \ldots, \boldsymbol{x}_{t-1})^\top \\ &= \left(M_{x,s}\right)_{t-\tau \le s \le t-1, x_0 \le x \le x_1} \in \mathbb{R}^{\tau \times (x_1 - x_0 + 1) \times 1} = \mathbb{R}^{5 \times 100 \times 1}, \end{aligned}$$

for a lookback period of $\tau = 5$. The LSTM cell encodes this tensor/matrix into a 20-dimensional vector which is then concatenated with the embeddings of the country code and the gender code (8.24). We use the same architecture here, only the LSTM part is replaced by a CN network in (8.25), the corresponding code is given on lines 14–17 of Listing 9.5.

Listing 9.5 CN network architecture to directly process the raw mortality rates $(M_{x,t})_{x,t}$

```
1  Tensor  = layer_input(shape=c(lookback,100,1), dtype='float32', name='Tensor')
2  Country = layer_input(shape=c(1), dtype='int32',    name='Country')
3  Gender  = layer_input(shape=c(1), dtype='int32',    name='Gender')
4  Time    = layer_input(shape=c(1), dtype='float32', name='Time')
5  #
6  CountryEmb = Country %>%
7   layer_embedding(input_dim=8,output_dim=1,input_length=1,name='CountryEmb') %>%
8   layer_flatten(name='Country_flat')
9  #
10 GenderEmb = Gender %>%
11  layer_embedding(input_dim=2,output_dim=1,input_length=1,name='GenderEmb') %>%
12  layer_flatten(name='Gender_flat')
13 #
14 CN = Tensor %>%
15  layer_conv_2d(filter = 10, kernel_size = c(5,5), activation = 'linear') %>%
16                 layer_max_pooling_2d(pool_size = c(1,8)) %>%
17                 layer_flatten()
18 #
19 Output = list(CN,CountryEmb,GenderEmb) %>% layer_concatenate() %>%
20  layer_dense(units=100, activation='linear',  name='scalarproduct') %>%
21  layer_reshape(c(1,100), name = 'Output')
22 #
23 model = keras_model(inputs = list(Tensor, Country, Gender),
24                     outputs = c(Output))
```

Line 15 maps the input tensor $5 \times 100 \times 1$ to a tensor $1 \times 96 \times 10$ having 10 filters, the max-pooling layer reduces this tensor to $1 \times 12 \times 10$, and the flatten layer encodes this tensor into a 120-dimensional vector. This vector is then concatenated with the embedding vectors of the country and the gender codes, and this provides us with $r = 12'570$ network parameters, thus, the LSTM architecture and the CN network architecture use roughly equally many network parameters that need to be fitted. We then use the identical partition in training, validation and test data as in Sect. 8.4.2, i.e., we use the data from 1950 to 2003 for fitting the network architecture, which is then used to forecast the calendar years 2004 to 2018. The results are presented in Table 9.3.

Table 9.3 Comparison of the out-of-sample mean squared losses for the calendar years $2004 \leq t \leq 2018$; the figures are in 10^{-4}

	Female			Male		
	LC	LSTM	CN	LC	LSTM	CN
Austria AUT	0.765	**0.312**	0.635	2.527	**1.169**	1.569
Belgium BE	0.371	0.311	**0.290**	2.835	**0.960**	1.100
Switzerland CH	0.654	**0.478**	0.772	1.609	**1.134**	2.035
Spain ESP	1.446	0.514	**0.199**	1.742	0.245	**0.240**
France FRA	**0.175**	1.684	0.309	**0.333**	0.363	0.770
Italy ITA	**0.179**	0.330	0.186	0.874	**0.320**	0.421
The Netherlands NL	0.426	0.315	**0.266**	1.978	**0.601**	0.606
Portugal POR	2.097	0.464	**0.416**	1.848	**1.239**	1.880

We observe that in our case the CN network architecture provides good results for the female populations, whereas for the male populations we rather prefer the LSTM architecture. At the current stage we rather see this as a proof of concept, because we have not really fine-tuned the network architectures, nor has the SGD fitting been perfected, e.g., often bigger architectures are used in combination with drop-outs, etc. We refrain from doing so, here, but refer to the relevant literature Perla et al. [301], Schnürch–Korn [330] and Wang et al. [375] for a more sophisticated fine-tuning.

Chapter 10
Natural Language Processing

Natural language processing (NLP) is a vastly growing field that is studying language, communication and text recognition. The purpose of this chapter is to present an introduction to NLP. Important milestones in the field of NLP are the work of Bengio et al. [28, 29] who have introduced the idea of word embedding, the work of Mikolov et al. [275, 276] who have developed word2vec which is an efficient word embedding tool, and the work of Pennington et al. [300] and Chaubard et al. [68] who provide the pre-trained word embedding model GloVe[1] and detailed educational material.[2] An excellent overview of the NLP working pipeline is provided by the tutorial of Ferrario–Nägelin [126]. This overview distinguishes three approaches: (1) the classical approach using bag-of-words and bag-of-part-of-speech models to classify text documents; (2) the modern approach using word embeddings to receive a low-dimensional representation of the dictionary, which is then further processed; (3) the contemporary approach uses a minimal amount of text pre-processing but directly feeds raw data to a machine learning algorithm. We discuss these different approaches and show how they can be used to extract the relevant information from claim descriptions to predict the claim types and the claim sizes; in the actuarial literature first papers on this topic have been published by Lee et al. [236] and Manski et al. [264].

10.1 Feature Pre-processing and Bag-of-Words

NLP requires an extensive feature pre-processing and engineering as different texts can be rather diverse in language, grammar, abbreviations, typos, etc. The current developments aim at automating this process, nevertheless, many of these steps

[1] https://nlp.stanford.edu/projects/glove/.

[2] https://nlp.stanford.edu/teaching/.

M. V. Wüthrich, M. Merz, *Statistical Foundations of Actuarial Learning and its Applications*, Springer Actuarial, https://doi.org/10.1007/978-3-031-12409-9_10

are still (tedious) manual work. Our goal here is to present the whole working pipeline to process language, perform text recognition and text understanding. As an example we use the claim data described in Chap. 13.3; this data has been made available through the book project of Frees [135], and it comprises property claims of governmental institutions in Wisconsin, US. An excerpt of the data is given in Listing 10.1; our attention applies to line 11 which provides a (very) short claim description for every claim.

Listing 10.1 Excerpt of the Wisconsin Local Government Property Insurance Fund (LGPIF) data set with short claim descriptions on line 11

```
1  'data.frame':     5424 obs. of  10 variables:
2   $ PolicyNum     : int 120002 120003 120003 120003 120003 120003 120003 ...
3   $ Year          : int 2010 2007 2008 2007 2009 2010 2007 2007 2009 2007 ...
4   $ Claim         : num 6839 2085 8775 600 34610 ...
5   $ Deduct        : int 1000 5000 5000 5000 5000 5000 5000 5000 5000 5000 ...
6   $ EntityType    : Factor w/ 6 levels "City","County",..: 2 2 2 2 2 2 2 2 2 2 ...
7   $ CoverageCode: Factor w/ 13 levels "CE","CF","CS",..: 12 12 11 11 11 12 ...
8   $ Fire5         : int 4 0 0 0 0 0 0 0 0 0 ...
9   $ CountyCode    : Factor w/ 72 levels "ADA","ASH","BAR",..: 2 3 3 3 3 3 3 3...
10  $ Hazard        : Factor w/ 9 levels "Fire","Hail",..: 3 3 5 5 9 6 3 3 3 3 ...
11  $ Description : chr  "lightning damage" "lightning damage at Comm. Center" ...
```

In a first step we need to pre-process the texts to make them suitable for predictive modeling. This first step is called *tokenization*. Essentially, tokenization labels the words with integers, that is, the used vocabulary is encoded by integers. There are several issues that one has to deal with in this first step such as upper and lower case, punctuation, orthographic errors and differences, abbreviations, etc. Different treatments of these issues will lead to different results, for more on this topic we refer to Sect. 1 in Ferrario–Nägelin [126]. We simply use the standard routine offered in R `keras` [77] called `text_tokenizer()` with its standard settings.

Listing 10.2 Tokenization within R `keras` [77]

```
1  library(keras)
2
3  ## initialize tokenizer and fit
4  tokenizer <- text_tokenizer() %>% fit_text_tokenizer(dat$Description)
5
6  ## number of tokens/words
7  length(tokenizer$word_index)
8
9  ## frequency of word appearances in each text
10 freq.text <- texts_to_matrix(tokenizer, dat$Description, mode = "count")
```

The R code in Listing 10.2 shows the crucial steps in tokenization. Line 4 extracts the relevant vocabulary from all available claim descriptions. In total the 5'424 claim

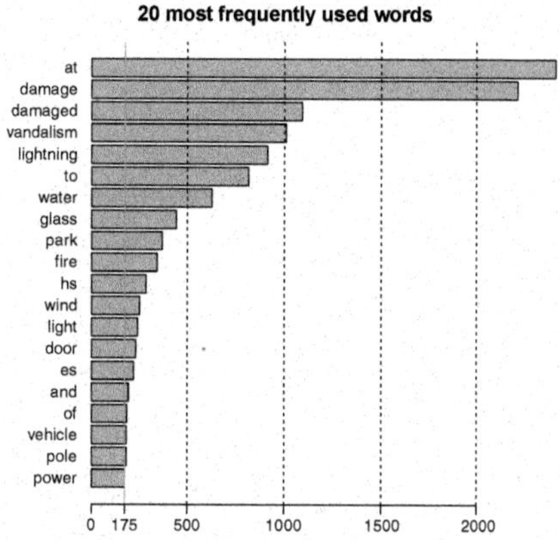

Fig. 10.1 Most frequently used words in the claim descriptions of Listing 10.1

descriptions of Listing 10.1 use $W = 2'237$ different words. This double counts different spellings, e.g., 'color' vs. 'colour'.

Figure 10.1 shows the most frequently used words in the claim descriptions of Listing 10.1. These are (in this order): 'at', 'damage', 'damaged', 'vandalism', 'lightning', 'to', 'water', 'glass', 'park', 'fire', 'hs', 'wind', 'light', 'door', 'es', 'and', 'of', 'vehicle', 'pole' and 'power'. We observe that many of these words are directly related to insurance claims, such as 'damage' and 'vandalism', others are frequent *stopwords* like 'at' and 'to', and then there are abbreviations like 'hs' and 'es' standing for high school and elementary school.

Listing 10.3 Word and text encoding

```
1  maxlen <- max(rowSums(freq.text))
2
3  ## encode the sentences
4  text.seq <- texts_to_sequences(tokenizer, dat$Description)
5
6  ## pad the sentences
7  text.seq.pad <- pad_sequences(text.seq, maxlen = maxlen, padding = "post")
8
9  ## examples
10 lightning/hail damage to equip at airport
11    5  48   2   6 196   1  40   0   0   0   0
12 ##
13 garage door damaged
14  36 14   3   0   0   0   0   0   0   0   0
```

The next step is to assign the (integer) labels $1 \leq w \leq W$ from the tokenization to the words in the texts. The maximal length over all texts/sentences is $T = 11$ words. This step and padding the sentences with zeros to equal length T is presented on lines 1–7 of Listing 10.3. Lines 11 and 14 of this listing give two explicit text examples

$$\texttt{text} = (w_1, \ldots, w_T)^\top \in \mathcal{W}_0^T,$$

where we set for the vocabulary $\mathcal{W}_0$ used

$$\mathcal{W} = \{1, \ldots, W\} \subset \mathbb{N} \qquad \text{and} \qquad \mathcal{W}_0 = \mathcal{W} \cup \{0\}.$$

The label 0 is used for padding shorter texts to the common length $T = 11$. The method of *bag-of-words* embeds $\texttt{text} = (w_1, \ldots, w_T)^\top$ into $\mathbb{N}_0^W$

$$\psi : \mathcal{W}_0^T \to \mathbb{N}_0^W, \qquad \texttt{text} \mapsto \psi(\texttt{text}) = \left(\sum_{t=1}^{T} \mathbb{1}_{\{w_t = w\}} \right)^\top_{w \in \mathcal{W}} . \tag{10.1}$$

The bag-of-words $\psi(\texttt{text})$ counts how often each word $w \in \mathcal{W}$ appears in a given $\texttt{text} = (w_1, \ldots, w_T)^\top$; the corresponding code is given on line 10 of Listing 10.2. The bag-of-words mapping ψ is not injective as the order of occurrence of the words gets lost, and, thus, also the semantics of the sentence gets lost. E.g., the bag-of-words of the following two sentences is the same 'The claim is expensive.' and 'Is the claim expensive?'. This is the reason for calling it a "bag of words" (which is unordered). This bag-of-words encoding resembles one-hot encoding, namely, if every text consists of a single word $T = 1$, then we receive the one-hot encoding with W describing the number of different levels, see (7.28). The bag-of-words $\psi(\texttt{text}) \in \mathbb{N}_0^W$ can directly be used as an input to a regression model. The disadvantage of this approach is that the input typically is high-dimensional (and likely sparse), and it is recommended that only the frequent words are considered.

Listing 10.4 Removal of stopwords and lemmatization

```
1  library(textstem)
2  library(tm)
3
4  text.clean <- removeWords(dat$Description, stopwords("english"))
5  text.clean <- lemmatize_strings(text.clean, dictionary = lexicon::hash_lemmas)
```

Additionally, stopwords can be removed. We perform this removal below because frequent stopwords like 'and' or 'to' may not essentially contribute to the understanding of the (short) claim descriptions; the code for the stopword removal is provided on line 4 of Listing 10.4. Moreover, stemming can be performed which means that inflectional forms are reduced to their stem by just truncating pre- and suffixes, conjugations, declensions, etc. Lemmatization is a more sophisticated form of reducing inflectional forms by using vocabularies and morphological analyses; an example is provided on line 5 of Listing 10.4. If we perform these two steps of removing stopwords and lemmatization to our example, the number of different words is reduced from 2'237 to 1'982.

Another step that can be performed is tagging words with part-of-speech (POS) attributes. These POS attributes indicate whether the corresponding words are used as nouns, adjectives, adverbs, etc., in the corresponding sentences. We then call the resulting encoding bag-of-POS. We refrain from doing this because we will present more sophisticated methods in the next sections.

10.2 Word Embeddings

The bag-of-words (10.1) can be interpreted as representing each word $w \in \mathcal{W} = \{1, \ldots, W\}$ by a one-hot encoding in $\{0, 1\}^W$, and then aggregating these one-hot encodings over all words that appear in the given $\texttt{text} = (w_1, \ldots, w_T)^\top$. Bengio et al. [28, 29] have introduced the technique of *word embedding* that maps words to a lower dimensional Euclidean space $\mathbb{R}^b$, $b \ll W$, such that proximity in $\mathbb{R}^b$ is associated with similarity in the meaning of the word, e.g., 'rain', 'water' and 'flood' should be more close to each other in $\mathbb{R}^b$ than to 'vandalism' (in an insurance context). This is exactly the idea promoted in the embedding mapping (7.31) using the embedding layers. Thus, we are looking for an embedding mapping

$$\boldsymbol{e} : \mathcal{W} \to \mathbb{R}^b, \qquad w \mapsto \boldsymbol{e}(w), \tag{10.2}$$

that maps each word w (or rather its tokenization) to a b-dimensional vector $\boldsymbol{e}(w)$, for a given embedding dimension $b \ll W$. The general idea now is that similarity in the meaning of words can be learned from the context in which the words are used in. That is, when we consider a text

$$\texttt{text} = (w_1, \ldots, w_{t-1}, w_t, w_{t+1}, \ldots, w_T)^\top,$$

then it might be possible to infer w_t from its neighbors w_{t-j} and w_{t+j}, $j \geq 1$. This explains the context of a word w_t, and using suitable learning tools it should also be possible to learn synonyms for w_t as these synonyms will stand in similar contexts.

More mathematically speaking, we assume that there exists a probability distribution p over the set of all texts of length T (using padding with zeros to common length)

$$\mathsf{T} = \left\{\texttt{text} = (w_1, \ldots, w_T)^\top\right\} \subseteq \mathcal{W}_0^T,$$

such that a randomly chosen $\texttt{text} \in \mathsf{T}$ appears with probability $p(w_1, \ldots, w_T) \in [0, 1)$. Inference of a word w_t from its context can then be obtained by studying the conditional probablity of w_t, given its context, that is

$$p\left(w_t \middle| w_1, \ldots, w_{t-1}, w_{t+1}, \ldots, w_T\right) = \frac{p(w_1, \ldots, w_T)}{p(w_1, \ldots, w_{t-1}, w_{t+1}, \ldots, w_T)}. \tag{10.3}$$

Since, typically, the probability distribution p is not known we aim at learning it from the available data. This idea has been taken up by Mikolov et al. [275, 276] who designed the word to vector (word2vec) algorithm. Pennington et al. [300] designed an alternative algorithm called global vectors (GloVe); we also refer to Chaubard et al. [68]. We describe these algorithms in the following sections.

10.2.1 Word to Vector Algorithms

There are two ways of estimating the probability p in (10.3). Either we can try to predict the *center word* w_t from its context as in (10.3) or we can try to predict the context from the center word w_t, which applies Bayes's rule to (10.3). The latter variant is called *skip-gram* and the former variant is called *continuous bag-of-words* (CBOW), if we neglect the order of the words in the context. These two approaches have been developed by Mikolov et al. [275, 276].

Skip-gram Approach

Typically, inferring a general probability distribution p over T is too complex. Therefore, we make a simplifying assumption. This simplifying assumption is not reasonable from a practical linguistic point of view, but it is sufficient to receive a reasonable word embedding map $\boldsymbol{e} : \mathcal{W} \to \mathbb{R}^b$. We assume conditional i.i.d. of the context words, given the center word w_t. Choosing a fixed context (window) size $c \in \mathbb{N}$, we try to maximize the log-likelihood over all probabilities p satisfying this conditional i.i.d. assumption

$$\begin{aligned}\ell_{\boldsymbol{W}} &= \sum_{i=1}^{n} \log p\left(w_{i,t-c}, \ldots, w_{i,t-1}, w_{i,t+1}, \ldots, w_{i,t+c} \middle| w_{i,t}\right) \\ &= \sum_{i=1}^{n} \sum_{-c \le j \le c, j \ne 0} \log p\left(w_{i,t+j} \middle| w_{i,t}\right), \end{aligned} \tag{10.4}$$

having n independent rows in the observed data matrix $\boldsymbol{W} = (w_{i,t-c}, \ldots, w_{i,t+c})_{1 \le i \le n} \in \mathcal{W}^{n \times (2c+1)}$. Thus, under the conditional i.i.d. of the context words, given the center word, the probabilities (10.4) infer the occurrence of (individual) context words of a given center word $w_{i,t}$ within a symmetric window of fixed size c. In the sequel we directly work with the log-likelihood (10.4), supposed that a context word $w_{i,t+j}$ exists for index j, otherwise the corresponding term is just dropped from the sum in (10.4).

The remaining step is to estimate the conditional probabilities $p(w_{t+j}|w_t)$ from the data matrix $\boldsymbol{W}$. This step will provide us with the embeddings (10.2). This estimation step is received by considering an approach similar to a GLM for categorical responses, see Sect. 5.7. We make the following ansatz for the context word w_s and the center word w_t (for all j)

$$p\left(w_s \middle| w_t\right) = \frac{\exp\left\langle \widetilde{e}(w_s), e(w_t)\right\rangle}{\sum_{w=1}^{W} \exp\left\langle \widetilde{e}(w), e(w_t)\right\rangle} \in (0, 1), \tag{10.5}$$

where e and $\widetilde{e}$ are two (different) embedding maps (10.2) that have the same embedding dimension $b \in \mathbb{N}$. Thus, we construct two different embeddings e and $\widetilde{e}$ for the center words and for the context words, respectively, and these embeddings (embedding weights) are chosen such that the log-likelihood (10.4) is maximized for the given observations W. These assumptions give us a minimization problem for the negative log-likelihood in the embedding mappings, i.e., we minimize over the embeddings e and $\widetilde{e}$

$$\begin{aligned} -\ell_W &= -\sum_{i=1}^{n} \sum_{-c\le j\le c, j\neq 0} \log\left(\frac{\exp\left\langle \widetilde{e}(w_{i,t+j}), e(w_{i,t})\right\rangle}{\sum_{w=1}^{W} \exp\left\langle \widetilde{e}(w), e(w_{i,t})\right\rangle}\right) \qquad (10.6) \\ &= -\sum_{i=1}^{n} \left(\sum_{-c\le j\le c, j\neq 0} \left\langle \widetilde{e}(w_{i,t+j}), e(w_{i,t})\right\rangle - 2c \log\left(\sum_{w=1}^{W} \exp\left\langle \widetilde{e}(w), e(w_{i,t})\right\rangle\right)\right). \end{aligned}$$

These optimal embeddings are learned using a variant of the gradient descent algorithm. This often results in a very high-dimensional optimization problem as we have $2bW$ parameters to learn, and the calculation of the last (normalization) term in (10.6) can be very expensive in gradient descent algorithms. For this reason we present the method of negative sampling below.

Continuous Bag-of-Words

For the CBOW method we start from the log-likelihood for a context size $c \in \mathbb{N}$ and given the observations W

$$\sum_{i=1}^{n} \log p\left(w_{i,t} \middle| w_{i,t-c}, \ldots, w_{i,t-1}, w_{i,t+1}, \ldots, w_{i,t+c}\right).$$

Again we need to reduce the complexity which requires an approximation to the above. Assume that the embedding map of the context words is given by $\widetilde{e} : \mathcal{W} \to \mathbb{R}^b$. We then average over the embeddings of the context words in order to predict the center word. Define the average embedding of the context words of $w_{i,t}$ (with a fixed window size c) by

$$\widetilde{e}_{i,t} = \frac{1}{2c} \sum_{-c\le j\le c, j\neq 0} \widetilde{e}(w_{i,t+j}).$$

Making an ansatz similar to (10.5), the full log-likelihood is approximated by

$$\sum_{i=1}^{n} \log p\left(w_{i,t} \middle| \widetilde{e}_{i,t}\right) = \sum_{i=1}^{n} \log \left(\frac{\exp \langle \widetilde{e}_{i,t}, \boldsymbol{e}(w_{i,t}) \rangle}{\sum_{w=1}^{W} \exp \langle \widetilde{e}_{i,t}, \boldsymbol{e}(w) \rangle} \right) \tag{10.7}$$

$$= \sum_{i=1}^{n} \langle \widetilde{e}_{i,t}, \boldsymbol{e}(w_{i,t}) \rangle - \log \left(\sum_{w=1}^{W} \exp \langle \widetilde{e}_{i,t}, \boldsymbol{e}(w) \rangle \right).$$

Again the gradient descent method is applied to the negative log-likelihood to learn the optimal embedding maps $\boldsymbol{e}$ and $\widetilde{\boldsymbol{e}}$.

Remark 10.1 In both cases, skip-gram and CBOW, we estimate two separate embeddings $\boldsymbol{e}$ and $\widetilde{\boldsymbol{e}}$ for the center word and the context words. Typically, CBOW is faster but skip-gram is better on words that are less frequent.

Negative Sampling

There is a computational issue in (10.6) and (10.7) because the probability normalizations in (10.6) and (10.7) aggregate over all available words $w \in \mathcal{W}$. This can be computationally demanding because we need to perform this calculation in each gradient descent step. For this reason, Mikolov et al. [276] turn the log-likelihood optimization problem (10.6) into a binary classification problem. Consider a pair $(w, \widetilde{w}) \in \mathcal{W} \times \mathcal{W}$ of center word w and context word $\widetilde{w}$. We introduce a binary response variable $Y \in \{1, 0\}$ that indicates whether an observation $(W, \widetilde{W}) = (w, \widetilde{w})$ is coming from a true center-context pair (from our texts) or whether we have a fake center-context pair (that has been generated randomly). Choosing the canonical link of the Bernoulli EF (logistic/sigmoid function) we make the following ansatz (in the skip-gram approach) to test for the authenticity of a center-context pair $(w, \widetilde{w})$

$$\mathbb{P}[Y = 1 | w, \widetilde{w}] = \frac{1}{1 + \exp\{-\langle \widetilde{\boldsymbol{e}}(\widetilde{w}), \boldsymbol{e}(w) \rangle\}}. \tag{10.8}$$

The recipe now is as follows: (1) Consider for a given window size c all center-context pairs $(w_i, \widetilde{w}_i) \in \mathcal{W} \times \mathcal{W}$ of our texts, and equip them with a response $Y_i = 1$. Assume we have N such observations. (2) Simulate N i.i.d. pairs $(W_{N+k}, \widetilde{W}_{N+k})$, $1 \leq k \leq N$, by randomly choosing W_{N+k} and $\widetilde{W}_{N+k}$, independent from each other (by performing independent re-sampling with or without replacements from the data $(w_i)_{1 \leq i \leq N}$ and $(\widetilde{w}_i)_{1 \leq i \leq N}$, respectively). Equip these (false) pairs with the response $Y_{N+k} = 0$. (3) Maximize the following log-likelihood as a function of the embedding maps $\boldsymbol{e}$ and $\widetilde{\boldsymbol{e}}$

$$\begin{aligned}\ell_Y &= \sum_{i=1}^{2N} \log \mathbb{P}[Y = Y_i |\, w_i, \widetilde{w}_i] \qquad (10.9)\\ &= \sum_{i=1}^{N} \log\left(\frac{1}{1+\exp\langle -\widetilde{\boldsymbol{e}}(\widetilde{w}_i), \boldsymbol{e}(w_i)\rangle}\right) + \sum_{k=N+1}^{2N} \log\left(\frac{1}{1+\exp\langle \widetilde{\boldsymbol{e}}(\widetilde{w}_k), \boldsymbol{e}(w_k)\rangle}\right).\end{aligned}$$

This approach is called *negative sampling* because we sample false or negative pairs $(W_{N+k}, \widetilde{W}_{N+k})$ that should not appear in our texts (as W_{N+k} and $\widetilde{W}_{N+k}$ have been generated independently from each other). The binary classification (10.9) aims at detecting the negative pairs be letting the scalar products $\langle \widetilde{\boldsymbol{e}}(\widetilde{w}_i), \boldsymbol{e}(w_i)\rangle$ be large for the true pairs and letting the scalar products $\langle \widetilde{\boldsymbol{e}}(\widetilde{w}_k), \boldsymbol{e}(w_k)\rangle$ be small for the false pairs. The former means that $\widetilde{\boldsymbol{e}}(\widetilde{w}_i)$ and $\boldsymbol{e}(w_i)$ should point into the same direction in the embedding space $\mathbb{R}^b$. The same should apply for a synonym of w_i and, thus, we receive the desired behavior that synonyms or words with similar meanings tend to cluster.

Example 10.2 (word2vec with Negative Sampling) We provide an example by constructing a word2vec embedding based on negative sampling. For this we aim at maximizing the log-likelihood (10.9) by finding optimal embedding maps $\boldsymbol{e}$ and $\widetilde{\boldsymbol{e}} : \mathcal{W} \to \mathbb{R}^b$. To construct these embedding maps we use the Wisconsin LGPIF data described in Sect. 13.3. The first decision (hyper-parameter) is the choice of the embedding dimension b. English language has millions of different words, and these words should be (in some sense) densely embedded into a b-dimensional Euclidean space. Typical choices of b vary between 50 and 300. Our LGPIF data vocabulary is much smaller, and for this example we choose $b = 2$ because this allows us to nicely illustrate the learned embeddings. However, apart from illustration, we should not choose such a small dimension as it does not allow for a sufficient flexibility in discriminating the words, as we will see.

We consider all available claim texts described in Sect. 13.3. These are 6'031 texts coming from the training and validation data sets (we include the validation data here to have more texts for learning the embeddings; this is different from Sect. 10.1). We extract the claim descriptions from these two data sets and we apply some pre-processing to the texts. This involves transforming all letters to lower case, removing the special characters like !"/&, and removing the stopwords. Moreover, we remove the words 'damage' and 'damaged' as these two words are very common in our insurance claim descriptions, see Fig. 10.1, but they do not further specify the claim type. Then we apply lemmatization, see Listing 10.4, and we adjust the vocabulary with the GloVe database,[3] see also Example 10.4. The latter step is

[3] https://nlp.stanford.edu/projects/glove/.

(tedious) manual work, and we do this step to be able to compare our results to pre-trained word2vec versions.

After this pre-processing we apply the tokenizer, see line 4 of Listing 10.2. This gives us $1'829$ different words. To construct our (illustrative) embedding we only consider the words that appear at least 20 times over all texts, these are $W = 142$ words. Thus, the following analysis is only based on the $W = 142$ most frequent words. Of course, we could increase our vocabulary by considering any text that can be downloaded from the internet. Since we would like to perform an insurance claim analysis, these texts should be related to an insurance context so that the learned embeddings reflect an insurance experience; we come back to this in Remark 10.4, below. We refrain here from doing so and embed these $W = 142$ words into the Euclidean plane ($b = 2$).

Listing 10.5 Tokenization of the most frequent words

```
1 ## applying the tokenizer to the cleaned texts
2 tokenizer <- text_tokenizer(num_words=142+1) %>% fit_text_tokenizer(dat$clean)
3
4 seqs <- texts_to_sequences(tokenizer, dat$clean)
5
6 ## skip-gram of text 1 using a window of size 2
7 skipgrams(sequence=unlist(seqs[[1]]),
8                       vocabulary_size=142, window_size=2, negative_samples=0)
```

Listing 10.5 shows the tokenization of the most frequent words, and on line 4 we build the (shortened) texts $w_1, w_2, \ldots$, only considering these most frequent words $w \in \mathcal{W} = \{1, \ldots, W\}$. In total we receive 4'746 texts that contain at least two words from $\mathcal{W}$ and, hence, can be used for the skip-gram building of center-context pairs $(w, \widetilde{w}) \in \mathcal{W} \times \mathcal{W}$. Lines 7–8 give the code for building these pairs for a window of size $c = 2$. In total we receive $N = 23'952$ center-context pairs $(w_i, \widetilde{w}_i)$ from our texts. We equip these pairs with a response $Y_i = 1$. For the false pairs, we randomly permute the second component of the true pairs $(W_{N+i}, \widetilde{W}_{N+i}) = (w_i, \widetilde{w}_{\tau(i)})$, where τ is a random permutation of $\{1, \ldots, N\}$. These false pairs are equipped with a response $Y_{N+i} = 0$. Thus, altogether we have $2N = 47'904$ observations $(Y_i, w_i, \widetilde{w}_i)$, $1 \le j \le 2N$, that can be used to learn the embeddings $\boldsymbol{e}$ and $\widetilde{\boldsymbol{e}}$.
Listing 10.6 shows the R code to perform the embedding learning using the negative sampling (10.9). This network has $2bW = 568$ embedding weights that need to be learned from the data. There are two more parameters involved on line 10 of Listing 10.6. These two parameters shift the scalar products by an intercept β_0 and scale them by a constant β_1. We could set $(\beta_0, \beta_1) = (0, 1)$, however, keeping these two parameters trainable has led to results that are better centered around the origin. Of course, these two parameters do not harm the arguments as they only

Listing 10.6 R code for negative sampling

```
center  = layer_input(shape = c(1), dtype = 'int32')
context = layer_input(shape = c(1), dtype = 'int32')
#
centerEmb = center %>%
     layer_embedding(input_dim=142,output_dim=2,input_length=1) %>% layer_flatten()
contextEmb = context %>%
     layer_embedding(input_dim=142,output_dim=2,input_length=1) %>% layer_flatten()
#
response = list(centerEmb, contextEmb) %>% layer_dot(axes = 1) %>%
                layer_dense(units=1, activation='sigmoid', name='response')
#
model = keras_model(inputs = c(center, context), outputs = c(response))
```

replace (10.8) by a slightly different model

$$\mathbb{P}[Y=1|\, w, \widetilde{w}] = \frac{1}{1+\exp\{-\beta_0 - \beta_1\langle \widetilde{\boldsymbol{e}}(\widetilde{w}), \boldsymbol{e}(w)\rangle\}} = \frac{e^{\beta_0}}{e^{\beta_0} + e^{-\beta_1\langle \widetilde{\boldsymbol{e}}(\widetilde{w}), \boldsymbol{e}(w)\rangle}},$$

and

$$\mathbb{P}[Y=0|\, w, \widetilde{w}] = 1 - \frac{e^{\beta_0}}{e^{\beta_0} + e^{-\beta_1\langle \widetilde{\boldsymbol{e}}(\widetilde{w}), \boldsymbol{e}(w)\rangle}} = \frac{e^{-\beta_0}}{e^{-\beta_0} + e^{\beta_1\langle \widetilde{\boldsymbol{e}}(\widetilde{w}), \boldsymbol{e}(w)\rangle}}.$$

We fit this model using the `nadam` version of the gradient descent algorithm, and the fitted embedding weights can be extracted with `get_weights(model)`. Figure 10.2 shows the learned embedding weights $\boldsymbol{e}(w) \in \mathbb{R}^2$ of all words $w \in \mathcal{W}$. We highlight the words that coincide with the insured hazards in red color, see line 10 of Listing 10.1. The word 'vehicle' is in the first quadrant and it is surrounded by 'pole', 'truck', 'garage', 'car', 'traffic'. The word 'vandalism' is in the third quadrant surrounded by 'graffito', 'window', 'pavilion', names of cites and parks, 'ms' for middle school. Finally, the words 'fire', 'wind', 'lightning' and 'hail' are in the first and fourth quadrant, close to 'water'; these words are surrounded by 'bldg' (building), 'smoke', 'equipment', 'alarm', 'safety', 'power', 'library', etc. We conclude that these embeddings make perfect sense in an insurance claim context. Note that we have applied some pre-processing, and embeddings could even be improved by further pre-processing, e.g., 'vandalism' and 'vandalize' or 'hs' and 'high school' are used.

Another nice observation is that the embeddings tend to build a circle around the origin, see Fig. 10.2. This is enforced by embedding $W = 142$ different words into a $b = 2$ dimensional space so that dissimilar words optimally repulse each other. ■

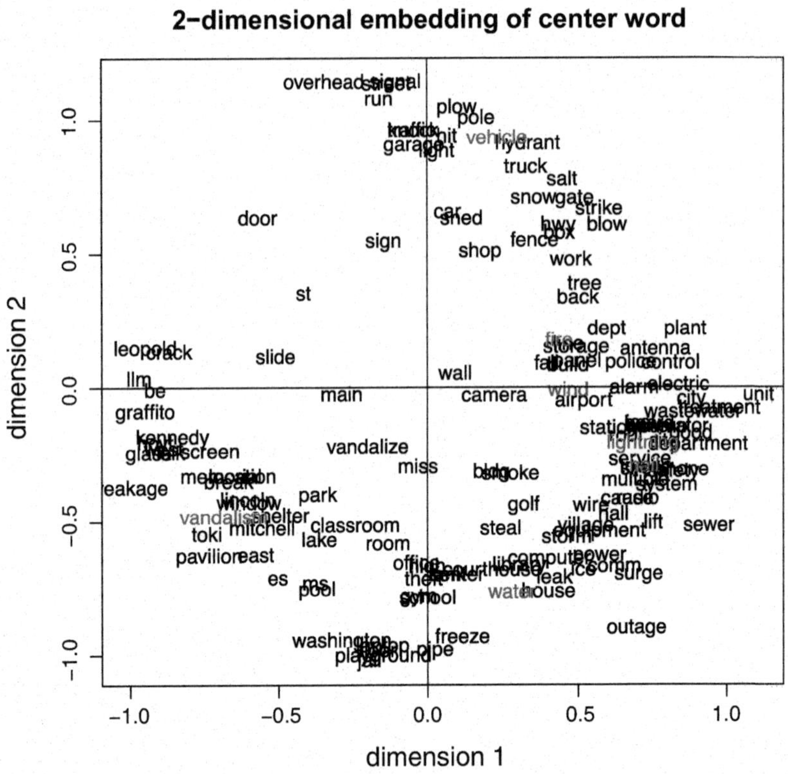

Fig. 10.2 Two-dimensional skip-gram embedding using negative sampling; in red color are the insured hazards 'vehicle', 'fire', 'lightning', 'wind', 'hail', 'water' and 'vandalism'

10.2.2 *Global Vectors Algorithm*

A second popular word embedding approach is global vectors (GloVe) developed by Pennington et al. [300], we also refer to Chaubard et al. [68]. GloVe is an unsupervised learning method that performs a word-word clustering (center-context pairs) over all available texts. Assume that the tokenization of all texts provides us with the words $w \in \mathcal{W}$. Choose a fixed context window size $c \in \mathbb{N}$ and define the matrix

$$\boldsymbol{C} = \left(C(w, \widetilde{w})\right)_{w,\widetilde{w} \in \mathcal{W}} \in \mathbb{N}_0^{W \times W},$$

with $C(w, \widetilde{w})$ counting the number of co-occurrences of w and $\widetilde{w}$ over all available texts where the word $\widetilde{w}$ appears as a context word of the center word w (for the given window size c). We note that $\boldsymbol{C}$ is a symmetric matrix that is typically sparse as many words do not appear in the context of other words (on finitely many texts). Figure 10.3 shows the center-context pairs $(w, \widetilde{w})$ co-occurrence matrix $\boldsymbol{C}$

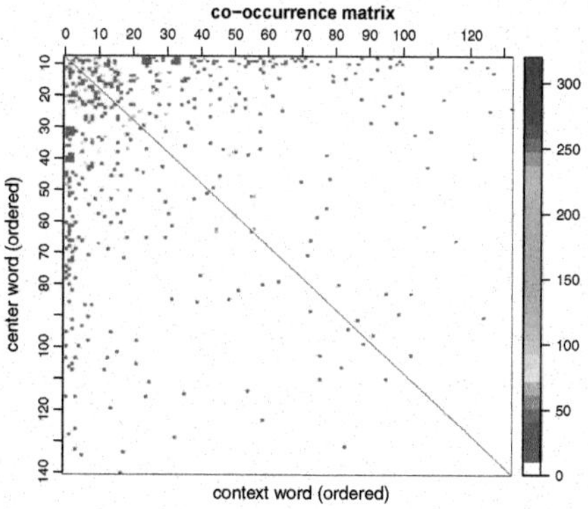

Fig. 10.3 Center-context pairs $(w, \widetilde{w})$ co-occurrence matrix $\boldsymbol{C}$ of Example 10.2; the color scale gives the observed frequencies

of Example 10.2 which is based on $W = 142$ words and 23'952 center-context pairs. The color pixels indicate the pairs that occur in the data, $C(w, \widetilde{w}) > 0$, and the white space corresponds to the pairs that have not been observed in the texts, $C(w, \widetilde{w}) = 0$. This plot confirms the sparsity of the center-context pairs; the words are ordered w.r.t. their frequencies in the texts.

In an empirical analysis Pennington et al. [300] have observed that the crucial quantities to be considered are the ratios for fixed context words. That is, for a context word $\widetilde{w}$ study a function of the center words w and v (subject to existence of the right-hand side)

$$(w, v, \widetilde{w}) \mapsto F(w, v, \widetilde{w}) = \frac{C(w, \widetilde{w}) / \sum_{\widetilde{u} \in \mathcal{W}} C(w, \widetilde{u})}{C(v, \widetilde{w}) / \sum_{\widetilde{u} \in \mathcal{W}} C(v, \widetilde{u})} = \frac{\widehat{p}(\widetilde{w}|w)}{\widehat{p}(\widetilde{w}|v)},$$

$\widehat{p}$ denoting the empirical probabilities. An empirical analysis suggests that such an approach seems to lead to a good discrimination of the meanings of the words, see Sect. 3 in Pennington et al. [300]. Further simplifications and assumptions provide the following ansatz, for details we refer to Pennington et al. [300],

$$\log C(w, \widetilde{w}) \approx \langle \widetilde{\boldsymbol{e}}(\widetilde{w}), \boldsymbol{e}(w) \rangle + \widetilde{\beta}_{\widetilde{w}} + \beta_w,$$

with intercepts $\widetilde{\beta}_{\widetilde{w}}, \beta_w \in \mathbb{R}$. There is still one issue, namely, that $\log C(w, \widetilde{w})$ may not be well-defined as certain pairs $(w, \widetilde{w})$ are not observed. Therefore, Pennington et al. [300] propose to solve a weighted squared error loss function problem to find the embedding mappings $\boldsymbol{e}, \widetilde{\boldsymbol{e}}$ and intercepts $\widetilde{\beta}_{\widetilde{w}}, \beta_w \in \mathbb{R}$. Their objective function is given by

$$\sum_{w, \widetilde{w} \in \mathcal{W}} \chi(C(w, \widetilde{w})) \left(\log C(w, \widetilde{w}) - \langle \widetilde{\boldsymbol{e}}(\widetilde{w}), \boldsymbol{e}(w) \rangle - \widetilde{\beta}_{\widetilde{w}} - \beta_w \right)^2, \tag{10.10}$$

with weighting function

$$x \geq 0 \mapsto \chi(x) = \left(\frac{x \wedge x_{\max}}{x_{\max}}\right)^{\alpha},$$

for $x_{\max} > 0$ and $\alpha > 0$. Pennington et al. [300] state that the model depends weakly on the cutoff point $x_{\max}$, they propose $x_{\max} = 100$, and a sub-linear behavior seems to outperform a linear one, suggesting, e.g., a choice of $\alpha = 3/4$. Under these choices the embeddings $\boldsymbol{e}$ and $\widetilde{\boldsymbol{e}}$ are found by minimizing the objective function (10.10) for the given data. Note that $\lim_{x \downarrow 0} \chi(x)(\log x)^2 = 0$.

Example 10.3 (GloVe Word Embedding) We provide an example using the GloVe embedding model, and we revisit the data of Example 10.2; we also use exactly the same pre-processing as in that example. We start from $N = 23'952$ center-context pairs.

In a first step we count the number of co-occurrences $C(w, \widetilde{w})$. There are only 4'972 pairs that occur, $C(w, \widetilde{w}) > 0$, this corresponds to the colors in Fig. 10.3. With these 4'972 pairs we have to fit 568 embedding weights (for the embedding dimension $b = 2$) and 284 intercepts $\widetilde{\beta}_{\widetilde{w}}, \beta_w$, thus, 852 parameters in total. The results of this fitting are shown in Fig. 10.4.

The general picture in Fig. 10.4 is similar to Fig. 10.2, e.g., 'vandalism' is surrounded by 'graffito', 'window', 'pavilion', names of cites and parks, 'ms' and 'es'; or 'vehicle' is surrounded by 'pole', 'traffic', 'street', 'signal'. However, the clustering of the words around the origin shows a crucial difference between GloVe and the negative sampling of word2vec. The problem here is that we do not have sufficiently many observations. We have 4'972 center-context pairs that occur, $C(w, \widetilde{w}) > 0$. 2'396 of these pairs occur exactly once, $C(w, \widetilde{w}) = 1$, this is almost half of the observations with $C(w, \widetilde{w}) > 0$. GloVe (10.10) considers these observations on the log-scale which provides $\log C(w, \widetilde{w}) = 0$ for the pairs that occur exactly once. The weighted square loss for these pairs is minimized by either setting $\widetilde{\boldsymbol{e}}(\widetilde{w}) = 0$ or $\boldsymbol{e}(w) = 0$, supposed that the intercepts are also set to 0. This is exactly what we observe in Fig. 10.4 and, thus, successfully fitting GloVe would require much more (frequent) observations. ■

Remark 10.4 (Pre-trained Word Embeddings) In practical applications we rely on pre-trained word embeddings. For GloVe there are pre-trained versions that can be downloaded.[4] These pre-trained versions comprise a vocabulary of 400K words, and they exist for the embedding dimensions $b = 50, 100, 200, 300$. These GloVe's have been trained on Wikipedia 2014 and Gigaword 5 which provided roughly 6B tokens. Another pre-trained open-source model that can be downloaded is spaCy.[5]

4 https://nlp.stanford.edu/projects/glove/.

5 https://spacy.io/models/en#en_core_web_md.

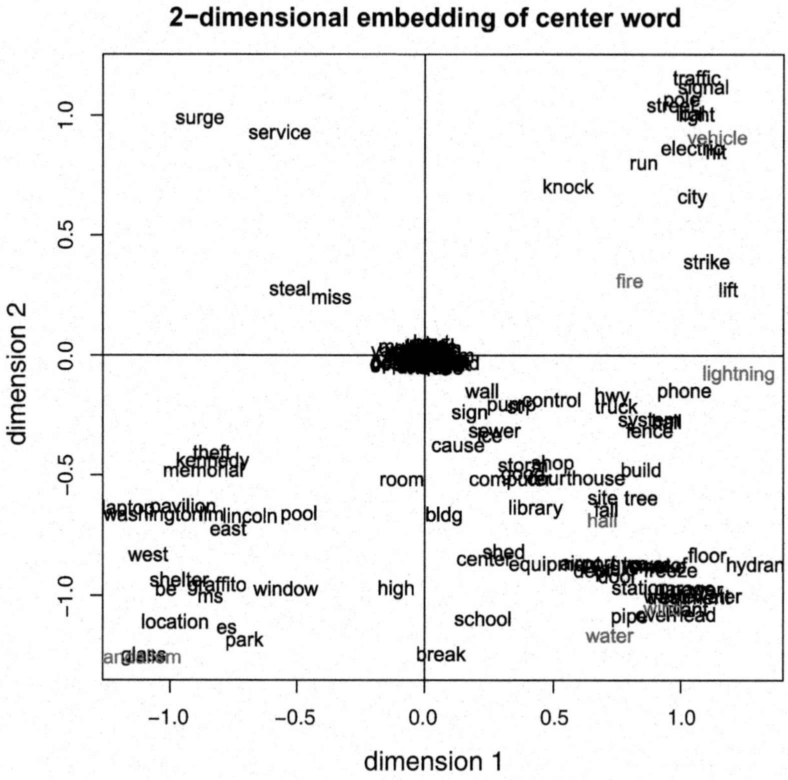

Fig. 10.4 Two-dimensional GloVe embedding; in red color are the insured hazards 'vehicle', 'fire', 'lightning', 'wind', 'hail', 'water' and 'vandalism'

Pre-trained embeddings can be problematic if we work in very specific settings. For instance, the Wisconsin LGPIF data contains the word 'Lincoln' in the claim descriptions. Now, Lincoln is a county in Wisconsin, it is town in Kewaunee County in Wisconsin, it is a former US president, there are Lincoln memorials, it is a common street name, it is a car brand and there are restaurants with this name. In our context, Lincoln is most commonly used w.r.t. the Lincoln Elementary and Middle Schools. On the other hand, it is likely that in pre-trained embeddings a different meaning of Lincoln is predominant, and therefore the embedding may not be reasonable for our insurance problem.

10.3 Lab: Predictive Modeling Using Word Embeddings

This section gives an example of applying the word embedding technique to a predictive modeling setting. This example is based on the Wisconsin LGPIF data set illustrated in Listing 10.1. Our goal is to predict the hazard types on line 10 of Listing 10.1 from the claim descriptions on line 11. We perform the same data cleaning process as in Example 10.2. This provides us with $W = 1'829$ different words, and the resulting (short) claim descriptions have a maximal length of $T = 9$. After padding with zeros we receive $n = 6'031$ claim descriptions given by texts $(w_1, \ldots, w_T)^\top \in \mathcal{W}_0^T$; we apply the padding to the left end of the sentences.

Word2vec Using Negative Sampling We start by the word2vec embedding technique using the negative sampling. We follow Example 10.2, and to successfully embed the available words $w \in \mathcal{W}$ we restrict the vocabulary to the words that are used at least 20 times. This reduces the vocabulary from 1'892 different words to 142 different words. The number of claim descriptions are reduced to 5'883 because 148 claim descriptions do not contain any of these 142 different words and, thus, cannot be classified as one of the hazard types (based on this reduced vocabulary).

In a first analysis we choose the embedding dimension $b = 2$, and this provides us with the word2vec embedding map that is illustrated in Fig. 10.2. Based on these embeddings we aim at predicting the hazard types from the claim descriptions. We have 9 different hazard types: Fire, Lightning, Hail, Wind, WaterW, WaterNW, Vehicle, Vandalism and Misc.[6] Therefore, we design a categorical classification model that has 9 different labels, we refer to Sect. 2.1.4.

Listing 10.7 R code for the hazard type prediction based on a word2vec embedding

```
1  input = layer_input(shape = list(T), name = "input")
2  #
3  word2vec = input %>%
4      layer_embedding(input_dim = W+1, output_dim = b, input_length = T,
5                    weights=list(wordEmb), trainable=FALSE) %>%
6      layer_flatten()
7  # response = word2vec %>%
8      layer_dense(units=20, activation='tanh', name='FNLayer1') %>%
9      layer_dense(units=15, activation='tanh', name='FNLayer2') %>%
10     layer_dense(units=9, activation='softmax', name='output')
11 #
12 model = keras_model(inputs = c(input), outputs = c(response))
```

The R code for the hazard type prediction is presented in Listing 10.7. The crucial part is shown on line 5. Namely, the embedding map $\boldsymbol{e}(w) \in \mathbb{R}^b$, $w \in \mathcal{W}$ is initialized with the embedding weights `wordEmb` received from Example 10.2, and

[6] WaterW relates to weather related water claims, and WaterNW relates to non-weather related water claims.

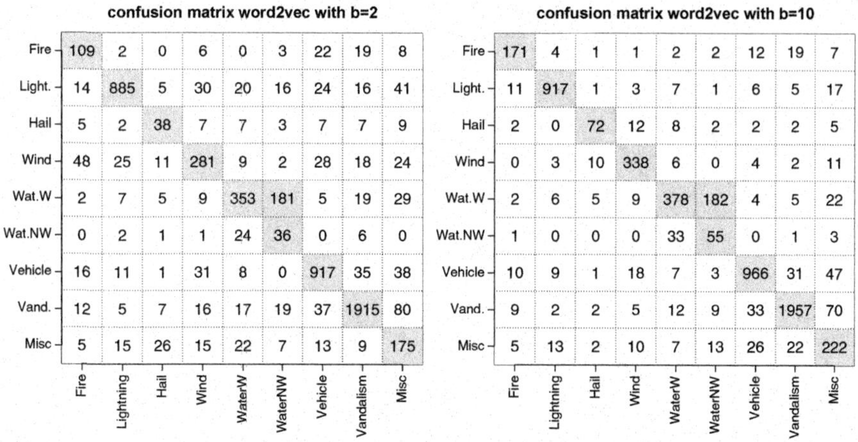

Fig. 10.5 Confusion matrices of the hazard type prediction using a word2vec embedding based on negative sampling (lhs) $b = 2$ dimensional embedding and (rhs) $b = 10$ dimensional embedding; columns show the observations and rows show the predictions

these embedding weights are declared to be non-trainable.[7] These features are then inputted into a FN network with two FN layers having $(q_1, q_2) = (20, 15)$ neurons, and as output activation we choose the softmax function. This model has 286 non-trainable embedding weights, and $r = (9 \cdot 2 + 1)20 + (20 + 1)15 + (15 + 1)9 = 839$ trainable parameters.

We fit this network using the nadam version of the gradient descent method, and we exercise an early stopping on a 20% validation data set (of the entire data). This network is fitted in a few seconds, and the results are presented in Fig. 10.5 (lhs). This figure shows the confusion matrix of prediction vs. observed (row vs. column). The general results look rather good, there are only difficulties to distinguish WaterN from WaterNW claims.

In a second analysis, we increase the embedding dimension to $b = 10$ and we perform exactly the same procedure as above. A higher embedding dimension allows the embedding map to better discriminate the words in their meanings. However, we should not go for a too high b because we have only 142 different words and 47'904 center-context pairs $(w, \widetilde{w})$ to learn these embeddings $\boldsymbol{e}(w) \in \mathbb{R}^b$. A higher embedding dimension also increases the number of network weights in the first FN layer on line 9 of Listing 10.7. This time, we need to train $r = (9 \cdot 10 + 1)20 + (20 + 1)15 + (15 + 1)9 = 2'279$ parameters. The results are presented in Fig. 10.5 (rhs). We observe an overall improvement compared to the 2-dimensional embeddings. This is also confirmed by Table 10.1 which gives the deviance losses and the misclassification rates.

[7] The zeros from padding are mapped to the origin.

Table 10.1 Hazard prediction results summarized in deviance losses and misclassification rates

	Number of parameters		Deviance	Misclassification
	Embedding	Network	loss	rate
word2vec negative sampling, $b = 2$	286	839	0.1442	19.9%
word2vec negative sampling, $b = 10$	1'430	2'279	0.0912	13.7%
FN GloVe using all words, $b = 50$	91'500	9'479	0.0802	11.7%
LSTM GloVe using all words, $b = 50$	91'500	3'369	0.0802	12.1%
Word similarity embedding, $b = 7$	12'810	1'739	0.1396	21.1%

Pre-trained GloVe Embedding In a next analysis we use the pre-trained GloVe embeddings, see Remark 10.4. This allows us to use all $W = 1'892$ words that appear in the $n = 6'031$ claim descriptions, and we can also classify all these claims. I.e., we can classify more claims, here, compared to the 5'883 claims we have classified based on the self-trained word2vec embeddings. Apart from that, all modeling steps are chosen as above. Only the higher embedding dimension $b = 50$ from the pre-trained `glove.6B.50d` increases the size of the network parameter to $r = (9 \cdot 50 + 1)20 + (20 + 1)15 + (15 + 1)9 = 9'479$ parameters; remark that the 91'500 embedding weights are not trained as they come from the pre-trained GloVe embeddings. Using the `nadam` optimizer with an early stopping provides us with the results in Fig. 10.6 (lhs). Using this pre-trained GloVe embedding leads to a further improvement, this is also verified by Table 10.1. Using the pre-trained GloVe is two-fold. On the one hand, it allows us to use all words of the claim descriptions, which improves the prediction accuracy. On the other hand, the embeddings are not adapted to insurance problems, as these have been trained on Wikipedia and Gigaword texts. The former advantage overrules the latter shortcoming in our example.

All the results above have been using the FN network of Listing 10.7. We made this choice because our texts have a maximal length of $T = 9$, which is very short. In general, texts should be understood as time-series, and RN networks are a canonical choice to analyze these time-series. Therefore, we study again the pre-trained GloVe embeddings, but we process the texts with a LSTM architecture, we refer to Sect. 8.3.1 for LSTM layers.
Listing 10.8 shows the LSTM architecture used. On line 9 we set the variable `return_sequences` to true which implies that all intermediate steps $z_t^{[1]}$, $1 \le t \le T$, are outputted to a time-distributed FN layer on line 10, see Sect. 8.2.4 for time-distributed layers. This LSTM network has $r = 4(50 + 1 + 10)10 + (10 + 1)10 + (90 + 1)9 = 3'369$ parameters. The flatten layer on line 11 of Listing 10.8 turns the $T = 9$ outputs $z_t^{[2]} \in \mathbb{R}^{q_2}$, $1 \le t \le T$, of dimension $q_2 = 10$ into a vector of size $Tq_2 = 90$. This vector is then fed into the output layer on line 12. At this stage, one could reduce the dimension of the parameter by setting a max-pooling layer in between the flatten and the output layer.

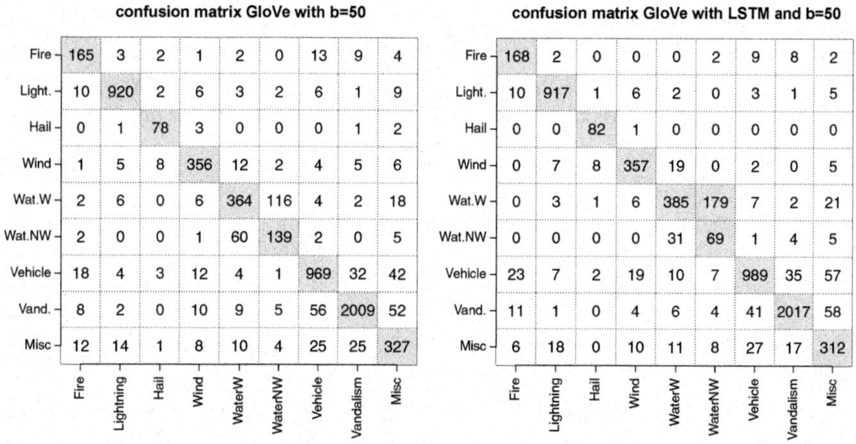

Fig. 10.6 Confusion matrices of the hazard type prediction using the pre-trained GloVe with $b = 50$ (lhs) FN network and (rhs) LSTM network; columns show the observations and rows show the predictions

Listing 10.8 R code for the hazard type prediction using a LSTM architecture

```
input = layer_input(shape = list(T), name = "input")
#
word2vec = input %>%
    layer_embedding(input_dim = W+1, output_dim = b, input_length = T,
                  weights=list(wordEmb), trainable=FALSE) %>%
    layer_flatten()
#
response = word2vec %>%
 layer_lstm(units=10, activation='tanh', return_sequences=TRUE,
                                        name='LSTM') %>%
 time_distributed(layer_dense(units=10, activation='tanh', name='FNLayer')) %>%
 layer_flatten() %>%
 layer_dense(units=9, activation='softmax', name='output')
#
model = keras_model(inputs = c(input), outputs = c(response))
```

We fit this LSTM architecture to the data using the pre-trained GloVe embeddings. The results are presented in Fig. 10.6 (rhs) and Table 10.1. We receive the same deviance loss, and the misclassification rate is slightly worse than in the FN network case (with the same pre-trained GloVe embeddings). Note that the deviance loss is calculated on the estimated classification probabilities $\widehat{\boldsymbol{p}}(\boldsymbol{x}) = (\widehat{p}_1(\boldsymbol{x}), \ldots, \widehat{p}_9(\boldsymbol{x}))^\top$, and the labels are received by

$$\widehat{Y} = \widehat{Y}(\boldsymbol{x}) = \underset{k=1,\ldots,9}{\arg\max}\ \widehat{p}_k(\boldsymbol{x}).$$

Thus, it may happen that the improvements on the estimated probabilities are not fully reflected on the predicted labels.

Word (Cosine) Similarity In our final analysis we work with the pre-trained GloVe embeddings $\boldsymbol{e}(w) \in \mathbb{R}^{50}$ but we first try to reduce the embedding dimension b. For this we follow Lee et al. [236], and we consider a *word similarity*. We can define the similarity of the words w and $w' \in \mathcal{W}$ by considering the scalar product of their embeddings

$$\text{sim}^{(u)}(w, w') = \langle \boldsymbol{e}(w), \boldsymbol{e}(w') \rangle \qquad \text{or} \qquad \text{sim}^{(n)}(w, w') = \frac{\langle \boldsymbol{e}(w), \boldsymbol{e}(w') \rangle}{\|\boldsymbol{e}(w)\|_2 \|\boldsymbol{e}(w')\|_2}. \tag{10.11}$$

The first one is an unweighted version and the second one is a normalized version scaling with the corresponding Euclidean norms so that the similarity measure is within $[-1, 1]$. In fact, the latter is also called cosine similarity. To reduce the embedding dimension and because we have a classification problem with hazard names, we can evaluate the (cosine) similarity of all used words $w \in \mathcal{W}$ to the hazards $h \in \mathcal{H} = \{\texttt{fire}, \texttt{lightning}, \texttt{hail}, \texttt{wind}, \texttt{water}, \texttt{vehicle}, \texttt{vandalism}\}$. Observe that `water` is further separated into weather related and non-weather related claims, and there is a further hazard type called `misc`, which collects all the rest. We could choose more words in $\mathcal{H}$ to more precisely describe these water and other claims. If we just use $\mathcal{H}$ we obtain a $b = |\mathcal{H}| = 7$ dimensional embedding mapping

$$w \in \mathcal{W}_0 \mapsto \boldsymbol{e}^{(a)}(w) = \left(\text{sim}^{(a)}(w, \texttt{fire}), \ldots, \text{sim}^{(a)}(w, \texttt{vandalism})\right)^\top \in \mathbb{R}^{b=7}, \tag{10.12}$$

for $a \in \{u, n\}$. This gives us for every $\texttt{text} = (w_1, \ldots, w_T)^\top \in \mathcal{W}_0^T$ the pre-processed features

$$\texttt{text} \mapsto \left(\boldsymbol{e}^{(a)}(w_1), \ldots, \boldsymbol{e}^{(a)}(w_T)\right)^\top \in \mathbb{R}^{T \times b}. \tag{10.13}$$

Lee et al. [236] apply a max-pooling layer to these embeddings which are then inputted into GAM classification model. We use a different approach here, and directly use the unweighted ($a = u$) text representations (10.13) as an input to a network, either of FN network type of Listing 10.7 or of LSTM type of Listing 10.8. If we use the FN network type we receive the results on the last line of Table 10.1 and Fig. 10.7.

Comparing the results of the word similarity through the embeddings (10.12) and (10.13) to the other prediction results, we conclude that this word similarity approach is not fully competitive compared to working directly with the word2vec or GloVe embeddings. It seems that the projection (10.12) does not discriminate sufficiently for our classification task.

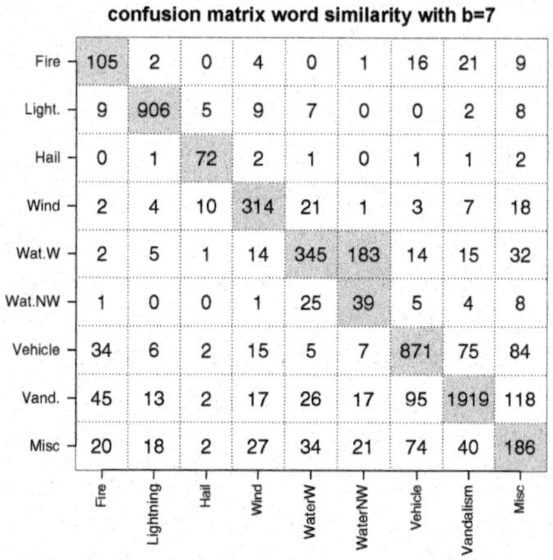

Fig. 10.7 Confusion matrix of the hazard type prediction using the word similarity (10.12)–(10.13) for $a = u$; columns show the observations and rows show the predictions

10.4 Lab: Deep Word Representation Learning

All examples above have been relying on embedding the words $w \in \mathcal{W}$ into a Euclidean space $\boldsymbol{e}(w) \in \mathbb{R}^b$ by performing a sort of unsupervised learning that provided word similarity clusters. The advantage of this approach is that the embedding is decoupled from the regression or classification task, this is computationally attractive. Moreover, once a suitable embedding has been learned, it can be used for several different tasks (in the spirit of transfer learning). The disadvantage of the pre-trained embeddings is that the embedding is not targeted to the regression task at hand. This has already been discussed in Remark 10.4 where we have highlighted that the meaning of some words (such as Lincoln) depends very much on its context.

Recent NLP aims at pre-processing a text as little as necessary, but tries to directly feed the raw sentences into RN networks such as LSTM or GRU architectures. Computationally this is much more demanding because we have to learn the embeddings and the network weights simultaneously, we refer to Table 10.1 to indicate the number of parameters involved. The purpose of this short section is to give an example, though our NLP database is rather small; this latter approach usually requires a huge database and the corresponding computational power. Ferrario–Nägelin [126] provide a more comprehensive example on the classification of movie reviews. For their analysis they evaluated approximately 50'000 movie reviews each using between 235 and 2'498 words. Their analysis was implemented on the ETH High Performance Computing (HPC) infrastructure Euler[8], and their run times have been between 20 and 30 minutes, see Table 8 of Ferrario–Nägelin [126].

[8] https://scicomp.ethz.ch/wiki/Euler

Since we neither have the computational power nor the big data to fit such a NLP application, we start the gradient descent fitting in the initial embedding weights $\boldsymbol{e}(w) \in \mathbb{R}^b$ that either come from the word2vec or the GloVe embeddings. During the gradient descent fitting, we allow these weights to change w.r.t. the regression task at hand. In comparison to Sect. 10.3, this only requires minor changes to the R code, namely, the only modification needed is to change from `FALSE` to `TRUE` on lines 5 in Listings 10.7 and 10.8. This change allows us to learn adapted weights during the gradient descent fitting. The resulting classification models are now very high-dimensional, and we need to carefully assess the early stopping rule, otherwise the model will (in-sample) over-fit to the learning data.

In Fig. 10.8 we provide the results that correspond to the self-trained word2vec embeddings given in Fig. 10.5, and the corresponding numerical results are given in Table 10.2. We observe an improvement in the prediction accuracy in both cases by letting the embedding weights being learned during the network fitting, and we receive a misclassification rate of 11.6% and 11.0% for the embedding dimensions $b = 2$ and $b = 10$, respectively, see Table 10.2.

Figure 10.8 (rhs) illustrates how the embeddings have changed from the initial (pre-trained) embeddings $\boldsymbol{e}^{(0)}(w)$ (coming from the word2vec negative sampling) to the learned embeddings $\widehat{\boldsymbol{e}}(w)$. We measure these changes in terms of the unweighted similarity measure defined in (10.11), and given by

$$\left\langle \boldsymbol{e}^{(0)}(w), \widehat{\boldsymbol{e}}(w) \right\rangle. \tag{10.14}$$

The upper horizontal line is a manually set threshold to identify the words w that experience a major change in their embeddings. These are the words 'vandalism', 'lightning', 'grafito', 'fence', 'hail', 'freeze', 'blow' and 'breakage'. Thus, these words receive a different embedding location/meaning which is more favorable for our classification task.

A similar analysis can be performed for the pre-trained GloVe embeddings. There we expected bigger changes to the embeddings since the GloVe embeddings have not been learned in an insurance context, and the embeddings will be adapted to the insurance prediction problem. We refrain from giving an explicit analysis, here, because to perform a thorough analysis we would need (much) more data.

We conclude this example with some remarks. We emphasize once more that our available data is minimal, and we expect (even much) better results for longer claim descriptions. In particular, our data is not sufficient to discriminate the weather related from the non-weather related water claims, as the claim descriptions seem to focus on the water claim itself and not on its cause. In a next step, one should use claim descriptions in order to predict the claim sizes, or to improve their predictions if they are based on classical tabular features, only. Here, we see some potential, in particular, w.r.t. medical claims, as medical reports may clearly indicate the severity of the claim as well as these reports may give some insight into the recovery process. Thus, our small example may only give some intuition of what is possible with

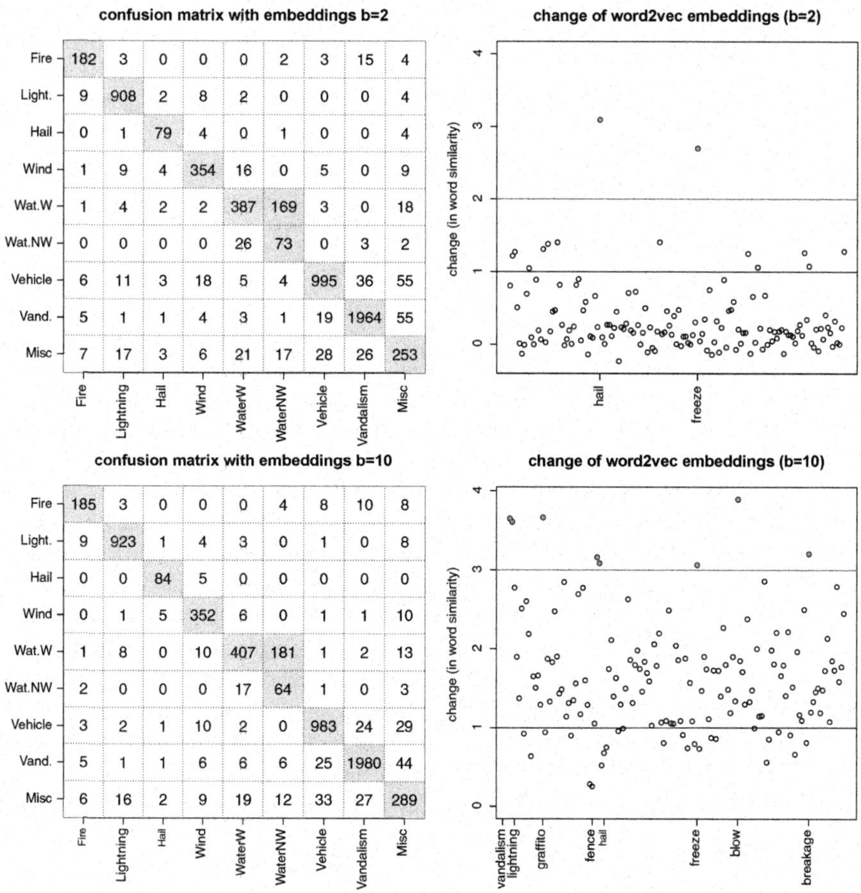

Fig. 10.8 Confusion matrices and the changes in the embeddings compared to the pre-trained word2vec embeddings of Fig. 10.5 for the dimensions $b = 2$ and $b = 10$

Table 10.2 Hazard prediction results summarized in deviance losses and misclassification rates: pre-trained embeddings vs. network learned embeddings

	Number of parameters		Deviance	Misclass.
	Non-trainable	Trainable	loss	rate
word2vec negative sampling, $b = 2$	286	839	0.1442	19.9%
word2vec improved embedding, $b = 2$		1'125	0.0814	11.7%
word2vec negative sampling, $b = 10$	1'430	2'279	0.0912	13.7%
word2vec improved embedding, $b = 10$		3'709	0.0714	10.5%

(unstructured) text data. Unfortunately, the LGPIF data of Listing 10.1 did not give us any satisfactory results for the claim size prediction, this for several reasons. Firstly, the data is rather heterogeneous ranging from small to very large claims and any member of the EDF struggles to model this data; we come back to a different modeling proposal of heterogeneous data in Sect. 11.3.2. Secondly, the claim descriptions are not very explanatory as they are too short for a more detailed information. Thirdly, the data has only 5'424 claims which seems small compared to the complexity of the problem that we try to solve.

10.5 Outlook: Creating Attention

In text recognition problems, obviously, not all the words in a sentence have the same importance. In the examples above, we have removed the stopwords as they may disturb the key understanding of our texts. Removing the stopwords means that we pay more attention to the remaining words. RN networks often face difficulty in giving the right recognition to the different parts of a sentence. For this reason, *attention layers* have gained more popularity recently. Attention layers are special modules in network architectures that allow the network to impose more weight on certain parts of the information in the features to emphasize their importance. The attention mechanism has been introduced in Bahdanau et al. [21]. There are different ways of modeling attention, the most popular one is the so-called *dot-product attention*, we refer to Vaswani et al. [366], and in the actuarial literature we mention Kuo–Richman [231] and Troxler–Schelldorfer [354].

We start by describing a simple attention mechanism. Consider a sentence $\texttt{text} = (w_1, \ldots, w_T) \in \mathcal{W}_0^T$ that provides, under an embedding map $\boldsymbol{e} : \mathcal{W}_0 \to \mathbb{R}^b$, the embedded sentence $(\boldsymbol{e}(w_1), \ldots, \boldsymbol{e}(w_T))^\top \in \mathbb{R}^{T\times b}$. We choose a weight matrix $U_Q \in \mathbb{R}^{b\times b}$ and an intercept vector $\boldsymbol{u}_Q \in \mathbb{R}^b$. Based on these choices we consider for each word w_t of our sentence the score, called *query*,

$$\boldsymbol{q}_t = \tanh\left(\boldsymbol{u}_Q + U_Q \boldsymbol{e}(w_t)\right) \in (-1, 1)^b. \tag{10.15}$$

Matrix $Q = (\boldsymbol{q}_1, \ldots, \boldsymbol{q}_T)^\top \in \mathbb{R}^{T\times b}$ collects all queries. It is obtained by applying a time-distributed FN layer with b neurons to the embedded sentence $(\boldsymbol{e}(w_1), \ldots, \boldsymbol{e}(w_T))^\top$.

These queries $\boldsymbol{q}_t$ are evaluated with a so-called *key* $\boldsymbol{k} \in \mathbb{R}^b$ giving us the *attention weights*

$$\alpha_t = \frac{\exp\langle \boldsymbol{k}, \boldsymbol{q}_t\rangle}{\sum_{s=1}^T \exp\langle \boldsymbol{k}, \boldsymbol{q}_s\rangle} \in (0, 1) \qquad \text{for } 1 \le t \le T. \tag{10.16}$$

Using these attention weights $\boldsymbol{\alpha} = (\alpha_1, \ldots, \alpha_T)^\top \in (0, 1)^T$ we encode the sentence `text` as

$$\texttt{text} = (w_1, \ldots, w_T) \mapsto \boldsymbol{w}^* = \sum_{t=1}^{T} \alpha_t \boldsymbol{e}(w_t) \tag{10.17}$$
$$= (\boldsymbol{e}(w_1), \ldots, \boldsymbol{e}(w_T))\, \boldsymbol{\alpha} \in \mathbb{R}^b.$$

Thus, to every sentence `text` we assign a categorical probability vector $\boldsymbol{\alpha} = \boldsymbol{\alpha}(\texttt{text}) \in \Delta_T$, see Sect. 2.1.4, (6.22) and (5.69), which is encoding this sentence `text` to a b-dimensional vector $\boldsymbol{w}^* \in \mathbb{R}^b$. This vector is then further processed by the network. Such a construction is called a *self-attention mechanism* because the text $(w_1, \ldots, w_T) \in \mathcal{W}_0^T$ is used to formulate the queries in (10.15), but, of course, these queries could also be coming from a completely different source. In the above set-up we have to learn the following parameters $U_Q \in \mathbb{R}^{b \times b}$ and $\boldsymbol{u}_Q, \boldsymbol{k} \in \mathbb{R}^b$, assuming that the embedding map $\boldsymbol{e} : \mathcal{W}_0 \to \mathbb{R}^b$ has already been specified.

There are several generalizations and modifications to this self-attention mechanism. The most common one is to expand the vector $\boldsymbol{w}^* \in \mathbb{R}^b$ in (10.17) to a matrix $W^* = (\boldsymbol{w}_1^*, \ldots, \boldsymbol{w}_q^*) \in \mathbb{R}^{b \times q}$. This matrix W^* can be interpreted as having q neurons $\boldsymbol{w}_j^* \in \mathbb{R}^b$, $1 \le j \le q$. For this, one replaces the key $\boldsymbol{k} \in \mathbb{R}^b$ by a matrix-valued key $K = (\boldsymbol{k}_1, \ldots, \boldsymbol{k}_q) \in \mathbb{R}^{b \times q}$. This allows one to calculate the attention weight matrix

$$A = \left(\alpha_{t,j}\right)_{1 \le t \le T, 1 \le j \le q} = \left(\frac{\exp\langle \boldsymbol{k}_j, \boldsymbol{q}_t \rangle}{\sum_{s=1}^{T} \exp\langle \boldsymbol{k}_j, \boldsymbol{q}_s \rangle} \right)_{1 \le t \le T, 1 \le j \le q}$$
$$= \operatorname{softmax}(QK) \in (0, 1)^{T \times q},$$

where the softmax function is applied column-wise. I.e., the attention weight matrix $A \in (0, 1)^{T \times q}$ has columns $\boldsymbol{\alpha}_j = (\alpha_{1,j}, \ldots, \alpha_{T,j})^\top \in \Delta_T$, $1 \le j \le q$, which are normalized to total weight 1, this is equivalent to (10.16). This is used to encode the sentence `text`

$$(\boldsymbol{e}(w_1), \ldots, \boldsymbol{e}(w_T)) \in \mathbb{R}^{b \times T} \mapsto W^* = (\boldsymbol{e}(w_1), \ldots, \boldsymbol{e}(w_T))\, A \tag{10.18}$$
$$= \left(\sum_{t=1}^{T} \alpha_{t,j} \boldsymbol{e}(w_t) \right)_{1 \le j \le q} \in \mathbb{R}^{b \times q}.$$

Mapping (10.18) is called an *attention layer*. Let us give some remarks.

Remarks 10.5

- Encoding (10.18) gives a natural multi-dimensional extension of (10.17). The crucial parts are the attention weights $\boldsymbol{\alpha}_j \in \Delta_T$ which weigh the different words

$(w_t)_{1\le t\le T}$. In the multi-dimensional case, we perform this weighting mechanism multiple times (in different directions), allowing us to extract different features from the sentences. In contrast, in (10.17) we only do this once. This is similar as going form one neuron to a layer of q neurons.

- The above structure uses a self-attention mechanism because the queries involve the words themselves, and the weight matrix $U_Q \in \mathbb{R}^{b\times b}$ and the intercept vector $\boldsymbol{u}_Q \in \mathbb{R}^b$ are learned with gradient descent. Concerning the key $K \in \mathbb{R}^{b\times q}$ one often chooses another self-attention mechanism by choosing a (non-linear) function $K = K(w_1, \ldots, w_T)$ to infer optimal keys.
- These attention layers are also the building blocks of *transformer models*. Transformer models use attention layers (10.18) of dimension $W^* \in \mathbb{R}^{b\times T}$ and skip connections to transform the input

$$W = (\boldsymbol{e}(w_1), \ldots, \boldsymbol{e}(w_T)) \in \mathbb{R}^{b\times T} \mapsto \frac{W + W^*}{2} \in \mathbb{R}^{b\times T}. \tag{10.19}$$

  Stacking multiple of these layers (10.19) transforms the original input W by weighing the important information in feature W for the prediction task at hand. Compared to LSTM layers this no longer sequentially screens the text but it directly acts on the part of the text that seems important.
- The attention mechanism is applied to a matrix $(\boldsymbol{e}(w_1), \ldots, \boldsymbol{e}(w_T))^\top \in \mathbb{R}^{T\times b}$ which presents a numerical encoding of the sentence $(w_1, \ldots, w_T)^\top \in \mathcal{W}_0^T$. Kuo–Richman [231] propose to apply this attention mechanism more generally to categorical feature components. Assume that we have T categorical feature components $x_1, \ldots, x_T$, after embedding them into b-dimensional Euclidean spaces we receive a representation $(\boldsymbol{e}(x_1), \ldots, \boldsymbol{e}(x_T))^\top \in \mathbb{R}^{T\times b}$, see (7.31). Naturally, this can now be further processed by putting different attention on the components of this embedding exactly using an attention layer (10.18), alternatively we can use transformer layers (10.19).

Example 10.6 We revisit the hazard type prediction example of Sect. 10.3. We select the $b = 10$ word2vec embedding (using negative sampling) and the pre-trained GloVe embedding of Table 10.1. These embeddings are then further processed by applying the attention mechanism (10.15)–(10.17) on the embeddings using one single attention neuron. Listing 10.9 gives the corresponding implementation. On line 9 we have the query (10.15), on lines 10–13 the key and the attention weights (10.16), and on line 15 the encodings (10.17). We then process these encodings through a FN network of depth $d = 2$, and we use the softmax output activation to receive the categorical probabilities. Note that we keep the learned word embeddings $\boldsymbol{e}(w)$ as non-trainable on line 5 of Listing 10.9.
Table 10.3 gives the results, and Fig. 10.9 shows the confusion matrix. We conclude that the results are rather similar, this attention mechanism seems to work quite well, and with less parameters, here. ■

Listing 10.9 R code for the hazard type prediction using an attention layer with $q = 1$

```
input = layer_input(shape = list(T), name = "input")
#
word2vec = input %>%
    layer_embedding(input_dim = W+1, output_dim = b, input_length = T,
                  weights=list(wordEmb), trainable=FALSE) %>%
    layer_flatten()
#
attention = word2vec %>%
 time_distributed(layer_dense(units=b, activation='tanh')) %>%
 time_distributed(layer_dense(units=1, activation='linear',
                                     use_bias=FALSE)) %>%
 layer_flatten() %>%
 layer_dense(unit=T, activation='softmax', weights=list(diag(T)),
                                         use_bias=FALSE, trainable=FALSE)
#
response = list(attention, word2vec) %>% layer_dot(axes=1) %>%
          layer_dense(units=20, activation='tanh') %>%
          layer_dense(units=15, activation='tanh') %>%
          layer_dense(units=9, activation='softmax')
#
model = keras_model(inputs = c(input), outputs = c(response))
```

Table 10.3 Hazard prediction results summarized in deviance losses and misclassification rates

	Number of parameters		Deviance	Misclassification
	Embedding	Network	loss	rate
word2vec negative sampling, $b = 10$	1'430	2'279	0.0912	13.7%
word2vec attention, $b = 10$	1'430	799	0.0784	12.0%
FN GloVe using all words, $b = 50$	91'500	9'479	0.0802	11.7%
GloVe attention, $b = 50$	91'500	4'079	0.0824	12.6%

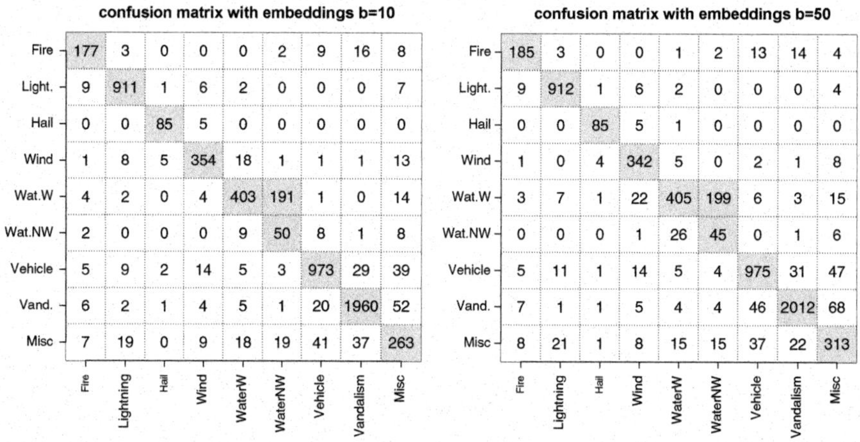

Fig. 10.9 Confusion matrices of the hazard type prediction (lhs) using an attention layer on the word2vec embeddings with $b = 10$, and (rhs) using an attention layer on the pre-trained GloVe embeddings with $b = 50$; columns show the observations and rows show the predictions

Chapter 11
Selected Topics in Deep Learning

11.1 Deep Learning Under Model Uncertainty

We revisit claim size modeling in this section. Claim size modeling is challenging because often there is no (simple) off-the-shelf distribution that allows one to appropriately describe all claim size observations. E.g., the main body of the claim size data may look like gamma distributed, and, at the same time, large claims seem to be more heavy-tailed (contradicting a gamma model assumption). Moreover, different product and claim types may lead to multi-modality in the claim size densities. In Sects. 5.3.7 and 5.3.8 we have explored a gamma and an inverse Gaussian GLM to model a motorcycle claims data set. In that example, the results have been satisfactory because this motorcycle data is neither multi-modal nor does it have heavy tails. These two GLM approaches have been based on the EDF (2.14), modeling the mean $\boldsymbol{x} \mapsto \mu(\boldsymbol{x})$ with a regression function and assuming a constant dispersion parameter $\varphi > 0$. There are two natural ways to extend this approach. One considers a double GLM with a dispersion submodel $\boldsymbol{x} \mapsto \varphi(\boldsymbol{x})$, see Sect. 5.5, the other explores multi-parameter extensions like the generalized inverse Gaussian model, which is a $k = 3$ vector-valued EF, see (2.10), or the GB2 family that involves 4 parameters, see (5.79). These extensions provide more complexity, also in MLE. In this section, we are not going to consider multi-parameter extensions, but in a first step we aim at robustifying (mean) parameter estimation within the EDF. In a second step we are going to analyze the resulting dispersion $\varphi(\boldsymbol{x})$. For these steps, we perform representation learning and parameter estimation under model uncertainty by simultaneously considering multiple models from Tweedie's family. These considerations are closely related to Tweedie's forecast dominance given in Definition 4.22.

M. V. Wüthrich, M. Merz, *Statistical Foundations of Actuarial Learning and its Applications*, Springer Actuarial, https://doi.org/10.1007/978-3-031-12409-9_11

We emphasize that we remain within a single distribution function choice in this section, i.e., we neither consider mixture distributions nor composite models in this section. Mixture density networks are going to be considered in Sect. 11.6, below, and a composite model approach is studied in Sect. 11.3, below. These mixture density networks and composite models allow us to model the body and the tail of the data with different distribution functions by either mixing or concatenating suitable distributions.

11.1.1 Recap: Tweedie's Family

Tweedie's family with power variance function $V(\mu) = \mu^p$, $p \geq 2$, provides us with a rich model class for claim size modeling if the claim sizes are strictly positive, a.s., and extending to $p \in (1, 2)$ allows us to model claims with a positive point mass in 0. This class of distribution functions contains the gamma case ($p = 2$) and the inverse Gaussian case ($p = 3$). In general, $p > 2$ provides us with positive stable generated distributions and $p \in (1, 2)$ gives Tweedie's CP models, see Table 2.1. Tweedie's family has cumulant function for $p > 1$

$$\kappa(\theta) = \kappa_p(\theta) = \begin{cases} \frac{1}{2-p}\left((1-p)\theta\right)^{\frac{2-p}{1-p}} & \text{for } p > 1 \text{ and } p \neq 2, \\ -\log(-\theta) & \text{for } p = 2, \end{cases} \tag{11.1}$$

on the effective domain $\theta \in \boldsymbol{\Theta} \in (-\infty, 0)$ for $p \in (1, 2]$, and $\theta \in \boldsymbol{\Theta} \in (-\infty, 0]$ for $p > 2$. The mean and the power variance function are for $p > 1$ given by

$$\theta \mapsto \mu = \mu(\theta) = \left((1-p)\theta\right)^{\frac{1}{1-p}} \qquad \text{and} \qquad \mu \mapsto V(\mu) = \mu^p.$$

The unit deviance takes the following form for $p > 1$ and $p \neq 2$, see (4.18),

$$\mathfrak{d}_p(y, \mu) = 2\left(y\frac{y^{1-p} - \mu^{1-p}}{1-p} - \frac{y^{2-p} - \mu^{2-p}}{2-p}\right) \geq 0, \tag{11.2}$$

and in the gamma case $p = 2$ we have, see Table 4.1,

$$\mathfrak{d}_2(y, \mu) = 2\left(\frac{y}{\mu} - 1 + \log\left(\frac{\mu}{y}\right)\right) \geq 0. \tag{11.3}$$

Figure 11.1 (lhs) shows the unit deviances $y \mapsto \mathfrak{d}_p(y, \mu)$ for fixed mean parameter $\mu = 2$ and power variance parameters $p \in \{0, 2, 2.5, 3, 3.5\}$; the case $p = 0$ corresponds to the symmetric Gaussian case $\mathfrak{d}_0(y, \mu) = (y - \mu)^2$. We observe that with an increasing power variance parameter p large claims $Y = y$ receive a smaller loss punishment (if we interpret the unit deviance as a loss function). This is the situation where we have a fixed mean μ and where we assess claim sizes

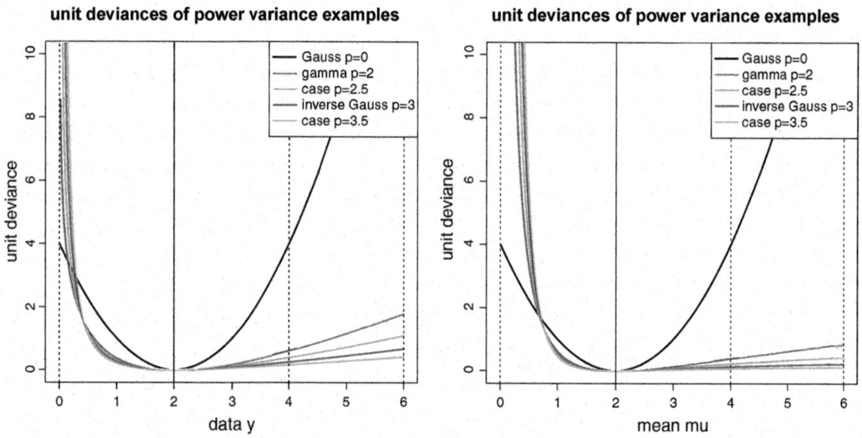

Fig. 11.1 (lhs) Unit deviances $y \mapsto \mathfrak{d}_p(y, \mu) \geq 0$ for fixed mean $\mu = 2$ and (rhs) unit deviances $\mu \mapsto \mathfrak{d}_p(y, \mu) \geq 0$ for fixed observation $y = 2$ for power variance parameters $p \in \{0, 2, 2.5, 3, 3.5\}$

$Y = y$ relative to this mean. For estimation purposes we have fixed observations $Y = y$ and we study the sensitivities in μ. Note that, in general, the unit deviances $\mathfrak{d}_p(y, \mu)$ are not symmetric in y and μ. This second case is shown in Fig. 11.1 (rhs), and the general behavior in p is similar. As a result, by selecting different hyper-parameters $p > 1$, we can control the influence of large (and small) claims on parameter estimation, because the unit deviances $\mathfrak{d}_p(y, \cdot)$ have different slopes for different p's. Basically, the choice of the loss function (unit deviance) determines the choice of the underlying distributional model, which then assesses the claim observations $Y = y$ according to their sizes and how these sizes match the model assumptions made.

In Lemma 2.22 we have seen that the unit deviances $\mathfrak{d}_p(y, \mu) \geq 0$ are zero if and only if $y = \mu$. The second derivatives given in Lemma 2.22 allow us to consider a second order Taylor expansion around a minimum $\mu_0 = y_0$

$$\mathfrak{d}_p(y_0 + \epsilon y, \mu_0 + \epsilon \mu) = \frac{\epsilon^2}{\mu_0^p}(y - \mu)^2 + o(\epsilon^2) \qquad \text{as } \epsilon \to 0.$$

Thus, locally around the minimum the unit deviances behave symmetric and like Gaussian squares, but this is only a local approximation around a minimum $\mu_0 = y_0$ as can be seen from Fig. 11.1. I.e., in general, model fitting turns out to be rather different from the Gaussian square loss if we have small and large claim sizes under choices $p > 1$.

Remarks 11.1

- Since unit deviances are Bregman divergences, we know that every unit deviance gives us a strictly consistent scoring function for the mean functional, see Theorem 4.19. Therefore, the specific choice of the power variance parameter p seems less relevant. However, strict consistency is an asymptotic statement, and choosing a unit deviance that matches the property of the data has better finite sample properties, i.e., a smaller variance in asymptotic normality; we come back to this in Sect. 11.1.4, below.
- A function $(y, \mu) \mapsto \psi(y, \mu)$ is called b-homogeneous if there exists $b \in \mathbb{R}$ such that for all (y, μ) and all $\lambda > 0$ we have $\psi(\lambda y, \lambda \mu) = \lambda^b \psi(y, \mu)$. Unit deviances $\mathfrak{d}_p$ are b-homogeneous with $b = 2 - p$. This b-homogeneity has the nice consequence that the decisions taken are independent of the scale, i.e., we have an invariance under changes of currencies. On the other hand, such a scaling influences the estimation of the dispersion parameter, i.e., if we scale the observation and the mean with λ we have unit deviance

$$\mathfrak{d}_p(\lambda y, \lambda \mu) = \lambda^{2-p}\, \mathfrak{d}_p(y, \mu). \tag{11.4}$$

  This influences the dispersion estimation for the cases different from the gamma case $p = 2$, see, e.g., saddlepoint approximation (5.60)–(5.62). This also relates to the different parametrizations in Sect. 5.3.8 where we study the inverse Gaussian model $p = 3$, which has a dispersion $\varphi_i = 1/\alpha_i$ in the reproductive form and $\varphi_i = 1/\alpha_i^2$ in parametrization (5.51).
- We only consider power variance parameters $p > 1$ in this section for non-negative claim size modeling. Technically, this analysis could be extended to $p \in \{0, 1\}$. We do not consider the Gaussian case $p = 0$ to exclude negative claims, and we do not consider the Poisson case $p = 1$ because this is used for claim counts modeling.

We recall that unit deviances of the EDF are equal to twice the corresponding KL divergences, which in turn are special cases of Bregman divergences. From Theorem 4.19 we know that Bregman divergences D_ψ are the only strictly consistent loss/scoring functions for mean estimation.

Lemma 11.2 *Choose $p > 1$. The scaled unit deviance $\mathfrak{d}_p(y, \mu)/2$ is a Bregman divergence $D_{\psi_p}(y, \mu)$ on $\mathbb{R}_+ \times \mathbb{R}_+$ with strictly decreasing and strictly convex* *function on* $\mathbb{R}_+$

$$\psi_p(y) = y h_p(y) - \kappa_p(h_p(y)) = \begin{cases} \frac{1}{(2-p)(1-p)} y^{2-p} & \text{for } p > 1 \text{ and } p \neq 2, \\ -1 - \log(y) & \text{for } p = 2, \end{cases}$$

for canonical link $h_p(y) = (\kappa_p')^{-1}(y) = y^{1-p}/(1-p)$.

Proof of Lemma 11.2 The Bregman divergence property follows from (2.29). For $p > 1$ and $y > 0$ we have the strictly decreasing property

$$\psi_p'(y) = h_p(y) = y^{1-p}/(1-p) < 0.$$

The second derivative is $\psi_p''(y) = h_p'(y) = y^{-p} = 1/V(y) > 0$ which provides the strict convexity. □

In the Gaussian case we have $\psi_0(y) = y^2/2$, and $\psi_0'(y) > 0$ on $\mathbb{R}_+$ implies that this is a strictly increasing convex function for positive claims $y > 0$. This is different to Lemma 11.2.

Assume we have independent observations $(Y_i, \boldsymbol{x}_i)$ following the same Tweedie's distribution, and with means given by $\mu_{\boldsymbol{\vartheta}}(\boldsymbol{x}_i)$ for some parameter $\boldsymbol{\vartheta}$. The M-estimator of $\boldsymbol{\vartheta}$ using this Bregman divergence is given by

$$\widehat{\boldsymbol{\vartheta}} = \underset{\boldsymbol{\vartheta}}{\arg\max}\, \ell_{\boldsymbol{Y}}(\boldsymbol{\vartheta}) = \underset{\boldsymbol{\vartheta}}{\arg\min} \sum_{i=1}^n \frac{v_i}{\varphi} D_{\psi_p}\left(Y_i, \mu_{\boldsymbol{\vartheta}}(\boldsymbol{x}_i)\right).$$

If we turn this M-estimator into a Z-estimator (supposed we have differentiability), the parameter estimate $\widehat{\boldsymbol{\vartheta}}$ is found as a solution of the score equations

$$\begin{aligned} 0 &\overset{!}{=} -\nabla_{\boldsymbol{\vartheta}} \sum_{i=1}^n \frac{v_i}{\varphi} D_{\psi_p}\left(Y_i, \mu_{\boldsymbol{\vartheta}}(\boldsymbol{x}_i)\right) \\ &= \sum_{i=1}^n \frac{v_i}{\varphi} \psi_p''(\mu_{\boldsymbol{\vartheta}}(\boldsymbol{x}_i)) \left(Y_i - \mu_{\boldsymbol{\vartheta}}(\boldsymbol{x}_i)\right) \nabla_{\boldsymbol{\vartheta}} \mu_{\boldsymbol{\vartheta}}(\boldsymbol{x}_i) \\ &= \sum_{i=1}^n \frac{v_i}{\varphi} \frac{Y_i - \mu_{\boldsymbol{\vartheta}}(\boldsymbol{x}_i)}{V(\mu_{\boldsymbol{\vartheta}}(\boldsymbol{x}_i))} \nabla_{\boldsymbol{\vartheta}} \mu_{\boldsymbol{\vartheta}}(\boldsymbol{x}_i) \qquad (11.5) \\ &= \sum_{i=1}^n \frac{v_i}{\varphi} \frac{Y_i - \mu_{\boldsymbol{\vartheta}}(\boldsymbol{x}_i)}{\mu_{\boldsymbol{\vartheta}}(\boldsymbol{x}_i)^p} \nabla_{\boldsymbol{\vartheta}} \mu_{\boldsymbol{\vartheta}}(\boldsymbol{x}_i). \end{aligned}$$

In the GLM case this exactly corresponds to (5.9). To determine the Z-estimator from (11.5), we scale the residuals $Y_i - \mu_i$ inversely proportional to the variances $V(\mu_i) = \mu_i^p$ of the chosen Tweedie's distribution. It is a well-known result that if we scale individual unbiased estimators inversely proportional to their variances, we receive the unbiased estimator with minimal variance, we come back to this in (11.16), below. This gives us the intuition behind a specific choice of the power variance parameter for mean estimation, as the sizes of the variances μ_i^p scale (weight) the observed residuals $Y_i - \mu_i$, and balance potential outliers in the observations correspondingly.

11.1.2 Lab: Claim Size Modeling Under Model Uncertainty

We present a proposal for deep learning under model uncertainty in this section. We explain this on an explicit example within Tweedie's distributions. We emphasize that this methodology can be applied in more generality, but it is beneficial here to have an explicit example in mind to illustrate the different phenomena.

Generalized Linear Models

We analyze a Swiss accident insurance claims data set. This data is illustrated in Sect. 13.4, and an excerpt of the data is given in Listing 13.7. In total we have 339'500 claims with positive payments. We choose this data set because it ranges from very small claims of 1 CHF to very large claims, the biggest one exceeding 1'300'000 CHF. These claims are supported by feature information such as the labor sector, the injury type or the injured body part, see Listing 13.7 and Fig. 13.25. For our analysis, we partition the data into a learning data set $\mathcal{L}$ and a test data set $\mathcal{T}$. We do this partition stratified w.r.t. the claim sizes and in a ratio of $9:1$. This results in a learning data set $\mathcal{L}$ of size $n = 305'550$ and in a test data set $\mathcal{T}$ of size $T = 33'950$.

We consider three Tweedie's distributions with power variance parameters $p \in \{2, 2.5, 3\}$, the first one is the gamma model, the last one the inverse Gaussian model, and the power variance parameter $p = 2.5$ gives a model in between. In a first step we consider GLMs, this requires feature engineering. We have three categorical features, one binary feature and two continuous ones. For the categorical and binary features we use dummy coding, and the continuous features `Age` and `AccQuart` are just included in its raw form. As link function g we choose the log-link which respects the positivity of the dual mean parameter space $\mathcal{M}$, see Table 2.1, but this is not the canonical link of the selected models. In the gamma GLM this leads to a convex minimization problem, but in Tweedie's GLM with $p = 2.5$

Table 11.1 In-sample and out-of-sample losses (gamma loss, power variance case $p = 2.5$ loss (in 10^{-2}) and inverse Gaussian (IG) loss (in 10^{-3})) and AIC values; the losses use unit dispersion $\varphi = 1$, AIC relies on the MLE of φ

	In-sample loss on $\mathcal{L}$			Out-of-sample loss on $\mathcal{T}$			AIC
	$\mathfrak{d}_{p=2}$	$\mathfrak{d}_{p=2.5}$	$\mathfrak{d}_{p=3}$	$\mathfrak{d}_{p=2}$	$\mathfrak{d}_{p=2.5}$	$\mathfrak{d}_{p=3}$	value
Null model	3.0094	10.2208	4.6979	3.0240	10.2420	4.6931	4'707'115 (IG)
Gamma GLM	**2.0695**	7.7127	3.9582	**2.1043**	7.7852	3.9763	4'741'472
$p = 2.5$ GLM	2.0744	**7.6971**	3.9433	2.1079	**7.7635**	3.9580	4'648'698
IG GLM	2.0865	7.7069	**3.9398**	2.1191	7.7730	**3.9541**	4'653'501

and in the inverse Gaussian GLM we have non-convex minimization problems, see Example 5.6. Therefore, we initialize Fisher's scoring method (5.12) in the latter two GLMs with the solution of the gamma GLM. The gamma and the inverse Gaussian cases can directly be fitted with the R command `glm` [307], for the power variance parameter case $p = 2.5$ we have coded our own MLE routine using Fisher's scoring method.

Table 11.1 shows the in-sample losses on the learning data $\mathcal{L}$ and the corresponding out-of-sample losses on the test data $\mathcal{T}$. The fitted GLMs (gamma, power variance parameter $p = 2.5$ and inverse Gaussian) are always evaluated on all three unit deviances $\mathfrak{d}_{p=2}(y, \mu)$, $\mathfrak{d}_{p=2.5}(y, \mu)$ and $\mathfrak{d}_{p=3}(y, \mu)$, respectively. We give some remarks. First, we observe that the in-sample loss is always minimized for the GLM with the same power variance parameter p as the loss $\mathfrak{d}_p$ studied (2.0695, 7.6971 and 3.9398 in bold face). This result simply states that the parameter estimates are obtained by minimizing the in-sample loss (or maximizing the corresponding in-sample log-likelihood). Second, the minimal out-of-sample losses are also highlighted in bold face. From these results we cannot give any preference to a single model w.r.t. Tweedie's forecast dominance, see Definition 4.20. Third, we calculate the AIC values for all models. The gamma and the inverse Gaussian cases have a closed-form solution for the normalizing term $a(y; v/\varphi)$ in the EDF density, and we can directly calculate AIC. The case $p = 2.5$ is more difficult and we use the saddlepoint approximation of Sect. 5.5.2. Considering AIC we give preference to Tweedie's GLM with $p = 2.5$. Note that the AIC values use the MLE for φ which is obtained from a general purpose optimizer, and which uses the saddlepoint approximation in the power variance case $p = 2.5$. Fourth, under a constant dispersion parameter φ, the mean estimation $\widehat{\mu}_i$ can be done without explicitly specifying φ because it cancels in the score equations. In fact, we perform this mean estimation in the additive form and not in the reproductive form, see (2.13) and the discussions in Sects. 5.3.7–5.3.8.

Figure 11.2 plots the deviance residuals (for unit dispersion) against the logged fitted means $\widehat{\mu}(\boldsymbol{x}_i)$ for $p \in \{2, 2.5, 3\}$ for 2'000 randomly selected claims; this is the Tukey–Anscombe plot. The green line has been obtained by a spline fit to the deviance residuals as a function of the fitted means $\widehat{\mu}(\boldsymbol{x}_i)$, and the cyan

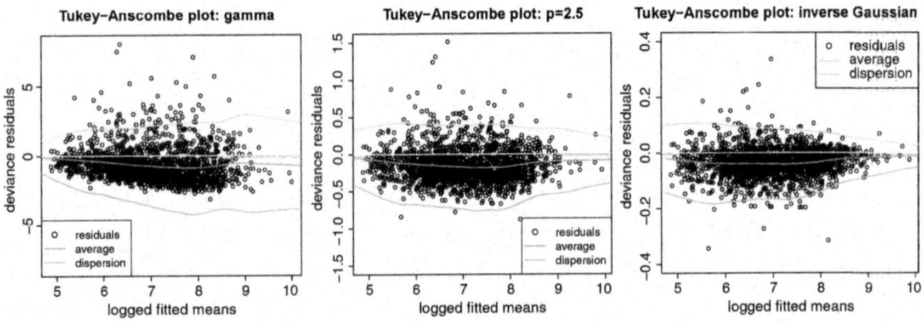

Fig. 11.2 Tukey–Anscombe plots showing the deviance residuals against the logged GLM fitted means $\widehat{\mu}(\boldsymbol{x}_i)$: (lhs) gamma GLM $p = 2$, (middle) power variance case $p = 2.5$, (rhs) inverse Gaussian GLM $p = 3$; the cyan lines show twice the estimated standard deviation of the deviance residuals as a function of the size of the logged estimated means $\widehat{\mu}$

lines give twice the estimated standard deviation of the deviance residuals as a function of the fitted means (also obtained from spline fits). This estimated standard deviation corresponds to the square-rooted deviance dispersion estimate $\widehat{\varphi}^{\mathrm{D}}$, see (5.30), however, in the additive form because we work with unscaled claim size observations. A constant dispersion assumption is supported by cyan lines of roughly constant size. In the gamma case the dispersion seems increasing in the mean estimate, and in the inverse Gaussian case it is decreasing, thus, the power variance parameters $p = 2$ and $p = 3$ do not support a constant dispersion in this example. Only the choice $p = 2.5$ may support a constant dispersion assumption (because it does not have an obvious trend). This says that the variance should scale as $V(\mu) = \mu^{2.5}$ as a function of the mean μ, see also (11.5).

Deep FN Networks

We compare the above GLMs to FN networks of depth $d = 3$ with $(q_1, q_2, q_3) = (20, 15, 10)$ neurons. The categorical features are modeled with embedding layers of dimension $b = 2$. We fit this network architecture with Tweedie's deviances losses having power variance parameters $p \in \{2, 2.5, 3\}$. Moreover, we use 20% of the learning data $\mathcal{L}$ as validation data $\mathcal{V}$ to explore the early stopping rule.[1] To reduce the randomness coming from early stopping with different seeds, we average the deviance losses over 20 runs (this is not the nagging predictor: we only average the deviance losses to have stable conclusions concerning forecast dominance). The results are presented in Table 11.2.

[1] In the standard implementation of SGD with early stopping, the learning and validation data partition is done non-stratified. If necessary, this can be changed manually.

Table 11.2 In-sample and out-of-sample losses (gamma loss, power variance case $p = 2.5$ loss (in 10^{-2}) and inverse Gaussian (IG) loss (in 10^{-3})) and average claim amounts; the losses use unit dispersion $\varphi = 1$ and the network losses are averaged deviance losses over 20 runs with different seeds

	In-sample loss on $\mathcal{L}$			Out-of-sample loss on $\mathcal{T}$			Average
	$\mathfrak{d}_{p=2}$	$\mathfrak{d}_{p=2.5}$	$\mathfrak{d}_{p=3}$	$\mathfrak{d}_{p=2}$	$\mathfrak{d}_{p=2.5}$	$\mathfrak{d}_{p=3}$	claim
Null model	3.0094	10.2208	4.6979	3.0240	10.2420	4.6931	1’774
Gamma GLM	**2.0695**	7.7127	3.9582	**2.1043**	7.7852	3.9763	1’701
$p = 2.5$ GLM	2.0744	**7.6971**	3.9433	2.1079	**7.7635**	3.9580	1’652
IG GLM	2.0865	7.7069	**3.9398**	2.1191	7.7730	**3.9541**	1’614
Gamma network	1.9738	7.4556	3.8693	**2.0543**	**7.6478**	3.9211	1’748
$p = 2.5$ network	**1.9712**	**7.4128**	**3.8458**	2.0654	7.6551	**3.9178**	1’739
IG network	1.9977	7.4568	3.8525	2.0762	7.6682	3.9188	1’712

First, we observe that the networks outperform the GLMs, saying that the feature engineering has not been done optimally for GLMs. Second, in-sample we no longer receive the lowest deviance loss in the model with the same p. This comes from the fact that we exercise early stopping, and, for instance, the gamma in-sample loss of the gamma network ($p = 2$) 1.9738 is bigger than the corresponding gamma loss of 1.9712 from the network with $p = 2.5$. Third, considering forecast dominance, preference is given either to the gamma network or to the power variance parameter $p = 2.5$. In general, it seems that fitting with higher power variance parameters leads to less stable results, but this statement needs more analysis. The disadvantage of this fitting approach is that we independently fit the models with the different power variance parameters to the observations, and, thus, the learned representations $z^{(d:1)}(x_i)$ are rather different for different p's. This makes it difficult to compare these models. This is exactly the point that we address next.

Robustified Representation Learning

To deal with the drawback of missing comparability of the network approaches with different power variance parameters, we can try to learn a representation that simultaneously fits different models. The implementation of this idea is rather straightforward in network modeling. We choose the above network of depth $d = 3$, which gives us the new (learned) representation $z_i = z^{(d:1)}(x_i)$ in the last FN layer. The general idea now is that we design multiple outputs for this learned representation to fit the different distributional models. That is, in the case of three Tweedie's loss functions with power variance parameters $p \in \{2, 2.5, 3\}$ we consider a three-dimensional output mapping

$$\begin{aligned} x \mapsto & \left(\mu_{p=2}(x), \mu_{p=2.5}(x), \mu_{p=3}(x)\right)^\top \qquad (11.6) \\ &= \left(g^{-1}\langle \beta_2, z^{(d:1)}(x)\rangle, g^{-1}\langle \beta_{2.5}, z^{(d:1)}(x)\rangle, g^{-1}\langle \beta_3, z^{(d:1)}(x)\rangle\right)^\top \in \mathbb{R}^3, \end{aligned}$$

for different output parameters $\boldsymbol{\beta}_2, \boldsymbol{\beta}_{2.5}, \boldsymbol{\beta}_3 \in \mathbb{R}^{q_d+1}$. These three expected responses (11.6) share the network parameters $\boldsymbol{w} = (\boldsymbol{w}_1^{(1)}, \ldots, \boldsymbol{w}_{q_d}^{(d)})$ in the FN layers, and the network fitting should learn these parameters such that $\boldsymbol{z}_i = \boldsymbol{z}^{(d:1)}(\boldsymbol{x}_i)$ gives a good representation for all considered loss functions. Choose positive weights $\eta_p > 0$, and define the combined deviance loss function

$$\mathfrak{D}\left(\boldsymbol{Y}, (\boldsymbol{w}, \boldsymbol{\beta}_2, \boldsymbol{\beta}_{2.5}, \boldsymbol{\beta}_3)\right) = \sum_{p\in\{2,2.5,3\}} \frac{\eta_p}{\varphi_p} \sum_{i=1}^{n} v_i \, \mathfrak{d}_p\left(Y_i, \mu_p(\boldsymbol{x}_i)\right), \tag{11.7}$$

for the given observations $(Y_i, \boldsymbol{x}_i, v_i)$, $1 \leq i \leq n$. Note that the unit deviances $\mathfrak{d}_p$ live on different scales for different p's. We use the (constant) weights $\eta_p > 0$ to balance these scales so that all power variance parameters p roughly equally contribute to the total loss, while setting $\varphi_p \equiv 1$ (which can be done for a constant dispersion). This approach is now fitted to the available learning data $\mathcal{L}$. The corresponding R code is given in Listing 11.1. Note that the fitting also requires that we triplicate the observations (Y_i, Y_i, Y_i) so that we can simultaneously evaluate the three chosen power variance deviance losses, see lines 18–21 of Listing 11.1. We fit this model to the Swiss accident insurance data, and the results are presented in Table 11.3 on the lines called 'multi-out'.

Listing 11.1 FN network with multiple output

```
1  Design   = layer_input(shape = c(q0), dtype = 'float32', name = 'Design')
2  #
3  Network = Design %>%
4             layer_dense(units=20, activation='tanh', name='FNLayer1') %>%
5             layer_dense(units=15, activation='tanh', name='FNLayer2') %>%
6             layer_dense(units=10, activation='tanh', name='FNLayer3')
7  #
8  Output1 = Network %>%
9             layer_dense(units=1, activation='exponential', name='Output1')
10 #
11 Output2 = Network %>%
12            layer_dense(units=1, activation='exponential', name='Output2')
13 #
14 Output3 = Network %>%
15            layer_dense(units=1, activation='exponential', name='Output3')
16
17 #
18 keras_model(inputs = c(Design), outputs = c(Output1, Output2, Output3))
19 #
20 model %>% compile(loss = list(loss1, loss2, loss3),
21                   loss_weights=list(eta1, eta2, eta3), optimizer = 'nadam')
```

This simultaneous representation learning across different loss functions leads to more stability in the results between the different loss function choices, i.e., there is less variability between the losses of the different outputs compared to fitting the three different models independently. The predictive performance seems slightly better in this robustified vs. the independent case (see bold face out-of-sample figures). The similarity of the results across the different loss functions (using the

Table 11.3 In-sample and out-of-sample losses (gamma loss, power variance case $p = 2.5$ loss (in 10^{-2}) and inverse Gaussian (IG) loss (in 10^{-3})) and average claim amounts; the losses use unit dispersion $\varphi = 1$ and the network losses are averaged deviance losses over 20 runs with different seeds

	In-sample loss on $\mathcal{L}$			Out-of-sample loss on $\mathcal{T}$			Average
	$\mathfrak{D}_{p=2}$	$\mathfrak{D}_{p=2.5}$	$\mathfrak{D}_{p=3}$	$\mathfrak{D}_{p=2}$	$\mathfrak{D}_{p=2.5}$	$\mathfrak{D}_{p=3}$	claim
Null model	3.0094	10.2208	4.6979	3.0240	10.2420	4.6931	1'774
Gamma network	1.9738	7.4556	3.8693	**2.0543**	**7.6478**	3.9211	1'748
$p = 2.5$ network	**1.9712**	**7.4128**	**3.8458**	2.0654	7.6551	**3.9178**	1'739
IG network	1.9977	7.4568	3.8525	2.0762	7.6682	3.9188	1'712
Gamma multi-output (11.6)	**1.9731**	**7.4275**	**3.8519**	2.0581	7.6422	3.9146	1'745
$p = 2.5$ multi-output (11.6)	1.9736	7.4281	3.8522	**2.0576**	7.6407	3.9139	1'732
IG multi-output (11.6)	1.9745	7.4295	3.8525	**2.0576**	**7.6401**	**3.9134**	1'705
Multi-loss fitting (11.8)	**1.9677**	**7.4118**	**3.8468**	**2.0580**	**7.6417**	**3.9144**	1'744

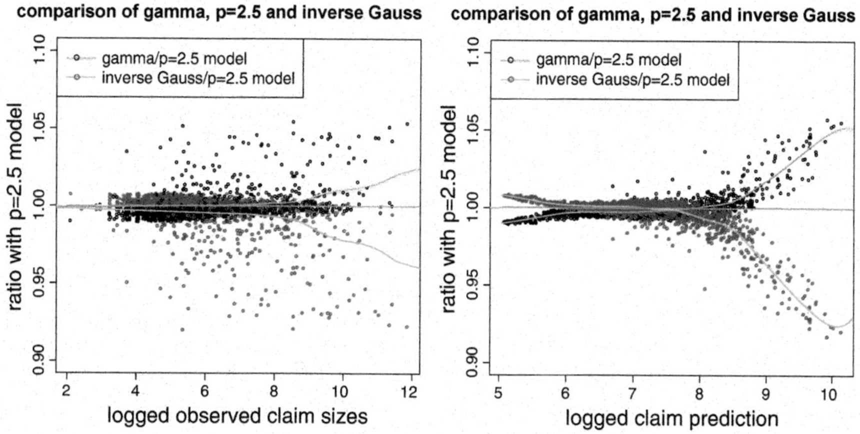

Fig. 11.3 Ratios $\widehat{\mu}_{p=2}(\boldsymbol{x}_i)/\widehat{\mu}_{p=2.5}(\boldsymbol{x}_i)$ (black color) and $\widehat{\mu}_{p=3}(\boldsymbol{x}_i)/\widehat{\mu}_{p=2.5}(\boldsymbol{x}_i)$ (blue color) of the three predictors (lhs) in-sample figures ordered on the x-axis w.r.t. the logged observed claims Y_i, darkgray and cyan lines give spline fits, (rhs) out-of-sample figures ordered on the x-axis w.r.t. the logged average size of the three predictors

jointly learned representation z_i) allows us to directly compare the corresponding predictors $\widehat{\mu}_p(\boldsymbol{x}_i)$ for the different p's.

Figure 11.3 compares the three predictors by considering the ratios $\widehat{\mu}_{p=2}(\boldsymbol{x}_i)/\widehat{\mu}_{p=2.5}(\boldsymbol{x}_i)$ in black color and $\widehat{\mu}_{p=3}(\boldsymbol{x}_i)/\widehat{\mu}_{p=2.5}(\boldsymbol{x}_i)$ in blue color, i.e., we divide by the (middle) predictor with power variance parameter $p = 2.5$. The figure on the left-hand side shows these ratios in-sample and ordered on the x-axis w.r.t. the observed claim sizes Y_i, and the darkgray and cyan lines give spline fits to these ratios. The figure on the right-hand side shows these ratios out-of-sample and ordered on the x-axis w.r.t. the average predictors $\bar{\mu}_i = (\widehat{\mu}_{p=2}(\boldsymbol{x}_i) + \widehat{\mu}_{p=2.5}(\boldsymbol{x}_i) + \widehat{\mu}_{p=3}(\boldsymbol{x}_i))/3$. In view of (11.5) we expect that the models with a smaller power variance parameter p over-fit more to large claims. From Fig. 11.3 (lhs) we can observe that, indeed, this is the case (see gray and cyan spline fits which bifurcate for large claims). That is, models with a smaller power variance parameter react more sensitively to large observations Y_i. The ratios in Fig. 11.3 provide differences of up to 7% for large claims.

Remark 11.3 The loss function (11.7) can also be interpreted as regularization. For instance, if we choose $\eta_2 = 1$, and if we assume that this is our preferred model, then we can regularize this model with further models, and their weights $\eta_p > 0$ determine the degree of regularization. Thus, in contrast to ridge and LASSO regularization of Sect. 6.2, regularization does not directly act on the model parameters, here, but rather on what we learn in terms of the representation $z_i = z^{(d:1)}(\boldsymbol{x}_i)$.

Using Forecast Dominance to Deal with Model Uncertainty

In GLMs, the power variance parameter p typically acts as a hyper-parameter, i.e., one fits different GLMs for different choices of p. Model selection is then done, e.g., by analyzing the Tukey–Anscombe plot, AIC, cross-validation or by studying out-of-sample forecast dominance. In networks we should not use AIC as we neither have a parsimonious network parameter nor do we use the MLE. Here, we focus on forecast dominance for the network predictors (based on the different chosen power variance parameters). If we are mainly interested in receiving a model that provides optimal forecast dominance, we should not consider three different outputs as in (11.7), but rather fit the same output to different loss functions; the required changes are minimal, see Listing 11.2. Namely, consider one FN network with one output $\mu(\boldsymbol{x}_i)$, but evaluate this output simultaneously on the different chosen loss functions

$$\mathfrak{D}(\boldsymbol{Y}, \boldsymbol{\vartheta}) = \sum_{p \in \{2, 2.5, 3\}} \frac{\eta_p}{\varphi_p} \sum_{i=1}^{n} v_i \, \mathfrak{d}_p \left(Y_i, \mu(\boldsymbol{x}_i)\right). \tag{11.8}$$

In contrast to (11.7), we only have one FN network regression function $\boldsymbol{x}_i \mapsto \mu(\boldsymbol{x}_i)$, here.

We present the results on the last line of Table 11.3, called 'multi-loss'. In our case, this approach is slightly less competitive (out-of-sample), however, it is less sensitive to outliers since we need to have a good regression function simultaneously for multiple loss functions. Of course, this multiple loss fitting approach is not restricted to different power variance parameters. As stated in Theorem 4.19, Bregman divergences are the only consistent loss functions for mean estimation, and the unit deviances are examples of Bregman divergences. Forecast dominance now suggests that we may choose any Bregman divergence as a loss function in Listing 11.2 as long as it reflects the expected properties of the model (and of

Listing 11.2 FN network with a single output for multiple losses

```
Design   = layer_input(shape = c(q0), dtype = 'float32', name = 'Design')
#
Network = Design %>%
          layer_dense(units=20, activation='tanh', name='FNLayer1') %>%
          layer_dense(units=15, activation='tanh', name='FNLayer2') %>%
          layer_dense(units=10, activation='tanh', name='FNLayer3')
#
Output = Network %>%
          layer_dense(units=1, activation='exponential', name='Output')
#
keras_model(inputs = c(Design), outputs = c(Output, Output, Output))
#
model %>% compile(loss = list(loss1, loss2, loss3),
                  loss_weights=list(eta1, eta2, eta3), optimizer = 'nadam')
```

the observed data), otherwise we will receive bad convergence properties, see also Sect. 11.1.4, below. For instance, we can robustify the Poisson claim counts model by additionally considering the deviance loss of the negative binomial model that also assesses over-dispersion.

Nagging Predictor

The loss figures in Table 11.3 are averaged deviance losses over 20 different runs of the gradient descent algorithm with different seeds (to receive stable results). Rather than averaging over the losses, we should improve the models by averaging over the predictors and, then, calculate the losses on these averaged predictors; this is exactly the proposal of the nagging predictor (7.44). We calculate the nagging predictor of the models that are simultaneously fit to the different loss functions (lines 'multi-output' and 'multi-loss' of Table 11.3). The resulting nagging predictors are reported in Table 11.4. This table shows that we give a clear preference to the nagging predictors. The simultaneous loss fitting (11.8) gives the best out-of-sample results for the nagging predictor, see the last line of Table 11.4.

Figure 11.4 shows the Tukey–Anscombe plot of the multi-loss nagging predictor for the different deviance losses (for unit dispersion). Again, the case $p = 2.5$ is closest to having a constant dispersion, and the other cases will require dispersion modeling $\varphi(\boldsymbol{x})$.

Figure 11.5 shows the empirical auto-calibration property of the multi-loss nagging predictor. This auto-calibration property is calculated as in Listing 7.8. We observe that the auto-calibration property holds rather accurately. Only for claim predictors $\widehat{\mu}(\boldsymbol{x}_i)$ above 10'000 CHF (vertical dotted line in Fig. 11.5) the fitted means under-estimate the observed average claim sizes. This affects (only) 1.7% of all claims and it could be corrected as described in Example 7.19.

Table 11.4 In-sample and out-of-sample losses (gamma loss, power variance case $p = 2.5$ loss (in 10^{-2}) and inverse Gaussian (IG) loss (in 10^{-3})) and average claim amounts; the losses use unit dispersion $\varphi = 1$

	In-sample loss on $\mathcal{L}$			Out-of-sample loss on $\mathcal{T}$			Average
	$\mathfrak{d}_{p=2}$	$\mathfrak{d}_{p=2.5}$	$\mathfrak{d}_{p=3}$	$\mathfrak{d}_{p=2}$	$\mathfrak{d}_{p=2.5}$	$\mathfrak{d}_{p=3}$	claim
Null model	3.0094	10.2208	4.6979	3.0240	10.2420	4.6931	1'774
Gamma multi-output (11.6)	**1.9731**	**7.4275**	**3.8519**	2.0581	7.6422	3.9146	1'745
$p = 2.5$ multi-output (11.6)	1.9736	7.4281	3.8522	**2.0576**	7.6407	3.9139	1'732
IG multi-output (11.6)	1.9745	7.4295	3.8525	**2.0576**	**7.6401**	**3.9134**	1'705
Multi-loss fitting (11.8)	**1.9677**	**7.4118**	**3.8468**	**2.0580**	**7.6417**	**3.9144**	1'744
Gamma multi-out & nagging	**1.9486**	**7.3616**	**3.8202**	**2.0275**	**7.5575**	**3.8864**	1'745
$p = 2.5$ multi-out & nagging	1.9496	7.3640	3.8311	2.0276	7.5578	**3.8864**	1'732
IG multi-out & nagging	1.9510	7.3666	3.8320	2.0281	7.5583	3.8865	1'705
Multi-loss with nagging	**1.9407**	**7.3403**	**3.8236**	**2.0244**	**7.5490**	**3.8837**	1'744

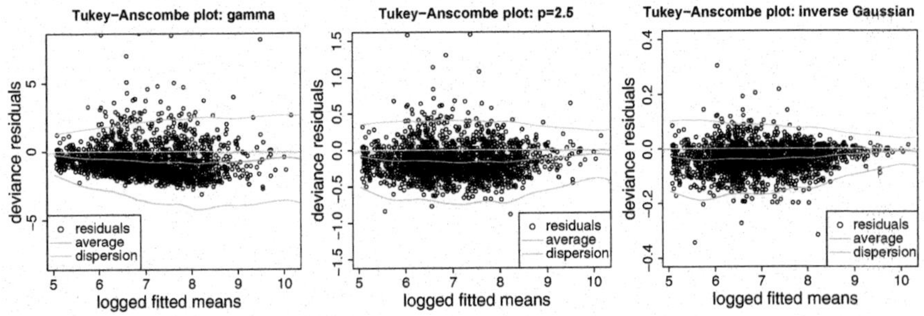

Fig. 11.4 Tukey–Anscombe plots giving the deviance residuals of the multi-loss nagging predictor of Table 11.4 for different power variance parameters: (lhs) gamma deviances $p = 2$, (middle) power variance deviances $p = 2.5$, (rhs) inverse Gaussian deviances $p = 3$; the cyan lines show twice the estimated standard deviation of the deviance residuals as a function of the size of the logged estimated means $\widehat{\mu}$

11.1.3 Lab: Deep Dispersion Modeling

From the Tukey–Anscombe plots in Fig. 11.4 we conclude that the dispersion requires regression modeling, too, as the dispersion does not seem to be constant over the whole range of the expected claim sizes. We therefore explore a *double FN network model*, in spirit this is similar to the double GLM of Sect. 5.5. We therefore assume to work within Tweedie's family with power variance parameters $p \geq 2$, and with unit deviances given by (11.2)–(11.3). The saddlepoint approximation (5.59) gives us

$$f(y; \theta, v/\varphi) \approx \left(\frac{2\pi\varphi}{v} V(y)\right)^{-1/2} \exp\left\{-\frac{1}{2\varphi/v}\,\mathfrak{d}_p(y, \mu)\right\},$$

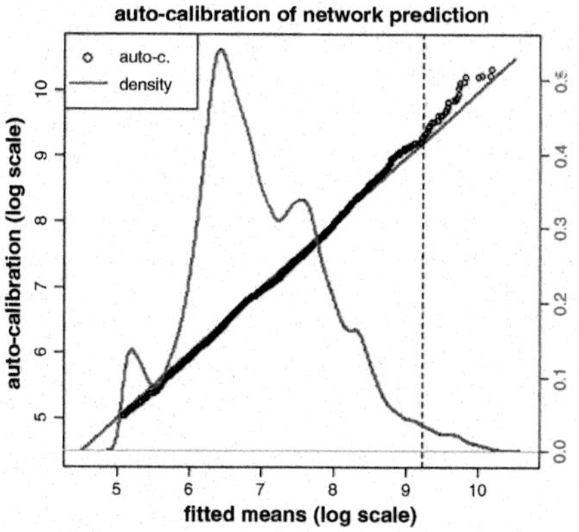

Fig. 11.5 Empirical auto-calibration property of the claim size predictor; the blue curve shows the empirical density of the multi-loss nagging predictor $\widehat{\mu}(\boldsymbol{x}_i)$

with power variance function $V(y) = y^p$. This saddlepoint approximation is formulated in the reproductive form for $Y = X/\omega = X\varphi/v$. This requires scaling of the observations X with the unknown φ to receive Y. In Sect. 5.5.4 we have shown how this problem can be solved. In this section we give a different proposal which is more robust in network fitting, and which benefits from the b-homogeneity of $\mathfrak{d}_p$, see (11.4).

We consider the variable transformation $y \mapsto x = y\omega = yv/\varphi$. In the absolutely continuous case $p \geq 2$ this gives us the approximation

$$f(x; \theta, v/\varphi) \approx \left(\frac{2\pi\varphi^{1+p}}{v^{1+p}} V(x)\right)^{-1/2} \exp\left\{-\frac{1}{2\varphi/v}\, \mathfrak{d}_p\left(\frac{x\varphi}{v}, \frac{\mu\varphi v}{\varphi v}\right)\right\} \frac{\varphi}{v}$$

$$= \left(\frac{2\pi\varphi^{p-1}}{v^{p-1}} V(x)\right)^{-1/2} \exp\left\{-\frac{1}{2\varphi^{p-1}/v^{p-1}}\, \mathfrak{d}_p\left(x, \mu_p\right)\right\},$$

with mean $\mu_p = \mu v/\varphi$ of $X = Yv/\varphi$. We set $\phi = -1/\varphi^{p-1} < 0$. This gives us the approximation

$$\ell_X(\mu_p, \phi) \approx \frac{v^{p-1}\mathfrak{d}_p(X, \mu_p)\phi - (-\log(-\phi))}{2} - \frac{1}{2}\log\left(\frac{2\pi}{v^{p-1}} V(X)\right). \qquad (11.9)$$

For given mean μ_p we again have a gamma approximation on the right-hand side, but we scale the dispersion differently. This gives us the approximate first moment

$$\mathbb{E}_\phi\left[v^{p-1}\mathfrak{d}_p(X, \mu_p)\middle|\, \mu_p\right] \approx \kappa_2'(\phi) = -1/\phi = \varphi^{p-1} \stackrel{\text{def.}}{=} \varphi_p.$$

The remainder of this modeling is similar to the residual MLE approach in Section 5.5.3. Namely, we set up two FN network regression functions

$$\boldsymbol{x} \mapsto \mu_p(\boldsymbol{x}) \qquad \text{and} \qquad \boldsymbol{x} \mapsto \varphi_p(\boldsymbol{x}) = \kappa_2'(\phi(\boldsymbol{x})) = -1/\phi(\boldsymbol{x}).$$

Parameter fitting is achieved by alternating the network parameter fitting of $\mu_p(\boldsymbol{x})$ and $\varphi_p(\boldsymbol{x})$ see also Section 5.5.4. We start the iteration by setting the dispersion constant to $\widehat{\varphi}_p^{(0)}(\boldsymbol{x}) \equiv$ const. In this case, the dispersion cancels in the score equations and the mean $\widehat{\mu}_p^{(1)}(\boldsymbol{x})$ can be estimated without the explicit knowledge of the (constant) dispersion parameter $\widehat{\varphi}_p^{(0)}$; this exactly provides the results of the previous Sect. 11.1.2. Then, we iterate this procedure for $t \geq 1$. For given mean estimate $\widehat{\mu}_p^{(t)}(\boldsymbol{x})$ we receive deviances $v^{p-1}\mathfrak{d}_p(X, \widehat{\mu}_p^{(t)}(\boldsymbol{x}))$, and this allows us to estimate $\widehat{\varphi}_p^{(t)}(\boldsymbol{x})$ from the approximate gamma model (11.9), and for given dispersion parameters $\widehat{\varphi}_p^{(t)}(\boldsymbol{x})$ we estimate $\widehat{\mu}_p^{(t+1)}(\boldsymbol{x})$ from the corresponding Tweedie's model for the observation X.

Example 11.4 We revisit the Swiss accident insurance data example of Sect. 11.1.2, and we use the robustified representation learning approach (11.7) that simultaneously fits Tweedie's models for the power variance parameters $p = 2, 2.5, 3$. The initial calibration step is done for constant dispersions $\widehat{\varphi}_p^{(0)}(\boldsymbol{x}) \equiv$ const, and it provides us with the estimated means $\widehat{\mu}_p^{(1)}(\boldsymbol{x})$ as illustrated in Fig. 11.3. For stability reasons we choose the nagging predictor averaging over 20 different SGD runs with 20 different seeds. These estimated means $\widehat{\mu}_p^{(1)}(\boldsymbol{x})$ give us the deviances $v^{p-1}\mathfrak{d}_p(X, \widehat{\mu}_p^{(1)}(\boldsymbol{x}))$.

Using these deviances allows us to alternate the dispersion and mean estimation for $t \geq 1$. For given means $\widehat{\mu}_p^{(t)}(\boldsymbol{x})$, $p = 2, 2.5, 3$, we set up a deep FN network $\boldsymbol{x} \mapsto \boldsymbol{z}^{(d:1)}(\boldsymbol{x})$ that allows for a robustified deep dispersion learning $\varphi_p(\boldsymbol{x})$, for $p = 2, 2.5, 3$. Under the log-link choice we consider the regression function with multiple outputs

$$\begin{aligned} \boldsymbol{x} \mapsto & \left(\varphi_{p=2}(\boldsymbol{x}), \varphi_{p=2.5}(\boldsymbol{x}), \varphi_{p=3}(\boldsymbol{x})\right)^\top \qquad (11.10) \\ &= \left(\exp\langle \boldsymbol{\alpha}_2, \boldsymbol{z}^{(d:1)}(\boldsymbol{x})\rangle, \exp\langle \boldsymbol{\alpha}_{2.5}, \boldsymbol{z}^{(d:1)}(\boldsymbol{x})\rangle, \exp\langle \boldsymbol{\alpha}_3, \boldsymbol{z}^{(d:1)}(\boldsymbol{x})\rangle\right)^\top \in \mathbb{R}_+^3, \end{aligned}$$

for different output parameters $\boldsymbol{\alpha}_2, \boldsymbol{\alpha}_{2.5}, \boldsymbol{\alpha}_3 \in \mathbb{R}^{q_d+1}$. These three dispersion responses (11.10) share the common network parameter $\widetilde{\boldsymbol{w}} = (\widetilde{\boldsymbol{w}}_1^{(1)}, \ldots, \widetilde{\boldsymbol{w}}_{q_d}^{(d)})$ in the FN layers of $\boldsymbol{z}^{(d:1)}$. The network fitting learns these parameters simultaneously for the different power variance parameters. Choose positive weights $\widetilde{\eta}_p > 0$, and define the combined deviance loss function (based on the gamma model κ_2 and having dispersion parameter 2)

$$\mathfrak{D}\left(\mathfrak{d}(\boldsymbol{X}, \widehat{\boldsymbol{\mu}}^{(t)}), (\widetilde{\boldsymbol{w}}, \boldsymbol{\alpha}_2, \boldsymbol{\alpha}_{2.5}, \boldsymbol{\alpha}_3)\right) = \sum_{p\in\{2,2.5,3\}} \frac{\widetilde{\eta}_p}{2} \sum_{i=1}^{n} \mathfrak{d}_2\left(v_i^{p-1}\mathfrak{d}_p(X_i, \widehat{\mu}_p^{(t)}(\boldsymbol{x}_i)), \varphi_p(\boldsymbol{x}_i)\right), \qquad (11.11)$$

where $\boldsymbol{X} = (X_1, \ldots, X_n)$ collects the unscaled observations $X_i = Y_i v_i / \varphi_i$. Thus, for all power variance parameters $p = 2, 2.5, 3$ we fit a gamma model $\mathfrak{d}_2(\cdot, \cdot)/2$ to the observed deviances (observations) $v_i^{p-1} \mathfrak{d}_p(X_i, \widehat{\mu}_p^{(t)}(\boldsymbol{x}_i))$ providing us with the estimated dispersions $\widehat{\varphi}_p^{(t)}(\boldsymbol{x}_i)$. This fitting step is received by the R code of Listing 11.1, where the losses on line 20 are all given by gamma deviance losses (11.11) and the deviances $v_i^{p-1} \mathfrak{d}_p(X_i, \widehat{\mu}_p^{(t)}(\boldsymbol{x}_i))$ play the role of the responses (observations).

In the next step we update the mean estimates $\widehat{\mu}_p^{(t+1)}(\boldsymbol{x}_i)$, given the estimated dispersions $\widehat{\varphi}_p^{(t)}(\boldsymbol{x}_i)$ from the previous step. This requires that we optimize the expected responses (11.6) for given heterogeneous dispersion parameters. We therefore consider the loss function for positive weights $\eta_p > 0$, see (11.7),

$$\mathfrak{D}\left(\boldsymbol{X}, \widehat{\boldsymbol{\varphi}}^{(t)}, (\boldsymbol{w}, \boldsymbol{\beta}_2, \boldsymbol{\beta}_{2.5}, \boldsymbol{\beta}_3)\right) = \sum_{p \in \{2,2.5,3\}} \eta_p \sum_{i=1}^{n} \frac{v_i^{p-1}}{\widehat{\varphi}_p^{(t)}(\boldsymbol{x}_i)} \mathfrak{d}_p\left(X_i, \mu_p(\boldsymbol{x}_i)\right). \tag{11.12}$$

We fit this model by iterating this approach for $t \geq 1$: we start from the predictors of Sect. 11.1.2 providing us with the first mean estimates $\widehat{\mu}_p^{(1)}(\boldsymbol{x}_i)$. Based on these mean estimates we iterate this robustified estimation of $\widehat{\varphi}_p^{(t)}(\boldsymbol{x}_i)$ and $\widehat{\mu}_p^{(t)}(\boldsymbol{x}_i)$. We give some remarks:

1. We use the robustified versions (11.11) and (11.12), respectively, where we simultaneously fit all power variance parameters $p = 2, 2.5, 3$ on the commonly learned representations $\boldsymbol{z}_i = \boldsymbol{z}^{(d:1)}(\boldsymbol{x}_i)$ in the last FN layer of the mean and the dispersion network, respectively.
2. For both FN networks of mean μ and dispersion φ modeling we use the same network architecture of depth $d = 3$ having $(q_1, q_2, q_3) = (20, 15, 10)$ neurons in the FN layers, the hyperbolic tangent activation function, and the log-link for the output. These two networks only differ in their network parameters $(\boldsymbol{w}, \boldsymbol{\beta}_2, \boldsymbol{\beta}_{2.5}, \boldsymbol{\beta}_3)$ and $(\widetilde{\boldsymbol{w}}, \boldsymbol{\alpha}_2, \boldsymbol{\alpha}_{2.5}, \boldsymbol{\alpha}_3)$, respectively.
3. For fitting we use the `nadam` version of SGD. For the early stopping we use a training data $\mathcal{U}$ to validation data $\mathcal{V}$ split of 8 : 2.
4. To ensure consistency within the individual SGD runs across $t \geq 1$, we use the learned network parameter of loop t as initial value for loop $t + 1$. This ensures monotonicity across the iterations in the log-likelihood and the loss function, respectively, up to the fact that the random mini-batches in SGD may distort this monotonicity.
5. To reduce the elements of randomness in SGD fitting we run this iteration procedure 20 times with different seeds, and we output the nagging predictors for $\widehat{\mu}_p^{(t)}(\boldsymbol{x}_i)$ and $\widehat{\varphi}_p^{(t)}(\boldsymbol{x}_i)$ averaged over the 20 runs for every t in Table 11.5.

We iterate this algorithm over two loops, and the results are presented in Table 11.5. We observe a decrease of $-2\ell_{\boldsymbol{X}}(\widehat{\mu}_p^{(t)}, \widehat{\varphi}_p^{(t)})$ by iterating the fitting algorithm for $t \geq 1$. For AIC, we would have to correct twice the negative log-likelihood by twice

Table 11.5 Iteration of mean $\widehat{\mu}_p^{(t)}$ and dispersion $\widehat{\varphi}_p^{(t)}$ estimation for the gamma model $p = 2$, the power variance parameter $p = 2.5$ model and the inverse Gaussian model $p = 3$: the numbers correspond to $-2\ell_X(\widehat{\mu}_p^{(t)}, \widehat{\varphi}_p^{(t)})$; the last line corrects $-2\ell_X(\widehat{\mu}_p^{(t)}, \widehat{\varphi}_p^{(t)})$ by $2 \cdot 2 \cdot 812 = 3'248$ (twice the number of parameters used in the mean and dispersion FN networks)

Iteration	$-2\cdot$ log-likelihood		
t	Gamma $p = 2$	Power variance $p = 2.5$	Inverse Gaussian $p = 3$
$(\widehat{\mu}^{(1)}, \widehat{\varphi}^{(0)})$	4'722'961	4'635'038	4'644'869
$(\widehat{\mu}^{(1)}, \widehat{\varphi}^{(1)})$	4'702'247	4'622'097	4'617'593
$(\widehat{\mu}^{(2)}, \widehat{\varphi}^{(1)})$	4'701'234	4'621'123	4'616'869
$(\widehat{\mu}^{(2)}, \widehat{\varphi}^{(2)})$	4'700'686	4'620'845	4'616'588
"AIC"	4'703'978	4'624'137	4'619'880

the number of MLE estimated parameters. We also adjust here correspondingly, though the correction is not justified by any theory, because we do not work with the MLE nor do we have a parsimonious model for mean and dispersion estimation. Nevertheless, we receive smaller values than in Table 11.1 which supports the use of this more complex double FN network model.

Comparing the three power variance parameter models, we now give preference to the inverse Gaussian model, as it has the biggest log-likelihood. Note that we directly compare all power variance models as the complexity is equal in all models (they only differ in the chosen power variance parameter) and the joint robustified fitting applies the same stopping rule to all power variance parameter models. The same result is obtained by comparing the out-of-sample log-likelihoods. Note that we do not compare the deviance losses, here, because the unit deviances are not designed to estimate parameters in vector-valued parameter families; we model dispersion as a second parameter.

Next, we study the estimated dispersions $\widehat{\varphi}_p(\boldsymbol{x}_i)$ as a function of the estimated means $\widehat{\mu}_p(\boldsymbol{x}_i)$. We fit a spline to $\widehat{\varphi}_p(\boldsymbol{x}_i)$ as a function of $\widehat{\mu}_p(\boldsymbol{x}_i)$, and we receive estimates that almost perfectly match the cyan lines in Fig. 11.4. This provides a proof of concept that the dispersion regression model finds the right level of dispersion as a function of the expected means.

Using the mean and dispersion estimates, we can calculate the dispersion scaled deviance residuals

$$r_i^{\mathrm{D}} = \operatorname{sign}(X_i - \widehat{\mu}_p(\boldsymbol{x}_i))\sqrt{v_i^{p-1}\mathfrak{d}\left(X_i, \widehat{\mu}_p(\boldsymbol{x}_i)\right)/\widehat{\varphi}_p(\boldsymbol{x}_i)}. \tag{11.13}$$

This then allows us to give the Tukey–Anscombe plots for the three considered power variance parameters.

The corresponding plots are given in Fig. 11.6; the difference to Fig. 11.4 is that the latter considers unit dispersion whereas the former scales the residuals with the rooted dispersion $\sqrt{\widehat{\varphi}_p(\boldsymbol{x}_i)}$; note that $v_i \equiv 1$ in this example. By scaling with the rooted dispersion the resulting deviance residuals r_i^{D} should roughly have unit standard deviation. From Fig. 11.6 we observe that indeed this is the case, the cyan

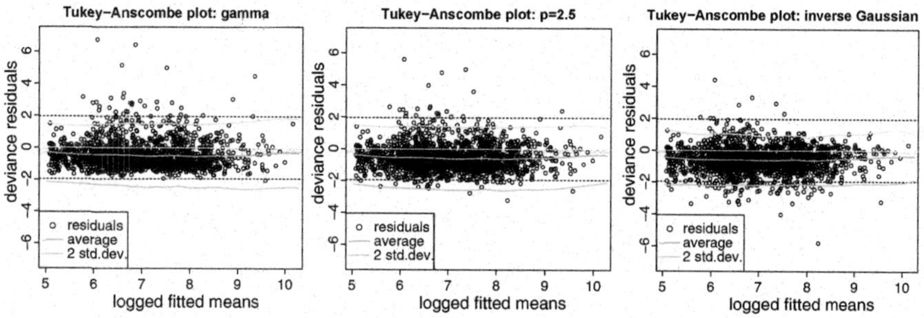

Fig. 11.6 Tukey–Anscombe plots giving the dispersion scaled deviance residuals r_i^{D} (11.13) of the models jointly fitting the mean parameters $\widehat{\mu}_p(\boldsymbol{x}_i)$ and the dispersion parameters $\widehat{\varphi}_p(\boldsymbol{x}_i)$: (lhs) gamma model, (middle) power variance parameter $p = 2.5$ model, and (rhs) inverse Gaussian models; the cyan lines correspond to 2 standard deviations

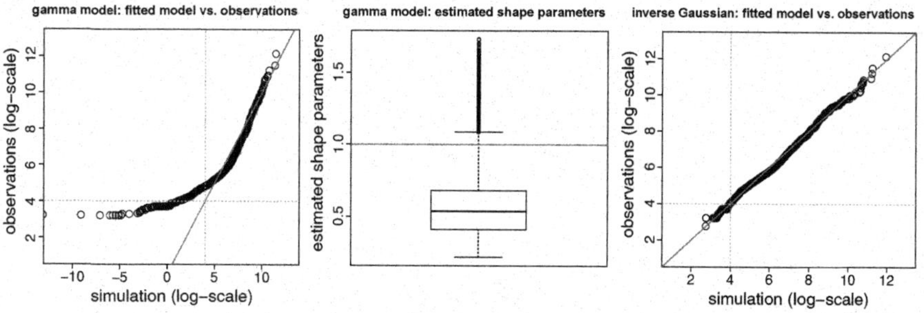

Fig. 11.7 (lhs) Gamma model: observations vs. simulations on log-scale, (middle) gamma model: estimated shape parameters $\widehat{\alpha}_t^\dagger = 1/\widehat{\varphi}_2(\boldsymbol{x}_t^\dagger) < 1$, $1 \le t \le T$, and (rhs) inverse Gaussian model: observations vs. simulations on log-scale

line shows a spline fit of twice the standard deviation of the deviance residuals r_i^{D}. These splines are of magnitude 2 which verifies the unit standard deviation property. Moreover, the cyan lines are roughly horizontal which indicates that the dispersion estimation and the scaling works across all expected claim sizes $\widehat{\mu}_p(\boldsymbol{x}_i)$. The three different power variance parameters $p = 2, 2.5, 3$ show different behaviors in the lower and upper tails in the residuals (centering around the orange horizontal zero line in Fig. 11.6) which corresponds to the different distributional properties of the chosen models.

We further analyze the gamma and the inverse Gaussian models. Note that the analysis of the power variance models for general power variance parameters $p \neq 0, 1, 2, 3$ is more difficult because neither the EDF density nor the EDF distribution function have a closed form. To analyze the gamma and the inverse Gaussian models we simulate observations X_t^{sim}, $t = 1, \ldots, T$, from the estimated models (using the out-of-sample features $\boldsymbol{x}_t^\dagger$ of the test data $\mathcal{T}$), and we compare them against the true out-of-sample observations $X_t^\dagger$. Figure 11.7 shows the results for the gamma model (lhs) and the inverse Gaussian model (rhs) on the log-scale. A good fit has been achieved if the black dots lie on the red diagonal line (in the colored version), because then the simulated data shares similar features as the observed data. The fit of the inverse Gaussian model seems reasonably good.

On the other hand, we see that the gamma model gives a poor fit, especially in the lower tail. This supports the AIC values of Table 11.5. The problem with the gamma model is that the data is more heavy-tailed than the gamma model can accomplish. As a consequence, the dispersion parameter estimates $\widehat{\varphi}_2(\boldsymbol{x}_t^\dagger)$ in the gamma model are compensating for this by taking values bigger than 1. A dispersion parameter bigger than 1 implies a shape parameter in the gamma model of $\widehat{\alpha}_t^\dagger = 1/\widehat{\varphi}_2(\boldsymbol{x}_t^\dagger) < 1$, and the resulting gamma density is strictly decreasing, see Fig. 2.1. If we simulate from this model we receive many observations X_t^{sim} close to zero (from the strictly decreasing density). This can be seen from the lower-left part of the graph in Fig. 11.7 (lhs), suggesting that we have many observations with $X_t^\dagger \in (0, 1)$, or on the log-scale $\log(X_t^\dagger) < 0$. However, the graph shows that this is not the case in the real data. Figure 11.7 (middle) shows the boxplot of the estimated shape parameters $\widehat{\alpha}_t^\dagger$ on the test data, $1 \le t \le T$, verifying that most insurance policies of the test data $\mathcal{T}$ receive a shape parameter $\widehat{\alpha}_t^\dagger$ less than 1.

We conclude that the inverse Gaussian double FN network model seems to work well for this data, and we give preference to this model. ■

11.1.4 Pseudo Maximum Likelihood Estimator

This short section gives a mathematical foundation to parameter estimation under model uncertainty and model misspecification. We summarize the results of Gourieroux et al. [168], and we refrain from giving any proofs in this section. Assume that the real-valued observations Y_i, $1 \le i \le n$, have been generated by the model

$$Y_i = \mu_{\zeta_0}(\boldsymbol{x}_i) + \varepsilon_i, \tag{11.14}$$

with (true) parameter $\zeta_0 \in \Lambda \subset \mathbb{R}^r$, feature $\boldsymbol{x}_i \in \mathcal{X} \subseteq \{1\} \times \mathbb{R}^q$, and where the conditional distribution of the noise random variables $(\varepsilon_i)_{1 \le i \le n}$ satisfies the conditional independence property $p_\varepsilon(\varepsilon_1, \ldots, \varepsilon_n | \boldsymbol{x}_1, \ldots, \boldsymbol{x}_n) = \prod_{i=1}^n p_\varepsilon(\varepsilon_i | \boldsymbol{x}_i)$. Denote by $p_{\boldsymbol{x}}(\boldsymbol{x})$ the portfolio distribution of the features $\boldsymbol{x}$. Thus, under (11.14), the claim Y of a randomly selected policy is generated by the joint probability measure $p_{\epsilon,\boldsymbol{x}}(\varepsilon, \boldsymbol{x}) = p_\varepsilon(\varepsilon | \boldsymbol{x}) p_{\boldsymbol{x}}(\boldsymbol{x})$. The technical assumptions under which the following statements hold are given in Assumption 11.9 at the end of this section.

Let $F_0(\cdot | \boldsymbol{x}_i)$ denote the true conditional distribution of Y_i, given $\boldsymbol{x}_i$. Typically, this (true) conditional distribution is unknown. It is assumed to provide the first two conditional moments

$$\mathbb{E}_{\zeta_0}[Y_i | \boldsymbol{x}_i] = \mu_{\zeta_0}(\boldsymbol{x}_i) \qquad \text{and} \qquad \text{Var}_{\zeta_0}(Y_i | \boldsymbol{x}_i) = \sigma_0^2(\boldsymbol{x}_i).$$

Thus, $\varepsilon_i|_{\boldsymbol{x}_i}$ is assumed to be centered with conditional variance $\sigma_0^2(\boldsymbol{x}_i)$, see (11.14). Our goal is to estimate the (true) parameter $\zeta_0 \in \Lambda$, based on the fact that the conditional distribution $F_0(\cdot|\boldsymbol{x})$ of the observations is unknown. Throughout we assume parameter identifiability, i.e., if $\mu_{\zeta_1}(\boldsymbol{x}) = \mu_{\zeta_2}(\boldsymbol{x})$, $p_{\boldsymbol{x}}$-a.s., then $\zeta_1 = \zeta_2$. The following estimator is called *pseudo maximum likelihood estimator* (PMLE)

$$\widehat{\zeta}_n^{\text{PMLE}} = \underset{\zeta\in\Lambda}{\arg\min}\ \frac{1}{n}\sum_{i=1}^{n} \mathfrak{d}(Y_i, \mu_\zeta(\boldsymbol{x}_i)), \tag{11.15}$$

where $\mathfrak{d}(y, \mu)$ is the unit deviance of a (pre-chosen) single-parameter linear EDF being parametrized by the same parameter space $\Lambda \subset \mathbb{R}^r$ as the original random variables (11.14); note that Λ is not the effective domain $\boldsymbol{\Theta}$ of the chosen EDF. $\widehat{\zeta}_n^{\text{PMLE}}$ is called PMLE because it is a MLE for $\zeta_0 \in \Lambda$, but not in the right model, because the pre-chosen EDF in (11.15) typically differs from the (unknown) true conditional distribution $F_0(\cdot|\boldsymbol{x})$. Nevertheless, we may hope to find the true parameter ζ_0, but possibly at a slower asymptotic rate. This is exactly what is going to be stated in the next theorems.

Theorem 11.5 (Theorem 1 of Gourieroux et al. [168]) *Denote by $\mathcal{M} = \kappa'(\mathring{\boldsymbol{\Theta}})$ the dual mean parameter space of the pre-chosen EDF (having cumulant function κ), and assume that $\mu_\zeta(\boldsymbol{x}) \in \mathcal{M}$ for all $\boldsymbol{x} \in \mathcal{X}$ and $\zeta \in \Lambda$. Let Assumption 11.9, below, hold. The PMLE $\widehat{\zeta}_n^{\text{PMLE}}$ is strongly consistent for ζ_0, i.e., it converges a.s. as $n \to \infty$.*

This theorem tells us that we can perform MLE in a pre-chosen EDF (which may differ from the true data model), and asymptotically we find the true parameter ζ_0 of the data model $F_0(\cdot|\boldsymbol{x})$. Of course, this uses the fact that any unit deviance $\mathfrak{d}$ is a strictly consistent loss function for mean estimation, see Theorem 4.19. We do not only receive consistency, but the following theorem also gives us the rate of convergence.

Theorem 11.6 (Theorem 3 of Gourieroux et al. [168]) *Set the same assumptions as in Theorem 11.5. The PMLE $\widehat{\zeta}_n^{\text{PMLE}}$ has the following asymptotic behavior*

$$\sqrt{n}\left(\widehat{\zeta}_n^{\text{PMLE}} - \zeta_0\right) \implies \mathcal{N}\left(0, \mathcal{I}^*(\zeta_0)^{-1}\Sigma(\zeta_0)\mathcal{I}^*(\zeta_0)^{-1}\right) \qquad \textit{for } n \to \infty,$$

with the following matrices evaluated in $\zeta = \zeta_0$

$$\mathcal{I}^*(\zeta) = \mathbb{E}_{\boldsymbol{x}}\left[\mathcal{I}^*(\zeta;\boldsymbol{x})\right] = \mathbb{E}_{\boldsymbol{x}}\left[J(\zeta;\boldsymbol{x})^\top \kappa''(h(\mu_\zeta(\boldsymbol{x})))J(\zeta;\boldsymbol{x})\right] \in \mathbb{R}^{r\times r},$$

$$\Sigma(\zeta) = \mathbb{E}_{\boldsymbol{x}}\left[J(\zeta;\boldsymbol{x})^\top \sigma_0^2(\boldsymbol{x})J(\zeta;\boldsymbol{x})\right] \in \mathbb{R}^{r\times r},$$

where $h = (\kappa')^{-1}$ is the canonical link of the pre-chosen EDF, and with the change of variable $\zeta \mapsto \theta = \theta(\zeta) = h(\mu_\zeta(\boldsymbol{x})) \in \boldsymbol{\Theta}$, for given feature $\boldsymbol{x}$, having Jacobian

$$J(\zeta;\boldsymbol{x}) = \left(\frac{\partial}{\partial \zeta_k} h(\mu_\zeta(\boldsymbol{x}))\right)_{1\le k\le r} = \frac{1}{\kappa''(h(\mu_\zeta(\boldsymbol{x})))} \left(\nabla_\zeta \mu_\zeta(\boldsymbol{x})\right)^\top \in \mathbb{R}^{1\times r}.$$

Remark that $\mathcal{I}^*(\zeta)$ averages Fisher's information $\mathcal{I}^*(\zeta;\boldsymbol{x})$ (of the chosen EDF) over the feature distribution $p_{\boldsymbol{x}}$. This theorem can be seen as a modification of (3.36) to the regression case. Theorem 11.6 gives us the asymptotic normality of the PMLE, and the resulting asymptotic variance depends on how well the pre-chosen EDF matches the true data distribution $F_0(\cdot|\boldsymbol{x})$. The following lemma corresponds to Property 5 in Gourieroux et al. [168].

Lemma 11.7 *The asymptotic variance in Theorem 11.6 has the lower bound, set $\zeta = \zeta_0$ and $\sigma^2(\boldsymbol{x}) = \sigma_0^2(\boldsymbol{x})$,*

$$\mathcal{I}^*(\zeta)^{-1}\Sigma(\zeta)\mathcal{I}^*(\zeta)^{-1} \;\ge\; \mathcal{H}(\zeta) = \mathbb{E}_{\boldsymbol{x}}\left[\nabla_\zeta \mu_\zeta(\boldsymbol{x})\sigma^{-2}(\boldsymbol{x})\left(\nabla_\zeta \mu_\zeta(\boldsymbol{x})\right)^\top\right]^{-1} \in \mathbb{R}^{r\times r}.$$

Proof We set $\tau^2(\boldsymbol{x}) = \kappa''(h(\mu_\zeta(\boldsymbol{x})))$. We have $J(\zeta;\boldsymbol{x})^\top = \nabla_\zeta \mu_\zeta(\boldsymbol{x})\tau^{-2}(\boldsymbol{x})$. The following matrix is positive semi-definite and it satisfies

$$\begin{aligned}
&\mathbb{E}_{\boldsymbol{x}}\Big[\left[\mathcal{I}^*(\zeta)^{-1}J(\zeta;\boldsymbol{x})^\top - \mathcal{H}(\zeta)J(\zeta;\boldsymbol{x})^\top\tau^2(\boldsymbol{x})\sigma^{-2}(\boldsymbol{x})\right]\sigma^2(\boldsymbol{x}) \\
&\qquad\qquad \times\left[\mathcal{I}^*(\zeta)^{-1}J(\zeta;\boldsymbol{x})^\top - \mathcal{H}(\zeta)J(\zeta;\boldsymbol{x})^\top\tau^2(\boldsymbol{x})\sigma^{-2}(\boldsymbol{x})\right]^\top\Big] \\
&= \mathcal{I}^*(\zeta)^{-1}\Sigma(\zeta)\mathcal{I}^*(\zeta)^{-1} - \mathcal{H}(\zeta)\mathcal{I}^*(\zeta)\mathcal{I}^*(\zeta)^{-1} - \mathcal{I}^*(\zeta)^{-1}\mathcal{I}^*(\zeta)\mathcal{H}(\zeta) + \mathcal{H}(\zeta)\mathcal{H}(\zeta)^{-1}\mathcal{H}(\zeta) \\
&= \mathcal{I}^*(\zeta)^{-1}\Sigma(\zeta)\mathcal{I}^*(\zeta)^{-1} - \mathcal{H}(\zeta).
\end{aligned}$$

This proves the claim. □

Theorem 11.6 and Lemma 11.7 tell us that if we estimate the parameter ζ_0 of the unknown model $F_0(\cdot|\boldsymbol{x})$ with PMLE based on a single-parameter linear EDF, we receive minimal asymptotic variance if we can match the variance $V(\mu_{\zeta_0}(\boldsymbol{x})) = \kappa''(h(\mu_{\zeta_0}(\boldsymbol{x})))$ of the chosen EDF with the variance $\sigma_0^2(\boldsymbol{x})$ of the true data model. E.g., if we know that the variance in the true model behaves as $\sigma_0^2(\boldsymbol{x}) = \mu_{\zeta_0}^3(\boldsymbol{x})$ we should select the inverse Gaussian model with variance function $V(\mu) = \mu^3$ for PMLE.

If the members of the single-parameter linear EDF do not fully match the variance structure of the true data, we can turn our attention to a dispersion submodel as in Sect. 5.5.1. Assume for the variance structure of the true data

$$\mathrm{Var}_{\zeta_0}(Y_i|\boldsymbol{x}_i) = \sigma_0^2(\boldsymbol{x}_i) = \frac{1}{v_i}s_{\alpha_0}^2(\boldsymbol{x}_i),$$

for a regression function $\boldsymbol{x} \mapsto s^2_{\alpha_0}(\boldsymbol{x})$ involving the (true) regression parameter α_0 and exposures $v_i > 0$. If we choose a fixed EDF, we have the log-likelihood function

$$(\mu, \varphi) \mapsto \ell_Y(\mu, \varphi; v) = \frac{v}{\varphi}\left[Yh(\mu) - \kappa(h(\mu))\right] + a(y; v/\varphi).$$

Equating the variance structure of the true data model with the variance in this pre-specified EDF, we obtain feature-dependent dispersion parameter

$$\varphi(\boldsymbol{x}_i) = \frac{s^2_{\alpha_0}(\boldsymbol{x}_i)}{V(\mu_{\zeta_0}(\boldsymbol{x}_i))}, \tag{11.16}$$

with variance function $V(\mu) = (\kappa'' \circ h)(\mu)$. The following theorem proposes a two-step procedure for this estimation problem.

Theorem 11.8 (Theorem 4 of Gourieroux et al. [168]) *Assume $\widetilde{\zeta}_n$ and $\widetilde{\alpha}_n$ are strongly consistent estimators for ζ_0 and α_0, as $n \to \infty$, such that $\sqrt{n}(\widetilde{\zeta}_n - \zeta_0)$ and $\sqrt{n}(\widetilde{\alpha}_n - \alpha_0)$ are bounded in probability. The quasi-generalized pseudo maximum likelihood estimator (QPMLE) of ζ_0 is obtained by*

$$\widehat{\zeta}_n^{\text{QPMLE}} = \underset{\zeta \in \Lambda}{\arg\max} \sum_{i=1}^n \ell_{Y_i}\left(\mu_\zeta(\boldsymbol{x}_i), \frac{s^2_{\widetilde{\alpha}_n}(\boldsymbol{x}_i)}{V(\mu_{\widetilde{\zeta}_n}(\boldsymbol{x}_i))}; v_i\right).$$

Under Assumption 11.9, below, $\widehat{\zeta}_n^{\text{QPMLE}}$ is strongly consistent and best asymptotically normal, i.e.,

$$\sqrt{n}\left(\widehat{\zeta}_n^{\text{QPMLE}} - \zeta_0\right) \implies \mathcal{N}(0, \mathcal{H}(\zeta_0)) \qquad \textit{for } n \to \infty.$$

This justifies the approach(es) in the previous chapters and sections, though, not fully, because we neither work with the MLE in FN networks nor do we care about identifiability in parameters. Nevertheless, this short section suggests to find strongly consistent estimators $\widetilde{\zeta}_n$ and $\widetilde{\alpha}_n$ for ζ_0 and α_0. This gives us a first model calibration step that allows us to specify the dispersion structure $\boldsymbol{x} \mapsto \varphi(\boldsymbol{x})$ via (11.16). Using this dispersion structure and the deviance loss function (4.9) for a variable dispersion parameter $\varphi(\boldsymbol{x})$, the QPMLE is obtained in the second step by, we replace the likelihood maximization by the deviance loss minimization,

$$\widehat{\zeta}_n^{\text{QPMLE}} = \underset{\zeta \in \Lambda}{\arg\min}\ \frac{1}{n}\sum_{i=1}^n \frac{v_i}{s^2_{\widetilde{\alpha}_n}(\boldsymbol{x}_i)/V(\mu_{\widetilde{\zeta}_n}(\boldsymbol{x}_i))}\ \mathfrak{d}(Y_i, \mu_\zeta(\boldsymbol{x}_i)).$$

This QPMLE is best asymptotically normal, thus, asymptotically optimal within the EDF. There might still be better estimators for ζ_0, but these are outside the EDF.

If we turn M-estimation into Z-estimation we have the requirement for ζ, see also (11.5),

$$\frac{1}{n}\sum_{i=1}^{n} v_i \, \frac{V(\mu_{\widetilde{\zeta}_n}(\boldsymbol{x}_i))}{s^2_{\widetilde{\alpha}_n}(\boldsymbol{x}_i)} \, \frac{Y_i - \mu_\zeta(\boldsymbol{x}_i)}{V(\mu_\zeta(\boldsymbol{x}_i))} \, \nabla_\zeta \mu_\zeta(\boldsymbol{x}_i) \stackrel{!}{=} 0.$$

Thus, it all boils down to find the right variance structure to receive the optimal asymptotic behavior.

The previous statements hold true under the following technical assumptions. These are taken from Appendix 1 of Gourieroux et al. [167], and they are an adapted version of the ones in Burguete et al. [61].

Assumption 11.9

(i) $\mu_\zeta(\boldsymbol{x})$ *and* $\mathfrak{d}(y, \mu_\zeta(\boldsymbol{x}))$ *are continuous w.r.t. all variables and twice continuously differentiable in* ζ*;*

(ii) $\Lambda \subset \mathbb{R}^r$ *is a compact set and the true parameter* ζ_0 *is in the interior of* Λ*;*

(iii) almost every realization of $(\varepsilon_i, \boldsymbol{x}_i)$ *is a Cesàro sum generator w.r.t. the probability measure* $p_{\epsilon,\boldsymbol{x}}(\varepsilon, \boldsymbol{x}) = p_\varepsilon(\varepsilon|\boldsymbol{x})p_{\boldsymbol{x}}(\boldsymbol{x})$ *and to a dominating function* $b(\varepsilon, \boldsymbol{x})$*;*

(iv) the sequence $(\boldsymbol{x}_i)_i$ *is a Cesàro sum generator w.r.t.* $p_{\boldsymbol{x}}$ *and* $b(\boldsymbol{x}) = \int_{\mathbb{R}} b(\varepsilon, \boldsymbol{x}) dp_\varepsilon(\varepsilon|\boldsymbol{x})$*;*

(v) for each $\boldsymbol{x} \in \{1\} \times \mathbb{R}^q$*, there exists a neighborhood* $N_{\boldsymbol{x}} \subset \{1\} \times \mathbb{R}^q$ *such that*

$$\int_{\mathbb{R}} \sup_{\boldsymbol{x}' \in N_{\boldsymbol{x}}} b(\varepsilon, \boldsymbol{x}')\, dp_\varepsilon(\varepsilon|\boldsymbol{x}) < \infty;$$

(vi) the functions $\mathfrak{d}(Y, \mu_\zeta(\boldsymbol{x}))$, $\partial\mathfrak{d}(Y, \mu_\zeta(\boldsymbol{x}))/\partial\zeta_k$, $\partial^2\mathfrak{d}(Y, \mu_\zeta(\boldsymbol{x}))/\partial\zeta_k\partial\zeta_l$ *are dominated by* $b(\varepsilon, \boldsymbol{x})$.

11.2 Deep Quantile Regression

So far, in network regression modeling, we have not addressed the question of prediction uncertainty. As mentioned in Remarks 4.2 on forecast evaluation, there are different sources that contribute to prediction uncertainty. There is the model and parameter estimation uncertainty, which may result in an inappropriate model choice, and there is the irreducible risk which comes from the fact that we forecast random variables which inherit a natural randomness that cannot be controlled.

We have discussed methods of evaluating model and parameter estimation error, such as the asymptotic normality of MLEs within GLMs, and we have discussed forecast dominance, the bootstrap method or the nagging predictor that allow one to assess the different sources of prediction uncertainty. However, we have not explicitly quantified these sources of uncertainty within the class of network regression models. We do an attempt in Sect. 11.4, below, by considering the fluctuations generated by bootstrap simulations. The irreducible risk can be assessed once we have a suitable statistical model; in Example 11.4 we have studied a gamma and an inverse Gaussian model on an explicit data set, and these models can be used, e.g., to calculate quantiles. In this section we consider a distribution-free approach that directly estimates these quantiles. Recall from Section 5.8.3 that quantiles are elicitable with the pinball loss as a strictly consistent loss function, see Theorem 5.33. This allows us to directly estimate the quantiles from the data.

11.2.1 Deep Quantile Regression: Single Quantile

In this section we present a way of assessing the irreducible risk which does not require a sophisticated model evaluation of distributional assumptions. Quantile regression is increasingly used in the machine learning community because it is a robust way of quantifying the irreducible risk, we refer to Meinshausen [270], Takeuchi et al. [350] and Richman [314]. We recall that quantiles are elicitable having the pinball loss as a strictly consistent loss function, see Theorem 5.33. We define a FN network regression model that allows us to directly estimate the quantiles based on the pinball loss. We therefore use an adapted version of the R code of Listing 9 in Richman [314], this adapted version has been proposed in Fissler et al. [130] to ensure that different quantiles respect monotonicity. For any two quantile levels $0 < \tau_1 < \tau_2 < 1$ we have

$$F^{-1}(\tau_1) \le F^{-1}(\tau_2), \tag{11.17}$$

where F^{-1} denotes the generalized inverse of distribution function F, see (5.80). If we simultaneously learn these quantiles for different quantile levels $\tau_1 < \tau_2$, we need to enforce the network to respect this monotonicity (11.17). This can be achieved by exploring a special network architecture in the output layer, and this is going to be presented in the next section.

We start by considering a single deep τ-quantile regression for a quantile level $\tau \in (0, 1)$. For datum $(Y, \boldsymbol{x})$ we consider the regression function

$$\boldsymbol{x} \mapsto F_{Y|\boldsymbol{x}}^{-1}(\tau) = g^{-1}\langle \boldsymbol{\beta}_\tau, \boldsymbol{z}^{(d:1)}(\boldsymbol{x})\rangle, \tag{11.18}$$

for a strictly monotone and smooth link function g, output parameter $\boldsymbol{\beta}_\tau \in \mathbb{R}^{q_d+1}$, and where $\boldsymbol{x} \mapsto \boldsymbol{z}^{(d:1)}(\boldsymbol{x})$ is a deep network. We add a lower index $Y|\boldsymbol{x}$ to the generalized inverse $F_{Y|\boldsymbol{x}}^{-1}$ to highlight that we consider the conditional distribution of Y, given feature $\boldsymbol{x} \in \mathcal{X}$. In the case of a deep FN network, (11.18) involves a network parameter $\boldsymbol{\vartheta} = (\boldsymbol{w}_1^{(1)}, \ldots, \boldsymbol{w}_{q_d}^{(d)}, \boldsymbol{\beta}_\tau)^\top$ that needs to be estimated. Of course, the deep network architecture $\boldsymbol{x} \mapsto \boldsymbol{z}^{(d:1)}(\boldsymbol{x})$ could also involve any other feature, such as CN or LSTM layers, embedding layers or a NLP text recognition feature. This would change the network architecture, but it would not change anything from a methodological viewpoint.

To estimate this regression parameter ϑ from independent data $(Y_i, \boldsymbol{x}_i)$, $1 \le i \le n$, we consider the objective function

$$\vartheta \mapsto \sum_{i=1}^{n} L_\tau \left(Y_i, g^{-1}\langle \boldsymbol{\beta}_\tau, \boldsymbol{z}^{(d:1)}(\boldsymbol{x}_i)\rangle \right),$$

with the strictly consistent pinball loss function L_τ for the τ-quantile. Alternatively, we could choose any other loss function satisfying Theorem 5.33, and we may try to find the asymptotically optimal one (similarly to Theorem 11.8). We refrain from doing so, but we mention Komunjer–Vuong [222]. Fitting the network parameter ϑ is then done in complete analogy to finding an optimal network parameter for network mean modeling. The only change is that we replace the deviance loss function by the pinball loss, e.g., in Listing 7.3 we have to exchange the loss function on line 5 correspondingly.

11.2.2 Deep Quantile Regression: Multiple Quantiles

We now turn our attention to the multiple quantile case that should satisfy the monotonicity requirement (11.17) for any quantile levels $0 < \tau_1 < \tau_2 < 1$. A separate deep quantile estimation for both quantile levels, as described in the previous section, may violate the monotonicity property, at least, in some part of the feature space $\mathcal{X}$, especially if the two quantile levels are close. Therefore, we enforce the monotonicity by a special choice of the network architecture.

For simplicity, in the remainder of this section, we assume that the response Y is positive, a.s. This implies for the quantiles $\tau \mapsto F_{Y|\boldsymbol{x}}^{-1}(\tau) \ge 0$, and we should choose a link function with $g^{-1} \ge 0$ in (11.18). To ensure the monotonicity (11.17) for the quantile levels $0 < \tau_1 < \tau_2 < 1$, we choose a second positive link function with $g_+^{-1} \ge 0$, and we set for multi-task forecasting

$$\begin{aligned} \boldsymbol{x} \mapsto & \left(F_{Y|\boldsymbol{x}}^{-1}(\tau_1),\ F_{Y|\boldsymbol{x}}^{-1}(\tau_2) \right)^\top \qquad (11.19) \\ &= \left(g^{-1}\langle \boldsymbol{\beta}_{\tau_1}, \boldsymbol{z}^{(d:1)}(\boldsymbol{x})\rangle,\ g^{-1}\langle \boldsymbol{\beta}_{\tau_1}, \boldsymbol{z}^{(d:1)}(\boldsymbol{x})\rangle + g_+^{-1}\langle \boldsymbol{\beta}_{\tau_2}, \boldsymbol{z}^{(d:1)}(\boldsymbol{x})\rangle \right)^\top \in \mathbb{R}_+^2, \end{aligned}$$

for a regression parameter $\vartheta = (\boldsymbol{w}_1^{(1)}, \ldots, \boldsymbol{w}_{q_d}^{(d)}, \boldsymbol{\beta}_{\tau_1}, \boldsymbol{\beta}_{\tau_2})^\top$. The positivity $g_+^{-1} \ge 0$ enforces the monotonicity in the two quantiles. We call (11.19) an *additive approach* as we start from a base level characterized by the smaller quantile $F_{Y|\boldsymbol{x}}^{-1}(\tau_1)$, and any bigger quantile is modeled by an additive increment. To ensure monotonicity for multiple quantiles we proceed recursively by choosing the lowest quantile as the initial base level.

We can also consider the upper quantile as the base level by multiplicatively lowering this upper quantile. Choose the (sigmoid) function $g_\sigma^{-1} \in (0, 1)$ and set for the *multiplicative approach*

$$
\begin{aligned}
\boldsymbol{x} \mapsto & \left(F_{Y|\boldsymbol{x}}^{-1}(\tau_1),\ F_{Y|\boldsymbol{x}}^{-1}(\tau_2)\right)^\top \qquad (11.20)\\
& = \left(g_\sigma^{-1}\langle \boldsymbol{\beta}_{\tau_1}, \boldsymbol{z}^{(d:1)}(\boldsymbol{x})\rangle\, g^{-1}\langle \boldsymbol{\beta}_{\tau_2}, \boldsymbol{z}^{(d:1)}(\boldsymbol{x})\rangle,\ g^{-1}\langle \boldsymbol{\beta}_{\tau_2}, \boldsymbol{z}^{(d:1)}(\boldsymbol{x})\rangle\right)^\top \in \mathbb{R}_+^2.
\end{aligned}
$$

Remark 11.10 In (11.19) and (11.20) we directly enforce the monotonicty by a corresponding regression function choice. Alternatively, we can also design a (plain-vanilla) multi-output network

$$
\begin{aligned}
\boldsymbol{x} \mapsto & \left(F_{Y|\boldsymbol{x}}^{-1}(\tau_1),\ F_{Y|\boldsymbol{x}}^{-1}(\tau_2)\right)^\top \qquad (11.21)\\
& = \left(g^{-1}\langle \boldsymbol{\beta}_{\tau_1}, \boldsymbol{z}^{(d:1)}(\boldsymbol{x})\rangle,\ g^{-1}\langle \boldsymbol{\beta}_{\tau_2}, \boldsymbol{z}^{(d:1)}(\boldsymbol{x})\rangle\right)^\top \in \mathbb{R}_+^2.
\end{aligned}
$$

If we just use a classical SGD fitting algorithm, we will likely result in a situation where the monotonicity will be violated in some part of the feature space. Kellner et al. [211] consider this problem. They add a penalization (regularization term) that punishes during SGD training network parameters that violate the monotonicity. Such a penalization can be constructed, e.g., with the ReLU function.

11.2.3 Lab: Deep Quantile Regression

We revisit the Swiss accident insurance data of Sect. 11.1.2, and we provide an example of a deep quantile regression using both the additive approach (11.19) and the multiplicative approach (11.20).
We select 5 different quantile levels $\mathcal{Q} = (\tau_1, \tau_2, \tau_3, \tau_4, \tau_5) = (10\%, 25\%, 50\%, 75\%, 90\%)$. We start with the additive approach (11.19). It requires to set $\tau_1 = 10\%$ as the base level, and the remaining quantile levels are modeled additively in a recursive way for $\tau_j < \tau_{j+1}$, $1 \le j \le 4$. The corresponding R code is given on lines 8–20 of Listing 11.3, and this compiles to the 5-dimensional output on line 22. For the multiplicative approach (11.20) we set $\tau_5 = 90\%$ as the base level, and the remaining quantile levels are received multiplicatively in a recursive way for $\tau_{j+1} > \tau_j$, $4 \ge j \ge 1$, see Listing 11.4. The additive and the multiplicative approaches take the extreme quantiles as initialization. One may also be interested in initializing the model in the median $\tau_3 = 50\%$, the smaller quantiles can then be received by the multiplicative approach and the bigger quantiles by the additive approach. We also explore this case and we call it the *mixed approach*.

Listing 11.3 Multiple FN quantile regression: additive approach

```
Design  = layer_input(shape = c(q0), dtype = 'float32', name = 'Design')
#
Network = Design %>%
            layer_dense(units=20, activation='tanh', name='FNLayer1') %>%
            layer_dense(units=15, activation='tanh', name='FNLayer2') %>%
            layer_dense(units=10, activation='tanh', name='FNLayer3')
#
q1  = Network %>% layer_dense(units=1, activation='exponential')
#
q20 = Network %>%  layer_dense(units=1, activation='exponential')
q2  = list(q1,q20) %>% layer_add()
#
q30 = Network %>%  layer_dense(units=1, activation='exponential')
q3  = list(q2,q30) %>% layer_add()
#
q40 = Network %>% layer_dense(units=1, activation='exponential')
q4  = list(q3,q40) %>% layer_add()
#
q50 = Network %>% layer_dense(units=1, activation='exponential')
q5  = list(q4,q50) %>% layer_add()
#
model = keras_model(inputs = list(Design), outputs = c(q1,q2,q3,q4,q5))
```

Listing 11.4 Multiple FN quantile regression: multiplicative approach

```
q5  = Network %>% layer_dense(units=1, activation='exponential')
#
q40 = Network %>% layer_dense(units=1, activation='sigmoid')
q4  = list(q5,q40) %>% layer_multiply()
#
q30 = Network %>% layer_dense(units=1, activation='sigmoid')
q3  = list(q4,q30) %>% layer_multiply()
#
q20 = Network %>% layer_dense(units=1, activation='sigmoid')
q2  = list(q3,q20) %>% layer_multiply()
#
q10 = Network %>% layer_dense(units=1, activation='sigmoid')
q1  = list(q2,q10) %>% layer_multiply()
```

Listing 11.5 Fitting a multiple FN quantile regression

```
Q_loss1 = function(y_true, y_pred){k_mean(k_maximum(y_true - y_pred, 0) * 0.1
                                 + k_maximum(y_pred - y_true, 0) * (1 - 0.1))}
Q_loss2 = function(y_true, y_pred){k_mean(k_maximum(y_true - y_pred, 0) * 0.25
                                 + k_maximum(y_pred - y_true, 0) * (1 - 0.25))}
Q_loss3 = function(y_true, y_pred){k_mean(k_maximum(y_true - y_pred, 0) * 0.5
                                 + k_maximum(y_pred - y_true, 0) * (1 - 0.5))}
Q_loss4 = function(y_true, y_pred){k_mean(k_maximum(y_true - y_pred, 0) * 0.75
                                 + k_maximum(y_pred - y_true, 0) * (1 - 0.75))}
Q_loss5 = function(y_true, y_pred){k_mean(k_maximum(y_true - y_pred, 0) * 0.9
                                 + k_maximum(y_pred - y_true, 0) * (1 - 0.9))}
#
model %>% compile(loss = list(Q_loss1,Q_loss2,Q_loss3,Q_loss4,Q_loss5),
                                                     optimizer = 'nadam')
```

These network architectures are fitted to the data using the pinball loss (5.81) for the quantile levels of $\mathcal{Q}$; note that the pinball loss requires the assumption of having a finite first moment. Listing 11.5 shows the choice of the pinball loss functions. We then fit the three architectures (additive, multiplicative and mixed) to our learning data $\mathcal{L}$, and we apply early stopping to prevent from over-fitting. Moreover, we consider the nagging predictor over 20 runs with different seeds to reduce the randomness coming from SGD fitting.

In Table 11.6 we give the out-of-sample pinball losses on the test data $\mathcal{T}$ of the three considered approaches, and illustrating the 5 quantile levels of $\mathcal{Q}$. The losses of the three approaches are rather close, giving a slight preference to the mixed approach, but the other two approaches seem to be competitive, too. We further analyze these quantile regression models by considering the empirical coverage ratios defined by

$$\widehat{\tau}_j = \frac{1}{T}\sum_{t=1}^{T} \mathbb{1}_{\left\{Y_t^\dagger \le \widehat{F}^{-1}_{Y|\boldsymbol{x}_t^\dagger}(\tau_j)\right\}}, \tag{11.22}$$

where $\widehat{F}^{-1}_{Y|\boldsymbol{x}_t^\dagger}(\tau_j)$ is the estimated quantile for level τ_j and feature $\boldsymbol{x}_t^\dagger$. Remark that the coverage ratios (11.22) correspond to the identification functions that are essentially the derivatives of the pinball losses, we refer to Dimitriadis et al. [106]. Table 11.7 reports these out-of-sample coverage ratios on the test data $\mathcal{T}$. From these results we conclude that on the portfolio level the quantiles are matched rather well.

In Fig. 11.8 we illustrate the estimated out-of-sample quantiles $\widehat{F}^{-1}_{Y|\boldsymbol{x}_t^\dagger}(\tau_j)$ for individual claims on the quantile levels $\tau_j \in \{10\%, 25\%, 50\%, 75\%, 90\%\}$ (cyan, blue, black, blue, cyan colors) using the mixed approach. The x-axis considers the logged estimated medians $\widehat{F}^{-1}_{Y|\boldsymbol{x}_t^\dagger}(50\%)$. We observe heteroskedasticity resulting in quantiles that are not ordered w.r.t. the median (black line). This supports the multiple deep quantile regression model because we cannot (simply) extrapolate the median to receive the other quantiles.

In the final step we compare the estimated quantiles $\widehat{F}^{-1}_{Y|\boldsymbol{x}}(\tau_j)$ from the mixed deep quantile regression approach to the ones that can be calculated from the fitted inverse Gaussian model using the double FN network approach of Example 11.4. In the latter model we estimate the mean $\widehat{\mu}(\boldsymbol{x})$ and the dispersion $\widehat{\varphi}(\boldsymbol{x})$ with two FN networks, which then allow us to calculate the quantiles using the inverse Gaussian distributional assumption. Note that we cannot calculate the quantiles in Tweedie's family with power variance parameter $p = 2.5$ because there is no

Table 11.6 Out-of-sample pinball losses of quantile regressions using the additive, the multiplicative and the mixed approaches; nagging predictors over 20 different seeds

	Out-of-sample losses on $\mathcal{T}$				
	10%	25%	50%	75%	90%
Additive approach	171.20	412.78	765.60	988.78	936.31
Multiplicative approach	171.18	412.87	766.04	988.59	936.57
Mixed approach	171.15	412.55	764.60	988.15	935.50

Table 11.7 Out-of-sample coverage ratios $\widehat{\tau}_j$ below the estimated deep FN quantile estimates $\widehat{F}^{-1}_{Y|x_t^\dagger}(\tau_j)$

	Out-of-sample coverage ratios				
	10%	25%	50%	75%	90%
Additive approach	10.27%	25.30%	50.19%	75.08%	90.03%
Multiplicative approach	10.18%	25.15%	49.64%	75.14%	90.22%
Mixed approach	10.13%	25.03%	50.32%	75.20%	90.08%

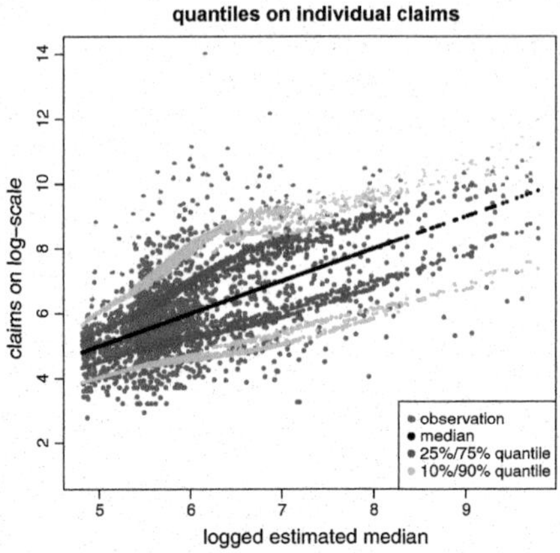

Fig. 11.8 Estimated out-of-sample quantiles $\widehat{F}^{-1}_{Y|x_t^\dagger}(\tau_j)$ of 2'000 randomly selected individual claims on the quantile levels $\tau_j \in \{10\%, 25\%, 50\%, 75\%, 90\%\}$ (cyan, blue, black, blue, cyan colors) using the mixed approach, the red dots are the out-of-sample observations $Y_t^\dagger$; the x-axis gives $\log \widehat{F}^{-1}_{Y|x_t^\dagger}(50\%)$ (also corresponding to the black diagonal line)

closed form of the distribution function. Figure 11.9 compares the two approaches on the quantile levels of $\mathcal{Q}$. Overall we observe a reasonably good match though it is not perfect. The small quantiles for level $\tau_1 = 10\%$ seem slightly under-estimated by the inverse Gaussian approach (see Fig. 11.9 (top-left)), whereas big quantiles $\tau_4 = 75\%$ and $\tau_5 = 90\%$ seem more conservative in the inverse Gaussian approach (see Fig. 11.9 (bottom)). This may indicate that the inverse Gaussian distribution does not fully fit the data, i.e., that one cannot fully recover the true quantiles from the mean $\widehat{\mu}(\boldsymbol{x})$, the dispersion $\widehat{\varphi}(\boldsymbol{x})$ and an inverse Gaussian assumption. There are two ways to further explore these issues. One can either choose other distributional assumptions which may better match the properties of the data, this further explores the distributional approach. Alternatively, Theorem 5.33 allows us to choose loss functions different from the pinball loss, i.e., one could consider different increasing functions G in that theorem to further explore the distribution-free approach. In general, any increasing choice of the function G leads to a strictly consistent quantile estimation (this is an asymptotic statement), but these choices may have different finite sample properties. Following Komunjer–Vuong [222], we can determine asymptotically efficient choices for G. This would require feature dependent choices $G_{\boldsymbol{x}_i}(y) = F_{Y|\boldsymbol{x}_i}(y)$, where $F_{Y|\boldsymbol{x}_i}$ is the (true) distribution of Y_i, conditionally given $\boldsymbol{x}_i$. This requires the knowledge of the true distribution, and Komunjer–Vuong [222] derive asymptotic efficiency when replacing this true

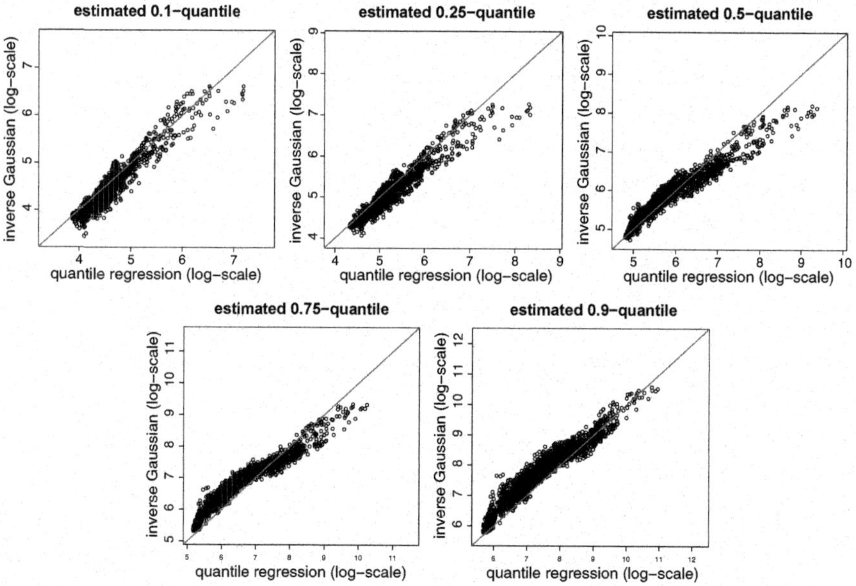

Fig. 11.9 Inverse Gaussian quantiles vs. deep quantile regression estimates of 2'000 randomly selected claims on the quantile levels of $\mathcal{Q} = (10\%, 25\%, 50\%, 75\%, 90\%)$

distribution by a non-parametric estimator, this is in spirit similar to Theorem 11.8. We refrain from giving more details but refer to the corresponding paper.

11.3 Deep Composite Model Regression

We have established a deep quantile regression in the previous section. Next we jointly estimate quantiles and conditional tail expectations (CTEs), leading to a composite regression model that has a splicing point determined by a quantile level; for composite models we refer to Sect. 6.4.4. This is exactly the proposal of Fissler et al. [130] which we are going to present in this section. Note that having a composite model allows us to have different distributions and regression structures below and above the splicing point, e.g., we can have a more heavy-tailed model in the upper tail using a different feature engineering from the main body of the data.

11.3.1 Joint Elicitability of Quantiles and Expected Shortfalls

In the previous examples we have seen that the distributional models may misestimate the true tail of the data because model fitting often pays more attention to an accurate model fit in the main body of the data. An idea is to directly estimate this tail in a distribution-free way by considering the (upper) CTE

$$\mathrm{CTE}_\tau^+(Y|\boldsymbol{x}) = \mathbb{E}\left[Y \,\middle|\, Y > F_{Y|\boldsymbol{x}}^{-1}(\tau),\, \boldsymbol{x}\right], \tag{11.23}$$

for a given quantile level $\tau \in (0, 1)$. The problem with (11.23) is that this is not an elicitable quantity, i.e., there is no loss/scoring function that is strictly consistent for the CTE functional.

If the distribution function $F_{Y|\boldsymbol{x}}$ is continuous, we can rewrite the upper CTE as follows, see Lemma 2.16 in McNeil et al. [268] and (11.35) below,

$$\mathrm{CTE}_\tau^+(Y|\boldsymbol{x}) = \mathrm{ES}_\tau^+(Y|\boldsymbol{x}) = \frac{1}{1-\tau}\int_\tau^1 F_{Y|\boldsymbol{x}}^{-1}(p)\,dp \;\geq\; F_{Y|\boldsymbol{x}}^{-1}(\tau). \tag{11.24}$$

This second object $\mathrm{ES}_\tau^+(Y|\boldsymbol{x})$ is called the upper expected shortfall (ES) of Y, given $\boldsymbol{x}$, on the security level τ. Fissler–Ziegel [131] and Fissler et al. [132] have proved that $\mathrm{ES}_\tau^+(Y|\boldsymbol{x})$ is *jointly* elicitable with the τ-quantile $F_{Y|\boldsymbol{x}}^{-1}(\tau)$. That is, there is a strictly consistent bivariate loss function that allows one to jointly estimate the τ-quantile and the corresponding ES. In fact, Corollary 5.5 of Fissler–Ziegel [131] give the full characterization of the strictly consistent bivariate loss functions for the joint elicitability of the τ-quantile and the ES; note that Fissler–Ziegel [131] use a different sign convention. This result is used in Guillén et al. [175] for the joint estimation of the quantile and the ES within a GLM. Guillén et al. [175] use a two-step approach to fit the quantile and the ES.

Fissler et al. [130] extend the results of Fissler–Ziegel [131], allowing for the joint estimation of the *composite triplet* consisting of the lower ES, the τ-quantile and the upper ES. This gives us a composite model that has the τ-quantile as splicing point. The beauty of this approach is that we can fit (in one step) a deep learning model to the upper and the lower ES, and perform a (potentially different) regression in both parts of the distribution. The lower CTE and the lower ES are defined by, respectively,

$$\mathrm{CTE}_\tau^-(Y|\boldsymbol{x}) = \mathbb{E}\left[Y \,\middle|\, Y \leq F_{Y|\boldsymbol{x}}^{-1}(\tau),\, \boldsymbol{x}\right],$$

and

$$\mathrm{ES}_\tau^-(Y|\boldsymbol{x}) = \frac{1}{\tau}\int_0^\tau F_{Y|\boldsymbol{x}}^{-1}(p)\,dp \;\leq\; F_{Y|\boldsymbol{x}}^{-1}(\tau).$$

Again, in case of a continuous distribution function $F_{Y|\boldsymbol{x}}$ we have the following identity $\mathrm{CTE}_\tau^-(Y|\boldsymbol{x}) = \mathrm{ES}_\tau^-(Y|\boldsymbol{x})$. From the lower and upper CTEs we receive the mean of Y, given $\boldsymbol{x}$, by

$$\mu(\boldsymbol{x}) = \mathbb{E}[Y|\boldsymbol{x}] = \tau\,\mathrm{CTE}_\tau^-(Y|\boldsymbol{x}) + (1-\tau)\,\mathrm{CTE}_\tau^+(Y|\boldsymbol{x}). \tag{11.25}$$

We introduce the auxiliary scoring functions

$$S_\tau^-(y,a) = \left(\mathbb{1}_{\{y\le a\}} - \tau\right) a - \mathbb{1}_{\{y\le a\}} y,$$
$$S_\tau^+(y,a) = \left(1 - \tau - \mathbb{1}_{\{y>a\}}\right) a + \mathbb{1}_{\{y>a\}} y \;=\; S_\tau^-(y,a) + y,$$

for $y, a \in \mathbb{R}$ and for $\tau \in (0,1)$. These auxiliary functions consider only the part of the pinball loss (5.81) that depends on action a, and we get the pinball loss as follows

$$L_\tau(y,a) \;=\; S_\tau^-(y,a) + \tau y \;=\; S_\tau^+(y,a) - (1-\tau)y.$$

Therefore, all three functions provide strictly consistent scoring functions for the τ-quantile, but only the pinball loss satisfies the calibration property (L0) on page 92.

For the following theorem we recall the general definition of the τ-quantile $Q_\tau(F_{Y|\boldsymbol{x}})$ of a distribution function $F_{Y|\boldsymbol{x}}$, see (5.82).

Theorem 11.11 (Theorem 2.8 of Fissler et al. [130], Without Proof) *Choose $\tau \in (0,1)$ and let $\mathcal{F}$ contain only distributions with a finite first moment, and being supported in the interval $\mathfrak{C} \subseteq \mathbb{R}$. The loss function $L : \mathfrak{C} \times \mathfrak{C}^3 \to \mathbb{R}_+$ of the form*

$$L(y; e^-, q, e^+) = (G(y) - G(q))\left(\tau - \mathbb{1}_{\{y\le q\}}\right) + \left\langle \nabla\Psi(e^-,e^+), \begin{pmatrix} e^- + \frac{1}{\tau} S_\tau^-(y,q) \\ e^+ - \frac{1}{1-\tau} S_\tau^+(y,q) \end{pmatrix} \right\rangle - \Psi(e^-,e^+) + \Psi(y,y), \tag{11.26}$$

is strictly consistent for the composite triplet $(\mathrm{ES}_\tau^-, Q_\tau, \mathrm{ES}_\tau^+)$ relative to the class $\mathcal{F}$, if Ψ is strictly convex with (sub-)gradient $\nabla\Psi$ such that for all $(e^-, e^+) \in \mathfrak{C}^2$ the function

$$q \mapsto G_{e^-,e^+}(q) = G(q) + \frac{1}{\tau}\frac{\partial}{\partial e^-}\Psi(e^-,e^+) q - \frac{1}{1-\tau}\frac{\partial}{\partial e^+}\Psi(e^-,e^+) q, \tag{11.27}$$

is strictly increasing, and if $\mathbb{E}_F[|G(Y)|] < \infty$, $\mathbb{E}_F[|\Psi(Y,Y)|] < \infty$ for all $Y \sim F \in \mathcal{F}$.

This opens the door for regression modeling of CTEs for continuous distribution functions $F_{Y|\boldsymbol{x}}$, $\boldsymbol{x} \in \mathcal{X}$. Namely, we can choose a regression function ξ_ϑ with a three-dimensional output

$$\boldsymbol{x} \in \mathcal{X} \;\mapsto\; \xi_\vartheta(\boldsymbol{x}) \in \mathfrak{C}^3,$$

depending on a regression parameter $\boldsymbol{\vartheta}$. This regression function is now used to describe the composite triplet $(\mathrm{ES}_\tau^-(Y|\boldsymbol{x}), F_{Y|\boldsymbol{x}}^{-1}(\tau), \mathrm{ES}_\tau^+(Y|\boldsymbol{x}))$. Having i.i.d. data $(Y_i, \boldsymbol{x}_i)$, $1 \le i \le n$, it can be fitted by solving

$$\widehat{\boldsymbol{\vartheta}} = \underset{\boldsymbol{\vartheta}}{\arg\min} \, \frac{1}{n} \sum_{i=1}^{n} L\left(Y_i; \xi_{\boldsymbol{\vartheta}}(\boldsymbol{x}_i)\right), \tag{11.28}$$

with loss function L given by (11.26). This then provides us with the estimates for the composite triplet

$$\boldsymbol{x} \mapsto \xi_{\widehat{\boldsymbol{\vartheta}}}(\boldsymbol{x}) = \left(\widehat{\mathrm{ES}}_\tau^-(Y|\boldsymbol{x}), \widehat{F}_{Y|\boldsymbol{x}}^{-1}(\tau), \widehat{\mathrm{ES}}_\tau^+(Y|\boldsymbol{x})\right).$$

There remains the choice of the functions G and Ψ, such that Ψ is strictly convex and G_{e^-,e^+}, defined in (11.27), is strictly increasing. Section 2.3 in Fissler et al. [130] discusses possible choices. A simple choice is to select the identity function $G(y) = y$ (which gives the pinball loss on the first line of (11.26)) and

$$\Psi(e^-, e^+) = \psi_1(e^-) + \psi_2(e^+),$$

with ψ_1 and ψ_2 strictly convex and with (sub-)gradients $\psi_1' > 0$ and $\psi_2' < 0$. Inserting this choice into (11.26) provides the loss function

$$L(y; e^-, q, e^+) = \left[1 + \frac{\psi_1'(e^-)}{\tau} + \frac{-\psi_2'(e^+)}{1-\tau}\right] L_\tau(y, q) + D_{\psi_1}(y, e^-) + D_{\psi_2}(y, e^+), \tag{11.29}$$

where $L_\tau(y, q)$ is the pinball loss (5.81) and D_{ψ_1} and D_{ψ_2} are Bregman divergences (2.28). There remains the choices of ψ_1 and ψ_2 which should be strictly convex, the first one being strictly increasing and the second one being strictly decreasing.

We restrict ourselves to strictly convex functions ψ on the positive real line $\mathbb{R}_+$, i.e., for positive claims $Y > 0$, a.s. For $b \in \mathbb{R}$, we consider the following functions on $\mathbb{R}_+$

$$\psi^{(b)}(y) = \begin{cases} \frac{1}{b(b-1)} y^b & \text{for } b \neq 0 \text{ and } b \neq 1, \\ -1 - \log(y) & \text{for } b = 0, \\ y\log(y) - y & \text{for } b = 1. \end{cases} \tag{11.30}$$

We compute the first and second derivatives. These are for $y > 0$ given by

$$\frac{\partial}{\partial y}\psi^{(b)}(y) = \begin{cases} \frac{1}{b-1}y^{b-1} & \text{for } b \neq 1, \\ \log(y) & \text{for } b = 1, \end{cases} \qquad \text{and} \qquad \frac{\partial^2}{\partial y^2}\psi^{(b)}(y) = y^{b-2} > 0.$$

Thus, for any $b \in \mathbb{R}$ we have a convex function, and this convex function is decreasing on $\mathbb{R}_+$ for $b < 1$ and increasing for $b > 1$. Therefore, we have to select $b > 1$ for ψ_1 and $b < 1$ for ψ_2 to get suitable choices in (11.29). Interestingly, these choices correspond to Lemma 11.2 with power variance parameters $p = 2 - b$, i.e., they provide us with Bregman divergences from Tweedie's distributions. However, (11.30) is more general, because it allows us to select any $b \in \mathbb{R}$, whereas for power variance parameters $p \in (0, 1)$ there do not exist any Tweedie's distributions, see Theorem 2.18.

In view of Lemma 11.2 and using the fact that unit deviances $\mathfrak{d}_p$ are Bregman divergences, we select a power variance parameter $p = 2 - b > 1$ for ψ_2 and we select the Gaussian model $p = 2 - b = 0$ for ψ_1. This gives us the special choice for the loss function (11.29) for strictly positive claims $Y > 0$, a.s.,

$$L(y; e^-, q, e^+) = \left[1 + \frac{\eta_1 e^-}{\tau} + \frac{\eta_2 (e^+)^{1-p}}{(1-\tau)(p-1)}\right] L_\tau(y, q) + \frac{\eta_1}{2}\mathfrak{d}_0(y, e^-) + \frac{\eta_2}{2}\mathfrak{d}_p(y, e^+), \tag{11.31}$$

with the Gaussian unit deviance $\mathfrak{d}_0(y, e^-) = (y - e^-)^2$ and Tweedie's unit deviance $\mathfrak{d}_p$ with power variance parameter $p > 1$, see Sect. 11.1.1. The additional constants $\eta_1, \eta_2 > 0$ are used to balance the contributions of the individual terms to the total loss. Typically, we choose $p \geq 2$ for the upper ES reflecting claim size models. This choice for ψ_2 implies that the residuals are weighted inversely proportional to the corresponding variances μ^p within Tweedie's family, see (11.5). Using this loss function (11.31) in (11.28) allows us to estimate the composite triplet $(\mathrm{ES}^-_\tau(Y|\boldsymbol{x}), F^{-1}_{Y|\boldsymbol{x}}(\tau), \mathrm{ES}^+_\tau(Y|\boldsymbol{x}))$ with a strictly consistent loss function.

11.3.2 Lab: Deep Composite Model Regression

The joint elicitability of Theorem 11.11 allows us to directly estimate these functionals for a fixed quantile level $\tau \in (0, 1)$. In a similar way to quantile regression we set up a FN network that respects the monotonicity $\mathrm{ES}^-_\tau(Y|\boldsymbol{x}) \leq$

$F^{-1}_{Y|\boldsymbol{x}}(\tau) \le \mathrm{ES}^+_\tau(Y|\boldsymbol{x})$. We set for the regression function in the additive approach for multi-task learning

$$\begin{aligned}
\boldsymbol{x} \mapsto &\left(\mathrm{ES}^-_\tau(Y|\boldsymbol{x}),\ F^{-1}_{Y|\boldsymbol{x}}(\tau),\ \mathrm{ES}^+_\tau(Y|\boldsymbol{x})\right)^\top \\
&= \Big(g^{-1}\langle \boldsymbol{\beta}_1, \boldsymbol{z}^{(d:1)}(\boldsymbol{x})\rangle,\ g^{-1}\langle \boldsymbol{\beta}_1, \boldsymbol{z}^{(d:1)}(\boldsymbol{x})\rangle + g_+^{-1}\langle \boldsymbol{\beta}_2, \boldsymbol{z}^{(d:1)}(\boldsymbol{x})\rangle, \\
&\qquad g^{-1}\langle \boldsymbol{\beta}_1, \boldsymbol{z}^{(d:1)}(\boldsymbol{x})\rangle + g_+^{-1}\langle \boldsymbol{\beta}_2, \boldsymbol{z}^{(d:1)}(\boldsymbol{x})\rangle + g_+^{-1}\langle \boldsymbol{\beta}_3, \boldsymbol{z}^{(d:1)}(\boldsymbol{x})\rangle\Big)^\top \in \mathbb{A},
\end{aligned} \tag{11.32}$$

for link functions g and g_+ with $g_+^{-1} \ge 0$, deep FN network $\boldsymbol{z}^{(d:1)} : \mathbb{R}^{q_0+1} \to \mathbb{R}^{q_d+1}$, regression parameters $\boldsymbol{\beta}_1, \boldsymbol{\beta}_2, \boldsymbol{\beta}_3 \in \mathbb{R}^{q_d+1}$, and with the action space $\mathbb{A} = \{(e^-, q, e^+) \in \mathbb{R}^3_+; e^- \le q \le e^+\}$ for positive claims. We also remind of Remark 11.10 for a different way of modeling the monotonicity.

Fitting this model is similar to the multiple deep quantile regression presented in Listings 11.3 and 11.5. There is one important difference though. Namely, we do not have multiple outputs and multiple loss functions, but we have a three-dimensional output with a single loss function (11.31) simultaneously evaluating all three components of the output (11.32). Listing 11.6 gives this loss for the inverse Gaussian case $p = 3$ in (11.31).

Listing 11.6 Loss function (11.31) for $p = 3$

```
Bregman_IG = function(y_true, y_pred){
  k_mean( (k_maximum(y_true[,1]-y_pred[,2],0)*tau0 +
                                   k_maximum(y_pred[,2]-y_true[,1],0)*(1-tau0) ) *
          ( 1 + eta1*y_pred[,1]/tau0 + eta2*y_pred[,3]^(-2)/(2*(1-tau0)) ) +
          eta1*(y_true[,1]-y_pred[,1])^2/2 +
          eta2*((y_true[,1]-y_pred[,3])^2/(y_pred[,3]^2*y_true[,1]))/2 )}
```

We revisit the Swiss accident insurance data of Sect. 11.2.3. We again use a FN network of depth $d = 3$ with $(q_1, q_2, q_3) = (20, 15, 10)$ neurons, hyperbolic tangent activation, two-dimensional embedding layers for the categorical features, exponential output activations for g^{-1} and g_+^{-1}, and the additive structure (11.32). We implement the loss function (11.31) for quantile level $\tau = 90\%$ and with power variance parameter $p = 3$, see Listing 11.6. This implies that for the upper ES estimation we scale residuals with $V(\mu) = \mu^3$, see (11.5). We then run an initial calibration of this FN network. Based on this initial calibration we can calculate the three loss contributions in (11.31) coming from the composite triplet. Based on these figures we choose the constants $\eta_1, \eta_2 > 0$ in (11.31) so that all three terms of the composite triplet contribute equally to the total loss. For the remainder of our calibration we hold on to these choices of η_1 and η_2.

We calibrate this deep FN architecture to the learning data $\mathcal{L}$, using the strictly consistent loss function (11.31) for the composite triplet $(\mathrm{ES}^-_{90\%}(Y|\boldsymbol{x}), F^{-1}_{Y|\boldsymbol{x}}(90\%), \mathrm{ES}^+_{90\%}(Y|\boldsymbol{x}))$, and to reduce the randomness in prediction we average over 20 early stopped SGD calibrations with different seeds (nagging predictor).

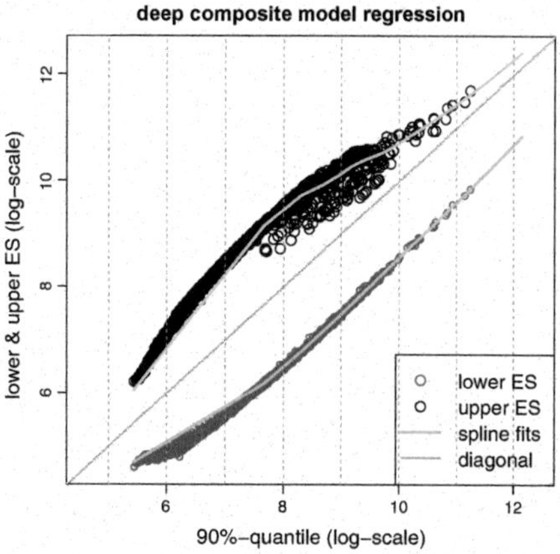

Fig. 11.10 Comparison of the estimated lower $\widehat{\mathrm{ES}}^-_{90\%}(Y|\boldsymbol{x}_t^\dagger)$ and the estimated upper $\widehat{\mathrm{ES}}^+_{90\%}(Y|\boldsymbol{x}_t^\dagger)$ against the estimated 90%-quantile $\widehat{F}^{-1}_{Y|\boldsymbol{x}_t^\dagger}(90\%)$ in the deep composite regression

Figure 11.10 shows the estimated lower and upper ES against the corresponding 90%-quantile estimates for 2'000 randomly selected insurance claims $\boldsymbol{x}_t^\dagger$. The diagonal orange line shows the estimated 90%-quantiles $\widehat{F}^{-1}_{Y|\boldsymbol{x}_t^\dagger}(90\%)$, and the cyan lines give spline fits to the estimated lower and upper ES. It is clearly visible that these respect the ordering

$$\widehat{\mathrm{ES}}^-_{90\%}(Y|\boldsymbol{x}_t^\dagger) \le \widehat{F}^{-1}_{Y|\boldsymbol{x}_t^\dagger}(90\%) \le \widehat{\mathrm{ES}}^+_{90\%}(Y|\boldsymbol{x}_t^\dagger),$$

for fixed features $\boldsymbol{x}_t^\dagger \in \mathcal{X}$.

The deep quantile regression has been back-tested using the coverage ratios (11.22). Back-testing the ES is more difficult, the standalone ES is not elicitable, and the ES can only be back-tested jointly with the corresponding quantile. The part of the joint identification function that corresponds to the ES is given by, see (4.2)–(4.3) in Fissler et al. [130],

$$\widehat{v}_- = \frac{1}{T}\sum_{t=1}^{T} \widehat{\mathrm{ES}}^-_\tau(Y|\boldsymbol{x}_t^\dagger) - \frac{Y_t^\dagger \mathbb{1}_{\left\{Y_t^\dagger \le \widehat{F}^{-1}_{Y|\boldsymbol{x}_t^\dagger}(\tau)\right\}} + \widehat{F}^{-1}_{Y|\boldsymbol{x}_t^\dagger}(\tau)\left(\tau - \mathbb{1}_{\left\{Y_t^\dagger \le \widehat{F}^{-1}_{Y|\boldsymbol{x}_t^\dagger}(\tau)\right\}}\right)}{\tau}, \tag{11.33}$$

and

$$\widehat{v}_+ = \frac{1}{T}\sum_{t=1}^{T} \widehat{\mathrm{ES}}^+_\tau(Y|\boldsymbol{x}_t^\dagger) - \frac{Y_t^\dagger \mathbb{1}_{\left\{Y_t^\dagger > \widehat{F}^{-1}_{Y|\boldsymbol{x}_t^\dagger}(\tau)\right\}} + \widehat{F}^{-1}_{Y|\boldsymbol{x}_t^\dagger}(\tau)\left(\mathbb{1}_{\left\{Y_t^\dagger \le \widehat{F}^{-1}_{Y|\boldsymbol{x}_t^\dagger}(\tau)\right\}} - \tau\right)}{1-\tau}. \tag{11.34}$$

These (empirical) identifications should be close too zero if the model fits the data.

Remark that the latter terms in (11.33)–(11.34) describe the lower and upper ES also in the case of non-continuous distribution functions because we have the identity

$$\mathrm{ES}_\tau^-(Y|\boldsymbol{x}) = \frac{1}{\tau}\left(\mathbb{E}\left[Y\mathbb{1}_{\{Y\le F_{Y|\boldsymbol{x}}^{-1}(\tau)\}}\middle|\boldsymbol{x}\right] + F_{Y|\boldsymbol{x}}^{-1}(\tau)\left(\tau - F_{Y|\boldsymbol{x}}\left(F_{Y|\boldsymbol{x}}^{-1}(\tau)\right)\right)\right), \tag{11.35}$$

the second term being zero for a continuous distribution $F_{Y|\boldsymbol{x}}$, but it is needed for non-continuous distribution functions.

We compare the deep composite regression results of this section to the deep gamma and inverse Gaussian models using a double FN network for dispersion modeling, see Sect. 11.1.3. This requires to calculate the ES in the gamma and the inverse Gaussian models. This can be done within the EDF, see Landsman–Valdez [233]. The upper ES in the gamma model $Y \sim \Gamma(\alpha, \beta)$ is given by, see (6.47),

$$\mathbb{E}\left[Y\middle|Y > F_Y^{-1}(\tau)\right] = \frac{\alpha}{\beta}\left(\frac{1-\mathcal{G}\left(\alpha+1, \beta F_Y^{-1}(\tau)\right)}{1-\tau}\right),$$

where $\mathcal{G}$ is the scaled incomplete gamma function (6.48) and $F_Y^{-1}(\tau)$ is the τ-quantile of $\Gamma(\alpha, \beta)$.

Example 4.3 of Landsman–Valdez [233] gives the inverse Gaussian case (2.8) with $\alpha, \beta > 0$

$$\begin{aligned}\mathbb{E}\left[Y\middle|Y > F_Y^{-1}(\tau)\right] = \frac{\alpha}{\beta}&\left(1+\frac{1/\alpha}{1-\tau}\sqrt{F_Y^{-1}(\tau)}\varphi(z_\tau^{(1)})\right)\\ &+\frac{\alpha}{\beta}\frac{1/\alpha}{1-\tau}e^{2\alpha\beta}\left(2\alpha\Phi(-z_\tau^{(2)}) - \sqrt{F_Y^{-1}(\tau)}\varphi(-z_\tau^{(2)})\right),\end{aligned}$$

where φ and Φ are the standard Gaussian density and distribution, respectively, $F_Y^{-1}(\tau)$ is the τ-quantile of the inverse Gaussian distribution and

$$z_\tau^{(1)} = \frac{\alpha}{\sqrt{F_Y^{-1}(\tau)}}\left(\frac{F_Y^{-1}(\tau)}{\alpha/\beta} - 1\right) \quad \text{and} \quad z_\tau^{(2)} = \frac{\alpha}{\sqrt{F_Y^{-1}(\tau)}}\left(\frac{F_Y^{-1}(\tau)}{\alpha/\beta} + 1\right).$$

This now allows us to calculate the identifications (11.33)–(11.34) in the fitted deep double networks using the gamma and the inverse Gaussian distributions of Sect. 11.1.3.

Table 11.8 shows the out-of-sample coverage ratios and the identifications of the deep composite regression and the two distributional approaches. These figures suggest that the gamma model is not competitive; the deep composite model has the most precise coverage ratio. In terms of the ES identification terms, the deep

Table 11.8 Out-of-sample coverage ratios $\widehat{\tau}$ and identifications $\widehat{\upsilon}_-$ and $\widehat{\upsilon}_+$ of the deep composite regression model and the deep double networks in the gamma and inverse Gaussian cases

	Coverage ratio $\tau = 90\%$	Lower ES identification $\widehat{\upsilon}_-$	Upper ES identification $\widehat{\upsilon}_+$
Deep composite model	90.12%	32.9	-143.5
Deep double network gamma	93.51%	356.6	-2'409.0
Deep double network inverse Gaussian	92.56%	−13.0	115.1

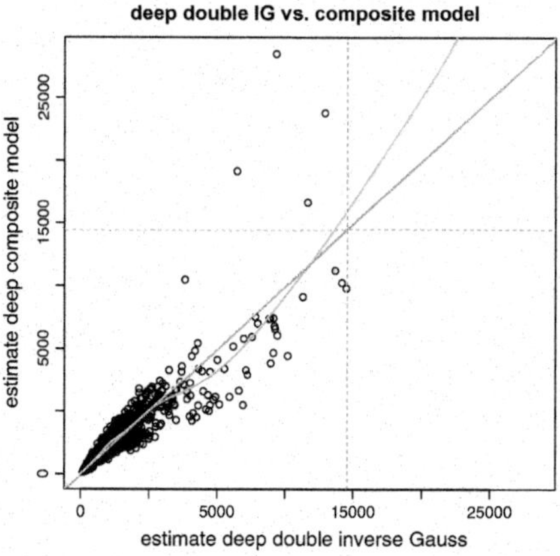

Fig. 11.11 Comparison of the estimated means from the deep double inverse Gaussian model and the deep composite model (11.25)

composite model and the double network with inverse Gaussian claim sizes are comparably accurate (out-of-sample) determining the lower and upper 90% ES. Finally, we paste the lower and upper ES from the deep composite regression model according to (11.25). This gives us an estimated mean (under a continuous distribution function)

$$\widehat{\mu}(\boldsymbol{x}) = \widehat{\mathbb{E}}[Y|\boldsymbol{x}] = \tau\, \widehat{\mathrm{ES}}_\tau^-(Y|\boldsymbol{x}) + (1-\tau)\, \widehat{\mathrm{ES}}_\tau^+(Y|\boldsymbol{x}).$$

Figure 11.11 compares these estimates of the deep composite regression model to the deep double inverse Gaussian model estimates. The black dots show 2'000 randomly selected claims $\boldsymbol{x}_t^\dagger$, and the cyan line gives a spline fit to all out-of-sample claims in $\mathcal{T}$. The body of the estimates is rather similar in both approaches but the deep composite approach provides more large estimates, the dotted orange lines show the maximum estimate from the deep double inverse Gaussian model.

We conclude that in the case where no member of the EDF reflects the properties of the data in the tail, the deep composite regression approach presented in this section provides an alternative method for mean estimation that allows for separate models in the main body and the tail of the data. Fixing the quantile level allows for a straightforward fitting in one step, this is in contrast to the composite models where we fix the splicing point. The latter approaches are more difficult in fitting, e.g., using the EM algorithm.

11.4 Model Uncertainty: A Bootstrap Approach

As described in Sect. 4, there are different sources of prediction uncertainty when forecasting random variables. There is the irreducible risk that comes from the fact that we try to predict random variables. This source of uncertainty is always present, even if we know the true data generating mechanism, i.e., it is irreducible. In most applied situations we do not know the true data generating mechanism which results in additional prediction uncertainty. Within GLMs this source of uncertainty has mainly been allocated to parameter estimation uncertainty deriving from the fact that we estimate the parameters from a finite sample, we refer to Sects. 3.4 and 11.1.4 on asymptotic results. In network modeling, the situation is more complicated. Firstly, we have seen that there is no best network regression model even if the architecture and the hyper-parameters are fully specified. In Fig. 7.18 we have seen that in a claim frequency context the different solutions from an early stopped SGD fitting can have a coefficient of variation of up to 40% on the individual policy level, on average these coefficients of variation were around 10%. This has led to the consideration of network ensembling and the nagging predictor in Sect. 7.4.4. These considerations have been based on a fixed learning data set $\mathcal{L}$. In this section, we assume that also the learning data set $\mathcal{L}$ may look differently by considering different realizations of the (randomly generated) observations Y_i. To reflect this source of randomness in outcomes we bootstrap new data from $\mathcal{L}$ by exploring a non-parametric bootstrap with random drawings with replacements from $\mathcal{L}$, see Sect. 4.3.1. This will allow us to study the volatility implied in estimation by considering a different set of observations, i.e., a different sample.

Ideally we would like to generate new observations from the true data generating mechanism, but, since this mechanism is not known, we can at best generate data from an estimated model. If we rely on a distributional model, we may suffer from model error, e.g., in Sect. 11.3 we have seen that it is rather difficult to specify a distributional regression model that has the right tail behavior. Therefore, we may give preference to a distribution-free approach. Non-parametric bootstrapping is such a distribution-free approach, the disadvantage being that we cannot enrich the existing observations by new observations, but we can only rearrange the available observations.

We revisit the robust representation learning approach of Sect. 11.1.2 on the same Swiss accident insurance data as explored in that section. In particular, we reconsider the deep multi-output models introduced in (11.6) and studied in Table 11.3 for power variance parameters $p = 2, 2.5, 3$ (and constant dispersion parameter). We perform exactly the same analysis, here, however we consider for this analysis bootstrapped data $\mathcal{L}^*$ for model fitting.

First, we fit 100 times the same deep FN network architecture as in (11.6) with different seeds (on identical learning data $\mathcal{L}$). From this we calculate the nagging predictor. Second, we generate 100 different bootstrap samples $\mathcal{L}^* = \mathcal{L}^{*(s)}$, $1 \leq s \leq 100$, from $\mathcal{L}$ (having an identical sample size) with random drawings with replacements, and we fit the same network architecture to these 100

Table 11.9 Out-of-sample losses (gamma loss, power variance case $p = 2.5$ loss (in 10^{-2}) and inverse Gaussian (IG) loss (in 10^{-3})) and average claim amounts; the losses use unit dispersion $\varphi = 1$

	Out-of-sample loss on $\mathcal{T}$			Average
	$\mathfrak{d}_{p=2}$	$\mathfrak{d}_{p=2.5}$	$\mathfrak{d}_{p=3}$	claim
Null model	4.6979	10.2420	4.6931	1'774
Gamma multi-output of Table 11.3	2.0581	7.6422	3.9146	1'745
$p = 2.5$ multi-output of Table 11.3	2.0576	7.6407	3.9139	1'732
IG multi-output of Table 11.3	2.0576	7.6401	3.9134	1'705
Gamma multi-output: nagging 100	2.0280	7.5582	3.8864	1'752
$p = 2.5$ multi-output: nagging 100	2.0282	7.5586	3.8865	1'739
IG multi-output: nagging 100	2.0286	7.5592	3.8865	1'711
Gamma multi-output: bootstrap 100	**2.0189**	**7.5301**	**3.8745**	1'803
$p = 2.5$ multi-output: bootstrap 100	2.0191	7.5305	3.8746	1'790
IG multi-output: bootstrap 100	2.0194	7.5309	3.8746	1'756

bootstrap samples. We then also average over these 100 predictors obtained from the different bootstrap samples. Table 11.9 provides the resulting out-of-sample deviance losses on the test data $\mathcal{T}$. We always hold on to the same test data $\mathcal{T}$ which is disjoint/independent from the learning data $\mathcal{L}$ and the bootstrap samples $\mathcal{L}^* = \mathcal{L}^{*(s)}$, $1 \le s \le 100$.

The nagging predictors over 100 seeds are roughly the same as over 20 seeds (see Table 11.3), which indicates that 20 different network fits suffice, here. Interestingly, the average bootstrapped version generally improves the nagging predictors. Thus, here the average bootstrap predictor provides a better balance among the observations to receive superior predictive power on the test data $\mathcal{T}$, compare lines 'nagging 100' vs. 'bootstrap 100' of Table 11.9.

The main purpose of this analysis is to understand the volatility involved in nagging and bootstrap predictors. We therefore consider the coefficients of variation Vco_t introduced in (7.43) on individual policies $1 \le t \le T$. Figure 11.12 shows these coefficients of variation on the individual predictors, i.e., for the individual claims $\boldsymbol{x}_t^\dagger$ and the individual network calibrations with different seeds. The left-hand side gives the coefficients of variation based on 100 bootstrap samples, the right-hand side gives the coefficients of variation of 100 predictors fitted on the same data $\mathcal{L}$ but with different seeds for the SGD algorithm; the y-scale is identical in both plots. We observe that the coefficients of variation are clearly higher under the bootstrap approach compared to holding on to the same data $\mathcal{L}$ for SGD fitting with different seeds. Thus, the nagging predictor averages over the randomness in different seeds for network calibrations, whereas bootstrapping additionally considers possible different samples $\mathcal{L}^*$ for model learning. We analyze the difference in magnitudes in more detail.

Figure 11.13 compares the two coefficients of variation for different claim sizes. The average coefficient of variation for fixed observations $\mathcal{L}$ is 15.9% (cyan columns). This average coefficient of variation is increased to 24.8% under bootstrapping

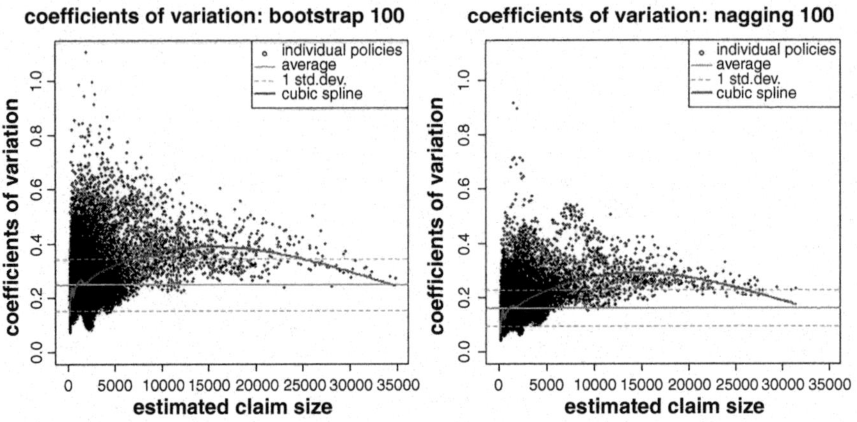

Fig. 11.12 Coefficients of variation in individual estimators (lhs) bootstrap 100, and (rhs) nagging 100; the y-scale is identical in both plots

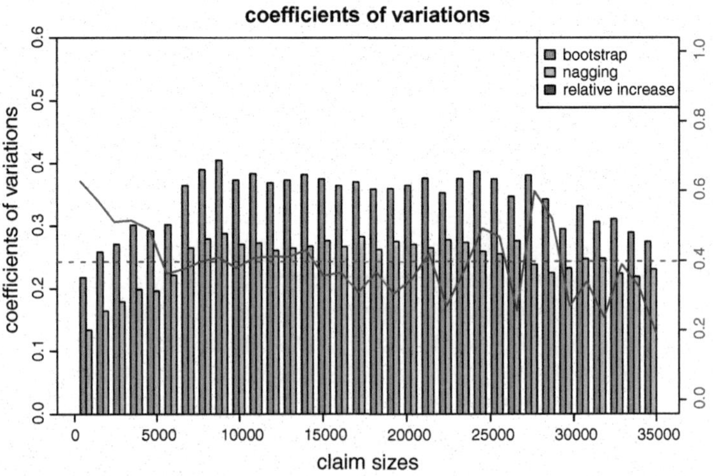

Fig. 11.13 Coefficients of variation in individual predictors of the bootstrap and the nagging approaches (ordered w.r.t. estimated claim sizes)

(orange columns). The blue line shows the average relative increase for the different claim sizes (right axis), and the blue dotted line is at a relative increase of 40%. From Fig. 11.13 we observe that this spread (relative increase) is rather constant across all claim predictions; we remark that 93.5% of all claim predictions are below 5'000. Thus, most claims are at the left end of Fig. 11.13.

From this small analysis we conclude that there is substantial model and estimation uncertainty involved, recall that we fit the deep network architecture to 305'550 individual claims having 7 feature components, this is a comparably large portfolio. On average, we have a coefficient of variation of 15% implied by SGD fitting with different seeds, and this coefficient of variation is increased to roughly 25% under additionally bootstrapping the observations. This is considerable, and it requires that we ensemble these predictors to receive more robust predictions. The results of Table 11.9 support this re-sampling and ensembling approach as we receive a better out-of-sample performance.

11.5 LocalGLMnet: An Interpretable Network Architecture

Network architectures are often criticized for not being (sufficiently) explainable. Of course, this is not fully true as we have gained a lot of insight about the data examples studied in this book. This criticism of non-explainability has led to the development of the post-hoc model-agnostic tools studied in Sect. 7.6. This approach has been questioned at many places, and it is not clear whether one should try to explain black box models, or whether one should rather try to make the models interpretable in the first place, see, e.g., Rudin [322]. In this section we take this different approach by working with a network architecture that is (more) interpretable. We present the LocalGLMnet proposal of Richman–Wüthrich [317, 318]. This approach allows for interpreting the results, and it allows for variable selection either using an empirical Wald test or LASSO regularization.

There are different other proposals that try to achieve similar explainability in specific network architectures. There is the explainable neural network of Vaughan et al. [367] and the neural additive model of Agarwal et al. [3]. These proposals rely on parallel networks considering one single variable at a time. Of course, this limits their performance because of a missing interaction potential. This has been improved in the Combined Actuarial eXplainable Neural Network (CAXNN) approach of Richman [314], which requires a manual specification of parallel networks for potential interactions. The LocalGLMnet, proposed in this section, does not require any manual engineering, and it still possesses the universal approximation property.

11.5.1 Definition of the LocalGLMnet

Starting point of the LocalGLMnet is a classical GLM. Choose a strictly monotone and smooth link function g. A GLM is received by considering the regression function

$$\boldsymbol{x} \mapsto g(\mu(\boldsymbol{x})) = \beta_0 + \langle \boldsymbol{\beta}, \boldsymbol{x} \rangle = \beta_0 + \sum_{j=1}^{q} \beta_j x_j, \tag{11.36}$$

for features $\boldsymbol{x} \in \mathcal{X} \subset \mathbb{R}^q$, intercept $\beta_0 \in \mathbb{R}$ and regression parameter $\boldsymbol{\beta} \in \mathbb{R}^q$. Compared to (5.5) we change the notation in this section by excluding the intercept component from the feature $\boldsymbol{x} = (x_1, \ldots, x_q)^\top$, because this will be more convenient for the LocalGLMnet proposal. The beauty of this GLM regression function is that we obtain a linear function after applying the link function g. This linear function is considered to be explainable as we can precisely quantify how much the expected response will change by slightly changing one of the feature components x_j. In particular, this holds true for the log-link which leads to a multiplicative structure in the expected response.

The idea is to hold on to this additive structure (11.36) as far as possible, still trying to benefit from the universal approximation property of network architectures. Richman–Wüthrich [317] propose the following regression structure.

Definition 11.12 (LocalGLMnet) Choose a FN network architecture $z^{(d:1)} : \mathbb{R}^q \to \mathbb{R}^q$ of depth $d \in \mathbb{N}$ with equal input and output dimensions to model the *regression attention*

$$\boldsymbol{\beta} : \mathbb{R}^q \to \mathbb{R}^q$$
$$\boldsymbol{x} \mapsto \boldsymbol{\beta}(\boldsymbol{x}) \stackrel{\text{def.}}{=} z^{(d:1)}(\boldsymbol{x}) = \left(z^{(d)} \circ \cdots \circ z^{(1)}\right)(\boldsymbol{x}).$$

The *LocalGLMnet* is defined by the generalized *additive decomposition*

$$\boldsymbol{x} \mapsto g\left(\mu(\boldsymbol{x})\right) = \beta_0 + \langle \boldsymbol{\beta}(\boldsymbol{x}), \boldsymbol{x} \rangle = \beta_0 + \sum_{j=1}^{q} \beta_j(\boldsymbol{x}) x_j,$$

for a strictly monotone and smooth link function g.

This architecture is called LocalGLMnet because locally, around a given feature value $\boldsymbol{x}$, it can be understood as a GLM, supposed that $\boldsymbol{\beta}(\boldsymbol{x})$ does not change too much in the environment of $\boldsymbol{x}$. In the GLM context $\boldsymbol{\beta}$ is called *regression parameter*, and in the LocalGLMnet context $\boldsymbol{\beta}(\boldsymbol{x})$ is called *regression attention* because the components $\beta_j(\boldsymbol{x})$ determine how much attention there should be given to a specific value x_j. We highlight this in the following discussion. Select one component $1 \le j \le q$ and study the individual term

$$\boldsymbol{x} \mapsto \beta_j(\boldsymbol{x}) x_j. \tag{11.37}$$

(1) If $\beta_j(\boldsymbol{x}) \equiv 0$, we should drop the term $\beta_j(\boldsymbol{x}) x_j$ from the regression function.
(2) If $\beta_j(\boldsymbol{x}) \equiv \beta_j$ ($\neq 0$) is not feature dependent (and different from zero), we receive a GLM term in x_j with regression parameter β_j.

(3) Property $\beta_j(\boldsymbol{x}) = \beta_j(x_j)$ implies that we have a term $\beta_j(x_j)x_j$ that does not interact with any other term $x_{j'}$, $j' \neq j$.
(4) Sensitivities of $\beta_j(\boldsymbol{x})$ in the components of $\boldsymbol{x}$ can be obtained by the gradient

$$\nabla_{\boldsymbol{x}} \beta_j(\boldsymbol{x}) = \left(\frac{\partial}{\partial x_1} \beta_j(\boldsymbol{x}), \ldots, \frac{\partial}{\partial x_q} \beta_j(\boldsymbol{x}) \right)^\top \in \mathbb{R}^q. \tag{11.38}$$

The j-th component of $\nabla_{\boldsymbol{x}} \beta_j(\boldsymbol{x})$ determines the (non-)linearity in term x_j, the components different from j describe the interactions of term x_j with the other components.
(5) These interpretations need some care because we do not have identifiability. For the special regression attention $\beta_j(\boldsymbol{x}) = x_{j'}/x_j$ we have

$$\beta_j(\boldsymbol{x})x_j = x_{j'}. \tag{11.39}$$

Therefore, we talk about *terms* in items (1)–(4), e.g., item (1) means that the term $\beta_j(\boldsymbol{x})x_j$ can be dropped, however, the feature component x_j may still play a significant role in some of the regression attentions $\beta_{j'}(\boldsymbol{x})$, $j' \neq j$.
In practical applications we have not experienced identifiability issue (11.39). Having already the linear terms in the LocalGLMnet regression structure and starting the SGD fitting in the GLM gives already quite pre-determined regression functions, and the LocalGLMnet is built around this initialization, hardly falling into a completely different model (11.39).
(6) The LocalGLMnet architecture has the universal approximation property discussed in Sect. 7.2.2, because networks can approximate any continuous function arbitrarily well on a compact support for sufficiently large networks. We can then select one component, say, x_1 and let $\beta_1(\boldsymbol{x}) = z_1^{(d:1)}(\boldsymbol{x})$ approximate a given continuous function $f(\boldsymbol{x})/x_1$, i.e., $f(\boldsymbol{x}) \approx \beta_1(\boldsymbol{x})x_1$ arbitrarily well on the compact support.

11.5.2 Variable Selection in LocalGLMnets

The LocalGLMnet allows for variable selection through the regression attentions $\beta_j(\boldsymbol{x})$. Roughly speaking, if the estimated regression attentions $\widehat{\beta}_j(\boldsymbol{x}) \approx 0$, then the term $\beta_j(\boldsymbol{x})x_j$ can be dropped. We can also explore whether the entire variable x_j should be dropped (not only the corresponding term $\beta_j(\boldsymbol{x})x_j$). For this, we have to refit the LocalGLMnet excluding the feature component x_j. If the out-of-sample performance on validation data does not change, then x_j also does not play an important role in any other regression attention $\beta_{j'}(\boldsymbol{x})$, $j' \neq j$, and it should be completely dropped from the model.

In GLMs we can either use the Wald test or the LRT to test a null hypothesis H_0 : $\beta_j = 0$, see Sect. 5.3. We explore a similar idea in this section, however, empirically.

We therefore first need to ensure that all feature components live on the same scale. We consider standardization with the empirical mean and the empirical standard deviation, see (7.30), and from now on we assume that all feature components are centered and have unit variance. Then, the main problem is to determine whether an estimated regression attention $\widehat{\beta}_j(\boldsymbol{x})$ is significantly different from 0 or not.

We therefore extend the features $\boldsymbol{x}^+ = (x_1, \ldots, x_q, x_{q+1})^\top \in \mathbb{R}^{q+1}$ by an additional independent and purely random component x_{q+1} that is also standardized. Since this additional component is independent of all other components it cannot have any predictive power for the response under consideration, thus, fitting this extended model should result in a regression attention $\widehat{\beta}_{q+1}(\boldsymbol{x}^+) \approx 0$. The estimate will not be exactly zero, because there is noise involved, and the magnitude of this fluctuation will determine the rejection/acceptance region of the null hypothesis of not being significant.

We fit the LocalGLMnet to the learning data $\mathcal{L}$ with features $\boldsymbol{x}_i^+ \in \mathbb{R}^{q+1}$ extended by the standardized i.i.d. component $x_{i,q+1}$ being independent of $(Y_i, \boldsymbol{x}_i)$. This gives us the estimated regression attentions $\widehat{\beta}_1(\boldsymbol{x}_i^+), \ldots, \widehat{\beta}_q(\boldsymbol{x}_i^+), \widehat{\beta}_{q+1}(\boldsymbol{x}_i^+)$. We compute the empirical mean and standard deviation of the attention weight of the additional component x_{q+1}

$$\bar{b}_{q+1} = \frac{1}{n}\sum_{i=1}^{n} \widehat{\beta}_{q+1}(\boldsymbol{x}_i^+) \qquad \text{and} \qquad \widehat{s}_{q+1} = \sqrt{\frac{1}{n-1}\sum_{i=1}^{n}\left(\widehat{\beta}_{q+1}(\boldsymbol{x}_i^+) - \bar{b}_{q+1}\right)^2}. \tag{11.40}$$

We expect approximate centering $\bar{b}_{q+1} \approx 0$ because this additional component x_{q+1} does not enter the true regression function, and the empirical standard deviation $\widehat{s}_{q+1}$ quantifies the expected fluctuation around zero of insignificant components.

We can now test the null hypothesis $H_0 : \beta_j(\boldsymbol{x}) = 0$ of component j on significance level $\alpha \in (0, 1/2)$. We define centered interval

$$I_\alpha = \left[\Phi^{-1}(\alpha/2) \cdot \widehat{s}_{q+1},\ \Phi^{-1}(1-\alpha/2) \cdot \widehat{s}_{q+1}\right], \tag{11.41}$$

where $\Phi^{-1}(p)$ denotes the standard Gaussian quantile for $p \in (0, 1)$. H_0 should be rejected if the coverage ratio of this centered interval I_α is substantially smaller than $1 - \alpha$, i.e.,

$$\frac{1}{n}\sum_{i=1}^{n} \mathbb{1}_{\{\widehat{\beta}_j(\boldsymbol{x}_i^+) \in I_\alpha\}} < 1 - \alpha.$$

This proposal is designed for continuous feature components, and categorical variables are discussed in Sect. 11.5.4, below. For x_{q+1} we can choose a standard Gaussian distribution, a normalized uniform distribution or we can randomly permute one of the feature components $x_{i,j}$ across the entire portfolio $1 \le i \le n$. Usually, the resulting empirical standard deviations $\widehat{s}_{q+1}$ are rather similar.

11.5.3 Lab: LocalGLMnet for Claim Frequency Modeling

We revisit the French MTPL data example. We compare the LocalGLMnet approach to the deep FN network considered in Sect. 7.3.2, and we benchmark with the results of Table 7.3; we benchmark with the crudest FN network from above because, at the current stage, we need one-hot encoding for the LocalGLMnet approach. The analysis in this section is the same as in Richman–Wüthrich [317].

The French MTPL data has 6 continuous feature components (we treat `Area` as a continuous variable), 1 binary component and 2 categorical components. We pre-process the continuous and binary variables to centering and unit variance using standardization (7.30). This will allow us to do variable selection as presented in (11.41). The categorical variables with more than two levels are more difficult. In a first attempt we use one-hot encoding for the categorical variables. We prefer one-hot encoding over dummy coding because this ensures that for all levels there is a component x_j with $x_j \neq 0$. This is important because the terms $\beta_j(\boldsymbol{x})x_j$ are equal to zero for the reference level in dummy coding (since $x_j = 0$). This does not allow us to study interactions with other variables for the term corresponding to the reference level. Remark that one-hot encoding and dummy coding do not lead to centering and unit variance.

This feature pre-processing gives us a feature vector $\boldsymbol{x} \in \mathbb{R}^q$ of dimension $q = 40$. For variable selection of the continuous and binary components we extend the feature $\boldsymbol{x}$ by two additional independent components x_{q+1} and x_{q+2}. We select two components to explore whether the particular distributional choice has some influence on the choice of the acceptance/rejection interval I_α in (11.41). We choose for policies $1 \le i \le n$

$$x_{i,q+1} \overset{\text{i.i.d.}}{\sim} \text{Uniform}\left[-\sqrt{3}, \sqrt{3}\right] \qquad \text{and} \qquad x_{i,q+2} \overset{\text{i.i.d.}}{\sim} \mathcal{N}(0,1),$$

these two sets of variables being mutually independent, and being independent from all other variables. We define the extended features $\boldsymbol{x}_i^+ = (x_{i,1}, \ldots, x_{i,q}, x_{i,q+1}, x_{i,q+2})^\top \in \mathbb{R}^{q_0}$ with $q_0 = q + 2$, and we consider the LocalGLMnet regression function

$$\boldsymbol{x}^+ \mapsto \log\left(\mu(\boldsymbol{x}^+)\right) = \beta_0 + \sum_{j=1}^{q_0} \beta_j(\boldsymbol{x}^+)x_j.$$

We choose the log-link for Poisson claim frequency modeling. The time exposure $v > 0$ can either be integrated as a weight to the EDF or as an offset on the canonical scale resulting in the same Poisson model, see Sect. 5.2.3.

Listing 11.7 LocalGLMnet architecture

```
1  Design   = layer_input(shape = c(42), dtype = 'float32', name = 'Design')
2  Vol      = layer_input(shape = c(1), dtype = 'float32', name = 'Vol')
3  #
4  Attention = Design %>%
5           layer_dense(units=20, activation='tanh', name='FNLayer1') %>%
6           layer_dense(units=15, activation='tanh', name='FNLayer2') %>%
7           layer_dense(units=10, activation='tanh', name='FNLayer3') %>%
8           layer_dense(units=42, activation='linear', name='Attention')
9  #
10 LocalGLM = list(Design, Attention) %>% layer_dot(name='LocalGLM', axes=1) %>%
11          layer_dense(units=1, activation='exponential', name='Balance')
12 #
13 Response = list(LocalGLM, Vol) %>% layer_multiply(name='Multiply')
14 #
15 keras_model(inputs = c(Design, Vol), outputs = c(Response))
```

We are now ready to define the LocalGLMnet architecture. We choose a network $z^{(d:1)} : \mathbb{R}^{q_0} \to \mathbb{R}^{q_0}$ of depth $d = 4$ with $(q_1, q_2, q_3, q_4) = (20, 15, 10, 42)$ neurons. The R code is given in Listing 11.7. We note that this is not much more involved than a plain-vanilla FN network. Slightly special in this implementation is the integration of the intercept β_0 on line 11. Naturally, we would like to add this intercept, however, there is no simple code for doing this. For that reason, we model the additive decomposition by

$$\boldsymbol{x}^+ \mapsto \log\left(\mu(\boldsymbol{x}^+)\right) = \alpha_0 + \alpha_1 \sum_{j=1}^{q_0} \beta_j(\boldsymbol{x}^+) x_j,$$

with real-valued parameters α_0 and α_1 being estimated on line 11 of Listing 11.7. Thus, in this implementation the regression attentions are obtained by $\alpha_1 \beta_j(\boldsymbol{x}^+)$. Of course, there are also other ways of implementing this. This LocalGLMnet architecture has 1'799 network weights to be fitted.

We fit this LocalGLMnet using a training to validation data split of $8:2$ and a batch size of 5'000. We initialize the gradient descent algorithm such that we exactly start in the GLM with $\beta_j(\boldsymbol{x}^+) \equiv \widehat{\beta}_j^{\text{MLE}}$. For this we set all weights in the last layer on line 8 of Listing 11.7 to zero, $w_{l,j}^{(d)} = 0$, and the corresponding intercepts to the MLEs of the GLM, i.e., $w_{0,j}^{(d)} = \widehat{\beta}_j^{\text{MLE}}$. This gives us the GLM initialization $\sum_{j=1}^{q_0} \widehat{\beta}_j^{\text{MLE}} x_j$ on line 10 of Listing 11.7. Moreover, on line 11 of that listing, we initialize $\alpha_1 = 1$ and $\alpha_0 = \widehat{\beta}_0^{\text{MLE}}$. This implies that the gradient descent algorithm starts in the MLE estimated GLM. The SGD fitting turns out to be faster than in the plain-vanilla FN case, probably, because we start in the GLM having already the reasonable linear terms x_j in the model, and we only need to find the regression attentions $\beta_j(\boldsymbol{x}^+)$ around these linear terms. The results are presented on the second last line of Table 11.10. The out-of-sample results are slightly worse than in the plain-vanilla FN case. There are many reasons for that, for instance, many levels in one-hot encoding may lead to more potential for over-fitting, and hence to an earlier

Table 11.10 Run times, number of parameters, in-sample and out-of-sample deviance losses (units are in 10^{-2}) and in-sample average frequency of the Poisson regressions, see also Table 7.3

	Run time	# param.	In-sample loss on $\mathcal{L}$	Out-of-sample loss on $\mathcal{T}$	Aver. freq.
Poisson null	–	1	25.213	25.445	7.36%
Poisson GLM3	15s	50	24.084	24.102	7.36%
One-hot FN $(q_1, q_2, q_3) = (20, 15, 10)$	51s	1'306	23.757	23.885	6.96%
LocalGLMnet on $\boldsymbol{x}^+$	20s	1'799	23.728	23.945	7.46%
LocalGLMnet on $\boldsymbol{x}^+$ bias regularized	–	–	23.727	23.943	7.36%

stopping, here. The same applies if we add too many purely random components x_{q+l}, $l \geq 1$. Since the balance property will not hold, in general, we apply the bias regularization step (7.33) to adjust α_0 and α_1, the results are presented on the last line of Table 11.10; in Remark 3.1 of Richman–Wüthrich [317] a more sophisticated balance property correction is presented. Our goal now is to analyze this solution.

Listing 11.8 Extracting the regression attentions from the LocalGLMnet architecture

```
zz       <- keras_model(inputs=model$input,
                   outputs=get_layer(model, 'Attention')$output)
beta     <- data.frame(zz %>% predict(list(Xlearn, Vlearn)))
alpha1   <- as.numeric(get_weights(model)[[9]])
beta     <- beta * alpha1
```

We start by analyzing the two additional components $x_{i,q+1}$ and $x_{i,q+2}$ being uniformly and Gaussian distributed, respectively. Listing 11.8 shows how to extract the estimated regression attentions $\widehat{\boldsymbol{\beta}}(\boldsymbol{x}_i^+)$. We calculate the means and standard deviations of the estimated regression attentions of the two additional components

$$\bar{b}_{q+1} = 0.0042 \qquad \text{and} \qquad \bar{b}_{q+2} = 0.0213,$$

and

$$\widehat{s}_{q+1} = 0.0516 \qquad \text{and} \qquad \widehat{s}_{q+2} = 0.0482.$$

From these numbers we see that the regression attentions $\widehat{\beta}_{q+2}(\boldsymbol{x}_i)$ are slightly biased, whereas $\widehat{\beta}_{q+1}(\boldsymbol{x}_i)$ are fairly centered compared to the magnitudes of the standard deviations. If we select a significance level of $\alpha = 0.1\%$, we receive a two-sided standard normal quantile of $|\Phi^{-1}(\alpha/2)| = 3.29$. This provides us for interval (11.41) with

$$I_\alpha = \left[\Phi^{-1}(\alpha/2) \cdot \widehat{s}_{q+1},\ \Phi^{-1}(1-\alpha/2) \cdot \widehat{s}_{q+1}\right] = [-0.17, 0.17].$$

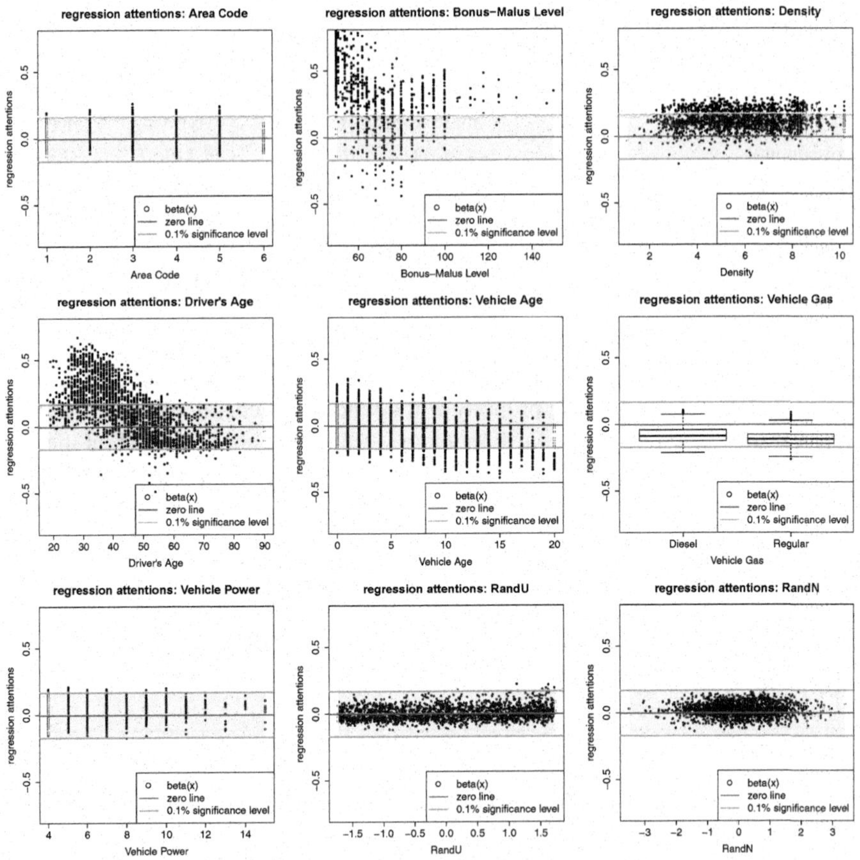

Fig. 11.14 Estimated regression attentions $\widehat{\beta}_j(\boldsymbol{x}_i^+)$ of the continuous and binary feature components `Area`, `BonusMalus`, log-`Density`, `DrivAge`, `VehAge`, `VehGas`, `VehPower` and the two random features $x_{i,q+1}$ and $x_{i,q+2}$ of 2'000 randomly selected policies $\boldsymbol{x}_i^+$; the orange area shows the interval I_α for dropping term $\beta_j(\boldsymbol{x})x_j$ on significance level $\alpha = 0.1\%$

Figure 11.14 shows the estimated regression attentions $\widehat{\beta}_j(\boldsymbol{x}_i^+)$ of the continuous and binary feature components for 2'000 randomly selected policies $\boldsymbol{x}_i^+$, and the orange area shows the acceptance region I_α on significance level $\alpha = 0.1\%$. Focusing on the figures of the two additional variables $x_{i,q+1}$ and $x_{i,q+2}$, Fig. 11.14 (bottom, middle and right), we observe that the estimated regression attentions are mostly within the confidence bounds of I_α. This says that we should drop these two terms (of course, this is clear since we have set the bounds according to these regression attentions). Focusing on the other variables, we question the inclusion of the term `VehPower` as it seems concentrated within I_α, and hence we cannot reject the null hypothesis $H_0 : \beta_{\text{VehPower}}(\boldsymbol{x}) = 0$. Moreover, the inclusion of the term `Area` needs further exploration.

Table 11.11 Run times, number of parameters, in-sample and out-of-sample deviance losses (units are in 10^{-2}) and in-sample average frequency of the Poisson regressions, see also Table 7.3

	Run time	# param.	In-sample loss on $\mathcal{L}$	Out-of-sample loss on $\mathcal{T}$	Aver. freq.
Poisson null	–	1	25.213	25.445	7.36%
Poisson GLM3	15s	50	24.084	24.102	7.36%
One-hot FN $(q_1, q_2, q_3) = (20, 15, 10)$	51s	1'306	23.757	23.885	6.96%
LocalGLMnet on $\boldsymbol{x}^+$	20s	1'799	23.728	23.945	7.46%
LocalGLMnet on $\boldsymbol{x}^+$ bias regularized	–	–	23.727	23.943	7.36%
LocalGLMnet on $\boldsymbol{x}^-$	20s	1'675	23.715	23.912	7.30%
LocalGLMnet on $\boldsymbol{x}^{\cdot}$ bias regularized	–	–	23.714	23.911	7.36%

We remind that dropping a term $\beta_j(\boldsymbol{x})x_j$ does not necessarily imply that we have to completely drop x_j because it may still play an important role in one of the other regression attentions $\beta_{j'}(\boldsymbol{x})$, $j' \neq j$. Therefore, we re-run the whole fitting procedure, but we drop the purely random feature components $x_{i,q+1}$ and $x_{i,q+2}$, and we also drop `VehPower` and `Area` to see whether we receive a model with a similar predictive power. This then would imply that we can drop these variables, in the sense of variable selection similar to the LRT and the Wald test of Sect. 5.3. We denote the feature where we drop these components by $\boldsymbol{x}^- \in \mathbb{R}^{q-2}$.
We re-fit the LocalGLMnet on the reduced features $\boldsymbol{x}_i^-$, and the results are presented in Table 11.11. We observe that the loss figures decrease. Indeed, this supports the null hypothesis of dropping `VehPower` and `Area`. The reason for being able to drop `VehPower` is that it does not contribute (sufficiently) to explain the systematic effects in the responses. The reason for being able to drop `Area` is slightly different: we have seen that `Area` and log-`Density` are highly correlated, see Fig. 13.12 (rhs), and it turns out that it is sufficient to only keep the `Density` variable (on the log-scale) in the model.

In a next step, we should analyze the robustness of these results by exploring the nagging predictor and/or bootstrapping as described in Sect. 11.4. We refrain from doing so, but we illustrate the LocalGLMnet solution of Table 11.11 in more detail. Figure 11.15 shows the feature contributions $\widehat{\beta}_j(\boldsymbol{x}_i^-)x_{i,j}$ of 2'000 randomly selected policies on the significant continuous and binary feature components. The magenta line gives a spline fit, and the more the black dots spread around these splines, the more interactions we have; for instance, higher bonus-malus levels interact with the age of driver which explains the scattering of the black dots. On average, frequencies are increasing in bonus-malus levels and density, decreasing in vehicle age, and for the driver's age variable it is important to understand the interactions. We observe that the spline fit for the log-`Density` is close to a linear function, this reflects that the regression attentions $\widehat{\beta}_{\texttt{Density}}(\boldsymbol{x}_i)$ in Fig. 11.14 (top-right) are more or less constant. This is also confirmed by the marginal plot in Fig. 5.4 (bottom-rhs) which has motivated the choice of a linear term for the log-`Density` in model Poisson GLM1 of Table 5.3.

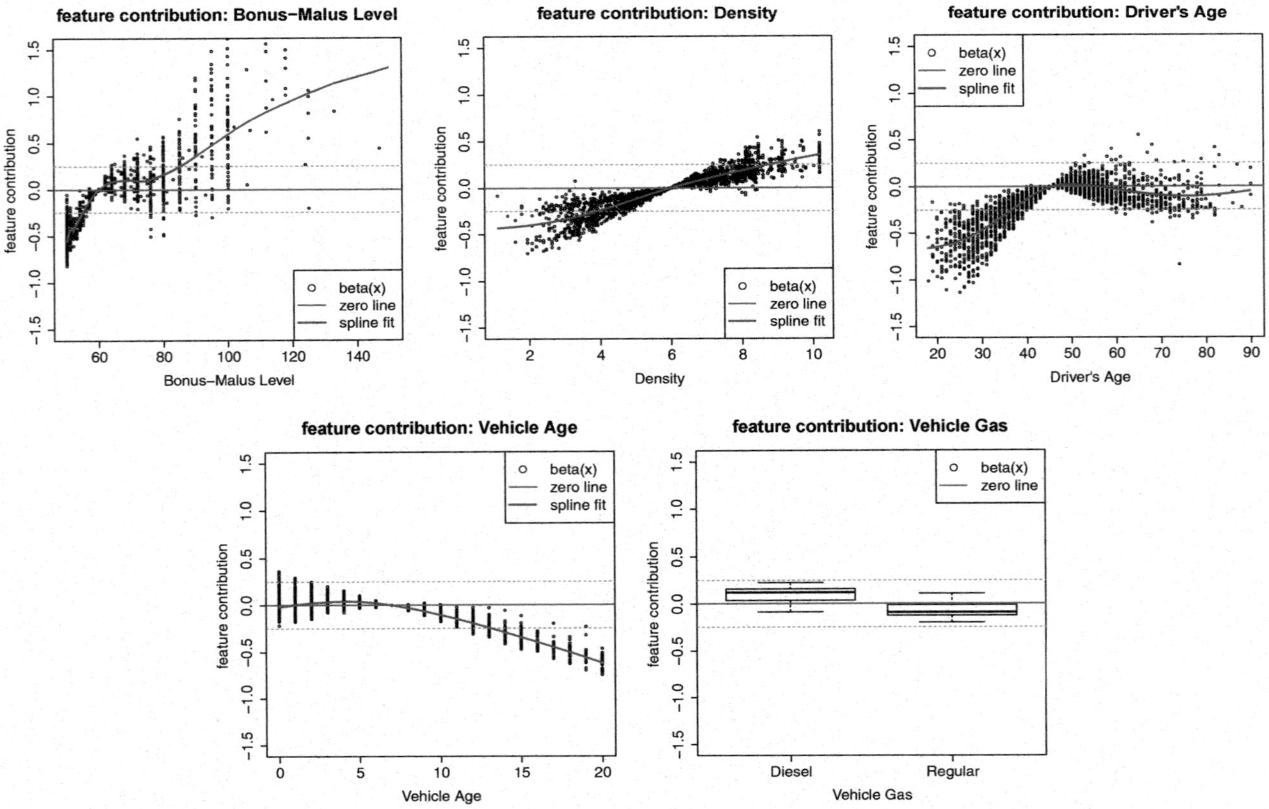

Fig. 11.15 Estimated feature contributions $\widehat{\beta}_j(\boldsymbol{x}_i^-)x_{i,j}$ of the significant continuous and binary components `BonusMalus`, log-`Density`, `DrivAge`, `VehAge` and `VehGas` of 2'000 randomly selected policies $\boldsymbol{x}_i^-$; the magenta line gives a spline fit

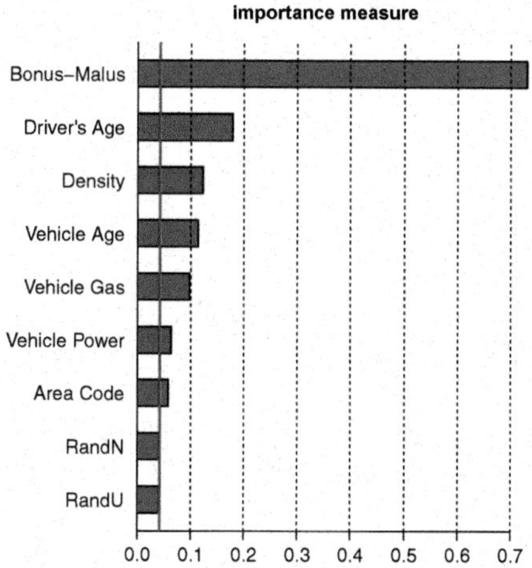

Fig. 11.16 Importance measure IM_j of the continuous and binary variables

Using the regression attentions we define an importance measure. We consider the extended features $\boldsymbol{x}^+$ in the following numerical analysis. We set

$$\mathrm{IM}_j = \frac{1}{n}\sum_{i=1}^{n} \left|\widehat{\beta}_j(\boldsymbol{x}_i^+)\right|,$$

for $1 \le j \le q+2$, and where we aggregate over all policies $1 \le i \le n$.
Figure 11.16 shows the importance measures IM_j of the continuous and binary variables j. The bars are ordered w.r.t. these importance measures. The graph confirms our previous conclusion, the least important variables are the two additional purely random components $x_{i,q+1}$ and $x_{i,q+2}$, followed by `Area` and `VehPower`. These are exactly the components that have been dropped going from the full model $\boldsymbol{x}^+$ to the reduced model $\boldsymbol{x}^-$.
Next, we analyze the interactions by studying the gradients (11.38). Figure 11.17 illustrates spline fits to the components $\partial\widehat{\beta}_j(\boldsymbol{x}_i^-)/\partial x_k$ w.r.t. x_j of the continuous variables `BonusMalus`, log-`Density`, `DrivAge` and `VehAge` over all policies $i = 1, \ldots, n$. The components $\partial\widehat{\beta}_j(\boldsymbol{x}_i^-)/\partial x_j$ show the non-linearity in x_j. We conclude that `BonusMalus`, `DrivAge` and `VehAge` should be non-linear, and log-`Density` is linear because $\partial\widehat{\beta}_j(\boldsymbol{x}_i^-)/\partial x_j \approx 0$. The components $\partial\widehat{\beta}_j(\boldsymbol{x}_i^-)/\partial x_k$, $k \neq j$, determine the interactions. We have the strongest interactions between `BonusMalus` and `DrivAge`, and `BonusMalus` has interactions with all variables. On the other hand, the log-`Density` only interacts with `BonusMalus`.
The reader will have noticed that we have excluded the categorical components `VehBrand` and `Region` from all model discussions. Firstly, these components are not standardized to zero mean and unit variance, and, secondly, we cannot study one level in isolation to be able to decide to keep or drop that variable. I.e., similar to group LASSO we need to study all levels simultaneously of each categorical feature component. We do this in the next section, and we conclude with the regression

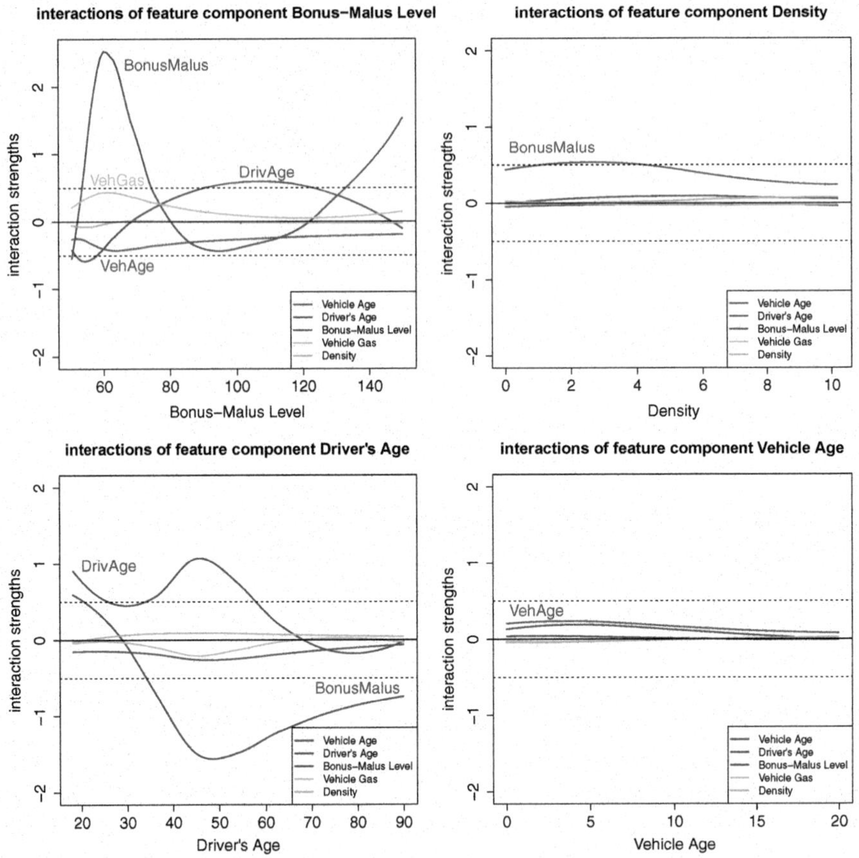

Fig. 11.17 Spline fits to the derivatives $\partial \widehat{\beta}_j(\boldsymbol{x}_i^-)/\partial x_k$ w.r.t. x_j of the continuous variables BonusMalus, log-Density, DrivAge and VehAge over all policies $i = 1, \ldots, n$

attentions $\widehat{\beta}_j(\boldsymbol{x})$ of the categorical feature components in Fig. 11.18, which seem to be significantly different from zero (VehBrands B10, B11, and Regions R22, R43, R82, R93), but which do not allow for variable selection as just described.

Remark 11.13 The bias regularization in Table 11.11 has simply been obtained by applying an additional MLE step to α_0 and α_1. Alternatively, we can also define the new features $\widehat{\boldsymbol{z}}_i = (\widehat{\alpha}_1 \widehat{\beta}_1(\boldsymbol{x}_i)x_{i,1}, \ldots, \widehat{\alpha}_1 \widehat{\beta}_{q_0}(\boldsymbol{x}_i)x_{i,q_0})^\top \in \mathbb{R}^{q_0}$, and then apply a proper GLM step to these newly (learned) features $\widehat{\boldsymbol{z}}_1, \ldots, \widehat{\boldsymbol{z}}_n$. Working with the canonical link will give us the balance property. This is discussed in more detail in Remark 3.1 of Richman–Wüthrich [317].

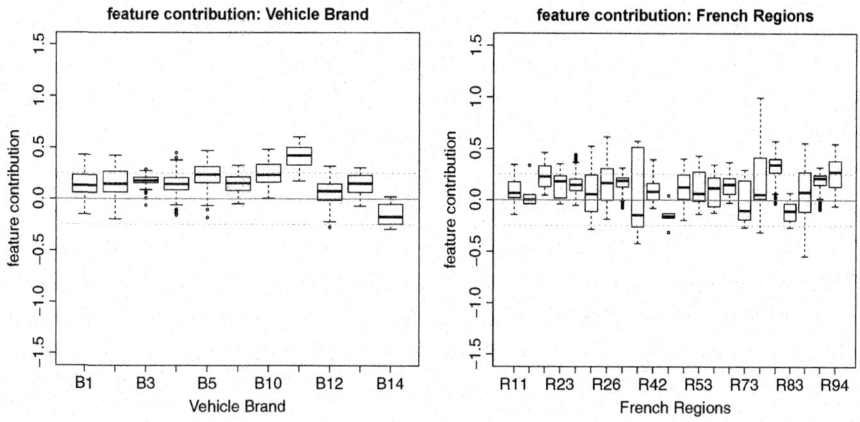

Fig. 11.18 Boxplot of the regression attentions $\widehat{\beta}_j(\boldsymbol{x})$ of the categorical feature components `VehBrand` and `Region`; the y-scale is the same as in Fig. 11.15

11.5.4 Variable Selection Through Regularization of the LocalGLMnet

A natural next step is to introduce regularization on the regression attentions $\boldsymbol{\beta}(\boldsymbol{x})$; this is the proposal suggested in Richman–Wüthrich [318]. We choose the LocalGLMnet architecture $\boldsymbol{x} \mapsto \mu(\boldsymbol{x})$ of Definition 11.12 having an intercept parameter $\beta_0 \in \mathbb{R}$ and the network weights $\boldsymbol{w}$. For fitting, we consider a loss function L and we add a regularization term to this loss function penalizing large regression attentions. That is, we aim at minimizing

$$\underset{\beta_0, \boldsymbol{w}}{\arg\min}\ \frac{1}{n}\sum_{i=1}^{n} L\left(Y_i, \mu(\boldsymbol{x}_i)\right) - \mathfrak{R}(\boldsymbol{\beta}(\boldsymbol{x}_i)), \tag{11.42}$$

with a penalty term (regularizer) $\mathfrak{R}(\cdot) \geq 0$. For the penalty term $\mathfrak{R}$ we can choose different forms, e.g., the elastic net regularizer of Zou–Hastie [409] is obtained by, see Remark 6.3,

$$\underset{\beta_0, \boldsymbol{w}}{\arg\min}\ \frac{1}{n}\sum_{i=1}^{n} L\left(Y_i, \mu(\boldsymbol{x}_i)\right) + \eta\left((1-\alpha)\|\boldsymbol{\beta}(\boldsymbol{x}_i)\|_2^2 + \alpha\|\boldsymbol{\beta}(\boldsymbol{x}_i)\|_1\right), \tag{11.43}$$

for a regularization parameter $\eta \geq 0$ and weight $\alpha \in [0, 1]$. For $\alpha = 0$ we receive ridge regularization, and for $\alpha = 1$ we get LASSO regularization of $\boldsymbol{\beta}(\cdot)$.

For variable selection of categorical feature components we should rather use the group LASSO penalization of Yuan–Lin [398], see also (6.5). Assume the features $\boldsymbol{x}$ have a natural group structure $\boldsymbol{x} = (\boldsymbol{x}_1^\top, \ldots, \boldsymbol{x}_K^\top)^\top \in \mathbb{R}^q$. We consider the optimization

$$\underset{\beta_0, \boldsymbol{w}}{\arg\min}\ \frac{1}{n}\sum_{i=1}^{n} L\left(Y_i, \mu(\boldsymbol{x}_i)\right) + \sum_{k=1}^{K} \eta_k \|\boldsymbol{\beta}_k(\boldsymbol{x}_i)\|_2, \tag{11.44}$$

for regularization parameters $\eta_k \geq 0$, and where $\boldsymbol{\beta}_k(\boldsymbol{x})$ collects all components $\beta_j(\boldsymbol{x})$ of $\boldsymbol{\beta}(\boldsymbol{x})$ that belong to the k-th group $\boldsymbol{x}_k$ of $\boldsymbol{x}$. Yuan–Lin [398] propose to scale the regularization parameters as $\eta_k = \sqrt{q_k}\eta \geq 0$, where q_k is the size of group k. Remark that if every group has size one we exactly obtain LASSO regularization.

Solving the optimization problem (11.44) poses some challenges because the regularizer is not differentiable in zero. In Sect. 6.2.5 we have presented the generalized projection operator (using the soft-thresholding operator) to solve the group LASSO regularization within GLMs. However, this proposal will not work here: the generalized projection operator may help to project the regression attentions $\boldsymbol{\beta}(\boldsymbol{x}_i)$ back to the constraint set $\mathcal{C}$. However, this does not tell us anything about how to choose the network parameters $\boldsymbol{w}$ and, therefore, will not work here. In a different setting, Oelker–Tutz [288] propose to use a differentiable ϵ-approximation to the terms in (11.44). Choose $\epsilon > 0$ and define for $\boldsymbol{\beta}_k \in \mathbb{R}^{q_k}$

$$\|\boldsymbol{\beta}_k\|_{2,\epsilon} = \sqrt{\|\boldsymbol{\beta}_k\|_2^2 + \epsilon} = \sqrt{\boldsymbol{\beta}_k^\top \boldsymbol{\beta}_k + \epsilon} \quad \to \quad \|\boldsymbol{\beta}_k\|_2 \quad \text{as } \epsilon \downarrow 0. \tag{11.45}$$

This motivates to study the optimization problem for a fixed (small) $\epsilon > 0$

$$\underset{\beta_0, \boldsymbol{w}}{\arg\min}\ \frac{1}{n}\sum_{i=1}^{n} L\left(Y_i, \mu(\boldsymbol{x}_i)\right) + \sum_{k=1}^{K} \eta_k \|\boldsymbol{\beta}_k(\boldsymbol{x}_i)\|_{2,\varepsilon}. \tag{11.46}$$

In Fig. 11.19 we plot these ϵ-approximations for $\epsilon \in \{10^{-1}, 10^{-2}, 10^{-3}, 10^{-4}, 10^{-5}\}$. The plot on the left-hand side gives $\beta \in \mathbb{R} \mapsto \|\beta\|_{2,\epsilon} = \sqrt{\beta^2 + \epsilon} \to |\beta|$ for $\epsilon \downarrow 0$, and the plot on the right-hand side gives the unit ball

$$\mathcal{B}_\epsilon = \left\{ \boldsymbol{\beta} = (\beta_1, \beta_2)^\top \in \mathbb{R}^2;\ \|\beta_1\|_{2,\epsilon} + \|\beta_2\|_{2,\epsilon} = 1 \right\}.$$

For the last two ϵ choices there is no visible difference to the ℓ_1-norm.

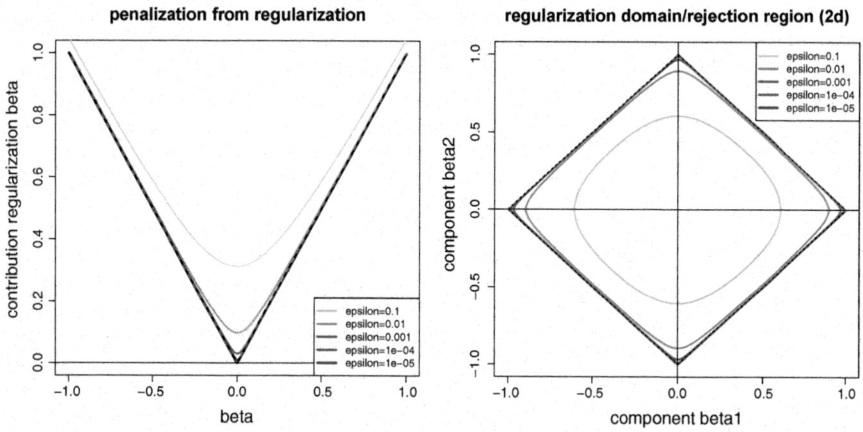

Fig. 11.19 (lhs) Comparison of $|\beta|$ and $\|\beta\|_{2,\epsilon} = \sqrt{\beta^2 + \epsilon}$ for $\beta \in \mathbb{R}$, and (rhs) unit balls $\mathcal{B}_\epsilon$ for $\epsilon \in \{10^{-1}, 10^{-2}, 10^{-3}, 10^{-4}, 10^{-5}\}$ compared to the Manhattan unit ball

The main disadvantage of the ϵ-approximation is that it does not shrink unimportant components $\beta_j(\boldsymbol{x})$ exactly to zero. But it allows us to identify unimportant (small) components, which can then be removed manually. As mentioned in Lee et al. [237], LASSO regularization needs a second model calibration step only fitting the model on the selected components (and without regularization) to receive an optimal predictive power and a minimal bias. Thus, we need a second calibration step after the removal of the unimportant components anyway.

11.5.5 Lab: LASSO Regularization of LocalGLMnet

We revisit the LocalGLMnet architecture applied to the French MTPL claim frequency data, see Sect. 11.5.3. The goal is to perform a group LASSO regularization so that we can also study the importance of the terms coming from the categorical feature components `VehBrand` and `Region`. We first pre-process all feature components as follows. We apply dummy coding to the categorical variables, and then we standardize all components to centering and unit variance, this includes the dummy coded components.

In a next step we need to define the natural groups $\boldsymbol{x} = (\boldsymbol{x}_1^\top, \ldots, \boldsymbol{x}_K^\top)^\top \in \mathbb{R}^q$. We have 7 continuous and binary components which give us dimensions $q_k = 1$ for $1 \le k \le 7$. `VehBrand` provides us with a group of size $q_8 = 10$, and `Region` gives us a group of size $q_9 = 21$. We set $K = 9$ and $q = \sum_{k=1}^{9} q_k = 38$. We code

Listing 11.9 Group LASSO regularization design

```
group.lasso.grouping <- function(xx){
   pp <- array(0, dim=c(length(xx),sum(xx)))
   for (k in 1:length(xx)){
       if (k==1){pp[k,1:xx[k]] <- 1
              }else{
              pp[k,(sum(xx[1:(k-1)])+1):sum(xx[1:k])] <-  1
              }}
   t(pp)
   }
#
ww <- group.lasso.grouping(c(rep(1,7),10,21)) 12 etaK <- eta
etaK <- eta * sqrt(c(rep(1,7),10,21))
```

a (sort of) regularization design matrix to encode the K groups and weights $\sqrt{q_k}$ for the q components of $\boldsymbol{x}$. This is done in Listing 11.9 providing us with a matrix of size 38×9 and the weights $\sqrt{q_k}$. This regularization design matrix enters the penalty term on lines 13 and 16 of Listing 11.10 which weights the penalizations $\|\cdot\|_{2,\epsilon}$.

Listing 11.10 LocalGLMnet with group LASSO regularization

```
1  Design   = layer_input(shape = c(38), dtype = 'float32')
2  LogVol   = layer_input(shape = c(1),  dtype = 'float32')
3  Bias1    = layer_input(shape = c(1),  dtype = 'float32')
4  #
5  Attention = Design %>%
6         layer_dense(units=15, activation='tanh') %>%
7         layer_dense(units=10, activation='tanh') %>%
8         layer_dense(units=38, activation='linear', name='Attention')
9  #
10 Penalty = Attention %>%
11        layer_lambda(function(x) k_square(x)) %>%
12        layer_dense(units=9, activation='linear',
13                    weights=list(ww), use_bias=FALSE, trainable=FALSE) %>%
14        layer_lambda(function(x) k_sqrt(x+epsilon)) %>%
15        layer_dense(units=1, activation='linear',
16        weights=list(array(etaK, dim=c(9,1))), use_bias=FALSE, trainable=FALSE)
17 #
18 LocalGLM = list(Design, Attention) %>% layer_dot(axes=1)
19 #
20 Bias = Bias1 %>%
21            layer_dense(units=1, activation='linear', use_bias=FALSE)
22 #
23 Response = list(LocalGLM, Bias, LogVol) %>% layer_add() %>%
24               layer_lambda(function(x) k_exp(x))
25 #
26 Output = list(Response, Penalty) %>% layer_concatenate()
27 #
28 keras_model(inputs = c(Design, LogVol, Bias1), outputs = c(Output))
```

The entire group LASSO regularized LocalGLMnet is depicted in Listing 11.10, showing the regression attentions on lines 5–8, the regularization on lines 10–16, and the output on line 26 returns the expected response $v_i \mu(\boldsymbol{x}_i)$ and the regularizer $\sum_{k=1}^K \eta_k \|\boldsymbol{\beta}_k(\boldsymbol{x}_i)\|_{2,\epsilon}$, we choose $\epsilon = 10^{-5}$ for our example.

Listing 11.11 Group LASSO regularized Poisson deviance loss

```
Poisson.reg <- function(y_true, y_pred){k_mean(
   y_pred[,1]-y_true[,1] + y_true[,1]*k_log((y_true[,1]/y_pred[,1]+.00000001))
                                         + y_pred[,2] )}
```

Finally, we need to code the loss function (11.42). This is done in Listing 11.11. We combine the Poisson deviance loss function with the group LASSO ϵ-approximation $\sum_{k=1}^K \eta_k \|\boldsymbol{\beta}_k(\boldsymbol{x}_i)\|_{2,\epsilon}$, the latter being outputted by Listing 11.10. We fit this network to the French MTPL data (as above) for regularization parameters $\eta \in \{0, 0.0025, 0.005\}$. Firstly, we note that the resulting networks are not fully competitive, this is probably due to the fact that the high-dimensional dummy coding leads to too much over-fitting potential which leads to a very early stopping in gradient descent fitting. Thus, this approach may not be useful to directly receive a good predictive model, but it may be helpful to select the right feature components to design a good predictive model.

Figure 11.20 gives the importance measures of the estimated regression attentions

$$\mathrm{IM}_j = \frac{1}{n}\sum_{i=1}^n \left|\widehat{\beta}_j(\boldsymbol{x}_i)\right|,$$

of all components $1 \le j \le q = 38$. The red color corresponds to regularization parameter $\eta = 0.005$, red + yellow colors to $\eta = 0.0025$, and red + yellow + green colors to $\eta = 0$ (no regularization). Figure 11.20 (lhs) shows the results on the original (standardized) features $\boldsymbol{x}$. By far the smallest red + yellow column among the continuous features is observed for `VehPower` which confirms the variable selection of Sect. 11.5.3. Among the categorical variables `Region` seems more important (on average) than `VehBrand` because the red and yellow columns are generally bigger for `Region`. All these red and yellow columns of `VehBrand` and `Region` are bigger than the ones of `VehPower` which supports the inclusion of the two categorical variables.

Figure 11.20 (rhs) verifies this decision of keeping the categorical variables. For this latter graph we randomly permute `Region` across the entire portfolio, and we run the same group LASSO regularized fitting procedure again on this modified data. The vertical black line shows the average importance of the permuted `Region` variable for $\eta = 0.0025$. We see that only `VehPower` has a smaller importance measure, and all other variables dominate the permuted `Region` variable. This confirms our conclusions above.

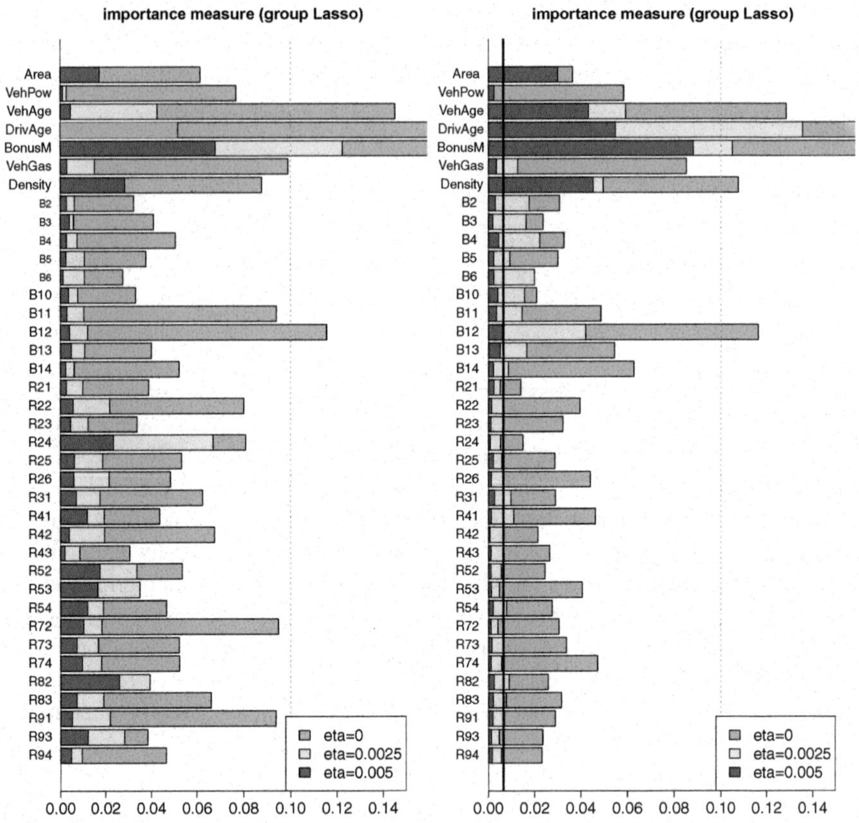

Fig. 11.20 Importance measures IM_j of the group LASSO regularized LocalGLMnet for variable selection with different regularization parameters $\eta \in \{0, 0.0025, 0.005\}$: (lhs) original data, and (rhs) randomly permuted `Region` labels; the x-scale is the same in both plots

We conclude that the LocalGLMnet architecture with a group LASSO regularization is helpful for variable selection, and, more generally, the LocalGLMnet architecture is useful for model interpretation, finding interactions and functional forms of the features entering the regression function. In examples that have categorical variables with many levels, the LocalGLMnet approach may not lead to a regression model that is fully competitive. In this case, the LocalGLMnet can be used for variable selection, and an other network architecture should then be fitted on the selected variables. Alternatively, we can embed the categorical variables in a preparatory network step, and then work with these embeddings of the categorical variables (kept fixed within the LocalGLMnet).

11.6 Selected Applications

11.6.1 Mixture Density Networks

In Sect. 6.3 we have introduced mixture distributions and we have presented the EM algorithm for fitting these mixture distributions. The EM algorithm considers two steps, an expectation step (E-step) and a maximization step (M-step). The E-step is motivated by (6.34). In this step the posterior distribution of the latent variable $\boldsymbol{Z}$ is determined, given the observation Y and the parameter estimates for the model parameters $\boldsymbol{\theta}$ and $\boldsymbol{p}$. The M-step (6.35) determines the optimal model parameters $\boldsymbol{\theta}$ and $\boldsymbol{p}$, based on the observation Y and the posterior distribution of $\boldsymbol{Z}$. Typically, we explore MLE in the M-step. However, for the EM algorithm to function it is not important that we really work with the maximum in the M-step, but monotonicity in (6.38) is sufficient. Thus, if at algorithmic time $t-1$ we have a parameter estimate $(\widehat{\boldsymbol{\theta}}^{(t-1)}, \widehat{\boldsymbol{p}}^{(t-1)})$, it suffices that the next estimate $(\widehat{\boldsymbol{\theta}}^{(t)}, \widehat{\boldsymbol{p}}^{(t)})$ increases the log-likelihood, without necessarily being the MLE; this latter approach is called generalized EM (GEM) algorithm. Exactly this point makes it feasible to also use the EM algorithm in cases where we model the parameters through networks which are fit using gradient descent (ascent) algorithms. These methods go under the name of mixture density networks (MDNs).

MDNs have been introduced by Bishop [35], who explores MDNs on Gaussian mixtures, and using SGD and quasi-Newton methods for model fitting. MDNs have also started to gain more popularity within the actuarial community, recent papers include Delong et al. [95], Kuo [230] and Al-Mudafer et al. [6], the latter two considering MDNs for claims reserving.

We recall the mixture density for a selected member of the EDF. The incomplete log-likelihood of the data $(Y_i, \boldsymbol{x}_i, v_i)_{1\le i\le n}$ is given by, see (6.24),

$$
\begin{aligned}
(\boldsymbol{\theta}, \boldsymbol{\varphi}, \boldsymbol{p}) \mapsto \ell_{\boldsymbol{Y}}(\boldsymbol{\theta}, \boldsymbol{\varphi}, \boldsymbol{p}) &= \sum_{i=1}^{n} \ell_{Y_i}(\boldsymbol{\theta}(\boldsymbol{x}_i), \boldsymbol{\varphi}(\boldsymbol{x}_i), \boldsymbol{p}(\boldsymbol{x}_i)) \\
&= \sum_{i=1}^{n} \log\left(\sum_{k=1}^{K} p_k(\boldsymbol{x}_i) f_k\left(Y_i; \theta_k(\boldsymbol{x}_i), \frac{v_i}{\varphi_k(\boldsymbol{x}_i)}\right)\right),
\end{aligned}
$$

for canonical parameter $\boldsymbol{\theta} = (\theta_1, \ldots, \theta_K)^\top \in \boldsymbol{\Theta} = \boldsymbol{\Theta}_1 \times \cdots \times \boldsymbol{\Theta}_K$, dispersion parameter $\boldsymbol{\varphi} = (\varphi_1, \ldots, \varphi_K)^\top \in \mathbb{R}_+^K$, mixture probability $\boldsymbol{p} \in \Delta_K$, and K denotes the number of mixture components. MDNs model these parameters with networks. Choose a FN network $\boldsymbol{z}^{(d:1)} : \mathbb{R}^{q+1} \to \{1\}\times\mathbb{R}^{q_d}$ of depth d, with input dimension q being equal to the dimension of the features $\boldsymbol{x} \in \mathcal{X} \subseteq \{1\}\times\mathbb{R}^q$ and output dimension $q_d + 1$. This gives us the learned representations $\boldsymbol{z}_i = \boldsymbol{z}^{(d:1)}(\boldsymbol{x}_i)$. These learned representations are used to model the parameters. For the mixture probability $\boldsymbol{p}$ we build a logistic categorical GLM, based on z_i. For the (canonical) link h, we set linear predictor, see (5.72),

$$h(\boldsymbol{p}(z_i)) = h\left(\boldsymbol{p}\left(z^{(d:1)}(\boldsymbol{x}_i)\right)\right) = \left(\langle \boldsymbol{\beta}_1^p, z_i\rangle, \ldots, \langle \boldsymbol{\beta}_K^p, z_i\rangle\right)^\top \in \mathbb{R}^K, \qquad (11.47)$$

with regression parameter $\boldsymbol{\beta}^p = ((\boldsymbol{\beta}_1^p)^\top, \ldots, (\boldsymbol{\beta}_K^p)^\top)^\top \in \mathbb{R}^{K(q_d+1)}$. For the canonical parameter $\boldsymbol{\theta}$, the mean parameter $\boldsymbol{\mu}$, respectively, and the dispersion parameter $\boldsymbol{\varphi}$ we proceed analogously. Choose strictly monotone and smooth link functions g_μ and g_φ, and consider the double GLMs, for $1 \le k \le K$, on the learned representations z_i

$$g_\mu(\mu_k(z_i)) = \langle \boldsymbol{\beta}_k^\mu, z_i\rangle \qquad \text{and} \qquad g_\varphi(\varphi_k(z_i)) = \langle \boldsymbol{\beta}_k^\varphi, z_i\rangle, \qquad (11.48)$$

with regression parameters $\boldsymbol{\beta}^\mu = ((\boldsymbol{\beta}_1^\mu)^\top, \ldots, (\boldsymbol{\beta}_K^\mu)^\top)^\top \in \mathbb{R}^{K(q_d+1)}$ for the mean parameters and $\boldsymbol{\beta}^\varphi = ((\boldsymbol{\beta}_1^\varphi)^\top, \ldots, (\boldsymbol{\beta}_K^\varphi)^\top)^\top \in \mathbb{R}^{K(q_d+1)}$ for the dispersion parameters. Thus, altogether this gives us a network parameter of dimension, set $q_0 = q$,

$$r = \sum_{m=1}^{d} q_m(q_{m-1} + 1) + 3K(q_d + 1).$$

Remarks 11.14

- The regression functions (11.47)–(11.48) use a slight abuse of notation, because, strictly speaking, these should be functions w.r.t. the features $\boldsymbol{x}_i \in \mathcal{X}$, i.e., we should understand the learned representations z_i as a short form for $\boldsymbol{x}_i \mapsto z^{(d:1)}(\boldsymbol{x}_i)$.
- It is not fully correct to say that (11.47) is the logistic categorical GLM of formula (5.72), because (11.47) does not lead to identifiable regression parameters. In fact, we should reduce the dimension of the categorical GLM to $K - 1$, by setting $\boldsymbol{\beta}_K^p = 0$, see (5.70), because the probability of the last label K is fully determined if we know the probabilities of all other labels; this would also justify to say that h is the canonical link. Since in FN network modeling we do not have identifiability anyway, we neglect this normalization (redundancy), see line 16 of Listing 11.12, below.
- The above proposal (11.47)–(11.48) suggests to use the same network $z^{(d:1)}$ for all mixture parameters involved. This requires that the chosen network is

sufficiently large, so that it can comply simultaneously with these different tasks. Alternatively, we could choose three separate (parallel) networks for $\boldsymbol{p}$, $\boldsymbol{\mu}$ and $\boldsymbol{\varphi}$, respectively. This second proposal does not (easily) allow for (non-trivial) interactions between the parameters, and it may also suffer from less robustness in fitting.

- Proposal (11.48) defines double GLMs for the mixture components f_k, $1 \le k \le K$. If we decide to not model the dispersion parameters feature dependent, i.e., if we set $\varphi_k(z) \equiv \varphi_k \in \mathbb{R}_+$, then the mixture components are modeled with GLMs on the learned representations $z_i = z^{(d:1)}(\boldsymbol{x}_i)$. Nevertheless, this latter approach still requires that the dispersion parameters φ_k are set to reasonable values, as they enter the score equations, this can be seen from (6.29) adapted to MDNs. Thus, in MDNs, the dispersion parameters do not cancel in the score equations, which is different from the single distribution case. The dispersion parameter can either be estimated (updated) during the M-step of the EM algorithm (supposed we use the EM algorithm), or it can be pre-specified as a given hyper-parameter.
- As mentioned in Sect. 6.3, mixture density fitting can be challenging because, in general, mixture density log-likelihoods are unbounded. Therefore, a suitable initialization of the EM algorithm is important for a successful model fitting. This problem is less pronounced in MDNs as we use early stopping in SGD fitting that prevents the fitted parameters to depend on a small set of observations. For instance, Example 6.13 cannot occur because an individual observation Y_1 enters at most one (mini-)batch of SGD, and the SGD algorithm will provide a good balance across all batches. Moreover, early stopping will imply that the selected parameters must also be good on the validation data being disjoint (and independent) from the training data.
- Delong et al. [95] present two different ways of fitting such MDNs. The crucial property in EM fitting is to preserve the monotonicity in the M-step. For MDNs this can either be achieved by using the parameters as offsets for the next EM iteration (this is called 'EM network boosting' in Delong et al. [95]) or to forward the network weights from one to the next loop (called 'EM forward network' in Delong et al. [95]). We are going to present the second option in the next example.

Example 11.15 (Gamma Claim Size Modeling and MDNs) We revisit Example 6.14 which models the claim sizes of the French MTPL data. For the modeling of these claim sizes we choose the mixture distribution (6.39) which has four gamma components $f_1, \ldots, f_4$ and one Lomax component f_5. In a first step we again model these five mixture components independent of the feature information $\boldsymbol{x}$, and the feature information only enters the mixture probabilities $\boldsymbol{p}(\boldsymbol{x}) \in \Delta_5$. This modeling approach has been motivated by Fig. 13.17 which suggests that the features mainly result in systematic effects on the mixture probabilities. We choose the same model and feature information as in Example 6.14. We only replace the logistic categorical GLM part (6.40) for modeling $\boldsymbol{p}(\boldsymbol{x})$ by a depth $d = 2$ FN network with $(q_1, q_2) = (20, 10)$ neurons. `Area`, `VehAge`, `DrivAge` and `BonusMalus` are modeled as continuous variables, and for the categorical variables `VehBrand` and `Region` we choose two-dimensional embedding layers.

Listing 11.12 R code of the MDN for modeling the mixture probability $\boldsymbol{p}(\boldsymbol{x})$

```
1  Design    = layer_input(shape = c(4), dtype = 'float32')
2  VehBrand = layer_input(shape = c(1), dtype = 'int32')
3  Region    = layer_input(shape = c(1), dtype = 'int32')
4  Bias      = layer_input(shape = c(1), dtype = 'float32')
5  #
6  BrandEmb  = VehBrand %>%
7        layer_embedding(input_dim = 11, output_dim = 2, input_length = 1) %>%
8        layer_flatten()
9  RegionEmb = Region %>%
10       layer_embedding(input_dim = 22, output_dim = 2, input_length = 1) %>%
11       layer_flatten()
12 #
13 pp = list(Design, BrandEmb, RegionEmb) %>% layer_concatenate() %>%
14           layer_dense(units=20, activation='tanh') %>%
15           layer_dense(units=10, activation='tanh') %>%
16           layer_dense(units=5, activation='softmax')
17 #
18 mu    = Bias %>% layer_dense(units=4, activation='exponential',
19                                       use_bias=FALSE)
20 #
21 tail  = Bias %>% layer_dense(units=1, activation='sigmoid',
22                                       use_bias=FALSE)
23 #
24 shape = Bias %>% layer_dense(units=4, activation='exponential',
25                                       use_bias=FALSE)
26 #
27 Response = list(pp, mu, tail, shape) %>% layer_concatenate()
28 #
29 keras_model(inputs = c(Design, VehBrand, Region, Bias), outputs = c(Response))
```

Listing 11.12 shows the chosen network. Lines 13–16 model the mixture probability $\boldsymbol{p}(\boldsymbol{x})$. We also integrate the modeling of the (homogeneous) parameters of the mixture densities $f_1, \ldots, f_5$. Lines 18 and 24 of Listing 11.12 consider the mean and shape parameter of the gamma components, and line 21 the tail parameter $1/\beta_5$ of the Lomax component. Note that we use the sigmoid activation for this Lomax parameter. This implies $1/\beta_5 \in (0, 1)$ and, thus, $\beta_5 > 1$, which enforces a finite mean model. The exponential activations on lines 18 and 24 ensure positivity of these parameters. The input `Bias` to these variables is simply the constant 1, which is the homogeneous case not differentiating w.r.t. the features.

Observe that in most of the networks so far, the output of the network was equal to an expected response of a random variable that we try to predict. In this MDN we output the parameters of a distribution function, see line 27 of Listing 11.12. In our case this output has dimension 14, which then enters the score in Listing 11.13. In a first attempt we fit this MDN brute-force by just implementing the incomplete log-likelihood received from (6.39). Since the gamma function $\Gamma(\cdot)$ is not easily available in `keras` [77], we replace the gamma density by its saddlepoint approximation, see Sect. 5.5.2. Listing 11.13 shows the negative log-likelihood of the mixture density that is used to perform the brute-force SGD fitting.

Listing 11.13 Mixture density negative incomplete log-likelihood

```
1  mixture_LogLikeli <- function(true, pred){ - k_mean(k_log(
2      pred[,1]*k_exp(-k_log(2*pi*true[,1]^2/pred[,11])/2 -
3              pred[,11]*(true[,1]/pred[,6]-1+k_log(pred[,6]/true[,1]))) +
4      pred[,2]*k_exp(-k_log(2*pi*true[,1]^2/pred[,12])/2 -
5              pred[,12]*(true[,1]/pred[,7]-1+k_log(pred[,7]/true[,1]))) +
6      pred[,3]*k_exp(-k_log(2*pi*true[,1]^2/pred[,13])/2 -
7              pred[,13]*(true[,1]/pred[,8]-1+k_log(pred[,8]/true[,1]))) +
8      pred[,4]*k_exp(-k_log(2*pi*true[,1]^2/pred[,14])/2 -
9              pred[,14]*(true[,1]/pred[,9]-1+k_log(pred[,9]/true[,1]))) +
10     pred[,5]*k_exp(k_log(1/(pred[,10]*M))-(1/pred[,10]+1)
11                                        *k_log(true[,1]/M+1))))
12     }
```

Lines 2–9 give the saddlepoint approximations to the four gamma components, and line 10 the Lomax component for the scale parameter M. Note that this brute-force approach is based only on the incomplete observation Y encoded in `true[,1]`, see Listing 11.13.

We fit this logistic categorical FN network of Listing 11.12 under the score function of Listing 11.13 using the `nadam` version of SGD. Moreover, we use a stratified training-validation split, otherwise we did not obtain a competitive model. The results are presented in Table 11.12 on line 'logistic FN network: brute-force fitting'. We observe a slightly worse performance (in-sample) than in the logistic GLM. This does not justify the use of the more complex network architecture. Or in other words, feature pre-processing seems to been done suitably in Example 6.14.

In a next step, we fit this MDN with the (generalized) EM algorithm. The E-step is exactly the same as in Example 6.14. For the M-step, having knowledge of the (latent mixture component) variables $\widehat{\boldsymbol{Z}}_i$, $1 \le i \le n$, implies that the mixture probability estimation and the mixture density estimation completely decouples. As a consequence, the parameters of the density components $f_1, \ldots, f_5$ can directly be estimated using univariate MLEs, this is the same as in Example 6.14. The only part that needs further explanation is the estimation of the logistic categorical FN network for $\boldsymbol{p}(\boldsymbol{x})$. In each loop of the EM iteration we would like to find the optimal network parameter for $\boldsymbol{p}(\boldsymbol{x})$, and at the same time we have to ensure the monotonicity (6.38). Following the 'EM forward network' approach of Delong et

Table 11.12 Mixture models for French MTPL claim size modeling; we set $M = 2'000$

	# Param.	$\ell_Y(\widehat{\boldsymbol{\theta}}, \widehat{\boldsymbol{p}})$	$\widehat{\mu} = \mathbb{E}_{\widehat{\boldsymbol{\theta}}, \widehat{\boldsymbol{p}}}[Y]$
Empirical			2'266
Null model	13	−199'306	2'381
Logistic GLM, Example 6.14	193	−198'404	2'176
Logistic FN network: brute-force fitting	520	−198'623	2'003
Logistic FN network: EM fitting	520	−198'449	2'119
MDN: brute-force fitting	825	−198'178	2'144
MDN: EM fitting	825	−198'085	2'240

al. [95], this is most easily achieved by just initializing the FN network in loop t of the algorithm with the optimal network parameter of the previous loop $t-1$. Thus, the starting parameter of SGD reflects the optimal parameter from the previous step, and since SGD generally decreases losses, the monotonicity (6.38) holds. The latter statement is not strictly true, SGD introduces additional randomness through the building of (mini-)batches, therefore, monotonicity should be traced explicitly (which also ensures that the early stopping rule is chosen suitably). We have implemented such an EM-SGD algorithm, essentially, we just have to drop lines 17–28 of Listing 11.12 and lines 13–16 provide the entire response. As loss function we choose the categorical (multi-class) cross-entropy loss, see (4.19). The results in Table 11.12 on line 'logistic FN network: EM fitting' indicate a superior fitting behavior compared to the brute-force fitting. Nevertheless, this network approach is still not outperforming the GLM approach, saying that we should stay with the simpler GLM.

In a final step, we also model the mean parameters $\mu_k(\boldsymbol{x})$, $1 \le k \le 4$, of the gamma components feature dependent, to see whether we can gain predictive power from this additional flexibility or whether our initial model choice is sufficient. For robustness reasons we neither model the shape parameters β_k , $1 \le k \le 4$, of the gamma components feature dependent nor the tail parameter β_5 of the Lomax component. The implementation only requires small changes to Listing 11.12, see Listing 11.14.

A brute-force fitting of the MDN architecture of Listing 11.14 can directly be based on the score function (negative incomplete log-likelihood) of Listing 11.13. In the case of the EM algorithm we need to change the score function to the complete log-likelihood accounting for the variables $\widehat{\boldsymbol{Z}}_i \in \Delta_5$. This is done in Listing 11.15 where $\widehat{\boldsymbol{Z}}_i$ is encoded in the variables `true[,2]` to `true[,6]`.

We fit this MDN using the two different fitting approaches, and the results are given on the last two lines of Table 11.12. Again the performance of the EM fitting is slightly better than the brute-force fitting, and the bigger log-likelihoods indicate that we can gain predictive power by also modeling the means of the gamma components feature dependent.

Figure 11.21 compares the QQ plot of the resulting MDN with EM fitting to the one received from the logistic categorical GLM of Example 6.14. These graphs are very similar. We conclude that in this particular example it seems that the simpler proposal of Example 6.14 is sufficient. ■

In a next step, we try to understand which feature components influence the mixture probabilities $\boldsymbol{p}(\boldsymbol{x}) = (p_1(\boldsymbol{x}), \ldots, p_K(\boldsymbol{x}))^\top$ most. Similarly to Examples 6.14 and 11.15, we therefore use a MDN where we only fit the mixture probability $\boldsymbol{p}(\boldsymbol{x})$ with a network and the mixture components $f_1, \ldots, f_K$ are assumed to be homogeneous.

Example 11.16 (MDN with LocalGLMnet) We revisit Example 11.15. We choose the mixture distribution (6.39) which has four gamma components $f_1, \ldots, f_4$ and a Lomax component f_5. We select their parameters independent of the features. The feature information $\boldsymbol{x}$ should only enter the mixture probability $\boldsymbol{p}(\boldsymbol{x}) \in \Delta_5$, similarly to the first part of Example 11.15. We replace the logistic FN network of

Listing 11.14 R code of the MDN for modeling the mixture probability $p(x)$ and the gamma means $\mu_k(x)$

```
Design    = layer_input(shape = c(4), dtype = 'float32')
VehBrand = layer_input(shape = c(1), dtype = 'int32')
Region    = layer_input(shape = c(1), dtype = 'int32')
Bias      = layer_input(shape = c(1), dtype = 'float32')
#
BrandEmb  = VehBrand %>%
      layer_embedding(input_dim = 11, output_dim = 2, input_length = 1) %>%
      layer_flatten()
RegionEmb = Region %>%
      layer_embedding(input_dim = 22, output_dim = 2, input_length = 1) %>%
      layer_flatten()
#
Network = list(Design, BrandEmb, RegionEmb) %>% layer_concatenate() %>%
          layer_dense(units=20, activation='tanh') %>%
          layer_dense(units=15, activation='tanh') %>%
          layer_dense(units=10, activation='tanh')
#
pp     = Network %>% layer_dense(units=5, activation='softmax')
#
mu     = Network %>% layer_dense(units=4, activation='exponential',
                                          use_bias=FALSE)
#
tail   = Bias %>% layer_dense(units=1, activation='sigmoid',
                                       use_bias=FALSE)
#
shape = Bias %>% layer_dense(units=4, activation='exponential',
                                       use_bias=FALSE)
#
Response = list(pp, mu, tail, shape) %>% layer_concatenate()
#
keras_model(inputs = c(Design, VehBrand, Region, Bias), outputs = c(Response))
```

Listing 11.15 Mixture density negative complete log-likelihood

```
mixture_LogLikeli_Complete <- function(true, pred){ - k_mean(
     true[,2]*(k_log(pred[,1])-k_log(2*pi*true[,1]^2/pred[,11])/2 -
               pred[,11]*(true[,1]/pred[,6]-1+k_log(pred[,6]/true[,1]))) +
     true[,3]*(k_log(pred[,2])-k_log(2*pi*true[,1]^2/pred[,12])/2 -
               pred[,12]*(true[,1]/pred[,7]-1+k_log(pred[,7]/true[,1]))) +
     true[,4]*(k_log(pred[,3])-k_log(2*pi*true[,1]^2/pred[,13])/2 -
               pred[,13]*(true[,1]/pred[,8]-1+k_log(pred[,8]/true[,1]))) +
     true[,5]*(k_log(pred[,4])-k_log(2*pi*true[,1]^2/pred[,14])/2 -
               pred[,14]*(true[,1]/pred[,9]-1+k_log(pred[,9]/true[,1]))) +
     true[,6]*(k_log(pred[,5])+k_log(1/(pred[,10]*M))-
                    (1/pred[,10]+1)*k_log(true[,1]/M+1)))
       }
```

Example 11.15 for modeling $p(x)$ by a LocalGLMnet such that we can analyze the importance of the variables, see Sect. 11.5.

For the feature information we choose the continuous variables `Area`, `VehPower`, `VehAge`, `DrivAge` and `BonusMalus`, the binary variable `VehGas` and the categorical variables `VehBrand` and `Region`, thus, we extend by `VehPower` and `VehGas` compared to Example 11.15. These latter two variables have not been included previously, because they did not seem to be important

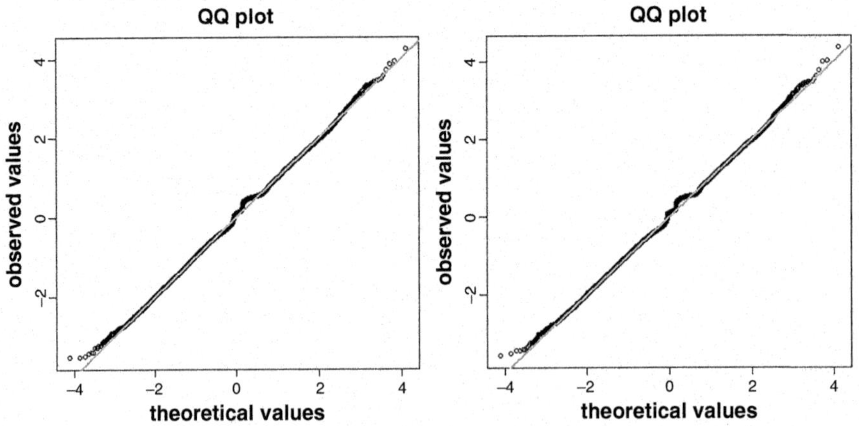

Fig. 11.21 QQ plots of mixture models: (lhs) logistic categorical GLM for mixture probabilities and (rhs) for MDN with EM fitting

w.r.t. Fig. 13.17. The continuous and binary variables are centered and normalized to unit variance. For the categorical variables we use two-dimensional embedding layers, and afterwards they are concatenated with the continuous variables with a subsequent normalization layer (to ensure that all components live on the same scale). This provides us with a 10-dimensional feature vector. This feature vector is complemented with an i.i.d. standard Gaussian component, called `Random`, to perform an empirical Wald type test. We call this pre-processed feature (after embedding and normalization of the categorical variables) $\boldsymbol{x} \in \mathbb{R}^{q_0}$ with $q_0 = 11$.

We design a LocalGLMnet that acts on this feature $\boldsymbol{x} \in \mathbb{R}^{q_0}$ for modeling a categorical multi-class output with $K = 5$ levels. Therefore, we choose the regression attentions

$$z^{(d:1)} : \mathbb{R}^{q_0} \to \mathbb{R}^{q_0 \times K}, \qquad \boldsymbol{x} \mapsto \boldsymbol{\beta}(\boldsymbol{x}) = \left(\boldsymbol{\beta}_1(\boldsymbol{x}), \ldots, \boldsymbol{\beta}_K(\boldsymbol{x})\right) = z^{(d:1)}(\boldsymbol{x}),$$

where $z^{(d:1)}$ is a network of depth d having a matrix-valued output of dimension $q_0 \times K$. For the (canonical) link h, this gives us the predictor, see (5.72),

$$h(\boldsymbol{p}(\boldsymbol{x})) = \left(\beta_{1,0} + \langle \boldsymbol{\beta}_1(\boldsymbol{x}), \boldsymbol{x}\rangle, \ldots, \beta_{K,0} + \langle \boldsymbol{\beta}_K(\boldsymbol{x}), \boldsymbol{x}\rangle\right)^\top \in \mathbb{R}^K, \tag{11.49}$$

with intercepts $\beta_{k,0} \in \mathbb{R}$, and where $\boldsymbol{\beta}_k(\boldsymbol{x}) \in \mathbb{R}^{q_0}$ is the k-th column of regression attention $\boldsymbol{\beta}(\boldsymbol{x}) = z^{(d:1)}(\boldsymbol{x}) \in \mathbb{R}^{q_0 \times K}$. We also refer to the second item of Remarks 11.14 concerning a possible dimension reduction in (11.49), i.e., in fact we apply the `softmax` activation function to the right-hand side of (11.49), neglecting the identifiability issue. Moreover, as in the introduction of the LocalGLMnet, we separate the intercept components from the remaining features in (11.49).

We fit this LocalGLMnet-MDN with the EM version presented in Example 11.15. We apply early stopping based on the same stratified training-validation split as in the aforementioned example, and this provides us with a log-likelihood of -198'290, thus, slightly bigger than the corresponding numbers in Table 11.12. More interestingly, our goal is to understand the regression attentions given by $\boldsymbol{\beta}(\boldsymbol{x}_i) = (\boldsymbol{\beta}_1(\boldsymbol{x}_i), \ldots, \boldsymbol{\beta}_5(\boldsymbol{x}_i)) \in \mathbb{R}^{11\times 5}$ over all claims $1 \le i \le n$. Figure 11.22 shows the resulting boxplots, where each of the five graphs corresponds to one mixture component $1 \le k \le 5$, and the different colors illustrate the 11 feature components providing the attention weights $\beta_{k,j}(\boldsymbol{x}_i)$, $1 \le j \le 11$. The red boxplots show the purely random component `Random` for $1 \le k \le 5$, which provides the acceptance region of an empirical Wald test for the null hypothesis that the corresponding term should be dropped. This is highlighted by the orange shaded area (at a significance level of 0.1%). Thus, whenever a boxplot lies within this orange shaded area we may consider dropping this term, e.g., for $k = 2$ (top-right), this is the case for `Area`, `VehPower` and `Region2` (being the second component of the two-dimensional region embedding). Note that this interpretation needs some care because we do not have identifiability in the class probabilities.

The first observation is that, indeed, `VehPower` is mostly in the orange confidence area and, thus, may be dropped. This does not apply to the other feature components, and, thus, we should keep them in the model. The three gamma mixture components f_1, f_2 and f_3 correspond to the three modes at 75, 600 and 1'175 in Fig. 13.17. Component f_4 is a gamma component covering the whole range of claims, and f_5 is the Lomax component modeling the regular variation in the tail. Interestingly, `DrivAge` and `BonusMalus` seem very important for mixture components $k = 1$, $k = 3$ and $k = 4$ (with different signs), this is supported by Fig. 13.17. The Lomax component seems mostly impacted by `DrivAge`, `VehBrand` and `Region`. Only mixture component $k = 2$ is more difficult to interpret. This component seems influenced by most the feature components, in particular, the combination of `VehAge`, `VehGas` and `VehBrand` seems important. This could mean that mixture component $k = 2$ belongs to a certain type of vehicle.

In a next step we could study interactions and their impact on the mixture components, and LASSO regularization would provide us with another method of variable selection, see Sect. 11.5.4. We refrain from doing so and close the example.

■

11.6.2 Estimation of Conditional Expectations

FN networks have also found their way into solving risk management problems. We briefly introduce a valuation problem and then describe a way of solving this problem. Assume we have a liability cash flow $Y_{1:T} = (Y_1, \ldots, Y_T)$ with (random) payments Y_t at time points $t = 1, \ldots, T$. We assume that this liability cash flow $Y_{1:T}$ is adapted to a filtration $(\mathcal{A}_t)_{1\le t\le T}$ on the underlying probability space $(\Omega, \mathcal{A}, \mathbb{P})$. Moreover, we assume to have a pricing kernel (state price deflator) $\psi_{1:T} = (\psi_1, \ldots, \psi_T)$ on that probability space which is an $(\mathcal{A}_t)_{1\le t\le T}$-adapted

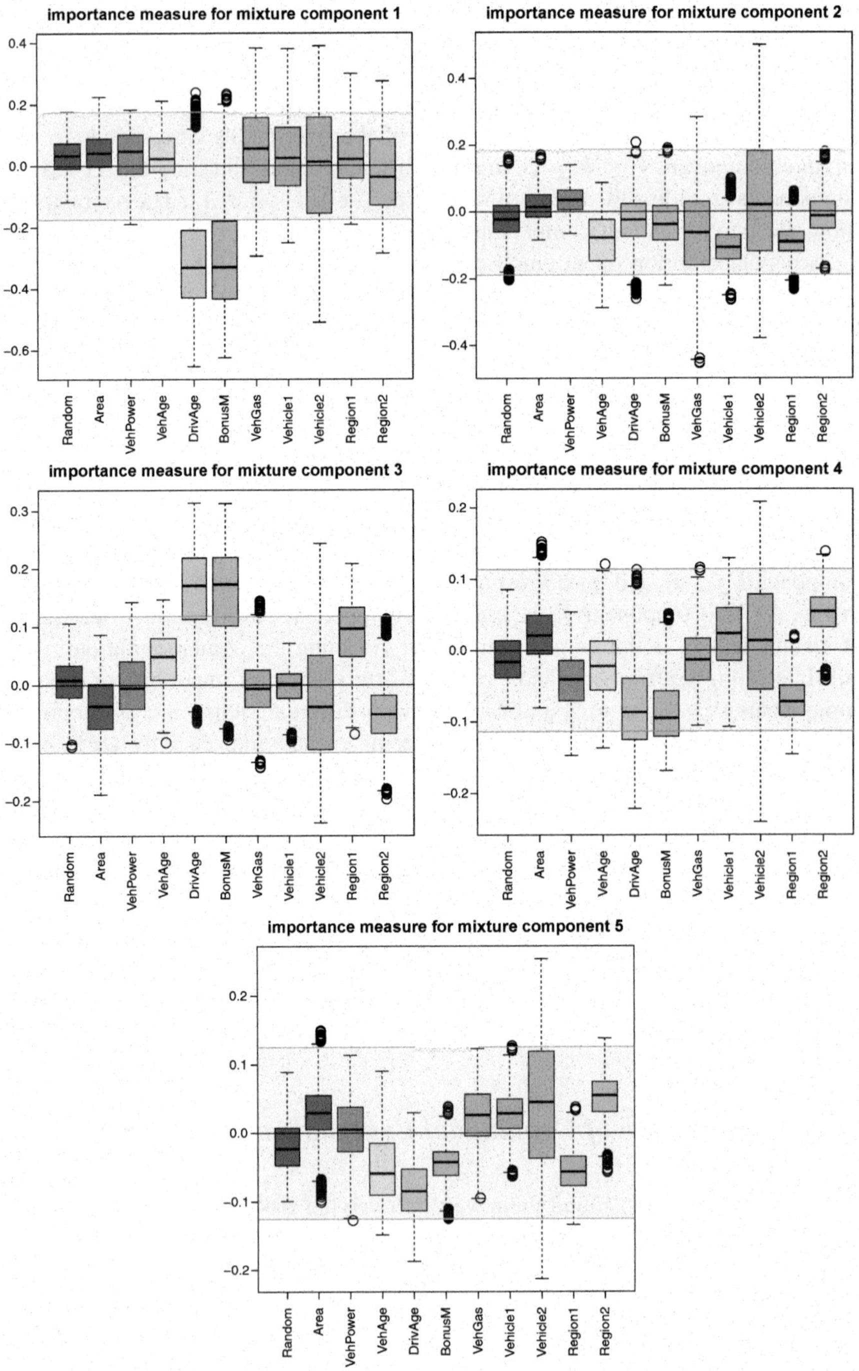

Fig. 11.22 Boxplot of regression attentions $\boldsymbol{\beta}(\boldsymbol{x}_i) = (\boldsymbol{\beta}_1(\boldsymbol{x}_i), \ldots, \boldsymbol{\beta}_5(\boldsymbol{x}_i)) \in \mathbb{R}^{11\times 5}$ over all claims $1 \le i \le n$ for the different mixture components $f_1, \ldots, f_5$

random vector with strictly positive components $\psi_t > 0$, a.s., for all $1 \le t \le T$. A no-arbitrage value of the outstanding liability cash flow at time $1 \le \tau < T$ can be defined by (we assume existence of all second moments)

$$\mathcal{R}_\tau = \sum_{s=\tau+1}^{T} \frac{1}{\psi_\tau} \mathbb{E}\left[\psi_s Y_s \middle| \mathcal{A}_\tau\right]. \tag{11.50}$$

For the mathematical background on no-arbitrage pricing using state price deflators we refer to Wüthrich–Merz [393]. The $\mathcal{A}_\tau$-measurable quantity $\mathcal{R}_\tau$ is called reserves of the outstanding liabilities at time τ. From a risk management and solvency point of view we would like to understand the volatility in the reserves $\mathcal{R}_\tau$ seen from time 0, i.e., we try to model the random variable $\mathcal{R}_\tau$ seen from time 0 (based on the trivial σ-algebra $\mathcal{A}_0 = \{\emptyset, \Omega\}$). In applied problems, the difficulty often is that the conditional expectations under the summation in (11.50) cannot be computed in closed form. Therefore the law of $\mathcal{R}_\tau$ cannot be determined explicitly.

We provide a numerical solution to the calculation of the conditional expectations in (11.50). Assume that the information set $\mathcal{A}_\tau$ can be described by a random vector $\boldsymbol{X}_\tau$, i.e., $\mathcal{A}_\tau = \sigma(\boldsymbol{X}_\tau)$. In that case we rewrite (11.50) as follows

$$\mathcal{R}_\tau = \sum_{s=\tau+1}^{T} \frac{1}{\psi_\tau} \mathbb{E}\left[\psi_s Y_s \middle| \boldsymbol{X}_\tau\right]. \tag{11.51}$$

The latter now indicates that we can determine the conditional expectations in (11.51) as regression functions in features $\boldsymbol{X}_\tau$, and we try to understand for $s > \tau$

$$\boldsymbol{x}_\tau \mapsto \mathbb{E}\left[\frac{\psi_s}{\psi_\tau} Y_s \middle| \boldsymbol{X}_\tau = \boldsymbol{x}_\tau\right]. \tag{11.52}$$

The random variable $\mathcal{R}_\tau$ can then be determined empirically by simulation. This requires two steps: (1) We have to be able to simulate $\psi_s Y_s / \psi_\tau$, conditionally given $\boldsymbol{X}_\tau = \boldsymbol{x}_\tau$. This allows us to estimate the conditional expectation (11.52) with a regression function. (2) We need to be able to simulate $\boldsymbol{X}_\tau$. This provides us with the empirical occurrence probabilities of specific choices $\boldsymbol{X}_\tau = \boldsymbol{x}_\tau$ in (11.52) which then gives an empirical version of $\mathcal{R}_\tau$.

In theory, this problem can be approached by nested simulations which is a two-stage procedure that first performs step (2), and then calculates step (1) empirically with Monte Carlo simulations for every realization of step (2), see, e.g., Lee [242] and Glynn–Lee [161]. The disadvantage of this two-stage nested simulation procedure is that it is computationally demanding. Building upon the work on valuation of American options by Carriere [65], Tsitsiklis–Van Roy [356] and Longstaff–Schwartz [257], the papers of Broadie et al. [55] and Ha–Bauer [177] propose to regress future cash flows on finitely many basis functions depending on the state variable $\boldsymbol{X}_\tau$. More recently, machine learning tools such as FN networks have been proposed to determine these basis and regression functions, see, e.g., Cheridito et al. [74] or Krah et al. [224].

In the following, we assume that all random variables considered are square-integrable and, thus, we can work in a Hilbert space with the scalar product $\langle X, Z\rangle = \mathbb{E}[XZ]$ for $X, Z \in \mathcal{L}^2(\Omega, \mathcal{A}, \mathbb{P})$. Moreover, for simplicity, we drop the time indices and we also drop the stochastic discounting in (11.52) by assuming $\psi_s/\psi_\tau \equiv 1$. These simplifications are not essential technically and simplify our outline. The conditional expectation $\mu(\boldsymbol{X}) = \mathbb{E}[Y|\boldsymbol{X}]$ can then be found by the orthogonal projection of Y onto the sub-space $\sigma(\boldsymbol{X})$, generated by $\boldsymbol{X}$, in the Hilbert space $\mathcal{L}^2(\Omega, \mathcal{A}, \mathbb{P})$. That is, the conditional expectation is the measurable function $\mu : \mathbb{R}^q \to \mathbb{R}$, $\boldsymbol{X} \mapsto \mu(\boldsymbol{X})$, that minimizes the mean squared error

$$\mathbb{E}\left[(Y - \mu(\boldsymbol{X}))^2\right] \stackrel{!}{=} \min, \tag{11.53}$$

among all measurable functions on $\boldsymbol{X}$. In Example 3.7, we have seen that $\mu(\cdot)$ is the minimizer of this problem if and only if

$$\mu(\boldsymbol{x}) = \underset{m\in\mathbb{R}}{\arg\min} \int_{\mathbb{R}} (y-m)^2 \, dF_{Y|\boldsymbol{x}}(y), \tag{11.54}$$

for $p_{\boldsymbol{x}}$-a.e. $\boldsymbol{x} \in \mathbb{R}^q$, where $p_{\boldsymbol{x}}$ is the distribution of $\boldsymbol{X}$, and where $F_{Y|\boldsymbol{x}}$ is the conditional distribution of Y, given feature $\boldsymbol{X} = \boldsymbol{x}$; we also refer to (3.6).

Under the assumption that we can simulate observations $(Y, \boldsymbol{X})$ under $\mathbb{P}$, we can solve (11.53)–(11.54) approximately by restricting to a sufficiently rich family of regression functions. Choose a FN network $\boldsymbol{z}^{(d:1)} : \mathbb{R}^q \to \mathbb{R}^{q_d}$ of depth d and the identity link $g(x) = x$. An optimal network parameter $\widehat{\boldsymbol{\vartheta}}$ is found by minimizing

$$\widehat{\boldsymbol{\vartheta}} = \underset{\boldsymbol{\vartheta}\in\mathbb{R}^r}{\arg\min} \frac{1}{n}\sum_{i=1}^n \left(Y_i - \left\langle \boldsymbol{\beta}, \boldsymbol{z}^{(d:1)}(\boldsymbol{X}_i)\right\rangle\right)^2, \tag{11.55}$$

where $(Y_i, \boldsymbol{X}_i)$, $1 \le i \le n$, are i.i.d. copies of $(Y, \boldsymbol{X})$. This provides us with the fitted FN network $\widehat{\boldsymbol{z}}^{(d:1)}(\cdot)$ and the fitted output parameter $\widehat{\boldsymbol{\beta}}$. These can be used to receive an approximation to the conditional expectation, solution of (11.54),

$$\boldsymbol{x} \mapsto \widehat{\mu}(\boldsymbol{x}) = \left\langle \widehat{\boldsymbol{\beta}}, \widehat{\boldsymbol{z}}^{(d:1)}(\boldsymbol{x})\right\rangle \approx \mu(\boldsymbol{x}) = \mathbb{E}\left[Y | \boldsymbol{X} = \boldsymbol{x}\right]. \tag{11.56}$$

This then allows us to approximate the random variable in (11.51) empirically by simulating features $\boldsymbol{X}$ and inserting them into left-hand side of (11.56).

Remarks 11.17

- There are different types of errors involved. First, there is an irreducible approximation error if the chosen family of FN networks is not sufficiently rich to approximate the conditional expectation well. For example, if we choose the hyperbolic tangent activation function, then, naturally, $\boldsymbol{z}^{(d:1)}(\cdot)$ is uniformly bounded for a fixed network parameter ϑ. This does not necessarily apply to the conditional expectation $\mathbb{E}[Y|\boldsymbol{X} = \cdot]$ and, thus, the approximation in the tail may be poor. Second, we consider an approximation based on a finite sample in (11.55). However, this error can be made arbitrarily small by letting $n \to \infty$. In-sample over-fitting should not be an issue as we may generate samples of arbitrary large sample sizes. Third, having the approximation (11.56), we still need to simulate i.i.d. samples $\boldsymbol{X}_k$, $k \geq 1$, having the same distribution as $\boldsymbol{X}$ to empirically approximate the distribution of the random variable $\mathcal{R}_\tau$ in (11.51). Also in this step we benefit from the fact that we can simulate infinitely many samples to mitigate this approximation error.

- To fit the network parameter ϑ in (11.55) we use i.i.d. copies $(Y_i, \boldsymbol{X}_i)$, $1 \leq i \leq n$, that have the same distribution as $(Y, \boldsymbol{X})$ under $\mathbb{P}$. However, to receive a good approximation to regression function $\boldsymbol{x} \mapsto \mu(\boldsymbol{x})$ we only need to simulate $Y_i|_{\{\boldsymbol{X}_i=\boldsymbol{x}_i\}}$ from $F_{Y|\boldsymbol{x}_i}(\cdot) = \mathbb{P}[\cdot|\boldsymbol{X}_i = \boldsymbol{x}_i]$, and $\boldsymbol{X}_i$ can be simulated from an arbitrary equivalent distribution to $p_{\boldsymbol{x}}$, and we still get the right conditional expectation in (11.54). This is worth mentioning because if we need a higher precision in some part of the feature space of $\boldsymbol{X}$, we can apply a sort of importance sampling by choosing a distribution for $\boldsymbol{X}$ that generates more samples in the corresponding part of the feature space compared to the original (true) distribution $p_{\boldsymbol{x}}$ of $\boldsymbol{X}$; this proposal has been emphasized in Cheridito et al. [74].

We study the example presented in Ha–Bauer [177] and Cheridito et al. [74]. This example considers a variable annuity (VA) with a guaranteed minimum income benefit (GMIB), and we revisit the network approach of Cheridito et al. [74].

Example 11.18 (Approximation of Conditional Expectations) We consider the VA example with a GMIB introduced and studied in Ha–Bauer [177]. This example involves a 3-dimensional stochastic process, for $t \geq 0$,

$$\boldsymbol{X}_t = (q_t, r_t, m_{x+t}),$$

with q_t being the log-value of the VA account at time t, r_t is the short rate at time t, and m_{x+t} is the force of mortality at time t of a person aged x at time 0. The payoff at fixed maturity date $T > 1$ of this insurance contract is given by

$$S = S(\boldsymbol{X}_T) = \max\left\{e^{q_T},\ b\, a_{x+T}(r_T, m_{x+T})\right\},$$

where e^{q_T} is the VA account value at time T, and $b\, a_{x+T}(r_T, m_{x+T})$ is the GMIB at time T consisting of a face value $b > 0$ and with $a_{x+T}(r_T, m_{x+T})$ being the value of an immediate annuity at time T of a person aged $x + T$. Our goal is to model the conditional expectation

$$\begin{aligned} \mu(\boldsymbol{X}_\tau) &= D(\tau, T; \boldsymbol{X}_\tau)\, \mathbb{E}\left[S(\boldsymbol{X}_T)|\, \boldsymbol{X}_\tau\right] \\ &= D(\tau, T; \boldsymbol{X}_\tau)\, \mathbb{E}\left[\max\left\{e^{q_T},\ b\, a_{x+T}(r_T, m_{x+T})\right\}\middle|\, \boldsymbol{X}_\tau\right], \end{aligned} \tag{11.57}$$

for a fixed valuation time point $0 < \tau < T$, and where $D(\tau, T) = D(\tau, T; \boldsymbol{X}_\tau)$ is a $\sigma(\boldsymbol{X}_\tau)$-measurable discount factor. This requires the explicit specification of the GMIB term as a function of (r_T, m_{x+T}), the modeling of the stochastic process $(\boldsymbol{X}_t)_{0 \le t \le T}$, and the specification of the discount factor $D(\tau, T; \boldsymbol{X}_\tau)$. In financial and actuarial valuation the regression function $\mu(\cdot)$ in (11.57) should reflect a no-arbitrage price. Therefore, $\mathbb{P}$ in (11.57) should be an equivalent martingale measure w.r.t. the selected numéraire. In our case, we choose a force of mortality $(m_{x+t})_t$-adjusted zero-coupon bond price as numéraire. This implies that $\mathbb{P}$ is a mortality-adjusted forward measure; for details and its explicit derivation we refer to Sect. 5.1 of Ha–Bauer [177]. In particular, Ha–Bauer [177] introduce a three-dimensional Brownian motion based model for $(\boldsymbol{X}_t)_t$ from which they deduce all relevant terms explicitly. We skip these calculations here, because, once the GMIB term and the discount factor are determined, everything boils down to knowing the distribution of the random vector $(\boldsymbol{X}_\tau, \boldsymbol{X}_T)$ under the corresponding probability measure $\mathbb{P}$. We choose initial age $x = 55$, maturity $T = 15$ and (solvency) time horizon $\tau = 1$. Under the model and parametrization of Ha–Bauer [177] we receive a multivariate Gaussian distribution under $\mathbb{P}$ given by

$$(\boldsymbol{X}_\tau, \boldsymbol{X}_T)^\top = (q_\tau, r_\tau, m_{x+\tau}, q_T, r_T, m_{x+T})^\top \tag{11.58}$$

$$\sim \mathcal{N}\left(\begin{pmatrix} 4.64 \\ 0.02 \\ 0.01 \\ 4.71 \\ 0.02 \\ 0.03 \end{pmatrix}, \begin{pmatrix} 3.2 \cdot 10^{-2} & -4.8 \cdot 10^{-4} & 1.3 \cdot 10^{-5} & 3.1 \cdot 10^{-2} & -1.4 \cdot 10^{-5} & 3.6 \cdot 10^{-5} \\ -4.8 \cdot 10^{-4} & 7.9 \cdot 10^{-5} & -4.4 \cdot 10^{-7} & -1.7 \cdot 10^{-4} & 2.4 \cdot 10^{-6} & -1.2 \cdot 10^{-6} \\ 1.3 \cdot 10^{-5} & -4.4 \cdot 10^{-7} & 1.5 \cdot 10^{-6} & 1.2 \cdot 10^{-5} & -1.3 \cdot 10^{-8} & 4.1 \cdot 10^{-6} \\ 3.1 \cdot 10^{-2} & -1.7 \cdot 10^{-4} & 1.2 \cdot 10^{-5} & 4.5 \cdot 10^{-1} & -1.3 \cdot 10^{-3} & 3.0 \cdot 10^{-4} \\ -1.4 \cdot 10^{-5} & 2.4 \cdot 10^{-6} & -1.3 \cdot 10^{-8} & -1.3 \cdot 10^{-3} & 2.0 \cdot 10^{-4} & -2.5 \cdot 10^{-6} \\ 3.6 \cdot 10^{-5} & -1.2 \cdot 10^{-6} & 4.1 \cdot 10^{-6} & 3.0 \cdot 10^{-4} & -2.5 \cdot 10^{-6} & 7.4 \cdot 10^{-5} \end{pmatrix}\right).$$

Under the model specification of Ha–Bauer [177], one can furthermore work out the discount factor and the annuity. Define for $t \ge 0$ and $k > 0$ the affine term structure

$$F(t, k; r_t, m_{x+t}) = \exp\left\{A(t, t+k) - B(t, t+k; \alpha) r_t - B(t, t+k; -\kappa) m_{x+t}\right\},$$

with deterministic functions

$$\begin{aligned} B(t, t+k; \alpha) &= \frac{1 - e^{-\alpha k}}{\alpha}, \\ A(t, t+k) &= \bar{\gamma}\left(B(t, t+k; \alpha) - k\right) + \frac{\sigma_r^2}{2\alpha^2}\left(k - 2B(t, t+k; \alpha) + B(t, t+k; 2\alpha)\right) \\ &\quad + \frac{\psi^2}{2\kappa^2}\left(k - 2B(t, t+k; -\kappa) + B(t, t+k; -2\kappa)\right) \\ &\quad + \frac{\varrho_{2,3}\sigma_r \psi}{\alpha\kappa}\left(B(t, t+k; -\kappa) - k + B(t, t+k; \alpha) - B(t, t+k; \alpha - \kappa)\right), \end{aligned}$$

with parameters for the short rate process $\alpha = 25\%$, $\sigma_r = 1\%$, for the force of mortality $\kappa = 7\%$, $\psi = 0.12\%$, the correlation between the short rate and the force of mortality $\varrho_{2,3} = -4\%$, and with market-price of the risk-adjusted mean reversion

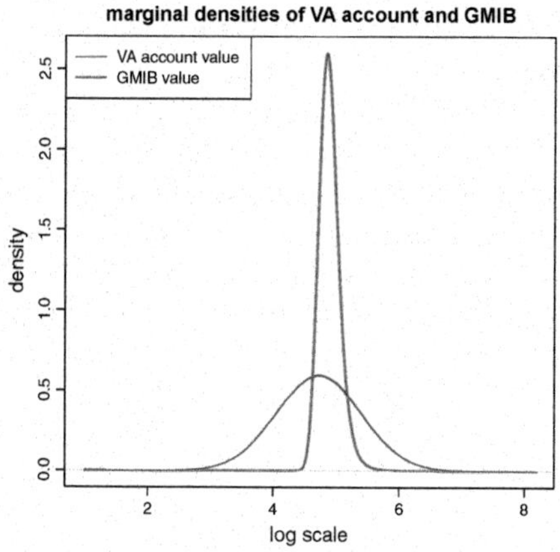

Fig. 11.23 Marginal densities of the VA account value e^{q_T} and the GMIB value $b\, a_{x+T}(r_T, m_{x+T})$

level $\bar{\gamma} = 1.92\%$ of the short rate process. These formulas can be retrieved because we work under an affine Gaussian structure. The discount factor is then given by

$$D(\tau, T; \boldsymbol{X}_\tau) = F(\tau, T - \tau; r_\tau, m_{x+\tau}),$$

and the annuity is determined by (we cap at age $55 + 50 = 105$)

$$a_{x+T}(r_T, m_{x+T}) = \sum_{k=1}^{50} F(T, k; r_T, m_{x+T}).$$

Moreover, we set for the face value $b = 10.79205$. This parametrization implies that the VA account value e^{q_T} exceeds the GMIB $b\, a_{x+T}(r_T, m_{x+T})$ with a probability of roughly 40%, i.e., in roughly 60% of the cases we exercise the GMIB option. Figure 11.23 shows the marginal densities of these two variables, moreover, their correlation is close to 0.

The model is now fully specified so that we can estimate the conditional expectation in (11.57) as a function of $\boldsymbol{X}_\tau$. We therefore simulate $n = 3'000'000$ i.i.d. Gaussian observations $(\boldsymbol{X}_\tau^{(i)}, \boldsymbol{X}_T^{(i)})$, $1 \leq i \leq n$, from (11.58). This provides us with the observations

$$\begin{aligned} Y_i &= D(\tau, T; \boldsymbol{X}_\tau^{(i)})\, S(\boldsymbol{X}_T^{(i)}) \\ &= F(\tau, T - \tau; r_\tau^{(i)}, m_{x+\tau}^{(i)}) \max\left\{ e^{q_T^{(i)}}, b \sum_{k=1}^{50} F(T, k; r_T^{(i)}, m_{x+T}^{(i)}) \right\}. \end{aligned}$$

The resulting data $(Y_i, \boldsymbol{X}_\tau^{(i)})_{1 \leq i \leq n}$ is used for determining the regression function $\mu(\cdot)$ in (11.57). We choose $n = 3'000'000$ samples in line with the least squares Monte Carlo approximation of Ha–Bauer [177].

We choose a FN network of depth $d = 3$ for approximating $\mu(\cdot)$. For the three FN layers we choose $(q_1, q_2, q_3) = (20, 15, 10)$ neurons with the hyperbolic tangent activation function, and as output activation we choose the identity function; we choose a more complex network compared to Cheridito et al. [74] because it seems that this gives us more accurate results. We fit this FN network using the square loss function. The square loss is motivated by (11.55). Furthermore, we average over 20 runs with different seeds. Thus, we receive 20 fitted FN networks $\widehat{\mu}_k(\cdot)$ for the 20 different seeds $1 \le k \le 20$ and the nagging predictor is obtained by averaging

$$\widehat{\mu}(\cdot) = \frac{1}{20} \sum_{k=1}^{20} \widehat{\mu}_k(\cdot).$$

We then generate new i.i.d. samples $\boldsymbol{X}_\tau^{(l)}$, $1 \le l \le L$, from the multivariate Gaussian distribution (11.58), where this time we only need the first 3 components. This gives us the empirical samples

$$\widehat{\mu}(\boldsymbol{X}_\tau^{(l)}) \qquad \text{for } 1 \le l \le L, \tag{11.59}$$

providing an empirical distribution $\widehat{F}_{\mu(\boldsymbol{X}_\tau)}$ that approximates the distribution of $\mu(\boldsymbol{X}_\tau)$, given in (11.57). In risk management and solvency analysis, this empirical distribution can be used to estimate the Value-at-Risk (VaR) and the (upper) conditional tail expectation (CTE) in valuation $\mu(\boldsymbol{X}_\tau)$, seen from time 0, on different safety levels $p \in (0, 1)$

$$\widehat{\text{VaR}}_p = \widehat{F}^{-1}_{\mu(\boldsymbol{X}_\tau)}(p) = \inf\left\{y \in \mathbb{R};\ \widehat{F}_{\mu(\boldsymbol{X}_\tau)}(y) \ge p\right\},$$

and

$$\widehat{\text{CTE}}_p = \mathbb{E}_{\widehat{F}_{\mu(\boldsymbol{X}_\tau)}}\left[\widehat{\mu}(\boldsymbol{X}_\tau) \,\middle|\, \widehat{\mu}(\boldsymbol{X}_\tau) > \widehat{\text{VaR}}_p\right].$$

We also refer to Sect. 11.3. The VaR and the CTE are two commonly used risk measures in insurance practice that determine the necessary risk bearing capital to run the corresponding insurance business. Typically, the VaR is evaluated on $p = 99.5\%$, i.e., we allow for a default probability of 0.5% of not being able to cover the changes in valuation over a $\tau = 1$ year time horizon. Alternatively, the CTE is considered on $p = 99\%$ which means that we need sufficient capital to cover on average the 1% worst changes in valuation over a 1 year time horizon.

Figure 11.24 shows our FN network approximations. The boxplots shows the individual results of the estimates $\widehat{\mu}_k(\cdot)$ with 20 different seeds, and the horizontal lines show the results of the nagging predictor (11.59). The red line at 140.97 gives the estimated VaR for $p = 99.5\%$, this value is slightly bigger than the best estimate of 139.47 (orange line) in Ha–Bauer [177] which is based on a functional approximation involving 37 monomials and 40'000'000 simulated samples. CTEs on $p = 99.5\%$ and $p = 99\%$ are given by 145.09 and 141.49. We conclude that in the present example $\widehat{\text{VaR}}_{99.5\%}$ (used in Europe) and $\widehat{\text{CTE}}_{99\%}$ (used in Switzerland) are approximately of the same size for this VA with a GMIB.

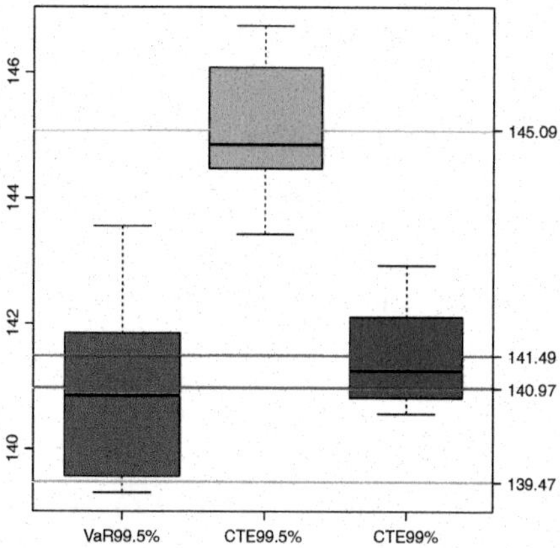

Fig. 11.24 Resulting $\widehat{\text{VaR}}_{99.5\%}$ (red), $\widehat{\text{CTE}}_{99.5\%}$ (green) and $\widehat{\text{CTE}}_{99\%}$ (blue); the orange line gives the result of Ha–Bauer [177] for the 99.5% VaR

This example shows how problems can be solved that require the computation of a conditional expectation. Alternatively, we could explore the LocalGLMnet architecture, which would allow us to explain the conditional expectation more explicitly in terms of the information $\boldsymbol{X}_\tau$ available at time τ. This may also be relevant in practice because it allows to determine the main risk drivers of the underlying insurance business.

Figure 11.25 shows the marginal densities of the components of $\boldsymbol{X}_\tau = (q_\tau, r_\tau, m_{x+\tau})$ in blue color. In red color we show the corresponding conditional densities of $\boldsymbol{X}_\tau$, conditioned on $\widehat{\mu}(\boldsymbol{X}_\tau) > \widehat{\text{VaR}}_{99.5\%}$, thus, these are the feature values $\boldsymbol{X}_\tau$ that lead to a shortfall beyond the 99.5% VaR of $\widehat{\mu}(\boldsymbol{X}_\tau)$. From this figure we conclude that the main driver of VaR is the VA account variable q_τ, whereas the short rate r_τ and the force of mortality $m_{x+\tau}$ are slightly lower beyond the VaR compared to their unconditioned counterparts. The explanation for these smaller values is that they lead to less discounting and, henceforth, to bigger GMIB values. This is useful information for exploring importance sampling as mentioned in Remarks 11.17. This closes the example. ∎

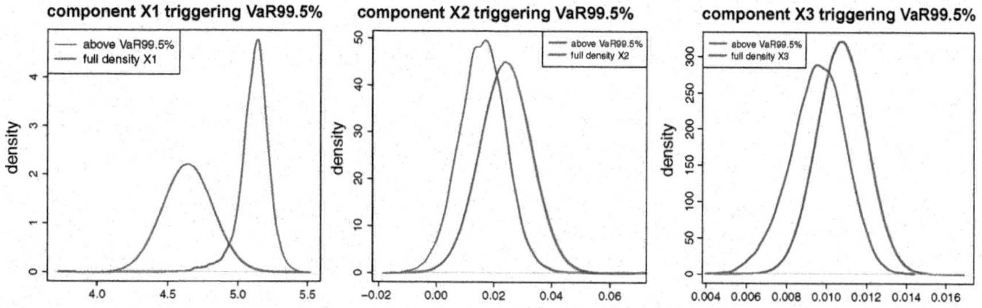

Fig. 11.25 Feature values $\boldsymbol{X}_\tau$ triggering VaR on the 99.5% level: (lhs) VA account log-value q_τ, (middle) short rate r_τ, and (rhs) force of mortality $m_{x+\tau}$, blue color shows the full density and red color shows the conditional density conditioned on being above the 99.5% VaR of $\widehat{\mu}(\boldsymbol{X}_\tau)$

11.6.3 Bayesian Networks: An Outlook

This section provides a short introduction to Bayesian networks and to variational inference. We see this section as a motivation for doing more research in that direction. In Sect. 11.4 we have assessed model uncertainty through bootstrapping. Alternatively, we could take a Bayesian viewpoint. We start from a fixed network architecture that involves a network parameter $\boldsymbol{\vartheta}$. The Bayesian approach considered in Section 6.1 selects a prior density $\pi(\boldsymbol{\vartheta})$ on the space of network parameters (w.r.t. a measure ν). For given data $(Y, \boldsymbol{x})$ we can then calculate the posterior density of $\boldsymbol{\vartheta}$ by

$$\pi\left(\boldsymbol{\vartheta} \mid Y, \boldsymbol{x}\right) \;\propto\; f\left(Y, \boldsymbol{\vartheta} \mid \boldsymbol{x}\right) = f\left(Y \mid \boldsymbol{\vartheta}, \boldsymbol{x}\right) \pi(\boldsymbol{\vartheta}). \tag{11.60}$$

A new data point $Y^\dagger$ with feature $\boldsymbol{x}^\dagger$ has conditional density, given observation $(Y, \boldsymbol{x})$,

$$f\left(y^\dagger \middle| \boldsymbol{x}^\dagger; Y, \boldsymbol{x}\right) = \int_{\boldsymbol{\vartheta}} f\left(y^\dagger \middle| \boldsymbol{\vartheta}, \boldsymbol{x}^\dagger\right) \pi\left(\boldsymbol{\vartheta} \mid Y, \boldsymbol{x}\right) d\nu(\boldsymbol{\vartheta}),$$

supposed that $(Y, \boldsymbol{x})$ and $(Y^\dagger, \boldsymbol{x}^\dagger)$ are conditionally independent, given $\boldsymbol{\vartheta}$. Thus, there only remains to determine the posterior density (11.60) of the network parameter $\boldsymbol{\vartheta}$. Unfortunately, this is a rather challenging problem because of the curse of dimensionality, and even advanced MCMC methods, such as HMC, often do not lead to satisfactory results (convergence), for MCMC we refer to Section 6.1. For this reason one often explores approximate inference methods, see, e.g., Chapter 10 of Bishop [36] or the tutorial of Jospin et al. [205]. A scalable version is to approximate the posterior density using the so-called method of variational inference. This is presented in the following.

Choose a family $\mathcal{F} = \{q(\cdot; \theta); \theta \in \boldsymbol{\Theta}\}$ of (more tractable) densities that have the same support as the prior $\pi(\cdot)$, and being parametrized by $\theta \in \boldsymbol{\Theta} \subset \mathbb{R}^K$. This family $\mathcal{F}$ is called the set of variational distributions, and the goal is to find the variational density $q(\cdot; \theta) \in \mathcal{F}$ that is closest to the posterior density (11.60).

To evaluate the similarity between two densities, we use the KL divergence which analyzes the divergence from $\pi\left(\cdot \mid Y, \boldsymbol{x}\right)$ to $q(\cdot; \theta)$ given by

$$D_{\mathrm{KL}}\left(q(\cdot; \theta) \middle|\middle| \pi\left(\cdot \mid Y, \boldsymbol{x}\right)\right) = \int_{\boldsymbol{\vartheta}} q(\boldsymbol{\vartheta}; \theta) \log\left(\frac{q(\boldsymbol{\vartheta}; \theta)}{\pi\left(\boldsymbol{\vartheta} \mid Y, \boldsymbol{x}\right)}\right) d\nu(\boldsymbol{\vartheta}).$$

The optimal approximation within $\mathcal{F}$, for given data $(Y, \boldsymbol{x})$, is found by solving

$$\widehat{\theta} = \widehat{\theta}(Y, \boldsymbol{x}) \;=\; \underset{\theta \in \boldsymbol{\Theta}}{\arg\min}\, D_{\mathrm{KL}}\left(q(\cdot; \theta) \middle|\middle| \pi\left(\cdot \mid Y, \boldsymbol{x}\right)\right);$$

for the moment we neglect existence and uniqueness questions. A main difficulty is the computation of this KL divergence because it involves the intractable posterior density of $\boldsymbol{\vartheta}$, given $(Y, \boldsymbol{x})$. We modify the optimization problem such that we can circumvent the explicit calculation of this KL divergence.

Lemma 11.19 *We have the following identity*

$$\log f(Y|\boldsymbol{x}) = \mathcal{E}(\theta|Y,\boldsymbol{x}) + D_{\mathrm{KL}}\left(q(\cdot;\theta)\middle|\middle|\pi\left(\cdot|\,Y,\boldsymbol{x}\right)\right),$$

for the (unconditional) density $f(y|\boldsymbol{x}) = \int_{\boldsymbol{\vartheta}} f(y|\boldsymbol{\vartheta},\boldsymbol{x})\pi(\boldsymbol{\vartheta})d\nu(\boldsymbol{\vartheta})$ *and the so-called evidence lower bound (ELBO)*

$$\mathcal{E}(\theta|Y,\boldsymbol{x}) = \int_{\boldsymbol{\vartheta}} q(\boldsymbol{\vartheta};\theta)\log\left(\frac{f\left(Y,\boldsymbol{\vartheta}\,|\,\boldsymbol{x}\right)}{q(\boldsymbol{\vartheta};\theta)}\right)d\nu(\boldsymbol{\vartheta}).$$

Observe that the left-hand side in the statement of Lemma 11.19 is independent of $\theta \in \boldsymbol{\Theta}$. Therefore, minimizing the KL divergence in θ is equivalent to maximizing the ELBO in θ. This follows exactly the same philosophy as the EM algorithm, see (6.32), in fact, the ELBO $\mathcal{E}$ plays the role of functional $\mathcal{Q}$ defined in (6.33).

Proof of Lemma 11.19 We start from the left-hand side of the statement

$$\begin{aligned}\log f(Y|\boldsymbol{x}) &= \int_{\boldsymbol{\vartheta}} q(\boldsymbol{\vartheta};\theta)\log f(Y|\boldsymbol{x})\,d\nu(\boldsymbol{\vartheta}) \;=\; \int_{\boldsymbol{\vartheta}} q(\boldsymbol{\vartheta};\theta)\log\left(\frac{f(Y,\boldsymbol{\vartheta}|\boldsymbol{x})}{\pi(\boldsymbol{\vartheta}|Y,\boldsymbol{x})}\right)d\nu(\boldsymbol{\vartheta})\\ &= \int_{\boldsymbol{\vartheta}} q(\boldsymbol{\vartheta};\theta)\log\left(\frac{f(Y,\boldsymbol{\vartheta}|\boldsymbol{x})/q(\boldsymbol{\vartheta};\theta)}{\pi(\boldsymbol{\vartheta}|Y,\boldsymbol{x})/q(\boldsymbol{\vartheta};\theta)}\right)d\nu(\boldsymbol{\vartheta})\\ &= \mathcal{E}(\theta|Y,\boldsymbol{x}) + D_{\mathrm{KL}}\left(q(\cdot;\theta)\middle|\middle|\pi\left(\cdot|\,Y,\boldsymbol{x}\right)\right).\end{aligned}$$

This proves the claim. □

The ELBO provides the lower bound (also called variational lower bound)

$$\log f(Y|\boldsymbol{x}) \;\geq\; \sup_{\theta\in\boldsymbol{\Theta}} \mathcal{E}(\theta|Y,\boldsymbol{x}).$$

Interestingly, the ELBO does not include the posterior density, but only the joint density of Y and $\boldsymbol{\vartheta}$, given $\boldsymbol{x}$, which is assumed to be known (available). It can be rewritten as

$$\begin{aligned}\mathcal{E}(\theta|Y,\boldsymbol{x}) &= \int_{\boldsymbol{\vartheta}} q(\boldsymbol{\vartheta};\theta)\log f\left(Y,\boldsymbol{\vartheta}\,|\,\boldsymbol{x}\right)d\nu(\boldsymbol{\vartheta}) - \int_{\boldsymbol{\vartheta}} q(\boldsymbol{\vartheta};\theta)\log q(\boldsymbol{\vartheta};\theta)\,d\nu(\boldsymbol{\vartheta})\\ &= \mathbb{E}_{q(\cdot;\theta)}\Big[\log f\left(Y,\boldsymbol{\vartheta}\,|\,\boldsymbol{x}\right)\Big|Y,\boldsymbol{x}\Big] - \mathbb{E}_{q(\cdot;\theta)}\Big[\log q(\boldsymbol{\vartheta};\theta)\Big],\end{aligned}$$

the first term being the expected joint log-likelihood of $(Y, \boldsymbol{\vartheta})$ under the variational density $\boldsymbol{\vartheta} \sim q(\cdot;\theta)$, and the second term being the entropy of the variational density.

The optimal approximation within $\mathcal{F}$ for given data $(Y, \boldsymbol{x})$ is then found by solving

$$\widehat{\theta} = \widehat{\theta}(Y, \boldsymbol{x}) = \underset{\theta \in \boldsymbol{\Theta}}{\arg\max}\ \mathcal{E}(\theta | Y, \boldsymbol{x}).$$

That is we try to simultaneously maximize the expected joint log-likelihood of $(Y, \boldsymbol{\vartheta})$ and the entropy over all variational densities $q(\cdot; \theta)$ in $\mathcal{F}$.

If we have multiple observations $\mathcal{D} = \{(Y_i, \boldsymbol{x}_i); 1 \le i \le n\}$, that are conditionally i.i.d., given $\boldsymbol{\vartheta}$, we have to solve (we use conditional independence)

$$\begin{aligned}
\widehat{\theta} &= \underset{\theta \in \boldsymbol{\Theta}}{\arg\max}\ \mathcal{E}(\theta | \mathcal{D}) \\
&= \underset{\theta \in \boldsymbol{\Theta}}{\arg\max}\ \mathbb{E}_{q(\cdot;\theta)}\left[\log \left(\pi(\boldsymbol{\vartheta}) \prod_{i=1}^{n} f\left(Y_i | \boldsymbol{\vartheta}, \boldsymbol{x}_i\right) \right) \middle| \mathcal{D} \right] - \mathbb{E}_{q(\cdot;\theta)}\left[\log q(\boldsymbol{\vartheta}; \theta) \right] \\
&= \underset{\theta \in \boldsymbol{\Theta}}{\arg\max} \left(\sum_{i=1}^{n} \mathbb{E}_{q(\cdot;\theta)}\left[\log f\left(Y_i | \boldsymbol{\vartheta}, \boldsymbol{x}_i\right) \middle| Y_i, \boldsymbol{x}_i \right] \right) - \mathbb{E}_{q(\cdot;\theta)}\left[\log \left(\frac{q(\boldsymbol{\vartheta}; \theta)}{\pi(\boldsymbol{\vartheta})} \right) \right] \\
&= \underset{\theta \in \boldsymbol{\Theta}}{\arg\max} \left(\sum_{i=1}^{n} \mathbb{E}_{q(\cdot;\theta)}\left[\log f\left(Y_i | \boldsymbol{\vartheta}, \boldsymbol{x}_i\right) \middle| Y_i, \boldsymbol{x}_i \right] \right) - D_{\mathrm{KL}}\left(q(\cdot; \theta) \| \pi \right).
\end{aligned}$$

Typically, one solves this problem with gradient ascent methods which requires calculation of the gradient ∇_θ of the objective function on the right-hand side. This is more difficult than plain vanilla gradient descent in network fitting because θ enters the expectation operator $\mathbb{E}_{q(\cdot;\theta)}$.

Kingma–Welling [217] propose to use the following reparametrization trick. Assume that we can receive the random variable $\boldsymbol{\vartheta} \sim q(\cdot; \theta)$ by a reparametrization $\boldsymbol{\vartheta} \overset{(d)}{=} t(\epsilon, \theta)$ for some smooth function t and where $\epsilon \sim p$ does not depend on θ. E.g., if $\boldsymbol{\vartheta}$ is multivariate Gaussian with mean $\boldsymbol{\mu}$ and covariance matrix $AA^\top$, then $\boldsymbol{\vartheta} \overset{(d)}{=} \boldsymbol{\mu} + A\epsilon$ for ϵ being standard multivariate Gaussian. Under the assumption that the reparametrization trick works for the family $\mathcal{F} = \{q(\cdot; \theta); \theta \in \boldsymbol{\Theta}\}$ we arrive at, for $\epsilon \sim p$,

$$\begin{aligned}
\widehat{\theta} &= \underset{\theta \in \boldsymbol{\Theta}}{\arg\max}\ \mathcal{E}(\theta | \mathcal{D}) \qquad (11.61) \\
&= \underset{\theta \in \boldsymbol{\Theta}}{\arg\max} \sum_{i=1}^{n} \left(\mathbb{E}_p\left[\log f\left(Y_i | t(\epsilon, \theta), \boldsymbol{x}_i\right) \middle| Y_i, \boldsymbol{x}_i \right] - \frac{1}{n} \mathbb{E}_p\left[\log \left(\frac{q(t(\epsilon, \theta); \theta)}{\pi(t(\epsilon, \theta))} \right) \right] \right) \\
&= \underset{\theta \in \boldsymbol{\Theta}}{\arg\max} \sum_{i=1}^{n} \mathbb{E}_p\left[\log \left(\frac{f\left(Y_i | t(\epsilon, \theta), \boldsymbol{x}_i\right) \pi\left(t(\epsilon, \theta)\right)^{1/n}}{q\left(t(\epsilon, \theta); \theta\right)^{1/n}} \right) \middle| Y_i, \boldsymbol{x}_i \right].
\end{aligned}$$

The gradient of the ELBO is then given by (supposed we can exchange $\mathbb{E}_p$ and ∇_θ)

$$\nabla_\theta\, \mathcal{E}(\theta|\mathcal{D}) = \sum_{i=1}^n \mathbb{E}_p\left[\nabla_\theta \log\left(\frac{f\,(Y_i\,|t(\epsilon,\theta),\boldsymbol{x}_i)\,\pi\,(t(\epsilon,\theta))^{1/n}}{q\,(t(\epsilon,\theta);\theta)^{1/n}}\right)\middle| Y_i,\boldsymbol{x}_i\right].$$

These expected gradients are calculated empirically using Monte Carlo methods. Sample i.i.d. observations $\epsilon^{(i,j)} \sim p$, $1 \le i \le n$ and $1 \le j \le m$, and consider the empirical approximation

$$\nabla_\theta \mathcal{E}(\theta|\mathcal{D}) \approx \sum_{i=1}^n \frac{1}{m}\sum_{j=1}^m \nabla_\theta \log\left(\frac{f\left(Y_i\,\middle|t(\epsilon^{(i,j)},\theta),\boldsymbol{x}_i\right)\pi\left(t(\epsilon^{(i,j)},\theta)\right)^{1/n}}{q\left(t(\epsilon^{(i,j)},\theta);\theta\right)^{1/n}}\right). \tag{11.62}$$

Using this empirical approximation we can use gradient ascent methods to estimate θ, known as stochastic gradient variational Bayes (SGVB) estimator, see Sect. 2.4.3 of Kingma–Welling [217], or as Bayes by Backprop, see Blundell et al. [41] and Jospin et al. [205].

Example 11.20 We consider the gradient (11.62) for an example from the EDF. First, if n is sufficiently large, it often suffices to set $m = 1$, and we still receive an accurate estimate. In that case we drop index j giving $\epsilon^{(i)}$. Assume that the (conditionally independent) observations Y_i belong to the same member of the EDF having cumulant function κ. Moreover, assume that the (conditional) mean of Y_i, given $\boldsymbol{x}_i$, can be described by a FN network and a link function g such that, see (7.8),

$$\mu_i = \mu(\boldsymbol{x}_i) = \mu_{\boldsymbol{\vartheta}}(\boldsymbol{x}_i) = g^{-1}\left\langle \boldsymbol{\beta}, \boldsymbol{z}_{\boldsymbol{w}}^{(d:1)}(\boldsymbol{x}_i)\right\rangle,$$

for network parameter $\boldsymbol{\vartheta} = (\boldsymbol{\beta},\boldsymbol{w}) \in \mathbb{R}^r$. In a Bayesian FN network this network parameter is not fixed but rather acts as a latent variable. In (11.62) this latent variable is for realization i given by (and using the reparametrization trick) $\boldsymbol{\vartheta} = t(\epsilon^{(i)};\theta) \in \mathbb{R}^r$; θ is not the canonical parameter, here. Thus, we receive conditional mean of Y_i, given $\epsilon^{(i)}$ and $\boldsymbol{x}_i$,

$$\mu_i = \mu_{t(\epsilon^{(i)};\theta)}(\boldsymbol{x}_i) = g^{-1}\left\langle \boldsymbol{\beta}(\epsilon^{(i)};\theta), \boldsymbol{z}_{\boldsymbol{w}(\epsilon^{(i)};\theta)}^{(d:1)}(\boldsymbol{x}_i)\right\rangle,$$

with network parameter $\boldsymbol{\vartheta}(\epsilon^{(i)};\theta) = (\boldsymbol{\beta}(\epsilon^{(i)};\theta), \boldsymbol{w}(\epsilon^{(i)};\theta)) = t(\epsilon^{(i)},\theta) \in \mathbb{R}^r$. Maximizing the ELBO implies that we need to calculate the gradients w.r.t. θ. First, we calculate the gradient w.r.t. the network parameter $\boldsymbol{\vartheta}$ of the data log-likelihood

$$\nabla_{\boldsymbol{\vartheta}} \log f\,(Y_i\,|\boldsymbol{\vartheta},\boldsymbol{x}_i) = \nabla_{\boldsymbol{\vartheta}}\, \ell_{Y_i}(\boldsymbol{\vartheta}) \in \mathbb{R}^r.$$

This gradient is calculated with back-propagation, we refer to (7.16) and Proposition 7.5. There remains the chain rule for evaluating the inner derivative coming from the reparametrization trick $\theta \in \boldsymbol{\Theta} \subset \mathbb{R}^K \mapsto \boldsymbol{\vartheta} = t(\epsilon^{(i)}; \theta) \in \mathbb{R}^r$. Consider the Jacobian matrix

$$J(\theta; \epsilon^{(i)}) = \left(\frac{\partial}{\partial \theta_k} t_j(\epsilon^{(i)}; \theta)\right)_{1 \le j \le r, 1 \le k \le K} \in \mathbb{R}^{r \times K}.$$

This gives us the gradient w.r.t. θ

$$\nabla_\theta \log f\left(Y_i \middle| t(\epsilon^{(i)}, \theta), \boldsymbol{x}_i\right) = J(\theta; \epsilon^{(i)})^\top \left(\nabla_{\boldsymbol{\vartheta}} \ell_{Y_i}(\boldsymbol{\vartheta})\Big|_{\boldsymbol{\vartheta} = t(\epsilon^{(i)}, \theta)}\right) \in \mathbb{R}^K. \tag{11.63}$$

The prior distribution is often taken to be the multivariate Gaussian with prior mean $\tau \in \mathbb{R}^r$ and (symmetric and positive definite) prior covariance matrix $T \in \mathbb{R}^{r \times r}$, thus,

$$\pi(\boldsymbol{\vartheta}) = ((2\pi)^{r/2} |T|^{1/2})^{-1} \exp\left\{-\frac{1}{2}(\boldsymbol{\vartheta} - \tau)^\top T^{-1} (\boldsymbol{\vartheta} - \tau)\right\}.$$

This implies for the gradient w.r.t. θ for the prior

$$\nabla_\theta \log \pi(t(\epsilon^{(i)}, \theta)) = -J(\theta; \epsilon^{(i)})^\top T^{-1} \left(t(\epsilon^{(i)}, \theta) - \tau\right) \in \mathbb{R}^K.$$

There remains the choice of the family $\mathcal{F} = \{q(\cdot; \theta); \theta \in \boldsymbol{\Theta}\}$ of variational densities such that the reparametrization trick works. This is discussed in the remainder. ■

We briefly discuss the most popular and simplest family chosen for the variational distributions $\mathcal{F}$. This family is the so-called mean field Gaussian variational family, meaning that all components of $\boldsymbol{\vartheta} \in \mathbb{R}^r$ are assumed to be independent Gaussian, that is,

$$q(\boldsymbol{\vartheta}; \theta) = \prod_{j=1}^r \frac{1}{\sqrt{2\pi}\sigma_j} \exp\left\{-\frac{1}{2\sigma_j^2}(\vartheta_j - \mu_j)^2\right\},$$

for $\theta = (\mu_1, \sigma_1, \ldots, \mu_r, \sigma_r)^\top \in \mathbb{R}^K$ with $K = 2r$ and with $\sigma_j > 0$ for all $1 \le j \le r$. This allows us to apply the reparametrization trick

$$\boldsymbol{\vartheta} \stackrel{(d)}{=} t(\epsilon, \theta) = \boldsymbol{\mu} + \mathrm{diag}(\sigma_1, \ldots, \sigma_r)\epsilon = \begin{pmatrix} \mu_1 + \sigma_1 \epsilon_1 \\ \vdots \\ \mu_r + \sigma_r \epsilon_r \end{pmatrix},$$

with r-dimensional standard Gaussian variable $\epsilon \sim \mathcal{N}(\mathbf{0}, \mathbb{1})$. The Jacobian matrix is

$$J(\theta;\epsilon) = \begin{pmatrix} 1 & \epsilon_1 & 0 & 0 & \cdots & 0 & 0 \\ 0 & 0 & 1 & \epsilon_2 & \cdots & 0 & 0 \\ \vdots & & & & \ddots & & \vdots \\ 0 & 0 & 0 & 0 & \cdots & 1 & \epsilon_r \end{pmatrix} \in \mathbb{R}^{r\times K}.$$

The mean field Gaussian case provides the entropy of the variational distribution

$$-\mathbb{E}_{q(\cdot;\theta)}\Big[\log q(\vartheta;\theta)\Big] = \sum_{j=1}^{r}\frac{1}{2}\log(2\pi\sigma_j^2) + \frac{1}{2} = \sum_{j=1}^{r}\log(\sqrt{2\pi e}\sigma_j).$$

This mean field Gaussian variational inference can be implemented with the R package `tfprobability` of Keydana et al. [212] and an explicit example is given in Kuo [230].

Example 11.20, Revisited Working under the assumptions of Example 11.20 and additionally assuming that the family of variational distributions $\mathcal{F}$ is multivariate Gaussian $q(\cdot;\theta) \overset{(d)}{=} \mathcal{N}(\boldsymbol{\mu}, \Sigma)$ leads us after some calculation to (the well-known formula)

$$D_{\mathrm{KL}}\left(q(\cdot;\theta)\middle|\middle|\pi\right) = \frac{1}{2}\left[\log\left(\frac{|T|}{|\Sigma|}\right) - r + \mathrm{trace}\left(T^{-1}\Sigma\right) + (\tau-\boldsymbol{\mu})^\top T^{-1}(\tau-\boldsymbol{\mu})\right].$$

This further simplifies if T and Σ are diagonal, the latter being the mean field Gaussian case. The remaining terms of the ELBO are treated empirically as in (11.63). ■

This section has provided a short introduction to uncertainty estimation in networks using Bayesian methods. We believe that this gives a promising outlook that certainly needs more theoretical and practical work to become useful in practical applications.

Chapter 12
Appendix A: Technical Results on Networks

The reader may have noticed that for GLMs we have developed an asymptotic theory that allowed us to assess the quality of predictors as well as it allowed us to validate the fitted models. For networks there does not exist such a theory, yet, and the purpose of this appendix is to present more technical results on the asymptotic behavior of FN networks and their estimators that may lead to an asymptotic theory. This appendix hopefully stimulates further research in this field of statistical modeling.

12.1 Universality Theorems

We present a specific version of the universality theorems for shallow FN networks; we refer to the discussion in Sect. 7.2.2. This section follows Hornik et al. [192]. Choose an input dimension $q_0 \in \mathbb{N}$ and consider the set of all affine functions

$$\mathcal{A}^{q_0} = \left\{ A : \{1\} \times \mathbb{R}^{q_0} \to \mathbb{R}; \quad \boldsymbol{x} \mapsto A(\boldsymbol{x}) = \langle \boldsymbol{w}, \boldsymbol{x} \rangle, \ \boldsymbol{w} \in \mathbb{R}^{q_0+1} \right\},$$

we add a 0th component in feature $\boldsymbol{x} = (x_0 = 1, x_1, \ldots, x_{q_0})^\top \in \{1\} \times \mathbb{R}^{q_0}$ for the intercept. Choose a measurable (activation) function $\phi : \mathbb{R} \to \mathbb{R}$ and define

$$\Sigma^{q_0}(\phi) = \left\{ f : \{1\} \times \mathbb{R}^{q_0} \to \mathbb{R}; \ \boldsymbol{x} \mapsto f(\boldsymbol{x}) = \sum_{j=0}^{q_1} \beta_j \phi(A_j(\boldsymbol{x})), \ A_j \in \mathcal{A}^{q_0}, \beta_j \in \mathbb{R}, q_1 \in \mathbb{N} \right\}.$$

M. V. Wüthrich, M. Merz, *Statistical Foundations of Actuarial Learning and its Applications*, Springer Actuarial, https://doi.org/10.1007/978-3-031-12409-9_12

This is the set of all shallow FN networks $f(\boldsymbol{x}) = \langle \boldsymbol{\beta}, \boldsymbol{z}^{(1:1)}(\boldsymbol{x})\rangle$ with activation function ϕ and the linear output activation, see (7.8); the intercept component of the output is integrated into the 0th component $j = 0$. Moreover, we define the networks

$$\Sigma\Pi^{q_0}(\phi) = \Big\{ f : \{1\} \times \mathbb{R}^{q_0} \to \mathbb{R}; \mapsto f(\boldsymbol{x}) = \sum_{j=0}^{q_1} \beta_j \prod_{k=1}^{l_j} \phi(A_{j,k}(\boldsymbol{x})),$$

$$A_{j,k} \in \mathcal{A}^{q_0}, \beta_j \in \mathbb{R}, l_j \in \mathbb{N}, q_1 \in \mathbb{N} \Big\}.$$

The latter networks contain the former $\Sigma^{q_0}(\phi) \subset \Sigma\Pi^{q_0}(\phi)$, by setting $l_j = 1$ for all $0 \le j \le q_1$. We are going to prove a universality theorem first for the networks $\Sigma\Pi^{q_0}(\phi)$, and afterwards for the shallow FN networks $\Sigma^{q_0}(\phi)$.

Definition 12.1 The function $\phi : \mathbb{R} \to [0, 1]$ is called a squashing function if it is non-decreasing with $\lim_{x\to-\infty} \phi(x) = 0$ and $\lim_{x\to\infty} \phi(x) = 1$.

Since squashing functions can have at most countably many discontinuities, they are measurable; a continuous and a non-continuous example are given by the sigmoid and by the step function activation, respectively, see Table 7.1.

Lemma 12.2 *The sigmoid activation function is Lipschitz with constant* $1/4$.

Proof The derivative of the sigmoid function is given by $\phi' = \phi(1 - \phi)$. This provides for the second derivative $\phi'' = \phi' - 2\phi\phi' = \phi'(1 - 2\phi)$. The latter is zero for $\phi(x) = 1/2$. This says that the maximal slope of ϕ is attained for $x = 0$ and it is $\phi'(0) = 1/4$. □

We denote by $C(\mathbb{R}^{q_0})$ the set of all continuous functions from $\{1\} \times \mathbb{R}^{q_0}$ to $\mathbb{R}$, and by $\mathcal{M}(\mathbb{R}^{q_0})$ the set of all measurable functions from $\{1\} \times \mathbb{R}^{q_0}$ to $\mathbb{R}$. If the measurable activation function ϕ is continuous, we have $\Sigma\Pi^{q_0}(\phi) \subset C(\mathbb{R}^{q_0})$, otherwise $\Sigma\Pi^{q_0}(\phi) \subset \mathcal{M}(\mathbb{R}^{q_0})$.

Definition 12.3 A subset $S \subset \mathcal{M}(\mathbb{R}^{q_0})$ is said to be uniformly dense on compacta in $C(\mathbb{R}^{q_0})$ if for every compact subset $K \subset \{1\}\times\mathbb{R}^{q_0}$ the set S is ρ_K-dense in $C(\mathbb{R}^{q_0})$ meaning that for all $\epsilon > 0$ and all $g \in C(\mathbb{R}^{q_0})$ there exists $f \in S$ such that

$$\rho_K(g, f) = \sup_{\boldsymbol{x}\in K} |g(\boldsymbol{x}) - f(\boldsymbol{x})| < \epsilon.$$

Theorem 12.4 (Theorem 2.1 in Hornik et al. [192]) *Assume ϕ is a non-constant and continuous activation function. $\Sigma\Pi^{q_0}(\phi) \subset C(\mathbb{R}^{q_0})$ is uniformly dense on compacta in $C(\mathbb{R}^{q_0})$.*

Proof The proof is based on the Stone–Weierstrass theorem. We briefly recall the Stone–Weierstrass theorem. Assume $\mathcal{A}$ is a family of real functions defined on a set E. $\mathcal{A}$ is called an *algebra* if it is closed under addition, multiplication and scalar multiplication. A family $\mathcal{A}$ *separates points* in E, if for every $x, z \in E$ with $x \neq z$ there exists a function $A \in \mathcal{A}$ with $A(x) \neq A(z)$. The family $\mathcal{A}$ does *not vanish at any point* of E if for all $x \in E$ there exists a function $A \in \mathcal{A}$ such that $A(x) \neq 0$.

Let $\mathcal{A}$ be an algebra of continuous real functions on a compact set K. The Stone–Weierstrass theorem says that if $\mathcal{A}$ separates points in K and if it does not vanish at any point of K, then $\mathcal{A}$ is ρ_K-dense in the space of all continuous real functions on K.

Choose any compact set $K \subset \{1\} \times \mathbb{R}^{q_0}$. For any activation function ϕ, $\Sigma\Pi^{q_0}(\phi)$ is obviously an algebra. So there remains to prove that this algebra separates points and does not vanish at any point. Firstly, choose $\boldsymbol{x}, \boldsymbol{z} \in K$ such that $\boldsymbol{x} \neq \boldsymbol{z}$. Since ϕ is non-constant we can choose $a, b \in \mathbb{R}$ such that $\phi(a) \neq \phi(b)$. Next choose $A \in \mathcal{A}^{q_0}$ such that $A(\boldsymbol{x}) = a$ and $A(\boldsymbol{z}) = b$. Then, $\phi(A(\boldsymbol{x})) \neq \phi(A(\boldsymbol{z}))$ and $\Sigma\Pi^{q_0}(\phi)$ separates points. Secondly, since ϕ is non-constant, we can choose $a \in \mathbb{R}$ such that $\phi(a) \neq 0$. Moreover, choose weight $\boldsymbol{w} = (a, 0, \ldots, 0)^\top \in \mathbb{R}^{q_0+1}$. Then for this $A \in \mathcal{A}^{q_0}$, $A(\boldsymbol{x}) = \langle \boldsymbol{w}, \boldsymbol{x} \rangle = a$ for any $\boldsymbol{x} \in K$. Henceforth, $\phi(A(\boldsymbol{x})) \neq 0$, therefore $\Sigma\Pi^{q_0}(\phi)$ does not vanish at any point of K. The claim then follows from the Stone–Weierstrass theorem and using that ϕ is continuous by assumption. □

For Theorem 12.4 to hold, the activation function ϕ can be any continuous and non-constant function, i.e., it does not need to be a squashing function. This is fairly general, but it rules out the step function activation as it is not continuous. However, for squashing functions continuity is not needed and one still receives the uniformly dense on compacta property of $\Sigma\Pi^{q_0}(\phi)$ in $\mathcal{C}(\mathbb{R}^{q_0})$, this has been proved in Theorem 2.3 of Hornik et al. [192]. The following theorem also does not need continuity, i.e., we do not require $\Sigma^{q_0}(\phi) \subset \mathcal{C}(\mathbb{R}^{q_0})$ as ϕ only needs to be measurable (and squashing).

Theorem 12.5 (Universality, Theorem 2.4 in Hornik et al. [192]) *Assume ϕ is a squashing activation function. $\Sigma^{q_0}(\phi)$ is uniformly dense on compacta in $\mathcal{C}(\mathbb{R}^{q_0})$.*

Sketch of Proof For the (continuous) cosine activation function choice $\cos(\cdot)$, Theorem 12.4 applies to $\Sigma\Pi^{q_0}(\cos)$. Repeatedly applying the trigonometric identity $\cos(a)\cos(b) = \cos(a+b) - \cos(a-b)$ allows us to rewrite any trigonometric polynomial $\prod_{k=1}^{l_j} \cos(A_{j,k}(\boldsymbol{x}))$ as $\sum_{t=1}^{T} \alpha_t \cos(A_t(\boldsymbol{x}))$ for suitable $A_t \in \mathcal{A}^{q_0}$, $\alpha_t \in \mathbb{R}$ and $T \in \mathbb{N}$. This allows us to identify $\Sigma^{q_0}(\cos) = \Sigma\Pi^{q_0}(\cos)$. As a consequence of Theorem 12.4, shallow FN networks $\Sigma^{q_0}(\cos)$ are uniformly dense on compacta in $\mathcal{C}(\mathbb{R}^{q_0})$.

The remaining part relies on approximating the cosine activation function. Firstly, Lemma A.2 of Hornik et al. [192] says that for any continuous squashing function ψ and any $\epsilon > 0$ there exists $H_\epsilon(x) = \sum_{j=1}^{q_1} \beta_j \phi(w_0^j + w_1^j x) \in \Sigma^1(\phi)$, $x \in \mathbb{R}$, such that

$$\sup_{x \in \mathbb{R}} |\psi(x) - H_\epsilon(x)| < \epsilon. \tag{12.1}$$

For the proof we refer to Lemma A.2 of Hornik et al. [192], it uses that ψ is a continuous squashing function, implying that for every $\delta \in (0, 1)$ there exists $m > 0$

such that $\psi(-m) < \delta$ and $\psi(m) > 1 - \delta$. Approximation $H_\epsilon \in \Sigma^1(\phi)$ of ψ is then constructed on $(-m, m)$ so that the error bound holds (and for δ sufficiently small).

Secondly, choose $\epsilon > 0$ and $M > 0$, there exists $\cos_{M,\epsilon} \in \Sigma^1(\phi)$ such that

$$\sup_{x \in [-M,M]} \left| \cos(x) - \cos_{M,\epsilon}(x) \right| < \epsilon. \tag{12.2}$$

This is Lemma A.3 of Hornik et al. [192]; to prove this, we consider the cosine squasher of Gallant–White [150], for $x \in \mathbb{R}$

$$\chi(x) = \frac{1}{2}\left(1 + \cos\left(x + \frac{3\pi}{2}\right)\right) \mathbb{1}_{\{-\pi/2 \le x \le \pi/2\}} + \mathbb{1}_{\{x > \pi/2\}} \in [0, 1].$$

This is a continuous squashing function. Adding, subtracting and scaling a *finite* number of affinely shifted versions of the cosine squasher χ can exactly replicate the cosine on $[-M, M]$. Claim (12.2) then follows from the fact that we need a finite number of cosine squashers χ to replicate the cosine on $[-M, M]$, the triangle equality, and the fact that the (continuous) cosine squasher can be approximated arbitrarily well in $\Sigma^1(\phi)$ using (12.1).

The final step is to patch everything together. Consider $\sum_{t=1}^T \alpha_t \cos(A_t(\boldsymbol{x}))$ which approximates on the compact set $K \subset \{1\} \times \mathbb{R}^{q_0}$ a given continuous function $g \in C(\mathbb{R}^{q_0})$ with a given tolerance $\epsilon/2$. Choose $M > 0$ such that $A_t(K) \subset [-M, M]$ for all $1 \le t \le T$. Note that this M can be found because K is compact, A_t are continuous and T is finite. Define $T' = T \sum_{t=1}^T |\alpha_t| < \infty$. By (12.2) we can then choose $\cos_{M,\epsilon/(2T')} \in \Sigma^1(\phi)$ such that

$$\sup_{\boldsymbol{x} \in K} \left| \sum_{t=1}^T \alpha_t \cos(A_t(\boldsymbol{x})) - \sum_{t=1}^T \alpha_t \cos_{M,\epsilon/(2T')}(A_t(\boldsymbol{x})) \right| < \epsilon/2.$$

This completes the proof. □

12.2 Consistency and Asymptotic Normality

Universality Theorem 12.5 tells us that we can approximate any compactly supported continuous function arbitrarily well by a sufficiently large shallow FN network, say, with sigmoid activation function ϕ. The next natural question is whether we can *learn* these approximations from data $(Y_i, \boldsymbol{x}_i)_{i \ge 1}$ that follow the true but unknown regression function $\boldsymbol{x} \mapsto \mu_0(\boldsymbol{x})$, or in other words whether we have consistency for a certain class of learning methods. This is the question addressed, e.g., in White [379, 380], Barron [26], Chen–Shen [73], Döhler–Rüschendorf [109] and Shen et al. [336]. This turns the algebraic universality question into a statistical question about consistency.

Assume that the true data model satisfies

$$Y = \mu_0(\boldsymbol{x}) + \varepsilon = \mathbb{E}[Y|\boldsymbol{x}] + \varepsilon, \tag{12.3}$$

for a continuous regression function $\mu_0 : \mathcal{X} \to \mathbb{R}$ on a compact set $\mathcal{X} \subset \{1\} \times \mathbb{R}^{q_0}$, and with a centered error ε satisfying $\mathbb{E}[|\varepsilon|^{2+\delta}] < \infty$ for some $\delta > 0$ and being independent of $\boldsymbol{x}$. The question now is whether we can learn this (true) regression function μ_0 from independent data $(Y_i, \boldsymbol{x}_i)$, $1 \le i \le n$, obeying (12.3). Throughout this section we use the square error loss function $L(y, a) = (y - a)^2$. For given data, this results in solving

$$\widetilde{\mu}_n = \underset{\mu \in \mathcal{C}(\mathcal{X})}{\arg\min} \frac{1}{n} \sum_{i=1}^{n} L\left(Y_i, \mu(\boldsymbol{x}_i)\right) = \underset{\mu \in \mathcal{C}(\mathcal{X})}{\arg\min} \frac{1}{n} \sum_{i=1}^{n} \left(Y_i - \mu(\boldsymbol{x}_i)\right)^2, \tag{12.4}$$

where $\mathcal{C}(\mathcal{X})$ denotes the set of continuous functions on the compact set $\mathcal{X} \subset \{1\} \times \mathbb{R}^{q_0}$. The main question is whether estimator $\widetilde{\mu}_n$ approaches the true regression function μ_0 for increasing sample size n.

Typically, the family of continuous functions $\mathcal{C}(\mathcal{X})$ is much too rich to be able to solve optimization problem (12.4), and the solution may have undesired properties. In particular, the solution to (12.4) will over-fit to the data for any sample size n, and consistency will not hold, see, e.g., Section 2.2.1 in Chen [72]. Therefore, the optimization needs to be done over (well-chosen) smaller sets $\mathcal{S}_n \subset \mathcal{C}(\mathcal{X})$. For instance, $\mathcal{S}_n$ can be the set of shallow FN networks having a maximal width $q_1 = q_1(n)$, depending on the sample size n of the data. Considering this regression problem in a non-parametric sense, we let grow these sets $\mathcal{S}_n$ with the sample size n. This idea is attributed to Grenander [172] and it is called the *method of sieve estimators* of μ_0. We define for $d \in \mathbb{N}$, $\Delta > 0$, $\widetilde{\Delta} > 0$ and activation function ϕ

$$\mathcal{S}(d, \Delta, \widetilde{\Delta}, \phi) = \left\{ f \in \Sigma^{q_0}(\phi);\ q_1 = d,\ \sum_{j=0}^{q_1} |\beta_j| \le \Delta,\ \max_{1 \le j \le q_1} \sum_{l=0}^{q_0} |w_{l,j}| \le \widetilde{\Delta} \right\}.$$

These sets $\mathcal{S}(d, \Delta, \widetilde{\Delta}, \phi)$ are shallow FN networks of a given width $q_1 = d$ and with some restrictions on the network parameters.[1] We then choose increasing sequences

[1] The bound $\sum_{j=0}^{q_1} |\beta_j| \le \Delta$ in $\mathcal{S}(d, \Delta, \widetilde{\Delta}, \phi)$ allows us to view this set of shallow FN networks as a symmetric convex hull of the family of functions $\mathcal{S}_0(\phi) = \{\boldsymbol{x} \mapsto \phi(A(\boldsymbol{x}));\ A \in \mathcal{A}^{q_0}\}$, see Sect. 2.6.3 in Van der Vaart–Wellner [364]. If we choose an increasing activation function ϕ, this family of functions $\phi \circ A$ is a composition of a fixed increasing function ϕ and a finite dimensional vector space $\mathcal{A}^{q_0}$ of functions A. This implies that $\mathcal{S}_0(\phi)$ is a VC-class saying that it has a finite Vapnik–Chervonenkis (VC) dimension [365]; see also Condition A and Theorem 2.1 in Döhler–Rüschendorf [109]. This VC-class is an important property in many proofs as it leads to a finite covering (metric entropy) of function spaces, and this allows to apply limit theorems to point processes, we refer to Van der Vaart–Wellner [364].

$(d_n)_{n\geq 1}$, $(\Delta_n)_{n\geq 1}$ and $(\widetilde{\Delta}_n)_{n\geq 1}$ which provides us with an increasing sequence of sieves (becoming finer as n increases)

$$\ldots \subseteq \mathcal{S}_n(\phi) \stackrel{\text{def.}}{=} \mathcal{S}(d_n, \Delta_n, \widetilde{\Delta}_n, \phi) \subseteq \mathcal{S}_{n+1}(\phi) \stackrel{\text{def.}}{=} \mathcal{S}(d_{n+1}, \Delta_{n+1}, \widetilde{\Delta}_{n+1}, \phi) \subseteq \ldots .$$

The following corollary is a simple consequence of Theorem 12.5.

Corollary 12.6 *Assume ϕ is a squashing activation function, and let the increasing sequences $(d_n)_{n\geq 1}$, $(\Delta_n)_{n\geq 1}$ and $(\widetilde{\Delta}_n)_{n\geq 1}$ tend to infinity for $n \to \infty$. Then $\bigcup_{n\geq 1} \mathcal{S}_n(\phi)$ is uniformly dense in $C(\mathcal{X})$.*

This corollary says that for any regression function $\mu_0 \in C(\mathcal{X})$ we can find $n \in \mathbb{N}$ and $\mu_n \in \mathcal{S}_n(\phi)$ such that μ_n is arbitrarily close to μ_0; remark that all functions are continuous on the compact set $\mathcal{X}$, and uniformly dense means $\rho_{\mathcal{X}}$-dense in that case. Corollary 12.6 does not hold true if $\Delta_n \equiv \Delta > 0$, for all n. In that case we can only approximate the smaller function class $\overline{\bigcup_{n\geq 1} \mathcal{S}_n(\phi)} \subset C(\mathcal{X})$. This is going to be used in one of the cases, below.

For increasing sequences $(d_n)_{n\geq 1}$, $(\Delta_n)_{n\geq 1}$ and $(\widetilde{\Delta}_n)_{n\geq 1}$ we define the sieve estimator $(\widehat{\mu}_n)_{n\geq 1}$ by

$$\widehat{\mu}_n = \underset{\mu \in \mathcal{S}_n(\phi)}{\arg\min} \frac{1}{n} \sum_{i=1}^{n} L\left(Y_i, \mu(\boldsymbol{x}_i)\right). \tag{12.5}$$

Under the following assumptions one can prove a consistency theorem.

Assumption 12.7 *Choose a complete probability space $(\Omega, \mathcal{A}, \mathbb{P})$*[2] *and $\mathcal{X} = \{1\} \times [0,1]^{q_0}$.*

(1) *Assume $\mu_0 \in C(\mathcal{X})$. Assume $(Y_i, \boldsymbol{X}_i)_{i\geq 1}$ are i.i.d. on $(\Omega, \mathcal{A}, \mathbb{P})$ following the regression structure* (12.3) *with ε_i being centered, having $\mathbb{E}[|\varepsilon_i|^{2+\delta}] < \infty$ for some $\delta > 0$ and being independent of $\boldsymbol{X}_i$. Set $\sigma^2 = \mathrm{Var}(\varepsilon_i) < \infty$.*
(2) *The activation function ϕ is the sigmoid function.*
(3) *The sequences $(d_n)_{n\geq 1}$, $(\Delta_n)_{n\geq 1}$ and $(\widetilde{\Delta}_n)_{n\geq 1}$ are increasing and tending to infinity as $n \to \infty$ with $d_n \Delta_n^2 \log(d_n \Delta_n) = o(n)$.*

Most results that we are going to present below hold for activation functions that are Lipschitz. The sigmoid activation function is Lipschitz, see Lemma 12.2.

The following considerations are based on the pseudo-norm, given $(\boldsymbol{X}_i)_{1\leq i\leq n}$,

$$\|\mu\|_n = \sqrt{\frac{1}{n}\sum_{i=1}^{n} (\mu(\boldsymbol{X}_i))^2} \qquad \text{for } \mu \in C(\mathcal{X}).$$

[2] A probability space $(\Omega, \mathcal{A}, \mathbb{P})$ is complete if for any $\mathbb{P}$-null set $B \in \mathcal{A}$ with $\mathbb{P}[B] = 0$ and every subset $A \subset B$ it follows that $A \in \mathcal{A}$.

This is a pseudo-norm because it is positive $\|\mu\|_n \geq 0$, absolutely homogeneous $\|a\mu\|_n = |a| \, \|\mu\|_n$ and the triangle inequality holds, but it is not definite because $\|\mu\|_n = 0$ does not imply that μ is the zero function (i.e. it is not point-separating). This pseudo-norm $\|\cdot\|_n$ depends on the (random) features $(\boldsymbol{X}_i)_{1\leq i\leq n}$ and, therefore, the subsequent statements involving this pseudo-norm hold in probability. The following result provides consistency, and that the true regression function μ_0, indeed, can be learned from i.i.d. data.

Theorem 12.8 (Consistency, Theorem 3.1 of Shen et al. [336]) *Under Assumption 12.7, the sieve estimator* $(\widehat{\mu}_n)_{n\geq 1}$ *in* (12.5) *exists. We have consistency* $\|\widehat{\mu}_n - \mu_0\|_n \to 0$ *in probability as* $n \to \infty$, *i.e., for all* $\epsilon > 0$

$$\lim_{n\to\infty} \mathbb{P}\left[\|\widehat{\mu}_n - \mu_0\|_n > \epsilon\right] = 0.$$

Remarks 12.9

- Such a consistency result for FN networks has first been proved in Theorem 3.3 of White [380], however, on slightly different spaces and under slightly different assumptions. Similar consistency results have been obtained for related point process situations by Döhler–Rüschendorf [109] and for time-series in White [380] and Chen–Shen [73].
- Item (3) of Assumption 12.7 gives upper complexity bounds on shallow FN networks as a function of the sample size n of the data, so that asymptotically they do not over-fit to the data. These bounds allow for much freedom in the choice of the growth rates, and different choices may lead to different speeds of convergence. The conditions of Assumption 12.7 are, e.g., satisfied for $\Delta_n = O(\log n)$ and $d_n = O(n^{1-\delta'})$, for any small $\delta' > 0$. Under these choices, the complexity d_n of the shallow FN network grows rather quickly. Table 1 of White [380] gives some examples, for instance, if for $n = 100$ data points we have a shallow FN network with 5 neurons, then these magnitudes support 477 neurons for $n = 10'000$ and 45'600 neurons for $n = 1'000'000$ data points (for the specific choice $\delta' = 0.01$). Of course, these numbers do not provide any practical guidance on the selection of the (shallow) FN network size.
- Theorem 12.8 requires that we can explicitly calculate the sieve estimator $\widehat{\mu}_n$, i.e., the global minimizer of the objective function in (12.5). In practical applications, relying on gradient descent algorithms, typically, this is not the case. Therefore, Theorem 12.8 is mainly of theoretical value saying that learning the true regression function μ_0 is possible within FN networks.

Sketch of Proof of Theorem 12.8 The proof of this theorem is based on a theorem in White–Woolridge [381] which states that if we have a sequence $(\mathcal{S}_n(\phi))_{n\geq 1}$ of compact subsets of $\mathcal{C}(\mathcal{X})$, and if $L_n : \Omega \times \mathcal{S}_n(\phi) \to \overline{\mathbb{R}}$ is a $\mathcal{A} \otimes \mathcal{B}(\mathcal{S}_n(\phi))/\mathcal{B}(\overline{\mathbb{R}})$-measurable sequence, $n \geq 1$, with $L_n(\omega, \cdot)$ being lower-semicontinuous on $\mathcal{S}_n(\phi)$ for all $\omega \in \Omega$. Then, there exists $\widehat{\mu}_n : \Omega \to \mathcal{S}_n(\phi)$ being $\mathcal{A}/\mathcal{B}(\mathcal{S}_n(\phi))$-measurable such that for each $\omega \in \Omega$, $L_n(\omega, \widehat{\mu}_n(\omega)) = \min_{\mu\in\mathcal{S}_n(\phi)} L_n(\omega, \mu)$. For the proof of the compactness of $\mathcal{S}_n(\phi)$ in $\mathcal{C}(\mathcal{X})$ we need that d_n and Δ_n are finite for any n. This then provides the existence of the sieve estimator, for details we refer Lemma 2.1 and Corollary 2.1 in Shen et al. [336]. The proof of the consistency result then uses the growth rates on $(d_n)_{n\geq 1}$ and $(\Delta_n)_{n\geq 1}$, for the details of the proof we refer to Theorem 3.1 in Shen et al. [336]. □

The next step is to analyze the rates of convergence of the sieve estimator $\widehat{\mu}_n \to \mu_0$, as $n \to \infty$. These rates heavily depend on (additional) regularity assumptions on the true regression function $\mu_0 \in \mathcal{C}(\mathcal{X})$; we refer to Remark 3 in Sect. 5 of Chen–Shen [73]. Here, we present some results of Shen et al. [336]. From the proof of Theorem 12.8 we know that $\mathcal{S}_n(\phi)$ is a compact set in $\mathcal{C}(\mathcal{X})$. This motivates to consider the closest approximation $\pi_n\mu \in \mathcal{S}_n(\phi)$ to $\mu \in \mathcal{C}(\mathcal{X})$. The uniform denseness of $\bigcup_{n\geq 1} \mathcal{S}_n(\phi)$ in $\mathcal{C}(\mathcal{X})$ implies that $\pi_n\mu$ converges to μ. The aforementioned rates of convergence of the sieve estimators will depend on how fast $\pi_n\mu_0 \in \mathcal{S}_n(\phi)$ converges to the true regression function $\mu_0 \in \mathcal{C}(\mathcal{X})$.

If one cannot determine the global minimum of (12.5), then often an accurate approximation is sufficient. For this one introduces an approximate sieve estimator. A sequence $(\widehat{\mu}_n)_{n\geq 1}$ is called an *approximate sieve estimator* if

$$\frac{1}{n}\sum_{i=1}^{n}(Y_i - \widehat{\mu}_n(\boldsymbol{X}_i))^2 \leq \inf_{\mu\in\mathcal{S}_n(\phi)} \frac{1}{n}\sum_{i=1}^{n}(Y_i - \mu(\boldsymbol{X}_i))^2 + O_P(\eta_n), \tag{12.6}$$

where $(\eta_n)_{n\geq 1}$ is a positive sequence converging to 0 as $n \to \infty$. The last term $O_P(\eta_n)$ denotes stochastic boundedness meaning that for all $\epsilon > 0$ there exits $K_\epsilon > 0$ such that for all $n \geq 1$

$$\mathbb{P}\left[\frac{1}{n}\sum_{i=1}^{n}(Y_i - \widehat{\mu}_n(\boldsymbol{X}_i))^2 - \inf_{\mu\in\mathcal{S}_n(\phi)} \frac{1}{n}\sum_{i=1}^{n}(Y_i - \mu(\boldsymbol{X}_i))^2 > K_\epsilon\eta_n\right] < \epsilon.$$

Theorem 12.10 (Theorem 4.1 of Shen et al. [336], Without Proof) *Set Assumption 12.7. If*

$$\eta_n = O\left(\min\left\{\|\pi_n\mu_0 - \mu_0\|_n^2,\ \frac{d_n\log(d_n\Delta_n)}{n},\ \frac{d_n\log n}{n}\right\}\right),$$

the following stochastic boundedness holds for $n \geq 1$

$$\|\widehat{\mu}_n - \mu_0\|_n = O_P\left(\max\left\{\|\pi_n\mu_0 - \mu_0\|_n,\ \sqrt{\frac{d_n\log n}{n}}\right\}\right).$$

Remarks 12.11

- Assumption 12.7 implies that $d_n\log(d_n\Delta_n) = o(n)$ as $n \to \infty$. Therefore, $\eta_n \to 0$ as $n \to \infty$.

- The statement in Theorem 4.1 of Shen et al. [336] is more involved because it is stated under slightly different assumptions. Our assumptions are sufficient for having consistency of the sieve estimator, see Theorem 12.8, and making these assumptions implies that the rate of convergence in Theorem 12.10 is determined by the rate of convergence of $\|\pi_n\mu_0 - \mu_0\|_n$ and $(n^{-1}d_n \log n)^{1/2}$, see Remark 4.1 in Shen et al. [336].
- The rate of convergence in Theorem 12.10 crucially depends on the rate $\|\pi_n\mu_0 - \mu_0\|_n$, as $n \to \infty$. If μ_0 lies in the (sub-)space of functions with finite first absolute moments of the Fourier magnitude distributions, denoted by $\mathcal{F}(\mathcal{X}) \subset \mathcal{C}(\mathcal{X})$, Makavoz [262] has shown that $\|\pi_n\mu_0 - \mu_0\|_n$ decays at least as $d_n^{-(q_0+1)/(2q_0)} = d_n^{-1/2-1/(2q_0)}$, this has improved the rate of $d_n^{-1/2}$ obtained by Barron [25]. This space $\mathcal{F}(\mathcal{X})$ allows for the choices $d_n = (n/\log n)^{q_0/(2+q_0)}$, $\Delta_n \equiv \Delta > 0$ and $\widetilde{\Delta}_n \equiv \widetilde{\Delta} > 0$ to receive consistency and the following rate of convergence, see Chen–Shen [73] and Remark 4.1 in Shen et al. [336],

$$\|\widehat{\mu}_n - \mu_0\|_n = O_P(r_n^{-1}),$$

for

$$r_n = \left(\frac{n}{\log n}\right)^{(q_0+1)/(4q_0+2)} \qquad n \geq 2. \tag{12.7}$$

Note that $1/4 \leq (q_0 + 1)/(4q_0 + 2) \leq 1/2$. Thus, this is a slower rate than the square root rule of typical asymptotic normality, for instance, for $q_0 = 1$ we get $1/3$. Interestingly, Barron [26] proposes the choice $d_n \sim (n/\log n)^{1/2}$ to receive an approximation rate of $(n/\log n)^{-1/4}$.

Also note that the space $\mathcal{F}(\mathcal{X})$ allows us to choose a finite $\Delta_n \equiv \Delta > 0$ in the sieves, thus, here we do not receive denseness of the sieves in the space of continuous functions $\mathcal{C}(\mathcal{X})$, but only in the space of functions with finite first absolute moments of the Fourier magnitude distributions $\mathcal{F}(\mathcal{X})$.

The last step is to establish the asymptotic normality. For this we have to define perturbations of shallow FN networks $\mu \in \mathcal{S}_n(\phi)$. Choose $\eta_n \in (0, 1)$ and define the function

$$\widetilde{\mu}_n(\mu) = (1 - \eta_n^{1/2})\mu + \eta_n^{1/2}(\mu_0 + 1).$$

This allows us to state the following asymptotic normality result.

Theorem 12.12 (Theorem 5.1 of Shen et al. [336], Without Proof) *Set Assumption 12.7. We make the following additional assumptions: suppose $\eta_n = o(n^{-1})$ and choose ϱ_n such that we have stochastic boundedness $\varrho_n\|\widehat{\mu}_n - \mu_0\|_n = O_P(1)$. Let the following conditions hold:*

(C1) $d_n\Delta_n \log(d_n\Delta_n) = o(n^{1/4})$;
(C2) $n\varrho_n^{-2}/\Delta_n^{\delta} = o(1)$;

(C3) $\sup_{\mu \in \mathcal{S}_n(\phi):\|\mu-\mu_0\|_n \le \varrho_n^{-1}} \|\pi_n \widetilde{\mu}_n(\mu) - \widetilde{\mu}_n(\mu)\|_n = O_P(\varrho_n \eta_n)$;

(C4) $\sup_{\mu \in \mathcal{S}_n(\phi):\|\mu-\mu_0\|_n \le \varrho_n^{-1}} \frac{1}{n} \sum_{i=1}^n \varepsilon_i \left(\pi_n \widetilde{\mu}_n(\mu)(\boldsymbol{X}_i) - \widetilde{\mu}_n(\mu)(\boldsymbol{X}_i)\right) = O_P(\eta_n)$.

We have the following asymptotic normality for $n \to \infty$

$$\frac{1}{\sqrt{n}} \sum_{i=1}^{n} \left(\widehat{\mu}_n(\boldsymbol{X}_i) - \mu_0(\boldsymbol{X}_i)\right) \Rightarrow \mathcal{N}\left(0, \sigma^2\right).$$

The assumptions of Theorem 12.12 require a slower growth rate d_n on the shallow FN network compared to the consistency results. Shen et al. [336] bring forward the argument that for the asymptotic normality result to hold, the shallow FN network should grow slower in order to get the Gaussian property, otherwise the sieve estimator may skew towards the true function μ_0. Conditions (C3)–(C4) on the other side give lower growth rates on the networks such that the approximation error decreases sufficiently fast.

If the variance parameter $\sigma^2 = \mathrm{Var}(\varepsilon_i)$ is not known, we can empirically estimate it

$$\widehat{\sigma}_n^2 = \frac{1}{n} \sum_{i=1}^{n} \left(Y_i - \widehat{\mu}_n(\boldsymbol{X}_i)\right)^2.$$

Theorem 5.2 in Shen et al. [336] proves that this estimator is consistent for σ^2, and the asymptotic normality result also holds true under this estimated variance parameter (using Slutsky's theorem), and under the same assumptions as in Theorem 12.12.

12.3 Functional Limit Theorem

Horel–Giesecke [190] push the above asymptotic results even one step further. Note that the asymptotic normality of Theorem 12.12 is not directly useful for variable selection, since the asymptotic result integrates over the feature space $\mathcal{X}$. Horel–Giesecke [190] prove a functional limit theorem which we briefly review in this section.

A q_0-tuple $\alpha = (\alpha_1, \ldots, \alpha_{q_0})^\top \in \mathbb{N}_0^{q_0}$ is called a multi-index, and we set $|\alpha| = \alpha_1 + \ldots + \alpha_{q_0}$. Define the derivative operator

$$\nabla^\alpha = \frac{\partial^{|\alpha|}}{\partial x_1^{\alpha_1} \cdots \partial x_{q_0}^{\alpha_{q_0}}}.$$

Consider the compact feature space $\mathcal{X} = \{1\} \times [0, 1]^{q_0}$ with $q_0 \geq 3$. Choose a distribution ν on this feature space $\mathcal{X}$ and define the L^2-space

$$L^2(\mathcal{X}, \nu) = \left\{ \mu : \mathcal{X} \to \mathbb{R} \text{ measurable; } \mathbb{E}_\nu[\mu(X)^2] = \int_{\mathcal{X}} \mu(x)^2 d\nu(x) < \infty \right\}.$$

Next, define the Sobolev space for $k \in \mathbb{N}$

$$W^{k,2}(\mathcal{X}, \nu) = \left\{ \mu \in L^2(\mathcal{X}, \nu);\ \nabla^\alpha \mu \in L^2(\mathcal{X}, \nu) \text{ for all } \alpha \in \mathbb{N}_0^{q_0} \text{ with } |\alpha| \leq k \right\},$$

where $\nabla^\alpha \mu$ is the weak derivative of μ. The motivation for studying Sobolev spaces is that for sufficiently large k and the existence of weak derivatives $\nabla^\alpha \mu \in L^2(\mathcal{X}, \nu)$, $|\alpha| \leq k$, we eventually receive a classical derivative of μ, see below. We define the Sobolev norm for $\mu \in W^{k,2}(\mathcal{X}, \nu)$ by

$$\|\mu\|_{k,2} = \left(\sum_{|\alpha| \leq k} \mathbb{E}_\nu \left[\left(\nabla^\alpha \mu(X) \right)^2 \right] \right)^{1/2}.$$

The normed Sobolev space $(W^{k,2}(\mathcal{X}, p), \|\cdot\|_{k,2})$ is a Hilbert space. Since we would like to consider gradient-based methods, we consider the following space

$$\mathcal{C}^1_B(\mathcal{X}, \nu) = \left\{ \mu : \mathcal{X} \to \mathbb{R} \text{ continuously differentiable; } \|\mu\|_{\lfloor q_0/2 \rfloor + 2, 2} \leq B \right\}, \tag{12.8}$$

for some positive constant $B < \infty$. We will assume that the true regression function $\mu_0 \in \mathcal{C}^1_B(\mathcal{X}, \nu)$, thus, the true regression function has a bounded Sobolev norm $\|\cdot\|_{\lfloor q_0/2 \rfloor + 2, 2}$ of maximal size B. Assume that $\mathring{\mathcal{X}} \subset \mathbb{R}^{q_0}$ is the open interior of $\mathcal{X}$ (excluding the intercept component), and that ν is absolutely continuous w.r.t. the Lebesgue measure with a strictly positive and bounded density on $\mathcal{X}$ (excluding the intercept component). The Sobolev number of the space $W^{\lfloor q_0/2 \rfloor + 2, 2}(\mathring{\mathcal{X}}, \nu)$ is given by $m = \lfloor q_0/2 \rfloor + 2 - q_0/2 \geq 1.5 > 1$. The Sobolev embedding theorem then tells us that for any function $\mu \in W^{\lfloor q_0/2 \rfloor + 2, 2}(\mathring{\mathcal{X}}, \nu)$, there exists an $\lfloor m \rfloor$-times continuously differentiable function on $\mathring{\mathcal{X}}$ that is equal to μ a.e., thus, the class of equivalent functions $\mu \in W^{\lfloor q_0/2 \rfloor + 2, 2}(\mathring{\mathcal{X}}, \nu)$ has a representative in $\mathcal{C}^1(\mathring{\mathcal{X}})$, $\lfloor m \rfloor = 1$, this motivates the consideration of the space in (12.8).

In practice, the bound B needs a careful consideration because the true μ_0 is unknown. Therefore, B should be sufficiently large so that μ_0 is contained in the space $\mathcal{C}^1_B(\mathcal{X}, \nu)$ and, on the other hand, it should not be too large as this will weaken the power of the tests, below.

We choose the sigmoid activation function for ϕ and we consider the approximate sieve estimators $(\widehat{\mu}_n)_{n\geq 1}$ for given data $(Y_i, \boldsymbol{X}_i)_i$ obtained by a solution to

$$\frac{1}{n}\sum_{i=1}^{n}(Y_i - \widehat{\mu}_n(\boldsymbol{X}_i))^2 \leq \inf_{\mu\in\mathcal{S}_n(\phi)} \frac{1}{n}\sum_{i=1}^{n}(Y_i - \mu(\boldsymbol{X}_i))^2 + o_P(1), \qquad (12.9)$$

where we allow for an error term $o_P(1)$ that converges in probability to zero as $n \to \infty$. In contrast to (12.6) we do not specify the error rate, here.

Assumption 12.13 *Choose a complete probability space* $(\Omega, \mathcal{A}, \mathbb{P})$ *and* $\mathcal{X} = \{1\} \times [0, 1]^{q_0}$.

(1) *Assume* $\mu_0 \in \mathcal{C}^1_B(\mathcal{X}, \nu)$ *for some* $B > 0$, *and* $(Y_i, \boldsymbol{X}_i)_{i\geq 1}$ *are i.i.d. on* $(\Omega, \mathcal{A}, \mathbb{P})$ *following regression structure* (12.3) *with* ε_i *being centered, having* $\mathbb{E}[|\varepsilon_i|^{2+\delta}] < \infty$ *for some* $\delta > 0$, *being absolutely continuous w.r.t. the Lebesgue measure, and being independent of* $\boldsymbol{X}_i$; *the features* $\boldsymbol{X}_i \sim \nu$ *are absolutely continuous w.r.t. the Lebesgue measure having a bounded and strictly positive density on* $\mathcal{X}$ *(excluding the intercept component). Set* $\sigma^2 = \text{Var}(\varepsilon_i) < \infty$.
(2) *The activation function* ϕ *is the sigmoid function.*
(3) *The sequence* $(d_n)_{n\geq 1}$ *is increasing and going to infinity satisfying* $d_n^{2+1/q_0}\log(d_n) = O(n)$ *as* $n \to \infty$, *and* $\Delta_n \equiv \Delta > 0$, $\widetilde{\Delta}_n \equiv \widetilde{\Delta} > 0$ *for* $n \geq 1$.
(4) *Define* $L_\mu(\boldsymbol{X}, \varepsilon) = -2\varepsilon(\mu(\boldsymbol{X}) - \mu_0(\boldsymbol{X})) + (\mu(\boldsymbol{X}) - \mu_0(\boldsymbol{X}))^2$, *and it holds for* $n \geq 2$

$$\frac{1}{\sqrt{n}}\sum_{i=1}^{n}\left(L_{\widehat{\mu}_n}(\boldsymbol{X}_i, \varepsilon_i) - \mathbb{E}_\nu\left[L_{\widehat{\mu}_n}(\boldsymbol{X}_1, \varepsilon_1)\right]\right)$$

$$\leq \inf_{h\in\mathcal{C}^1_B(\mathcal{X},\nu)} \frac{1}{\sqrt{n}}\sum_{i=1}^{n}\left(L_{\mu_0+h/r_n}(\boldsymbol{X}_i, \varepsilon_i) - \mathbb{E}_\nu\left[L_{\mu_0+h/r_n}(\boldsymbol{X}_1, \varepsilon_1)\right]\right) + o_P(r_n^{-1}),$$

for r_n *being the rate defined in* (12.7).

The first three items of this assumption are rather similar to Assumption 12.7 which provides consistency in Theorem 12.8 and the rates of convergence in Theorem 12.10. Item (4) of Assumption 12.13 needs to be compared to (C3)–(C4) of Theorem 12.12 which is used for getting the asymptotic normality. $(r_n)_n$ is the rate that provides convergence in probability of the sieve estimator to the true regression function, and this magnitude is used for the perturbation, see also (C3)–(C4) in Theorem 12.12.

Theorem 12.14 (Asymptotics, Theorem 1 of Horel–Gisecke [190], Without Proof) *Under Assumption 12.13 the approximate sieve estimator* $(\widehat{\mu}_n)_{n\geq 1}$ (12.9) *converges weakly in the metric space* $(\mathcal{C}^1_B(\mathcal{X}, \nu), d_\nu)$ *with* $d_\nu(\mu, \mu') = \mathbb{E}_\nu[(\mu(\boldsymbol{X}) - \mu'(\boldsymbol{X}))^2]$:

$$r_n\left(\widehat{\mu}_n - \mu_0\right) \Rightarrow \mu^\star \qquad \text{as } n \to \infty,$$

where $\mu^\star$ *is the* arg max *of the Gaussian process* $\{G_\mu;\ \mu \in \mathcal{C}_B^1(\mathcal{X}, \nu)\}$ *with mean zero and covariance function* $\text{Cov}(G_\mu, G_{\mu'}) = 4\sigma^2 \mathbb{E}_\nu[\mu(\boldsymbol{X})\mu'(\boldsymbol{X})]$.

Remarks 12.15 We highlight the differences between Theorems 12.12 and 12.14.

- Theorem 12.12 provides a convergence in distribution to a Gaussian random variable, whereas the limit in Theorem 12.14 is a random function $\boldsymbol{x} \mapsto \mu^\star(\boldsymbol{x}) = \mu^\star_\omega(\boldsymbol{x})$, $\omega \in \Omega$, thus, the former convergence result integrates over the (empirical) feature distribution, whereas the latter also allows for a point-wise consideration in feature $\boldsymbol{x}$.
- The former theorem does not allow for variable selection in $\boldsymbol{X}$ whereas the latter does because the limiting function still discriminates different feature values.
- For the proof of Theorem 12.14 we refer to Horel–Giesecke [190]. It is based on asymptotic results on empirical point processes; we refer to Van der Vaart–Wellner [364]. The Gaussian process $\{G_\mu;\ \mu \in \mathcal{C}_B^1(\mathcal{X}, \nu)\}$ is parametrized by the (totally bounded) space $\mathcal{C}_B^1(\mathcal{X}, \nu)$, and it is continuous over this compact index space. This implies that it takes its maximum. Uniqueness of the maximum then gives us the random function $\mu^\star$ which exactly describes the limiting distribution of $r_n(\widehat{\mu}_n - \mu_0)$ as $n \to \infty$.

12.4 Hypothesis Testing

Theorem 12.14 can be used to provide a significance test for feature component selection, similarly to the LRT and the Wald test presented in Sect. 5.3.2 on GLMs. We define gradient-based test statistics, for $1 \le j \le q_0$, and w.r.t. the approximate sieve estimator $\widehat{\mu}_n \in \mathcal{S}_n(\phi)$ given in (12.9),

$$\Lambda_j^{(n)} = \int_{\mathcal{X}} \left(\frac{\partial \widehat{\mu}_n(\boldsymbol{x})}{\partial x_j}\right)^2 d\nu(\boldsymbol{x}) \qquad \text{and} \qquad \widehat{\Lambda}_j^{(n)} = \frac{1}{n}\sum_{i=1}^n \left(\frac{\partial \widehat{\mu}_n(\boldsymbol{X}_i)}{\partial x_j}\right)^2.$$

The test statistics $\Lambda_j^{(n)}$ integrates the squared partial derivative of the sieve estimator $\widehat{\mu}_n$ w.r.t. the distribution ν, whereas $\widehat{\Lambda}_j^{(n)}$ can be considered as its empirical counterpart if $\boldsymbol{X} \sim \nu$. Note that both test statistics depend on the data $(Y_i, \boldsymbol{X}_i)_{1 \le i \le n}$ determining the sieve estimator $\widehat{\mu}_n$, see (12.9). These test statistics are used to test the following null hypothesis H_0 against the alternative hypothesis H_1 for the true regression function $\mu_0 \in \mathcal{C}_B^1(\mathcal{X}, \nu)$

$$H_0 : \lambda_j = \mathbb{E}_\nu\left[\left(\frac{\partial \mu_0(\boldsymbol{X})}{\partial x_j}\right)^2\right] = 0 \qquad \text{against} \qquad H_1 : \lambda_j \neq 0. \tag{12.10}$$

We emphasize that the expression λ_j in (12.10) is a deterministic number, for this reason we use the expected value notation $\mathbb{E}_\nu[\cdot]$. This in contrast to $\Lambda_j^{(n)}$, which is only a conditional expectation, conditionally given the data $(Y_i, \boldsymbol{X}_i)_{1\le i\le n}$.

Proposition 12.16 (Theorem 2 and Proposition 3 of Horel–Giesecke [190], Without Proof) *Under Assumption 12.13 and under the null hypothesis H_0 we have for $n \to \infty$*

$$r_n^2 \Lambda_j^{(n)}, r_n^2 \widehat{\Lambda}_j^{(n)} \Rightarrow \Psi_j \stackrel{\text{def.}}{=} \int_{\mathcal{X}} \left(\frac{\partial \mu^\star(\boldsymbol{x})}{\partial x_j} \right)^2 d\nu(\boldsymbol{x}). \tag{12.11}$$

In order to use this proposition we need to be able to calculate the limiting distribution characterized by random variable Ψ_j. The maximal argument $\mu^\star$ of the Gaussian process $\{G_\mu;\ \mu \in \mathcal{C}_B^1(\mathcal{X}, \nu)\}$ is given by a random function such that for all $\omega \in \Omega$, $\mu_\omega^\star(\cdot)$ fulfills

$$G_{\mu_\omega^\star(\cdot)}(\omega) \ge G_\mu(\omega) \qquad \text{for all } \mu \in \mathcal{C}_B^1(\mathcal{X}, \nu).$$

A discretization and simulation approach can be explored to approximate this maximal argument $\mu^\star$ for different $\omega \in \Omega$, see Section 5.7 in Horel–Giesecke [190].

1. Sample random functions f_k from $\mathcal{C}_B^1(\mathcal{X}, \nu)$, $k \ge 1$. The universality theorems suggest that we sample these random functions f_k from the sieves $(\mathcal{S}_n \cap \mathcal{C}_B^1(\mathcal{X}, \nu))_{n\ge 1}$. This requires sampling dimension q_1 of the shallow FN network and the corresponding network weights. This provides us with candidate functions $f_1, \ldots, f_K \in \mathcal{C}_B^1(\mathcal{X}, \nu)$, these candidate functions can be understood as a random covering of the (totally bounded) index space $\mathcal{C}_B^1(\mathcal{X}, \nu)$.
2. Simulate K-dimensional multivariate Gaussian random variables $\boldsymbol{G}^{(t)}$ (i.i.d.) with mean zero and (empirical) covariance matrix

$$\widehat{\Sigma} = \left(\frac{1}{n} \sum_{i=1}^{n} f_k(\boldsymbol{X}_i) f_l(\boldsymbol{X}_i) \right)_{1\le k,l\le K}.$$

   These random variables $\boldsymbol{G}^{(1)}, \ldots, \boldsymbol{G}^{(T)}$ play the role of discretized random samples of the Gaussian process $\{G_\mu;\ \mu \in \mathcal{C}_B^1(\mathcal{X}, \nu)\}$.
3. The empirical arg max of the sample $\boldsymbol{G}^{(t)}$, $1 \le t \le T$, is obtained by

$$\widehat{\mu}_t^\star = \underset{f_k:\ 1\le k\le K}{\arg\max}\ G_{f_k}^{(t)},$$

   where $G_{f_k}^{(t)}$ is the k-th component of $\boldsymbol{G}^{(t)}$.

4. The empirical distribution of the following sample $\widehat{\Psi}_j^{(t)}$, $1 \le t \le T$, gives us an approximation to the limiting distribution in Proposition 12.16

$$\widehat{\Psi}_j^{(t)} = \frac{1}{n}\sum_{i=1}^{n}\left(\frac{\partial \widehat{\mu}_t^{\star}(\boldsymbol{X}_i)}{\partial x_j}\right)^2,$$

i.e., under the null hypothesis H_0 we approximate the right-hand side of (12.11) by the empirical distribution of $(\widehat{\Psi}_j^{(t)})_{1\le t\le T}$.

We close this section we some remarks.

Remarks 12.17

- The quality of the empirical approximation $(\widehat{\Psi}_j^{(t)})_{1\le t\le T}$ to the limiting distribution of Ψ_j will depend on how well we cover the index set $\mathcal{C}_B^1(\mathcal{X}, \nu)$. We could try to use covering theorems to control the accuracy. However, this is often too challenging. The simulation approach presented above suffers from not giving us any control on the quality of this covering, nor is it clear how the Sobolev norm condition for B in (12.8) can efficiently be checked during the simulation approach. We highlight that this Sobolev norm bound $\|f_k\|_{\lfloor q_0/2\rfloor+2,2} \le B$ is crucial when we want to empirically estimate the distribution of Ψ_j; under special assumptions Horel–Giesecke [190] prove in their Theorem 4 that Ψ_j scales as B^2. Thus, if we do not have any control over the Sobolev norm of the sampled shallow FN networks f_k, the above simulation algorithm is not useful to approximate the limiting distribution in Proposition 12.16.
- The assumptions of Proposition 12.16 require that $\boldsymbol{X} \sim \nu$ has a strictly positive density over the entire feature space $\mathcal{X}$ (excluding the intercept component). This is necessary to be able to capture any non-zero partial derivative $\partial\mu_0(\boldsymbol{x})/\partial x_j$ over the entire feature space $\mathcal{X}$. In practical applications, where we rely on a finite sample $(\boldsymbol{X}_i)_{1\le i\le n}$, this may be problematic and needs some care. For instance, there may be the situation where the samples cluster in two disjoint regions, say $C_1 \subset \mathcal{X}$ and $C_2 \subset \mathcal{X}$, because we may have $\nu(C_1 \cup C_2) \approx 1$. That is, in that case we rarely have observations $\boldsymbol{X}_i$ not lying in one of these two clusters. If $\partial\mu_0(\boldsymbol{x})/\partial x_j = 0$ on these two clusters $\boldsymbol{x} \in C_1 \cup C_2$, but if μ_0 has a very steep slope between the two clusters (i.e., if they are really different in terms of μ_0), then the test on this finite sample will not find the significant slope.
- The distribution $\boldsymbol{X} \sim \nu$ of the features is assumed to be absolutely continuous on the hypercube $[0, 1]^{q_0}$, this is not fulfilled for binary and categorical features.
- Another question is how the test of Proposition 12.16 is affected by collinearity in feature components. Note that we only test one component at a time. Moreover, we would like to highlight the j-dependency in the limiting random variable Ψ_j. This dependency is induced by the properties of the feature distribution ν that may not be exchangeable in the components of $\boldsymbol{x}$.

Chapter 13
Appendix B: Data and Examples

This appendix presents and describes the data sets used.

13.1 French Motor Third Party Liability Data

We consider a French motor third party liability (MTPL) claims data set. This data set is available through the R library CASdatasets[1] being hosted by Dutang–Charpentier [113]. The specific data sets chosen from CASdatasets are called FreMTPL2freq and FreMTPL2sev, the former contains the insurance policy and claim frequency information and the latter the corresponding claim severity information.[2]

Before we can work with this data set we perform data cleaning. It has been pointed out by Loser [259] that the claim counts on the insurance policies with policy IDs $\leq$ 24500 in FreMTPL2freq do not seem to be correct because these claims do not have claim severity counterparts in FreMTPL2sev. For this reason we work with the claim counts extracted from the latter file. In Listing 13.1 we give the code used for data cleaning.[3] In this code we merge FreMTPL2freq with the aggregated severities on each insurance policy and the corresponding claim counts are received from FreMTPL2sev, this is done on lines 2–11 of Listing 13.1. A

[1] CASdatasets website: http://cas.uqam.ca/.

[2] We use CASdatasets version 1.0–8 which has been packaged on 2018-05-20. This version uses for the 22 French regions the labels R11, ..., R94. In later versions of CASdatasets these labels have been replaced by the region names, in this transformation the labels R31 (Nord-Pas-de-Calais) and R41 (Lorraine) have been merged to one region called Nord-Pas-de-Calais. We believe that this is an error and therefore prefer to work with an older version of CASdatasets. This older version can be downloaded in R with library(OpenML), library(farff), freMTPL2freq <- getOMLDataSet(data.id = 41214)$data

[3] The code in Listing 13.1 is a modified version of the R code provided by Loser [259].

M. V. Wüthrich, M. Merz, *Statistical Foundations of Actuarial Learning and its Applications*, Springer Actuarial, https://doi.org/10.1007/978-3-031-12409-9_13

further inspection of the data indicates that policies with more than 5 claims may be data error because they all seem to belong to the same driver (and they have very short exposures).[4] For this reason we drop these records on line 12. On line 13 we censor exposures at one accounting year (since these policies are active within one calendar year). Finally, on lines 15–16 we re-level the `VehBrands`.[5] All subsequent analysis is based on this cleaned data set.

Listing 13.1 Data cleaning applied to the French MTPL data set

```
1  #
2  data(freMTPL2freq)
3  dat <- freMTPL2freq[, -2]
4  dat$VehGas <- factor(dat$VehGas)
5  data(freMTPL2sev)
6  sev <- freMTPL2sev
7  sev$ClaimNb <- 1
8  dat0 <- aggregate(sev, by=list(IDpol=sev$IDpol), FUN = sum)[c(1,3:4)]
9  names(dat0)[2] <- "ClaimTotal"
10 dat <- merge(x=dat, y=dat0, by="IDpol", all.x=TRUE)
11 dat[is.na(dat)] <- 0
12 dat <- dat[which(dat$ClaimNb <=5),]
13 dat$Exposure <- pmin(dat$Exposure, 1)
14 sev <- sev[which(sev$IDpol %in% dat$IDpol), c(1,2)]
15 dat$VehBrand <- factor(dat$VehBrand, levels=c("B1","B2","B3","B4","B5","B6",
16                                          "B10","B11","B12","B13","B14"))
```

Listing 13.2 Excerpt of the French MTPL data set

```
1  'data.frame':    678007 obs. of  13 variables:
2   $ IDpol      : num  1 3 5 10 11 13 15 17 18 21 ...
3   $ Exposure   : num  0.1 0.77 0.75 0.09 0.84 0.52 0.45 0.27 0.71 0.15 ...
4   $ Area       : Factor w/ 6 levels "A","B","C","D",..: 4 4 2 2 2 5 5 3 3 2 ...
5   $ VehPower   : int  5 5 6 7 7 6 6 7 7 7 ...
6   $ VehAge     : int  0 0 2 0 0 2 2 0 0 0 ...
7   $ DrivAge    : int  55 55 52 46 46 38 38 33 33 41 ...
8   $ BonusMalus: int  50 50 50 50 50 50 50 68 68 50 ...
9   $ VehBrand   : Factor w/ 11 levels "B1","B2","B3",..: 9 9 9 9 9 9 9 9 9 9 ...
10  $ VehGas     : Factor w/ 2 levels "Diesel","Regular": 2 2 1 1 1 2 2 1 1 1 ...
11  $ Density    : int  1217 1217 54 76 76 3003 3003 137 137 60 ...
12  $ Region     : Factor w/ 22 levels "R11","R21","R22",..: 18 18 3 15 15 8 8 ...
13  $ ClaimTotal: num  0 0 0 0 0 0 0 0 0 0 ...
14  $ ClaimNb    : num  0 0 0 0 0 0 0 0 0 0 ...
15 ####
16 'data.frame':    26383 obs. of  2 variables:
17  $ IDpol      : int  1552 1010996 4024277 4007252 4046424 4073956 4012173 ...
18  $ ClaimAmount: num  995 1128 1851 1204 1204 ...
```

Listing 13.2 gives an excerpt of the cleaned French MTPL data set, lines 2–14 give the insurance policy and claim counts information, and lines 17–18

[4] Short exposure policies may also belong to a commercial car rental company.

[5] The data set `FreMTPLfreq` of `CASdatasets` is a subset of `FreMTPL2freq` with slightly changed feature components, for instance, the former data set contains car brand names in a more aggregated version than the latter, see Table 13.2, below.

display the individual claim amounts. We have 9 feature components on lines 4–12 (1 component is binary, 3 components are categorical, and 5 components are continuous), an exposure variable on line 3, and claim information on lines 13–14 and 18. In total we have 26'383 claims on 678'007 insurance policies.

We start by giving a descriptive analysis of the data, this closely follows Noll et al. [287]. We have the following insurance policy information:

1. `IDpol`: policy number (unique identifier);
2. `Exposure`: total exposure in yearly units (years-at-risk) and within $(0, 1]$;
3. `Area`: area code (categorical, ordinal with 6 levels);
4. `VehPower`: power of the car (continuous);
5. `VehAge`: age of the car in years;
6. `DrivAge`: age of the (most common) driver in years;
7. `BonusMalus`: bonus-malus level between 50 and 230 (with entrance level 100);
8. `VehBrand`: car brand (categorical, nominal with 11 levels), see also Table 13.2;
9. `VehGas`: diesel or regular fuel car (binary);
10. `Density`: density of population per km^2 at the location of the living place of the driver;
11. `Region`: regions in France (prior to 2016), see also Fig. 13.1 (categorical).

We start by describing the `Exposure`. The `Exposure` measures the duration of an insurance policy in yearly units; sometimes it is also called *years-at-risk*. The shortest exposure in our data set is 0.0027 which corresponds to 1 day, and the longest exposure is 1 which corresponds to 1 year. Figure 13.2 (lhs, middle) shows a histogram and a boxplot of these exposures. In view of the histogram we conclude that roughly 1/4 of all policies have a full exposure of 1 calendar year, and all other policies are only partly exposed during the calendar year. From a practical insurance point of view this high ratio of partly exposed policies seems rather

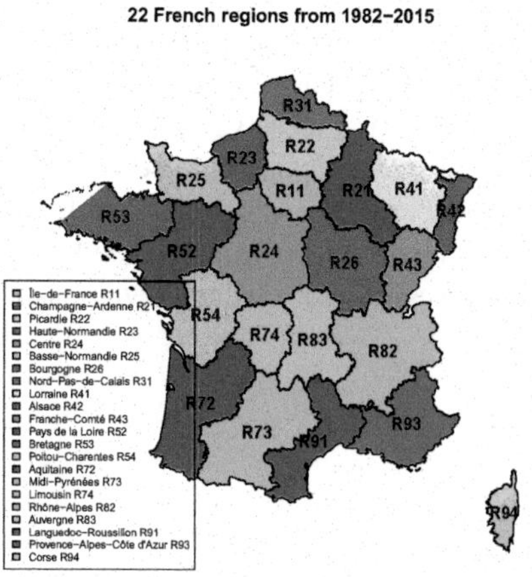

Fig. 13.1 The 22 regions in France between 1982 and 2015

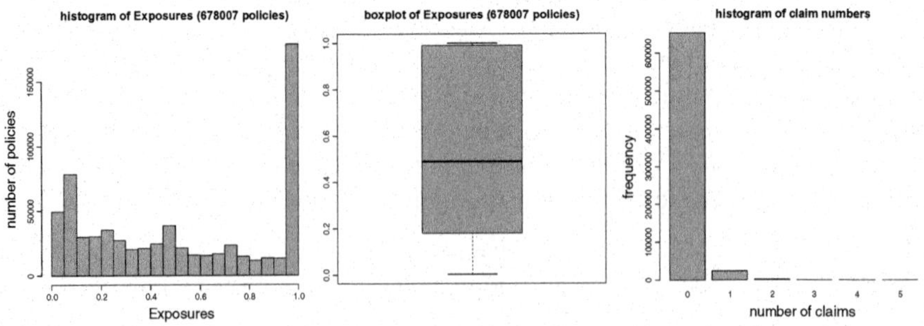

Fig. 13.2 (lhs) Histogram of Exposure, (middle) boxplot of Exposure, (rhs) number of observed claims ClaimNb of the French MTPL data

Table 13.1 Split of the portfolio w.r.t. the number of claims

Number of claims	0	1	2	3	4	5
Number of policies	653'069	23'571	1'298	62	5	2
Total exposure	341'090	16'315	909	42	2	1

unusual. A further inspection of the data indicates that policy renewals during the year account for two separate records in the data set. Of course, such split policies should be merged to one yearly policy. Unfortunately, we do not have the necessary information to perform this merger, therefore, we need to work with the data as it is. In Table 13.1 and Fig. 13.2 (rhs) we split the portfolio w.r.t. the number of claims. On 653'069 insurance policies (amounting to a total exposure of 341'090 years-at-risk) we do not have any claim, and on the remaining 24'938 policies (17'269 years-at-risk) we have at least one claim. The overall portfolio claim frequency (w.r.t. Exposure) is $\overline{\lambda} = 7.35\%$.

We study the split of this overall frequency $\overline{\lambda} = 7.35\%$ across the different feature levels. This empirical analysis is crucial for the model choice in regression modeling.[6] For the empirical analysis we provide 3 different types of graphs for each feature component (where applicable), these are given in Figs. 13.3, 13.4, 13.5, 13.6, 13.7, 13.8, 13.9, 13.10, and 13.11. The first graph (lhs) gives the split of the total exposure to the different feature levels, the second graph (middle) gives the average feature value in each French region (green meaning low and red meaning high),[7] and the third graph (rhs) gives the observed average frequency per feature level. This observed frequency is obtained by dividing the total number of claims by the total exposure per feature level. The frequencies are complemented by confidence bounds of two standard deviations (shaded area). These confidence bounds correspond to twice the estimated standard deviations. The standard deviations are estimated under

[6] The empirical analysis in these notes differs from Noll et al. [287] because data cleaning has been done differently here, we refer to Listing 13.1.

[7] We acknowledge the use of UNESCO (1987) database through UNEP/GRID-Geneva for the French map.

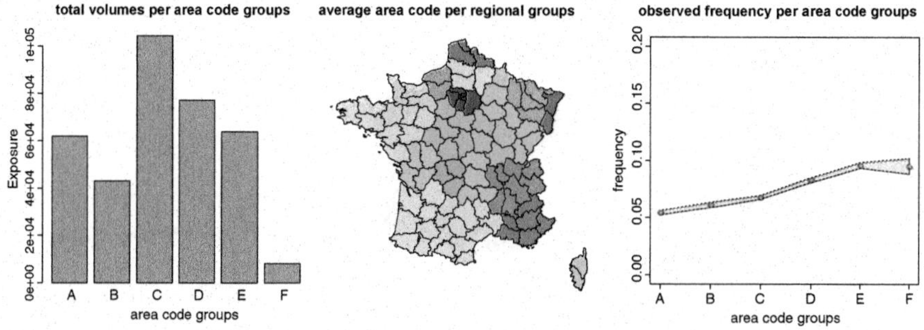

Fig. 13.3 (lhs) Histogram of exposures per `Area` code, (middle) average `Area` code per `Region`, we map $(A, \dots, F) \mapsto (1, \dots, 6)$, (rhs) observed frequency per `Area` code

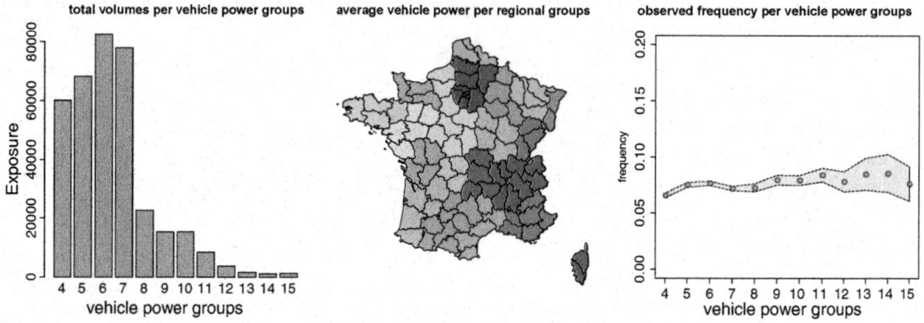

Fig. 13.4 (lhs) Histogram of exposures per `VehPower`, (middle) average `VehPower` per `Region`, (rhs) observed frequency per `VehPower`

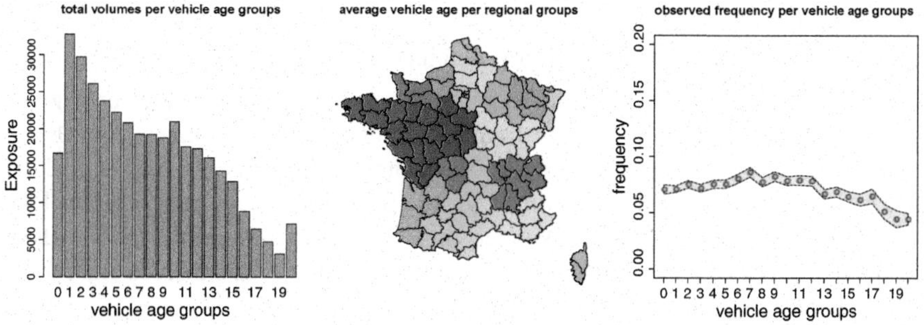

Fig. 13.5 (lhs) Histogram of exposures per `VehAge` (censored at 20), (middle) average `VehAge` per `Region`, (rhs) observed frequency per `VehAge`

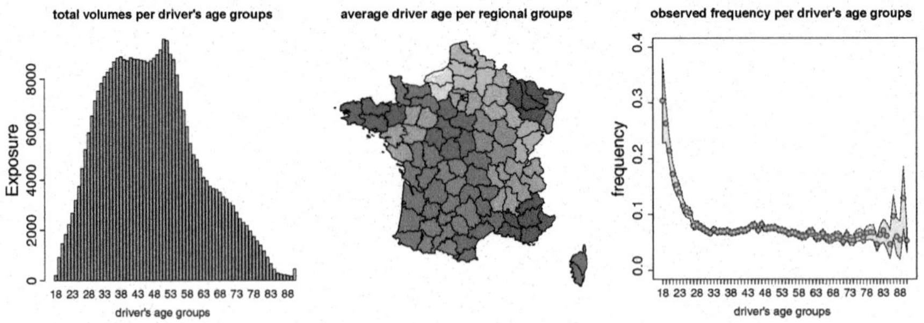

Fig. 13.6 (lhs) Histogram of exposures per DrivAge (censored at 90), (middle) average DrivAge per Region, (rhs) observed frequency per DrivAge (y-scale is different compared to the other frequency plots)

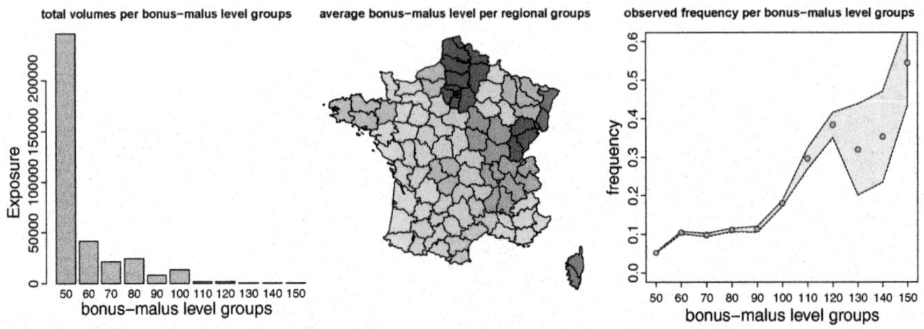

Fig. 13.7 (lhs) Histogram of exposures per BonusMalus level (censored at 150), (middle) average BonusMalus level per Region, (rhs) observed frequency per BonusMalus level (y-scale is different compared to the other frequency plots)

a Poisson assumption, thus, they are obtained by $\pm 2\sqrt{\bar{\lambda}_k/\texttt{Exposure}_k}$, where $\bar{\lambda}_k$ is the observed frequency and $\texttt{Exposure}_k$ is the total exposure for a given feature level k. We note that in all frequency plots the y-axis ranges from 0% to 20%, except in the BonusMalus plot where the maximum is set to 60%, and the DrivAge plot where the maximum is set to 40%. From these plots we conclude that some levels have only a small underlying Exposure; BonusMalus leads to the highest variability in frequencies followed by DrivAge; and there is quite some heterogeneity.

Table 13.2 gives the assignment of the different VehBrand levels to car brands. This list has been compiled from the two data sets FreMTPLfreq and FreMTPL2freq contained in the R package CASdatasets [113], see Footnote 5.

Next, we analyze collinearity between the feature components. For this we calculate Pearson's correlation and Spearman's Rho for the continuous feature components, see Table 13.3. In general, these correlations are low, except for DrivAge vs. BonusMalus. Of course, the latter is very sensible because a BonusMalus

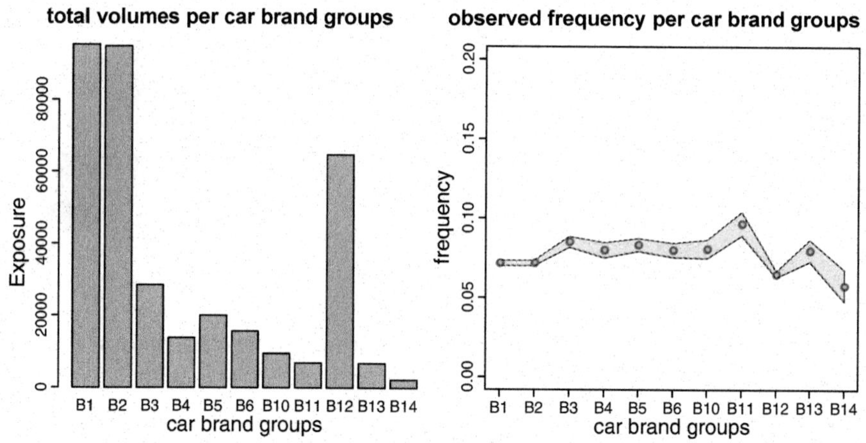

Fig. 13.8 (lhs) Histogram of exposures per `VehBrand`, (rhs) observed frequency per `VehBrand`; for `VehBrand` assignment we refer to Table 13.2

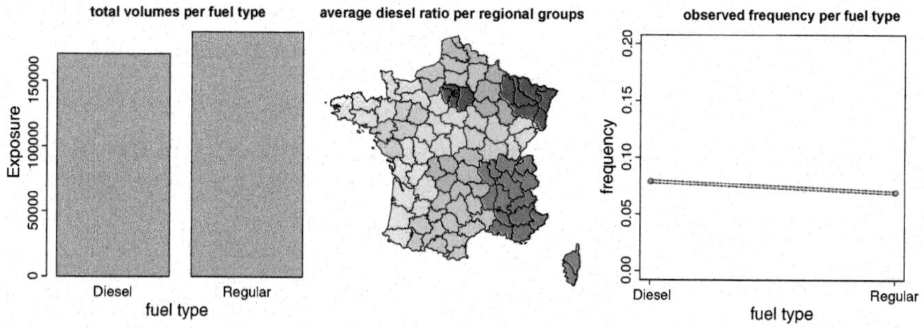

Fig. 13.9 (lhs) Histogram of exposures per `VehGas`, (middle) average `VehGas` per `Region` (diesel is green and regular red), (rhs) observed frequency per `VehGas`

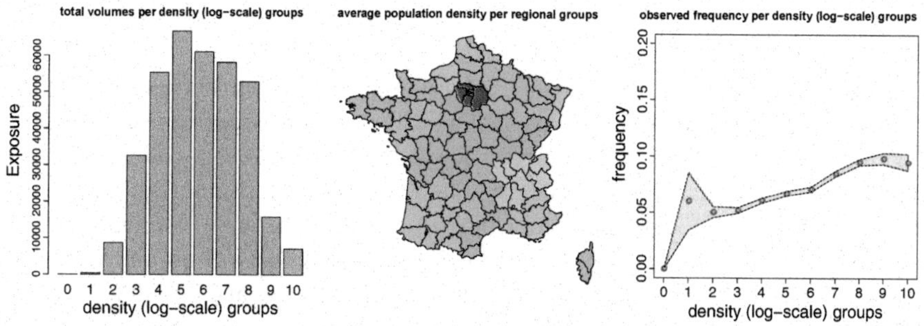

Fig. 13.10 (lhs) Histogram of exposures per population `Density` (on log-scale), (middle) average population `Density` per `Region`, (rhs) observed frequency per population `Density`; in general, we always consider `Density` on the log-scale

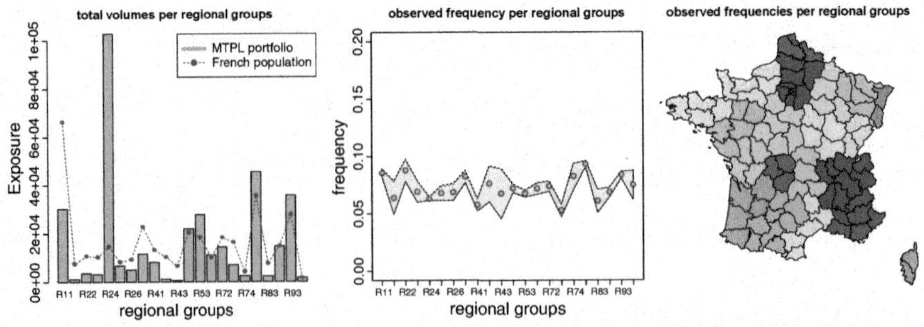

Fig. 13.11 (lhs) Histogram of exposures `Exposure`, and (middle, rhs) observed claim frequencies per `Region` in France (prior to 2016)

Table 13.2 `VehBrand` assignment

Renault, Nissan and Citroën	B1 / B2
Volkswagen, Audi, Skoda and Seat	B3
Opel, General Motors and Ford	B4 / B5
Fiat	B6
Mercedes, Chrysler and BMW	B10 / B11
Japanese (except Nissan) and Korean cars	B12
Other cars	B13 / B14

Table 13.3 Correlations in feature components: top-right shows Pearson's correlation; bottom-left shows Spearman's Rho; `Density` is considered on the log-scale; significant correlations are boldface

	`VehPower`	`VehAge`	`DrivAge`	`BonusMalus`	`Density`
`VehPower`		−0.01	0.03	−0.08	0.01
`VehAge`	0.00		−0.06	0.08	−0.10
`DrivAge`	0.04	−0.08		**−0.48**	−0.05
`BonusMalus`	−0.07	0.08	**−0.57**		0.13
`Density`	−0.01	−0.10	−0.05	0.14	

level below 100 needs a certain number of driving years without claims. We give the corresponding boxplot in Fig. 13.12 (lhs) which confirms this negative correlation. Figure 13.12 (rhs) gives the boxplot of log-`Density` vs. `Area` code. From this plot we conclude that the area code has likely been set w.r.t. the log-`Density`. For our regression models this means that we can drop the area code information, and we should only work with `Density`. Nevertheless, we will use the area code to show what happens in case of collinear feature components, i.e., if we replace $(A, \dots, F) \mapsto (1, \dots, 6)$.

Figure 13.13 illustrates each continuous feature component w.r.t. the different `VehBrand`s. Vehicle brands `B10` and `B11` (Mercedes, Chrysler and BMW) have more `VehPower` than other cars, `B10` being more likely a diesel car, and vehicle brand `B12` (Japanese and Korean cars) has comparably new cars in more densely populated French regions.

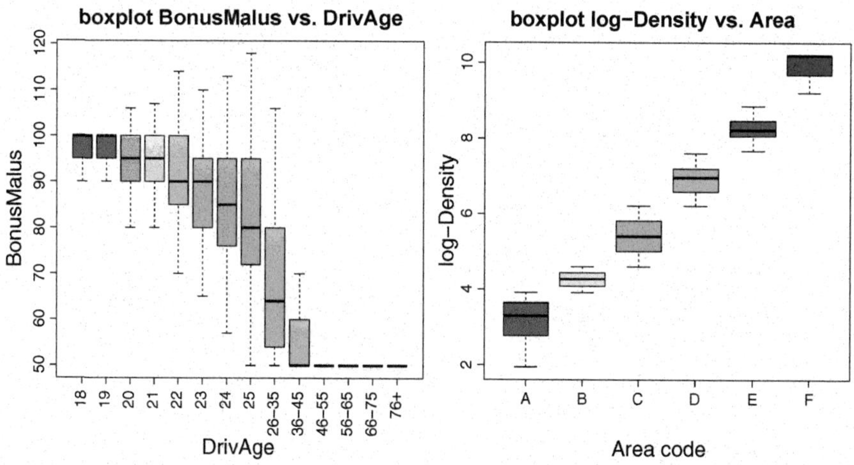

Fig. 13.12 Boxplots (lhs) `BonusMalus` vs. `DrivAge`, (rhs) log-`Density` vs. `Area` code; these plots are inspired by Fig. 2 in Lorentzen–Mayer [258]

More formally, the strength of dependence between categorical variables can be measured by Cramér's V. Cramér's V is based on the χ^2-test of independence on contingency tables. We briefly explain this. Assume we have two-dimensional categorical features $\boldsymbol{x} = (x_1, x_2) \in \mathcal{X}$ having m_1 and m_2 levels, respectively. Let $p_{\boldsymbol{x}}$ describe the probability on $\mathcal{X}$ that a randomly chosen insurance policy takes feature $\boldsymbol{x}$, and let p_{x_1} and p_{x_2} be the marginal distributions of $p_{\boldsymbol{x}}$. If the two components of $\boldsymbol{x}$ are independent with these two marginals, then we have special (independence) distribution

$$\pi_{\boldsymbol{x}} = p_{x_1} p_{x_2} \qquad \text{for all } \boldsymbol{x} = (x_1, x_2) \in \mathcal{X}.$$

The χ^2-test for independence now analyzes $p_{\boldsymbol{x}}$ vs. $\pi_{\boldsymbol{x}}$. Assume we have n observations. Denote by $n_{\boldsymbol{x}} = n_{x_1,x_2}$ the number of instances that have feature $\boldsymbol{x} = (x_1, x_2)$, and let $n_{x_1,\cdot}$ and $n_{\cdot,x_2}$ be the corresponding marginal observations. The χ^2-test statistics is given by

$$\chi^2 = \sum_{\boldsymbol{x}=(x_1,x_2)\in\mathcal{X}} \frac{\left(n_{\boldsymbol{x}} - \frac{n_{x_1,\cdot}\, n_{\cdot,x_2}}{n}\right)^2}{\frac{n_{x_1,\cdot}\, n_{\cdot,x_2}}{n}}.$$

Under the null hypothesis of having independence between the components of $\boldsymbol{x}$, the test statistics χ^2 converges in distribution to a χ^2-distribution with $(m_1 m_2 - 1)$ degrees of freedom if we let the number of independently drawn instances go to infinity. Seven different proofs of this statement are given in Benhamou–Melot [30].

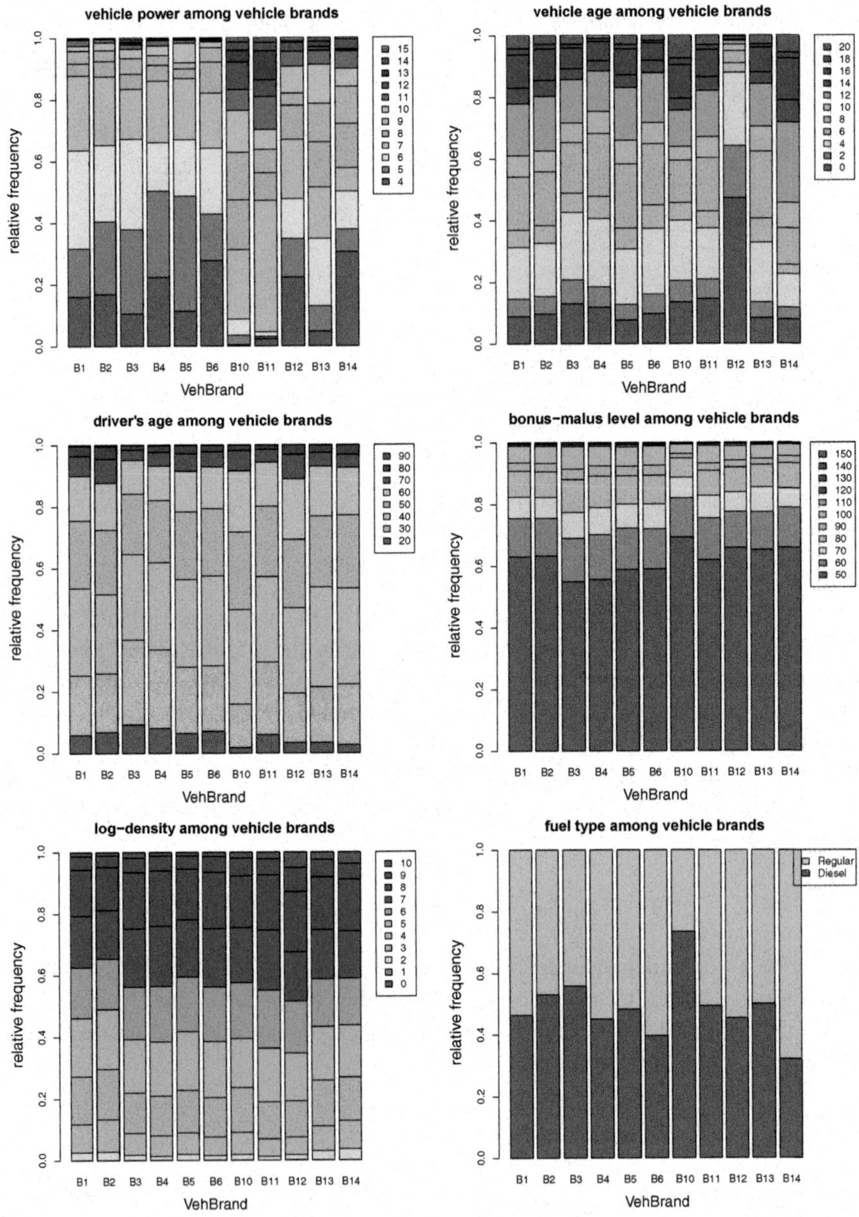

Fig. 13.13 Distribution of the variables `VehPower`, `VehAge`, `DrivAge`, `BonusMalus`, log-`Density`, `VehGas` for each car brand `VehBrand`, individually

Table 13.4 Cramér's V for the categorical feature components vs. the categorized continuous components

	VehPower	VehAge	DrivAge	BonusMalus	log-Density	VehGas	Region
VehBrand	0.16	0.17	0.06	0.03	0.05	0.12	0.13
Region	0.04	0.09	0.05	0.04	0.24	0.09	
Area					0.87		

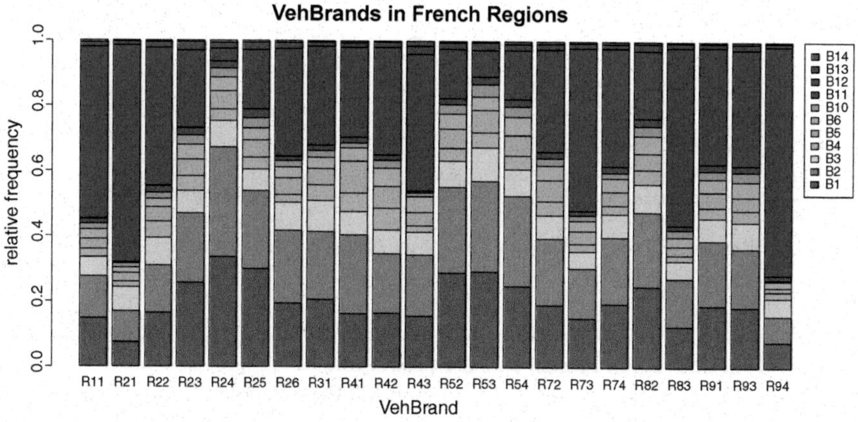

Fig. 13.14 VehBrands in the different French Regions

We scale the test statistics to the interval [0, 1] by dividing it by the comonotonic (maximal dependent) case and by the sample size n. This motivates Cramér's V

$$V = \sqrt{\frac{\chi^2/n}{\min\{m_1 - 1, m_2 - 1\}}} \in [0, 1].$$

Section 7.2.3 of Cohen [78] gives a rule of thumb for small, medium and large dependence. Cohen [78] calls the association between x_1 and x_2 small if $V\sqrt{\min\{m_1 - 1, m_2 - 1\}}$ is less 0.1, it is of medium strength for $V\sqrt{\min\{m_1 - 1, m_2 - 1\}}$ of size 0.3, and it is a large effect if this value is around 0.5. Our results are presented in Table 13.4. Clearly, there is some association between VehBrand and both VehPower and VehAge, this can also be seen from Fig. 13.13, for the remaining variables the dependence is somewhat weaker. Not surprisingly, Cramér's V shows the largest value between Region and log-Density.

In Fig. 13.14 we show the VehBrands in the different French Regions, Cramér's V is 0.13 for these two categorical variables, multiplying with $\sqrt{11 - 1}$ gives a value bigger than 0.4 which is a considerable association according to Cohen [78]. We note that in some regions the French car brands B1 and B2 are very dominant, whereas on the Isle of Corse (R94) 80% of the cars in our portfolio are Japanese

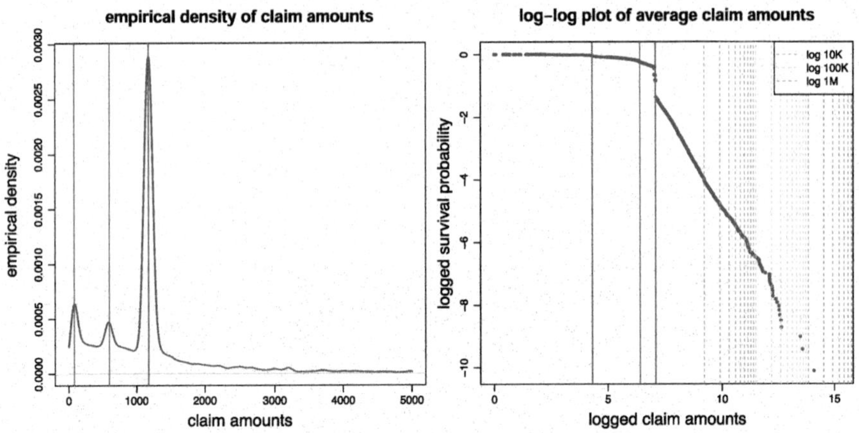

Fig. 13.15 Empirical density and log-log plots of the observed claim amounts

or Korean cars `B12`. Our portfolio has its biggest exposure in Region `R24`, see Fig. 13.11, in this region French cars are predominant.

Next, we study the claim sizes of this French MTPL example. Figure 13.15 shows the empirical density plot and the log-log plot. These two plots already illustrate the main difficulty we often face in claim size modeling. From the empirical density plot we observe that there are many payments of fixed size (red vertical lines) which do not match any absolutely continuous distribution function assumption. The log-log plot shows heavy-tailedness because we observe asymptotically a straight line with negative slope on the log-scale, this indicates regularly varying tails and, thus, the EDF is not a suitable model on the original observation scale.

Figure 13.16 gives the boxplots of the claim sizes per feature level (we omit the claims outside the whiskers because heavy-tailedness would distort the picture). The empirical mean in orange is much bigger than the median in red color, which also expresses the heavy-tailedness. From these plots we conclude that the claim sizes seem less sensitive in feature values which may question the use of a regression model for claim sizes.

Figure 13.17 shows the density plots for different feature levels. Interestingly, it seems that the features determine the sizes of the modes, for instance, if we focus on `Area`, Fig. 13.17 (top-left), we see that the area codes mainly influence the sizes of the modes. This may be interpreted by modes corresponding to different claim types which occur at different frequencies among the area codes.

13.2 Swedish Motorcycle Data

Our second example considers the Swedish motorcycle data which originally has been used in Ohlsson–Johansson [290]. It is available through the R library

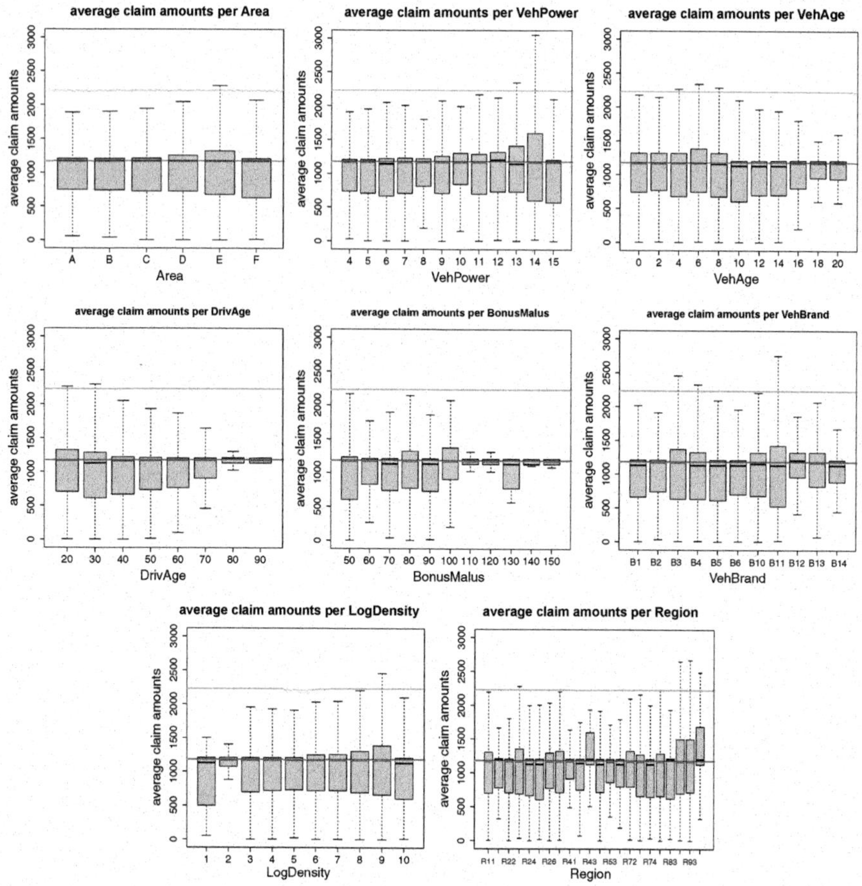

Fig. 13.16 Boxplots of claim sizes per feature level: these plots omit the claims outside the whiskers; red color shows the median and orange color the empirical mean

`CASdatasets` [113], and it is called `swmotorcycle`. Listing 13.3 shows the data cleaning that we have used, and Listing 13.4 gives an excerpt of the cleaned data.

We briefly describe the data. The data considers comprehensive insurance for motorcycles. This covers loss or damage of motorcycles other than collision, e.g., caused by theft, fire or vandalism. The data considers aggregated claims on feature levels for years 1994–1998. We have claims on 656 out of the 62'036 different features, thus, only slightly more than 1% of all feature combinations suffer a claim in the considered period.

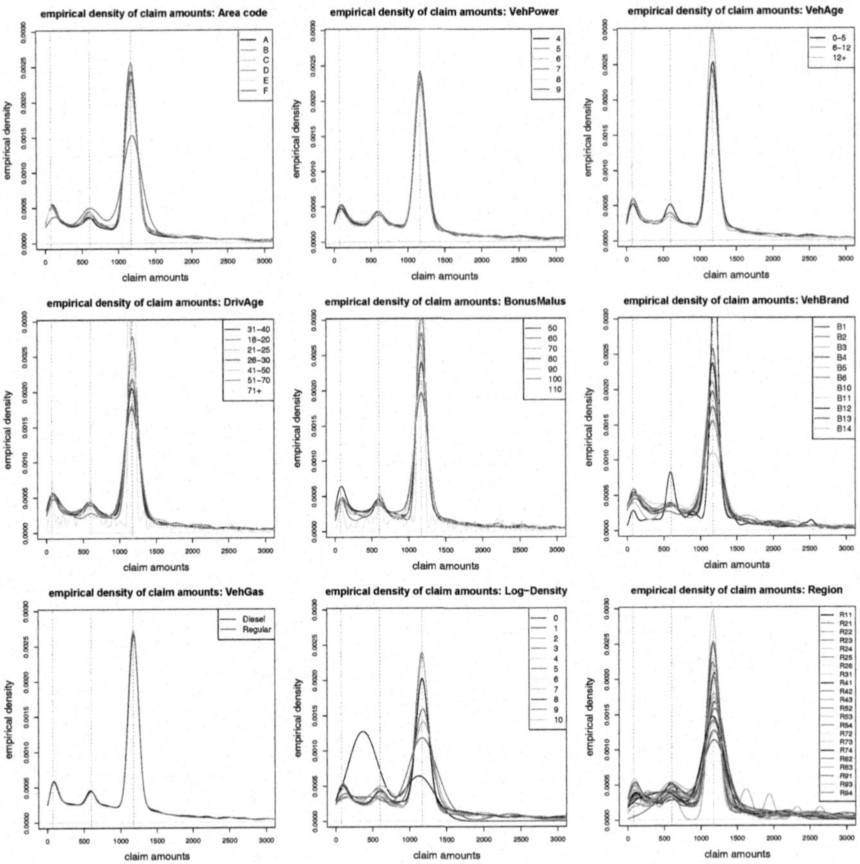

Fig. 13.17 Empirical claim size densities split w.r.t. the different levels of the feature components

We start by describing the available variables on lines 2–10 of Listing 13.4:

1. `OwnerAge`: age of motorcycle owner in $\{18, \ldots, 70\}$ years (we censor at 70 because of scarcity of data above);
2. `Gender`: gender of motorcycle owner either being `Female` or `Male`;
3. `Area`: 7 geographical Swedish zones being (1) central parts of Sweden's three largest cities, (2) suburbs and middle-sized towns, (3) lesser towns except those in zones (5)–(7), (4) small towns and countryside except those in zones (5)–(7), (5) Northern towns, (6) Northern countryside, and (7) Gotland (Sweden's largest island);
4. `RiskClass`: 7 ordered motorcycle classes received from the so-called EV ratio defined as (Engine power in kW $\times$ 100) / (Vehicle weight in kg + 75kg);
5. `VehAge`: age of motorcycle in $\{0, \ldots, 30\}$ years (we censor at 30);
6. `BonusClass`: ordered bonus-malus class from 1 to 7, entry level is 1;

Listing 13.3 Data cleaning applied to the Swedish motorcycle data set

```
library(CASdatasets)
data(swmotorcycle)
mcdata <- swmotorcycle
mcdata$Gender <- as.factor(mcdata$Gender)
mcdata$Area   <- as.factor(mcdata$Area)
mcdata$Area   <- factor(mcdata$Area,levels(mcdata$Area)[c(1,7,3,6,5,4,2)])
mcdata$Area   <- c("Zone 1","Zone 2","Zone 3","Zone 4","Zone 5",
                             "Zone 6","Zone 7")[as.integer(mcdata$Area)]
mcdata$Area        <- as.factor(mcdata$Area)
mcdata$RiskClass   <- as.factor(mcdata$RiskClass)
mcdata$RiskClass   <- factor(mcdata$RiskClass,
                          levels(mcdata$RiskClass)[c(1,6,7,3,4,5,2)])
mcdata$RiskClass   <- as.integer(mcdata$RiskClass)
mcdata$BonusClass <- as.integer(as.factor(mcdata$BonusClass))
#
mcdata   <- mcdata[which(mcdata$OwnerAge>=18),]   # only minimal age 18
mcdata$OwnerAge <- pmin(70, mcdata$OwnerAge)      # set maximal age 70
mcdata$VehAge <- pmin(30, mcdata$VehAge)          # set maximal motorcycle age 30
mcdata <- mcdata[which(mcdata$Exposure>0),]       # only positive exposures
```

Listing 13.4 Excerpt of the Swedish motorcycle data set

```
'data.frame':   62036 obs. of  9 variables:
 $ OwnerAge    : num  18 18 18 18 18 18 18 18 18 18 ...
 $ Gender      : Factor w/ 2 levels "Female","Male": 1 1 1 1 1 1 1 1 1 1 ...
 $ Area        : Factor w/ 7 levels "Zone 1","Zone 2",..: 1 1 1 1 2 2 2 3 ...
 $ RiskClass   : int  1 2 3 3 1 1 3 1 1 1 ...
 $ VehAge      : num  8 11 9 9 11 12 24 4 6 6 ...
 $ BonusClass  : int  2 2 3 4 1 1 2 1 1 2 ...
 $ Exposure    : num  1 0.778 0.499 0.501 0.929 ...
 $ ClaimNb     : int  0 0 0 0 0 0 0 0 0 0 ...
 $ ClaimAmount: int  0 0 0 0 0 0 0 0 0 0 ...
```

7. `Exposure`: total exposure in yearly units, these exposures are aggregated for given feature combinations, resulting in total exposures $[0.0274, 31.3397]$, the shortest entry referring to 10 days and the longest one to more than 31 years;
8. `ClaimNb`: number of claims N_i for a given feature;
9. `ClaimAmount`: total claim amount for a give feature (aggregated over all claims).

We start with a descriptive and exploratory analysis of the Swedish motorcycle data of Listing 13.4. We have $n = 62'036$ different feature combinations with positive `Exposure`. This `Exposure` is aggregated over individual policies with a fixed feature combination. We denote by N_i the number of claims on feature i, this corresponds to `ClaimNb`, and the total claim amount `ClaimAmount` is denoted by $S_i = \sum_{j=1}^{N_i} Z_{i,j}$, where $Z_{i,j}$ are the individual claim sizes on feature i (in case of claims). The empirical claim frequency is $\bar{\lambda} = \sum_{i=1}^n N_i / \sum_{i=1}^n v_i = 1.05\%$, and the average claim size is $\bar{\mu} = \sum_{i=1}^n S_i / \sum_{i=1}^n N_i = 24'641$ Swedish crowns SEK.

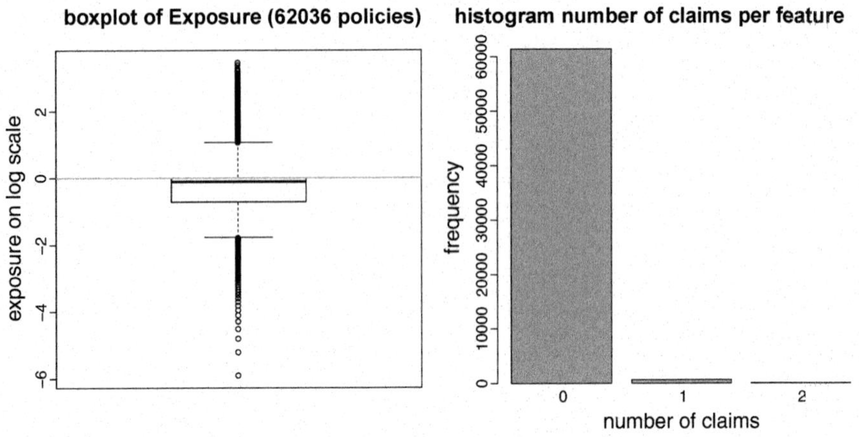

Fig. 13.18 (lhs) Boxplot of `Exposure` on the log-scale (the horizontal line corresponds to 1 accounting year), (rhs) histogram of the number of observed claims `ClaimNb` per feature of the Swedish motorcycle data

Figure 13.18 shows the boxplot over all `Exposures` and the claim counts on all insurance policies. We note that insurance claims are rare events for this product, because the empirical claim frequency is only $\bar{\lambda} = 1.05\%$.
Figures 13.19 and 13.20 give the marginal total exposures (split by gender), the marginal claim frequencies and the marginal average claim amounts for the covariate components `OwnerAge`, `Area`, `RiskClass`, `VehAge` and `BonusClass`. We observe that we have a very imbalanced portfolio between genders, only 11% of the total exposure is coming from females. The empirical claim frequency of females is 0.86% and the one of males is 1.08%. We note that the female claim frequency comes from (only) 61 claims (based on an exposure for female of 7'094 accounting years, versus 57'679 for male). Therefore, it is difficult to analyze females separately, and all marginal claim frequencies and claim sizes in Figs. 13.19 and 13.20 (middle and rhs) are analyzed jointly for both genders. If we run a simple Poisson GLM that only involves `Gender` as feature component, it turns out that the female frequency is 20% lower than the male frequency (remember we have the balance property on each dummy variable, see Example 5.12), but this variable should not be kept in the model on a 5% significance level. The same holds for claim amounts.

The empirical marginal frequencies in Figs. 13.19 and 13.20 (middle) are complemented with confidence bands of ± 2 standard deviations. From the plots we conclude that we should keep the explanatory variables `OwnerAge`, `Area`, `RiskClass` and `VehAge`, but the variable `BonusClass` does not seem to have any predictive power. At the first sight, this seems surprising because the bonus class encodes the past claims history. The reason that the bonus class is not needed for our claims is that we consider comprehensive insurance for motorcycles covering loss or damage of motorcycles other than collision (for instance, caused by theft, fire or vandalism), and the bonus class encodes collision claims.

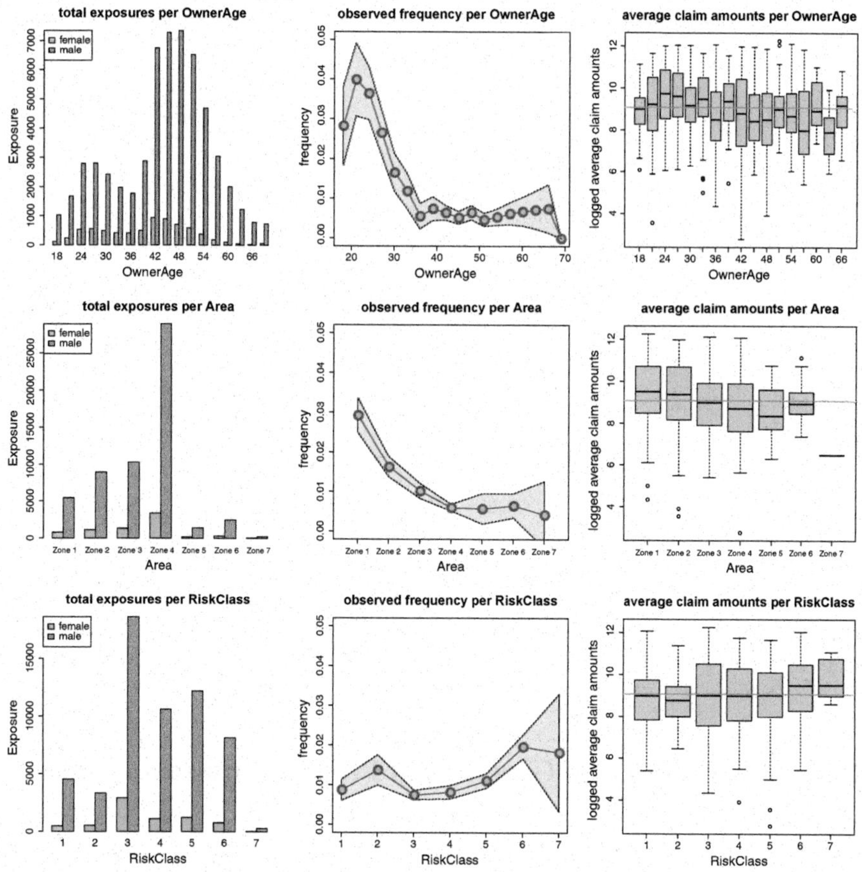

Fig. 13.19 (Top, middle and bottom rows) `OwnerAge`, `Area`, `RiskClass`: (lhs) histogram of exposures (split by gender), (middle) observed claim frequency, (rhs) boxplot of observed average claim amounts $\bar{\mu}_i = S_i/N_i$ of features with $N_i > 0$ (on log-scale)

For a regression analysis Zones 5 to 7 should be merged because of small exposures and a similar behavior, the same applies to `RiskClass` 6 and 7, and `VehAge` above 20.

Figure 13.21 shows the correlations between the features: (top) correlations between continuous features, (bottom), dependence between continuous features and the categorical `Area` features. We have some dependence, for instance, in `Zone 1` (three largest Swedish cities) the motorcycles are more light (`RiskClass`) and less old. Older people drive less heavy motorcycles that are more old, and older motorcycles are less heavy.

Figure 13.22 gives the empirical density, empirical distribution and log-log plot of average claim amounts $\bar{\mu}_i = S_i/N_i$. From the log-log plot we conclude that the average claim amounts are not heavy-tailed for this motorcycle insurance product.

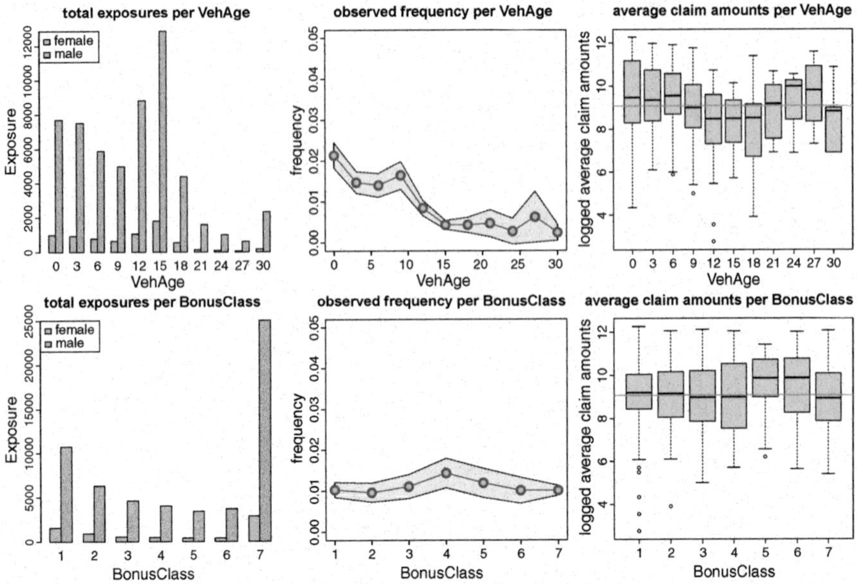

Fig. 13.20 (Top and bottom rows) `VehAge`, `BonusClass`: (lhs) histogram of exposures (split by gender), (middle) observed claim frequency, (rhs) boxplot of observed average claim amounts $\bar{\mu}_i = S_i/N_i$ of features with $N_i > 0$ (on log-scale)

13.3 Wisconsin Local Government Property Insurance Fund

The third example considers property insurance claims of the Wisconsin Local Government Property Insurance Fund (LGPIF). This data[8] has been made available through the book project of Frees [135],[9] and is also used in Lee et al. [236]. The Wisconsin LGPIF is an insurance pool that is managed by the Wisconsin Office of the Insurance Commissioner. This fund provides insurance protection to local governmental institutions such as counties, schools, libraries, airports, etc. It insures property claims for buildings and motor vehicles, and it excludes certain natural and man made perils like flood, earthquakes or nuclear accidents. We give a description of the data (we have applied some data cleaning to the original data).

The special feature of this data is that we have a short claim description on line 11 of Listing 13.5. This description will allow us to better understand the claim type beyond just knowing the hazard type that has been affected.

Figure 13.23 gives the empirical density (upper-truncated at 50'000) and the log-log plot of the observed LGPIF claim amounts. Most claims are below 10'000, however, the log-log plot shows clearly that the data is heavy-tailed, the largest claim being

[8] https://github.com/OpenActTexts/Loss-Data-Analytics/tree/master/Data.

[9] https://ewfrees.github.io/Loss-Data-Analytics/.

	OwnerAge	RiskClass	VehAge
OwnerAge		-11%	7%
RiskClass	-10%		-19%
VehAge	6%	-12%	

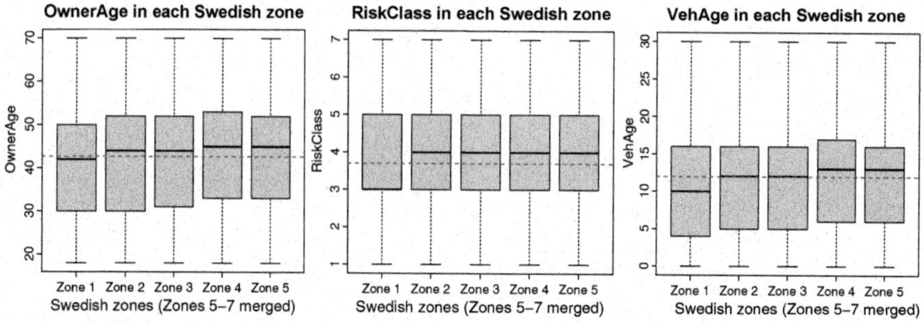

Fig. 13.21 (Top) Correlations: top-right shows Pearson's correlation; bottom-left shows Spearman's Rho; (bottom) boxplots of `OwnerAge`, `RiskClass`, `VehAge` versus `Area` (where Zones 5–7 have been merged)

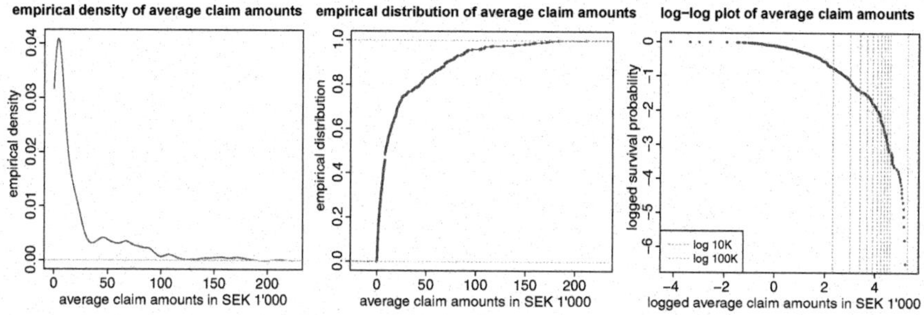

Fig. 13.22 (lhs) Empirical density (middle) empirical distribution and (rhs) log-log plot of average claim amounts $\bar{\mu}_i = S_i/N_i$ of features with $N_i > 0$

12'922'218 and 13 claims being above 1 million. These claims are further described by the features given in Listing 13.5.

In our example we will not focus on modeling the claim sizes, but we rather aim at predicting the hazard types from the claim descriptions. There are 9 different hazard types: Fire, Lightning, Hail, Wind, WaterW, WaterNW, Vehicle, Vandalism and Misc. The last label contains all claims that cannot be allocated to one of the previous hazard types, and WaterW refers to weather related water claims and WaterNW to the non-weather related ones. If we only focus on this latter problem we have more data available as there is a training data set and a validation data

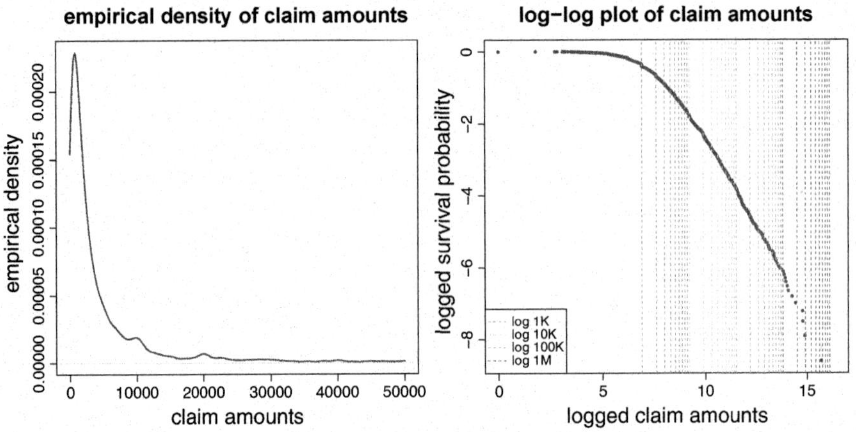

Fig. 13.23 (lhs) Empirical density (upper-truncated at 50'000), (rhs) log-log plot of the observed LGPIF claim amounts

Listing 13.5 Excerpt of the Wisconsin LGPIF data set

```
'data.frame':    5424 obs. of  10 variables:
 $ PolicyNum   : int 120002 120003 120003 120003 120003 120003 120003 ...
 $ Year        : int 2010 2007 2008 2007 2009 2010 2007 2007 2009 2007 ...
 $ Claim       : num 6839 2085 8775 600 34610 ...
 $ Deduct      : int 1000 5000 5000 5000 5000 5000 5000 5000 5000 5000 ...
 $ EntityType  : Factor w/ 6 levels "City","County",..: 2 2 2 2 2 2 2 2 2 2 ...
 $ CoverageCode: Factor w/ 13 levels "CE","CF","CS",..: 12 12 11 11 11 12 ...
 $ Fire5       : int 4 0 0 0 0 0 0 0 0 0 ...
 $ CountyCode  : Factor w/ 72 levels "ADA","ASH","BAR",..: 2 3 3 3 3 3 3 3...
 $ Hazard      : Factor w/ 9 levels "Fire","Hail",..: 3 3 5 5 9 6 3 3 3 3 ...
 $ Description : chr  "lightning damage" "lightning damage at Comm. Center" ...
```

set with hazard types and claim descriptions.[10] In total we have 6'031 such claim descriptions, see Listing 13.6, which are studied in our text recognition Chap. 10.

Listing 13.6 Excerpt of the Wisconsin LGPIF claim descriptions

```
'data.frame':    6031 obs. of  2 variables:
 Hazard     : Factor w/ 9 levels "Fire","Hail",..: 1 3 3 5 5 9 3 6 ...
 Description: chr "fire damage at Town Hall"
                  "lightning damage at water tower" ...
```

[10] https://github.com/OpenActTexts/Loss-Data-Analytics/tree/master/Data.

13.4 Swiss Accident Insurance Data

Our next example considers Swiss accident insurance data.[11] This data set is not publicly available. Swiss accident insurance is compulsory for employees, i.e., by law each employer has to sign an insurance contract to protect the employees against accidents. This insurance cover includes both work and leisure accidents, and it covers medical expenses and daily allowance. Listing 13.7 gives an excerpt of the data. Line BU indicates whether we have a workplace or a leisure accident, line 10 gives the medical expenses and line 12 shows the allowance expenses. In the subsequent analysis we only consider medical expenses.

Listing 13.7 Excerpt of the Swiss accident insurance data set

```
1  'data.frame':    339500 obs. of   11 variables:
2   $ Id         : int  1 2 3 4 5 6 7 8 9 10 ...
3   $ BU         : Factor w/ 2 levels "1","2": 1 1 2 2 2 1 2 2 2 1 ...
4   $ Sector     : Factor w/ 24 levels "5","12","13",..: 5 10 13 7 12 13 4 21 1 ...
5   $ AccQuart : int  3 2 1 3 4 4 1 2 1 3 ...
6   $ RepDel     : num  0 0 0 0 1 0 0 0 0 0 ...
7   $ Age        : num  45 20 20 20 60 55 30 25 20 20 ...
8   $ InjType    : Factor w/ 19 levels "1","2","3","4",..: 7 6 4 13 16 2 6 4 4 ...
9   $ InjPart    : Factor w/ 35 levels "1","2","3","4",..: 20 28 28 20 14 23 2 ...
10  $ Claim      : num  562 6675 700 57 2382 ...
11  $ NumbPaym : num  2 2 2 1 1 3 1 1 1 1 ...
12  $ Allowance: num  2345 5554 21 0 395 ...
```

Sector indicates the labor sector of the insured company, AccQuart gives the accident quarter since leisure claims have a seasonal component, RepDel gives the reporting delay in yearly units, Age is the age of the injured (in 5 years buckets), and InjType and InjPart denote the injury type and the injured body part.

Figure 13.24 gives the empirical density (upper-truncated at 10'000) and the log-log plot of the observed Swiss accident insurance claim amounts. Most claims are below 5'000, however, the log-log plot shows some heavy-tailedness, the largest claim exceeding 1'300'000 CHF.

Figure 13.25 shows the average claim amounts split w.r.t. the different feature components (top) Sector, AccQuart, RepDel, (bottom) Age, InjType, InjPart, and moreover, split by work and leisure accidents (in cyan and gray in the colored version). Typically, leisure accidents are more numerous and more expensive on average than accidents at the work place. From Fig. 13.25 (top, left) we observe considerable variability in average claim sizes between the different labor sectors (cyan bars), whereas average leisure claim sizes (gray bars) are similar

[11] https://www.unfallstatistik.ch/.

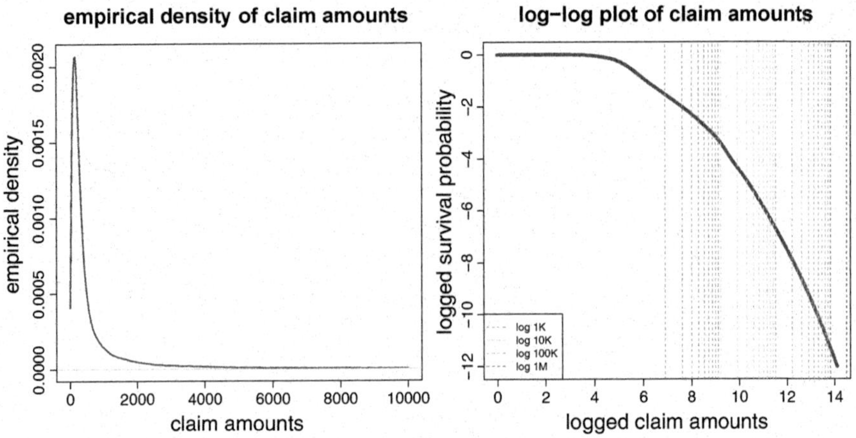

Fig. 13.24 (lhs) Empirical density (upper-truncated at 10'000), (rhs) log-log plot of the observed Swiss accident insurance claim amounts

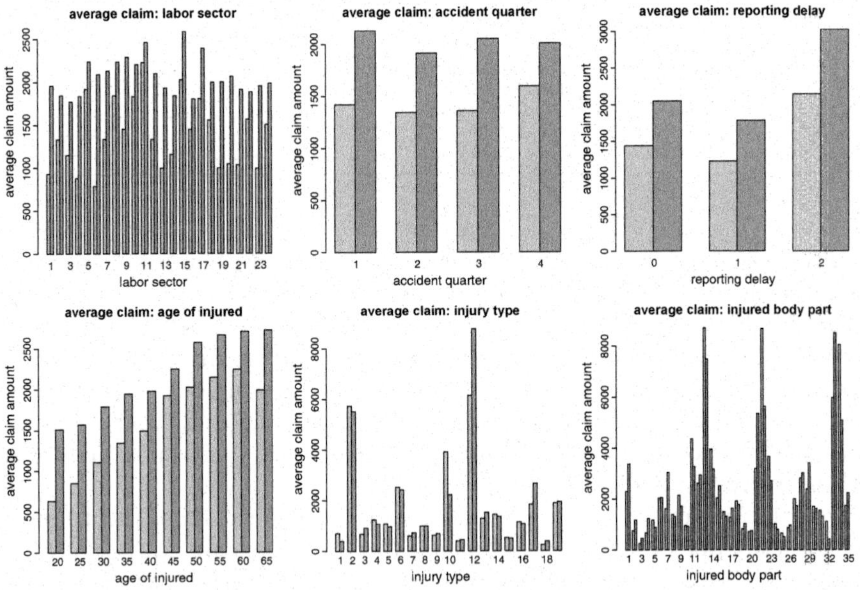

Fig. 13.25 Average claim amounts split w.r.t. the different feature components (top) `Sector`, `AccQuart`, `RepDel`, (bottom) `Age`, `InjType`, `InjPart`, and split by work and leisure accidents (cyan/gray in the colored version)

across the different labor sectors. Average claim sizes considerably differ between injury types and injured body parts (bottom, middle and right), but they do not differ between work and leisure claims.

Bibliography

1. Aas, K., Jullum, M., & Løland, A. (2021). Explaining individual predictions when features are dependent: More accurate approximations to Shapley values. *Artificial Intelligence, 298*. Article 103502.
2. Abadi, M., et al. (2015). *TensorFlow: Large-scale machine learning on heterogeneous systems*. https://www.tensorflow.org/
3. Agarwal, R., Melnick, L., Frosst, N., Zhang, X., Lengerich, B., Caruana, R., & Hinton, G. E. (2021). *Neural additive models: Interpretable machine learning with neural nets*. arXiv:2004.13912v2.
4. Ágoston, K. C., & Gyetvai, M. (2020). Joint optimization of transition rules and the premium scale in a bonus-malus system. *ASTIN Bulletin, 50/3*, 743–776.
5. Akaike, H. (1974). A new look at the statistical model identification. *IEEE Transactions on Automatic Control, 19/6*, 716–723.
6. Al-Mudafer, M. T., Avanzi, B., Taylor, G., & Wong, B. (2022). Stochastic loss reserving with mixture density neural networks. *Insurance: Mathematics & Economics, 105*, 144–147.
7. Albrecher, H., Beirlant, J., & Teugels, J. L. (2017). *Reinsurance: Actuarial and statistical aspects*. Hoboken: Wiley.
8. Albrecher, H., Bladt, M., & Yslas, J. (2022). Fitting inhomogeneous phase-type distributions to data: The univariate and the multivariate case. *Scandinavian Journal of Statistics, 49/1*, 44–77.
9. Alzner, H. (1997). On some inequalities for the gamma and psi functions. *Mathematics of Computation, 66/217*, 373–389.
10. Amari, S. (2016). *Information geometry and its applications*. New York: Springer.
11. Améndola, C., Drton, M., & Sturmfels, B. (2016). Maximum likelihood estimates for Gaussian mixtures are transcendental. In I. S. Kotsireas, S. M. Rump, & C. K. Yap (Eds.), *6th International Conference on Mathematical Aspects of Computer and Information Sciences. Lecture notes in computer science* (Vol. 9582, pp. 579–590). New York: Springer.
12. Ancona, M., Ceolini, E., Öztireli, C., & Gross, M. (2019). Gradient-based attribution methods. In W. Samek, G. Montavon, A. Vedaldi, L. K. Hansen, & K.-R. Müller (Eds.), *Explainable AI: Interpreting, explaining and visualizing deep learning. Lecture notes in artificial intelligence* (Vol. 11700, pp. 169–191). New York: Springer.
13. Apley, D. W., & Zhu, J. (2020). Visualizing the effects of predictor variables in black box supervised learning models. *Journal of the Royal Statistical Society, Series B, 82/4*, 1059–1086.
14. Asmussen, S., Nerman, O., & Olsson, M. (1996). Fitting phase-type distributions via the EM algorithm. *Scandinavian Journal of Statistics, 23/4*, 419–441.

M. V. Wüthrich, M. Merz, *Statistical Foundations of Actuarial Learning and its Applications*, Springer Actuarial, https://doi.org/10.1007/978-3-031-12409-9

15. Awad, Y., Bar-Lev, S. K., & Makov, U. (2022). A new class of counting distributions embedded in the Lee–Carter model of mortality projections: A Bayesian approach. *Risks, 10/6*. Article 111.
16. Ay, N., Jost, J., Lê, H. V., & Schwachhöfer, L. (2017). *Information geometry*. New York: Springer.
17. Ayuso, M., Guillén, M., & Nielsen, J. P. (2019). Improving automobile insurance ratemaking using telematics: Incorporating mileage and driver behaviour data. *Transportation, 46/3*, 735–752.
18. Ayuso, M., Guillén, M., & Pérez-Marín, A. M. (2016). Telematics and gender discrimination: Some usage-based evidence on whether men's risk of accidents differs from women's. *Risks, 4/2*. Article 10.
19. Ayuso, M., Guillén, M., & Pérez-Marín, A. M. (2016). Using GPS data to analyse the distance travelled to the first accident at fault in pay-as-you-drive insurance. *Transportation Research Part C: Emerging Technologies, 68*, 160–167.
20. Bachelier, L. (1900). The theory of speculation. English translation by May, D. R. (2011). *Annales Scientifiques de l'École Normale Supérieure, 3/17*, 21–89.
21. Bahdanau, D., Cho, K., & Bengio, Y. (2016). *Neural machine translation by jointly learning to align and translate*. arXiv:1409.0473v7.
22. Bailey, R. A. (1963). Insurance rates with minimum bias. *Proceedings of the Casualty Actuarial Society, 50*, 4–11.
23. Barndorff-Nielsen, O. (2014). *Information and exponential families: In statistical theory*. New York: Wiley.
24. Barndorff-Nielsen, O., & Cox, D. R. (1979). Edgeworth and saddlepoint approximations with statistical applications. *Journal of the Royal Statistical Society, Series B, 41/3*, 279–299.
25. Barron, A. R. (1993). Universal approximation bounds for superpositions of a sigmoidal function. *IEEE Transactions of Information Theory, 39/3*, 930–945.
26. Barron, A. R. (1994). Approximation and estimation bounds for artificial neural networks. *Machine Learning, 14*, 115–133.
27. Bengio Y., Courville A., & Vincent P. (2013). Representation learning: A review and new perspectives. *IEEE Transactions on Pattern Analysis and Machine Learning Intelligence, 35/8*, 1798–1828.
28. Bengio, Y., Ducharme, R., Vincent, P., & Jauvin, C. (2003). A neural probabilistic language model. *Journal of Machine Learning Research, 3/Feb*, 1137–1155.
29. Bengio, Y., Schwenk, H., Senécal, J.-S., Morin, F., & Gauvain, J.-L. (2006). Neural probabilistic language models. In D. E. Holmes & L. C. Jain (Eds.), *Innovations in machine learning. Studies in fuzziness and soft computing* (Vol. 194, pp. 137–186). New York: Springer.
30. Benhamou, E., & Melot, V. (2018). *Seven proofs of the Pearson Chi-squared independence test and its graphical interpretation*. arXiv:1808.09171.
31. Berger, J. O. (1985). *Statistical decision theory and Bayesian analysis* (2nd ed.). New York: Springer.
32. Bichsel, F. (1964). Erfahrungstarifierung in der Motorfahrzeug-Haftpflicht-Versicherung. *Bulletin of the Swiss Association of Actuaries, 1964*, 119–130.
33. Bickel, P. J., & Doksum, K. A. (2001). *Mathematical statistics: Basic ideas and selected topics* (Vol. I, 2nd ed.). Hoboken: Prentice Hall.
34. Billingsley, P. (1995). *Probability and measure* (3rd ed.). New York: Wiley.
35. Bishop, C. M. (1994). *Mixture Density Networks*. Technical Report. Aston University, Birmingham.
36. Bishop, C. M. (2006). *Pattern recognition and machine learning*. New York: Springer.
37. Bladt, M. (2022). Phase-type distributions for insurance pricing. *ASTIN Bulletin, 52/2*, 417–448.
38. Blæsild, P., & Jensen, J. L. (1985). Saddlepoint formulas for reproductive exponential models. *Scandinavian Journal of Statistics, 12/3*, 193–202.

39. Blier-Wong, C., Cossette, H., Lamontagne, L., & Marceau, E. (2022). Geographic ratemaking with spatial embeddings. *ASTIN Bulletin, 52/1*, 1–31.
40. Blostein, M., & Miljkovic, T. (2019). On modeling left-truncated loss data using mixture distributions. *Insurance: Mathematics & Economics, 85*, 35–46.
41. Blundell, C., Cornebise, J., Kavukcuoglu, K., & Wiersta, D. (2015). Weight uncertainty in neural network. *Proceedings of Machine Learning Research, 37*, 1613–1622.
42. Boucher, J. P., Côté, S., & Guillén, M. (2017). Exposure as duration and distance in telematics motor insurance using generalized additive models. *Risks, 5/4*. Article 54.
43. Boucher, J. P., Denuit, M., & Guillén, M. (2007). Risk classification for claim counts: A comparative analysis of various zeroinflated mixed Poisson and hurdle models. *North American Actuarial Journal, 11/4*, 110–131.
44. Boucher, J. P., Denuit, M., & Guillén, M. (2008). Modelling of insurance claim count with hurdle distribution for panel data. In B. C. Arnold, N. Balakrishnan, J. M. Sarabia, & R. Mínguez (Eds.), *Advances in mathematical and statistical modeling. Statistics for industry and technology* (pp. 45–59). Boston: Birkhäuser.
45. Boucher, J. P., Denuit, M., & Guillén, M. (2009). Number of accidents or number of claims? An approach with zero-inflated Poisson models for panel data. *Journal of Risk and Insurance, 76/4*, 821–846.
46. Boucher, J. P., & Inoussa, R. (2014). A posteriori ratemaking with panel data. *ASTIN Bulletin, 44/3*, 587–612.
47. Boucher, J. P., & Pigeon, M. (2018). *A claim score for dynamic claim counts modeling*. arXiv:1812.06157.
48. Box, G. E. P., & Cox, D. R. (1964). An analysis of transformations. *Journal of the Royal Statistical Society, Series B, 26/2*, 211–243.
49. Box, G. E. P., & Jenkins, G. M. (1976). *Time series analysis: Forecasting and control.* San Francisco: Holden-Day.
50. Bregman, L. M. (1967). The relaxation method of finding the common point of convex sets and its application to the solution of problems in convex programming. *USSR Computational Mathematics and Mathematical Physics, 7/3*, 200–217.
51. Breiman, L. (1996). Bagging predictors. *Machine Learning, 24/2*, 123–140.
52. Breiman, L. (2001). Random forests. *Machine Learning, 45/1*, 5–32.
53. Breiman, L. (2001). Statistical modeling: The two cultures. *Statistical Science, 16/3*, 199–215.
54. Breiman, L., Friedman, J. H., Olshen, R. A., & Stone, C. J. (1984). *Classification and regression trees. Wadsworth statistics/probability series*. Monterey: Brooks/Cole Publishing.
55. Broadie, M., Du, Y., & Moallemi, C. (2011). Efficient risk estimation via nested sequential estimation. *Management Science, 57/6*, 1171–1194.
56. Brouhns, N., Denuit, M., & Vermunt, J. K. (2002). A Poisson log-bilinear regression approach to the construction of projected lifetables. *Insurance: Mathematics & Economics, 31/3*, 373–393.
57. Brouhns, N., Guillén, M., Denuit, M., & Pinquet, J. (2003). Bonus-malus scales in segmented tariffs with stochastic migration between segments. *Journal of Risk and Insurance, 70/4*, 577–599.
58. Bühlmann, H., & Gisler, A. (2005). *A course in credibility theory and its applications*. New York: Springer.
59. Bühlmann, P., & Mächler, M. (2014). *Computational statistics. Lecture notes*. ETH Zurich: Department of Mathematics.
60. Bühlmann, P., & Yu, B. (2002). Analyzing bagging. *Annals of Statistics, 30/4*, 927–961.
61. Burguete, J., Gallant, R., & Souza, G. (1982). On unification of the asymptotic theory of nonlinear econometric models. *Economic Review, 1/2*, 151–190.
62. Calderín-Ojeda, E., Gómez-Déniz, E., & Barranco-Chamorro, I. (2019). Modeling zero-inflated count data with a special case of the generalised Poisson distribution. *ASTIN Bulletin, 49/3*, 689–708.

63. Cameron, A., & Trivedi, P. (1986). Econometric models based on count data: Comparisons and applications of some estimators and tests. *Journal of Applied Econometrics, 1*, 29–54.
64. Cantelli, F. P. (1933). Sulla determinazione empirica delle leggi di probabilità. *Giornale Dell'Istituto Italiano Degli Attuari, 4*, 421–424.
65. Carriere, J. F. (1996). Valuation of the early-exercise price for options using simulations and nonparametric regression. *Insurance: Mathematics & Economics, 19/1*, 19–30.
66. Chan, J. S. K., Choy, S. T. B., Makov, U. E., & Landsman, Z. (2018). Modelling insurance losses using contaminated generalised beta type-II distribution. *ASTIN Bulletin, 48/2*, 871–904.
67. Charpentier, A. (2015). *Computational actuarial science with R*. Boca Raton: CRC Press.
68. Chaubard, F., Mundra, R., & Socher, R. (2016). *Deep learning for natural language processing. Lecture notes*. Stanford: Stanford University.
69. Chen, A., Guillén, M., & Vigna, E. (2018). Solvency requirement in a unisex mortality model. *ASTIN Bulletin, 48/3*, 1219–1243.
70. Chen, A., & Vigna, E. (2017). A unisex stochastic mortality model to comply with EU Gender Directive. *Insurance: Mathematics & Economics, 73*, 124–136.
71. Chen, T., & Guestrin, C. (2016). *XGBoost: A scalable tree boosting system*. arXiv:1603.02754v3.
72. Chen, X. (2007). Large sample sieve estimation of semi-parametric models. In J. J. Heckman & E. E. Leamer (Eds.), *Handbook of econometrics* (Vol. 6B, Chap. 76, pp. 5549–5632). Amsterdam: Elsevier.
73. Chen, X., & Shen, X. (1998). Sieve extremum estimates for weakly dependent data. *Econometrica, 66/2*, 289–314.
74. Cheridito, P., Ery, J., & Wüthrich, M. V. (2020). Assessing asset-liability risk with neural networks. *Risks, 8/1*. Article 16.
75. Cheridito, P., Jentzen, A., & Rossmannek, F. (2022). Efficient approximation of high-dimensional functions with neural networks. *IEEE Transactions on Neural Networks and Learning Systems, 33/7*, 3079–3093.
76. Cho, K., van Merrienboer, B., Gulcehre, C., Bahdanau, D., Bougares, F., Schwenk, H., & Bengio, Y. (2014). *Learning phrase representations using RNN encoder-decoder for statistical machine translation*. arXiv:1406.1078.
77. Chollet, F., Allaire, J. J., et al. (2017). *R interface to Keras*. https://github.com/rstudio/keras
78. Cohen, J. (1988). *Statistical power analysis for the behavioral sciences* (2nd ed.) New York: Lawrence Erlbaum Associates.
79. Congdon, P. (2014). *Applied Bayesian modelling* (2nd ed.). New York: Wiley.
80. Cook, D. R., & Croos-Dabrera, R. (1993). Partial residual plots in generalized linear models. *Journal of the American Statistical Association, 93/442*, 730–739.
81. Cooray, K., & Ananda, M. M. A. (2005). Modeling actuarial data with composite lognormal-Pareto model. *Scandinavian Actuarial Journal, 2005/5*, 321–334.
82. Corradin, A., Denuit, M., Detyniecki, M., Grari, V., Sammarco, M., & Trufin, J. (2022). Joint modeling of claim frequencies and behavior signals in motor insurance. *ASTIN Bulletin, 52/1*, 33–54.
83. Cragg, J. G. (1971). Some statistical models for limited dependent variables with application to the demand for durable good. *Econometrica, 39/5*, 829–844.
84. Craven, P., & Wahba, G. (1978). Smoothing noisy data with spline functions. *Numerische Mathematik, 31*, 377–403.
85. Creal, D. (2012). A survey of sequential Monte Carlo methods for economics and finance. *Econometric Reviews, 31/3*, 245–296.
86. Cybenko, G. (1989). Approximation by superpositions of a sigmoidal function. *Mathematics of Control, Signals and Systems, 2*, 303–314.
87. Daniels, H. E. (1954). Saddlepoint approximations in statistics. *Annals of Mathematical Statistics, 25*, 631–650.
88. Darmois, G., (1935). Sur les lois de probabilité à estimation exhaustive. *Comptes Rendus de l'Académie des Sciences Paris, 260*, 1265–1266.

89. De Jong, P., & Heller, G. Z. (2008). *Generalized linear models for insurance data*. Cambridge: Cambridge University Press.
90. De Jong, P., Tickle, L., & Xu, J. (2020). A more meaningful parameterization of the Lee–Carter model. *Insurance: Mathematics & Economics, 94*, 1–8.
91. De Pril, N. (1978). The efficiency of a bonus-malus system. *ASTIN Bulletin, 10/1*, 59–72.
92. Del Moral, P., Doucet, A., & Jasra, A. (2006). Sequential Monte Carlo samplers. *Journal of the Royal Statistical Society, Series B, 68/3*, 411–436.
93. Del Moral, P., Peters, G. W., & Vergé, C. (2012). An introduction to stochastic particle integration methods: With applications to risk and insurance. In J. Dick, F. Y. Kuo, G. W. Peters, & I. H. Sloan (Eds.), *Monte Carlo and Quasi-Monte Carlo Methods 2012. Proceedings in Mathematics & Statistics* (Vol. 65, pp. 39–81). New York: Springer.
94. Delong, Ł., Lindholm, M., & Wüthrich, M. V. (2021). Making Tweedie's compound Poisson model more accessible. *European Actuarial Journal, 11/1*, 185–226.
95. Delong, Ł., Lindholm, M., & Wüthrich, M. V. (2021). Gamma mixture density networks and their application to modeling insurance claim amounts. *Insurance: Mathematics & Economics, 101/B*, 240–261.
96. Dempster, A. P., Laird, N. M., & Rubin, D. B. (1977). Maximum likelihood for incomplete data via the EM algorithm. *Journal of the Royal Statistical Society, Series B, 39/1*, 1–22.
97. Denuit, M., Charpentier, A., & Trufin, J. (2021). Autocalibration and Tweedie-dominance for insurance pricing in machine learning. *Insurance: Mathematics & Economics, 101/B*, 485–497.
98. Denuit, M., Guillén, M., & Trufin, J. (2019). Multivariate credibility modelling for usage-based motor insurance pricing with behavioural data. *Annals of Actuarial Science, 13/2*, 378–399.
99. Denuit, M., Hainaut, D., & Trufin, J. (2019). *Effective statistical learning methods for actuaries I: GLMs and extensions*. New York: Springer.
100. Denuit, M., Hainaut, D., & Trufin, J. (2020). *Effective statistical learning methods for actuaries II: Tree-based methods and extensions*. New York: Springer.
101. Denuit, M., Hainaut, D., & Trufin, J. (2019). *Effective statistical learning methods for actuaries III: Neural networks and extensions*. New York: Springer.
102. Denuit, M., Maréchal, X., Pitrebois, S., & Walhin, J.-F. (2007). *Actuarial modelling of claim counts: Risk classification, credibility and bonus-malus systems*. New York: Wiley.
103. Denuit, M., & Trufin, J. (2021). Generalization error for Tweedie models: Decomposition and error reduction with bagging. *European Actuarial Journal, 11/1*, 325–331.
104. Devriendt, S., Antonio, K., Reynkens, T., & Verbelen, R. (2021). Sparse regression with multi-type regularized feature modeling. *Insurance: Mathematics & Economics, 96*, 248–261.
105. Dietterich, T. G. (2000). Ensemble methods in machine learning. In J. Kittel & F. Roli (Eds.), *Multiple classifier systems. Lecture notes in computer science* (Vol. 1857, pp. 1–15). New York: Springer.
106. Dimitriadis, T., Fissler, T., & Ziegel, J. F. (2020). *The efficiency gap*. arXiv:2010.14146.
107. Dobson, A. J. (2001). *An introduction to generalized linear models*. Boca Raton: Chapman & Hall/CRC.
108. Döhler, S., & Rüschendorf, L. (2001). An approximation result for nets in functional estimation. *Statistics & Probability Letters, 52/4*, 373–380.
109. Döhler, S., & Rüschendorf, L. (2003). Nonparametric estimation of regression functions in point process models. *Statistics Inference for Stochastic Processes, 6*, 291–307.
110. Dong, Y., Huang, F., Yu, H., & Haberman, S. (2020). Multi-population mortality forecasting using tensor decomposition. *Scandinavian Actuarial Journal, 2020/8*, 754–775.
111. Doucet, A., & Johansen, A. M. (2011). A tutorial on particle filtering and smoothing: Fifteen years later. In D. Crisan & B. Rozovsky (Eds.), *Handbook of nonlinear filtering* (pp. 656–670). Oxford: Oxford University Press.
112. Dunn, P. K., & Smyth, G. K. (2005). Series evaluation of Tweedie exponential dispersion model densities. *Statistics and Computing, 15*, 267–280.

113. Dutang, C., & Charpentier, A. (2018). *CASdatasets R package vignette*. Reference manual. Version 1.0-8, packaged 2018-05-20.
114. Eckart, G., & Young, G. (1936). The approximation of one matrix by another of lower rank. *Psychometrika, 1*, 211–218.
115. Efron, B. (1979). Bootstrap methods: Another look at the jackknife. *Annals of Statistics, 7/1*, 1–26.
116. Efron, B. (2020). Prediction, estimation, and attribution. *Journal of the American Statistical Association, 115/530*, 636–655.
117. Efron, B., & Hastie, T. (2016). *Computer age statistical inference: Algorithms, evidence, and data science*. Cambridge: Cambridge University Press.
118. Efron, B., & Tibshirani, R. J. (1993). *An introduction to the bootstrap*. New York: Chapman & Hall.
119. Ehm, W., Gneiting, T., Jordan, A., & Krüger, F. (2016). Of quantiles and expectiles: Consistent scoring functions, Choquet representations and forecast rankings. *Journal of the Royal Statistical Society, Series B, 78/3*, 505–562.
120. Elbrächter, D., Perekrestenko, D., Grohs, P., & Bölcskei, H. (2021). Deep neural network approximation theory. *IEEE Transactions on Information Theory, 67/5*, 2581–2623.
121. Embrechts, P., Klüppelberg, C., & Mikosch, T. (2003). *Modelling extremal events for insurance and finance* (4th printing). New York: Springer.
122. Embrechts, P., & Wüthrich, M. V. (2022). Recent challenges in actuarial science. *Annual Review of Statistics and Its Applications, 9*, 119–140.
123. Fahrmeir, L., & Tutz, G. (1994). *Multivariate statistical modelling based on generalized linear models*. New York: Springer.
124. Fan, J., & Li, R. (2001). Variable selection via nonconcave penalized likelihood and its oracle properties. *Journal of the American Statistical Association, 96/456*, 1348–1360.
125. Ferrario, A., & Hämmerli, R. (2019). *On boosting: Theory and applications*. SSRN Manuscript ID 3402687. Version June 11, 2019.
126. Ferrario, A., & Nägelin, M. (2020). *The art of natural language processing: Classical, modern and contemporary approaches to text document classification*. SSRN Manuscript ID 3547887. Version March 1, 2020.
127. Ferrario, A., Noll, A., & Wüthrich, M. V. (2018). *Insights from inside neural networks*. SSRN Manuscript ID 3226852. Version April 23, 2020.
128. Fisher, R. A. (1934). Two new properties of mathematical likelihood. *Proceeding of the Royal Society A, 144/852*, 285–307.
129. Fissler, T., Lorentzen, C., & Mayer, M. (2022). *Model comparison and calibration assessment: User guide for consistent scoring functions in machine learning and actuarial practice*. arXiv:2202.12780.
130. Fissler, T., Merz, M., & Wüthrich, M. V. (2021). *Deep quantile and deep composite model regression*. arXiv:2112.03075.
131. Fissler, T., & Ziegel, J. F. (2016). Higher order elicitability and Osband's principle. *The Annals of Statistics, 4474*, 1680–1707.
132. Fissler, T., Ziegel, J. F., & Gneiting, T. (2015). *Expected shortfall is jointly elicitable with value at risk - Implications for backtesting*. arXiv:1507.00244v2.
133. Fortuin, C. M., Kasteleyn, P. W., & Ginibre, J. (1971). Correlation inequalities on some partially ordered sets. *Communication Mathematical Physics, 22/2*, 89–103.
134. Frees, E. W. (2010). *Regression modelling with actuarial and financial applications*. Cambridge: Cambridge University Press.
135. Frees, E. W. (2020). *Loss data analytics. An open text authored by the Actuarial Community*. https://ewfrees.github.io/Loss-Data-Analytics/
136. Frees, E. W., & Huang, F. (2021). The discriminating (pricing) actuary. *North American Actuarial Journal* (in press).
137. Frees, E. W., Lee, G., & Yang, L. (2016). Multivariate frequency-severity regression models in insurance. *Risks, 4/1*. Article 4.

138. Frei, D. (2021). *Insurance Claim Size Modelling with Mixture Distributions.* MSc Thesis. Department of Mathematics, ETH Zurich.
139. Freund, Y. (1995). Boosting a weak learning algorithm by majority. *Information and Computation, 121/2*, 256–285.
140. Freund, Y., & Schapire, R. E. (1997). A decision-theoretic generalization of on-line learning and an application to boosting. *Journal of Computer and System Sciences, 55/1*, 119–139.
141. Friedman, J. H. (2001). Greedy function approximation: A gradient boosting machine. *Annals of Statistics, 29/5*, 1189–1232.
142. Friedman, J. H., Hastie, T., & Tibshirani, R. (2010). Regularization paths for generalized linear models via coordinate descent. *Journal of Statistical Software, 33/1*, 1–22.
143. Friedman, J. H., & Popescu, B. E. (2008). Predictive learning via rule ensembles. *Annals of Applied Statistics, 2/3*, 916–954.
144. Fritsch, S., Günther, F., Wright, M. N., Suling, M., & Müller, S. M. (2019). *neuralnet: Training of neural networks.* https://github.com/bips-hb/neuralnet
145. Fukushima, K. (1980). Neocognitron: A self-organizing neural network model for a mechanism of pattern recognition unaffected by shift in position. *Biological Cybernetics, 36/4*, 193–202.
146. Fung, T. C., Badescu, A. L., & Lin, X. S. (2019). A class of mixture of experts models for general insurance: Application to correlated claim frequencies. *ASTIN Bulletin, 49/3*, 647–688.
147. Fung, T. C., Badescu, A. L., & Lin, X. S. (2022). Fitting censored and truncated regression data using the mixture of experts models. *North American Actuarial Journal* (in press).
148. Fung, T. C., Tzougas, G., & Wüthrich, M. V. (2022). Mixture composite regression models with multi-type feature selection. *North American Actuarial Journal* (in press).
149. Gabrielli, A., Richman, R., & Wüthrich, M. V. (2020). Neural network embedding of the over-dispersed Poisson reserving model. *Scandinavian Actuarial Journal, 2020/1*, 1–29.
150. Gallant, A. R., & White, H. (1988). There exists a neural network that does not make avoidable mistakes. In *IEEE 1988 International Conference on Neural Networks* (pp. I657–664).
151. Gao, G., Meng, S., & Wüthrich, M. V. (2019). Claims frequency modeling using telematics car driving data. *Scandinavian Actuarial Journal, 2019/2*, 143–162.
152. Gao, G., Meng, S., & Wüthrich, M. V. (2022). What can we learn from telematics car driving data: A survey. *Insurance: Mathematics & Economics, 104*, 185–199.
153. Gao, G., & Shi, Y. (2021). Age-coherent extensions of the Lee–Carter model. *Scandinavian Actuarial Journal, 2021/10*, 998–1016.
154. Gao, G., Wang, H., & Wüthrich, M. V. (2022). Boosting Poisson regression models with telematics car driving data. *Machine Learning, 111/1*, 243–272.
155. Gao, G., & Wüthrich, M. V. (2018). Feature extraction from telematics car driving heatmaps. *European Actuarial Journal, 8/2*, 383–406.
156. Gao, G., & Wüthrich, M. V. (2019). Convolutional neural network classification of telematics car driving data. *Risks, 7/1*. Article 6.
157. Gelman, A., Carlin, J. B., Stern, H. S., Dunson, D. B., Vehtari, A., & Rubin, D. B. (2013). *Bayesian data analysis* (3rd ed.). Boca Raton: Chapman & Hall/CRC.
158. Gilks, W. R., Richardson, S., & Spiegelhalter, D. J. (1995). *Markov chain Monte Carlo in practice*. Boca Raton: Chapman & Hall.
159. Glivenko, V. (1933). Sulla determinazione empirica delle leggi di probabilità. *Giornale Dell'Istituto Italiano Degli Attuari, 4*, 92–99.
160. Glorot, X., & Bengio, Y. (2010). Understanding the difficulty of training deep feedforward neural networks. In *Proceedings of the Thirteenth International Conference on Artificial Intelligence and Statistics. Proceedings of Machine Learning Research* (Vol. 9, pp. 249–256).
161. Glynn, P., & Lee, S. H. (2003). Computing the distribution function of a conditional expectation via Monte Carlo: Discrete conditioning spaces. *ACM Transactions on Modeling and Computer Simulation, 13/3*, 238–258.

162. Gneiting, T. (2011). Making and evaluating point forecasts. *Journal of the American Statistical Association, 106/494*, 746–762.
163. Gneiting, T., & Raftery, A. E. (2007). Strictly proper scoring rules, prediction, and estimation. *Journal of the American Statistical Association, 102/477*, 359–378.
164. Goldstein, A., Kapelner, A., Bleich, J., & Pitkin, E. (2015). Peeking inside the black box: Visualizing statistical learning with plots of individual conditional expectation. *Journal of Computational and Graphical Statistics, 24/1*, 44–65.
165. Golub, G., & Van Loan, C. (1983). *Matrix computations*. Baltimore: John Hopkins University Press.
166. Goodfellow, I., Bengio, Y., & Courville, A. (2016). *Deep learning*. Cambridge: MIT Press. http://www.deeplearningbook.org
167. Gourieroux, C., Laurent, J. P., & Scaillet, O. (2000). Sensitivity analysis of values at risk. *Journal of Empirical Finance, 7/3–4*, 225–245.
168. Gourieroux, C., Montfort, A., & Trognon, A. (1984). Pseudo maximum likelihood methods: Theory. *Econometrica, 52/3*, 681–700.
169. Green, P. J. (1995). Reversible jump Markov chain Monte Carlo computation and Bayesian model determination. *Biometrika, 82/4*, 711–732.
170. Green, P. J. (2003). Trans-dimensional Markov chain Monte Carlo. In P. J. Green, N. L. Hjort, & S. Richardson (Eds.), *Highly structured stochastic systems. Oxford statistical science series* (pp. 179–206). Oxford: Oxford University Press.
171. Greene, W. (2008). Functional forms for the negative binomial model for count data. *Economics Letters, 99*, 585–590.
172. Grenander, U. (1981). *Abstract inference*. New York: Wiley.
173. Grün, B., & Miljkovic, T. (2019). Extending composite loss models using a general framework of advanced computational tools. *Scandinavian Actuarial Journal, 2019/8*, 642–660.
174. Guillén, M. (2012). Sexless and beautiful data: From quantity to quality. *Annals of Actuarial Science, 6/2*, 231–234.
175. Guillén, M., Bermúdez, L., & Pitarque, A. (2021). Joint generalized quantile and conditional tail expectation for insurance risk analysis. *Insurance: Mathematics & Economics, 99*, 1–8.
176. Guo, C., & Berkhahn, F. (2016). *Entity embeddings of categorical variables*. arXiv:1604.06737.
177. Ha, H., & Bauer, D. (2022). A least-squares Monte Carlo approach to the estimation of enterprise risk. *Finance and Stochastics, 26*, 417–459.
178. Hainaut, D. (2018). A neural-network analyzer for mortality forecast. *ASTIN Bulletin, 48/2*, 481–508.
179. Hainaut, D., & Denuit, M. (2020). Wavelet-based feature extraction for mortality projection. *ASTIN Bulletin, 50/3*, 675–707.
180. Hampel, F. R., Ronchetti, E. M., Rousseeuw, P. J., & Stahel, W. A. (1986). *Robust statistics*. New York: Wiley.
181. Hastie, T., & Tibshirani, R. (1986). Generalized additive models (with discussion). *Statistical Science, 1*, 297–318.
182. Hastie, T., & Tibshirani, R. (1990). *Generalized additive models*. New York: Chapman & Hall.
183. Hastie, T., Tibshirani, R., & Friedman, J. (2009). *The elements of statistical learning: Data mining, inference, and prediction* (2nd ed.). New York: Springer.
184. Hastie, T., Tibshirani, R., & Wainwright, M. (2015). *Statistical learning with sparsity: The Lasso and generalizations*. Boca Raton: CRC Press.
185. Hastings, W. K. (1970). Monte Carlo sampling methods using Markov chains and their applications. *Biometrika, 57/1*, 97–109.
186. Hinton, G. E., & Salakhutdinov, R. R. (2006). Reducing the dimensionality of data with neural networks. *Science, 313/5786*, 504–507.
187. Hinton, G., Srivastava, N., & Swersky, K. (2012). *Neural networks for machine learning. Lecture slides*. Toronto: University of Toronto.

188. Hochreiter, S., & Schmidhuber, J. (1997). Long short-term memory. *Neural Computation, 9/8*, 1735–1780.
189. Hong, L. J. (2009). Estimating quantile sensitivities. *Operations Research, 57/1*, 118–130.
190. Horel, E., & Giesecke, K. (2020). Significance tests in neural networks. *Journal of Machine Learning Research, 21/227*, 1–29.
191. Hornik, K. (1991). Approximation capabilities of multilayer feedforward networks. *Neural Networks, 4/2*, 251–257.
192. Hornik, K., Stinchcombe, M., & White, H. (1989). Multilayer feedforward networks are universal approximators. *Neural Networks, 2/5*, 359–366.
193. Huang, Y., & Meng, S. (2019). Automobile insurance classification ratemaking based on telematics driving data. *Decision Support Systems, 127*. Article 113156.
194. Huber, P. J. (1981). *Robust statistics*. Hoboken: Wiley.
195. Human Mortality Database (2018). University of California, Berkeley (USA), and Max Planck Institute for Demographic Research (Germany). www.mortality.org
196. Hyndman, R. J., Booth, H., & Yasmeen, F. (2013). Coherent mortality forecasting: The product-ratio method with functional time series models. *Demography, 50/1*, 261–283.
197. Hyndman, R. J., & Ullah, M. S. (2007). Robust forecasting of mortality and fertility rates: A functional data approach. *Computational Statistics & Data Analysis, 51/10*, 4942–4956.
198. Isenbeck, M., & Rüschendorf, L. (1992). Completeness in location families. *Probability and Mathematical Statistics, 13/2*, 321–343.
199. Johansen, A. M., Evers, L., & Whiteley, N. (2010). *Monte Carlo methods. Lecture notes.* Bristol: Department of Mathematics, University of Bristol.
200. Jørgensen, B. (1981). *Statistical properties of the generalized inverse Gaussian distribution. Lecture notes in statistics*. New York: Springer.
201. Jørgensen, B. (1986). Some properties of exponential dispersion models. *Scandinavian Journal of Statistics, 13/3*, 187–197.
202. Jørgensen, B. (1987). Exponential dispersion models. *Journal of the Royal Statistical Society, Series B, 49/2*, 127–145.
203. Jørgensen, B. (1997). *The theory of dispersion models*. Boca Raton: Chapman & Hall.
204. Jørgensen, B., & de Souza, M. C. P. (1994). Fitting Tweedie's compound Poisson model to insurance claims data. *Scandinavian Actuarial Journal, 1994/1*, 69–93.
205. Jospin, L. V., Buntine, W., Boussaid, F., Laga, H., & Bennamoun, M. (2020). *Hands-on Bayesian neural networks - A tutorial for deep learning users*. arXiv: 2007.06823.
206. Jung, J. (1968). On automobile insurance ratemaking. *ASTIN Bulletin, 5/1*, 41–48.
207. Kalman, R. E. (1960). A new approach to linear filtering and prediction problems. *Journal of Basic Engineering, 82/1*, 35–45.
208. Karush, W. (1939). *Minima of Functions of Several Variables with Inequalities as Side Constraints*. MSc Thesis. Department of Mathematics, University of Chicago.
209. Kearns, M., & Valiant, L. G. (1988). *Learning Boolean Formulae or Finite Automata is Hard as Factoring*. Technical Report TR-14–88. Aiken Computation Laboratory, Harvard University.
210. Kearns, M., & Valiant, L. G. (1994). Cryptographic limitations on learning Boolean formulae and finite automata. *Journal of the Association for Computing Machinery ACM, 41/1*, 67–95.
211. Kellner, R., Nagl, M., & Rösch, D. (2022). Opening the black box - Quantile neural networks for loss given default prediction. *Journal of Banking & Finance, 134*, 1–20.
212. Keydana, S., Falbel, D., & Kuo, K. (2021). *R package 'tfprobability': Interface to 'TensorFlow Probability'*. Version 0.12.0.0, May 20, 2021.
213. Khalili, A. (2010). New estimation and feature selection methods in mixture-of-experts models. *Canadian Journal of Statistics, 38/4*, 519–539.
214. Khalili, A., & Chen, J. (2007). Variable selection in finite mixture of regression models. *Journal of the American Statistical Association, 102/479*, 1025–1038.
215. Kidger, P., & Lyons, T. (2020). Universal approximation with deep narrow networks. *Proceedings of Machine Learning Research, 125*, 2306–2327.
216. Kingma, D., & Ba, J. (2014). *Adam: A method for stochastic optimization*. arXiv:1412.6980.

217. Kingma, D. P., & Welling, M. (2019). An introduction to variational autoencoders. *Foundations and Trends in Machine Learning, 12/4*, 307–392.
218. Kleinow, T. (2015). A common age effect model for the mortality of multiple populations. *Insurance: Mathematics & Economics, 63*, 147–152.
219. Knyazev, B., Drozdzal, M., Taylor, G. W., & Romero-Soriano, A. (2021). *Parameter prediction of unseen deep architectures*. arXiv:2110.13100.
220. Koenker, R., & Bassett, G., Jr. (1978). Regression quantiles. *Econometrica, 46/1*, 33–50.
221. Kolmogoroff, A. (1933). *Grundbegriffe der Wahrscheinlichkeitsrechnung*. New York: Springer.
222. Komunjer, I., & Vuong, Q. (2010). Efficient estimation in dynamic conditional quantile models. *Journal of Econometrics, 157*, 272–285.
223. Koopman, B. O. (1936). On distributions admitting a sufficient statistics. *Transactions of the American Mathematical Society, 39*, 399–409.
224. Krah, A.-S., Nikolić, Z., & Korn, R. (2020). Least-squares Monte Carlo for proxy modeling in life insurance: neural networks. *Risks, 8/4*. Article 116.
225. Kramer, M. A. (1991). Nonlinear principal component analysis using autoassociative neural networks. *AIChE Journal, 37/2*, 233–243.
226. Krizhevsky, Al., Sutskever, I., & Hinton, G. E. (2017). ImageNet classification with deep convolutional neural networks. *Communications of the Association for Computing Machinery ACM, 60/6*, 84–90.
227. Krüger, F., & Ziegel, J. F. (2021). Generic conditions for forecast dominance. *Journal of Business & Economics Statistics, 39/4*, 972–983.
228. Kuhn, H. W., & Tucker, A. W. (1951). Nonlinear programming. *Proceedings of 2nd Berkeley Symposium* (pp. 481–492). Berkeley: University of California Press.
229. Künsch, H. R. (2005). *Mathematische Statistik. Lecture notes*. ETH Zurich: Department of Mathematics.
230. Kuo, K. (2020). *Individual claims forecasting with Bayesian mixture density networks*. arXiv:2003.02453.
231. Kuo, K., & Richman, R. (2021). *Embeddings and attention in predictive modeling*. arXiv:2104.03545v1.
232. Lambert, D. (1992). Zero-inflated Poisson regression, with an application to defects in manufacturing. *Technometrics, 34/1*, 1–14.
233. Landsman, Z., & Valdez, E. A. (2005). Tail conditional expectation for exponential dispersion models. *ASTIN Bulletin, 35/1*, 189–209.
234. LeCun, Y., Boser, B., Denker, J. S., Henderson, D., Howard, R. E., Hubbard, W., & Jackel, L. D. (1989). Backpropagation applied to handwritten zip code recognition. *Neural Computation, 1/4*, 541–551.
235. LeCun, Y., Bottou, L., Bengio, Y., & Haffner, P. (1998). Gradient-based learning applied to document recognition. *Proceedings of the IEEE, 86/11*, 2278–2324.
236. Lee, G. Y., Manski, S., & Maiti, T. (2020). Actuarial applications of word embedding models. *ASTIN Bulletin, 50/1*, 1–24.
237. Lee, J. D., Sun, D. L., Sun, Y., & Taylor, J. E. (2016). Exact post-selection inference, with application to the lasso. *Annals of Statistics, 44/3*, 907–927.
238. Lee, R. D., & Carter, L. R. (1992). Modeling and forecasting U.S. mortality. *Journal of the American Statistical Association, 87/419*, 659–671.
239. Lee, S. C. K. (2021). Addressing imbalanced insurance data through zero-inflated Poisson regression boosting. *ASTIN Bulletin, 51/1*, 27–55.
240. Lee, S. C. K., & Lin, X. S. (2010). Modeling and evaluating insurance losses via mixtures of Erlang distributions. *North American Actuarial Journal, 14/1*, 107–130.
241. Lee, S. C. K., & Lin, X. S. (2018). Delta boosting machine with application to general insurance. *North American Actuarial Journal, 22/3*, 405–425.
242. Lee, S. H. (1998). *Monte Carlo Computation of Conditional Expectation Quantiles*. PhD Thesis, Stanford University.
243. Lehmann, E. L. (1959). *Testing statistical hypotheses*. New York: Wiley.

244. Lehmann, E. L. (1983). *Theory of point estimation*. New York: Wiley.
245. Lemaire, J. (1995). *Bonus-malus systems in automobile insurance*. Dordrecht: Kluwer Academic Publisher.
246. Lemaire, J., Park, S. C., & Wang, K. (2016). The use of annual mileage as a rating variable. *ASTIN Bulletin, 46/1*, 39–69.
247. Leshno, M., Lin, V. Y., Pinkus, A., & Schocken, S. (1993). Multilayer feedforward networks with a nonpolynomial activation function can approximate any function. *Neural Networks, 6/6*, 861–867.
248. Li, H., & Lu, Y. (2017). Coherent forecasting of mortality rates: a sparse vector-autoregression approach. *ASTIN Bulletin, 47/2*, 563–600.
249. Li, N., & Lee, R. (2005). Coherent mortality forecasts for a group of populations: An extension of the Lee–Carter method. *Demography, 42/3*, 575–594.
250. Li, N., Lee, R., & Gerland, P. (2013). Extending the Lee–Carter method to model the rotation of age patterns of mortality decline for long-term projections. *Demography, 50/6*, 2037–2051.
251. Li, Z., Wang, F., & Zhao, Z. (2022). *A new class of composite GBII regression models with varying threshold for modelling heavy-tailed data*. arXiv:2203.11469v2.
252. Lindholm, M., & Palmborg, L. (2022). Efficient use of data from LSTM mortality forecasting. *European Actuarial Journal* (in press).
253. Lindholm, M., Richman, R., Tsanakas, A., & Wüthrich, M. V. (2022). Discrimination-free insurance pricing. *ASTIN Bulletin, 52/1*, 55–89.
254. Loader, C., Sun, J., Lucent Technologies, & Liaw, A. (2022). locfit: Local regression, likelihood and density estimation. https://cran.r-project.org/web/packages/locfit/index.html
255. Loimaranta, K. (1972). Some asymptotic properties of bonus systems. *ASTIN Bulletin, 6/3*, 233–245.
256. Lomax, K. S. (1954). Business failures: Another example of the analysis of failure data. *Journal of the American Statistical Association, 49/268*, 847–852.
257. Longstaff, F., & Schwartz, E. (2001). Valuing American options by simulation: A simple least-squares approach. *The Review of Financial Studies, 14/1*, 113–147.
258. Lorentzen, C., & Mayer, M. (2020). *Peeking into the black box: An actuarial case study for interpretable machine learning*. SSRN Manuscript ID 3595944. Version May 7, 2020.
259. Loser, F. (2020). Private communication.
260. Lu, J., Shen, Z., Yang, H., & Zhang, S. (2021). Deep network approximation for smooth functions. *SIAM Journal on Mathematical Analysis, 53/5*, 5465–5506.
261. Lundberg, S. M., & Lee, S.-I. (2017). A unified approach to interpreting model predictions. In I. Guyon, U. V. Luxburg, S. Bengio, H. Wallach, R. Fergus, S. Vishwanathan, & R. Garnett (Eds.), *Advances in neural information processing systems* (Vol. 30, pp. 4765–4774). New York: Curran Associates.
262. Makavoz, Y. (1996). Random approximants and neural networks. *Journal of Approximation Theory, 85/1*, 98–109.
263. Mallat, S. (2012). Group invariant scattering. *Communication in Pure and Applied Mathematics, 65/10*, 1331–1398.
264. Manski, S., Yang, K., Lee, G. Y., & Maiti, T. (2021). Extracting information from textual descriptions for actuarial applications. *Annals of Actuarial Science, 15/3*, 605–622.
265. McCullagh, P., & Nelder, J. A. (1983). *Generalized linear models*. Boca Raton: Chapman & Hall.
266. McGrayne, S. B. (2011). *The theory that would not die*. New Haven: Yale University Press.
267. McLachlan, G. J., & Krishnan, T. (2008). *The EM algorithm and extensions* (2nd ed.). New York: Wiley.
268. McNeil, A. J., Frey, R., & Embrechts, P. (2015). *Quantitative risk management: Concepts, techniques and tools* (revised edition). Princeton: Princeton University Press.
269. Meier, D., & Wüthrich, M. V. (2020). *Convolutional neural network case studies: (1) anomalies in mortality rates (2) image recognition*. SSRN Manuscript ID 3656210. Version July 19, 2020.

270. Meinshausen, N. (2006). Quantile regression forests. *Journal of Machine Learning Research, 7*, 983–999.
271. Meng, S., Wang, H., Shi, Y., & Gao, G. (2022). Improving automobile insurance claims frequency prediction with telematics car driving data. *ASTIN Bulletin, 52/2*, 363–391.
272. Mercer, J. (1909). Functions of positive and negative type and their connection with the theory of integral equations. *Philosophical Transactions of the Royal Society A, 209/441–458*, 415–446.
273. Merz, M., Richman, R., Tsanakas, A., & Wüthrich, M. V. (2022). Interpreting deep learning models with marginal attribution by conditioning on quantiles. *Data Mining and Knowledge Discovery, 36*, 1335–1370.
274. Metropolis, N., Rosenbluth, A. W., Rosenbluth, M. N., Teller, A. H., & Teller, E. (1953). Equation of state calculations by fast computing machines. *Journal of Chemical Physics, 21/6*, 1087–1092.
275. Mikolov, T., Chen, K., Corrado, G. S., & Dean, J. (2013). *Efficient estimation of word representations in vector space.* arXiv:1301.3781.
276. Mikolov, T., Sutskever, I., Chen, K., Corrado, G. S., & Dean, J. (2013). Distributed representations of words and phrases and their compositionality. *Advances in Neural Information Processing Systems, 26*, 3111–3119.
277. Mikosch, T. (2006). *Non-life insurance mathematics*. New York: Springer.
278. Miljkovic, T., & Grün, B. (2016). Modeling loss data using mixtures of distributions. *Insurance: Mathematics & Economics, 70*, 387–396.
279. Mirsky, L. (1960). Symmetric gauge functions and unitarily invariant norms. *Quarterly Journal of Mathematics, 11/1*, 50–59.
280. Montúfar, G., Pascanu, R., Cho, K., & Bengio, Y. (2014). On the number of linear regions of deep neural networks. *Neural Information Processing Systems Proceedings, 27*, 2924–2932.
281. Neal, R. M. (1996). *Bayesian learning for neural networks*. New York: Springer.
282. Nelder, J. A., & Pregibon, D. (1987). An extended quasi-likelihood function. *Biometrika, 74/2*, 221–232.
283. Nelder, J. A., & Wedderburn, R. W. M. (1972). Generalized linear models. *Journal of the Royal Statistical Society, Series A, 135/3*, 370–384.
284. Nesterov, Y. (2007). *Gradient Methods for Minimizing Composite Objective Function.* Technical Report 76. Center for Operations Research and Econometrics (CORE), Catholic University of Louvain.
285. Nielsen, F. (2020). An elementary introduction to information geometry. *Entropy, 22/10*, 1100.
286. Nigri, A., Levantesi, S., Marino, M., Scognamiglio, S., & Perla, F. (2019). A deep learning integrated Lee–Carter model. *Risks, 7/1*. Article 33.
287. Noll, A., Salzmann, R., & Wüthrich, M. V. (2018). *Case study: French motor third-party liability claims*. SSRN Manuscript ID 3164764. Version March 4, 2020.
288. Oelker, M.-R., & Tutz, G. (2017). A uniform framework for the combination of penalties in generalized structured models. *Advances in Data Analysis and Classification, 11*, 97–120.
289. O'Hagan, W., Murphy, B. T., Scrucca, L., & Gormley, I. C. (2019). Investigation of parameter uncertainty in clustering using a Gaussian mixture model via jackknife, bootstrap and weighted likelihood bootstrap. *Computational Statistics, 34/4*, 1779–1813.
290. Ohlsson, E., & Johansson, B. (2010). *Non-life insurance pricing with generalized linear models*. New York: Springer.
291. Paefgen, J., Staake, T., & Fleisch, E. (2014). Multivariate exposure modeling of accident risk: Insights from pay-as-you-drive insurance data. *Transportation Research Part A: Policy and Practice, 61*, 27–40.
292. Parikh, N., & Boyd, S. (2013). Proximal algorithms. *Foundations and Trends in Optimization, 1/3*, 123–231.
293. Park, J., & Sandberg, I. (1991). Universal approximation using radial-basis-function networks. *Neural Computation, 3/2*, 246–257.

294. Park, J., & Sandberg, I. (1993). Approximation and radial-basis-function networks. *Neural Computation, 5/2*, 305–316.
295. Parodi, P. (2020). A generalised property exposure rating framework that incorporates scale-independent losses and maximum possible loss uncertainty. *ASTIN Bulletin, 50/2*, 513–553.
296. Paszke, A., et al. (2019). PyTorch: An imperative style, high-performance deep learning library. In *Advances in Neural Information Processing Systems* (Vol. 32, pp. 8024–8035).
297. Patton, A. J. (2020). Comparing possibly misspecified forecasts. *Journal of Business & Economic Statistics, 38/4*, 796–809.
298. Pearl, J. (2009). Causal inference in statistics: An overview. *Statistics Surveys, 3*, 96–146.
299. Pearl, J., Glymour, M., & Jewell, N. P. (2016). *Causal inference in statistics: A primer*. Chichester: Wiley.
300. Pennington, J., Socher, R., & Manning, C. D. (2014). GloVe: Global vectors for word representation. *Proceedings of the 2014 Conference on Empirical Methods in Natural Language Processing (EMNLP)* (pp. 1532–1543).
301. Perla, F., Richman, R., Scognamiglio, S., & Wüthrich, M. V. (2021). Time-series forecasting of mortality rates using deep learning. *Scandinavian Actuarial Journal, 2021/7*, 572–598.
302. Petrushev, P. (1999). Approximation by ridge functions and neural networks. *SIAM Journal on Mathematical Analysis, 30/1*, 155–189.
303. Pinkus, A. (1999). Approximation theory of the MLP model in neural networks. *Acta Numerica, 8*, 143–195.
304. Pinquet, J. (1998). Designing optimal bonus-malus systems from different types of claims. *ASTIN Bulletin, 28/2*, 205–220.
305. Pinquet, J., Guillén, M., & Bolance, C. (2001). Long-range contagion in automobile insurance data: estimation and implications for experience rating. *ASTIN Bulletin, 31/2*, 337–348.
306. Pitman, E. J. G. (1936). Sufficient statistics and intrinsic accuracy. *Proceedings of the Cambridge Philosophical Society, 32/4*, 567–579.
307. R Core Team (2021). R: a language and environment for statistical computing. *R Foundation for Statistical Computing*, Vienna, Austria. https://www.R-project.org/
308. Renshaw, A. E., & Haberman, S. (2003). Lee–Carter mortality forecasting with age-specific enhancement. *Insurance: Mathematics & Economics, 33/2*, 255–272.
309. Renshaw, A. E., & Haberman, S. (2006). A cohort-based extension to the Lee–Carter model for mortality reduction factors. *Insurance: Mathematics & Economics, 38/3*, 556–570.
310. Rentzmann, S., & Wüthrich, M. V. (2019). *Unsupervised learning: What is a sports car?* SSRN Manuscript ID 3439358. Version October 14, 2019.
311. Ribeiro, M. T., Singh, S., & Guestrin, C. (2016). "Why should I trust you?": Explaining the predictions of any classifier. In *Proceedings of the 22nd ACM SIGKDD International Conference on Knowledge Discovery and Data Mining, KDD '16* (pp. 1135–1144). New York: Association for Computing Machinery.
312. Richman, R. (2021). AI in actuarial science - A review of recent advances - Part 1. *Annals of Actuarial Science, 15/2*, 207–229.
313. Richman, R. (2021). AI in actuarial science - A review of recent advances - Part 2. *Annals of Actuarial Science, 15/2*, 230–258.
314. Richman, R. (2021). *Mind the gap - Safely incorporating deep learning models into the actuarial toolkit*. SSRN Manuscript ID 3857693. Version April 2, 2021.
315. Richman, R., & Wüthrich, M. V. (2020). Nagging predictors. *Risks, 8/3*. Article 83.
316. Richman, R., & Wüthrich, M. V. (2021). A neural network extension of the Lee-Carter model to multiple populations. *Annals of Actuarial Science, 15/2*, 346–366.
317. Richman, R., & Wüthrich, M. V. (2022). LocalGLMnet: Interpretable deep learning for tabular data. *Scandinavian Actuarial Journal* (in press).
318. Richman, R., & Wüthrich, M. V. (2021). *LASSO regularization within the LocalGLMnet architecture*. SSRN Manuscript ID 3927187. Version June 1, 2022.
319. Robert, C. P. (2001). *The Bayesian choice* (2nd ed.). New York: Springer.
320. Rolski, T., Schmidli, H., Schmidt, V., & Teugels, J. (1999). *Stochastic processes for insurance and finance*. New York: Wiley.

321. Ruckstuhl, N. (2021). *Multi-Population Mortality Modeling Using Tensor Decomposition.* MSc Thesis. Department of Mathematics, ETH Zurich.
322. Rudin, C. (2019). Stop explaining black box machine learning models for high stakes decisions and use interpretable models instead. *Nature Machine Intelligence, 1*, 206–215.
323. Rüger, S. M., & Ossen, A. (1997). The metric structure of weight space. *Neural Processing Letters, 5/2*, 1–9.
324. Rumelhart, D. E., Hinton, G. E., & Williams, R. J. (1986). Learning representations by back-propagating errors. *Nature, 323/6088*, 533–536.
325. Russolillo, M., Giordano, G., & Haberman, S. (2010). Extending the Lee-Carter model: A three-way decomposition. *Scandinavian Actuarial Journal, 2011/2*, 96–117.
326. Saerens, M. (2000). Building cost functions minimizing to some summary statistics. *IEEE Transactions on Neural Networks, 11*, 1263–1271.
327. Savage, L. J. (1971). Elicitable of personal probabilities and expectations. *Journal of the American Statistical Association, 66/336*, 783–810.
328. Schapire, R. E. (1990). The strength of weak learnability. *Machine Learning, 5/2*, 197–227.
329. Schelldorfer, J., & Wüthrich, M. V. (2019). *Nesting classical actuarial models into neural networks*. SSRN Manuscript ID 3320525. Version January 25, 2019.
330. Schnürch, S., & Korn, R. (2022). Point and interval forecasts of death rates using neural networks. *ASTIN Bulletin, 52/1*, 333–360.
331. Schwarz, G. E. (1978). Estimating the dimension of a model. *Annals of Statistics, 6/2*, 461–464.
332. Scollnik, D. P. M. (2007). On composite lognormal-Pareto models. *Scandinavian Actuarial Journal, 2007/1*, 20–33.
333. Shang, H. L. (2019). Dynamic principal component regression: Application to age-specific mortality forecasting. *ASTIN Bulletin, 49/3*, 619–645.
334. Shang, H. L., & Haberman, S. (2020). Forecasting multiple functional time series in a group structure: an application to mortality. *ASTIN Bulletin, 50/2*, 357–379.
335. Shapley, L. S. (1953). A value for n-person games. In H. W. Kuhn, & A. W. Tucker (Eds.), *Contributions to the theory of games (AM-28)* (Vol. II, pp. 307–318). Princeton: Princeton University Press.
336. Shen, X., Jiang, C., Sakhanenko, L., & Lu, Q. (2019). *Asymptotic properties of neural network sieve estimators*. arXiv:1906.00875v2.
337. Shlens, J. (2014). *A tutorial on principal component analysis*. arXiv:1404.1100.
338. Shmueli, G. (2010). To explain or to predict? *Statistical Science, 25/3*, 289–310.
339. Shrikumar, A., Greenside, P., & Kundaje, A. (2017). Learning important features through propagating activation differences. In *Proceedings of the 34th International Conference on Machine Learning, Proceedings of Machine Learning Research, PMLR* (Vol. 70, pp. 3145–3153). Sydney: International Convention Centre.
340. Shrikumar, A., Greenside, P., Shcherbina, A., & Kundaje, A. (2016). *Not just a black box: Learning important features through propagating activation differences*. arXiv:1605.01713.
341. Smyth, G. K. (1989). Generalized linear models with varying dispersion. *Journal of the Royal Statistical Society, Series B, 51/1*, 47–60.
342. Smyth, G. K., & Jørgensen, B. (2002). Fitting Tweedie's compound Poisson model to insurance claims data: dispersion modeling. *ASTIN Bulletin, 32/1*, 143–157.
343. Smyth, G. K., & Verbyla, A. P. (1999). Double generalized linear models: Approximate REML and diagnostics. In H. Friedl, A. Berghold, & G. Kauermann (Eds.), *Proceedings of the 14th International Workshop on Statistical Modelling* (pp. 66–80). Technical University, Graz.
344. So, B., Boucher, J.-P., & Valdez, E. A. (2021). Cost-sensitive multi-class AdaBoost for understanding behavior based on telematics. *ASTIN Bulletin, 51/3*, 719–751.
345. Srivastava, N., Hinton, G., Krizhevsky, A. Sutskever, I., & Salakhutdinov, R. (2014). Dropout: A simple way to prevent neural networks from overfitting. *Journal of Machine Learning Research, 15/56*, 1929–1958.

346. Strassen, V. (1965). The existence of probability measures with given marginals. *Annals of Mathematical Statistics, 36/2*, 423–439.
347. Sun, S., Bi, J., Guillén, M., & Pérez-Marín, A. M. (2020). Assessing driving risk using internet of vehicles data: An analysis based on generalized linear models. *Sensors, 20/9*. Article 2712.
348. Sundberg, R. (1974). Maximum likelihood theory for incomplete data from an exponential family. *Scandinavian Journal of Statistics, 1/2*, 49–58.
349. Sundberg, R. (1976). An iterative method for solution of the likelihood equations for incomplete data from exponential families. *Communication in Statistics - Simulation and Computation, 5/1*, 55–64.
350. Takeuchi, I., Le, Q. V., Sears, T. D., & Smola, A. J. (2006). Nonparametric quantile estimation. *Journal of Machine Learning Research, 7*, 1231–1264.
351. Thomson, W. (1979). Eliciting production possibilities from a well-informed manager. *Journal of Economic Theory, 20*, 360–380.
352. Tibshirani, R. (1996). Regression shrinkage and selection via the LASSO. *Journal of the Royal Statistical Society, Series B, 58/1*, 267–288.
353. Tikhonov, A. N. (1943). On the stability of inverse problems. *Doklady Akademii Nauk SSSR, 39/5*, 195–198.
354. Troxler, A., & Schelldorfer, J. (2022). *Actuarial applications of natural language processing using transformers: Case studies for using text features in an actuarial context.* arXiv:2206.02014.
355. Tsanakas, A., & Millossovich, P. (2016). Sensitivity analysis using risk measures. *Risk Analysis, 36/1*, 30–48.
356. Tsitsiklis, J., & Van Roy, B. (2001). Regression methods for pricing complex American-style options. *IEEE Transactions on Neural Networks, 12/4*, 694–703.
357. Tukey, J. W. (1977). *Exploratory data analysis.* Reading: Addison-Wesley.
358. Tweedie, M. C. K. (1984). An index which distinguishes between some important exponential families. In J. K. Ghosh, & J. Roy (Eds.) *Statistics: Applications and new directions. Proceeding of the Indian Statistical Golden Jubilee International Conference* (pp. 579–604). Calcutta: Indian Statistical Institute.
359. Tzougas, G., & Karlis, D. (2020). An EM algorithm for fitting a new class of mixed exponential regression models with varying dispersion. *ASTIN Bulletin, 50/2*, 555–583.
360. Tzougas, G., Vrontos, S., & Frangos, N. (2014). Optimal bonus-malus systems using finite mixture models. *ASTIN Bulletin, 44/2*, 417–444.
361. Uribe, J. M., & Guillén, M. (2019). *Quantile regression for cross-sectional and time series data applications in energy markets using R.* New York: Springer.
362. Valiant, L. G. (1984). A theory of learnable. *Communications of the Association for Computing Machinery ACM, 27/11*, 1134–1142.
363. Van der Vaart, A. W. (1998). *Asymptotic statistics.* Cambridge: Cambridge University Press.
364. Van der Vaart, A. W., & Wellner, J. A. (1996). *Weak convergence and empirical processes: With applications to statistics.* New York: Springer.
365. Vapnik, V., & Chervonenkis, A. (1974). *The theory of pattern recognition.* Moscow: Nauka.
366. Vaswani, A., Shazeer, N., Parmar, N., Uszkoreit, J., Jones, L., Gomez, A. N., Kaiser, Ł., & Polosukhin, I. (2017). *Attention is all you need.* arXiv:1706.03762v5.
367. Vaughan, J., Sudjianto, A., Brahimi, E., Chen, J., & Nair, V. N. (2018). *Explainable neural networks based on additive index models.* arXiv:1806.01933v1.
368. Venables, W. N., & Ripley, B. D. (2002). *Modern applied statistics with S.* New York: Springer.
369. Venter, G. C. (1983). Transformed beta and gamma functions and losses. *Proceedings of the Casualty Actuarial Society, 71*, 289–308.
370. Verbelen, R., Antonio, K., & Claeskens, G. (2018). Unraveling the predictive power of telematics data in car insurance pricing. *Journal of the Royal Statistical Society: Series C, 67/5*, 1275–1304.

371. Verbelen, R., Gong, L., Antonio, K., Badescu, A., & Lin, S. (2015). Fitting mixtures of Erlangs to censored and truncated data using the EM algorithm. *ASTIN Bulletin, 45/3*, 729–758.
372. Verschuren, R. M. (2021). Predictive claim scores for dynamic multi-product risk classification in insurance. *ASTIN Bulletin, 51/1*, 1–25.
373. Wager, S., Wang, S., & Liang, P. S. (2013). Dropout training as adaptive regularization. In C. Burges, L. Bottou, M. Welling, Z. Ghahramani, & K. Weinberger (Eds.), *Advances in neural information processing systems* (Vol. 26, pp. 351–359). Red Hook: Curran Associates.
374. Wald, A. (1949). Note on the consistency of the maximum likelihood estimate. *Annals of Mathematical Statistics, 20/4*, 595–601.
375. Wang, C.-W., Zhang, J., & Zhu, W. (2021). Neighbouring prediction for mortality. *ASTIN Bulletin, 51/3*, 689–718.
376. Wedderburn, R. W. M. (1974). Quasi-likelihood functions, generalized linear models and the Gauss–Newton method. *Biometrika, 61/3*, 439–447.
377. Weidner, W., Transchel, F. W. G., & Weidner, R. (2016). Classification of scale-sensitive telematic observables for riskindividual pricing. *European Actuarial Journal, 6/1*, 3–24.
378. Weidner, W., Transchel, F. W. G., & Weidner, R. (2017). Telematic driving profile classification in car insurance pricing. *Annals of Actuarial Science, 11/2*, 213–236.
379. White, H. (1989). Learning in artificial neural networks: a statistical perspective. *Neural Computation, 1/4*, 425–464.
380. White, H. (1990). Connectionist nonparametric regression: multilayer feedforward networks can learn arbitrary mappings. *Neural Networks, 3/5*, 535–549.
381. White, H., & Woolridge, J. M. (1991). Some results on sieve estimation with dependent observations. In W. Barnett, J. Powell, & G. Tauchen (Eds.), *Nonparametric and semiparametric in econometrics and statistics* (pp. 459–493). Cambridge: Cambridge University Press.
382. Wiatowski, T., & Bölcskei, H. (2018). A mathematical theory of deep convolutional neural networks for feature extraction. *IEEE Transactions on Information Theory, 64/3*, 1845–1866.
383. Wilson, E. B., & Hilferty, M. M. (1931). The distribution of chi-square. *Proceedings of National Academy of Science, 17/12*, 684–688.
384. Wood, S. N. (2017). *Generalized additive models: An introduction with* R (2nd ed.). Boca Raton: CRC Press.
385. Wu, C. F. J. (1983). On the convergence properties of the EM algorithm. *Annals of Statistics, 11/1*, 95–103.
386. Wu, C. F. J. (1986). Jackknife, bootstrap and other resampling methods in regression analysis. *Annals of Statistics, 14/4*, 1261–1295.
387. Wüthrich, M. V. (2013). *Non-life insurance: Mathematics & statistics*. SSRN Manuscript ID 2319328. Version February 7, 2022.
388. Wüthrich, M. V. (2017). Covariate selection from telematics car driving data. *European Actuarial Journal, 7/1*, 89–108.
389. Wüthrich, M. V. (2017). Sequential Monte Carlo sampling for state space models. In V. Kreinovich, S. Sriboonchitta, & V.-N. Huynh (Eds.), *Robustness in econometrics. Studies in computational intelligence* (Vol. 592, pp. 25–50). New York: Springer.
390. Wüthrich, M. V. (2020). Bias regularization in neural network models for general insurance pricing. *European Actuarial Journal, 10/1*, 179–202.
391. Wüthrich, M. V. (2022). *Model selection with Gini indices under auto-calibration*. arXiv:2207.14372.
392. Wüthrich, M. V., & Buser, C. (2016). *Data analytics for non-life insurance pricing*. SSRN Manuscript ID 2870308. Version of October 27, 2021.
393. Wüthrich, M. V., & Merz, M. (2013). *Financial modeling, actuarial valuation and solvency in insurance*. New York: Springer.
394. Wüthrich, M. V., & Merz, M. (2019). Editorial: Yes, we CANN! *ASTIN Bulletin, 49/1*, 1–3.
395. Yan, H., Peters, G. W., & Chan, J. S. K. (2020). Multivariate long-memory cohort mortality models. *ASTIN Bulletin, 50/1*, 223–263.

396. Yin, C., & Lin, X. S. (2016). Efficient estimation of Erlang mixtures using iSCAD penalty with insurance application. *ASTIN Bulletin, 46/3*, 779–799.
397. Yu, B., & Barter, R. (2020). The data science process: One culture. *International Statistical Review, 88/S1*, S83–S86.
398. Yuan, X. T., & Lin, Y. (2007). Model selection and estimation in regression with grouped variables. *Journal of the Royal Statistical Society, Series B, 68/1*, 49–67.
399. Yukich, J., Stinchcombe, M., & White, H. (1995). Sup-norm approximation bounds for networks through probabilistic methods. *IEEE Transactions on Information Theory, 41/4*, 1021–1027.
400. Zaslavsky, T. (1975). *Facing up to arrangements: Face-count formulas for partitions of space by hyperplanes* (Vol. 154). Providence: Memoirs of the American Mathematical Society.
401. Zeileis, A., Kleiber C., & Jackman, S. (2008). Regression models for count data in R. *Journal of Statistical Software, 27/8*, 1–25.
402. Zhang, C., Ren, M., & Urtasun, R. (2020). *Graph hypernetworks for neural architecture search.* arXiv:1810.05749v3.
403. Zhang, W., Itoh, K., Tanida, J., & Ichioka, Y. (1990). Parallel distributed processing model with local space-invariant interconnections and its optical architecture. *Applied Optics, 29/32*, 4790–4797.
404. Zhang, W., Tanida, J., Itoh, K., & Ichioka, Y. (1988). Shift invariant pattern recognition neural network and its optical architecture. *Proceedings of the Annual Conference of the Japan Society of Applied Physics, 6p-M-14*, 734.
405. Zhao, Q., & Hastie, T. (2021). Causal interpretations of black-box models. *Journal of Business & Economic Statistics, 39/1*, 272–281.
406. Zhou, Z.-H., Wu, J., & Tang, W. (2002). Ensembling neural networks: Many could be better than all. *Artificial Intelligence, 137/1–2*, 239–263.
407. Zhu, R., & Wüthrich, M. V. (2021). Clustering driving styles via image processing. *Annals of Actuarial Science, 15/2*, 276–290.
408. Zou, H. (2006). The adaptive LASSO and its oracle properties. *Journal of the American Statistical Assocation, 101/476*, 1418–1429.
409. Zou, H., & Hastie, T. (2005). Regularization and variable selection via the elastic net. *Journal of the Royal Statistical Society, Series B, 67/2*, 301–320.

Index

M. V. Wüthrich, M. Merz, *Statistical Foundations of Actuarial Learning and its Applications*, Springer Actuarial, https://doi.org/10.1007/978-3-031-12409-9

F

G

N

T